Hormones and Reproduction in Fishes, Amphibians, and Reptiles

Hormones and Reproduction in Fishes, Amphibians, and Reptiles

Edited by

David O. Norris

and

Richard E. Jones

University of Colorado
Boulder, Colorado

Plenum Press • New York and London

Library of Congress Cataloging in Publication Data

Hormones and reproduction in fishes, amphibians, and reptiles.

Includes bibliographies and index.
1. Fishes—Reproduction—Endocrine aspects. 2. Amphibians—Reproduction—Endocrine aspects. 3. Reptiles—Reproduction—Endocrine aspects—Congresses. 4. Endocrinology, Comparative. I. Norris, David O. II. Jones, Richard E. (Richard Evan), 1940– .
QL639.2.H67 1987 597 87-6944
ISBN 0-306-42551-3

A Division of Plenum Publishing Corporation
233 Spring Street, New York, N.Y. 10013

Printed in the United States of America

We dedicate this volume to our teachers and mentors, Professor Howard A. Bern and Professor Aubrey Gorbman, and to all those to whom we have taught the excitement of comparative endocrinology that Professors Bern and Gorbman instilled in us.

PREFACE

Comparative endocrinology is one of the most rapidly developing subdisciplines within the field of endocrinology, and it is having a significant impact on research at the molecular, cellular, organismal and environmental levels. Much of the current ferment in endocrinology is in reproductive endocrinology. The purpose of this volume on hormones and reproduction in fishes, amphibians and reptiles is to summarize our present understandings and to identify important research problems to be addressed in the area of comparative reproductive endocrinology. It was inspired by the gathering at Copper Mountain, Colorado, of eminent endocrine scientists from around the world on the occasion of the Tenth International Symposium on Comparative Endocrinology in July, 1985. While preparing for that meeting, we decided that a special volume on reproductive endocrinology was needed to summarize what is known and to stimulate research in particular directions.

Why do we emphasize fishes, amphibians and reptiles? First, knowledge about the reproductive endocrinology of these ectothermic vertebrates can provide a clearer picture of the evolution of reproductive hormones and their effects on target organs. This comparative approach can lead to new theories about the evolution of reproductive control mechanisms.

Second, studies concerning the reproductive endocrinology of "lower" vertebrates can result in development of "model systems" for application to studies of birds and mammals. Indeed, information about the patterns of reproductive control in ectothermic vertebrates can tell us which are evolutionarily stable and which are labile. This historical perspective then can lead to predictions about reproductive control mechanisms in birds and mammals which have separate evolutionary origins from ectothermic reptiles. Even without enough information to form an evolutionary history, specific discoveries from, for example, a frog or lizard species can be tested in mammals. Contributions from such approaches are being made at all levels. Professor Howard A. Bern (1984) summarized this for us:

> . . . in recent years, has come the recognition that comparative endocrinology can break new ground: can discover new phenomena

and offer new concepts, which are then applicable to an understanding of endocrine physiology generally--in mammals and in humans.

Third, some ectothermic vertebrates are important food sources. This is especially true for fishes but also for some amphibians and reptiles. And, with the present state of human population growth, these ectothermic vertebrates will become more important economically in the future. In this regard, reproductive endocrinology has, and will continue to have, important benefits related to the propagation of certain food species (e.g., salmon, bullfrogs, green sea turtles).

Fourth, several ectothermic vertebrate species are endangered and could become extinct. Besides protection and habitat management, it will become necessary to house breeding stocks of endangered species as captive populations, and the utilization of hormonal manipulations may be a valuable tool for propagating these species for eventual release into nature. For example, synthetic gonadotropin-releasing hormone is now being used to induce ovulation in captive, endangered anuran amphibians.

Lastly, the reproductive biology of a species plays a vital role in its natural history and life cycle. Therefore, knowledge about the reproductive endocrinology of any species can fill an important gap in our understanding of its particular adaptations. Such knowledge will be essential for future management of natural habitats and for perpetuation of natural populations.

The topics chosen for this work include development and differentiation, gonadal cycles, oocyte maturation, vitellogenesis, ovulation, corpus luteum function, male gonaducts and sex accessory glands, steroid receptors and binding proteins, behavior, evolution of viviparity, aging of the neuroendocrine-reproductive axis, and the roles of gonadotropin-releasing hormones, gonadotropins, prolactin, thyroid hormones, steroid hormones, stress hormones, the pineal complex and environmental factors. The contributors are some of the current leaders in comparative endocrinology of ectothermic vertebrates, and they represent six countries evenly distributed among the three continents: Asia, Europe and North America.

It is the editors' hope and intention that this volume will not only serve as an introduction to specific topics in comparative reproductive endocrinology but also will stimulate new research in the field. We have attempted to coordinate the chapters contributed by the various authors through the use of cross-references. Furthermore, we have adhered as much as possible to a uniform terminology as well as a standard format for each chapter, beginning with mammalian (and where pertinent, avian) knowledge followed by discussions of fishes, amphibians and reptiles.

We wish to thank all of the contributors for their hard work, tolerance and patience. Frank Moore kindly supplied us with the cover photograph. Finally, special thanks go to Ann Underwood for typing the camera-ready copy and to Plenum Publishing Corporation for their support.

David O. Norris

Boulder, Colorado

Richard E. Jones

CONTENTS

CHAPTER 1

HORMONES AND SEXUAL DIFFERENTIATION

Elizabeth Adkins-Regan

CHAPTER 2

GONADOTROPIN-RELEASING HORMONES IN FISHES

Nancy Sherwood

CHAPTER 3

ROLE OF GONADOTROPIN-RELEASING HORMONE IN REGULATION OF GONADOTROPIN SECRETION FROM AMPHIBIAN AND REPTILIAN PITUITARIES

Paul Licht and David Porter

CHAPTER 4

PROLACTIN AND REPRODUCTION

Valdo Mazzi and Camillo Vellano

CHAPTER 5

REGULATION OF OVARIAN STEROIDOGENESIS

G. Chieffi and R. Pierantoni

CHAPTER 6

ENDOCRINOLOGY OF VITELLOGENESIS

Shuk-mei Ho

CHAPTER 7

ENDOCRINE CONTROL OF OOCYTE MATURATION

Yoshitaka Nagahama

CHAPTER 8

OVULATION: INSIGHTS ABOUT THE MECHANISMS BASED ON A COMPARATIVE APPROACH

Richard E. Jones

CHAPTER 9

FUNCTIONAL MORPHOLOGY AND REGULATION OF THE CORPUS LUTEUM

Françoise Xavier

CHAPTER 10

TESTICULAR FUNCTION

Brian Lofts

CHAPTER 11

REGULATION OF MALE GONADUCTS AND SEX ACCESSORY STRUCTURES

David O. Norris

CHAPTER 12

SEX STEROID RECEPTORS AND NON-RECEPTOR BINDING PROTEINS

Ian P. Callard and Gloria V. Callard

CHAPTER 13

SEASONAL REPRODUCTION; PATTERNS AND CONTROL

J. M. Whittier and David Crews

CHAPTER 14

THYROID HORMONES AND REPRODUCTION

John F. Leatherland

CHAPTER 15

PINEAL INVOLVEMENT IN SEASONALITY OF REPRODUCTION

William A. Gern, Jeanne M. Nervina, and Shelley S. Greenhouse

CHAPTER 16

STRESS AND REPRODUCTION: RECIPROCAL RELATIONSHIPS

Neil Greenberg and John Wingfield

CHAPTER 17

REGULATION OF REPRODUCTIVE BEHAVIORS

Frank L. Moore

CHAPTER 18

THE EVOLUTION OF VIVIPARITY IN FISHES, AMPHIBIANS AND REPTILES: AN ENDOCRINE APPROACH

Louis J. Guillette, Jr.

CHAPTER 19

AGING OF THE NEUROENDOCRINE SYSTEM

Martin P. Schreibman, Henrietta Margolis-Nunno,
and Leslie Halpern-Sebold

HORMONES AND SEXUAL DIFFERENTIATION

Elizabeth Adkins-Regan

Cornell University
Ithaca, New York 14853

I. INTRODUCTION

One of the most profound questions that can be asked about the animals around us concerns the developmental mechanisms that enable a single fertilized egg to become a functional adult with all its diverse tissues and behavioral capabilities. Sexual differentiation, in which initially undifferentiated embryos come to realize one of two distinct morphological and behavioral outcomes, is a particularly interesting reflection of the basic developmental phenomenon of differentiation, especially because of the involvement of sex hormones.

This chapter will discuss the endocrinology of sexual differentiation in fishes, amphibians, and reptiles. Because hormonal contributions to the differentiation of reproductive morphology have been extensively reviewed recently,[2,46,47,99] I will focus disproportionately on recent findings and on the differentiation of behavior and nervous systems.

A. Sex Determination

In theory, sexual differentiation and determination are distinct processes. In practice, however, they are not (see Austin et al.[6] for an interesting discussion of this point), particularly for fishes and amphibians, where the fate of the germ cells is not determined entirely by their genetic sex. In addition, the sex-determining mechanisms of animals might have some important consequences for sexual differentiation or place significant constraints on it. For these reasons, a few words about genetic sex-determining mechanisms in teleostean fishes, amphibians, and reptiles will be helpful (almost nothing is known in other kinds of fishes). Detailed reviews can be found in Bull,[15], Kallman,[54] and Price.[73]

As Table 1 indicates, heteromorphic sex chromosomes (paired chromosomes that look similar in one sex and different in the other sex) do occur in some reptiles and fishes, though not in most. Animals in the same class can differ with respect to the presence or absence of heteromorphic chromosomes, and animals in the same class or even genus can differ with respect to which sex is heterogametic (has the two dissimilar sex chromosomes). Furthermore, heteromorphic sex chromosomes are not necessary in order to have genetic sex determination.

In several poeciliid fish such as guppies (*Poecilia reticulata*), swordtails (*Xiphophorus helleri*), and platys (*Xiphophorus maculatus*), sex-linked color patterns have made it possible to analyze the genetic mechanisms of sex determination. Wild populations of platys are polymorphic for sex determination. There are three possible female chromosome types (WY, WX, and XX) and two possible male types (XY and YY).[53] In this and some other piscine sex determination systems, individuals of the same gonadal sex do not necessarily have the same sex genotype, and offspring sex ratios of different male-female pairs of the same species can differ dramatically. A common interpretation of the poeciliid fish data is that multiple sex genes or "genetic sex factors" determine sex by additively combined effects.[73] Kallman[54] rejects such a polygenic model, proposing instead that there is one major male-determining gene present on all the sex chromosomes that interacts with control elements on the sex chromosomes. The latter in turn interact with an autosomal regulatory locus. He points out that the involvement of one or more autosomal loci does not make the system polygenic (in the sense of additive effects of many minor genes) any more than the proven existence of autosomal regulatory genes for sex determination in mammals makes those systems polygenic.

H-Y antigen, the histocompatibility antigen first discovered in the tissues of male mammals, has now been found in several fishes, amphibians, and reptiles (see Table 1). In each case where the heterogametic sex is known through other methods, the H-Y+ sex (the one with more of the antigen) is always the heterogametic sex.

Sex is not always determined genetically in vertebrates, nor are vertebrates always gonochorists (animals with separate male and female individuals). Some fishes (omitted from Table 1) are successive hermaphrodites, with environmental control of sex change. For some animals, particularly reptiles, incubation temperature appears to be a major determinant of sex (see Table 1). Yet other vertebrates (also omitted from Table 1) are

Table 1. Sex Determination in Some Fishes, Amphibians and Reptiles

	Hetermorphic sex chromosomes?	Genetic sex determination?	Heterogametic Sex	$H\text{-}Y^{+}$ sex	Temperature sex determination?
Fishes (Osteichthyes)					
27 families	-	+	M (most)		+ (1 sp)
Platyfish (*Xiphophorus maculatus*)	-	+	F(M)**	M	
Medaka (*Oryzias latipes*)	-	+	M	M	
European eel (*Anguilla* anguilla)	-			F	
Mozambique cichlid (*Sarotherodon mossambicus*)	-	+	M		
Cichlid (*Sarotherodon hornorum*)	-	+	F		
Rainbow trout (*Salmo gairdneri*)	+				
Amphibians					
African clawed frog (*Xenopus laevis*)	-	+	F	F	
Leopard frog (*Rana pipiens*)	-	+	M	M	
Other frogs	-	+	M		
Ribbed salamander (*Pleurodeles waltli*)	-	+	F		+
Other salamanders and newts	-	+	F		
Reptiles					
Most turtles	-			F	+
Softshell turtles (Trionychidae) and Chelidae	-				-
3 turtle species	+		M		
Alligators and crocodiles	-				+
Snakes	+		F		-?
Agamid lizards (Agamidae)	-				+?
Geckos (Gekkonidae)	-		F		+?
Iguanid lizards (Iguanidae)	- and +		M		-?
Lacertid lizards (Lacertidae)	- and +		F		-?
Skinks (Scincidae)			M		
Whiptail lizards (Teiidae)	+ (1 sp)		M		

Reproduced with permission from Adkins-Regan.[3]
Based in part on Bull[15] and Wachtel and Koo[100]
(+) means yes; (-) means no; (M) = male; (F) = female. A blank means that no information is available.
Presumably at least some cases that are (+) for temperature sex determination are (-) for genetic sex determination, and most reptiles with heteromorphic sex chromosomes have genetic sex determination.
If the entry is based on information from several or many species and there are a small number of exceptions, these exceptions have not necessarily been indicated.
* The XY or ZW sex, based on chromosome heteromorphism or breeding experiments. In the case of fish, there may be minor sex factors acting in addition, but the system approximately resembles XX/XY or ZZ/ZW.
** This species is polymorphic with respect to sex determination (see text).

unisexual (a few fishes and lizards) or simultaneously hermaphroditic (a few fishes) and thus do not exhibit sexual differentiation in the usual sense.

II. THEORETICAL AND CONCEPTUAL CONTEXT

One convenient, if not totally correct, way to think about sexual differentiation is to assume that the gonads differentiate first, and that their secretions then influence the differentiation of gonaducts, other internal reproductive organs, and external genitalia plus secondary sex characters, in roughly that order.[74] In mammals and birds, behavior and the neural substrates for behavior differentiate at the same time as or after the internal and external genitalia.[2,42]

Some reproductive structures, such as phalli, differentiate from a single primordium that either enlarges greatly (in males) or enlarges only slightly (in females). Other structures, such as the gonaducts of amphibians, reptiles, birds, and mammals, differentiate from dual primordia. The mullerian ducts become the oviducts of females, whereas the wolffian ducts become the sperm ducts of males.

Behavioral researchers have found it valuable to ask whether this "structure" seems to develop from a single or a dual "primordium." For example, does masculinization always involve simultaneous defeminization (as would be the case with a single primordium) or can animals be masculinized but retain the capacity for female behavior (dual primordia)? In mammals and birds the evidence clearly indicates that there are two independent behavioral dimensions ("primordia").[2,42]

The gonads of vertebrates other than fishes originate from dual primordia. The region (cortex or medulla) of the indifferent germ cell-containing gonad that proliferates determines which type of gonad develops.

A. Organization vs. Activation: Critical Periods

The term "organizational" usually refers to permanent actions of hormones that serve to establish the fundamental form (male vs. female) of some organ or behavioral capacity (see Goy and McEwen[42]). Because most organs can be affected in such a profound way only during the period of most rapid cell differentiation, these effects are generally limited to critical periods during embryonic, larval, or neonatal development, at least in birds and mammals.[61] Activational effects of hormones, on the other hand, are not permanent (reversing if hormone exposure ends) and are not limited to any critical period. They cause quantitative changes in organs or systems that already have been organized previously and typically occur at and after puberty or sexual maturity. The distinction between organization and activation is important for interpreting many sexual differentiation experiments, particularly those focusing on behavior.

B. Adult Sexual Dimorphism

Because sexual differentiation is the process that produces adult sexual dimorphism, it only makes sense to look at the developmental effect of sex hormone treatment on characters that actually are dimorphic in adults. But some characters are dimorphic only because the sexes differ in the activational hormones that are normally present in adulthood. If adult males treated with estrogen become completely female-like, or adult females treated with androgen become completely male-like, then the adult dimorphism is simply a consequence of gonadal dimorphism, and there is no dimorphism in the substrate (soma) for the character itself. This point is particularly important in studies of behavioral and neural sexual differentiation. Neural sexual dimorphism and organizational effects of sex hormones on behavior and the nervous system are only likely to exist if animals treated with heterologous hormones are _not_ completely sex-reversed behaviorally.

C. The Nature of the Inducer(s)

It can be proven that differentiation of morphological structures, including gonads, is under the control of a hormone or hormone-like factor by showing that tissues fail to differentiate according to their sex genotype when removed from the embryo and allowed to develop *in vitro* or as a graft in an individual of the opposite sex.[16] Determining the identity of the hormone or other factor is less straightforward, particularly for gonadal induction. Two theories have traditionally dominated the field. The corticomedullary inductor model, first outlined and later updated by Witschi,[108] proposes that there are two non-steroidal inducers, one that stimulates development of the cortical tissue, causing ovarian differentiation, and another that stimulates development of the medullary tissue, causing testicular differentiation. There has been considerable interest recently in the possibility that H-Y antigen might be the non-steroidal inducer of the heterogametic gonad;[100] but the fact that the antigen is steroid-inducible, at least in chickens,[63] means that the antigen could be a result, rather than a cause, of gonadal differentiation. The steroid model, first outlined by Yamamoto,[111] proposes that steroidal estrogens and androgens are the gonadal inducers, and that in some species only one sex (the "dominant" sex) requires hormones for differentiation, whereas the other develops automatically without hormones. Such a dominant-neutral system of sexual dimorphism[48] is also theoretically possible for non-gonadal characters.

What is the nature of the substance that induces sexual differentiation of the other reproductive organs or of the nervous system and behavior? Again, steroid sex hormones are one possibility. But evidence is accumulating that the testicular factor that causes regression of the mullerian ducts in male birds and mammals is not a steroid,[51] raising the possibility of non-steroidal inducers in other vertebrates as well. This mullerian-inhibiting substance (MIS) recently was identified as a glycoprotein in mammals.[30]

III. DIFFERENTIATION OF THE GONADS

A. Fishes

Differentiation of the gonads of cyclostomes has been reviewed recently.[40] It is unfortunate that so little work has been done with cyclostomes, for their gonads differ from those of other vertebrates in ways that ought to have important consequences for differentiation. Both lampreys and hagfish have a single gonad, and neither has a corticomedullary dual primordium. Hagfish have an unusual anterior-posterior arrangement; the anterior gonad is the potentially ovarian tissue, whereas the posterior gonad is the potentially testicular tissue. The one reported attempt to alter lamprey gonadal differentiation by sex steroid treatment was unsuccessful.[43]

Cartilaginous fish have the cortico-medullary, dual gonadal primordium typical of all other vertebrates _except_ teleosts and cyclostomes.[29] If undifferentiated gonads of _Scyliorhinus canicula_ are transplanted, their differentiation is independent of the host's sex[97] as if gonadal differentiatio is not under hormonal control. Yet in this same species, injections of sex steroids prior to differentiation feminize the gonads.[21]

Most of what is known about differentiation of gonads in fishes is base on work with teleosts. Some of these, such as trout, salmon, and African cichlids, are commercially or agriculturally significant. There is consider able interest in using sex steroids added to the water to reverse or sterilize the gonads of fry, either because one sex has greater economic value or because high population densities produce smaller, less valuable fish.[48,75,88]

Teleostean fishes usually, but not always, have paired gonads, but thei gonads differ from those of amphibians and reptiles in two important respects. First, some undergo natural sex reversal in adulthood. These will be discussed later; this section will focus on gonochoristic species. Second, fish do not seem to have a dual cortico-medullary primordium; instea all the primordial tissue resembles cortical tissue.[48] This makes it difficult to tell when testicular vs. ovarian differentiation has begun. It also has engendered a long-standing controversy over whether the germ cells themselves are also bipotential, and, if so, for how long (see Reinboth[83] for a more extensive discussion). Can future female germ cells be distinguished from future male germ cells or is the sex of the primordial germ cells determined by inducers (sex steroids?) from the gonadal tissue? Using light microscopy, Satoh and Egami[85] could not tell male and female primordial germ cells apart in the medaka (*Oryzias latipes*), but using electron microscopy, Bruslé and Bruslé[14] could see ultrastructural sex differences in germ cells of the golden-grey mullet (*Mugil auratus*). Yet in some species sex steroids do alter germ cell sex (see below).

The normal course of development of the gonads has been described for several fish (see Nagahama,[65] Jensen and Shelton[50] for reviews). Ovaries often differentiate before testes, judging from the onset of germ cell differentiation.[65] Gonadal differentiation begins surprisingly late and takes remarkably long to complete. For example, in grass carp (*Ctenopharyngodon idella*) and rainbow trout (*Salmo gairdneri*) it begins 50 days or more after hatching,[50,98] and in *Mugil auratus* it begins at the end of the first year.[14] In some species the indifferent stage can last for several years,[18] raising the possibility that differentiation and sexual maturity (production of mature gametes) occur relatively closely together in time. With the exception of salmon (*Salmo salar*),[56] species with relatively early gonadal differentiation also reach sexual maturity early. Thus *Oreochromis* (*Sarotherodon*, *Tilapia*) gonads differentiate 15-30 days post-fertilization.[49]

The effects of sex steroid treatment early in development on gonadal differentiation have been studied extensively in fishes (see Hunter and Donaldson,[48] Schreck,[88] and Yamamoto[111] for detailed reviews). Hormones are usually administered in the water in which the fry develop, or in their food. Partial sex reversal often occurs, and in over a dozen species permanent complete functional sex reversal has occurred. These sex-reversed individuals produce offspring in matings with untreated individuals of the same genetic sex. The critical period for gonadal reversal is the period of initial gonadal differentiation. The most detailed information about complete experimental sex reversal comes from a series of experiments with *Oryzias latipes*.[111] When fry were given food containing androgens such as methyltestosterone or androstenedione, genetic females were completely sex-reversed, whereas if the food contained estrogens such as estradiol, estrone, or diethylstilbestrol, genetic males were reversed completely. These effects were permanent, and treatment after gonadal differentiation had occurred never caused complete reversal. The percentage of sex-reversed fish was dose-related, and about 15-20 μg steroid per gram of food reversed half the fish. Subsequent experiments in which gonadal grafts underwent reversal after being placed in the anterior chamber of the eye of an oppositely-sexed fish[85] indicate that physiological levels of hormones are adequate.

Are sex steroids the gonadal sex-inducers during normal development? One requirement of a steroid model of induction is that piscine gonads must actually produce and take up sex steroids prior to or at the beginning of gonadal differentiation. Hishida[45] showed that physiological levels of androgens and estrogens are accumulated only by actively differentiating *Oryzias* gonads. Takahashi and Iwasaki[96] obtained enzymatic evidence for steroidogenesis by the gonads of guppies, but only after the germ cells had already differentiated. Van den Hurk et al.,[98] using a variety of biochemical and histological methods in rainbow trout (*Salmo gairdneri*), failed to obtain evidence of steroid production in indifferent gonads and in differentiating ovaries. Differentiating testes could synthesize androgens.

Thus it is not clear that piscine gonads are producing steroids early enough for those steroids to be a cause, rather than a result, of differentiation. Further evidence against the steroid model includes the frequent experimental occurrence of paradoxical effects (masculinization caused by estrogens or feminization caused by androgens)[48] in the absence of any direct evidence that the differentiating testes aromatize androgens.[36,98]

H-Y antigen has been detected in the males of six species (three poeciliids, one cichlid, one eel, and the medaka),[64,70,90,104] and in both sexes of yet another species.[64] Its role, if any, in gonadal differentiation has not been studied experimentally in fishes. An H-Y theory would not necessarily be incompatible with a steroid model, for the following causal sequence of events is conceivable:[54] Male sex genes; steroid production; expression of H-Y antigen; induction of gonadal differentiation.

Two gonochoristic teleosts exhibit experimental sex reversal in adulthood. If adult female *Betta splendens* or *Macropodus opercularis* are ovariectomized, some of the females slowly turn into males with functional testes.[11,60] Either gonadal differentiation is not limited to a critical period in these species, or normal adult females have some testicular tissue that is inhibited by ovarian tissue.

B. Amphibians

Amphibian gonads are paired and originate from dual cortico-medullary primordia. Gonadal differentiation occurs mainly or entirely during the post-hatching larval stage. Several kinds of experiments using different techniques (grafts, parabiosis) show that hormonal factors can modify or reverse gonadal differentiation.[34,39] Male-produced hormones seem to predominate, for in cross-sex grafts or parabionts ovaries are masculinized, but testes are not feminized.

Administration of sex steroids to larvae has caused complete functional sex reversal in three *Rana* species, the tree frog *Pseudacris*, the African reed frog *Hyperolius*, *Ambystoma*, *Xenopus*, and four salamanders and newts of the genera *Hynobius*, *Triturus*, and *Pleurodeles*.[28,34,39,84,109] Genetic males have been turned into egg-laying individuals, or genetic females have been turned into sperm-producing males that then fathered offspring. The ease with which each sex can be reversed depends on the species. In anurans, female-to-male reversal occurs more readily (although tadpoles of *Rana pipiens* can be reversed in both directions)[19] whereas in *Xenopus* and the urodeles, male-to-female reversal is characteristic. In a recent experiment, testosterone treatment of *Hyperolius* larvae resulted in 100% males, whereas estradiol treatment had no effect on the sex ratio.[84] Experimental sex reversal occurs only during a critical period, a fact that has been most clearly demonstrated in *Xenopus*.[109] *Xenopus* gonads differentiate during the third week after hatching. A week or 10 days of estrogen treatment at this time will produce 100% females, but earlier or later treatment is much less or not at all effective.

The spectacular success of these experiments, together with biochemical evidence for steroid secretion by larval gonads,[67] suggests that sex steroids could be the natural gonadal inducers. This model has been tested by examining the effect of antihormones on gonadal differentiation in *Rana esculenta*.[76] Cyproterone acetate, an antiandrogen, failed to prevent androgens from masculinizing gonads, and tamoxifen, an antiestrogen, failed to prevent estrogens from feminizing gonads. When given alone, the antihormones failed to prevent normal differentiation. These results suggest that either the natural inducers are not steroids or the steroidal inducers do not act through the same mechanisms (receptors?) as those in adult target tissues.

H-Y antigen has been detected in male *Rana pipiens* and female *Xenopus laevis*[100] (see Table 1), but its role, if any, in gonadal differentiation is unknown.

Castrated male toads (*Bufo*) will occasionally become functional females.[38] The Bidder's organ contains the cortical primordium, which develops after the testes are removed. This adult sexual lability is all the more surprising given that larval hormone treatment does not sex reverse *Bufo*.

C. Reptiles

Our knowledge of all ontogenetic processes in reptiles, including sexual differentiation, is rather poor, due to slow developmental time scales and difficulties with captive rearing. This is a most unfortunate situation, for it is likely that reptiles gave rise to modern birds and mammals, and are therefore a crucial evolutionary "keystone."

In spite of the long incubation or gestation periods, gonadal differentiation is not always complete at hatching. Differentiation occurs earlier and faster in smaller, more rapidly maturing species such as the lizard *Lacerta vivipara*.[31]

Most experimental investigations of gonadal differentiation in reptiles have treated incubating eggs with sex steroids. Both estrogens and androgens have caused partial sex reversal in at least some, but not all, species.[28,31] Feminization occurs more readily than masculinization (see for example, Pieau[71]).

Completely permanent experimental sex reversal has never been observed in reptiles, and so the primary gonadal inducer, if any, is unlikely to be a sex steroid. Little is known about the secretions of embryonic or juvenile reptilian gonads.[68,69]. However, Pieau et al.[72] reported that androstenedione, dihydrotestosterone and estrone are synthesized by embryonic gonads of turtles, *Emys orbicularis*, and production is greater in testes than in ovaries. Progesterone, testosterone and dehydroepiandrosterone were not detected. There has been some interest in, but little evidence for, the possibility that pituitary hormones might contribute to induction.[77,78]

IV. DIFFERENTIATION OF OTHER REPRODUCTIVE STRUCTURES

A. Fishes

Chieffi[21] examined the effect of embryonic sex steroid treatment on the differentiation of reproductive ducts and claspers in the shark *Scyliorhinus canicula*. At low dosages, testosterone caused oviduct atrophy, but at higher dosages the oviduct hypertrophied and the wolffian duct was feminized, effects that could be due to aromatization. Estradiol caused oviduct hypertrophy in both sexes. The claspers differentiated early, at the same time as the gonad, but were not affected by hormone treatment.

In teleostean fishes there is good evidence that sex steroids induce gonaduct differentiation. Not only is duct differentiation preceded by gonadal steroidogenesis, but in addition steroid treatment can reverse the sex of the ducts.[36] (See Chapter 11 for more information about duct regulation in fishes and other animals.)

Teleostean fishes have a sizeable and varied array of secondary sex characters (gonopodia, body colors, etc.), and yet surprisingly little is known about the role of hormones in their development. Is this because the substrates for them are not actually differentiated? Androgen treatment of adult females or estrogen treatment of adult males often causes complete sex reversal of these sex characters,[58] as if dimorphism is entirely a product of activational, rather than organizational, hormonal influences.

B. Amphibians

In amphibians the wolffian ducts normally persist in females and function as ureters; thus they are not highly sexually differentiated and are probably not affected by physiological levels of sex steroids.[16,37] Mullerian ducts, on the other hand, seem to require estrogen for normal female differentiation. Androgens also can stimulate mullerian duct growth (see Chapter 11).

Cloacal glands are well-developed in adult male urodeles, and are important for the production of the spermatophore (see Chapter 11). As described earlier in the section on gonadal differentiation, some genetic male urodeles treated early in development with estrogens develop functional ovaries.[28,34] These animals also have feminized cloacal glands. It is not clear, however, whether estrogen treatment of *adult* males would have the same effect; i.e., whether an organizational effect is involved in the response.

Adult male *Xenopus* have nuptial pads. If adult gonadectomized females are implanted with pellets of testosterone, they form impressive nuptial pads.[55] Male *Xenopus* also have a larger larynx, and male laryngeal muscle binds more androgen.[89] These sex differences underlie the male courtship vocalization. The sex difference in androgen binding disappears if females are treated with testosterone.[89] Thus neither nuptial pads nor androgen binding by laryngeal muscle are sexually differentiated in the organizational sense.

C. Reptiles

In normal development both sets of ducts are retained in females, sometimes until early postnatal life, but mullerian ducts regress in males. As in amphibians, hormones have a minimal role in the development of the wolffian ducts.[74] Testicular hormones are necessary for mullerian duct regression in males (*Lacerta vivipara* is a possible exception[31]). If embryos are castrated prior to differentiation, males retain the mullerian ducts,[110] and testosterone treatment prevents mullerian duct development.[79] Furthermore, the mullerian ducts of males stop growing shortly after the onset of testicular differentiation. The role of the testis in gonaduct differentiation thus seems well established, but it would be premature to assume that the inducer is a steroid,[74] especially because the mullerian-inhibiting testicular hormone of mammals and birds is clearly not a steroid.[103,106]

Numerous experiments have found effects of embryonic or juvenile sex steroid treatments on reptilian reproductive organs of the cloacal region (particularly the penis or hemipenes). Both feminization and masculinization have been observed.[35] But it is difficult to assess the significance of these results for the normal process of differentiation, because it is not known if the effects are permanent or if effects of the same magnitude would occur with treatment of adults. Feminization predominates in several species of lizards,[2] which may explain why extensive testosterone treatment of embryonic or juvenile *Cnemidophorus uniparens* had relatively minor masculinizing effects on renal sex segments and secondary sex characters, in spite of the fact that it caused significant masculinization of the gonads.[12] This last experiment is nonetheless unique, for the species is parthenogenetic, and only females occur under normal conditions.

Some reptiles are highly size dimorphic with females being larger. A recent experiment looked at the contribution of sex hormones to this dimorphism in garter snakes (*Thamnophis sirtalis*).[27] No clear organizational effects were obtained. Instead, activational effects of testosterone in adulthood could account for most of the sex difference.

V. DIFFERENTIATION OF BEHAVIOR AND THE NERVOUS SYSTEM

A. Fishes

Through the efforts of many ethologists, the reproductive behavior of several species (all teleosts) has been thoroughly described (see Baerends[7] and references therein), yet little is known about sex differences in the nervous system or sexual differentiation of behavior. (See Chapter 17 for activational effects of sex hormones on fish behavior.)

Some information can be gleaned from research that does not focus directly on behavior or neurobiology. Experiments were described previously (see section on gonadal differentiation) in which several species of fish were completely sex reversed by treating the fry or embryos with sex steroids. Even though behavior was not observed systematically (see Clemens et al.[22] for an exception), the fact that these animals did produce offspring is evidence that some behavior patterns must have been sex-reversed as well.[57] But adult fish of many species are also capable of extensive behavioral sex reversal. In particular, females treated with androgens often show extensive male-typical sexual behavior,[58,82] as if the neural substrate for behavior is not sexually differentiated. In some species, female-typical behavior is activated easily in males with estrogen,[57] leading to the same conclusion.

An unusually interesting example of androgen-induced masculinization of adult females has been observed in fish of the family Mormyridae.[8,9] African mormyrids give off pulsatile electric organ discharges (EODs) which are used for species recognition and object localization. The EODs indicate that *Brienomyrus brachyistius* actually comprises three separate species, and the EODs of two of these species, *B. brachyistius* (long biphasic) and *B. brachyistius* (triphasic) are dimorphic in form and duration. In *B. brachyistius* (long biphasic) testosterone completely masculinizes the EOD of intact and gonadectomized adult females. The effect reverses when treatment ends. Androgen-induced masculinization does not occur in *B. brachyistius* (biphasic), a species with monomorphic EODs. Interestingly, in *B. brachyistius* (triphasic) androgen-induced masculinization is surprisingly long-lasting. EODs are still male-like 25 days after androgen treatment of females is terminated and 28 days after castration of a male.[10] It will be important to see if adult androgen treatment actually causes *permanent* masculinization, for such effects are quite rare. All told, these electric fish phenomena provide an excellent opportunity to delve into the molecular and electrochemical mechanisms of steroid action.

Are there behavior patterns that do seem to have a dimorphic substrate? Wai and Hoar[101] treated adult gonadectomized male and female sticklebacks with methyltestosterone. Under long photoperiods both sexes became aggressive and territorial, but only males exhibited nest-building behavior (a male-typical characteristic) or male-typical sexual behavior.

Very few experiments have been explicitly designed to determine whether early exposure to sex hormones alters behavioral sexual differentiation. In one preliminary study[59] guppies were treated with 0.5 ppm methyltestosterone either 1 to 7 days or 28 to 35 days after birth and later were tested as adults. Males were unaffected, and females were masculinized only to a very small extent (on two test days the proportion of early-treated females showing male-typical courtship was higher). The small effect is not surprising given that female-controls receiving the androgen as adults behaved very much like males.

Another study[13] examined the effect of early methyltestosterone treatment on the adult behavior of *Oreochromis* (*Sarotherodon*) *mossambicus*, a mouthbrooding cichlid. Animals were treated either before hatching, for 21 days after hatching, from 21 to 42 days after hatching, or as adults. The effectiveness of the treatment depended not only on the time of treatment, but also on the route of administration (in the water vs. in the food). Some sex-reversed genetic females were detected. All of them exhibited the complete normal complement of male behavior patterns and were not different from males receiving the same early treatment. Responsiveness of early androgen-treated but non-sex-reversed females to androgen was examined by giving adults methyltestosterone in the diet for 40 days and testing them on days 24-40. Only early-treated females tilted (a male-typical territorial defense display) and females treated before hatching gave more lateral displays (another aggressive behavior) than controls. All of the gonadally sex-reversed females exhibited the complete, normal complement of male behaviors and were not different from control males. Thus early androgen treatment masculinized the responsiveness of adult females to androgen. But early androgen also masculinized aggressive behavior in males tested as adults without further androgen treatment. Because such "hyper-masculinization" could mean that the treatment dosage exceeded the physiological range, it will be important in future studies to try different dosages to see if females can be affected without also masculinizing males.

Several examples of neuroanatomical sexual dimorphism have been discovered in birds and mammals.[5] Thus far, the only reported example in fish occurs in the oyster toadfish (*Opsanus tau*).[32] Males emit a courtship vocalization, and the swim-bladder muscle complex (the sound producing organ) is larger in males than in females. These muscles are innervated by the sonic motor nucleus (SMN), a possible homologue of the hypoglossal nucleus of birds and mammals. The male SMN contains more large motor neurons than the female SMN. It will be interesting to see whether early hormones have a role in the development of this sex difference.

B. Amphibians

Until quite recently, all that was known about the differentiation of behavior in amphibians was that animals that had been gonadally sex-reversed by larval steroid treatment could produce offspring, and therefore behaviorally must have been reversed. Whether the substrate for adult sexual behavior was actually sexually differentiated was not known, in part because researchers were unable to hormonally activate sexual behavior in gonadectomized adults. Even now that such activation has been achieved (see Chapter 17), the only direct assessment of sex differences in hormonal responsiveness is Kelley and Pfaff's study of *Xenopus laevis*.[55] Gonadectomized adults were given testosterone pellets; both sexes then clasped stimulus-females, and only minor and insignificant sex differences were observed. Nothing seems to be known about sex differences in the display of female-typical behavior in estrogen-treated amphibians.

Several behaviorally significant neural sex differences have been discovered in anurans. *Eleutherodactylus coqui*, a neotropical tree frog, has a sexually dimorphic auditory system.[66] The distributions of the best excitatory frequencies of primary auditory neurons differ in the two sexes, and can account for the fact that males respond mainly to the lower frequency syllable of the two-syllable male call whereas females respond mainly to the higher frequency syllable. The developmental origins of this sex difference are not yet known.

The neural mechanisms for vocalization in American toads (*Bufo americanus*) include the magnocellular portion of the pretrigeminal nucleus, a midbrain structure. Cells in this area are larger in males than in females, but are completely sex-reversed (masculinized) in size by injecting intact females with androgens,[86] indicating that adult, rather than embryonic or larval, hormones are responsible for the sex difference.

In *Rana pipiens*, the laryngeal nerves of isolated brainstems of males produce call-like firing patterns, provided that the preoptic area is stimulated first electrically.[87] This response is much less likely to occur in female brainstems, but does occur if the females are injected with androgens before preparation. Masculinization is extensive but not complete.

C. Reptiles

Most investigations of sex-hormone effects on behavior have focused on one of two species: *Anolis carolinensis*, the green or common anole, and *Thamnophis sirtalis*, the garter snake (see Chapter 13). These two animals are quite different reproductively. *Anolis* is oviparous but has many mammal-like endocrine characteristics, such as an estrous cycle.[24] *Thamnophis*, on the other hand, bears live young and exhibits the "dissociated reproductive tactic"[25] (see also Chapter 13). Gamete production and mating behavior occur at different times of the year, with profound consequences for the endocrine regulation of reproduction and sexual behavior.

Anolis is a promising candidate for differentiation experiments, for it develops relatively rapidly, reaching sexual maturity in a year or less.[41] No such experiments have been reported yet, but sex differences in the responsiveness of adults to gonadal hormones have been delineated.[1,62] When adult gonadectomized animals are injected or implanted with testosterone propionate, females and males exhibit equal levels of male-typical courtship and copulation. But if animals are treated with estradiol benzoate, males exhibit much less female-typical sexual receptivity than females do. Thus the primary sex difference lies in female-typical behavior, and early hormone treatment would be predicted to affect receptivity but not male-typical behavior.

A recent and particularly interesting study provides evidence for an unusual type of organizational effect of androgen in garter snakes.[26] Males undergo spermatogenesis and elevated androgen levels during the summer, then hibernate during the winter, storing the sperm. Mating occurs upon emergence from hibernation, when androgen levels are low. Male-typical courtship (chin-rubbing) and copulation are independent of activational hormones from the testes, adrenals, and pituitary, and cannot be stimulated in adult males with testosterone. Instead, cold temperature for several weeks (mimicking hibernation) followed by warm temperature activates male sexual behavior. But if neonatal males are treated with testosterone, then the males chin-rub (court). Therefore, some post-natal process changes the activational stimulus for male courtship from testosterone to temperature.

Table 2 summarizes these and other results from Crews' experiment.[26] Treatment with estradiol did not mimic these effects. The following interpretation is consistent with these results. In order for a male to court upon emergence, he must be exposed to elevated testosterone before hibernation. In the field the first such exposure would occur as a yearling during summer spermatogenesis, rather than during the neonatal period. Thus, testosterone organizes the behavior so that it will later be activated by temperature, and this organization can occur at any age. This interpretation does not explain why the females failed to court upon emergence, and it will be important to see if another organizational hormone effect is responsible for this sex difference. But the success of this first attempt to study sexual differentiation of behavior in a reptile should help overcome the misconception that such work is impossible.

In mammals sex hormones can organize the capacity for a behavior pattern that does not require activational hormones for expression (e.g., rough-and-tumble play in rhesus monkeys).[42] The organization of male sexual behavior in garter snakes may be similar with respect to the absence of an activational hormone requirement, but it is different in that organization is not limited to a critical period in early life.

VI. TEMPERATURE AND SEXUAL DIFFERENTIATION

In some vertebrates the temperature during development is a major influence on sex determination and differentiation.[15] In at least one fish species (the Atlantic silverside *Menidia menidia*) and in many species of reptiles and amphibians (see Bull[15] and Table 1) egg temperature has a dramatic

Table 2. Effect of Testosterone (T) Treatment on the Courtship Behavior of Garter Snakes

Sex	Neonatal treatment	Courted as neonates?	Courted at first emergence?	Yearling treatment	Courted as yearlings?	Courted at second emergence?
Males	Sham surgery	No	No	T	Yes*	Yes
Males	Castration	No	No	T	Yes	Yes
Males	Castr + T	Yes	Yes	T	Yes	Yes
Females	Sham surgery	No	No	T	Yes	No
Females	T	Yes	No	T	Yes	No

Based on data in Crews.[26] Testosterone was administered via silastic implants.

*Only 1 out of 4 males in this group courted, however.

effect on the sex ratios of the hatchlings or juveniles, an effect that cannot be accounted for by differential mortality. In *Menidia*, alligators, and lizards, cool temperatures produce more females and warm temperatures produce more males, whereas in many turtles the relationship between temperature and sex ratio is reversed.[15] In snapping turtles and crocodiles a U-shaped function best describes the relationship between temperature and sex ratio.

Temperature-influenced sex determination occurs under natural conditions,[23] but it is not yet known whether temperature-determined sex is stable and lasts into reproductive adulthood. Some species exhibiting the phenomenon may not have genetic sex determination at all. In the turtle *Emys orbicularis* H-Y antigen measurements taken in different tissues suggest that temperature actually overrides and reverses genetic sex.[113]

The evolutionary and ecological questions raised by this phenomenon are fascinating, but it is also important to determine the mechanisms underlying temperature-influenced sex determination. Unfortunately, little is known about these mechanisms. In the snapping turtle *Chelydra*, the ability of temperature to determine the sex ratio is limited to a well-defined critical period between embryonic stages 14 and 19.[112] But these stages are defined by the appearance of the forelimbs, and the gonadal condition corresponding to this period has not been described. Even species that are not affected by temperature variations within a normal range (*Xenopus*, *Salmo*, *Betta*, *Tilapia*) become intersexes or are sex-reversed if exposed to extreme temperatures during development.[17,18] A particularly interesting example of this phenomenon occurs in *Rivulus marmoratus*. This freshwater fish is a synchronous hermaphrodite, and in the field (at least in Florida) male or female individuals are rare. But in the laboratory low temperatures during embryonic development cause some pure males to appear.[44] After hatching, high temperatures cause a sex change from hermaphrodite to male. The fact that extreme temperatures can cause abnormal sexual differentiation, and that the abnormalities look like those produced by treating embryos with sex steroids, raises the possibility that hormones are part of the mechanism of temperature-influenced sex determination.

VII. SUCCESSIVE HERMAPHRODITISM AND SOCIAL INFLUENCES ON SEXUAL DIFFERENTIATION

A. Fishes

Naturally occuring hermaphroditism is very rare in all vertebrates except teleostean fishes. In these fishes the most common form of hermaphroditism is successive or sequential hermaphroditism in which individuals function reproductively as adults of one sex, then change to the other sex. The change can be from male to female (protandry) or female to male (protogyny). Successive hermaphroditism occurs in over 50 species of fish in several families living in a wide variety of habitats.[20,81,95] Like temperature-influenced sex determination, this phenomenon requires both physiological and ecological explanations in order to be understood. Size-advantage models have predominated in ecological theorizing.[20,102] According to these models the relationship between reproductive success and size differs for the two sexes such that the greatest lifetime reproductive success is achieved by reproducing as one sex at small sizes (young ages) and as the other sex at larger sizes (older ages). Some are skeptical that this explanation can account for the field data.[93]

Studies of the mechanisms of sex change include descriptions of gonadal histology for several species.[18] The gonads of juveniles contain both ovarian and testicular tissue, and primordial germ cells settle in both regions. During reproductive life first one and then the other tissue type predominates, but more comparative studies are needed before it can be concluded that adult sex reversal always requires an adult gonad that contains opposite sex tissue.

Sparus aurata studied during captive rearing illustrate one pattern seen in protandrous species.[114] At 5 months the future ovarian tissue is located dorsally in the gonad and the future testicular tissue is located ventrally. At 9-12 months, the testicular tissue develops as the ovarian tissue regresses. Between 13 and 16 months, the testis is latent, while the ovary grows and oogenesis occurs. Finally, at 17-24 months, two divergent paths are taken. In 80% of the fish, vitellogenesis proceeds and the testis degenerates. But in 20%, the testis grows again and the ovary degenerates, making sex change a probabilistic event.

All fishes, regardless of sex or reproductive mode, formerly were thought to pass through a juvenile stage of ovarian development.[28] Even if not true for all species, this observation suggests that successive hermaphroditism might have evolved through changes in the developmental timing of originally transitional gonadal intersex conditions.[94] The conceptual advantage of this scenario is that there are known genetic mechanisms for producing changes in developmental timing.

There has been great interest in the possibility that sex steroids cause or induce sex reversal in successive hermaphrodites, but so far the evidence is equivocal.[18] When studied _in vitro_, the gonads of these fish do change their steroidogenic profiles as sex change occurs.[18] As the ricefield eel (_Monopterus albus_) changes from female to male between age 2 and 4 years (the gonads take 3-5 months to reverse), there is an increase in androgen production.[17]

But such results do not distinguish between sex steroids as a cause and as a result of sex change. More direct evidence would come from successful experimental steroid induction of sex reversal. Such induction has been attempted in only a few species, and in most cases the results are difficult to interpret, because the gonadal sex of the fish was not confirmed before treatment.[18] Some protogynous females of _Coris julis_ (rainbow wrasse) treated with testosterone did reverse.[80] _Anthias squammipinnis_ females treated with high dosages of testosterone took on male behavior, color, and size, but at lower (more physiological?) dosages they reversed back to females.[33]

Monopterus albus, a protogynous eel, has been subjected to the most systematic efforts to induce sex reversal with hormones.[17,18] In spite of the fact that androgen production increases during reversal, neither testosterone, methyltestosterone, 11-ketotestosterone, nor androgens combined with a steroidogenetic inhibitor (to suppress possible endogenous inhibitory steroids) induce reversal when given to animals in the female phase. Androgens only facilitate sex change that had already begun. LH is more effective, but even it stimulates Leydig cells without reversing the germ cells. Thus the sex-reversing factor is still unknown. Furthermore, the effect of sex-steroid treatment during embryonic development on the _initial_ sex of successive hermaphrodites does not seem to have been investigated.

In over 16 species of protogynous and protandrous fishes the sex change is influenced or controlled by the social environment. This phenomenon has been studied most extensively in the coral reef fish _A. squammipinnis_.[91]

Groups normally contain one or a very few males with several females. Removal of the male or males causes the behaviorally dominant female to turn into a male. This sex change is triggered by a change in the behavioral interactions of the group members, not by simple removal of male visual cues.[92] The physiological change triggered by these behavioral interactions is not yet known, and as mentioned above, sex steroids cannot mimic this effect.

The discovery of H-Y antigen in fishes raises the possibility that the expression of this antigen might be part of the mechanism of sex change.[18] Recently H-Y antigen was assayed in spleen cells from females and from males that previously had been females. Males had a higher level of H-Y antigen than females, suggesting that sex reversal is similar to sexual differentiation in gonochoristic species (Shapiro, personal communication).

That successive hermaphroditism occurs at all in vertebrates under natural conditions raises some important questions. Why does it only occur in teleosts? The traditional explanation is that these fishes have very simple internal reproductive anatomy that is not highly sexually differentiated and that the teleostean gonad is more bipotential from earliest development. How satisfying this answer is depends on just how important anatomical and physiological constraints really are in limiting evolutionary possibilities. Where do the germ cells come from that function after sex change?[83] Did they differentiate earlier and wait in reserve, or do the germ cells remain bipotential well into adulthood?[17] It would be premature to state definitely as a general vertebrate law that germ cells are only bipotential during embryonic development.

VIII. SUMMARY AND CONCLUSIONS

A. The Nature of the Inducer(s)

The ease with which reproductive structures other than gonads can be sex-reversed by sex-steroid treatment strongly implies that gonadal steroids are the natural inducers during differentiation. The major exception is mullerian duct regression in male amphibians and reptiles, which could be caused by a non-steroidal testicular hormone. A dominant-neutral model, rather than one with two different opposite-sex inducers, seems adequate to account for development, again with the ducts as exceptions. In all cases where the relevant information is available, the gonads have begun steroidogenesis prior to the differentiation of the other reproductive structures, as must happen if gonadal steroids are the natural inducers. The very few experiments on behavioral differentiation also are consistent with the proposal that sex steroids are organizing hormones, a proposal that has been abundantly confirmed for birds and mammals.[42]

The inducers for gonadal differentiation remain more elusive, particularly for reptiles in which steroids never cause sex reversal. Even in amphibians and fishes in which complete reversal can be produced or occurs in adults under natural conditions, it is not clear that steroids are the natural inducers. Yet no non-steroidal inducers have ever been identified.[48] One can rule out pituitary hormones as candidates, for gonadal differentiation proceeds normally in hypophysectomized embryos.[99] H-Y antigen is currently a popular candidate, but to the extent that it is steroid-inducible itself, it may not be the initial causal event. There is no direct evidence that H-Y antigen induces gonadal differentiation in reptiles, amphibians, or fish.

B. How Useful is the Organizational Theory?

Distinct organizational and activational hormone actions are most likely to occur in stable gonochorists and for reproductive characters that exhibit adult sex differences in activation. Thus far the following characters seem to develop in this manner: gonads of amphibians and gonochoristic fish (but bearing in mind that the organizing factors may not be conventional hormones), gonaducts of fishes, amphibians and reptiles, courtship behavior in garter snakes, and possibly aggressive behavior in a mouth-brooding cichlid. There is unlikely to be a true organizational period of hormone action for many secondary sex characters and behavioral patterns in amphibians and fishes, because activational hormones can so easily sex-reverse these endpoints in adulthood. Organizational hormone actions on the nervous system have not been demonstrated directly, but several neural sex differences have been discovered that should be investigated. How successively hermaphroditic fish fit into the organization vs. activation framework is not yet clear. Do they have two organizational periods, or none?

C. Relationship Between Sex Determination and Differentiation

Aspects of vertebrate sexuality such as the type of sex determination, the homogametic sex, if any, the neutral sex (in the sense of a dominant-neutral model), and the sex that can more easily be reversed by steroid treatment (usually the neutral sex) all depend on the class, order, or even family of the animal. The likelihood that some meaningful pattern is contained in these species differences has been explored by several authors[16,34,39,52,99,107] and more recently has been extended to behavioral and neural development.[2,4] (See these last two sources for detailed summaries.) While there are significant exceptions, there is a marked tendency across all vertebrate classes for the homogametic sex to be the neutral sex for both morphological and behavioral development.

This correlation between genetic sex determination and sexual differentiation may come about because the heterogametic sex has the major sex gene(s) that express the inducer for the dominant gonad. Genetic modeling has shown that changes from one mode (e.g., male homogamety) to another (e.g., female homogamety) could easily have occurred during vertebrate evolution.[15] What is more puzzling is why such changes should have occurred, that is, why all gonochoristic vertebrates do not share the same mode. Are there some adaptive advantages to one mode vs. another?

D. Differentiation and Epigenesis

The difficulties involved in identifying the gonadal inducers or understanding successive hermaphroditism should not allow us to overlook an emerging and profound principle. Sexual differentiation in all vertebrates that have been studied systematically, regardless of reproductive mode, is controlled in at least one sex by hormones or hormone-like factors, as opposed to more direct genetic control. This is also true for many invertebrates (some insects are an exception).[2,17] Why is sexual differentiation so epigenetic? While this question cannot be answered empirically, nonetheless it is worth reflecting upon. The advantage may lie in the ease with which environmental sex determination can evolve, and in fact environmental sex determination may have been the original vertebrate condition.[15] The fact that in the laboratory environmental conditions such as temperature can alter sexual differentiation even in species with genetic sex determination is further testimony to the influence of epigenetic factors in this developmental process. An additional advantage might be that fewer genes are required to produce separate males and females when hormones (internal environmental factors) control their differentiation. In current models, as few as four genes can account for mammalian sexual differentiation.[105] Whatever the answer, sexual differentiation remains one of the best known examples of gene-environment interactions in the regulation of vertebrate development.

IX. ACKNOWLEDGMENTS

I thank Andy Bass, Allen Billy, David Crews, Robin Liley, Robert Schmidt, and Norm Stacey for sharing preprints and manuscripts with me.

X. REFERENCES

1. Adkins, E.K. and Schlesinger, L. (1979). Androgens and the social behavior of male and female lizards (*Anolis carolinensis*). Horm. Behav. 13, 139-152.
2. Adkins-Regan, E. (1981). Early organizational effects of hormones: An evolutionary perspective. In "Neuroendocrinology of Reproduction: Physiology and Behavior" (N. T. Adler, ed.), pp. 159-228. Plenum, New York.
3. Adkins-Regan, E. (1985). Mechanisms of sex determination in vertebrates. Sci. Prog. 69, 553-568.
4. Adkins-Regan, E. (1985). Non-mammalian psychosexual differentiation. In "Handbook of Behavioral Neurobiology: Neurobiology of Reproduction" (N. T. Adler, ed.), pp. 43-76. Plenum, New York.
5. Arnold, A.P. (1980). Sexual Differences in the Brain. Am. Sci. 68, 165-174.
6. Austin, C.R., Edwards, R. G., and Mittwoch, U. (1981). Introduction. In "Mechanisms of Sex Differentiation in Animals and Man" (C. R. Austin and R. G. Edwards, eds), pp. 1-54. Academic Press, London.
7. Baerends,G.P. (1957). Behavior: The ethological analysis of fish behavior. In "Physiology of Fishes" Vol II (M. Brown, ed.), pp. 229-269. Academic Press, New York.
8. Bass, A.H., and Hopkins, C.D. (1982). Gonadal steroids modulate sex differences in an electric organ discharge. Neurosci. Abstr. 8, 931.
9. Bass, A.E., and Hopkins, C.D. (1983). Hormonal control of sexual differentiation: Changes in electric organ discharge waveform. Science 220, 971-973
10. Bass, A.H., and Hopkins, C.D. (1985). Hormonal control of sex differences in the electric organ discharge (EOD) of mormyrid fishes, J. Comp. Physiol. 156, 587-604.
11. Becker, P., Roland, H., and Reinboth, R. (1975). An unusual approach to experimental sex inversion in the teleost fish, *Betta* and *Macropodus*, in "Intersexuality in the Animal Kingdom" (R. Reinboth, ed.), pp. 236-242. Springer-Verlag, New York.
12. Billy, A.J., and Crews, D. (1985). The effects of sex steroid treatments on sexual differentiation in a unisexual lizard, *Cnemidophorus uniparens* (Teiidae), unpublished manuscript.
13. Billy, A.J., and Liley, N.R. (1985). The effects of early and late androgen treatments on the behaviour of *Sarotherodon mossambicus* (Pisces:Cichlidae). Horm. Behav. 19, 311-330.
14. Bruslé, S., and Bruslé, J. (1978). Early sex differentiation in *Mugil* (*Liza*) *auratus Risso*, 1810 (Teleost Mugilidae). An ultrastructural study. Ann. Biol. Anim. Bioch. Biophys. 18, 871-875.
15. Bull, J.J. (1983). "Evolution of Sex Determining Mechanisms." Cummings, Menlo Park, California.
16. Burns, R.K. (1961). Role of hormones in the differentiation of sex. In "Sex and Internal Secretions" (W. C. Young, ed.), pp. 76-160. Williams & Wilkins, Baltimore.
17. Chan, S.T.H., and O, W.-S. (1981). Environmental and non-genetic mechanisms in sex determination. In "Mechanisms of Sex Differentiation in Animals and Man" (C.R. Austin and R.G. Edwards, eds.), pp. 55-112. Academic Press, London.
18. Chan, S.T.H., and Yeung, W.S.B. (1983). Sex control and sex reversal in fish under natural conditions. In "Fish Physiology," Vol. IXB (W.S. Hoar, D.J. Randall and E.M. Donaldson, eds.), pp. 171-222. Academic Press, New York.

19. Chang, C.Y., and Witschi, E. (1955). Genic control and hormonal reversal of sex differentiation in Xenopus. Proc. Soc. Exp. Biol. Med. 93, 140-144.
20. Charnov, E.L. (1982). "The Theory of Sex Allocation." Princeton University Press, Princeton.
21. Chieffi, G. (1967). The reproductive system of elasmobranchs: Developmental and endocrinological aspects. In "Sharks, Skates, and Rays" (P.W. Gilbert, R.F. Mathewson and D.P. Rall, eds.), pp. 553-580. The Johns Hopkins Press, Baltimore.
22. Clemens, H.P., McDermitt, C., and Inslee, T. (1966). The effect of feeding methyl testosterone to guppies for sixty days after birth. Copeia 1966, 280.
23. Conover, D.O. (1984). Adaptive significance of temperature dependent sex determination in a fish. Amer. Nat. 123, 297-313.
24. Crews, D. (1980). Interrelationships among ecological, behavioral, and neuroendocrine processes in the reproductive cycle of Anolis carolinensis and other reptiles. In "Advances in the Study of Behavior," Vol. XI (J.S. Rosenblatt, R.A. Hinde, C. Beer and M.-C Busnel, eds.), pp. 1-74. Academic Press, New York.
25. Crews, D. (1984). Gamete production, sex hormone secretion, and mating behavior uncoupled. Horm. Behav. 18, 22-28.
26. Crews, D. (1985). Effects of early sex steroid hormone treatment on courtship behavior and sexual attractivity in the red-sided garter snake, Thamnophis sirtalis parietalis, Physiol. Behav. 35, 569-575.
27. Crews, D., Diamond, M.A., Whittier, J., and Mason, R. (1985). Small male body size in the garter snake depends on the testes. Amer. J. Physiol. 249, R62-R66.
28. Dodd, J.M. (1960). Genetic and environmental aspects of sex determination in cold-blooded vertebrates. Mem. Soc. Endocrinol. 7, 17.
29. Dodd, J.M. (1983). Reproduction in cartilaginous fishes (Chondrichthyes). In "Fish Physiology," Vol. IXA (W.S. Hoar, D.J. Randall and E.M. Donaldson, eds.), pp. 31-95. Academic Press, New York.
30. Donahoe, P.K., Budzik, G.P., Trelstad, R., Mudgett-Hunter, M., Fuller, A., Jr., Hutson, J.M., Ikawa, N., Nayashi, A., and McLaughlin, D. (1982). Mullerian-inhibiting substance: an update. Rec. Prog. Horm. Res. 38, 279-330.
31. Dufaure, J.-P. (1966). Recherches descriptives et expérimentales sur les modalités et facteurs du développement de l'appareil génital chez le lézard vivipare (Lacerta vivipara Jacquin), Arch. Anat. Microsc. Morphol. Exp. 55, 437.
32. Fine, M.L., Economos, D., Radtke, R., and McClung, J.R. (1984). Ontogeny and sexual dimorphism of the sonic motor nucleus in the oyster toadfish. J. Comp. Neurol. 225, 105-110.
33. Fishelson, L. (1975). Ecology and physiology of sex reversal in Anthias squamipinnis (Peters), Teleostei: Anthiidae. In "Intersexuality in the Animal Kingdom (R. Reinboth, ed.), pp. 284-294. Springer-Verlag, New York.
34. Foote, C.L. (1964). Intersexuality in amphibians. In "Intersexuality in Vertebrates Including Man" (C. N. Armstrong and A.J. Marshall, eds.), pp. 233-272. Academic Press, London.
35. Forbes, T.R. (1964). Intersexuality in reptiles. In "Intersexuality in Vertebrates Including Man" (C.N. Armstrong and A.J. Marshall, eds.), pp. 273-284. Academic Press, London.
36. Fostier, A., Jalabert, B., Billard, R., Breton, B., and Zohar, Y. (1983). The gonadal steroids. In "Fish Physiology," Vol. IX A (W.S. Hoar, D.J. Randall and E.M. Donaldson, eds.), pp. 277-372. Academic Press, New York.
37. Gallien, L. (1955). The action of sex hormones on the development of sex in amphibia. Mem. Soc. Endocrinol. 4, 188.

38. Gallien, L.G. (1965). Genetic control of sexual differentiation in vertebrates. In "Organogenesis" (R. L. DeHaan and H. Ursprung, eds.), pp. 583-610. Holt, Rinehart and Winston, New York.
39. Gallien, L. (1967). Developments in sexual organogenesis. In "Advances in Morphogenesis," Vol. VI (M. Abercrombie and J. Brachet, eds.), pp. 259-317. Academic Press, New York.
40. Gorbman, A. (1983). Reproduction in cyclostome fishes and its regulation. In "Fish Physiology," Vol. IX A (W.S. Hoar, D.J. Randall and E.M. Donaldson, eds.),pp. 1-29. Academic Press, New York.
41. Gordon, R.E. (1956). The biology and biodemography of *Anolis carolinensis* Voigt. Unpublished doctoral dissertation, Tulane University.
42. Goy, R.W., and McEwen, B.S. (1980). "Sexual Differentiation of the Brain." MIT Press, Cambridge.
43. Hardisty, M.W., and Taylor, B.J. (1965). The effects of sex hormones on the ammocoete larva. Life Sci. 4, 743-747.
44. Harrington, R.W. (1975). Sex determination and differentiation among uniparental homozygotes of the hermaphroditic fish *Rivulus marmoratus* (Cyprinondontidae: Atheriniformes). In "Intersexuality in the Animal Kingdom" (R. Reinboth, ed.), pp. 249-262. SpringerVerlag, New York.
45. Hishida, T. (1965). Accumulation of estrone-16-C^{14} and diethylstilbestrol--(Monoethyl-1-C^{14}) in larval gonads of the medaka, *Oryzias latipes*, and determination of the minimum dosage of estrogen for sex reversal. Gen. Comp. Endocrinol. 5, 137-144.
46. Hoar, W.S., Randall, D.J., and Donaldson, E.M. (1983). "Fish Physiology," Vol. IX A. Academic Press, New York.
47. Hoar, W.S., Randall, D.J., and Donaldson, E.M. (1983). "Fish Physiology," vol. IX B. Academic Press, New York.
48. Hunter, G.A., and Donaldson, E.M. (1983). Hormonal sex control and its application to fish culture. In "Fish Physiology," Vol. IX B (W. S. Hoar, D. J. Randall, and E. M. Donaldson, eds.), pp. 223-303. Academic Press, New York.
49. Jalabert, B., and Zohar, Y. (1982). Reproductive physiology in cichlid fishes, with particular reference to *Tilapia* and *Sarotherodon*. In "The Biology and Culture of Tilapias." (R.S.V. Pullin and R.H. Lowe-McConnell, eds.), pp. 129-140. International Center for Living Aquatic Resources Management, Manila, Philippines.
50. Jensen, G.L., and Shelton, W.L. (1983). Gonadal differentiation in relation to sex control of grass carp, *Ctenopharyngodon idella* (Pisces: Cyprinidae), Copeia 1983, 749-755.
51. Josso, N. (1981). Differentiation of the genital tract: Stimulators and inhibitors. In "Mechanisms of Sex Differentiation in Animals and Man" (C.R. Austin and R.G. Edwards, eds.), pp. 165-203. Academic Press, London.
52. Jost, A. (1979). Basic sexual trends in the development of vertebrates. In "Sex, Hormones and Behaviour" (Ciba Foundation), pp. 5-13. Excerpta Medica, Amsterdam.
53. Kallman, K.D. (1973). The sex-determining mechanisms of the platyfish, *Xiphophorus maculatus*. In "Genetics and Mutagenesis of Fish" (J.H. Shroder, ed.), pp.19-28. Springer-Verlag, Berlin.
54. Kallman, K.D. (1984). A new look at sex determination in poeciliid fishes. In "Evolutionary Genetics of Fishes" (B.J. Turner, ed.), pp. 95-171. Plenum, New York.
55. Kelley, D.B., and Pfaff, D.W. (1976). Hormone effects on male sex behavior in adult south African clawed frogs, *Xenopus laevis*. Horm. Behav. 7, 159.

56. Laird, L.M., Ellis, A.E., Wilson, A.R., and Holliday, F.G.T. (1978). The development of the gonadal and immune systems in the Atlantic salmon (Salmo salar L.) and a consideration of the possibility of inducing autoimmune destruction of the testis. Ann. Biol. anim. Bioch. Biophys. 18, 1101-1106.
57. Liley, N.R. (1969). Hormones and reproductive behavior in fishes. In "Fish Physiology," vol. 3 (W.S. Hoar and D.J. Randall, eds.), pp. 73-116. Academic Press, New York.
58. Liley, N.R., and Stacey, N.E. (1983). Hormones, pheromones, and reproductive behavior in fish. In "Fish Physiology," Vol. 9B (W.S. Hoar, D.J. Randall, and E.M. Donaldson, eds.), pp. 1-63. Academic Press, New York.
59. Lindsay, W.K. (1974). The effects of neonatal androgen treatment on androgen-induced masculinization in the adult female guppy, Poecilia reticulata, unpublished B.Sc. Thesis, University of British Columbia, Vancouver, B.C., Canada.
60. Lowe, T.P., and Larkin, J.R. (1975). Sex reversal in Betta splendens Regan with emphasis on the problems of sex determination. J. Exp. Zool. 191, 25.
61. McClusky, N.J., and Naftolin, F. (1981). Sexual differentiation of the central nervous system. Science 211, 1924-1303
62. Mason, P., and Adkins, E.K. (1976). Hormones and social behavior in the lizard Anolis carolinensis. Horm. Behav. 7, 75.
63. Müller, U., Zenzes, M.T., and Wolf, U. (1979). Appearance of H-W (H-Y) antigen in the gonads of estradiol sex-reversed male chicken embryos. Nature 280, 142-144.
64. Müller, U., and Wolf, U. (1979). Cross-reactivity to mammalian anti H-Y antiserum in teleostean fish. Differentiation 14, 185-187.
65. Nagahama, Y. (1983). The functional morphology of teleost gonads. In "Fish Physiology," Vol. IX A (W.S. Hoar, D.J. Randall and E.M. Donaldson, eds), pp. 223-275. Academic Press, New York.
66. Narins, P.M., and Capranica, R.R. (1976). Sexual differences in the auditory system of the tree frog Eleutherodactylus coqui. Science 192, 378-380.
67. Ozon, R. (1969). Steroid biosynthesis in larval and embryonic gonads of lower vertebrates. Gen. Comp. Endocrinol. Suppl 2, 135-140.
68. Ozon, R. (1972). Androgens in fishes, amphibians, reptiles, and birds. In "Steroids in Nonmammalian Vertebrates" (D.R. Idler, ed.), pp. 329-389. Academic Press, New York.
69. Ozon, R. (1972). Estrogens in fishes, amphibians, reptiles, and birds. In "Steroids in Nonmammalian Vertebrates" (D.R. Idler, ed.), pp. 390-414. Academic Press, New York.
70. Pechan, P., Wachtel, S.S., and Reinboth, R. (1979). H-Y antigen in the teleost. Differentiation 14, 189-192.
71. Pieau, C. (1974). Différenciation du sexe en fonction de la température chez les embryons d'Emys orbicularis L. (Chélonien), effets des hormones sexuelles. Ann. Embryol. Morphol. 7, 365.
72. Pieau, C. Mignot, Th-M., Dorizzi, M. and Guichard, A. (1982). Gonadal steroid levels in the turtle Emys orbicularis L.: A preliminary study in embryos, hatchlings, and young as a function of the incubation temperature of eggs. Gen. Comp. Endocrinol. 47, 392-398.
73. Price, D.J. (1984). Genetics of sex determination in fishes: A brief review. In "Fish Reproduction: Strategies and Tactics" (G.W. Potts and R.J. Wooton, eds.), pp. 77-89. Academic Press, New York.
74. Price, D., Zaaijer, J.J.P., Ortiz, E., and Brinkmann, A.C. (1975). Current views on embryonic sex differentiation in reptiles, birds and mammals. In "Trends in Comparative Endocrinology" (E.J.W. Barrington, ed.), pp. 173-196. Amer. Zool. 15, Suppl.

75. Pullin, R.S.V., and Lowe-McConnell, R.H. (1982). "The Biology and Culture of Tilapias." International Center for Living Aquatic Resources Management, Manila, Philippines.
76. Rastogi, R.K., and Chieffi, G. (1975). The effects of antiandrogens and antiestrogens in nonmammalian vertebrates. Gen. Comp. Endocrinol. 26, 79-91.
77. Raynaud, A. (1962). Le développement de l'embryon d'Orvet (Anguis fragilis L.) décapité à un stade précoce. C.R. Acad. Sci. [D] (Paris) 255, 3041.
78. Raynaud, A. (1963). Hormones in development. In "Techniques in Endocrine Research" (P. Eckstein and F. Knowles, eds.), pp. 261-288. Academic Press, London.
79. Raynaud, A., and Pieau, C. (1973). Nouvelles observations relatives à l'action de la testostérone sur les conduits génitaux, explantés in vitro, d'embryons de Reptiles. C. R. Acad. Sci. [D] (Paris) 277, 2545.
80. Reinboth, R. (1962). Morphologische und funktionelle Zweigeschlechtlichkeit bei marinen Teleostiern (Serranidae, Sparidae, Centracanthidae, Labridae). Zool. Jb. (Physiol) 69, 405-480.
81. Reinboth, R. (1970). Intersexuality in fishes. Mem. Soc. Endocrinol. 18, 515-543.
82. Reinboth, R. (1972). Some remarks on secondary sex characters, sex, and sexual behavior in teleosts. Gen Comp. Endocrinol. Suppl. 3, 565-570.
83. Reinboth, R. (1982). The problem of sexual bipotentiality as exemplified by teleosts. Reprod. Nutr. Develop. 22, 397-403.
84. Richards, C.M. (1982). The alteration of chromatophore expression by sex hormones in the Kenyan reedfrog, Hyperolius viridiflavus. Gen. Comp. Endocrinol 45, 59-67.
85. Satoh, N., and Egami, N. (1973). Preliminary report on sex differentiation in germ cells of normal and transplanted gonads in the fish, Oryzias latipes. In "Genetics and Mutagenesis of Fish" (J.H. Schroder, ed.), pp. 29-32. Springer-Verlag, Berlin.
86. Schmidt, R.S. (1982). Masculinization of toad pretrigeminal nucleus by androgens. Brain Res. 244, 190-192.
87. Schmidt, R.S. (1983). Neural correlates of frog calling: Masculinization by androgens. Horm. Behav. 17, 94-102.
88. Schreck, C.B. (1974). Hormonal treatment and sex manipulation in fishes. In "Control of Sex in Fishes" (C.B. Schreck, ed.), Virginia Polytechnic Institute, Blacksburg.
89. Segil, N., Silverman, L., Kelley, D., and Rainbow, T. (1983). Androgen binding in the laryngeal muscle of Xenopus laevis: Sex differences and hormonal regulation. Neurosci. Abstr. 9, 1093.
90. Shalev, A., and Huebner, E. (1980). Expression of H-Y antigen in the guppy (Lebistes reticulatus). Differentiation 16, 81-83.
91. Shapiro, D.Y. (1979). Social behavior, group structure, and the control of sex reversal in hermaphroditic fish. In "Advances in the Study of Behavior," Vol. 19 (J.S. Rosenblatt, R.A. Hinde, C. Beer, and M.-C Busnel, eds.), pp. 43-103. Academic Press, New York.
92. Shapiro, D.Y. (1983). Distinguishing behavioral interactions from visual cues as causes of adult sex change in a coral reef fish. Horm. Behav. 17, 424-432.
93. Shapiro, D.Y. (1984). Sex reversal and sociodemographic processes in coral reef fishes. In "Fish Reproduction: Strategies and Tactics" (G.W. Potts and R.J. Wootton, eds.), pp. 103-118. Academic Press, London.
94. Shapiro, D.Y. (1985). "Sexual Differentiation, Social Behavior, and Sex Reversal in Coral Reef Fishes. Bioscience, in press.

95. Smith, C.L. (1975). The evolution of hermaphroditism in fishes. In "Intersexuality in the Animal Kingdom" (Reinboth, ed.), pp. 295-310. Springer-Verlag, New York.
96. Takahashi, H., and Iwasaki, Y. (1973). The occurrence of histochemical activity of 3β-hydroxysteroid dehydrogenase in the developing testes of Poecilia reticulata. Devel. Growth Different. 15, 241-253.
97. Thiebold, J.J. (1964). Contribution à l'étude de l'organogénèse uro-génitale et de son déterminisme chez un poisson élasmobranche: la petite roussete Scyliorhinus caniculus L. Bull. Biol. Fr. Belg. 98, 253-347.
98. van den Hurk, R., Lambert, J.G.D., and Peute, J. (1982). Steroidogenesis in the gonads of rainbow trout fry (Salmo gairdneri) before and after the onset of gonadal sex differentiation. Reprod. Nutr. Develop. 22, 413-425.
99. van Tienhoven, A. (1983). "Reproductive Physiology of Vertebrates," 2nd ed. Comstock, Ithaca, New York.
100. Wachtel, S.S., and Koo, G.C. (1981). H-Y antigen in gonadal differentiation. In "Mechanisms of Sex Differentiation in Animals and Man" (C.R. Austin and R.G. Edwards, eds.), pp. 255-300. Academic Press, London.
101. Wai, E.H., and Hoar, W.S. (1963). The secondary sex characters and reproductive behavior of gonadectomized sticklebacks treated with methyl testosterone. Canadian J. Zool. 41, 611-628.
102. Warner, R.R. (1984). Mating behavior and hermaphroditism in coral reef fishes. Am. Sci. 72, 128-136.
103. Weniger, J.-P., Mack, G., and Holder, F. (1975). L'hormone responsable de la régression des canaux de Muller chez l'embryon de poulet mâle n'est pas un androgène. C.R. Acad. Sci. [D] (Paris) 280, 1889-1891.
104. Wiberg, U. (1982). Serological cross-reactivity to rat anti H-Y antiserum in the female European eel (Anguilla anguilla). Differentiation 21, 206-208.
105. Wiberg, U., Mayerová, A., Müller, U., Fredga K., and Wolf, U. (1982). X-linked genes of the H-Y antigen system in the wood lemming (Myopus schisticolor). Hum. Genet. 60, 63-166.
106. Wilson, J.D., George, F. W., and Griffin, J.E. (1981). The hormonal control of sexual development. Science 211, 1278-1284.
107. Witschi, E. (1959). Age of sex-determining mechanisms in vertebrates. Science 130, 372.
108. Witschi, E. (1967). Biochemistry of sex differentiation in vertebrate embryos. In "The Biochemistry of Animal Development," Vol. II (R. Weber, ed.), pp. 193-223. Academic Press, New York.
109. Witschi, E. (1971). Mechanisms of sexual differentiation. In "Hormones in Development" (M. Hamburgh and E.J.W. Barrington, eds.), pp. 601-618. Appleton-Century-Crofts, New York.
110. Wolff, E. (1959). Endocrine function of the gonad in developing vertebrates. In "Comparative Endocrinology" (A. Gorbman, ed.), pp. 568-581. Wiley and Sons, New York.
111. Yamamoto, T.-O. (1969). Sex differentiation. In "Fish Physiology" (W.S. Hoar and D.J. Randall, eds.), pp. 117-163. Academic Press, New York.
112. Yntema, C.L. (1979). Temperature levels and periods of sex determination during incubation of eggs of Chelydra serpentina. J. Morphology 159, 17027.
113. Zaborski, P., Dorizzi, M., and Pieau, C. (1982). H-Y antigen expression in temperature sex-reversed turtles (Emys orbicularis). Differentiation 22, 73-78.

114. Zohar, Y., Abraham, M., and Gordin, H. (1978). The gonadal cycle of the captivity-reared hermaphroditic teleost *Sparus aurata* (L.) during the first two years of life. Ann. Biol. Anim. Bioch. Biophys. 18, 877-882.

GONADOTROPIN-RELEASING HORMONES IN FISHES

Nancy Sherwood

Department of Biology
University of Victoria
Victoria BC, Canada V8W 2Y2

Most vertebrates appear to synthesize in their brains a factor(s) which controls certain aspects of reproduction. Of the vertebrates studied to date, only the primitive hagfishes appear to lack this factor, gonadotropin-releasing hormone (GnRH). The most studied, but not the sole, function of GnRH is stimulation of the synthesis and release of pituitary gonadotropin(s) which subsequently affects the gonads. In this role GnRH is released into the brain-pituitary portal vessels in tetrapods, but in fishes GnRH is released in various ways from nerve terminals: near the pituitary gonadotropic cells, near connective tissue layers proximal to the anterior pituitary, or into a primitive portal system (see Ball[5]).

The primary structure of four GnRH molecules is known. In each case, GnRH is a peptide with 10 amino acids and common N and C terminal residues. Considerable homology exists among these molecules, suggesting the GnRH peptides in different vertebrates are members of a phylogenetic family.

I. THE PRIMARY STRUCTURE OF KNOWN GnRH FAMILY MEMBERS

A. Overview of Mammals and Birds

The structure of mammalian GnRH is shown in Table 1. Originally the structure was determined in both pigs and sheep;[16,68] the molecules, isolated from hypothalamic tissue, were the same in the two mammals. Later, the structure of GnRH in human placenta was found to be identical to that in pigs and sheep.[110] The sequence of GnRH in the rat brain is not determined, but the amino acid composition matches that in other mammals.[8] Miyamoto and coworkers[76] state that more than one form of

Table 1. Known Peptides of GnRH Family

	1	2	3	4	5	6	7	8	9	10
Piscine GnRH (salmon)	pGlu-	His-	Trp-	Ser-	Tyr-	Gly-	Trp-	Leu-	Pro-	Gly-NH_2
Avian GnRH I. (chicken)	pGlu-	His-	Trp-	Ser-	Tyr-	Gly-	Leu-	Gln-	Pro-	Gly-NH_2
II.	pGlu-	His-	Trp-	Ser-	His-	Gly-	Trp-	Tyr-	Pro-	Gly-NH_2
Mammalian GnRH II. (pig, sheep, human)	pGlu-	His-	Trp-	Ser-	Tyr-	Gly-	Leu-	Arg-	Pro-	Gly-NH_2

GnRH exists in mammals, but the evidence has not yet been reported. Synthetic forms of mammalian GnRH have been used extensively to study the role of GnRH in a variety of vertebrates.

The primary structure of GnRH in birds is known only for the chicken. Table 1 shows the two forms of GnRH which have been isolated from the chicken brain.[61,62,74-76] Chicken-I GnRH, which differs from the mammalian molecule by only one amino acid in position 8, has little ability to release gonadotropins (Gn) or bind to mammalian anterior pituitary cells in culture; the binding and potency of chicken-I GnRH is less than 5% compared with synthetic mammalian GnRH.[51,70,71,73,76] However, chicken-II GnRH has greater potency in rat pituitary cells; the chicken molecule released 32% LH and 41% FSH compared with mammalian GnRH.[76] The chicken-II peptide differs from the mammalian GnRH in three amino acid positions (5, 7 and 8) but is still clearly a member of the same family. Other functional differences in the two forms of chicken GnRH remain to be elucidated. It is not known if chicken-I and chicken-II GnRH are also present in other birds.

A difference in the mammalian and avian GnRH receptors may account for the potency effects. Milton and coworkers[73] have suggested that mammalian GnRH receptors are more "stringent" in that binding was strongest with the endogenous mammalian molecule; chicken GnRH receptors were termed "promiscuous" in that chicken-I and mammalian GnRH were equipotent in releasing LH from cultured chicken pituitary cells.[70,71,73] However, the difference in potency is less dramatic in other studies: chicken-I was 2.7 times more potent compared with mammalian GnRH in releasing LH from cultured chicken pituitary cells;[51] mammalian GnRH was only three times more potent compared with chicken-II in releasing LH from cultured mammalian pituitary cells.[76]

B. Fishes

To date, the primary structure of only one piscine GnRH molecule has been published.[98] Salmon-I GnRH differs compared with mammalian GnRH in amino acids 7 and 8 (Table 1). Again, the salmon form is clearly a family member; the length of the molecule, termini and 80% of the amino acid sequence are identical to the mammalian peptide. The native form is biologically active in salmon as detailed below (Section IV.C). In addition, immunological and chromatographic evidence suggests that a second. form of GnRH exists in salmon (Fig. 1). Cross-reactivity studies suggest that salmon-II GnRH is the same length as salmon-I, and both differ from mammalian GnRH in the C-terminal residues. Chromatographic studies (high pressure liquid chromatography, HPLC) show the salmon-II molecule is more hydrophobic than mammalian, but less than salmon-I, GnRH. Functional differences between salmon-I and II GnRH can be studied more easily when

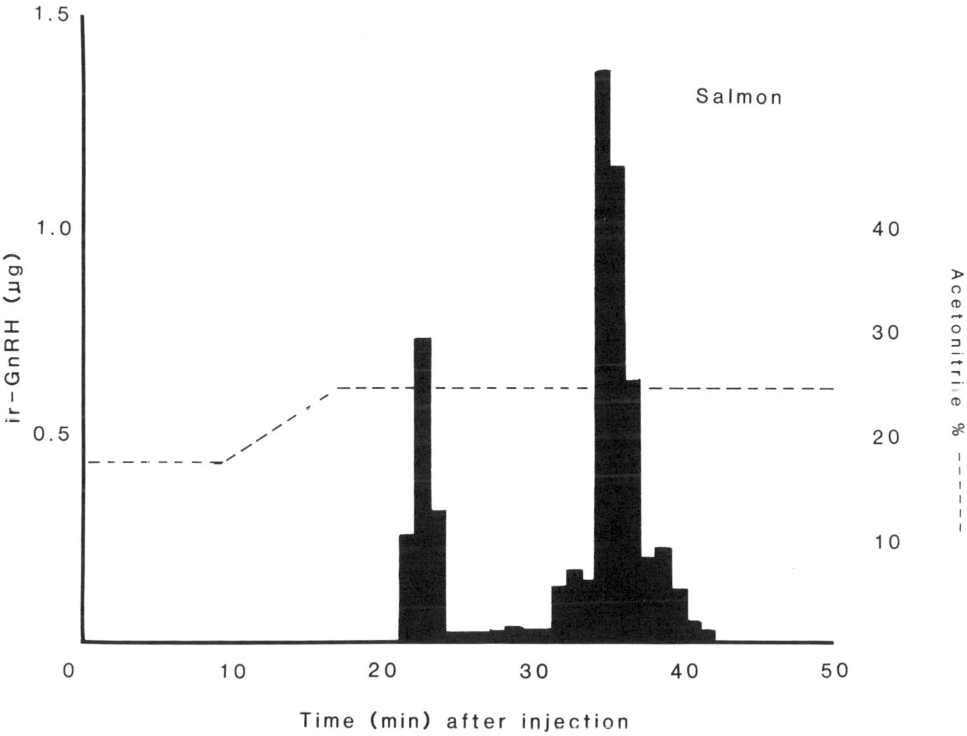

Fig. 1. HPLC of salmon brain fractions containing immunoreactive (ir) gonadotropin-releasing hormone (GnRH). Flow rate was 1 ml/min, and 1-ml fractions were collected. The amount of GnRH for each fraction was determined by RIA with antibody R-42. The mobile phase was composed of an increasing percentage of acetonitrile (----) diluted with triethylammonium phosphate at pH 6.5. Salmon II GnRH eluted at fractions 22-24; salmon I GnRH eluted at fractions 35-37.

synthetic salmon-II becomes available. A recent report shows that cod brains may have two forms: the first behaves like salmon-I GnRH; the second elutes early from a Sephadex G-25 column.[118] Another form of fish GnRH, distinguished by its HPLC elution pattern, has been reported for tilapia and winter flounder (see Crim[27]).

C. Amphibians and Reptiles

The primary structure of GnRH in amphibian and reptilian brains has not been determined. However, considerable evidence supports the presence of GnRH in the brains of both classes. The dominant form of GnRH in the adult bullfrog (*Rana catesbeiana*) has the same amino acid composition as mammalian GnRH.[91]

Although mammalian GnRH is the dominant form in the adult amphibian brain, small amounts of the two salmon-like forms of GnRH have been indirectly demonstrated by immunological and chromatographic methods.[100] This pattern was evident in both newts (*Ambystoma gracile*; *Taricha granulosa*) and frogs (*Hyla regilla* and *Rana pipiens*). Branton and associates[10] found that tadpole brains contain a fish-like GnRH molecule, whereas the adult brain (*Rana catesbeiana*) contains in addition the mammalian form of GnRH. However, in larval and neotenic newts (*Ambystoma gracile*), both salmon-like and mammalian-like GnRH are present.[100] Certain sympathetic ganglia in amphibians also contain only the fish-like GnRH molecule.[43]

In reptiles, certain antibodies cross-react with GnRH-like molecules in the brains of snakes, turtles, tortoises and lizards.[59,60,77,80,81] The primary structure of the reptilian GnRH molecules has not been reported. However, Millar (personal communication, 1984) found salmon-like GnRH in the brain of a South African lizard and identified chicken-I-like GnRH in the alligator brain.

II. CHARACTERIZATION OF NEW GnRH MOLECULES IN FISHES

It is now clear that salmon-I GnRH is not the only form in fishes. Although the primary structure is not known, indirect evidence from chromatographic and immunological studies suggests that at least lamprey, dogfish, ratfish and sturgeon have distinct GnRH molecules compared with salmon-I GnRH. Another distinct GnRH molecule may occur in certain teleosts (tilapia and winter flounder; see Crim[27]).

A. Agnatha

The only living representatives of the class Agnatha are hagfishes and lampreys. Most cross-reactivity studies have not revealed a GnRH-like molecule in hagfish brains. Workers did not detect GnRH activity in hagfish brains using immunocytochemistry,[25,26,80,81] although some have detected picogram amounts in whole brains using radioimmunoassay (RIA).[55,60] We used six antisera directed against different parts of the mammalian GnRH molecule, but did not find detectable levels by RIA in *Eptatretus stouti*.[97] One of the antisera, R-42, detects GnRH in all other vertebrates studied to date. Several possibilities exist. Hagfishes may have a GnRH molecule which is fundamentally different from the other vertebrates, or hagfishes may have lost their GnRH during degeneration of the brain. Gorbman[48] argues that, for certain hagfishes (e.g., *Eptatretus stouti*), environmental conditions of constant temperature and darkness may alleviate the need for a neural stimulus to coordinate the reproductive cycle with seasons. Finally, hagfishes may never have evolved a GnRH molecule; they continue to release eggs for months after hypophysectomy.[69]

The lamprey (*Petromyzon marinus*), on the other hand, has two distinct GnRH molecules. One is dominant; the amount of the other form is only 2.5% of the first. Lamprey-I GnRH is clearly a family member with the same length and termini compared with the other four molecules. It has

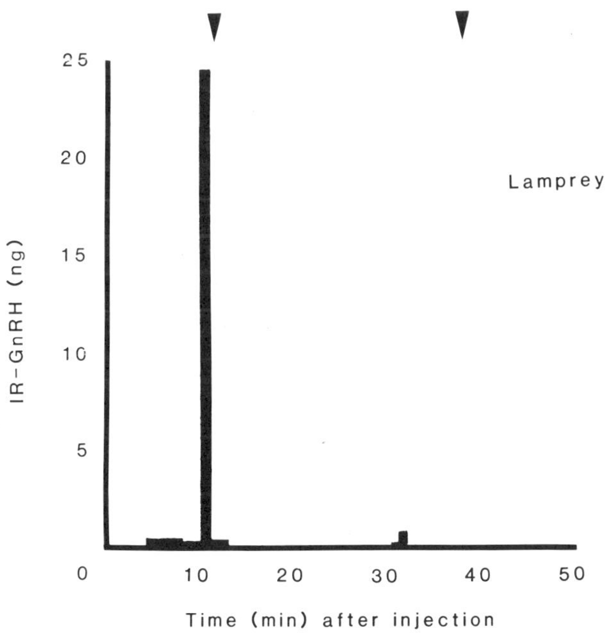

Fig. 2. HPLC of lamprey (*Petromyzon marinus*) brain fractions containing immunoreactive (ir) GnRH. The other details are the same as in Fig. 1. Lamprey I GnRH eluted at fraction 11; lamprey II GnRH eluted at fraction 32. Reproduced from *Neuropeptides* 6:205-214, 1985 with permission of journal and Churchill Livingstone.

the same hydrophobicity on a reverse phase C-18 column as mammalian GnRH, but not necessarily the same amino acid composition (Figure 2); lamprey-II GnRH is more hydrophobic.[97] This suggests the GnRH family of molecules extends throughout the vertebrate kingdom with the exception of hagfishes.

B. Chondrichthyes

Dogfish sharks (*Squalus acanthias*) appear to have a dominate GnRH-like peak with a small quantity of a second form; ratfish (*Hydrolagus colliei*) has a single peak that elutes at the same time compared with the dominant dogfish peak from a C-18 HPLC column (Figure 3; Sherwood and Carolsfeld, unpublished). Likewise, the dogfish and ratfish GnRH immunological profile is the same: positive for antiserum R-42, negative for B-6. The latter is directed against the C-terminal region of the mammalian molecule. Deery[35] also noted earlier that a mammalian antiserum did not detect dogfish GnRH. The primary structure is not known, but the shark and ratfish GnRH material also elute at the same time as salmon-II GnRH. However, the identity of the three molecules has not been established directly.

C. Osteichthyes

1. *Non-teleosts*. Ball[5] states that "it is probably correct to interpret the ancestors of teleosts as having been related remotely to sturgeons and more closely to *Amia* and *Lepisostesus*." Unlike teleosts, these ganoid fishes have a portal blood system between the hypothalamus and pituitary, although the system is less developed than in tetrapods. Sturgeons also are known to secrete gonadotropin.[46,53]

The presence of a unique form of GnRH in sturgeon (*Acipenser transmontanus*) has been observed (Sherwood, Carolsfeld & Doroshoff, unpublished). Sturgeon GnRH is detected by mammalian antisera B-6 and R-42, unlike the negative response to B-6 in other fishes (hagfish, lamprey, goldfish, salmon, trout, herring, mullet, milkfish, siganids and sea bass). Sturgeon GnRH is also chromatographically distinct from synthetic mammalian, chicken-I and II and salmon-I GnRH (Figure 3).

2. *Teleosts*. Herring (*Clupea harengus pallasi*) are teleosts, but not neoteleosts. However, the pattern of immunoreactive-GnRH material is similar to that of salmon and trout: two forms of GnRH can be separated with HPLC; elution times and cross-reactivities are the same (Sherwood and Carolsfeld, unpublished). Likewise milkfish (*Chanos chanos*), mullet (*Mugil cephalus*), trout (*Salmo gairdneri*),[99] goldfish (*Carassius*

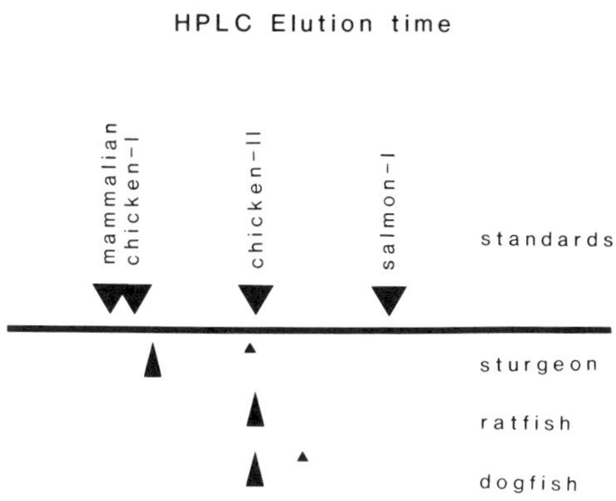

Cross-reactivity

Antisera	mammalian	chicken-I	chicken-II	salmon-I	sturgeon	ratfish	dogfish
R-42	+	+	+	+	+	+	+
B6	+	O	O	O	+	O	O

Fig. 3. Top. Diagram comparing elution times from high pressure liquid chromatography (HPLC) for the synthetic GnRH peptides on the top of the line and for the immunoreactive GnRH-like peaks from three fish shown below the line. Bottom. Cross-reactivity of mammalian, avian and piscine forms of GnRH with two antisera, R-42 and B-6. R-42 is a conformational antiserum which requires the termini and folding of GnRH, but allows conservative amino acid substitutions within the molecule. B-6 is directed toward the C-terminal region of mammalian GnRH and does not recognize GnRH molecules with substitutions in this region. Data from unpublished studies by Sherwood, Carolsfeld and Doroshov.

auratus),[96] siganid (*Siganus guttatus*) and sea bass (*Lates calcarifer*) (Sherwood, Nacario, Crim & Harvey, unpublished) have two forms of GnRH which have the same chromatographic and immunological profile as salmon-I and II. This indirect evidence suggests the eight teleosts tested to date have the same forms of GnRH. The second form of teleostean (or Salmon II) GnRH-like material also elutes with chicken-II GnRH, but no direct evidence establishes their identity. Tilapia and winter flounder are reported to have a distinct form of GnRH (see Crim[27]), but this form needs to be compared in the same HPLC system with the other teleosts listed above.

III. ANATOMICAL LOCATION OF GnRH IN FISH BRAIN

Immunocytochemistry is the main method used to locate GnRH cell bodies in fish brains. Success depends on using an antibody that is directed against the native GnRH molecule of the fish. To date most studies in fishes have used antisera made against mammalian GnRH. These antisera are likely to detect fish GnRH under two conditions: (1) the antiserum is a sequential type directed against a region of the GnRH molecule shared by fish and mammals or (2) the antiserum is a conformational-type directed against the common three-dimensional configuration. In the latter case, conservative amino acid substitutions can occur without interfering with detection.[23]

A. Agnatha

In the hagfish *E. stouti*, staining for immunoreactive (ir) GnRH cells was absent in all parts of the central nervous system.[25,26] Negative staining also occurred in *E. burgeri*.[80] However, ir-GnRH was detected by RIA in *Heptatretus hexatrema*[60] and in very low quantities in *E. stouti*.[55] None was found in whole brains of *E. stouti* by RIA.[97]

In lampreys, GnRH immunoreactive cells stained intensely in reproductively mature individuals (*Lampetra richardsoni* and *Entosphenus tridentata*,[25,26] *Entosphenus japonica*,[80] and *Petromyzon marinus*.[81,82] These neurons were located mainly in the preoptic nucleus, with processes extending to the CSF or to the infundibulum and neurohypophysis.[25,26,80] GnRH may reach the pars distalis by diffusion across the connective tissue layer. This anatomical arrangement seems inefficient compared with either the hypothalamo-hypophysial portal system or direct innervation of the anterior pituitary by GnRH axons. However, the transport of lamprey GnRH in the systemic circulation seems unlikely, as the concentration in the brain is at the same low level found in other vertebrates.[97]

B. Chondrichthyes

In both ratfish (*H. colliei*) and dogfish shark (*S. acanthias*), the form of GnRH is chromatographically distinct from mammalian and salmon-I GnRH; the elasmobranch GnRH-like molecule is not detected by all antisera made against mammalian GnRH (see section II.B.; Sherwood[95]). The distinctness of elasmobranch GnRH may explain the variability of detection in different studies. Some workers did not detect GnRH by RIA in other sharks (*Scyliorhinus canicula*,[35] *Triakis scyllia*[80]). Later Nozaki and associates[82] did find GnRH in *Triakis* brains using immunocytochemistry with any of three GnRH antisera. Others found low levels of GnRH-like material in mature female sharks (*S. acanthias*,[55] *Poroderma africanum*[59,60]) and ratfish.[55] We found substantial amounts of immunoreactive GnRH (2-6 ng/brain) in *S. acanthias* and *H. colliei* brains using a conformational-type antiserum (Sherwood & Carolsfeld, unpublished). Immunocytochemistry has revealed GnRH in some of the cell bodies of the dogfish terminal nerve, an olfactory-related system. The GnRH cells were concentrated near the olfactory epithelium; fibers projected rostrally toward, but were not shown to terminate in, the epithelium. The GnRH fibers also projected caudally to terminate in medial telencephalic nuclei. Dogfish terminal nerve GnRH, unlike that in teleosts, was neither co-localized with FMRFamine nor present in a major projection to the retina;[107] (Stell, personal communication). Another unusual feature in sharks (*T. scyllia*) was that GnRH fibers did not terminate in the median eminence.[82] But location of the gonadotropes in the ventral lobe of the pituitary, far removed from the median eminence, suggests an alternate pathway for GnRH release may occur in elasmobranchs.

C. Osteichthyes

In nonteleosts the bowfin fish is considered to be "mainline" in terms of evolution; this fish (*Amia calva*) contains GnRH fibers in the median eminence and beneath the preoptic recess.[24]

GnRH also has been detected by immunohistochemistry in the brains of a number of teleosts. Four populations of GnRH (LHRH) neurons have been identified in the brain: (1) the olfactory bulbs or transition area of the olfactory bulbs and ventral telencephalon, (2) preoptic region, (3) diencephalon including the hypothalamus and (4) midbrain.[49,77,81,82,92]

Munz and coworkers[77,78] found that in certain teleosts about 50-100 immunoreactive GnRH neurons were located in the olfactory-telencephalic area (platyfish, *Xiphophorus maculatus*; poecilid fish, *Xiphophorus helleri*; hybrids, *Xiphophorus sp*). In other teleosts the GnRH neurons were in the caudal part of the olfactory bulbs. These fishes include cichlids (*Cichlasoma biocellatum*) and centrarchids (*Lepomis machrochirus*). The authors did not find ir-GnRH neurons in a similar location in goldfish (*C. auratus*) or in catfish (*Ictalurus punctatus*; *Kryptopterus bicirrhis*). Subsequently, Stell and colleagues[107] were able to demonstrate these GnRH-positive neurons in goldfish. Similarly, the presence of GnRH in the rostral and caudal olfactory bulb and in the ventral telencephalon was reported for eel (*Anguilla japonica*);[81,82] twenty-one percent of total brain ir-GnRH was present in the telencephalon of eel (*Anguilla anguilla*).[42] It was also shown that the olfactory-telencephalic neurons are the first of the GnRH cells to develop in the brain.[49]

The olfactory-related GnRH neurons were originally termed the nucleus olfactoretinalis;[77,78] later, these GnRH neurons and fibers were shown to be identical to the terminal nerve.[107] The processes of these GnRH neurons project to the retina, olfactory bulb, telencephalon and optic tectum.[78,107] The function of the terminal nerve is not clear, but it has been suggested that sex-related olfactory stimuli, possibly pheromonal, may act through the GnRH-containing terminal nerve to affect the retina and other neural centers; sexual and reproductive behavior may be affected. For example, antidromic stimulation of the terminal nerve in a location distant from the olfactory tract activates sperm release in goldfish.[37]

A second cluster of about 20 LHRH-containing cell bodies were located in the diencephalon, lateral to the preoptic nucleus (basolateral preoptic nucleus).[57,77] Their axons extended toward the ventricle, anterior commissure, hypothalamus and pituitary. The location of this second clump of ir-GnRH cells also has been described as the periventricular preoptic nucleus; it receives GnRH-staining fibers from the olfactoretinal nucleus.[49]

Another group of GnRH cells was present in the hypothalamus, in the posterior part of the lateral tuberal nucleus (NLT);[49,57,92] "processes from the NLT form a network without apparent direction."[49] In addition, GnRH cells are in the thalamus of the three-spined stickleback, *Gasterosteus aculeatus* L.[9]

The fourth population of GnRH neurons was in the "dorsal midbrain, caudal to the posterior commissure and rostral to the nucleus of the third cranial nerve"; their fibers extend toward the third cranial nerve.[77] Although the cell body of origin was not always identified, GnRH processes were present in the retina, olfactory bulb, telencephalon, dorsal diencephalon, optic tectum, torus semicircularis, midbrain, cerebellum and medulla oblongata.[57,77]

In several teleosts, GnRH cell bodies or fibers were not stained in the brain.[35,47,80] However, it seems likely a different antiserum or reproductive stage might produce positive results. For example, subsequent to the above studies, trout[47] and goldfish[57,107] were shown to have GnRH-stained cells.

IV. REPRODUCTIVE RESPONSES TO GnRH

GnRH appears to be important in controlling reproduction in fishes, with the possible exception of hagfishes. The synthetic form of any of the four GnRH family members can produce a reproductive response, at least in the fish tested, but the analogs are considerably more effective. Positive reproductive responses to GnRH appear to depend on the correct reproductive stage at the time of treatment and the proper form of GnRH or analog given at the correct dose, acting for a sufficient time. The effect of GnRH has been studied primarily by assaying gonadotropin or steroid levels in the blood or by observation of ovulation, spawning or spermiation.

A. Agnatha

1. Gonadotropins (Gns). In agnathans, response to GnRH has been judged by steroidal and ovulatory changes because Gn(s) in lampreys and hagfishes have neither been identified nor detected in a heterologous RIA.

2. Steroids. An unusual pattern of steroids exists in the two sexes of lampreys compared with other vertebrates. Estradiol, estrone and androstenedione plasma levels are similar in male and female adult sea lampreys (*P. marinus*); testosterone and 5α-dihydrotestosterone are very low or not detectable in the sexes.[58] An injection of a mammalian GnRH analog (10 μg; D-Ala^{6}, des Gly^{10} GnRH ethylamide) in both male and female sea lampreys resulted in a significant elevation in plasma estradiol concentration. Plasma androgens were not affected in females and were elevated in males at 24 h, but not 8 or 48 h, after injection of the analog.[104] Thus, the estradiol response in both sexes during gonadal maturation is similar, but in general the physiological function of the sex steroids in lampreys is not clear.

3. Ovulation. Hypothalamic control over reproduction in lampreys was shown by injection of a mammalian GnRH analog; ovulation was advanced and steroidogenesis stimulated. A single or double injection of the analog (50 μg/kg) resulted in 56-64% ovulation in treated lampreys compared with 18% in controls on the 12th day after injection.[103]

B. Chondrichthyes

1. Gonadotropins. In dogfish (*S. canicula*) Gn is secreted primarily (98.8%) from the ventral lobe of the pituitary and has been partially purified.[108,109] It is not yet clear if only one Gn hormone is present in dogfish, but the material is biologically active and distinct from the Gns identified to date in teleosts.[65,108,109] The location of Gn in the ventral lobe of the pituitary is unusual in that the lobe is vascularized by the internal carotid arteries rather than by the hypothalamic portal vessels, which terminate in the medial lobe.[5,38] Synthetic mammalian GnRH injections in dogfish resulted in elevated plasma Gn, measured by a heterologous RIA.[38]

2. Steroids. Injection of synthetic mammalian GnRH induced a significant increase in androgen plasma levels in male dogfish (*S. canicula* L); estradiol was not detectable. In female dogfish, both estradiol and androgen increased after injection of mammalian GnRH in November or of dogfish hypothalamic extract in July; only plasma estradiol levels increased after an injection in February.[56]

3. Ovulation. In ratfish (*H. colliei*), an injection of 20 μg of mammalian GnRH resulted in ovulation (in Jenkins & Dodd[56]). An injection of mammalian GnRH in November caused all the female dogfish to release eggs. No response occurred in February. Also, dogfish hypothalamic extract resulted in a greater number of female fish ovulating within 72 h of injection compared with controls.[56]

C. Osteichthyes: Nonteleosts

1. Gonadotropin. All ganoids have at least a primitive median eminence-portal complex similar to the tetrapods. Sturgeons secrete Gn,[46,53] and it appears to be a single form.[18] The Gn purified from sturgeon (*Acipenser stellatus* Pallas) shows homology in amino acid composition of α and β subunits compared with mammalian Gns.[17] An injection of synthetic mammalian GnRH resulted in an increase in serum Gn; an initial rise occurred at 30 min and "was followed by a more significant rise lasting 4-15 hours."[6]

2. Steroids. In untreated sturgeon estradiol was elevated at vitellogenesis, but was low after spawning; GnRH-treated sturgeon were not studied for steroidal response.[6]

3. Ovulation. Ovulation in sturgeons was induced by treatment with synthetic mammalian GnRH.[6]

D. Osteichthyes: Teleosts

Many studies now confirm that synthetic GnRH is effective in causing a reproductive response in teleosts. Most of the testing has been with mammalian GnRH or its analogs; salmon GnRH, Trp^7-Leu^8 GnRH, is a newcomer and has not been tested extensively. However, analogs of either mammalian or salmon GnRH are more effective than the synthetic native form of GnRH.

1. Gonadotropins. In fishes as in other vertebrates, the pituitary is thought to be the main target for GnRH in reproduction. Gn released

from the pituitary into the blood is often used as a measure of the effectiveness of GnRH. Gns have been partially characterized from several teleosts: salmonids, carp, tilapia, winter flounder, American plaice and pike eel (see review[17,54]). Of the piscine Gns identified to date, none are identical with those in other vertebrates. However, amino acid composition of the α and β subunits of Gns purified from carp (*Cyprinus carpio*) show homology with α and β subunits of mammalian Gns. In the parts of the amino acid sequence of carp α and β subunits which are known, subunit is more closely related to mammalian LHβ than to FSHβ.[17] The number of Gns in a species is not yet clear. A single "carbohydrate-rich" Gn has been identified in the teleosts listed above; a second "carbohydrate-poor" Gn has been studied in salmon, carp, winter flounder and American plaice.[54] Some reports suggest a single Gn stimulates the entire process of gonadal development; others suggest "maturational" and "vitellogenic" functions are present in two Gn molecules (see[17,54]).

The measurement of blood Gn levels in response to GnRH is usually by RIA. However, several problems exist in Gn assays. The distinctness of Gn in certain teleostean species means that a purified form of each Gn is needed for antiserum preparation. Heterologous assays may not give an accurate answer. Also, circulating Gn may be neither the pituitary form nor the active form; and measured Gn may be only one of multiple forms. Given these problems, a number of studies have nonetheless reported a rise in Gn after GnRH administration.

Hypothalamic extracts have been used to demonstrate that a gonadotropic factor in the brain releases Gn from pituitaries. The advantage of extracts is that the correct native form(s) of GnRH is administered, although other substances in the extract may change the response. Injections of carp hypothalamic extract into mature carp (*Cyprinus carpio* L.) increased the plasma level of Gn.[11,115] Such extracts also released Gn from carp pituitaries *in vitro*.[12,14] Further, carp or trout hypothalamic extracts stimulated release of LH from sheep pituitaries *in vitro*.[13] In goldfish, another member of the carp family, injection of an extract of hypothalamus induced an increase in serum Gn 20 minutes later;[32] injection of plaice, winter flounder, salmon or trout hypothalamic extracts stimulated Gn release *in vitro* from pituitaries of testosterone-primed juvenile rainbow trout (*Salmo gairdneri*).[30] Hence the native, but unidentified, factors in brain extracts released Gn.

In 1983, two forms of salmon GnRH were noted by immunological and HPLC chromatographic studies; one of the forms was isolated and identified.[95,98,99] The synthetic salmon-I GnRH and its analogs were tested in salmon, goldfish and trout. In salmon, a single injection of salmon-I

GnRH increased Gn levels.[41]. In goldfish, Peter and coworkers[89] found that salmon-I, chicken-I and mammalian GnRH were similarly effective in stimulating Gn release. In perifused goldfish pituitary fragments, salmon GnRH was more potent compared with mammalian GnRH or its analog if increasing doses were tested; the three peptides were equipotent if decreasing doses were used.[67] Plasma Gn also was elevated by salmon-I, chicken-I or mammalian GnRH in juvenile trout primed with testosterone. However, certain salmon or mammalian "superactive" GnRH analogs were about 10-100 times more effective in releasing Gn compared with native salmon, mammalian or chicken-I GnRH.[28] The three analogs that produced the longest Gn response after a single injection were (1) D-Arg^{6}, des Gly^{10} salmon GnRH ethylamide, (2) D-Ala^{6}, des Gly^{10} mammalian GnRH ethylamide and (3) D-h Arg$(Et_2)^{6}$, des Gly^{10} mammalian GnRH ethylamide. The same order of potency applied for Gn release in adult male landlocked salmon. A variety of other salmon, mammalian or chicken GnRH analogs were less effective in males and females in releasing Gn.[28] In salmon, a single injection of one of two salmon or two mammalian analogs also produced a more prolonged rise of Gn. D-Arg^{6}, des Gly^{10} salmon GnRH ethylamide was also the most potent salmon analog in producing increases in serum Gn in goldfish.[89] Donaldson and colleagues[41] concluded that "teleost GnRH analogues are similar in potency to mammalian GnRH analogues under the conditions described."

Before the structure of salmon GnRH was known, a number of studies demonstrated that native synthetic mammalian GnRH was effective in releasing GnRH in some fishes under certain conditions, but mammalian GnRH analogs were effective in most studies. Native mammalian GnRH elevated plasma levels of Gn in Coho salmon (*Oncorhynchus kisutch*),[111] carp (*C. carpio*),[11,15,115,117] brown trout (*Salmo trutta*)[29] and in goldfish.[32] The rise occurred 15-30 min after injection. In trout, increased plasma Gn lasted at least 6 h.[29] In carp, elevated Gn remained high for 5 h before it fell, but two injections, 3 h apart were used.[117] In carp, a single injection of GnRH produced a rapid rise and fall of Gn.[11,15] Also, synthetic mammalian GnRH released Gn *in vitro* from pituitaries of testosterone-primed juvenile rainbow trout[30] or from organ-cultured, steroid-treated pituitaries of immature trout.[45]

Mammalian GnRH analogs produce a more consistent and prolonged elevation of Gn at lower doses compared with native GnRH. A single injection of a mammalian GnRH analog elevated plasma Gn in Coho salmon,[111] male goldfish[83] and carp.[15] Multiple injections or slow-release preparations of mammalian GnRH analogs produced a Gn release of greater magnitude or prolonged duration in salmon,[113] carp[15] and male goldfish.[83] A sustained (10-11 days) rise in Gn followed treatment of male Atlantic salmon (*Salmo salar*) with a mammalian GnRH analog given in several ways: multiple injections in saline or propylene glycol or implantation in a pellet made of silicone rubber or cholesterol.[114] Likewise in female trout, a GnRH analog was administered embedded in a pellet; Gn levels were elevated two weeks later in a dose-dependent way and remained far above the controls for at least two months.[34] The short half-life (under 15 min) of mammalian GnRH in fishes may account for the need to use multiple injections or prolonged release preparations.[96]

A true priming effect of GnRH occurs when two equal doses result in greater Gn release after the second dose compared with the first. If Gn levels rise and fall quickly as in the carp study,[15] priming like that seen in rats[3] is demonstrated. The prolonged Gn response in some studies makes it more difficult to demonstrate priming, but two injections 12 h apart were more effective than one injection of the same total dose in goldfish;[83] priming also may have occurred in brown trout[29] and male Atlantic salmon.[114]

In fishes, the Gn response to GnRH is unusual compared with mammals. Prolonged administration of GnRH in mammals is suppressive[90] unless it is given in a pulsatile fashion.[7] Even then, too frequent or large quantities of GnRH suppress Gn.[4] In fishes, it is not clear whether pulsatile stimulation is necessary. A prolonged Gn response to a GnRH injection occurs in some fishes.[29,83] There is as yet only a small body of evidence that high or prolonged levels of GnRH in teleosts suppress the reproductive system. Peter[83] found a reduced Gn response in goldfish after high doses of mammalian GnRH analog. Crim and coworkers[33] noted inhibition of expected testicular weight increases after long-term treatment with mammalian GnRH analog in male landlocked Atlantic salmon during a period of rapid reproductive development; another mammalian GnRH analog also suppressed the reproductive system in recrudescing goldfish.[66] *In vitro*, a 1000-fold dose of mammalian GnRH for 24 h was less effective compared with the lower dose in stimulating Gn synthesis in trout pituitaries.[45]

2. Steroids. Not all fishes appear to show the same steroidal pattern in response to GnRH. In 1980, Weil and colleagues[117] reported that double injections of GnRH in carp elevated plasma estradiol beginning 5-7 h postinjection and continuing to the end of sampling at 12 h postinjection. In coho salmon, however, double injections of mammalian GnRH analog advanced ovulation associated with a decrease in plasma estradiol levels and an increase in plasma 17α,20β-dihydroxy-4-pregnen-3-one (17α,20β-P) levels. In females, testosterone levels remained high throughout the preovulatory period, with a small peak during the rapid rise of

$17\alpha,20\beta$-P, which in turn determines the time of ovulation.[112,113] In goldfish 17α-hydroxyprogesterone (17α-P) and $17\alpha,20\beta$-P are released as a surge 6 h after the second injection of mammalian GnRH analog plus pimozide (a dopamine inhibitor). GnRH analog without pimozide produced a smaller and later peak of 17α-P. This preovulatory surge of secretion of progestogens is speculated to aid in the induction of oocyte maturation and ovulation.[88]

3. Ovulation and Spawning. The same principles that apply to Gn release tend to apply to ovulation: GnRH analogs are more effective compared with native GnRH; multiple injections or slow-release preparations of GnRH are sometimes necessary to induce a response. Ovulation and spawning tend to be all-or-none events; hence "priming" and "suppression" as a result of GnRH application are more difficult to measure. However, ovulation studies with GnRH clearly show that a proper stage of maturation must be reached before GnRH is effective.

Salmon-I GnRH induced advanced ovulation in coho salmon; 75-100% of the salmon treated with salmon-I ("teleost") GnRH or one of four analogs (two salmon and two mammalian) ovulated by day 14 compared with 33% of the controls.[41] Native mammalian GnRH induced ovulation in goldfish,[64] ayu (*Plecoglossus altivelis*),[1,52] plaice (*Limanda yokohamae*)[2] and goby (*Acanthogobius flavimanus*).[2] Mammalian GnRH was not always effective: two injections[117] or even five injections[101] induced oocyte maturation, but not ovulation. Twelve injections of GnRH over 6 weeks were required to induce ovulation in medaka (*Oryzias latipes*).[19]

An analog of mammalian GnRH had 50 times the biological potency of GnRH in its ability to accelerate oocyte maturation in coho salmon.[111] This explains why native mammalian GnRH did not advance oocyte development in coho salmon, whereas an equal or smaller amount of analog induced germinal vesicle breakdown.[112]. Double injections of mammalian GnRH analogs in coho salmon resulted in advanced ovulation.[40] Mammalian GnRH analog (D Ala^6, des Gly^{10} GnRH ethylamide) has been used extensively in China to induce ovulation in Chinese carp: silver, bighead, grass and black.[64] Slow-release GnRH (D Trp^6, des Gly^{10} mammalian GnRH ethylamide in a cholesterol matrix pellet) implanted into rainbow trout advanced ovulation 3-4 weeks; the eggs, however, had high rates of mortality at 15-21 days after fertilization.[34] Also, silastic matrix pellets with a mammalian GnRH analog induced spawning 1-2 days early in a once-a-month cycle in siganids (*Siganus guttatus*).[50] An emulsion, but not an aqueous solution, of mammalian GnRH was effective in inducing ovulation in flounder and goby.[2]

Fishes must mature naturally to a certain reproductive stage before GnRH treatment is successful in most studies. In sea bass (*L. calcarifer*) only 1 in 10 fish partially responded in May, but in July and August, as the fish became more reproductively mature, a greater response occurred.[50] Maturing, but not immature, ayu could be stimulated to ovulate with mammalian GnRH.[1] In trout and carp, the pituitary response to GnRH was low or absent in early maturational stages.[29,115,116] In landlocked Atlantic salmon, the maturational stage was important in the response to a GnRH analog: (1) in sexually-regressed males, some testicular stimulation occurred, (2) in recrudescing fish, the ratio of gonad/body weight was increased in females, but decreased in males, and (3) in prespawning fish, ovulation and spermiation were accelerated.[33]

Self-suppression of GnRH effects has not been documented fully. However, two reports show that a high dose of GnRH may be less effective compared with a low dose. Sokolowska[101] showed a dose of 1 μg/kg, but not 25 μg/kg mammalian GnRH advanced oocyte maturation in carp. Another report also showed that 0.2, but not 1.0 mg/kg of salmon-I GnRH advanced ovulation in coho salmon.[41] Likewise, a mammalian GnRH analog in pellet form suppressed testicular weight increases in developing salmon;[33] ten daily injections of 1 μg/g inhibited the reproductive system in goldfish undergoing recrudescence.[66] Evidence of self-suppression is counterbalanced by many experiments in which high doses and prolonged release preparations did not inhibit reproduction.

Spermiation is affected also by GnRH. The quantity of collectible milt was increased by a number of GnRH molecules and analogs. The effectiveness was highest for certain salmon and mammalian GnRH analogs, least for the native salmon, mammalian, chicken-I GnRHs and two other mammalian GnRH analogs.[28] A mammalian analog (10-30 μg/kg) increased milt production in black carp at 12 to 24 h and in bighead carp at 6-1/2 h (see Lam[64]). Finally, in landlocked Atlantic salmon[114] or in testosterone-primed juvenile rainbow trout,[31] onset of spermiation was accelerated by an analog of mammalian GnRH administered by one of several methods: injected in various diluents or implanted as silicone or cholesterol pellets.[114]

V. SUMMARY AND FUTURE RESEARCH DIRECTIONS

A. Evolution of the GnRH Family

A family of GnRH molecules exists for vertebrates. The characteristics shared by the family members are: constant length of 10 amino acids, common N-terminal of pyroglutamic acid, common C-terminal of glycine amide and stability in positions 2 and 3 (His^2-Trp^3). The variability in the GnRH molecules reported to date occurs in positions 5, 7 and 8. There is some evidence that positions 2 and 3 are important for the physiological function of releasing Gns, whereas positions 5-9 may be more important for binding to receptors (see Sherwood,[95] Sherwood et al.[98]). Of piscine GnRHs, only the primary structure of salmon-I GnRH is known. However, immunological and chromatographic evidence provides indirect support that lamprey, dogfish, ratfish, sturgeon and salmon-II GnRH-like molecules belong in the same family; their length and termini appear to be the same. Figure 4 presents a working diagram for the structural relationships of the GnRH molecules. The similarity in elution times and cross-reactivity (Figure 3) of several forms of GnRH including salmon-II does not prove their structures are identical; their grouping under salmon-II-like GnRH provides a working relationship only. Future studies need to determine the primary structure of GnRH in other fishes to help "flesh out" our skeletal conception of GnRH evolution. Further comparative studies also are needed to test the effectiveness of identified members of the GnRH family in various fishes and, vice versa, the effectiveness of the piscine GnRHs in other vertebrates. However, it is probable this set of related peptide hormones is derived from a common ancestral protein; these molecules are related by structure, immunochemistry and function.

B. The Primary Structure of GnRH in Other Fishes.

Determination of the amino acid sequence of GnRH depends on the availability of a large number of brains. This is possible in some fish species; in other fishes, different methods will be necessary. A combination of immunological and chromatographical (HPLC) methods requires only a few brains and provides indirect evidence for the presence of a unique form of GnRH in a species. Probes of complementary DNA might be useful in

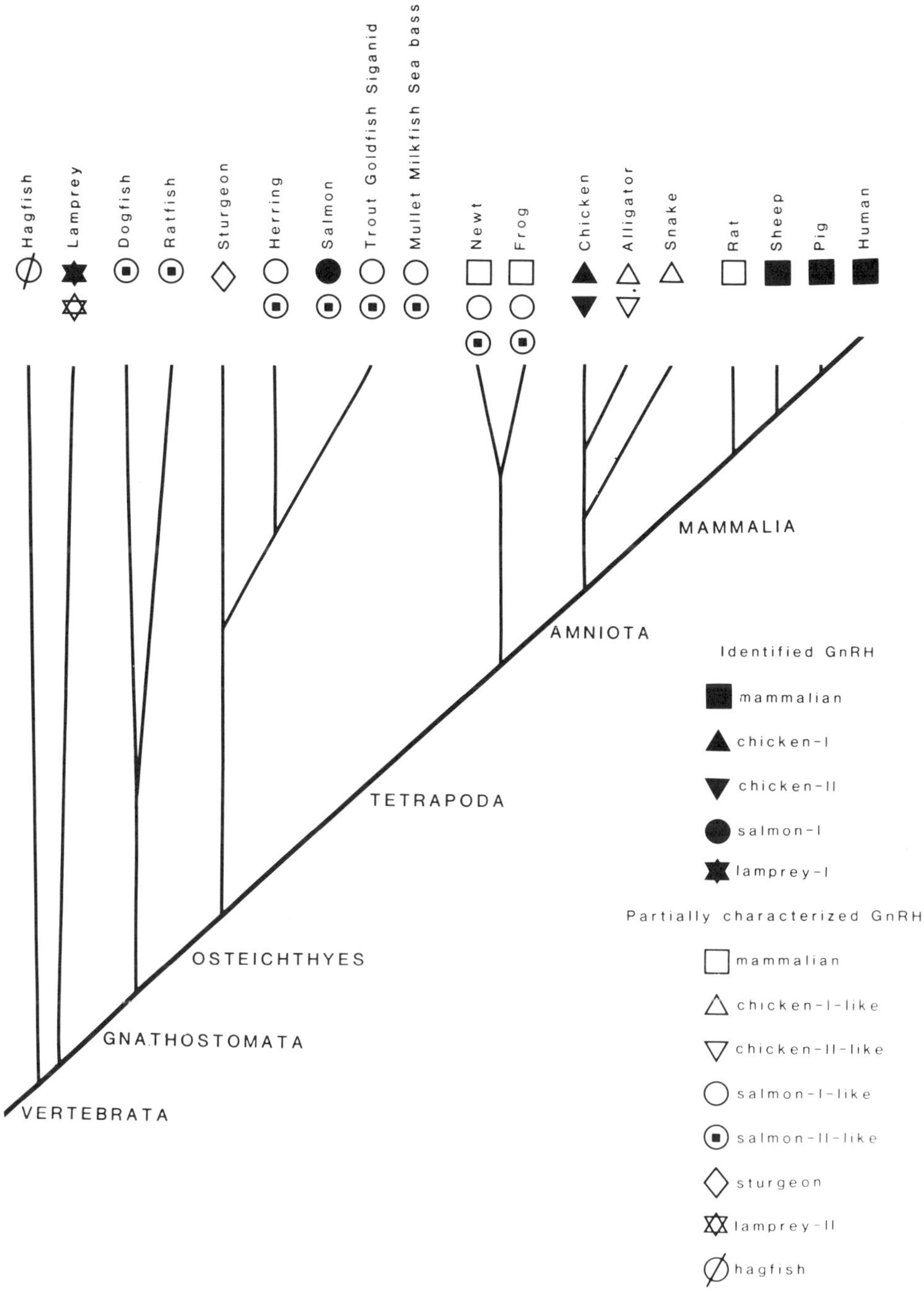

Fig. 4. A simplified cladogram to illustrate the identification of several forms of GnRH. The solid symbols show GnRH molecules for which the primary sequence is known. Rat GnRH has been identified by DNA sequencing; frog and lamprey II GnRH are known by amino acid composition. The other GnRH molecules have been only tentatively identified by HPLC and immunological methods. Hagfish GnRH appears not to be present or only in small quantities. Adapted from a figure in American Zoologist (as yet unpublished) and reproduced with permission.

determining if genetic material codes for GnRH-like material in fishes. Seeburg and Adelman[93] were able to identify the structure of GnRH precursor isolated from human placenta even though the encoded sequence for GnRH was short and the probe was not a perfect match to the genetic material. Complementary DNA may be used in the future also for in-situ hybridization cytochemistry. This technique could be used to detect mRNA for GnRH. Hence the cells which synthesize the molecules would be identified.

C. Multiple Forms of GnRH in Single Species of Fish.

In fishes, unlike mammals, more than one form of GnRH appears to be present in most brains. But the primary structure of two forms in one species is known only for chicken (Table 1). The evidence for multiple forms of GnRH in one species of fish is based on chromatographic and immunological data: lamprey, dogfish, sturgeon, salmon, trout, herring, goldfish, mullet, milkfish, siganid and sea bass each appear to have at least two forms of GnRH. Only a small quantity of a second form appears to exist in lamprey, dogfish and sturgeon. It is not known if the quantity of the second form of GnRH changes in any fishes during different stages of maturity or reproduction.

In amphibians (see also Chapter 3), the first or dominant form of GnRH has the same amino acid composition as mammalian GnRH,[91] but small amounts of GnRH in the brain, sympathetic ganglia and adrenal medulla have a cross-reactivity and HPLC elution pattern similar to salmon GnRH.[43,44,100] Mammalian chromaffin tissue (adrenal medulla) did not contain GnRH in rat, guinea pig or monkey, but it is not known if piscine chromaffin tissue contains GnRH. Likewise, amphibian retinae have more than one form of GnRH; one is chromatographically similar to mammalian and the other to piscine GnRH. Retinae of fish and amphibians contain GnRH; those of rat, guinea pit, monkey, turtle or chicken do not.[44]

The most intriguing question concerns the function of the two forms of GnRH. The two forms may have: (1) the same physiological role, but different potencies, (2) separate, but related roles in reproduction, (3) different functions, one of which is not related to reproduction. Furthermore, only one form may be functional. An understanding of the function of the two forms of GnRH probably depends on future identification of the primary structure of the second form and hence availability of the synthetic forms for physiological testing.

These multiple forms of peptides in one species may result from genetic reduplication. But it remains to be shown whether peptides such as GnRH have two forms with closely related amino acids in the same precursor molecule.

GnRH is not the only hormone that is present in more than one form in some species. Insulin, for example, is a single form in most species, but in tuna, toadfish and three species of rodents, two distinct forms of insu-in exist[72].

D. Other Roles for GnRH in Fishes

The main role ascribed to GnRH is the release of Gns from the pituitary. New evidence suggests other functions also may exist because (1) GnRH cell bodies are found near the olfactory bulbs with axons terminating in several brain areas in teleosts and dogfish; axons also end in the retina in teleosts, and possibly in dogfish, (2) GnRH also is located in brain areas distinct from the olfactory and hypothalamic regions, and (3) multiple forms appear to exist in the brains of most species examined.

The GnRH neurons near the olfactory bulbs may mediate certain environmental cues that activate the reproductive system, including behavior. For example, the GnRH-containing terminal nerve may be important in triggering sexual responses to pheromones;[37] damage to the terminal system both reduces sperm release after electrical stimulation of the olfactory tract (see Demski[36]) and reduces courtship behavior.[105]

GnRH-like material also has been identified in the retinae of carp, goldfish, rainbow trout and bullfrogs, but not turtles, chickens, rats, guinea pigs or monkey.[44] In goldfish, certain retinal ganglionic cells show a modified electrophysiological response after synthetic salmon-I GnRH was applied. One possible explanation is that the change in the visual system may be modulated by sex-related olfactory stimuli that act through the GnRH terminal nerve.[107] In this system, GnRH may be acting as a neurotransmitter in the retina.

The proximity of GnRH cells to steroid-concentrating neurons outside the olfactory or hypothalamic areas has led to the suggestion that the GnRH cells are involved in sexual motivation or behavior.[36]

In tetrapod vertebrates, GnRH-like material has been identified tentatively in tissues outside the CNS: autonomic nervous system (amphibians)[43,44] and testes (mammals).[94] GnRH definitely has been identified in human placenta;[93,110] it may act as a "local hormone" (paracrine). There is evidence that GnRH acts as a neurotransmitter after it is released from preganglionic nerves in the sympathetic nervous system. GnRH has not been reported as yet in the autonomic nervous system or gonads of fishes. However, Stell (personal communication) has noted evidence that the terminal nerve in dogfish resembles a typical autonomic nerve. This may be a connecting thread for the presence of GnRH both in the terminal nerve of dogfish and autonomic ganglia of amphibians.

E. Inhibitory Effects of GnRH

Suppression of the pituitary-gonadal axis with high, sustained doses of GnRH is well documented in mammals;[90] pulsatile administration of GnRH is more effective compared with sustained administration for stimulation of Gn release and other reproductive responses.[7] In fishes, there is only limited evidence as yet that prolonged exposure to GnRH inhibits the reproductive system (see section IV.D.1,3.). A basic question concerning the use of GnRH in fish culture is whether the hormone should be applied at a continuous level or in discrete doses separated by several hours or days. Sustained administration from slow-release pellets or constant-release pumps has two advantages: (1) hormone levels could be maintained during slow maturational processes or for repetitive spawning and (2) fish would not require repeated, stressful handling. It is important to determine whether sustained administration will suppress the reproductive system in fishes for aquaculture purposes.

F. A Possible GnRIF

The existence of a gonadotropin release-inhibiting factor (GnRIF) in goldfish has been proposed.[79,86,87] The nature of the factor is not known, but dopamine has been suggested as a GnRIF. Dopamine or an agonist, administered by injection (ip, but not intraventricular), resulted in a decrease of serum Gn. An antagonist of dopamine, pimozide, enhanced the ability of GnRH to induce ovulation[20,102] or release Gn in goldfish.[89] Also several pharmacological agents that interfere with dopamine resulted in an increase in serum Gn.[84] Peter[84] concludes that "dopamine acts at the level of the pituitary to block spontaneous release of Gn, and also to block, or modulate, the actions of GnRH." *In-vitro* experiments support this conclusion.[21] However, a number of criteria must be met[63] in order to prove that a factor, possibly dopamine, acts in the normal cycle as a chemical mediator to inhibit Gn synthesis or release.

G. The Nature of Fish GnRH Receptors

It has been assumed that GnRH receptors occur in fishes as in mammals in the pituitary and possibly in the ovary and testis. However, to date, high specific binding of radioactive GnRH or its agonists to receptors in fish tissue has not been reported. This is an important area for future research to determine the nature of the co-evolution of the GnRH molecule and its receptor.

H. The Fish GnRH Precursor(s)

One GnRH precursor in human placenta has been identified.[93] The precursor contained the code for a signal peptide, GnRH and a 53-amino acid associated peptide. It remains to be shown if vertebrates with multiple forms of GnRH, as occurs in some fish, have one precursor which codes for both forms as in enkephalin precursor in chromaffin tissue[22] or contains two or more precursor molecules which code for each GnRH form.

VI. ACKNOWLEDGMENTS

Canadian IDRC, MRC and Fisheries and Oceans supported the unpublished work reported here.

VII. REFERENCES

1. Aida, K. (1983). Effect of LH-releasing hormone on gonadal development in a salmonoid fish, the ayu. Bull. Japan. Soc. Sci. Fish. 49, 711-718.
2. Aida, K., Izumo, R.S., Satoh, H. and Hibiya, Y. (1978). Induction of ovulation in plaice and goby with synthetic LH-releasing hormone. Bull. Japan. Soc. Sci. Fish. 44, 445-450.
3. Aiyer, M.S., Chiappa, S.A. and Fink, G. (1974). A priming effect of luteinizing hormone releasing factor on the anterior pituitary gland in the female rat. J. Endocr. 62, 573-588.
4. Baird, C.J., Tharandt, L. and Tamarkin, L. (1984). Regulation of luteinizing hormone release by pulsatile and continuous administration of gonadotropin-releasing hormone to superfused rat and hamster pituitary cells. Endocrinology 114, 1041-1047.
5. Ball, J.N. (1981). Hypothalamic control of the pars distalis in fishes, amphibians, and reptiles. Gen. Comp. Endocr. 44, 135-170.
6. Barannikova, I.A., Bukovskaya, O.S., and Efimova, N.A. (1982). Hormonal control of the reproductive function of sturgeons (Chondrostei). In "Reproductive Physiology of Fish" (C.J.J. Richter and H.J.Th. Goos, eds.), p. 49. Pudoc, Wageningen, Netherlands.
7. Belchetz, P.E., Plant, T.M., Nakai, Y., Keogh, E.J. and Knobil, E. (1978). Hypophysial responses to continuous and intermittent delivery of hypothalmic gonadotropin-releasing hormone. Science 202, 631-633.
8. Bohlen, P., Castillo, F., Yin, S.Y., Brazeau, P., Baird, A., and Guillemin, R. (1981). A general approach to the microisolation of peptides. In "Peptides: synthesis-structure-function" (D.H. Rich and E. Gross, eds.), Proc. 7th American Symposium, pp. 777-780. Pierce Chemical Co., Rockford, Illinois.
9. Borg, B., Goos, H.J.Th., and Terlou, M. (1982) LHRH-immunoreactive cells in the brain of the three-spined stickleback, *Gasterosteus aculeatus* L. (Gasterosteidae). Cell Tissue Res. 226, 695-699.
10. Branton, W.D., Jan, L.Y. and Jan, Y.N. (1982). Non-mammalian luteinizing hormone-releasing factor (LRF) in tadpole and frog brain. Program of the 12th Annual meeting of Society for Neuroscience, p. 14. Minneapolis, Minnesota (Abstract).
11. Breton, B. and Weil, C. (1973). Effets du LH/FSH-RH synthetique et d'extraits hypothalamiques de carpe sur la secretion d'hormone gonadotrope in vivo chez la carpe (*Cyprinus carpio* L). C.R. Acad. Sci. 277, 2061-2064.

12. Breton, B., Jalabert, B., Billard, R. and Weil, C. (1971). Stimulation in vitro de la liberation d'hormone gonadotrope hypophysaire par un facteur hypothalamique chez la carpe (Cyprinus carpio L.) C.R. Acad. Sci. 273, 2591-2594.
13. Breton, B., Weil, C., Jalabert, B. and Billard, R. (1972). Activité réciproque des facteurs hypothalamiques de bélier (Ovis aries) et de poissons téléostéens sur la sécrétion in vitro des hormones gonadotropes c-HG et LH respectivement par des hypophyses de carpe et de bélier. C.R. Acad. Sci. Paris, 274, 2530-2533.
14. Breton, B., Jalabert, B. and Weil, C. (1975). Caracterisation partielle d'un facteur hypothalamique de liberation des hormones gonadotropes chez la carpe (Cyprinus carpio L.). Etude in vitro. Gen. Comp. Endocrinol. 25, 405-415.
15. Breton, B., Jalabert, B., Bieniarz, K., Sokolovska, M. and Epler, P. (1983). Effects of synthetic LH-RH and analog on plasma gonadotropin levels and maturational response to 17α-hydroxy-20β-dihydroprogesterone. Aquaculture 32, 105-114.
16. Burgus, R., Butcher, M., Amoss, M., Ling, N., Monahan, M., Rivier, J., Fellows, R., Blackwell, R., Vale, W. and Guillemin, R. (1972). Primary structure of the ovine hypothalamic luteinizing hormone-releasing factor (LRF). Proc. Natl. Acad. Sci. USA 69, 278-282.
17. Burzawa-Gérard, E. (1982). Chemical data on pituitary gonadotropins and their implication to evolution. Can. J. Fish Aquat. Sci. 39, 80-91.
18. Burzawa-Gérard, E., Goncharov, B.F. and Fontaine, Y.A. (1975). L'hormone gonadotrope hypophysaire d'un poisson chondrostéen, l'esturgeon (Acipenser stellatus Pall.). I. Purification. Gen. Comp. Endocrinol. 27, 289-295.
19. Chan, K.S. (1977). Effect of synthetic luteinizing hormone-releasing hormone (LH-RH) on ovarian development in Japanese medaka, Oryzias latipes. Can. J. Zool. 55, 155-160.
20. Chang, J.P. and Peter, R.E. (1983). Effects of pimozide and des Gly^{10}, $(D\text{-}Ala^{6})$ luteinizing hormone-releasing hormone ethylamide on serum gonadotropin concentrations, germinal vesicle migration, and ovulation in female goldfish, Carassius auratus. Gen. Comp. Endocrinol. 52, 30-37.
21. Chang, J.P., MacKenzie, D.S., Gould, D.R. and Peter, R.E. (1984). Effects of dopamine and norepinephrine on in vitro spontaneous and gonadotropin-releasing hormone-induced gonadotropin release by dispersed cells or fragments of the goldfish pituitary. Life Sci. 35, 2027-2033.
22. Comb, M., Seeburg, P.H., Adelman, J., Eiden, L. and Herbert, E. (1982). Primary structure of the human Met- and Leu-enkephalin precursor and its mRNA. Nature 295, 663-666..
23. Copeland, K.C., Aubert, M.L., Rivier, J. and Sizonenko, P.C. (1979). Luteinizing hormone-releasing hormone: sequential versus conformational specificity of antiluteinizing hormone-releasing hormone sera. Endocrinology 104, 1504-1512.
24. Crim, J.W. (1983). Immunoreactive luteinizing hormone-releasing hormone (LH-RH) in the brain of a holostean fish, Amia calva. Amer. Zool. 23, 908 (Abstract).
25. Crim, J.W., Urano, A. and Gorbman, A. (1979). Immunocytochemical studies of luteinizing hormone-releasing hormone in brains of agnathan fishes. I. Comparisons of adult Pacific lamprey (Entosphenus tridentata) and the Pacific hagfish (Eptatretus stouti). Gen. Comp. Endocrinol. 37, 294-305.
26. Crim, J.W., Urano, A. and Gorbman, A. (1979). Immunocytochemical studies of luteinizing hormone-releasing hornome in brains of agnathan fishes. II. Patterns of immunoreactivity in larval and maturing Western brook lamprey (Lampetra richardsoni). Gen. Comp. Endocrinol. 38, 290-299.

27. Crim, L.W. (1984). Actions of LHRH and its analogs in lower vertebrates. In "LHRH and Its Analogs. Contraceptive and Therapeutic Applications" (B.H. Vickery, J.J. Nestor, Jr. and E.S.E. Hafez, eds.), pp. 377-384, MTP Press Ltd., Boston.
28. Crim, L.W. (1984). A variety of synthetic LHRH peptides stimulate gonadotropin hormone secretion in rainbow trout and the landlocked Atlantic salmon. J. Steroid Biochem. 20, B1390 (Abstract).
29. Crim, L.W. and Cluett, D.M. (1974). Elevation of plasma gonadotropin concentration in response to mammalian gonadotropin releasing hormone (GRH) treatment of the male brown trout as determined by radioimmunoassay. Endocr. Res. Commun. 1, 101-110.
30. Crim, L.W. and Evans, D.M. (1980). LH-RH-stimulated gonadotropin release from the rainbow trout pituitary gland: An *in vitro* assay for detection of teleost gonadotropin releasing factor(s). Gen. Comp. Endocrinol. 40, 283-290.
31. Crim, L.W. and Evans, D.M. (1983). Influence of testosterone and/or luteinizing hormone-releasing hormone analogue on precocious sexual development in the juvenile rainbow trout. Biol. Reprod. 29, 137-142.
32. Crim, L.W., Peter, R.E. and Billard, R. (1976). Stimulation of gonadotropin secretion by intraventricular injection of hypothalamic extracts in the goldfish, *Carassius auratus*. Gen. Comp. Endocrinol. 30, 77-82.
33. Crim, L.W., Evans, D.M. and Vickery, B.H. (1983). Manipulation of the seasonal reproductive cycle of the landlocked Atlantic salmon (*Salmo salar*) by LHRH analogues administered at various stages of gonadal development. Can. J. Fish Aquat. Sci. 40, 61-67.
34. Crim, L.W., Sutterlin, A.M., Evans, D.M. and Weil, C.(1983). Accelerated ovulation by pelleted LHRH analogue treatment of spring-spawning rainbow trout (*Salmo gairdneri*) held at low temperature. Aquaculture 35, 299-307.
35. Deery, D.J. (1974). Determination by radioimmunoassay of the luteinising hormone-releasing hormone (LHRH) content of the hypothalamus of the rat and some lower vertebrates. Gen. Comp. Endocrinol. 24,280-285.
36. Demski, L.S. (1984). The evolution of neuroanatomical substrates of reproductive behavior: Sex steroid and LHRH-specific pathways including the terminal nerve. Amer. Zool. 24, 809-830.
37. Demski, L.S. and Northcutt, R.G. (1983). The terminal nerve: A new chemosensory system in vertebrates? Science 220, 435-437.
38. Dodd, J.M. (1975). The hormones of sex and reproduction and their effect in fish and lower chordates: Twenty years on. Amer. Zool. 15 (Suppl. 1), 137-171.
39. Donaldson, E.M., Hunter, G.A. and Dye, H.M. (1981/1982). Induced ovulation in coho salmon (*Oncorhynchus kisutch*). II. Preliminary study of the use of LH-RH and two high potency LH-RH analogues. Aquaculture 26, 129-141.
40. Donaldson, E.M., Hunter, G.A., Van Der Kraak, G. and Dye, H.M. (1982). Application LH-RH and LH-RH analogues to the induced final maturation and ovulation of coho salmon (*Oncorhynchus kisutch*) In "Reproductive Physiology of Fish" (C.J.J. Richter and H.J.Th. Goos, eds.), pp. 177-180. Pudoc, Wageningen, Netherlands.
41. Donaldson, E.M., Van Der Kraak, G., Hunter, G.A., Dye, H.M., Rivier, J. and Vale, W. (1984). Teleost GnRH and analogues: Effect on plasma GtH concentration and ovulation in coho salmon (*Oncorhynchus kisutch*). Gen. Comp. Endocrinol. 53, 458-459, (Abstract).
42. Dufour, S., Pasqualini, C., Kerdelhue, B. and Fontaine, Y.A. (1982). Presence and distribution of radioimmunoassayable LHRH in the European eel, *Anguilla anguilla*. Neuropeptides 3, 159-171.

43. Eiden, L.E. and Eskay, R.L. (1980). Characterization of LRF-like immunoreactivity in the frog sympathetic ganglia: Non-identity with LRF decapeptide. Neuropeptides 1, 29-37.
44. Eiden, L.E., Loumaye, E., Sherwood, N. and Eskay, R.L. (1982). Two chemically and immunologically distinct forms of luteinizing hormone-releasing hormone are differentially expressed in frog neural tissues. Peptides 3, 323-327.
45. Fahraeus-van Ree, G.E., van Vlaardingen, M. and Gielen, J.Th. (1983). Effect of 17α-methyltestosterone, estradiol-17β and synthetic LHRH on production of gonadotropic hormone in pituitaries of rainbow trout (organ culture). Cell Tissue Res. 232, 157-176.
46. Fontaine, M. and Olivereau, M. (1975). Some aspects of the organization and evolution of the vertebrate pituitary. Amer. Zool. 15 (Suppl. 1), 61-79.
47. Goos, H.J. Th. (1978). Hypophysiotropic centers in the brain of amphibians and fish. Amer. Zool. 18, 401-410.
48. Gorham, A. (1980). Evolution of the brain-pituitary relationship: Evidence from the Agnatha. Can. J. Fish. Aquat. Sci. 37, 1680-1686.
49. Halpern-Sebold, L.R. and Schreibman, M.P. (1983). Ontogeny of centers containing luteinizing hormone-releasing hormone in the brain of platyfish (*Xiphophorus maculatus*) as determined by immunocytochemistry. Cell Tissue Res. 229, 75-84.
50. Harvey, B., Nacario, J., Crim, L.W., Juario, J.V. and Marte, C.L. (1985). Induced spawning of sea bass *Lates calcarifer* and rabbitfish *Siganus guttatus* after implantation of pelleted LHRH analogue. Aquaculture, 47, 53-59.
51. Hasegawa, Y., Miyamoto, K., Igarashi, M., Chino, N., and Sakakibara,S. (1984). Biological properties of chicken luteinizing hormone-releasing hormone: gonadotropin release from rat and chicken cultured anterior pituitary cells and radioligand analysis. Endocrinology 114, 1441-1447.
52. Hirose, K. and Ishida, R. (1974). Induction of ovulation in the ayu, *Plecoglossus altivelis* and LH-releasing hormone (LH-RH). Bull. Japan. Soc. Sci. Fish. 40, 1235-1240.
53. Holmes, R.L. and Ball, J.N. (1974). "The Pituitary Gland. A Comparative Account," Cambridge Univ. Press, London.
54. Idler, D.R. (1982). Some perspectives on fish gonadotropins. *In* "Reproductive Physiology of Fish" (C.J.J. Richter and H.J.Th. Goos, eds.), pp. 4-13. Pudoc, Wageningen, Netherlands.
55. Jackson, I.M.D. (1980). Distribution and evolutionary significance of the hypophysiotropic hormones of the hypothalamus. Front. Horm. Res. 6, 35-69.
56. Jenkins, N. and Dodd, J.M. (1980). Effects of synthetic mammalian gonadotropin releasing hormone and dogfish hypothalamic extracts on levels of androgens and oestradiol in the circulation of the dogfish (*Scyliorhinus canicula* L.). J. Endocrinol. 86, 171-177.
57. Kah, O., Chambolle, P., Dubourg, P. and Dubois, M.P. (1982). Distribution of immunoreactive LH-RH in the brain of the goldfish. *In* "Reproductive Physiology of Fish" (C.J.J. Richter and H.J.Th. Goos, eds.), p. 56. Pudoc, Wageningen, Netherlands.
58. Katz, Y., Dashow, L. and Epple, A. (1982). Circulating steroid hormones of anadromous sea lampreys under various experimental conditions. Gen. Comp. Endocrinol. 48, 261-268.
59. King, J.A. and Millar, R.P. (1979). Heterogeneity of vertebrate luteinizing hormone-releasing hormone. Science 206, 67-69.
60. King, J.A. and Millar, R.P. (1980). Comparative aspects of

luteinizing hormone-releasing hormone structure and function in vertebrate phylogeny. Endocrinology 106, 707-717.

61. King, J.A. and Millar, R.P. (1982). Structure of chicken hypothalamic luteinizing hormone-releasing hormone. I. Structural determination on partially purified material. J. Biol. Chem. 257, 10722-10728.

62. King, J.A. and Millar, R.P. (1982). Structure of chicken hypothalamic luteinizing hormone-releasing hormone. II. Isolation and characterization. J. Biol. Chem. 257, 10729-10732.

63. Krieger, D.T., Brownstein, M.J. and Martin, J.B. (1983). Introduction. In "Brain Peptides" (D.T. Krieger, M.J. Brownstein and J.B. Martin, eds.), pp. 1-12. John Wiley & Sons, New York.

64. Lam, T.J. (1982). Applications of endocrinology to fish culture. Can. J. Fish. Aquat. Sci. 39, 111-137.

65. Lance, V. and Callard, I.P. (1978). Gonadotrophic activity in pituitary extracts from an elasmobranch (*Squalus acanthias* L.) J. Endocrinol. 78, 149-150.

66. Lin, H.-R., Peter, R.E., Nahorniak, C.S. and Bres, O. (1985). Actions of the superactive analogue des-Gly^{10}(D-Ala^{6})LRH ethylamide (LRH-A) on gonadotropin secretion in goldfish. In "Current Trends in Comparative Endocrinology," Vol 1 (B. Lofts and W. N. Holmes, eds.), pp. 77-79. University of Hong Kong Press, Hong Kong.

67. MacKenzie, D.S., Gould, D.R., Peter, R.E., Rivier, J. and Vale, W.W. (1984). Response of superfused goldfish pituitary fragments to mammalian and salmon gonadotropin-releasing hormones. Life Sci. 35, 2019-2026.

68. Matsuo, H., Baba, Y., Nair, R.M.G., Arimura, A. and Schally, A.V. (1971). Structure of the porcine LH- and FSH-releasing hormone. I. The proposed amino acid sequence. Biochem. Biophys. Res. Communic. 43, 1334-1339.

69. Matty, A.J., Tsuneki, K., Dickhoff, W.W. and Gorbman, A. (1976). Thyroid and gonadal function in hypophysectomized hagfish, *Eptatretus stouti*. Gen. Comp. Endocrinol. 30, 500-516.

70. Millar, R.P. and King, J.A. (1983). Synthesis, luteinizing hormone-releasing activity, and receptor binding of chicken hypothalamic luteinizing hormone-releasing hormone. Endocrinology 113, 1364-1369.

71. Millar, R.P. and King, J.A. (1983). Synthesis and biological activity of (D-Trp^{6}) chicken luteinizing hormone-releasing hormone. Peptides 4, 425-429.

72. Miller, W.L., Baxter, J.D. and Eberhardt, N.L. (1983). Peptide hormone genes: structure and evolution. In "Brain Peptides" (D.T. Krieger, M.J. Brownstein and J.B. Martin, eds.), pp. 15-74. John Wiley & Sons, New York.

73. Milton, R. de L., King, J.A., Badminton, M.N. Tobler, C.J., Lindsey, G.G., Fridkin, M. and Millar, R.P. (1983). Comparative structure-activity studies on mammalian (Arg^{8}) LH-RH and chicken (Gln^{8}) LH-RH by fluorimetric titration. Biochem. Biophys. Res. Communic. 111, 1082-1088.

74. Miyamoto, K., Hasegawa, Y., Minegishi, T., Nomura, M., Takahashi, Y., Igarashi, M., Kangawa, K. and Matsuo, H. (1982). Isolation and characterization of chicken hypothalamic luteinizing hormone-releasing hormone. Biochem. Biophys. Res. Communic. 107, 820-827.

75. Miyamoto, K., Hasegawa, Y., Igarashi, M., Chino, N., Sakakibara, S., Kangawa, K. and Matsuo, H. (1983). Evidence that chicken hypothalamic luteinizing hormone-releasing hormone is (Gln^{8})-LH-RH. Life Sci. 32, 1341-1347.

76. Miyamoto, K., Hasegawa, Y., Nomura, M., Igarashi, M., Kangawa, K., Matsuo, H. (1984). Identification of the second gonadotropin-releasing hormone in chicken hypothalamus: Evidence that gonadotropin secretion is probably controlled by two distinct gonadotropin-releasing hormones in avian species. Proc. Natl. Acad. Sci. USA 81, 3874-3878.
77. Munz, H., Stumpf, W. E. and Jennes, L. (1981). LHRH systems in the brain of platyfish. Brain Research 221, 1-13.
78. Munz, H., Claas, B., Stumpf, W.E. and Jennes, L. (1982). Centrifugal innervation of the retina by luteinizing hormone releasing hormone (LHRH)-immunoreactive telencephalic neurons in teleostean fishes. Cell Tissue Res. 222, 313-323.
79. Nagahama, Y. and Peter, R.E. (1982). Effects of brain lesions on gonadotrop ultrastructure and serum gonadotropin levels in goldfish. Cell Tissue Res. 225, 259-265.
80. Nozaki, M. and Kobayashi, H. (1979). Distribution of LHRH-like substance in the vertebrate brain as revealed by immunohistochemistry. Arch. Histol. Jap. 42, 201-219.
81. Nozaki, M., Tsukahara, T. and Kobayashi, H. (1984). Neuronal systems producing LHRH in vertebrates. _In_ "Endocrine Correlates of Reproduction" (K. Ochiai, Y. Arai, T. Shioda, and M. Takahashi, eds.), pp. 3-27. Japan Sci. Soc. Press, Tokyo/Springer-Verlag, Berlin.
82. Nozaki, M., Tsukahara, T. and Kobayashi, H. (1984). An immunocytochemical study on the distribution of neuropeptides in the brain of certain species of fish. Biomedical Res. 4 (suppl), 135-145.
83. Peter, R.E. (1980). Serum gonadotropin levels in mature male goldfish in response to luteinizing hormone-releasing hormone (LH-RH) and des-Gly^{10}-(D-Ala^{6})-LH-RH ethylamide. Can. J. Zool. 58, 1100-1104.
84. Peter, R.E. (1982). Nature, localization and actions of neurohormones regulating gonadotropin secretion in teleosts. _In_ "Reproductive Physiology of Fish" (C.J.J. Richter and H.J.Th. Goos, eds.), pp. 30-39. Pudoc, Wageningen, Netherlands.
85. Peter, R.E. (1983). Evolution of neurohormonal regulation of reproduction in lower vertebrates. Amer. Zool. 23, 685-695.
86. Peter, R.E. and Paulencu, C.R. (1980). Involvement of the preoptic region in gonadotropin release-inhibition in goldfish, _Carassius auratus_. Neuroendocrinology 31, 133-141.
87. Peter, R.E., Crim, L.W., Goos, H.J.Th. and Crim, J.W. (1978). Lesioning studies on the gravid female goldfish: Neuroendocrine regulation of ovulation. Gen. Comp. Endocrinol. 35, 391-401.
88. Peter, R.E., Sokolowska, M., Truscott, B., Walsh, J. and Idler, D.R. (1984). Secretion of progestogens during induced ovulation in goldfish. Can. J. Zool. 62, 1946-1949.
89. Peter, R.E. Nahorniak, C.S., Sokolowska, M., Chang, J.P., Rivier, J.E.,Vale, W.W., King, J.A. and Millar, R.P. (1985). Structure-activity relationships of mammalian, chicken, and salmon gonadotropin releasing hormone _in vivo_ in goldfish. Gen. Comp. Endocrinol. 58, 231-242.
90. Piper, E.L., Perkins, J.L., Tugwell, D.R. and Vaught, W.G. (1975). Inhibition of LH release as influenced by LRF infusion. Proc. Soc. Exp. Biol. Med. 148, 880-882.
91. Rivier, J., Rivier, C., Branton, D., Millar, R., Spiess, J., Vale, W. (1981). HPLC purification of ovine CRF, rat extra hypothalamic brain somatostatin and frog brain GnRH. _In_ "Peptides: Synthesis-structure-function," Proc. 7th American Peptide Symposium (D.H. Rich and E. Gross, eds.), pp. 771-776. Pierce Chemical Co., Rockford, Illinois.

92. Schreibman, M.P., Halpern, L.R., Goos, H.J.Th. and Margolis-Kazan, H. (1979). Identification of luteinizing hormone-releasing hormone (LH-RH) in the brain and pituitary gland of a fish by immunocytochemistry. J. Exp. Zool. 210, 153-160.
93. Seeburg, P.H. and Adelman, J.P. (1984). Characterization of cDNA for precursor of human luteinizing hormone releasing hormone. Nature, 311, 666-668.
94. Sharpe, R.M., Fraser, H.M., Cooper, I., and Rommerts, F.F.G. (1981). Sertoli-Leydig cell communication via an LHRH-like factor. Nature 290, 785-787.
95. Sherwood, N.M. (in press). Evolution of a neuropeptide family: Gonadotropin-releasing hormone. Amer. Zool.
96. Sherwood, N.M. and Harvey, B. (1986). Topical absorption of gonadotropin-releasing hormone (GnRH) in goldfish. Gen. Comp. Endocrinol. 61, 13-19.
97. Sherwood, N.M. and Sower, S.A. (1985). A new family member for gonadotropin-releasing hormone. Neuropeptides. 6, 205-214
98. Sherwood, N.M., Eiden, L., Brownstein, M., Spiess, J., Rivier, J. and Vale, W. (1983). Characterization of a teleost gonadotropin-releasing hormone. Proc. Natl. Acad. Sci.USA 80, 2794-2798.
99. Sherwood, N.M., Harvey, B., Brownstein, M.J. and Eiden, L.E. (1984). Gonadotropin-releasing hormone (Gn-RH) in striped mullet (*Mugil cephalus*), milkfish (*Chanos chanos*), and rainbow trout (*Salmo gairdneri*): Comparison with salmon Gn-RH. Gen. Comp. Endocrinol. 55, 174-181.
100. Sherwood, N.M., Zoeller, R.T., and Moore, F.L. (1986). Multiple forms of gonadotropin-releasing hormone in amphibian brains. Gen. Comp. Endocrinol. 61, 313-322.
101. Sokolowska, M. (1982). The influence of luteinizing hormone-releasing hormone (LH-RH) on serum gonadotropin levels and oocyte maturation in female carp (*Cyprinus carpio* L.). Comp. Biochem. Physiol. 73, 273-278.
102. Sokolowska, M., Peter, R. E., Nahorniak, C.S., Pan, C.H., Chang, J.P., Crim, L.W. and Weil, C. (1984). Induction of ovulation in goldfish, *Carassius auratus*, by pimozide and analogues of LH-RH. Aquaculture 36, 71-83.
103. Sower, S. A., Dickhoff, W.W., Gorbman, A., Rivier, J.E. and Vale, W.W. (1983). Ovulatory and steroidal responses in the lamprey following administration of salmon gonadotropin and agonistic and antagonistic analogues of gonadotropin-releasing hormone. Can. J. Zool. 61, 2653-2659.
104. Sower, S.A., Plisetskaya, E. and Gorbman, A. (1985). Steroid and thyroid hormone profiles following a single injection of partly purified salmon gonadotropin or GnRH analogues in male and female sea lamprey. J. Exp. Zool. 235, 403-408.
105. Stacey, N.E. and Kyle, A.L. (1983). Effects of olfactory tract lesions on sexual and feeding behavior in goldfish. Physiol. and Behav. 30, 621-628.
106. Stell, W.K. (1984). Luteinizing hormone-releasing hormone (LHRH)- and pancreatic polypeptide (PP)-immunoreactive neurons in the terminal nerve of the spiny dogfish, *Squalus acanthias*. Anat. Record 208, 173A-174A.
107. Stell, W.K., Walker, S.E., Chohan, K.S. and Ball, A.K. (1984). The goldfish nervus terminalis: A luteinizing hormone-releasing hormone and molluscan cardioexcitatory peptide immunoreactive olfactoretinal pathway. Proc. Natl. Acad. Sci. USA, 81, 940-944.
108. Sumpter, J.P., Follett, B.K., Jenkins, N. and Dodd, J.M. (1978). Studies on the purification and properties of gonadotropin from ventral lobes of the pituitary gland of the dogfish (*Scyliorhinus canicula* L.). Gen. Comp. Endocrinol. 36, 264-274.

109. Sumpter, J.P., Jenkins, N. and Dodd, J.M. (1978). Gonadotrophic hormone in the pituitary gland of the dogfish (Scyliorhinus canicula L.): Distribution and physiological significance. Gen. Comp. Endocrinol. 36, 275-285.

110. Tan, L. and Rousseau, P. (1982). The chemical identity of the immunoreactive LHRH-like peptide biosynthesized in the human placenta. Biochem. Biophys. Res. Communic. 109, 1061-1071.

111. Van Der Kraak, G., Lin, H.R., Donaldson, E.M., Dye, H.M. and Hunter, G.A. (1983). Effects of LH-RH and des-Gly^{10}($D-Ala^{6}$) LH-RH-ethylamide on plasma gonadotropin levels and oocyte maturation in adult female coho salmon (Oncorhynchus kisutch). Gen. Comp. Endocrinol. 49, 470-476.

112. Van Der Kraak, G., Dye, H.-M. and Donaldson, E.M. (1984). Effects of LH-RH and des Gly^{10}($D-Ala^{6}$)LH-RH ethylamide on plasma sex steroid profiles in adult female coho salmon (Oncorhynchus kisutch). Gen. comp. Endocrinol. 55,36-45.

113. Van Der Kraak, G., Dye, H., Donaldson, E.M. and Hunter G.A. (1985). Plasma gonadotropin, 17β-estradiol and 17α,20β-dihydroxy-4-pregnen-3-one levels during LH-RH analogue and gonadotropin induced ovulation in coho salmon (Oncorhynchus kisutch). Can. J. Zool. 63, 824-833.

114. Weil, C. and Crim, L.W. (1983). Administration of LHRH analogues in various ways: effect on the advancement of spermiation in prespawning landlocked salmon, Salmo salar. Aquaculture 35, 103-115.

115. Weil, C., Breton, B. and Reinaud, P. (1975). Etude de la réponse hypophysaire á l'administration de Gn-RH exogéne au cours du cycle reproducteur annuel chez la carpe Cyprinus carpio L. C.R. Acad. Sci., Paris 280, 2469-2472.

116. Weil, C., Billard, R., Breton, B. and Jalabert, B. (1978). Pituitary response to LH-RH at different stages of gametogenesis in the rainbow trout (Salmo gairdneri). Ann. Biol. Anim. Biochim. Biophys. 18, 863-869.

117. Weil, C., Fostier, A., Horvath, L., Marlot, S. and Berscenyi, M. (1980). Profiles of plasma gonadotropin and 17β-estradiol in the common carp, Cyprinus carpio L., as related to spawning induced by hypohysation or LH-RH treatment. Reprod. Nutr. Develop. 20, 1041-1050.

118. Wu, P. and Jackson, I.M.D. (1985). LHRH in cod-fish brain: Evidence of identity with salmon LHRH. 67th Annual Meeting of the Endocrine Society, 1985 (Abstract).

ROLE OF GONADOTROPIN-RELEASING HORMONE IN REGULATION OF GONADOTROPIN SECRETION FROM AMPHIBIAN AND REPTILIAN PITUITARIES

Paul Licht and David A. Porter

Department of Zoology
University of California
Berkeley, CA 94720

The ubiquity of a brain peptide involved in regulating pituitary gonadotropin (Gn) secretion in vertebrates has been confirmed by a variety of direct and indirect studies.[57] This gonadotropin-releasing hormone (GnRH) [also known as luteinizing hormone-releasing hormone (LHRH)] has been extensively characterized biochemically and physiologically in mammals, but data for nonmammalian species are still very limited. Indeed, our use of the term GnRH, unless specified otherwise, will refer to the mammalian decapeptide:

$$pGlu^{1}\text{-}His^{2}\text{-}Trp^{3}\text{-}Ser^{4}\text{-}Tyr^{5}\text{-}Gly^{6}\text{-}Leu^{7}\text{-}Arg^{8}\text{-}Pro^{9}\text{-}Gly^{10}\text{-}NH_2.$$

Considerable information has been amassed on GnRH physiology of amphibians in recent years, but the reptiles probably represent the least understood of the tetrapods. The difficulty in measuring directly the levels of Gns in nonmammalian species has certainly contributed to this problem. However, it is now established that a pair of gonadotropins, similar to the follicle-stimulating hormone (FSH) and luteinizing hormone (LH) of mammals, occurs in at least two orders of amphibians and in crocodilian and chelonian reptiles; squamates (lizards and snakes) may have only a single Gn whose homologies to FSH and LH remain uncertain (see review[32]). The present review focuses on recent evidence pertaining to whether and how the synthesis and secretion of these pituitary Gns are regulated by GnRH-like materials in the brain; the chemical nature of GnRHs, and how they modulate secretion of individual pituitary Gns are also considered.

I. IDENTIFICATION AND CHEMICAL CHARACTERIZATION OF GnRH

A. Amphibians

The potential importance of GnRH in the brain as a regulator of pituitary Gn release in anurans and urodeles has been inferred from results of experimental manipulations, including lesions and electrical stimulation, and immunocytochemistry; no data exist for caecilians. These data have been reviewed extensively[5,13,16] and will be surveyed only briefly here. In general, the combined evidence offers strong support for the view that several potential GnRH regulating 'centers' occur in different regions of the brain and that the peptide is released into the portal circulation from nerve endings in the median emience (ME) as in other tetrapods. However, immunological localization studies have revealed that GnRH-reactive tracts are generally widespread and variable in distribution among species, and no simple description can be given for either anurans or urodeles.

The concept of a bipartite control system (i.e., two separate centers exhibiting at least partial independence) has been especially attractive in light of data for mammals.[5] This concept in amphibians initially received support from the lesioning studies in _Rana temporaria_ (reviewed by Dierickx[17]). These studies implicated two distinct brain areas in pituitary Gn regulation: one included the ME and adjacent anterior medial-based hypothalamus behind the optic chiasm (the _nucleus infundibularis ventralis_, NIV), and the second included the preoptic area (POA) and possibly more rostral regions of the telencephalon. While data support the importance of both areas in anurans and urodeles, the relative importance and exact role of each site are not resolved, and species or sexual differences may exist. Lesioning in _R. temporaria_ indicated that the NIV has a high degree of autonomy in regard to the 'tonic' Gn secretion required for seasonal gametogenesis, since normal ovarian growth occurred even when the NIV was fully isolated from the POA. Input from the POA appeared essential for 'phasic' LH release since isolation of the NIV from this region in summer blocked ovulation in the following spring. However, one could argue that the NIV-ME might be capable of full autonomy in this species, since females similarly lesioned in February did ovulate normally six weeks later. In male toads, _Bufo bufo_, the maintenance of thumb pad development (indicative of testicular steroid secretion) after hypothalamic sections between the NIV and POA was consistent with the hypothesis that the basal hypothalamus was at least partially autonomous; but similar lesions in female toads blocked ovarian development.[25] Moreover, in both male[43] and female[41] newts, full stimulation of Gn secretion (as judged by gametogenesis or ovulation) apparently requires input from the telencephalon anterior to the POA. The indispensability of the hypothalamus in a urodele may be inferred from data on an eyeless mutant of the axolotl, Ambystoma mexicanum, in which sterility appears to arise from failure of normal hypothalamic development,[69] but these data do not clarify the role of different brain regions.

Responses to electrical stimulation of brain areas further implicate the POA as a site of pituitary regulation in anurans and urodeles. In newts, Triturus cristatus, stimulation of the POA areas promoted ovulation,[42] and in the bullfrog, Rana catesbeiana, graded electrical stimulation of the medial POA was clearly associated with graded increases in plasma FSH and LH.[46] Responsive areas corresponded to some of those identified as containing GnRH by immunological staining in this species.[2] These studies do not resolve whether the NIV can function autonomously.

Immunostaining of nerve tracts with antiserum to GnRH in brains of diverse anurans and urodeles not only confirms the presence of a molecule structurally related to the mammalian peptide (see Crim and Vigna[13]), but reveals the complexity of the distribution of this material (see especially Nozaki et al.;[55] Demski[16]). Crim and Vigna[13] argue that the variable staining and even absence of staining in some species implies structural divergence in the molecule, since typically such studies rely on a single anti-GnRH serum. Unfortunately, these immunocytochemical studies have not been combined with other more complete types of quantification of brain GnRH, and the functional corollaries of the variations in the distributional network of immunoreactive fibers are unknown. It is possible that only part of this distributed neuronal system is involved with pituitary regulation and that the function of other tracts or neurons relates to extra-pituitary target sites, e.g., direct neural behavioral effects (see Demski[16] and below).

1. Changes in Endogenous GnRH. Limited quantitative data on brain GnRH immunoreactivity lend support to the functional significance of this material in anurans. For example, seasonal variation in the hypothalamic content of immunoreactive GnRH paralleled ovarian development in Xenopus laevis.[27] Similar seasonal correlations for both sexes were inferred from qualitative changes in the intensity of immunostaining of GnRH fibers in the forebrain and median eminence of the toad, Bufo japonicus;[24] artificial 'hibernation' also resulted in decreased staining. In the newt, Taricha granulosa, seasonal changes in immuno-GnRH content of infundibulum, rostral hypothalamus and POA paralleled the testis cycle.[73] Interestingly, all three regions showed a sharp drop in content shortly before the end of the breeding season but then increased again coincident with

recrudescence. Both studies suggest that the termination and onset of breeding might be related to cessation and restoration, respectively, of brain GnRH output. Orchidectomy of these newts during the breeding season resulted in a rise in GnRH in the POA at 30 days.[72] Unfortunately, independent measures of circulating Gn are not available for this species. Age-related changes in GnRH content (as measured by immunostaining) are consistent with the general maturation of the reproductive system in the bullfrog: The intensity and number of immunocytochemically stained cells containing GnRH localized in the ME and the anterior POA increase as the bullfrog matures; and no staining was observed in prometamorphic tadpoles.[12] These changes parallel ontogenetic changes in sensitivity of the bullfrog pituitary to exogenous GnRH.[49,62]

In the only experimental manipulation involving measurements of GnRH, Moore and Zoeller[54] found that handling stress and injection of corticosterone in the breeding season caused a rapid depression of plasma testosterone and an increase in GnRH content of infundibulum and rostral hypothalamus, but not the POA of male newts. Stress had different effects in the fall, since the POA content was altered. The time course of these changes indicates a rapid depression (within 1 h) and recovery (within 3 h) of GnRH release which might account for stress-related changes in gonadal activity; these data also help explain declines in plasma Gn and steroids in captured bullfrogs.[35] Results were interpreted in terms of GnRH synthesis and release on the assumption that the peptide was produced in the POA and stored in the ME. The validity of this assumption depends on the extent to which the basal medial hypothalamus is independent of the POA.

2. Biochemistry of GnRH: Deery[15] observed parallelism between GnRH and hypothalamic extracts (HE) from a *X. laevis* in GnRH-RIA. In a more extensive comparative immunological study, King and Millar[28] employed four antisera directed against different regions of the GnRH molecule. Consistent parallelism between displacement curves for GnRH and amphibian HE argued for the identity of the amphibian and mammalian molecules. The effectiveness of the antisera for discerning differences in molecular structure was indicated by the non-parallel displacement curves observed with teleostean, reptilian and avian HE. The RIA studies from Moore's lab (see above) also demonstrate similarity between the newt and mammalian GnRH. Identity between bullfrog and mammalian GnRH was further demonstrated by direct isolation studies,[64] which examined behavior of bullfrog GnRH on HPLC, activity in several RIA systems and amino acid sequence. Eiden et al.[20] similarly found immunological identity between hypothalamic

GnRH from frogs and GnRH, but also identified a second form of GnRH in the bullfrog that occurred in sympathetic ganglia, retinas and adrenals. While the second GnRH was also biologically active (induced LH release from mammalian pituitaries), it seems unlikely from its localization that this GnRH plays a role in pituitary Gn secretion. Rather, it may function as a neurotransmitter (e.g., Jan and Jan[23]).

It now seems likely that multiple forms of GnRH are characteristic of both anurans and urodeles, and both the forms discussed above may also occur in the brains in members of the two orders.[66] The presence of the mammalian-type GnRH in both argues for the primitive nature of this molecule. These data do not support the suggestion of differences in GnRH forms based on immunostaining techniques.[13] However, the relative proportions of each may vary with the age of the animal. For example, Branton et al.[6] were only able to identify the major adult (i.e., mammalian type) GnRH in the brains of postmetamorphic bullfrogs; tadpoles possessed only the nonmammalian form.

B. Reptiles

No direct studies on isolation and analysis have been completed for a reptilian GnRH. Indirect immunological evidence confirms the presence of a GnRH molecule in various parts of the brain, but suggests that it is structurally distinct from that of GnRH. Tentative evidence for the presence of a GnRH-like regulatory system in reptiles came from demonstrations of GnRH-reactive materials in hypothalamic nerve tracts. However, information on the reptile is limited in this respect and little can be said about the details of the distribution of GnRH or its interspecific variation. Doerr-Schott and Dubois[18] were the first to report GnRH-reactive neurons in the ME of a reptile (lizard). Although no clear neural pathways were visualized, immunoreaction was observed in perikarya of the dorsal cortex of the telencephalon near the anterior limits of the ventricles. Immuno-staining of a variety of reptiles with two antisera (against the Gly^{10} and Try^{5} regions) was of only limited success except for one snake.[55] In an Elaphe, the antisera stained perikarya in the medial septal nucleus (septo-preopticus) and adjacent medial basal hypothalamus with fibers traceable to the ME. Other lizards and snakes (seven spp.) showed only weak reactions in the ME and nowhere else. The turtle differed in possessing an additional group of perikarya in the telencephalic region. Reptiles are either unusual in containing very low amounts of GnRH or their GnRH is immunochemically distinct from the mammalian molecule. Nonetheless, the few positive results suggest that the reptilian GnRH tracts are fundamentally like those of other species, at least with regard to the ME.

Assuming functional parallelism to other vertebrates, GnRH-like centers involved in pituitary regulation have been inferred from the existence of sex steroid sensitive areas in reptilian brains (see Ball,[5] Crews,[11], Demski[16] for reviews). More important, lesions in or rostral to the anterior hypothalamic-POA of lizards, but not dorsal or caudal, caused testicular regression, suggesting interference with a stimulatory GnRH regulating center as in other vertebrates.[71] Since this region has been associated with sex steroid binding it may represent a major locus of GN regulation. On the other hand, lesions to the telencephalic septal area, implicated in Gn regulation in some other vertebrates, was apparently without effect in the lizard (reviewed by Crews[11]). Thus, there is no evidence for a bipartite GnRH control system in reptiles. No electrical stimulation studies have been reported for reptiles.

The presence of bioactive GnRH in reptiles was suggested by the demonstration that turtle HE stimulated release of LH from dispersed turtle pituitary cells.[21] However, it is not clear from their brief abstract whether controls were performed to account for the relatively high LH content likely to occur in such extracts due to contamination with pars tuberalis.[56]

1. Biochemical Characterization. Information on the structure of reptilian GnRH comes largely from immunological characterization. In their first comparative survey of vertebrate hypothalamic GnRH using a combination of chromatographic and radioimmunological techniques, King and Millar[27,28] concluded that GnRH in a tortoise and lizard was distinct from the mammalian molecule but indistinguishable from that of birds and teleosts. Major differences appeared to be in the region of Leu^7, with similarities to all other species at the NH_2- and COOH-termini. More recently, King and Millar[29] cited unpublished data indicating that lizard brain GnRH is identical to that of salmon GnRH ($[Try^7, Leu^8]$-GnRH) while alligator brain GnRH was reported to resemble chicken GnRH-I.[52] Additional HPLC and immunological studies revealed at least four different GnRH molecules in the brain of the lizard, *Cordylis nigra*, with the major fraction showing properties much like that of the salmon GnRH, but material resembling chicken GnRH-II also was detected.[63] Thus, these data are only partially consistent with their earlier conclusion regarding similarities between reptilian and avian hypothalamic GnRH. On the basis of RIAs with two distinct antisera, Lance et al.[30] also concluded that GnRH in alligator

hypothalami differed from the mammalian GnRH at position 8 and possibly also 7; i.e., it was considered to resemble the avian molecule. Direct isolation and characterization is clearly required but structural divergence from the molecule found in mammals and amphibians is certainly suggested. It is unclear why the several antisera used by Nozaki et al.[55] failed to provide better resolution of GnRH tracts in reptiles (see above).

Immunological studies of King and Millar[28] showed that hypothalamic GnRH content in reptiles was at least an order of magnitude greater than that of the brain as predicted from immunocytochemical studies. Vale and Licht (unpubl.) also found that hypothalami of sea turtles contain about the same amount of immunoreactive GnRH as the remainder of the brain combined. Aside from the report of Hall et al.[21] regarding seasonal variations in the GnRH activity of turtle HE, quantitative information on variations in hypothalamic GnRH content in reptiles is unavailable.

Results of work in progress on the isolation of GnRH-like factors from hypothalami and brains of the green turtle, *Chelonia mydas* (Vale, Rivier, Joachim and Licht, unpublished) have not been completely consistent with the results of indirect immunological approaches. Thus far, two distinct GnRH-like factors have been identified in the turtle. One of them appears very closely related to mammalian GnRH, and the other is not clearly like any of the other GnRHs identified so far. Both exist in relatively low amounts. Their exact distributions and relative activities on pituitary secretion remain to be determined.

II. REGULATION OF PITUITARY GONADOTROPIN RELEASE: *IN-VIVO* STUDIES

A. Amphibians

While lesion studies implicated the brain in the control of Gn secretion, they constitute weak evidence for the involvement of GnRH *per se* and do not provide evidence for the separate control of FSH and LH. The first indication that exogenous GnRH was biologically active in amphibians came from indirect studies of gonadal activity. Studies on urodeles deal largely with newts, especially *Triturus cristatus*. Chronic daily administration of synthetic GnRH (0.5 μg/8 g) directly upon the pituitary autograft of hypophysectomized newts for 1 month resulted in variable but increased testicular volumes and spermatogenesis.[45] Stimulation of both FSH and LH was inferred from previous data suggesting that testicular recrudescence required the presence of both Gns. In female newts, GnRH also stimulated some oocyte growth[44] and ovulation.[70] In the latter case, GnRH was effective only when administered daily for a month directly into the region of the *in situ* pituitary, and not when given intraperitoneally (ip) or into the ventricle, suggesting a high clearance rate. In the same species, similar *in situ* perfusion with GnRH elevated plasma testosterone and mobilized spermatozoa[3] and a GnRH analog (buserelin) increased testicular androgen content.[60] Intracerebroventricular injection of GnRH into another newt, *Taricha granulosa*, activated sexual behavior and also elevated plasma androgen levels, but only when administered early in the breeding season.[53] Administration of a GnRH antagonist suppressed sexual behavior in these males.[53] Daily ip treatment of axolotls, *Ambystoma mexicanum*, with long-acting GnRH agonists can induce courtship and spawning among otherwise sexually recalcitrant animals (Brothers, pers. comm.), and spawning has been induced in Spanish newts, *Pleurodeles waltl*, with this agonist (R. Barnes, pers. comm.). Direct measurements of the effects of GnRH on FSH and LH have not been made in urodeles. Thus, there is no information on relative effects on the two Gns or on the dynamics of the GnRH response in this order.

Data for anuran amphibians is more extensive than for urodeles. The first evidence for a GnRH action in this group came from the demonstration of a graded spermiation response to increasing doses of GnRH in the tree-frog, *Hyla regilla*.[31] Spermiation responses also have been reported for several *Rana*.[14,19,51,67] GnRH appears to be effective for inducing rapid spermiation in a wide array of anurans (R. Barnes, pers. comm.). Because this response is typically rapid (within 30-60 min) and relatively sensitive to GnRH, it has proven a valuable method for obtaining viable sperm for breeding purposes since it circumvents the need to sacrifice males. Acute responses to GnRH in female anurans, especially with regard to ovulation, have been less predictable than in males. GnRH treatment typically results in rapid increases in plasma sex steroids in both sexes of frog (e.g., Daniels and Licht,[14] McCreery et al.[51]). Thornton and Geschwind[68] observed ovulation in a few *X. laevis* after an injection of GnRH but only after extensive priming with Gns (hCG and PMSG). In bullfrogs, ovulation could not be obtained even with multiple daily doses of GnRH, but a long-acting agonist (described below) was effective after 3-4 daily injections;[51] this situation apparently reflects the requirement for prolonged pituitary stimulation by GnRH in this species (see McCreery and Licht[47]). However, single injections of long-acting GnRH agonist frequently induce ovulation with 12-24 h in other anurans; e.g., *R.* (*Pixiecephalus*) *adspersa* and *Ceratophrys ornatus*, *Scaphiopus hammondi*, *Latoria* (R. Barnes, personal communication). The effectiveness of these analogs has made them valuable for artificial breeding purposes. Spawning behavior also is induced frequently when pairs of frogs are treated with the GnRH agonists.

Several problems exist for interpreting results of GnRH treatments such as those discussed above in which Gns were not measured directly. First, with regard to induction of sexual behavior, GnRH may have direct neural effects independently of pituitary Gn release. Moore et al.[53] ascribed the behavioral actions of GnRH in male newts to such direct central effects, because other studies in this species indicated that sex behavior could not be explained by changes in circulating androgens. Studies in gonadectomized and hypophysectomized animals are clearly required to evaluate this hypothesis. However, Kelley[26] demonstrated that 2-days treatment with GnRH at a high dose (100 μg) stimulated receptivity in both intact and ovariectomized _X. laevis_. Although the effectiveness of GnRH in the latter required steroid (estradiol and progesterone) pretreatment, the results support the view that it might act independently of Gn release. The possibility that GnRH also may effect the gonad directly poses a second problem in interpreting _in-vivo_ responses to GnRH. In _R. pipiens_[65] and _R. esculenta_[59,60] GnRH or an analog (buserelin) stimulated _in-vitro_ testicular testosterone secretion either alone or in combination with Gn or pituitary extracts. Moreover, Pierantoni et al.[59] reported that GnRH stimulated elevation of plasma testosterone in hypophysectomized males (in our experience, such conclusions require rigorous evidence against the possibility of pituitary remnants). Thus, _in-vivo_ gonadal responses to GnRH do not necessarily imply elevated pituitary Gn secretion. On the other hand, failure to find GnRH in bullfrog testes and demonstration of very low circulating levels (<15 pg/ml) compared to those used _in vitro_[20] casts doubt on the physiological significance of the direct gonadal actions of GnRH. Moreover, we have been unable to demonstrate any action of GnRH, either alone or in combination with Gn, on steroid production or oocyte maturation in ovaries of _R. catesbeiana_ or _R. pipiens_.[22]

Direct evidence for _in-vivo_ responses of FSH and LH to GnRH in an amphibian is limited to the bullfrog, _R. catesbeiana_. The first study[14] revealed that the basic response to a single injection of GnRH was comparable to that of mammals: the response was rapid (plasma Gn peaked within 10-15 min); transitory (plasma Gn returned to baseline in 60 min); and involved both FSH and LH. Age and sex dependent variations were also evident--juveniles were less responsive than adults, and adult males were most responsive (in terms of sensitivity and magnitude of Gn increments). This same pattern was observed in subsequent studies,[51] which further demonstrated a similarity between frogs and rats in their response to a potent GnRH agonist [(imBzl-D-His^6, Pro^9-NEt)-GnRH]. The agonist was more potent and its effects more long-lasting, but again adult males were most responsive. Inhibition of GnRH actions by an antagonist [(Ac-dehydro-Pro^{1}, pCl-D-Phe^{2}, D-Trp^{3,6})-N -MeLeu^{7}-GnRH] provided further evidence of the similarity between receptor characteristics of mammal and frog.[51]

Similarities in basic pituitary response characteristics are consistent with the structural homology between the mammalian and *Rana* GnRH molecules but do not imply identity of GnRH receptors, since the frog and rat differed in their sensitivity to a heterologous (chicken) GnRH.[37] Chicken- and mammalian-GnRH were equipotent in the frog, but the former was only about 1% as active in mammals. GnRH receptors on the frog pituitary apparently exhibit less stringent requirements for structural features of the molecule than those of mammals (see also Millar and King[29]).

Another fundamental difference in GnRH physiology between the frog and mammal, namely, resistance to GnRH desensitization, became apparent during chronic treatment of bullfrogs with the long-acting GnRH agonist. Although such treatment effectively induces sterility in mammals due to development of pituitary refractoriness, female bullfrogs continued to respond to daily injections and, indeed, ovulation occurred in preovulatory females after 3-4 days.[51] Resistance to desensitization was apparent especially when frogs were infused continuously with GnRH via in-dwelling arterial cannulae.[47,48]. Unlike the situation in mammals, such chronic infusion of *R. catesbeiana* produced elevated plasma Gn (and sex steroid) levels for at least 4 days. The response pattern was typically biphasic, with a rapid increase in plasma Gns (within 0.5-1 h) followed ca. 24 h later by a second larger increase in plasma Gns indicating self-priming.[48]

The generally low sensitivity of female bullfrogs to acute GnRH treatment led to the hypothesis that pituitary responsiveness might vary with ovarian stage. However, individual responses to acute injection did not correlate with either follicular size or plasma steroid levels (e.g., estradiol or testosterone) during the ovarian cycle,[36] and early responses to GnRH infusion did not vary significantly among females in different ovarian stages,[47] but striking differences were observed during chronic infusion. Females in early follicular stages remained relatively unresponsive during 2-4 days of continuous infusion with 1-10 μg/h, whereas preovulatory females exhibited markedly elevated levels of Gn starting after about 12 hours of infusion. This enhanced sensitivity culminated in a pronounced 'surge' in plasma Gn coincident with a rise in progesterone and ovulation; these were the only conditions in which females were more responsive to GnRH than males. Combined with results for long-acting agonists, it appears that the

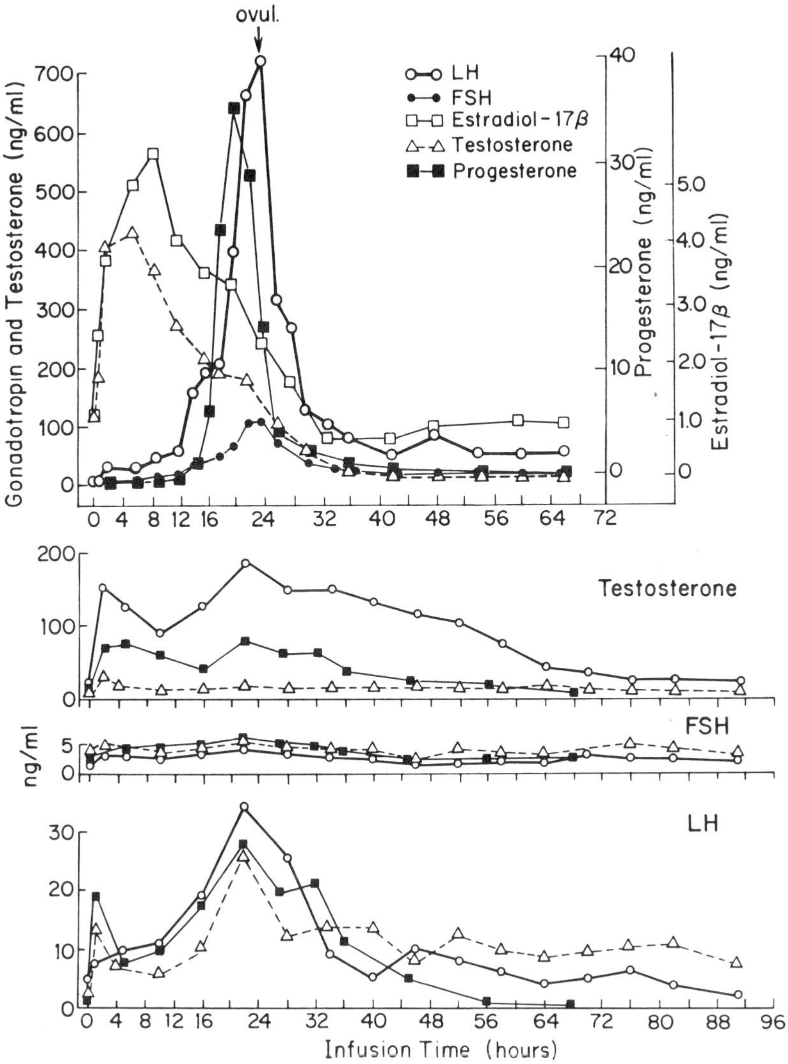

Fig. 1. In-vivo response to continuous infusion of 1 μg GnRH/hr in a representative adult female bullfrog with preovulatory ovarian follicles (top panel) and 3 individuals with ovaries in early follicular phases (bottom 3 panels). Plasma levels of steroids and gonadotropins were measured over a 3-4 day period via indwelling cannulae. Note that in each case final levels of gonadotropins were still elevated to about the same levels as were attained in the first few hours of the test although secretion rates were maximal at intermediate times. Reproduced from McCreery and Licht (1983) with permission.

frog pituitary may require prolonged exposure to GnRH to attain the sensitivity required for the periovulatory LH surge.

Ontogenetic differences in GnRH sensitivity also have been observed between the sexes in _R. catesbeiana_.[50] Tadpoles were wholly unresponsive to

GnRH, and males respond more to acute GnRH treatment than do females at metamorphosis; in fact, females remain relatively insensitive until after sexual maturation. The basis for these changes may be related in large part to the influence of androgen, notably the non-aromatizable androgen, 5α-dihydrotestosterone (DHT). Unlike mammals, estradiol appears to be relatively ineffective for GnRH 'priming' of the pituitary in frogs. Gonadectomy resulted in a marked decrease in responses of male but not female frogs, and these effects were reversed by implantation of silastic capsules containing DHT, but not estradiol.[50] Since both DHT and estradiol were equally effective in inhibiting the post-gonadectomy rise in plasma Gns, their negative feedback effects at the hypothalamic level are presumably distinct from sensitizing actions at the level of the pituitary. The rise in plasma DHT in late stages of ovarian development[35] could thus account for the increased GnRH responsiveness required for the ovulatory surge.

The gradual increase in plasma DHT commencing at metamorphosis in male bullfrogs can account for the ontogenetic changes in sexual differentiation in GnRH responsiveness.[49] Support for this conclusion comes also from the induction of GnRH sensitivity in DHT-implanted tadpoles.[49] The potentiating action of DHT may result in part from increased pituitary Gn content.[49] These results complement ontogenetic changes in GnRH staining in the bullfrog brain.[12]

B. Reptiles

In-vivo data for the actions of GnRH in reptiles based on tests with mammalian and avian GnRH molecules as well as a potent GnRH agonist (see above) have been equivocal or inconsistent compared to the amphibian data. Callard and Lance[8] first reported that a single injection of 4 μg GnRH elevated (somewhat erratically) plasma LH and progesterone in female painted turtles, *C. picta*. We were unable to repeat these results on females (even with 40 μg GnRH), but males responded to a single intracardiac injection of GnRH with increased plasma LH and testosterone; responses were considerably more prolonged than in the amphibians--LH continued to increase for at least 1 h.[39] Early tests of GnRH in green turtles, *C. mydas* also indicated a response of pituitary LH secretion (but not steroids),[33] but later tests with GnRH or agonist, even at high doses, failed to alter either plasma LH or gonadal steroids in this or another sea turtle, *Lepidochelys olivacea*.[38] Particularly noteworthy was the insensitivity of the sea turtles even when injected only a few hours before the normal ovulatory LH surge.[38] Similar unresponsiveness in both sexes of stinkpot turtle, *Sternotherus odoratus*, and cobra, *Naja naja*, was observed with injections of GnRH, GnRH agonist and avian GnRH (chicken GnRH-I).[37]

Three separate *in-vivo* studies provide at least indirect evidence for an action of GnRH in lizards. First, Arslan et al.[4] reported that injection of GnRH increased testicular androgen in a lizard, but dose and protocol were not given. Ciarci et al.[10] reported that daily injection of an agonist (buserelin) blocked spermatogenic recrudescence in the lizard, *Podarcis s. sicula*, whereas, injections every 4 days tended to prolong the period of maximal testicular activity; the former results suggest some desensitization due to constant treatment. Phillips et al.[58] reported that daily pulsatile treatment (exact doses were difficult to define) of female iguanas, *Iguana iguana*, with chicken GnRH-II induced a transient rise in plasma estradiol after several days and elicited sexual behavior in cohabiting males. Subsequent studies indicated that constant infusion with the peptide had similar effects (Phillips and Lasley, pers. comm.). Also, Lance et al.[30] observed a rise in plasma testosterone within 2 hr following injection of GnRH into mature and immature alligators (500 μg into a 30-50 kg animal); surprisingly, levels remained elevated for at least 24 h.

In general, results of direct tests on pituitary Gn release led us to postulate that either the reptilian pituitary might exhibit a high degree of species specificity to GnRH (consistent with their immunochemically distinctive GnRH) or that they differed from other vertebrates in their dependence on this peptide. Subsequent *in-vitro* studies failed to support either contention (see below).

Alderete et al.[1] presented evidence for a direct neural action of GnRH on sexual behavior in a female lizard, *Anolis carolinensis*, but there is otherwise little information on such extra-pituitary actions of the peptide in reptiles. It is of interest that we have consistently failed to observe pituitary response (by indirect measurement of testicular growth or androgen secretion) in males of this species. Evidence for direct effects of GnRH on the reptilian gonad (e.g., steroid secretion) as described for male frogs is lacking. Increased androgen in the lizard testis[4] parallels the situation that was attributed to a direct gonadal effect of GnRH in newts (see above).

III. *IN-VITRO* CHARACTERIZATION OF PITUITARY RESPONSIVENESS TO GnRH

A. Amphibians

In-vitro studies provide evidence for the direct action of GnRH on pituitary tissues and allow analysis of the mechanisms involved in these actions. The first such data were derived from bioassay (using the LH specific *Xenopus* oocyte maturation test) of *in-vitro* secretions by glands of *R. pipiens*. Thornton and Geschwind[68] demonstrated that HE prepared from *R. pipiens* (but not neurointermediate or cerebral cortical tissue) stimulated LH secretion from *R. pipiens* hemipituitaries in static incubations. Although dynamics of the response were not fully evaluated by the static incubation, this study indicated that LH secretion declined after about 4 h of GnRH stimulation. These glands appeared to be less sensitive to GnRH than mammalian pituitaries.

The development of an *in-vitro* superfusion system for single hemipituitaries combined with availability of sensitive and specific RIAs for each Gn allowed more detailed analyses of individual variability and the temporal response patterns to GnRH. The first study[62] confirmed that mammalian GnRH stimulated parallel secretion of both LH and FSH from *R. catesbeiana* and *R. pipiens* hemipituitaries, and showed that the sensitivity to GnRH was not appreciably different from that of mammals; e.g., 0.2 ng/ml GnRH stimulated a two-fold or greater elevation in Gn secretion from *R. pipiens* hemipituitaries.

Sexual differences in pituitary sensitivity to GnRH were verified also by *in-vitro* pituitary superfusion with *R. catesbeiana* glands.[62] Tissues from juvenile females ($<$ 100 g body wt.) consistently were less responsive to GnRH than those from males. Adult *R. pipiens* hemipituitaries did not exhibit consistent sexual differences, but two preovulatory females did show a much larger increase in LH and FSH secretion after 10-12 h of continuous superfusion with 20 ng GnRH/ml than was seen in any male. Thus, *in-vitro* data suggest that at least some of the sex differences and effects of reproductive stage on GnRH sensitivity involve properties of the anterior pituitary *per se*.

The maintenance of elevated Gn secretion during chronic superfusion with GnRH in *Rana* contrasted with results of static incubations with *R. pipiens* glands[68] and supported *in-vivo* results on the resistance to GnRH desensitization in the frog gland (see above). Pituitaries from either *R. pipiens* or *R. catesbeiana* superfused with minimal or high doses of GnRH retained responsiveness for at least 48 h.[62] Moreover, a second peak in FSH and LH secretion was observed after 7-12 h of continuous superfusion with GnRH as was observed *in vivo*.

B. Reptiles

The most consistent evidence for the sensitivity of the reptilian pituitary to GnRH and information on the role of gonadal feedback in controlling

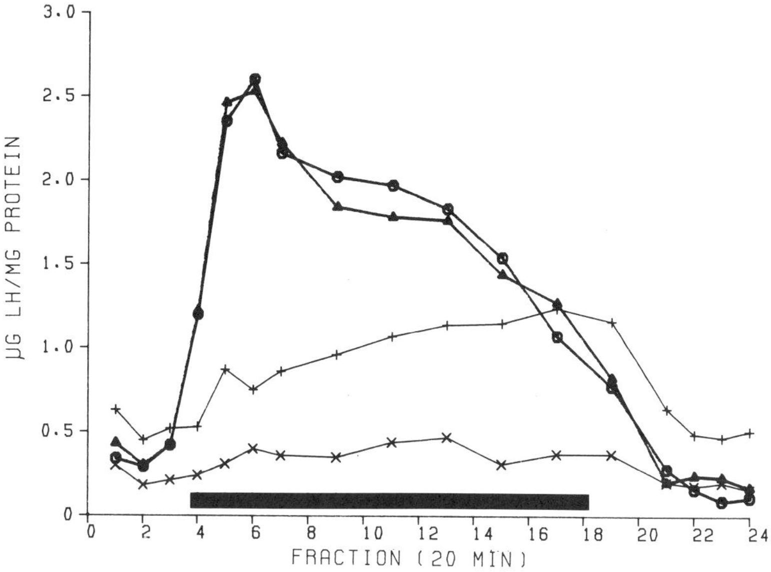

Fig. 2. In-vitro secretion of LH by hemipituitaries from male and female painted turtles Chrysemys picta superfused with 10 ng GnRH/ml for 5 hrs. (indicated by heavy black bar at bottom). Data represent averages for groups of 5 animals each that had been maintained at warm (O, males; +, females) or cold temperatures (Δ, males; X, females). Reproduced from Licht and Porter (1985) with permission.

pituitary GnRH sensitivity in reptiles come from in-vitro studies involving three species of turtle, including those that failed to respond in vivo.[34,39,40] Callard and Scanes[9] reported a small but dose-dependent LH response to GnRH in dispersed pituitary cells from C. picta in a static incubation system. Superfusion of individual turtle glands provided details on the dynamics of GnRH action and helped elucidate the basis for individual variability in GnRH responsiveness.

The first of the superfusion studies unexpectedly revealed that the turtle pituitary was capable of large and rapid increases in LH secretion in response to low doses of GnRH; but high individual variability suggested important seasonal, sexual and species differences.[40] For example, pituitaries from female P. scripta responded (albeit only briefly) in spring and late fall but not in summer, whereas all male glands were unresponsive during several hours of superfusion. Female C. picta also exhibited seasonality in their response, but the variation was more in the temporal aspects than in magnitude of response: in spring the LH response was very transitory (<1 h) compared to fall (at least 9 h), suggesting seasonal differences in resistance to 'desensitization.' Unlike *P. scripta*, glands from male *C. picta* were consistently more responsive to GnRH than those from females, although the sexes showed a different temporal pattern of response.[39] Gn secretion rates by male glands peaked within an hour and then showed progressive attenuation over the next 5 h of continuous GnRH superfusion, whereas female glands showed a progressive increment in the rate of Gn secretion during the entire period of treatment, suggesting 'self-priming.' Since the male glands secreted a very high percentage of their total Gn content, the gradual attenuation of Gn secretion may have represented an 'exhaustion' of stores rather than loss of sensitivity as in mammalian glands.

Manipulation of ovarian stage by temperature in *C. picta* suggested that ovarian feedback might underlie GnRH sensitivity.[39] Pituitaries from warm-treated females with regressed ovaries responded to GnRH consistently, whereas glands from females held at lower temperatures with enlarged ovaries were wholly unresponsive to GnRH (all incubations were at the same temperature). In contrast, pituitaries from males held at both temperatures showed similar responses to GnRH despite a pronounced difference in stage of testicular recrudescence (Fig. 2). Sexual differences in pituitary responsiveness could not be explained by gland content, since females had considerably higher LH concentrations (about six fold) than males. Interestingly, individual *in-vitro* responsiveness did not correlate well with results obtained *in vivo*: all females were unresponsive and only a few of the males responded to GnRH *in vivo*.[39]

The superfusion studies on *C. picta* also provided the first evidence for FSH secretion in response to GnRH in a reptile.[39] FSH and LH secretion were generally parallel in both sexes; i.e., like LH, FSH was more sensitive to GnRH in males than females, and FSH did not respond in cold-treated females. These studies also confirmed the specificity of the response to GnRH; thyrotropin secretion was unaffected by GnRH. Thus, the reptile is like the amphibian (and mammal) in that both Gns respond similarly to a single GnRH; this is presumably a primitive feature of the tetrapod hypothalamo-pituitary system.

More direct evidence for gonadal feedback on pituitary GnRH responsiveness came from superfusion studies on a third turtle, *S. odoratus.* Although unresponsive to various GnRH preparations *in vivo*,[37] glands from both sexes showed significant increases in LH secretion in response to GnRH superfusion.[34] Unlike the two species discussed above, there was no consistent sexual difference in either GnRH sensitivity or response pattern. Although gonadectomy did not consistently alter basal or GnRH-stimulated secretion, it tended to augment responsiveness in terms of increments in LH secretion. However, there was a striking inhibitory effect of androgen and especially estradiol on the GnRH-mediated LH secretion. In-vivo implants of estradiol reduced pituitary LH content and completely abolished GnRH responsiveness of gonadectomized turtles (both sexes) and reduced responsiveness of intact females. These results are consistent with the suggestion of an inverse correlation between sensitivity to GnRH and ovarian stage indicated by studies on intact females.

Contrary to expectations from in-vivo tests, the dose sensitivity of the three species of turtle glands was not appreciably different from that of other tetrapods (i.e., as little as 0.1-1 nM GnRH was effective); thus, conclusions regarding phylogenetic specificity in GnRH receptors based on in-vivo data must be reevaluated. The failure to obtain consistent responses in vivo may be related to extra-pituitary factors such as clearance rates. It is noteworthy in this context that in newts, GnRH was effective only if injected directly into the pituitary area (see above). Arguing against this explanation is the relatively prolonged response of plasma LH following intracardiac injection of GnRH in male C. picta,[39] and the prolonged elevation of plasma testosterone in alligators.[30] The high degree of individual variability, the relatively small increment in LH secretion, and the transitoriness of the response frequently observed in vitro may account in part for the equivocal results obtained from acute in-vivo injections in reptiles.

IV. MECHANISM OF ACTION OF GnRH

Details on the mechanisms of action of GnRH on the pituitary of nonmammalian species are fragmentary. There is still no information on reptiles, but limited progress on amphibians suggests that the physiology of the peptide is similar in many respects to that described for mammals. In-vitro studies utilizing mammalian pituitaries indicated the existence of two pools of gonadotropins. The first, or 'rapidly releasable' pool that is involved at initiation of GnRH treatment does not require protein synthesis for its release. Release from the second, or 'storage' pool, which is responsible for Gn secretion beyond the first peak is dependent on protein synthesis (e.g., Bourne and Baldwin[7].) In-vitro superfusion of R. pipiens hemipituitaries with a relatively high dose (100 ng/ml) of GnRH $\pm$ 71 μM cycloheximide revealed the presence of two similar pools of Gns in this anuran.[61] The first pool, which did not require protein synthesis for its release, accounted for about the first 2 h of elevated Gn secretion in response to GnRH. Further maintenance of elevated Gn secretion and self priming was reduced or blocked by cycloheximide (Fig. 3).

Superfusion of hemipituitaries from *R. pipiens* with medium containing 10X normal KCl induced a slight initial rise in FSH and LH secretion followed by steadily increasing Gn output (but not to levels achieved with 100 ng GnRH/ml) (Fig. 3). Treatment with GnRH or elevated KCl was ineffective in medium containing low Ca^{++} (Fig. 3). Similar results were obtained by the earlier static incubations of glands from this same species.[68] In medium containing normal Ca^{++} along with 2.5 mM cobalt chloride, the response to GnRH was greatly reduced (Co^{++} competes with Ca^{++} for calcium channels). Although elevated K^+ levels can stimulate Gn secretion, its elimination from the medium does not alter responsiveness to GnRH. Thus, elevated K^+ probably activates some element of the secretory mechanism associated with that activated by GnRH, resulting in Ca^{++} influx (Porter, unpublished).

V. FUTURE RESEARCH DIRECTIONS

Data for the physiology of GnRH in reptiles and amphibians is fragmentary from a comparative standpoint and in terms of the details of the integration of the hypothalamo-hypophysial gonadal axis. From the comparative perspective, it should be noted that no data exist for caecilian amphibians, and direct measurements of Gns are limited to a few anurans, a few species of turtles and one snake. Even these few data point to the need for a broader perspective; e.g., sexual differences varied among the three turtles studied. Discrepancies between *in-vivo* and *in-vitro* responsiveness must be resolved, especially for reptiles. This raises the important issue of the 'fate' of circulating GnRH in these nonmammalian species; e.g., there are no data on clearance rates. The problem of species specificity in GnRH receptors also exists; the nature of the reptilian and urodelean GnRH and whether structural differences are important for pituitary responsiveness remain to be resolved.

While the limited available data provide convincing evidence that GnRH may be the fundamental regulator of Gn secretion in both amphibians and reptiles, we are far from understanding the integration of this control. Particularly striking is the high degree of individual variability observed which may relate to age, sex, season, and reproductive stage to name a few candidates. It seems certain that gonadal feedback modifies pituitary sensitivity, but there is still no information on the extent to which such feedback involves direct effects at the pituitary as opposed to neural levels or whether it involves receptors or gland content. Amphibians and possibly also reptiles appear to differ from mammals in regard to the action of particular steroids. Induction of the Gn 'surge' required for ovulation is of

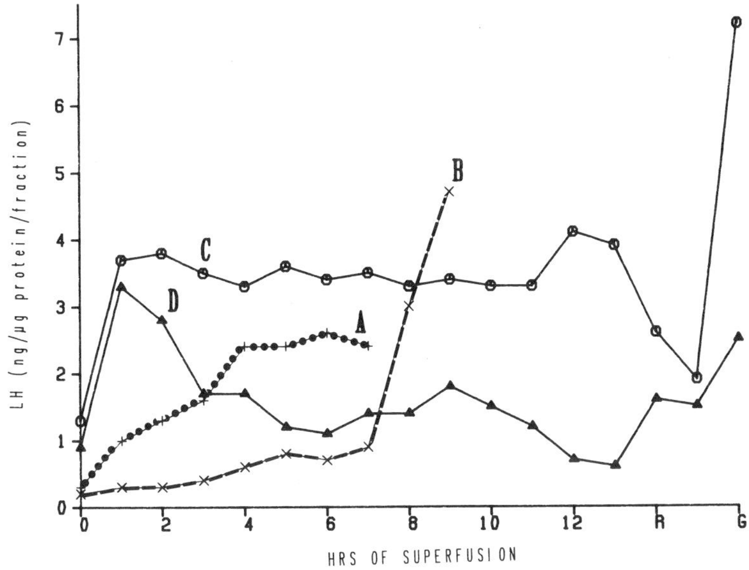

Fig. 3. Effects of various treatments on LH secretion in superfused hemipituitaries from <u>Rana pipiens</u>. Each curve represents averages from several replicates. Group A (dotted-line) shows the response of tissues incubated for 7 hrs in medium containing 10X normal KCl. All other groups were exposed to 100 mg GnRH/ml starting at time 0. In Group B (dashed line), the medium contained low CA^{++} + EGTA for 7 hr and Ca^{++} was then restored to normal. In C, normal medium was used (=controls) throughout; and in D, medium contained 71 µM cycloheximide. In the last two groups, glands were 'rinsed' in medium without GnRH for 2 hrs following 13 hrs. of superfusion (time="R") and were exposed to GnRH again for a final hour (time="G") to test for a priming effect. Data represent the composite of several experiments and curves represent averages of multiple replicates.
Data taken from Porter (1985) and unpublished data.

special interest in this regard; data are especially limited for reptiles. In mammals, gonadal factors other than steroids, e.g., inhibin, also may play a major role in regulating the differential release of FSH and LH. Since GnRHs in amphibians and reptiles show the same dual action on the release of the two gonadotropins, inhibin-like peptides also may berequired to explain differential regulation of the two gonadotropins; this is a completely unexplored area.

Although experience has taught us that temperature plays a major role in regulating endocrine activity (especially in reproductive events) in the ectothermal tetrapods, there is as yet no information regarding direct thermal effects on the GnRH-pituitary axis. Such considerations are key for our understanding of sensitivity, responsiveness, resistance to desensitization and self-priming. A key feature of the GnRH-pituitary axis in mammals is the pulsatile or episodic nature of GnRH and the attendant requirement for such stimulation at the pituitary level. This important issue remains to be explored for these nonmammalian species. Available data for amphibians could be used to argue that they are fundamentally different from mammals in lacking this requirement for temporal variability, but there are no data for reptiles, and nothing is known about normal short-term temporal patterns of circulating Gn or GnRH in either group. Until further details on the nature of the GnRH actions are known, further progress on identifying the neural centers of regulation may be difficult. An important issue for interpreting much of the existing data regards the different extrapituitary sites of action of GnRH; this is probably of special importance for behavioral responses.

Many aspects of the mechanisms of action of GnRH remain to be explored in amphibians, and there are no data for reptiles. Future work must distinguish between Gn synthesis and release from the pituitary and GnRH synthesis and release from the brain. While Ca^{++} appears to represent a potentially important "second messenger" for GnRH, its actions need more detailed analysis of receptor-dependent and receptor-independent interactions. A better understanding of the mechanisms of GnRH action may provide insights into the basis of apparent differences in responsiveness among amphibians, reptiles and mammals.

VI. ACKNOWLEDGMENT

Preparation of this manuscript and support for some of the unpublished results was provided by grant PCM 8406751 from the National Science Foundation.

VII. REFERENCES

1. Alderete, M.R., Tokarz, R.R. and Crews, D. (1980). Luteinizing hormone-releasing hormone (LHRH) and thyrotropin releasing hormone (TRH) induction of female sexual receptivity in the lizard, *Anolis carolinensis*. Neuroendocrinology 30, 200-205.
2. Alpert, L.C., Brawer, J.R., Jackson, I.M.D., and Reichlin, S. (1976). Localization of LH-RH in neurons in frog brain (*Rana pipiens* and *Rana catesbeiana*). Endocrinology 98, 910-921.
3. Andreoletti, G.E., Mazzi, V., Vellano, C. and Sacerdote, M. (1980). Effects of LHRH on spermiation and serum testosterone levels in the winter newt, *Triturus cristatus carnifex* (Laur.). Monitore zool. ital. (N.S.) 14, 255-262.

4. Arslan, M., Saulat, T., Saeed, S.A. and Qazi, M.H. (1985). Regulation of gonadal function in a lizard. IXth Intern. Congr. Comp. Endocrinol., Hong Kong. P58 (Abstr).
5. Ball, J.N. (1981). Hypothalamic control of the pars distalis in fishes, amphibians, and reptiles. Gen. Comp. Endocrinol. 44, 135-170.
6. Branton, W. D., Jan, L. Y., Jan, Y. N. (1982). Non-mammalian luteinizing hormone-releasing factor (LRF) in tadpole and frog brain. 12th Ann. Meeting, Soc. Neurosc., Minn. 1082, p. 14 (Abstr).
7. Bourne, G. A. and Baldwin, D. M. (1980). Extracellular CA^{++}-independent and dependent components of the biphasic release of LH in response to luteinizing hormone-releasing hormone in vitro. Endocrinology 107, 780-788.
8. Callard, I.P., and Lance, V. (1977). The control of reptilian follicular cycles. In "Reproduction and Evolution" (Calaby, J.H. and Tyndale-Biscoe, C.H. ed.), pp. 199-210. Australian Acad. Sci., Canberra City.
9. Callard, I.P. and Scanes, C.G. (1975). Synthesis of radioimmunoassayable LH by suspended pituitary cells derived from the turtle, Chrysemys picta. Proc. 57th Annu. Meet. Endocrine Soc., New York. Abstr. 307.
10. Ciarcia, G., Angelini, F., Botte, V. (1983). Effects of a chronic treatment with the luteinizing hormone-releasing hormone agonist buserelin, on the gonads of the lizard Podarcis s. sicular Raf. Atti Accad. Nazion. dei Lincei, Rend Cl. Sci. fis., mat. Series VIII, 74, 425-429.
11. Crews, D. (1979). The neuroendocrinology of reproduction in reptiles. Biol. Reprod. 20, 51-73.
12. Crim, J.W. (1985). Immunocytochemistry of luteinizing hormone-releasing hormone and sexual maturation of the frog brain: comparisons of juvenile and adult bullfrogs (Rana catesbeiana). Gen. Comp. Endocrinol. 59, 424-433.
13. Crim, J.W. and Vigna, S.R. (1983). Brain, gut and skin peptide hormones in lower vertebrates. Amer. Zool. 23, 621-638.
14. Daniels, E. and Licht, P. (1980). Effects of gonadotropin-releasing hormone on the levels of plasma gonadotropins in the bullfrog, Rana catesbeiana. Gen. Comp. Endocrinol. 42, 455-463.
15. Deery, D.J. (1974). Determination by radioimmunoassay of the luteinizing hormone-releasing hormone (LH-RH) content of the hypothalamus of the rat and some lower vertebrates. Gen. Comp. Endocrinol. 24, 280-285.
16. Demski, L.S. (1984). The evolution of neuroanatomical substrates of reproductive behavior: sex steroid and LHRH-specific pathways including the terminal nerve. Amer. Zool. 24, 809-830.
17. Dierickx, K. (1974). Identification of adenohypophysiotropic neurohormone producing cells in Rana temporaria. In "Neurosecretion --The final neuroendocrine pathway." (Knowles, F. and Vollrath, L., eds.) pp. 170-181. Springer, Berlin.
18. Doerr-Schott, J. and Dubois M.P. (1978). Immunohistochemical localization of different peptidergic substances in the brain of amphibians and reptiles. In "Comparative Endocrinology" (Gaillard, P.J. and Boer, H.H., eds.), pp. 367-370. Elsevier, Amsterdam.
19. Easley, K.A., Culley, D.D., Jr., Horseman, N.D. and Penkala, J.E. (1978). Environmental influences on hormonally induced spermiation of the bullfrog, Rana catesbeiana. J. Exp. Zool. 207, 407-416.
20. Eiden, L.E., Loumaye, E., Sherwood, N., Eskay, R.L. (1982). Two chemically and immunologically distinct forms of luteinizing hormone-releasing hormone are differentially expressed in frog neural tissues. Peptides 3, 323-327.

21. Hall, T.R., Chadwick, A., Scanes, C.G. and Callard, I.P. (1978). Effects of hypothalamic extract and steroids on secretion of prolactin, growth hormone and luteinizing hormone from pituitaries of Chrysemys picta cultured in vitro. J. Endocrinol. 76, 169-170.
22. Hubbard, G.M. and Licht, P. (1985). In vitro study of the direct ovarian effects of gonadotropin releasing hormone (GnRH) in the frogs, Rana pipiens and Rana catesbeiana. Gen. Comp. Endocrinol. 60, 154-161.
23. Jan, L.Y. and Jan, Y.N. (1981). Role of an LHRH-like peptide as a neurotransmitter in sympathetic ganglia of the frog. Fed. Proc. 40, 2560-2564.
24. Jokura, Y. and Urano, A. (1985). An immunohistochemical study of seasonal changes in luteinizing hormone-releasing hormone and vasotocin in the forebrain and the neurohypophysis of the toad, Bufo japonicus. Gen. Comp. Endocrinol. 59, 238-245.
25. Jorgensen, C. B. (1968). Central nervous control of adenohypophysial functions. In "Perspectives in Endocrinology" (Barrington, E.J.W. and Jorgensen, C.B., eds.), pp. 469-541. Academic Press, London.
26. Kelley, D.B. (1982). Female sex behaviors in the South African clawed frog, Xenopus laevis: Gonadotropin-releasing, gonadotropic, and steroid hormones. Horm. Behav. 16, 158-174.
27. King, J.A. and Millar, R.P. (1979). Hypothalamic luteinizing hormone-releasing hormone content in relation to the seasonal reproductive cycle of Xenopus laevis. Gen. Comp. Endocrinol. 39, 309-312.
28. King, J.A. and Millar, R.P. (1980). Comparative aspects of vertebrate luteinizing hormone-releasing hormone structure and function in vertebrate phylogeny. Endocrinology 106, 707-717.
29. King, J.A. and Millar, R.P. (1984). Isolation and structural characterization of chicken hypothalamic luteinizing releasing hormone. J. Exp. Zool. 232, 419-423.
30. Lance, V.A., Vilet, K. and Bolaffi, J.L. (1985). Effect of mammalian luteinizing hormone-releasing hormone on plasma testosterone in male alligators, with observations on the nature of alligator hypothalamic gonadotropin releasing hormone. Gen. Comp. Endocrinol. 60, 138-143.
31. Licht, P. (1974). Induction of sperm release in frogs by mammalian gonadotropin-releasing hormone. Gen. Comp. Endocrinol. 23, 352-354.
32. Licht, P. (1979). Reproductive endocrinology of reptiles and amphibians. Ann. Rev. Physiol. 41, 337-351.
33. Licht, P. (1980). Evolutionary and functional aspects of pituitary gonadotropins in the green turtle, Chelonia mydas. Amer. Zool. 20, 565-574.
34. Licht, P., Khorrami-Yaghoobi, P. and Porter, D.A. (1985). Effects of gonadectomy and steroid treatment on plasma gonadotropins and the response of superfused pituitaries to gonadotropin releasing hormone (GnRH) in the turtle Sternotherus odoratus. Gen. Comp. Endocrinol. 60, 441-449.
35. Licht, P., McCreery, B.R., Barnes, R. and Pang, R. (1983). Seasonal and stress related changes in plasma gonadotropins, sex steroids and corticosterone in the bullfrog, Rana catesbeiana. Gen. Comp. Endocrinol. 50, 124-145.
36. Licht, P., McCreery, B.R. and Barnes, R. (1983). Relation between acute pituitary responsiveness to gonadotropin releasing hormone (GnRH) and the ovarian cycle in the bullfrog, Rana catesbeiana. Gen. Comp. Endocrinol. 51, 148-153.
37. Licht, P., Millar, R., King, J.A., McCreery, B.R., Mondonca, M.T.,

Bona-Gallo, A. and Lofts, B. (1984). Effects of chicken and mammalian gonadotropin-releasing hormones (GnRH) on in vivo pituitary gonadotropin release in amphibians and reptiles. Gen. Comp. Endocrinol. 54, 89-96.

38. Licht, P., Owen, D., Cliffton, K. and Penaflores, C. (1982). Changes in LH and progesterone associated with the nesting cycle of the olive ridley sea turtle, Lepidochelys olivacea. Gen. Comp. Endocrinol., 48, 247-253.

39. Licht, P. and Porter, D.A. (1985). In vivo and in vitro responses to gonadotropin releasing hormone in the turtle, Chrysemys picta, in relation to sex and reproductive stage. Gen. Comp. Endocrinol. 60, 75-85.

40. Licht, P. and Porter, D.A. (1985). LH secretion in response to gonadotropin releasing hormone (GnRH) by superfused pituitaries from two species of turtles. Gen. Comp. Endocrinol. 59, 442-448.

41. Malacarne, G. and Giacomoa, C. (1978). Lesions to the CNS and ovariectomy in Triturus cristatus carnifex Laur. Effects on sexual behavior. Atti. Accad. Sci. Torino Cl. Sci. Mat. Fis. Natur. 112, 49-61.

42. Mazzi, V. (1970). The hypothalamus as a thermodependent neuroendocrine center in urodeles. In "The hypothalamus" (Martini, L., Motta, M. and Frashchini, F. eds.) pp. 663-676. Academic Press, New York.

43. Mazzi, V. (1978). Effects on spermatogenesis of permanent lesions of the rostral preoptic area in the crested newt (Triturus cristatus carnifex, Laur.). Gen. Comp. Endocrinol. 34, 247-250.

44. Mazzi, V., Vellano, C. and Bona, A. (1974). Oocyte recruitment in the ovaries of aestivating pituitary-autografted hypophysectomized female newts treated with synthetic LH-RH. Atti. Accad. Sci. Torino (I Cl., Sci. mat. fis. nat.) 108, 405-411.

45. Mazzi, V., Vellano, C., Colluci, D., and Merlo, A. (1974). Gonadotropin stimulation by chronic administration of synthetic luteinizing hormone releasing hormone in hypophysectomized pituitary grafted male newts. Gen. Comp. Endocrinol. 24, 1-9.

46. McCreery, B.R. (1984). Pituitary gonadotropin release by graded electrical stimulation of the preoptic area in the male bullfrog, Rana catesbeiana. Gen. Comp. Endocrinol. 55, 367-372.

47. McCreery, B.R. and Licht, P. (1983). Induced ovulation and changes in pituitary responsiveness to continuous infusion of gonadotropin-releasing hormone (GnRH) during the ovarian cycle in the bullfrog, Rana catesbeiana. Biol. Reprod. 29, 863-87.

48. McCreery, B.R. and Licht, P. (1983). Pituitary and gonadal responses to continuous infusion of gonadotropin releasing hormone (GnRH) in the male bullfrog, Rana catesbeiana. Biol. Reprod. 29, 129-136.

49. McCreery, B.R. and Licht, P. (1984). The role of androgen in the development of sexual differences in pituitary responsiveness to gonadotropin releasing hormone (GnRH) agonist in the bullfrog, Rana catesbeiana. Gen. Comp. Endocrinol. 54, 350-359.

50. McCreery, B.R. and Licht, P. (1984). Effects of gonadectomy and sex steroids on pituitary gonadotropin release and response to gonadotropin releasing hormone (GnRH) agonist in the bullfrog, Rana catesbeiana. Gen. Comp. Endocrinol. 54, 283-295.

51. McCreery, B.R., Licht, P., Barnes, R., Rivier, J. and Vale, W. (1982). Actions of agonistic and antagonistic analogs of gonadotropin releasing hormone (Gn-RH) in the bullfrog, Rana catesbeiana. Gen. Comp. Endocrinol. 46, 511-520.

52. Millar, R.P. and King, J.A. (1984). Structure-activity relations of LHRH in birds. J. Exp. Zool. 232, 425-430.

53. Moore, F.L., Miller, L.J., Spielvogel, S.P., Kubiak, T. and Folkers, K. (1982). Luteinizing hormone-releasing hormone involvement in the reproductive behavior of a male amphibian. Neuroendocrinology 35, 212-216.
54. Moore, F.L. and Zoeller, R.T. (1985). Stress-induced inhibition of reproduction. Evidence of suppressed secretion of LHRH in an amphibian. Gen. Comp. Endocrinol. 60, 252-258.
55. Nozaki, M., Tsukahara, T. and Kobayashi, H. (1984). Neuronal systems producing LHRH in vertebrates. In "Endocrine correlates of reproduction" (Ochiai, K., Arai, Y., Shioda, T. and Takahashi, M., eds.), pp. 3-27. Japan Sci. Soc. Press, Tokyo.
56. Pearson, A.K. and Licht, P. (1982). Morphology and immunocytochemistry of the turtle pituitary gland with special reference to the pars tuberalis. Cell Tissue Res. 22, 81-100.
57. Peter, R.E. (1983). Evolution of neurohormonal regulation of reproduction in lower vertebrates. Amer. Zool. 23, 685-695.
58. Phillips, J.A., Alexander, N., Karesh, W.B., Millar, R. and Lasley, B.L. (1985). Stimulating male sexual behavior with repetitive pulses of GnRH in female green iguanas, Iguana iguana. J. Exp. Zool. 234.
59. Pierantoni, R., Fasano, S., DiMatteo, L., Minucci, S., Varriale, B., Chieffi, G. (1984). Stimulatory effect of a GnRH agonist (buserelin) in in vitro and in vivo testosterone production by the frog (Rana esculenta) testis. Mol. Cell. Endocr. 38, 215-219.
60. Pierantoni, R., Iela, L. D'Istria, M., Fasano, S., Rastogi, R.K., Delrio, G. (1984). Seasonal testosterone profile and testicular responsiveness to pituitary factors and gonadotrophin releasing hormone during two different phases of the sexual cycle of the frog (Rana esculenta). J. Endocr. 102, 387-392.
61. Porter, D.A. (1985). The effects of cycloheximide on in vitro response of Rana pipiens pituitaries to continuously superfused gonadotropin-releasing hormone (GnRH). Biol. Reprod. 33, 393-400.
62. Porter, D.A. and Licht, P. (1985). Pituitary responsiveness to superfused gonadotropin releasing hormone in two species of ranid frogs. Gen. Comp. Endocrinol. 59, 308-315.
63. Powell, R.C., King, J.A. and Millar, R.P. (1985). [Trp^7, Leu^8] LH-RH in reptilian brain. Peptides 6, 223-227.
64. Rivier, J., Rivier, C., Branton, D., Millar, R., Spiess, J. and Vale, W. (1981). HPLC purification of ovine CRF, rat extrahypothalmic brain somatostatin and frog brain GnRH. In "Peptides: Synthesis-structure-function" (Rich, D.H. and Gross, E., eds.), pp. 771-776. Proc. Seventh Amer. Peptide Symposium, Pierce Chem. Co., Rock
65. Segal, S.J. and Adejuwon, C.A. (1979). Direct effect of LHRH on testicular steroidogenesis in Rana pipiens. Biol. Bull. 157, 393.
66. Sherwood, N.M., Zoeller, R.T. and Moore, F.L. (1986). Multiple forms of gonadotropin-releasing hormone in amphibian brains. Gen. Comp. Endocrinol. 61, 313-322.
67. Smith-Gill, S.J. and Berven, K.A. (1980). In vitro fertilization and assessment of male reproductive potential using mammalian gonadotropin-releasing hormone to induce spermiation in Rana sylvatica. Copeia 1980, 723-728.
68. Thornton, V.F. and Geschwind, I.I. (1974). Hypothalamic control of gonadotropin release in amphibians: Evidence from studies of gonadotropin release in vitro and in vivo. Gen. Comp. Endocrinol. 23, 294-301.

69. Van Deusen, E. (1973). Experimental studies on a mutant gene (e) preventing the differentiation of eye and normal hypothalamus primordia in the axolotl. Dev. Biol. 34, 135-158.
70. Vellano, C., Bona, A., Mazzi, V. and Colucci, D. (1974). The effect of synthetic luteinizing hormone releasing hormone on ovulation in the crested newt. Gen. Comp. Endocrinol. 24, 338-340.
71. Wheeler, J.M. and Crews, D. (1978). The role of the anterior hypothalamus preoptic area in the regulation of male reproductive behavior in the lizard, *Anolis carolinensis*: Lesion studies. Horm. Behav. 11, 42-60.
72. Zoeller, R. T. (1982). Changes in LHRH content of infundibulum, preoptic and septal areas following castration of male newts. Amer. Zool. 22, 954 (Abstr. 579).
73. Zoeller, R.T. and Moore, F. L. (1985). Seasonal changes in LHRH concentrations in microdissected brain regions of male rough-skinned newts (*Taricha granulosa*). Gen. Comp. Endocrinol. 58, 222-230.

PROLACTIN AND REPRODUCTION

Valdo Mazzi and Camillo Vellano

Dipartimento di Biologia Animale
Università di Torino
Via Giolitti 34
10123 Torino
Italia

In this review we set out to examine the state-of-the-art research into the effects of prolactin (PRL) on various aspects of reproduction in anamniote vertebrates and reptiles. As an introduction, we will examine our knowledge of PRL receptors from a comparative viewpoint (see also Ensor[61]).

I. PROLACTIN RECEPTORS

A. Overview of Mammals

Mammalian PRL receptors have been identified in numerous PRL target tissues including ovary and testis (see later) and have been isolated in particularly pure form from mammary gland and liver tissue.[58] The receptor is protein in nature and has a molecular weight of some 32,000 daltons (affinity labelling techniques combined with SDS-polyacrylamide gel electrophoresis were used to denature and dissociate solubilized PRL receptors). Higher molecular weight forms have also been reported. These probably contain PRL receptor aggregates consisting of two or more M_r = 31,000 binding components, although it is not clear whether other components, which do not contribute to the hormone binding subunit, exist in the overall molecule.[91]. In rat liver, the receptors differ from other liver receptors in involving reactive sulphydryl groups.[146]

After irreversibly binding to the receptor,[81] PRL is probably, like insulin,[68] internalized in the form of ligand-receptor complexes and transported (at least in the liver) to preferential sites in the Golgi apparatus, lysosomes and prelysosomes (see Kelly et al.,[91] Ferland et al.[69] for references). The lysosome- or prelysosome-bound hormone binds irreversibly to its receptor, and the hormone-receptors become partially solubilized.[69] The intracellular receptor may represent an intermediate step in new receptor synthesis before incorporation into the membrane.[1]

PRL interacts with the cell membrane receptor to induce the release of a second messenger, probably a small peptide molecule derived from the receptor molecule itself. In the mammary gland, this second messenger stimulates the transcription of milk protein genes from isolated mammary nuclei. When the messenger (mediator) interacts with the nucleus, it probably dephosphorylates one of the nuclear proteins. Milk protein gene stimulation appears to involve initiation of mRNA sequences.[91] For the first time, Shiu and Friesen[138] succeeded in purifying a PRL receptor antiserum from rabbit mammary gland tissue. The serum inhibited PRL action thus providing evidence for an obligatory functional role of the PRL receptors in mediating the action of this hormone. This antiserum was found to be capable of:

(1) Supporting the synthesis of casein and the uptake of α-isoaminobutyric acid by rabbit mammary explants in culture.[138]

(2) Reducing the effect of injected PRL on rat mammary glands and ovaries.[21]

(3) Both mimicking and inhibiting the action of prolactin on rabbit mammary explants[56a] and rat livers.[131]

(4) Inhibiting the binding of labelled PRL in various rabbit, pig and rat tissues.[90]

(5) Inducing the release of the mediator.[91]

Partially purified PRL antisera have been obtained for several species: binding studies suggest that the receptor molecule is more or less exposed to interaction with the antisera or that the receptor protein differs somewhat between species[90] and perhaps even between different organs in the same species. Indeed, the tacit assumption that there is only one receptor type was apparently contradicted by recent research when significant differences were found in antigenic structure and hormone affinity between PRL receptors from different tissues of the rabbit.[90] The kidney cortex receptor seems to be derived from a primitive form of the PRL receptor concerned with the regulation of electrolyte balance.[151]

B. Fishes

Data are available only for teleostean fishes, and this information on PRL receptor distribution amongst different organs and species is scanty and inconsistent. According to Fryer,[74] the specific binding of iodinated ovine prolactin (oPRL) to kidney preparations from freshwater tilapia was negligible. This finding was confirmed by Edery et al.[60] who also found a relatively high specific binding rate (5-10%) in liver, ovary and testis and lower specific binding in intestine, gill and bladder tissue. The problem of identifying PRL receptors in tilapia has been complicated by the recent finding of two separate PRL molecules with distinct but overlapping physiological effects.[143] In _Colis lalia_, only the kidney tubules were found to be sites of specific mammalian PRL localization.[78]

C. Amphibians

For amphibians, PRL receptors were identified for the first time in adult _Rana catesbeiana_ kidney membranes[128] and physically characterized in bullfrog tadpole livers and tail fins. The PRL-receptor complex has a molecular weight of 114,000 and 103,000 in the liver and tail fin respectively. The two receptor types differ in their sensitivity to PRL binding, to neuraminidase post-treatment of the membranes and in having different charges.[37,38] In some organs of _R. catesbeiana_, PRL receptor concentration has been found to vary between the larval and adult forms, appearing first in the gills and later in the kidneys[155] where the receptors proliferate during the later stages of metamorphosis.[77] Subsequent research on numerous anuran and urodelan species showed that few PRL receptors are present in the liver, although they are easily identifiable in the urinary bladder.[154]

In anuran kidneys, the degree of binding to renal microsomal fractions is generally higher in more advanced terrestial forms (Bufonidae, Ranidae). In totally aquatic and more primitive _Xenopus laevis_, where the receptors have been shown to reside in the proximal convoluted tubule by histoautoradiographic techniques,[80] the binding is low.[154] Amongst the urodeles, renal binding is low in _Siren lacertina_ and _Cryptobranchus alleganiensis_ and undetectable in _Necturus maculosus_, the most larval-like aquatic species studied.[154]

All of these investigations employed oPRL. In view of recent findings on molecular evolution of PRL (see Nicoll et al.[122]), and a difference between solubilized receptors obtained from amphibian and mammalian tissues,[38] it would appear more appropriate to use PRL isolated from the actual species upon which the investigations are being conducted. This approach was adopted by D'Istria et al.[56] They used PRL purified from the crested newt (_Triturus cristatus carnifex_) to study the effects of thyroidectomy on PRL and androgen receptors in the dorsal skin and tail fin of this species. It was found that PRL receptors decrease in the presence of T_4 and increase in the presence of PRL in the dorsal skin and tail fin.[56]. In contrast PRL receptors increase in kidney microsomal membrane fractions of R. catesbeiana tadpole,[156] probably due to delay in the normal development of the mesonephron. We thus cannot exclude the possibility that hormones determining the number of PRL receptors vary from species to species and from organ to organ.

D. Reptiles

PRL receptors have not been reported in reptiles. In birds PRL binding sites have been demonstrated and characterized in pigeon crop sac mucosa using biochemical and autoradiographic techniques (see Shain et al.[137] for references). In the liver of Streptopelia risoria receptors demonstrate a high affinity but a relatively low binding capacity; they are probably regulated by sex hormones.[32]

II. EFFECT OF PROLACTIN ON GONADS, SEXUAL ACCESSORIES AND SECONDARY SEXUAL CHARACTERS

A. Overview of Mammals

Prolactin receptors have been identified in testicular Sertoli cell of the rat[35,42,43,46] and hamster.[94] When PRL interacts with these receptors, it: (1) stimulates steroid production, interfering with acetate incorporation,[82,83] cholesterol metabolism and testicular dehydrogenase activity;[6] (2) stimulates testosterone production.[7,23,82,83,132] Blood testosterone levels fall after treatment with ergocryptine.[23] Prolactin may regulate its own receptors (although contrasting data are reported; see Klemcke et al.[94]) and those of LH, at least in rodents.[4] It has been known for some time that the prostate and seminal vesicle both contain PRL receptors and PRL target tissues. Prolactin brings about development of these organs and, acting synergistically with androgen, stimulates secretion (see de Vlaming,[53] Clarke and Bern[45] for references).

Prolactin receptors also have been found in the ovary, localized in the cells of the granulosa, the corpus luteum and the thecae. During follicular growth and development, granulosa PRL receptors are induced and maintained by a combination of luteinizing hormone (LH), follicle-stimulating hormone (FSH) and estradiol. Prolactin alone is not capable of inducing the formation of new LH receptors even though it is indispensable in the maintenance of these receptors in luteinized granulosa cells and the corpus luteum.

Prolactin also acts on the hypothalamus to regulate sensitivity to the testosterone feedback on pituitary FSH and LH secretion. This effect is gonadotropic in the golden hamster and antigonadotropic in the rat.[107]

Information on the role of prolactin in gonadotropin secretion and control of the ovarian follicle at different stages of maturity, ovulation and corpus luteum functions is briefly summarized according to McNeilly's recent survey.[116] A condition of hyperprolactemia hastens the onset of puberty in immature rats since it makes the granulosa cell LH receptors more sensitive to LH production. Prolactin plays in important part in maintaining the corpus luteum in the rat. In other species PRL does not act directly on steroid production but, in small doses, has a permissive role to provide receptors for the stimulation of or precursor for the production of steroids. In the rat and mouse it has been found that PRL plays a part in luteolysis, which is in turn dependent on gonadotropin and uterine luteolysins. The factors determining whether PRL acts as a luteotropic or a luteolytic agent have not, however, been identified clearly.

Prolactin's main effect on the follicle is the promotion of progesterone synthesis and secretion by the granulosa cells, probably through interaction with ovarian steroid hormones. Prolactin also inhibits secretion of estrogens by reducing aromatase activity. In vitro, it antagonizes LH-stimulated estrogen production in rat ovarian interstitial cells and isolated cow theca. Before weaning, the suckling stimulus and PRL suppress ovarian activity. Recovery of activity is delayed since suckling and/or prolactin bring about a reduction in hypothalamic gonadotropin-releasing hormone (GnRH) production with a resultant decrease in release of LH. Prolactin could also have a direct effect on the action of gonadotropins at the ovarian level. Finally, PRL could modulate estrogen secretion by altering gonadal steroid receptor levels in the hypothalamus.

Prolactin acts in synergism with estradiol and progesterone to promote the secretion of a sialic acid-rich mucus in the rat vagina. The effect is strengthened by growth hormone (GH). PRL also, together with the estrogens, brings about the production of specific antigens by the vaginal cells and helps maintain uterine quiescence during pregnancy by inhibiting oxytocin-induced myometrial contractions in several species (see de Vlaming,[53] Clarke and Bern[45] for references). The hormone also stimulates, together with other hormones (insulin, estrogen, progesterone, thyroid hormones, GH, placental lactogen), the growth, differentiation and functions of mammary glands. The mode of action differs, however, from species to species. After binding to a receptor in a mammary gland primed by cortisol and insulin, PRL alone can stimulate secretion of milk proteins and activity of enzymes which take part in lipid metabolism and regulation of ion flow (see Part III A). It also increases the capacity of the gland to attract and retain IgA-lymphoblasts (see Shiu and Friesen[139]).

B. Fishes

The environmental temperature and time of day at which the hormone is administered could affect hormone response in fishes.[55] Exogenous PRL is reported to exert an antigonadal effect on the testis in *Notemigonus*,[55] while oPRL stimulates the activity of steroid-3 -dehydrogenases in the cichlid *Aequidens pulcher*.[20] In females PRL induced the appearance of corpora atretica in two cichlids,[17] and regression of the ovary was reported for *Heteropneustes fossilis*[54] and *Notemigonus crysoleucas*.[55] Estradiol, LH and GnRH all stimulate PRL secretion in male and female carp. Prolactin stimulates ^{32}P uptake by the ovary and increases plasma calcium and phosphoprotein levels, suggesting an increase in vitellogenesis. No effect of PRL was recorded for *Rutilus rutilus*, and only a stimulation of plasma calcium levels was seen in *Leuciscus rutilus*. In dogfish sharks, PRL depresses vitellogenesis. This effect can be reversed by bromocryptine treatment. No changes in serum calcium levels were noted in the male.[63].

In the viviparous cyprinodont *Poecilia latipinna*, PRL cells do not change appreciably during the monthly cycle of vitellogenesis, intraovarian gestation and parturition.[158] On the basis of these results it may be concluded, as stated by Ensor et al.,[63] that PRL may have a reproductive role in some fish, but the effect is not constant and shows some degree of species specificity.

As regards the effects of PRL on sexual accessories and secondary sexual characters (see also Part V and Chapter 11) in both the Asian catfish *H. fossilis* and the gobiid fish *Gillichthys mirabilis* after hypophysectomy, it was found that PRL synergizes with testosterone (and GH in the former) to promote seminal vesicle growth.[54,144]

C. Amphibians

Mazzi et al.[115] found that PRL has an antigonadic effect on the crested newt *T. cristatus carnifex*, since the administration of high doses of oPRL to the summer male stops spermatogenesis, probably due to a concomitant reduction in blood FSH levels.[113] Indeed, it was found that simultaneous administration of FSH counteracted the effects of PRL. Deafferentation of the preoptic area produces similar results to those obtained following the administration of exogenous PRL;[110] i.e., a decrease in gonadotropin secretion and an increase in PRL production, indicated by an increase in tail fin height, an ambisexual PRL-dependent character.[150]

In the crested newt, PRL, testosterone and estradiol all follow more or less the same annual cycle (a summer minimum and an autumn-winter peak).

The administration of testosterone to the summer male brings about an increase in PRL secretion.[114] This effect is also noted following *in situ* perfusion of the hypophysis with GnRH and following oLH administration.[149] Conversely, in the same species, small doses of oPRL cause a significant drop in serum testosterone levels in both summer and winter males. It has not, however, been established whether this effect is due to interference with mechanisms regulating gonadotropin production within the hypothalamus and/or hypophysis, or with testicular LH receptors.[2]

So far, no antigonadal effects have been reported for either male or female in other urodelan species. Norris and Duvall[122] observed that *in-vivo* treatment of female tiger salamanders (*Ambystoma tigrinum*) with oPRL blocked the inhibitory action of *in-vivo* thyroxine treatment on LH-induced ovulation *in vitro*. These observations suggest a possible progonadal effect of PRL related to ovarian development. In anuran species Billeter and Barker Jørgensen[12] noted that PRL impedes compensatory hypertrophy in the ovaries of young toads. In *X. laevis*, prolactin does not modify the activity of 17β-hydroxysteroid oxidoreductase (an enzyme which catalyzes the conversion of 4-androstene-3,17-dione to testosterone) in fully grown follicles cultured *in vitro*. If, however, PRL is administered at the same time as human chorionic gonadotropin (hCG), it completely blocks the negative action of the gonadotropic hormone on this enzyme.[145]

No definite effects of PRL on the oviducts have been demonstrated for amphibians. Indeed, a reported increase in oviducal jelly secretion in *Bufo arenarum*[87] was not confirmed by investigations into *Rana pipiens* and *R. catesbeiana*.[8] Prolactin is implicated in the development of the cloacal cornified plates in the female and of nuptial pads in male red-spotted newt.[140,160] It also is involved in cloacal gland hypertrophy and secretion, acting in synergy with testosterone in *Cynops pyrrhogaster*.[93] In the male crested newt PRL causes turgescence of the cloacal swelling by acting, together with testosterone, on the connective tissue frame and blood vessels.[111] In *Discoglossus pictus*, however, PRL antagonized the action of androgen on the seminal vesicles.[120] In many Salamandridae species,[92,119,147,150] it has been shown that an ambisexual character, tail fin height, important in both osmoregulation (part III. C) and in sexual behavior (part IV. C), is exclusively PRL-dependent. In *Notophthalmus viridescens* cyclical variations of tail height occur only in the male.[140] In this species oxytocin (100 μU) antagonized the stimulatory action of 0.3U of PRL.[52] Tail height in neotenic tiger salamanders, although not a sexually dimorphic character, also is stimulated by PRL.[125] Furthermore, osmoregula--tory effects of PRL on tail tissue are antagonized by oxytocin in this species, too.[126] Oxytocin had no effect on the antimetamorphic activity of PRL on the gills of these animals, however.[126]

D. Reptiles

Information on the effect of PRL on sexual characters in reptiles is fragmentary and contradictory. A correlation between testicular activity and PRL could be assumed from the observation that in *Cnemidophorus sexlineatus* acidophilic cellular activity of lactotropes and somatotropes falls off in August in concomitance with testicular regression and a decrease in physical activity and growth rate.[89] In *Lacerta sicula campestris* either PRL or FSH, but neither LH or testosterone, stimulate development of the epididymis (Fisher and Fisher, 1971, cited by Clarke and Bern[45]). In *Anolis carolinensis*, though, PRL is without effect on the testis and accessory organs.[96]

Ovarian growth, stimulated by exogenous gonadotropins, was inhibited by oPRL in the lizards *Dipsosaurus dorsalis* and *Phrynosoma cornutum*.[33,36] Prolactin injections also depressed the ability of oFSH to stimulate ovarian growth only in vitellogenic females of *A. carolinensis*.[85] In *Chrysemys picta* mammalian and avian prolactin brought about a decrease in LH-stimulated progesterone synthesis in *in-vitro* cultured ovaries[35] but was without effect *in vivo*.[95] Whereas in *P. cornutum* PRL acting synergistically with steroid hormones stimulates oviduct development[33] in *A. carolinensis* oPRL depressed oviducal weight.[85]

III. PROLACTIN AND ADAPTATION TO THE REPRODUCTIVE ENVIRONMENT

A. Overview of Mammals

Of the five different kinds of physiological functions attributed to PRL (control of water and electrolyte balance, regulation of growth and development, metabolic effects, control of reproductive functions[11]), it seems that the regulation of water and electrolyte balance may have been prolactin's ancestral function in early vertebrates.[122] This role has retained its prominence in fishes (see part IIIA) and is not without importance in mammals, not so much as an adaptation to the reproductive environment but because two activities closely linked to reproduction and survival depend on PRL; i.e., the regulation of mammary gland functions (see part II A) and ion transport across some fetal membranes. Lactation "represents a substantial osmoregulatory stress to most mammalian species, particularly the smaller ones . . .";[121] for this reason prolactin's osmoregulatory capacities are particularly useful in water and electrolyte conservation.

It has been shown that lactating female rats are more resistant to dehydration than non-lactating rats;[62] that ouabain inhibits some PRL-induced activities in the mammary glands (Na^+-K^+ ATPase transport, lactation and fatty acid synthesis);[65-67] that in an *in-vitro* cultured monolayer of mammary gland cells PRL increased the transport of Na^+ and of an unidentified ion.[13-15] Recently, many workers have found evidence that placental lactogen and, in some species, PRL regulate exchanges between mother, fetus and amniotic fluid (see Nicoll,[121] Loretz and Bern[102] for references). As Loretz and Bern[102] emphasized, "The potential osmoregulatory effects of prolactin on the fetus are interesting in as much as mammalian amniotic fluids may be hypoosmotic or hyperosmotic to body fluids and may resemble the environments experienced by fresh-water and marine aquatic vertebrates respectively."

B. Fishes

Present knowledge of osmoregulatory mechanisms in fish is summarized here on the basis of information given in recent review of the subject[45,102,129] as well as the latest research findings. For an evolutionary approach to the problem of ionic extraction systems, see Evans.[64]

Information on the effects exercised by PRL on adaptation to the reproductive environment is practically restricted to the Teleostei. Indeed, neither the Holocephali (Chondrichthyes), nor the Dipnoi and the Coelacanthini among the Sarcopterygii have been investigated. It would undoubtedly be interesting to find out whether PRL has a role in the adaptation of migratory Cyclostomata to different environments, e.g., *Petromyzon marinus*, even though it has been found in other species (*Lampetra fluviatilis*) that hypophysectomy does not prevent survival in fresh water and that changes in external salinity do not change pituitary cytology. In exclusively marine elasmobranchs, PRL effects have been reported indicating that the hormone is able to influence water secretion by altering body surface permeability and Na^+ excretion. Among non-teleostean Actinopterygii such as *Amia*, hypophysectomy brings about sodium loss which may be prevented by administration of mammalian PRL or of pituitary homogenates.

We owe the first observation of an effect for prolactin on osmoregulatory adaptation in euryhaline Teleostei to Pickford and Phillips.[123a] They demonstrated that administration of exogenous PRL permitted hypophysectomized killifish to survive in fresh water. Subsequently, once it was known that PRL can keep plasma Na^+ and Cl^- concentrations constant, researchers turned their attention to elucidating the physiological effect of PRL on individual organs involved in the promotion of freshwater survival in these teleosts. To recapitulate, PRL reduces Na^+ efflux through the gills (a passive process) and increases the influx of Na^+ (an active process). Reports of both increased and decreased permeability of the gills to water can, on the other hand, probably be put down to different methods. The operculum was chosen as a model for *in-vitro* research into ion transport since it, like the branchial filaments, contains "chloride (salt) cells." In this membrane it was found that PRL reduces active Cl^- extrusion (probably an adaptive mechanism for freshwater survival) and, in addition, it can significantly decrease electrical conductance in seawater-adapted tilapia.[73]

During adaptation to sea water, the number of branchial chloride cells increases and an intracellular tubular reticulum develops. In addition, newly formed cells send out arms to existing salt cells. These arms form leaky junctions (low resistance pathways) with the salt cells. This permits the tubular network to contact external milieu. In freshwater-adapted fish, on the other hand, the chloride cell tubular reticulum is separated from the external environment by tight apical junctions. A transjunctional flow model was proposed on the basis of these observations.[135] This model is not mutually exclusive with respect to the vesicular transport model based mainly on ultrastructural observations.[24] As far as we know, however, the effects of prolactin on the ultrastructure of salt cells has not been investigated (see the recent review by Zadunaisky[159]).

It seems that PRL is capable of mediating osmoregulation by altering cutaneous mucus production and hence controlling epithelial surface permeability only in a few teleostean species. Indeed, while hypophysectomy brings about a decrease in mucous cell density and PRL brings about an increase in these cells and increases mucus production in some species, in others, such as the medaka (*Oryzias*), the tilapia and *Poecilia*, it has no effect whatsoever. Moreover, experiments measuring the diffusion coefficients of labelled sodium and chloride ions through mucus and the ion exchange capacity of sculpin (*Leptocottus armatus*) mucus showed that mucus could not limit the passive movement of ions sufficiently to account for the sodium-retaining effect of PRL. Experiments carried out on isolated strips of *Gillichthyes* cutis showed the existence of a functional chloride-secreting pump. Transplanting of rostral pars distalis (rich in PRL-producing cells) into hypophysectomized animals produced results which supported the view that PRL suppresses the ion extrusion mechanism by reducing the trans-epithelial potential and short-circuit current (SCC).

Teleosts adapted to live in sea water, a dehydrating environment, absorb water and NaCL through their intestines. Gravimetric experiments carried out on the eel and the trout have shown that this is reduced by PRL. This hormone, then, promotes an important adaptive effect since it is instrumental in preventing passive overhydration in spite of a drop in active NaCL absorption. In freshwater animals, PRL probably helps to reduce the intestinal reabsorption of "accidentally" ingested water.

Prolactin reduces reabsorption of water by the kidneys and increases urinary excretion of Na^+. This may depend on changes in the glomerular filtration rate (GFR) or gill permeability to water. The presence of renal tubule PRL receptors and the observation that PRL may be toxic to some teleosts (sometimes bringing about the degeneration of tubule cells and stimulating Na-K-ATPase activity) suggest a primary renal tubular site for PRL action. Prolactin also acts on the urinary bladder in freshwateradapted teleosts to restrict water-loss (by reducing osmotic permeability) in the urine and maintains urinary bladder ion reabsorption (so far inconclusive results have been obtained). During salt water adaptation in the goby, PRL brings about an increase in the reabsorption of Na^+ and Cl^- by the urinary bladder. As Loretz and Bern noted, "In those species demonstrating PRL-stimulated urinary bladder ion reabsorption, the effects are doubly adaptive for freshwater osmoregulation since urinary Na^+ and Cl^- reabsorption is enhanced while water reabsorption is inhibited, resulting in an increase in urinary water clearance."

Prolactin probably also plays a part in aiding survival in water exhibiting high acidity levels due to acid rain. Indeed, tilapia subjected to prolonged acid stress increase their PRL production while a restoration of plasma electrolytes and branchial osmotic water permeability takes place.[153]

The role played by PRL in stenohaline teleosts has attracted less research attention. In general, however, it seems to be of less importance than in euryhaline species. In various species (goldfish, bowfin, bullhead), it was found that after hypophysectomy the animal survives for only a very short time in fresh water but that survival could be prolonged if the animal were kept in physiological solution or if PRL were administered. Prolactin only partially restores plasma sodium and chloride levels which, like osmolarity, decrease in the hypophysectomized animal. In *Ictalurus melas*, PRL alters blood Na^+ and Cl^- levels in intact animals but only Na+ levels in hypophysectomized animals. Moreover, it also significantly lowers blood pH. Hypophysectomy brings about a drop in plasma Na^+ levels in *I. punctatus* though this is not due to an increased loss either from gills or kidney. Survival is longer (see Fortner and Pickford[72] for references).

C. Amphibians

There is no doubt that the part played by PRL in adaptation to the reproductive environment varies considerably, not only between the Anura and the Urodela but also within the two orders. The extent of participation depends on the osmoregulative capacity of individual species to live in a wide range of habitats: fresh water, brackish water, dry land. Since a detailed examination of the literature is beyond the range of this review, we will restrict ourselves here to a general summary of our current understanding of the best known species. Loretz and Bern[102] summarized the osmoregulatory effects of PRL on the larval and adult forms of Anura and Urodela (see Table 1). They also analyzed the most recent results reported in the literature on this topic. This analysis revealed that the part played by PRL in neotenic and larval urodeles and anuran larvae closely resembles the role played by PRL in eurhyaline teleosts, i.e., it increases plasma Na^+, stimulates Na^+ uptake by the gills and increases *in-vivo* water permeability. The antimetamorphic effect of PRL could also be osmoregulatory in nature.[127] It seems that hydromineral homeostasis shifts from the gills to the kidneys as the number of kidney PRL receptors increases in the adult animal. This increase of PRL receptors in the kidneys is already taking place in bullfrog tadpoles at the climax stage of metamorphosis.[155] In anuran amphibians, which do not drink, PRL plays a primary role in an event of fundamental importance: the transport of water through the skin. In particular, in *X. laevis*, a species adapted to live in brackish water, PRL exerts marked effects on the skin and mucous glands, on interrenal permeability, on the activity of liver enzymes governing the NH_3/urea balance, on muscular Na^+ and K^+ concentrations, on renal alkaline phosphatase activity, on liver glycogen and on blood glucose levels (see Mazzi et al.[112] for references). Moreover, as has been observed on isolated skin sections, PRL is able to increase SCC in the intact animal in water, the same response being evoked by exposure to a dry environment. Prolactin brings about a bodily Na^+ accumulation in animals raised in fresh water.[76] This effect also was observed in *Rana temporaria* in which species the hormone induces an elevation of potential difference and of SCC in the skin.[88] In urodeles, such as the red-spotted newt whose land-dwelling juvenile form (red eft) returns to the water to undergo a second metamorphosis, PRL induces water-drive behavior and an increase in tail fin height (see later). These effects are not antagonized by thyroxine[52] but, as in bullfrog tadpoles and plethodontid salamanders,[136] PRL has little effect on plasma Na^+ levels depressed by hypophysectomy. Prolactin, however, restores plasma Na^+ levels depressed

Table 1. Reported Actions of Prolactin on Amphibian Osmoregulation

Group	Plasma Na	Gill		Skin		Urinary bladder reabsorption		Muscle tissue content	
		H_2O permeability	Na uptake	H_2O permeability	Na uptake	H_2O	Na	H_2O	Na
Anurans									
Larval		↑(P_d)		↑(P_d)	↑, 0				
Adult	↑, ↓				↑		↑	↑, ↓	↑, 0
Urodeles									
Larval/neotenic	↑, 0		↑					↑	↑
Adult	↑, 0, ↓			↓(P_o)	↓, 0	0	↑		

↑ - Increase; 0 = no effect; ↓ = decrease; P_d = diffusional permeability; P_o = osmotic permeability.

(From Loretz and Bern[102] with permission of the authors and of Karger A. G. Basel.)

by thyroxine to normal[27,28] and significantly reduces in-vivo transepithelial potential.[31]

Season- and temperature-dependent changes in osmoregulatory performance have been observed in many Anura. These changes were not as marked and not accompanied by such dramatic integumental changes as those described for some salamandrid urodeles, such as *T. cristatus carnifex*, *Taricha granulosa* (rough-skinned newt) and *Hynobius retardatus*, which are characterized by a temperature-dependent circennial sexual cycle. These species pass the non-breeding period on land (terrestrial phase) and spend the cold months of the year (breeding period) in water, where they undergo a second seasonal metamorphosis. During the aquatic phase, in addition to water drive, the animals are characterized by: smooth, mucus-covered skin; an increase in body weight and tail fin height; edema in *H. retardatus* due to accumulation between skin and muscle of a hydrophilic tissue rich in polysaccharides also responsible for an increase in specific gravity which permits the animal to sink in water;[119] low integument osmotic permeability; scanty urine production and reduced transintegumental potential and SCC (see Dent,[51] de Vlaming,[53] Clarke and Bern[45] for references). These conditions are reversed in terrestrial phase newts (non-breeding). It therefore appears obvious, and also has been shown experimentally in the crested newt and the rough-skinned newt, that low temperature decreases integumental osmotic permeability and electrogenic ion transport. Thus, the skin may be an important site of seasonal shift in amphibian osmoregulatory physiology.[30,101] When rough-skinned or crested newts captured in the breeding season are subjected to an increase in temperature they change over to the terrestrial phase and undergo all the physiological and morphological changes that distinguish the aquatic animal from the terrestrial animal.

Morphological and physiological differences between summer and winter animals are due to changes in endocrine balance which also involve PRL. That this hormone is primarily responsible for water drive was first proved by Chadwick[40] for *N. viridescens* and subsequently confirmed for numerous other species (see De Vlaming,[53] Clarke and Bern[45] for references). The administration of PRL to hypophysectomized crested newts is alone sufficient to restore at least partially integumentary characteristics to within normal limits, perhaps also with the aid of the thyroid.[148] When the temperature is kept constant, administration of PRL to the summer animal brings about, in addition to water drive, changes in the integument, the thyroid, the testis, some secondary sexual characters, the interrenal, heart muscle glycogen content, fat-body weight and locomotor activity (also reported by Duvall and Norris[59] for *A. tigrinum*) characteristic of the winter animal (see Mazzi[108,109] for references).

In the crested newt, the abrupt drop in plasma sodium levels in the hypophysectomized animal (or animal treated with 2-Br-α-ergocryptine[99]) is due partly to changes in skin permeability with a considerable increase in its active transport and passive flow components. Prolactin[97] or PRL plus aldosterone[134] are able to restore these parameters to normal. Prolactin's restrictive action on salt loss is therefore an advantage during the aquatic phase.[101] In the rough-skinned newt, on the other hand, it seems that corticotropin (ACTH), presumably by causing secretion of interrenal hormones, is indispensable for the stimulation of the rapid weight and salt loss which occur when newts in the aquatic breeding phase are transferred to warm temperature conditions.[29] Administration of ovine PRL (2μg/g daily for one week) to intact terrestrial-phase newts caused their skin to become electrophysiologically indistinguishable from that of aquatic phase animals.[31]

In amphibians, not only the integument but also the renal system contributes significantly to osmoregulation. Experiments carried out at different times of the year on the crested newt show that under equilibrium conditions, the renal system is essentially dependent on the operation of the skin mechanisms since it filters and eliminates liquid volumes and salt quantities comparable to those absorbed through the skin.[98] It therefore follows that GFR and urinary flow undergo cyclic variations mirroring changes in skin permeability.

As far as we know, it seems that neurohypophysial and interrenal hormones are capable of influencing kidney function. The fact that PRL receptors are abundant in amphibian kidneys suggests that, just as in euryhaline teleosts, PRL plays a part in kidney osmoregulation in these animals. So far it has been found that PRL does not alter the GFR in crested newts but significantly increases tubular water reabsorption (Lodi and Biciotti, personal communication).

A role for PRL in urinary bladder function was proposed in the case of the marine toad,[49,141] and confirmed when PRL receptors were found in the organ.[154] In the context of adaptation to reproductive environments, it also would be rewarding to study the part played by PRL in helping to survive acidic conditions. This ability varies enormously among amphibian species.[124]

D. Reptiles

More is known about the role played by PRL in osmoregulation in birds than in reptiles. Indeed, in the former it has been demonstrated that, although PRL does not reduce Na^+ reabsorption by the jejunum and the rectum (in the hen), it exercises a wide range of dehydration-preventing actions and rapidly stimulates salt gland secretion in the duck (see Loretz and Bern[102] for references). Much less is known about the effect of PRL on reptilian salt glands although it is quite possible that these glands are stimulated by PRL as in the case of birds. Hypophysectomy lowers serum Na^+ levels in *Chrysemys picta*[34] and in *Testudo graeca*, but raises them in *D. dorsalis*.[41] Simultaneous administration of PRL and aldosterone, however, brings about a further drop in plasma Na^+ levels in *Testudo*[26] but restores them to normal in *Dipsosaurus*.[41] Prolactin plays a decisive adaptive role in the case of aquatic chelonians: it acts on the kidney to restore GFR to normal in hypophysectomized animals and increases GFR in intact animals. Prolactin also induces an increase in urinary Na^+ reabsorption. In *Testudo*, a terrestrial species, the effect of PRL on the volume of urine produced varies in relation to the state of dehydration, bringing about a decrease in intact turtles, and an increase in water-loaded turtles.[25,26]

Some mention also should be made of a possible adaptational role for PRL in the regulation of epidermal sloughing, which appears to be stimulated by this hormone in *A. carolinensis*[103] and also in *Gekko gecko* where PRL exercises an additive effect on thyroxine.[44]

IV. PROLACTIN AND REPRODUCTIVE BEHAVIOR

This section will deal with the effects of PRL on reproductive behavior in the widest sense of the term: i.e., not only behavior which directs male and females towards their breeding grounds but also sexual and parental behavior. By the term sexual behavior we mean any form of interaction between the two sexes that promotes union between the gametes: namely prespawning and spawning behavior (or premating and mating behavior for internally fertilizing species). The term parental behavior is used to describe nest building activities and any postspawning or postmating care of eggs or young.

A. Overview of Mammals

Prolactin acts on various central nervous system neurotransmitters (dopamine, norepinephrine, acetylcholine, GABA) to determine numerous behavioral effects, namely: induction of maternal behavior, stimulation of food-seeking drive, facilitation of conditioned response learning, reduction of sensitivity to pain, inhibition of morphine-like substance dependence, strengthening of the analgesic effect of morphine, stimulation of grooming behavior, strengthening of drug-induced stereotypes and facilitation of sexual behavior (see Drago et al.[57] for references).

B. Fishes

1. Migrations. There is little doubt that PRL plays a part, thanks to its ancestral osmoregulative properties, in the adaptation of euryhaline teleosts to different salt concentrations (Part IIIB) and therefore also in the great migrations. It is not known, however, whether the hormone triggers the migratory stimulus (as in birds) or whether it simply prepares the animal for survival in the new environment. Indeed, in closely related species such as the sockeye salmon (*Oncorhynchus nerka*) and the chum salmon (*O. keta*) signs of PRL cell activation were noted before the start of migration and when the animal reached fresh water, respectively (see Clarke and Bern[45] for references). In a small migratory species such as *Gasterosteus aculeatus* (three-spined stickleback), which spends autumn and winter in the sea, the PRL cells start to become more active at the beginning of the spring when the animal migrates to its freshwater breeding grounds, thus inducing the osmoregulatory adaptations (including cytological changes in kidney tubules[152]) which permit survival in the new environment.

2. Sexual Behavior. The many experiments carried out on teleosts appear to show that, at least in this group, sexual behavior does not depend on PRL but on sex hormones.[86]

3. Parental Behavior. Some aspects of teleostean parental behavior appear to be governed by PRL, or at least influenced by this hormone. Anabantid fishes such as *Macropodus* or *Betta* build a nest of air bubbles in which the female lays her eggs. Prolactin facilitates air-bubble production by increasing mucus secretion in the oral cavity. At the same time, in *Macropodus*, PRL calms the animal down and attenuates its normal aggressive behavior. The action of PRL is indispensable; testosterone alone induces the animal to build a nest but this could not be achieved successfully without an increase in mucus production.[70] Prolactin also specifically stimulates nest-building in *Pterophyllum scolare* and *Symphysodon*.[16,17]

In many teleosts parental care also includes fanning behavior; a current of fresh water is directed towards the eggs by the pectoral fins. When mammalian, or perch or tilapia PRL or PRL plus gonadal steroids are injected into teleosts, in many cases (*Lepomis*, *Acquidens*, *Pterophyllum*) high level fanning behavior is stimulated.[17] Reducing PRL production with L-dopa suppresses fanning behavior in *Cichlocosa nigrofasciatum*.[71] The literature contains conflicting reports concerning *G. aculeatus*. Recent electron microscopic observations carried out by Slijkhuis et al.[141] seem to show, however, that endogenous PRL is involved in the regulation of parental fanning behavior in this species, too. In addition, in certain cichlids (*Acquidens*, *Cichlosoma* and *Astronotus*) PRL treatment does not induce fanning but only a decrease in aggression.[18]

It only remains to establish the way in which PRL acts. One possibility is that the act of fertilization or the perception of the egg increases PRL production and consequently fanning. According to this theory, as the eggs grow PRL production increases and fanning behavior is intensified. Another possibility is that PRL plays only a permissive role, i.e., it permits the male to react to the increasing stimulus presented by the growing eggs.

In the sea horse, incubation in the brood pouch is governed by synergism between interrenal corticoids (control of connective tissue structures) and PRL (responsible for proliferation of the epithelium lining the brood pouch).[22]

Electrophysiological experiments have shown that in the bluegill (*Lepomis macrochirus*) stimulation of the dorsal area of the forebrain brings about the type of behavior associated with nest construction and cleaning.[50]

It therefore seems interesting that PRL produces intense phasal activity in a neuron population of this area in those species that show obvious fanning behavior (*Lepomis gibbosus*, *Astronotus ocellatus*, *Tilapia mariae*), but only a slight increase of activity in those species not displaying parental behavior (*Carassius gibelio*, *Idus idus*).[19]

The epidermal cells in many teleosts produce large quantities of mucus. This effect may be related to osmoregulation (see Part IIIB). This activity becomes particularly intense during the breeding period in some species, and the mucus is used to feed the young. This is what happens in the case of the cichlid *Symphysodon discus*: after hatching, the young feed for about five weeks on the abundant mucus produced by the "discus" (discus milk) of the parents.[17,18] It has been shown that teleostean "milk" production in *Symphysodon* is stimulated by PRL and also by paralactin, a "prolactin" of teleostean origin.[17] Experiments on hypophysectomized animals may be able to establish whether this effect is due only to PRL or whether other hormones play a part. The histotropic effect on mucous cell production in fishes seems to be analogous to the secretion of crop milk in pigeons and milk production in mammals and could represent a manifestation of the integumental effect characteristic of PRL in all classes of vertebrates and more or less directly linked to reproductive and maternal activity.[18,39]

C. Amphibians

1. *Water Drive*. For a review of this important PRL-dependent behavioral character, an exclusive feature of some urodele salamandrids, and for a discussion of its functional implications, see Part IIIC.

2. *Sexual Behavior*. Information on the role of PRL in the determination of sexual behavior is restricted essentially to the males of some urodele salamandrids. On the basis of the results of a series of experiments carried out on three European newts (*T. cristatus*, *T. helveticus*, *T. vulgaris*) Grant[79] concluded that, ". . . prolactin is implicated as an important factor in eliciting sexual behavior independently of most gonadal influences." Although more recent research into the crested newt[106] confirms this activating effect of prolactin on male mating behavior (also reported for *C. pyrrhogaster*),[92] it also highlights the important permissive role of the androgens, a function already reported for this same species.[105] In the rough-skinned newt, the permissive role played by androgens in sexual behavior does not relate to PRL, which only has a minor effect on amplectic clasping,[117] but rather to arginine-vasotocin (AVT).[118] The fact that ". . . two different hypophyseal principles (PRL and AVT) are active in

stimulating two taxonomically different species . . . can be extremely useful for the understanding of the evolutionary aspects of hormonal control of amphibian behavior."[106] Prolactin probably does not act directly on the central nervous system of the crested newt but through signals given out by directly stimulated peripheral structures, namely the tail (important in sexual behavior since courtship behavior consists primarily of tail vibrations and rapid lashings) or the secretions of mucous glands, particularly those involved in spermatophore production (cloacal glands[111] and pelvic glands[133]). Male libido is probably determined by spermatophore availability.[84,104]

3. Parental Care. Many examples of parental care have been observed for both urodeles and anurans.[130] Few experiments have been carried out, though, into the endocrine nature of their control, and none of these have been related to PRL.

D. Reptiles

Some evidence exists, albeit often conflicting, that PRL plays a part in endocrine control of reproductive activity in reptiles.[47,85] Very little information, and that almost always indirect or negative, is available regarding the influence of PRL on reproductive behavior.[48] It is known, for example, that the courtship behavior of the male red-sided garter snake (*Thamnophis sirtalis parietalis*) is stimulated neither by testicular hormones nor by interrenal hormones, but then again neither by a whole series of brain hormones including catecholamine neurotransmitters linked to PRL secretion.[75]

V. SUMMARY AND FUTURE RESEARCH DIRECTIONS

The foregoing analysis of our current understanding of the role of PRL in piscine, amphibian and reptilian reproduction essentially confirms Bern's[10] often-stated view that PRL does not have a major influence on sexual activity. It should, however, also be pointed out that the subject has received little attention from research workers and little progress has been made in the years following the last comprehensive reviews of Ensor,[61] DeVlaming,[53] Clark and Bern[45] and, on the subject of osmoregulation, Loretz and Bern.[102] One reason for this apparent lack of interest in what amounts to an important topic could be the difficulties involved in obtaining PRL of the specific species under investigation and in assaying the concentrations of PRL and other hormones in blood and culture media when the vertebrates used in the experiments are often small and difficult to obtain in sufficient numbers. We do consider, however, that some aspects of PRL endocrinology deserve closer scrutiny. The following approaches are suggested:

1. Analysis of the proganadal or antigonadal effects of PRL on spermatogenesis and oogenesis, using physiological doses of the hormone and simultaneously monitoring any changes in blood Gn levels.
2. A wide-ranging comparative analysis of PRL receptors in target organs; the testis and ovary in particular, using if possible homologous PRL and heterologous or homologous, polyclonal or, better still, monoclonal antibodies to PRL receptors.
3. Analysis of the effects of PRL on its own receptors and Gn receptors.
4. Analysis of the effects of PRL on ovarian and testicular steroid production and of changes in blood androgen and estrogen levels. The experimental conditions should be designed to permit the effect of PRL, or of a lack of PRL, to be distinguished from the effect of variations in Gn content, also on secondary sexual characters.
5. Analysis of the effects of PRL on central nervous system androgen and estrogen feedback mechanisms and also on mechanisms for neural regulation of Gn.
6. A more precise identification of the neural centers responsible for the different aspects of sexual behavior and an analysis of the effects of PRL on the electrical activity of neurones or of groups of neurons in these centers

VI. REFERENCES

1. Amit, T., Barkey,R.J., Gavish, M. and Youdim, M.B.H. (1984). Induction of prolactin (PRL) receptors by PRL in the rat lung and liver. Demonstration and characterization of a soluble receptor. Endocrinology 114, 545-552.

1a. Andreoletti, G.E., Colucci, D., Vellano, C. and Mazzi, V. (1984). Annual cycle of testosterone and estradiol in the crested newt (<u>Triturus</u> <u>cristatus</u> <u>carnifex</u>). Atti Accad. Sci. Torino, I Cl. Sci. Mat. Fis. Natur. 118, 157-164.

2. Andreoletti, G.E., Colucci, D., Vellano, C. and Mazzi V. (1985). Prolactin lowers serum testosterone level in summer and winter newts. Monitore Zool. Ital. (N.S.) 19, 129-135.

3. Aragona, C. and Friesen, H.G. (1975). Specific prolactin binding sites in the prostate and testis of rats. Endocrinology 97, 677-684.

4. Barenton, B. (1981). Failure of short-term hyperprolactinemia to increase the number of gonadotrophin receptors in the ram testis. J. Endocr. 88, 451-454.

5. Barkey, R.J., Shani, J., Amit, T. and Barzilai, D. (1977). Specific binding of prolactin to seminal vesicle, prostate and testicular homogenates of immature, mature and aged rats. J. Endocr. 74, 163-173.

6. Bartke, A. (1976). Pituitary testis relationships. Role of prolactin in the regulation of testicular function. In "Spermiation" (P.O. Hubinont, M. L'Hermite and J. Schwers, eds.), pp. 138-152. Karger, Basel.
7. Bartke, A., Croft, B.T. and Dalterio, S. (1975). Prolactin restores plasma testosterone levels and stimulates testicular growth in hamsters exposed to short day length. Endocrinology, 97, 1601-1604.
8. Basu, R.W., Bern, H.A. and Chan, H. (1965). Effects of prolactin, growth hormone and gonadal steroids on the oviduct of Salientia. Anat. Rec. 151, 441-442.
9. Bentley, P.J. and Yorio, T. (1979). Do frogs drink? J. Exptl. Biol. 79, 41-46.
10. Bern, H.A. (1975). On two possibly primary activities of prolactins: osmoregulatory and developmental. Verh. Dtsch. Zool. Ges. 1975, 40-46.
11. Bern, H.A. and Nicoll, C.S. (1968). The comparative endocrinology of prolactin. Rec. Progr. Horm. Res. 24, 681-720.
12. Billeter, E. and Jørgensen, C.B. (1976). Ovarian development in young toads; Effects of unilateral ovariectomy, hypophysectomy, treatment with gonadotropin (HCG), GH and prolactin, and importance of body growth. Gen. Comp. Endocrinol. 29, 531-544.
13. Bisbee, C.A. (1981a). Prolactin effects on ion transport across cultured mouse mammary epithelium. Am. J. Physiol. 240, C110-C115.
14. Bisbee, C. A. (1981b). Transepithelial electrophysiology of cultured mouse mammary epithelium: sensitivity to prolactins. Am. J. Physiol. 241, E410-E413.
15. Bisbee, C.A., Machen, T.A. and Bern, H.A. (1979). Mouse mammary epithelial cells on floating collagen gels: transepithelial ion transport and effects of prolactin. Proc. Natl. Acad. Sci. USA, 75, 536-540.
16. Blum, V. (1973). Experiments mit Teleosteer-Prolactin. Zool. Jb. Physiol. 77,335-347.
17. Blum, V. (1974) Die Rolle des Polaktins bei der Cichlidenbrutpflege. Fortschr. Zool. 22, 310-333.
18. Blum, V. and Fielder, K. (1965). Hormonal control of reproductive behavior in some cichlid fish. Gen. Comp. Endocrinol. 5, 186-196.
19. Blum, V. and Fiedler, K. (1974). Prolaktin empfindliche Strukturen im Fischgehirn. Fortschr. Zool. 22, 155-166.
20. Blum, V. and Weber, K.M. (1968). The influence of prolactin on the activity of steroid-3 -ol dehydrogenase in the ovaries of the cyclid fish Aequidens pulcher. Experientia 24, 1259-1260.
21. Bohnet, H.G., Shiu, R.P.C., Grinwich, D. and Friesen, H.G. (1978). In vivo effects of antisera to prolactin receptors in female rats. Endocrinology 102, 1657-1661.
22. Boisseau, J.P. (1968). Prolactine et incubation chez l'hippocampe. Collq. Int. C.N.R.S. no. 177, 205-214.
23. Boyns, A.R., Cole, E.N., Golder, M.P., Danutra, V., Harper, M.E., Brownsey, B., Cowley, T., Jones, G.E. and Griffiths, K. (1972). Prolactin studies with the prostate. In "Prolactin and carcinogenesis, Proceedings of the 4th Tenovus Workshop" (A.R. Boyns and K. Griffith, eds.), pp. 204-216, Alpha Omega Alpha Publishing, Cardiff.
24. Bradley, T.J. (1981). Improved visualization of apical vesicles in chloride cells of fish gills using an osmium quick-fix technique. J. Exp. Zool. 217, 185-198.
25. Brewer, K.J. and Ensor, D.M. (1980). Hormonal control of osmoregulation in the Chelonia. I. The effects of prolactin and interrenal steroids in freshwater chelonians. Gen. Comp. Endocrinol. 42, 304-309.

26. Brewer, K.J. and Ensor, D.M. (1980). Hormonal control of osmoregulation in the Chelonia. II. The effects of prolactin and corticosterone on Testudo graeca. Gen. Comp. Endocrinol. 42, 310-314.
27. Brown, P.S. and Brown, S.C. (1973). Prolactin and thyroid hormone interactions in salt and water balance in the newt Notophthalmus viridiscens. Gen. Comp. Endocrinol. 20, 456-466.
28. Brown, P.S., Brown, S.C. and Frye, B.E. (1979). The effects of hypophysectomy and thyroidectomy on salt balance in the plethodontid salamanders, Desmognathus fuscus and Desmognathus monticola. J. Comp. Physiol. 132, 357-361.
29. Brown, P.S., Brown, S.C. and Specker, J.L. (1984). Osmoregulatory changes during the aquatic to terrestrial transition in the rough-skinned newt, Taricha granulosa: the roles of temperature and ACTH. Gen. Comp. Endocrinol. 56, 130-139.
30. Brown, P.S., Brown, S.C., Bisaglio, I.T. and Lemke, S.M. (1983). Breeding condition temperature and the regulation of salt and water by pituitary hormones in the red spotted newt, Notophthalmus viridiscens, Gen. Comp. Endocrinol. 51, 292-302.
31. Brown, S.C., Brown, P.S., Bania, T.C. and Hayner, A.M. (1984). Integumental electrophysiology in newts: Effects of prolactin. Gen. Comp. Endocrinol. 53, 446-447 (Abstract).
32. Buntin, J.D., Keskey, T.S. and Janik, D.S. (1984). Properties of hepatic binding sites for prolactin in ring doves (Streptophelia risoria). Gen. Comp. Endocrinol. 55, 418-428.
33. Burns, J.M. and Richards, J.S. (1974). Effects of mammalian and host gonadotrophins on the ovaries and oviducts of female Texas horned lizard, Phrynosoma cornutum. Comp. Biochem. Physiol. 47A, 655-661.
34. Butler, D.G. and Knox, W.H. (1970). Adrenalectomy of the painted turtle (Chrysemys picta belli): Effect on ionoregulation and tissue glycogen. Gen. Comp. Endocrinol. 23, 65-76.
35. Callard, I.P., McChesney, I., Scanes, C. and Callard, G.V. (1976). The influence of mammalian and avian gonadotrophins on in vitro ovarian steroid synthesis in the turtle (Chrysemys picta). Gen. Comp. Endocrinol. 28, 2-9.
36. Callard, I.P. and Zeigler, H. (1970). Inhibitory effect of prolactin on gonadotrophin stimulated ovarian growth in the iguanid lizard, Dipsosaurus dorsalis. J. Endocr. 47, 131-132.
37. Carr, F.E. and Jaffe, R.C. (1980). Prolactin and tadpole metamorphosis. Evidence of prolactin receptors in premetamorphic Rana catesbeiana liver and tail fin. Mol. Cell. Endocrinol. 17, 145-155.
38. Carr, F.E. and Jaffe, R.C. (1981). Solubilization and molecular weight estimation of prolactin receptors from Rana catesbeiana tadpole liver and tail fin. Endocrinology 108, 943-949.
39. Chadwick, A. (1977). Comparison of milk-like secretions found in non-mammals. Symp. Zool. Soc. Lond. 41, 341-358.
40. Chadwick, C.S. (1941). Further observations on the water drive in Triturus viridescens. II. Induction of water drive which the lactogenic hormone. J. Exp. Zool. 86, 175-187.
41. Chan, D.R.O., Callard, I.P. and Chester Jones, I. (1970). Observations on the electrolyte composition of the iguanid lizard Dipsosaurus dorsalis (Baird and Girard), with special reference to the control by the pituitary gland and the adrenal cortex. Gen. Comp. Endocrinol. 15, 374-387.
42. Chan, V., Katikineni, M., Davies, T.F. and Catt, K.J. (1981). Hormonal regulation of testicular luteinizing hormone and prolactin receptors. Endocrinology 108, 1607-1612.
43. Charreau, E.H., Attramadal, A., Torjesen, P.A., Purvis, K., Calandra, R. and Hansson, V. (1977). Prolactin binding in rat testis: specific receptors in interstitial cells. Mol. Cell. Endocrinol. 6, 303-309.

44. Chiu, K.W. and Phillips, J.G. (1972). The effect of prolactin, thyroxine and thyrotropin on the sloughing cycle of the lizard Gekko gecko L. Gen. Comp. Endocrinol. 19, 592-593.
45. Clarke, W.C. and Bern, H.A. (1980). Comparative endocrinology of prolactin. Horm. Proteins Peptides 8, 105-197.
46. Costlow, M.E. and McGuire, W.V. (1977). Autoradiographic localization of the binding of ^{125}I-labelled prolactin to rat tissues in vitro. J. Endocr. 75,221-226.
47. Crews, D. (1979). Endocrine control of reptilian reproductive behavior. In "Endocrine control of sexual behavior" (C. Beyer, ed.), pp. 167-209. Raven Press, New York.
48. Crews, D. (1983). Alternative reproductive tactics in reptiles. Bioscience, 33,562-567.
49. Debnam, E.S. and Snart, R.S. (1975). Water transport response of toad bladder to prolactin. Comp. Biochem. Physiol. 52A, 75-76.
50. Demski, L.S. and Knigge, K.M. (1970). The telencephalon and hypothalamus of the bluegill (Lepomis macrochirus). Evoked feeding, aggressive and reproductive behavior in the represental frontal sections. J. Comp. Neurol. 143, 1-16.
51. Dent, J.N. (1975). Integumentary effects of prolactin in the lower vertebrates. Amer. Zool. 15, 923-935.
52. Dent, J.N. (1982). On the interaction of thyroxine and of oxytocin with prolactin on the growth of tail fin in the red-spotted newt. Gen. Comp. Endocrinol. 47, 269-275.
53. de Vlaming, V.L. (1979). Action of prolactin among the vertebrates. In "Hormones and evolution" (E.J.W. Barrington, ed.), vol. 2, pp. 561-642. Academic Press, New York, San Francisco, London.
54. de Vlaming, V.L. and Sundararaj, B.I. (1972). Endocrine influences on seminal vesicles in the estuarine gobiid fish, Gillichthys mirabilis. Biol. Bull. 142, 243-250.
55. de Vlaming, V.L. and Vodicnik, K.M. (1977). Diurnal variations in pituitary gonadotropin content and gonadal response to exogenous gonadotropin and prolactin in Notemigonus crysoleucas. J. Fish Biol. 10, 371-383.
56. D'Istria, M., Pierantoni, R., Citarella, F., Fasano, F., Vellano, C., Peyrot, A. and Delrio, G. (1984). The effects of thyroidectomy on androgen and prolactin receptors in the dorsal skin and caudal fin of Triturus cristatus carnifex. Gen. Comp. Endocrinol. 53, 153-158.
56a. Djiane, J., Delouis, C. and Kelly P.A. (1979). Prolactin induced in vivo and in vitro loss of prolactin receptors in rabbit mammary gland. 61st Annual Meeting of the Endocrine Society, Anaheim, CA (Abstract 29).
57. Drago, F., Bohns, B., Gispen, W.H., Ree, J.M. van, Scapagnini, V. and Wied, D. de (1982). Behavioral changes in short-term and long-term hyperprolactinaemic rats. In "Integrative neurohumoral mechanisms" (E. Endroczi, L. Angelucci, D. de Wied and V. Scapagnini, eds.), pp. 417-427, Elsevier Science Publishers, Amsterdam.
58. Dube, D., Kelly, P.A. and Pelletier, G. (1980). Comparative localization of prolactin binding sites in different rat tissues by immunohistochemistry, radioautography and radioreceptor assay. Mol. Cell. Endocrinol. 18, 109-122.
59. Duvall, D. and Norris, D.O. (1977). Prolactin and substrate stimulation of locomotor activity in adult tiger salamanders (Ambystoma tigrinum). J. Exp. Zool. 200, 103-106.
60. Edery, M., Young, G., Bern, H.A. and Steiney, S. (1984). Prolactin receptors in tilapia (Sarotherodon mossambicus) tissues: binding studies using ^{125}I labeled ovine prolactin. Gen. Comp. Endocrinol. 56, 19-23.
61. Ensor, D.M. (1978). "Comparative Endocrinology of Prolactin." Chapman and Hall, London.

62. Ensor, D.M., Edmondson, M.R. and Phillips, J.G. (1972). Prolactin and dehydration in rats. J. Endocr. 53, LIX-LX.
63. Ensor, D.M., Sahai, S., Lee, A.C. and Tinsley, D. (1984). Reproductive role of prolactin in fish. Gen. Comp. Endocrinol. 53, 435 (Abstract).
64. Evans, D.H. (1984). The roles of gill permeability and transport mechanism in euryhalinity. In "Fish Physiology" (W.S. Hoar, D. J. Randall, eds.), vol. X B, pp. 239-283. Academic Press, San Diego, San Francisco, New York, London, Toronto, Montreal, Sydney, Tokyo, São Paulo.
65. Falconer, I.R. (1980). Biochemical action of prolactin in the mammary gland. In "Endocrinology 1980" (I.A. Cumming, J.W. Funder and F.A.O. Mendelsohn, eds.), pp. 327-330. Australian Academy of Science, Canberra.
66. Falconer, I.R. and Rowe, J.M. (1975). Possible mechanism for action of prolactin on mammary cell sodium transport. Nature (London) 256, 327-328.
67. Falconer, I.R. and Rowe, J.M. (1977). Effects of prolactin on sodium and potassium concentrations in mammary alveolar tissue. Endocrinology 101, 181-186.
68. Fehlman, M., Carpentier, J.L., Le Cam, A., Thamm, P., Saunders, D., Brandemburg, D., Orci, L. and Freychet, P. (1982). Biochemical and morphological evidence that insulin receptor is internalized with insulin in hepatocytes. J. Cell. Biol. 93, 82-87.
69. Ferland, L.H., Djiane, J., Houdebine, L.M. and Kelly, P.A. (1984). Intracellular transformation of prolactin following internalization into rat liver. Mol. Cell. Endocrinol. 35,25-31.
70. Fiedler, K. (1974). Hormonale Kontrolle des Verhaltens bei Fischen. Fortscrh. Zool. 22, 269-309.
71. Fiedler, K., Bruss, R., Christ, H. and Lotz-Zoller, R. (1979). Behavioural effects of peptide hormones in fishes. In "Brain and pituitary peptides (W. Wuttke, A. Weindl, K.H. Vogt and R.R. Dries, eds), pp. 65-70. Karger, Basel.
72. Fortner, N.A. and Pickford, G.E. (1982). The effects of hypophysectomy and replacement therapy with prolactin, cortisone or their combination on the blood of the black bullhead. *Ictalurus melas*. Gen. Comp. Endocrinol. 47, 111-119.
73. Foskett, J.K., Machen, T.E. and Bern, H.A. (1982). Chloride secretion and conductance of teleost opercular membrane. Effects of prolactin. Am. J. Physiol. 242, R 380-R 389.
74. Fryer, J.N. (1979). Prolactin binding sites in tilapia (*Sarotherodon mossambicus*) kidney. Gen. Comp. Endocrinol. 3, 397-402.
75. Garstka, W.R., Camazine, B. and Crews, D. (1982). Interaction of behavior and physiology during the annual reproductive cycle of the red-sided garter snake (*Thamnophis sirtalis parietalis*). Herpetologica 38, 104-123.
76. Giunta, C., Campantico, E., Bianchero, F., Ciancia Peron, A. and Vietti, M. (1979). Biochemical data on some organs from normal or prolactin-treated *Xenopus laevis* (Daudin) adults reared in fresh and salt water. Gen. Comp. Endocrinol. 20, 183-189.
77. Gona, O. (1982). Uptake of ^{125}I labelled prolactin by bullfrog kidney tubules: an autoradiografic study. J. Endocr. 93, 133-137.
78. Gona, O. (1984). Localization of prolactin uptake by the kidney of the teleostean, *Colisa lalia*. Gen. Comp. Endocrinol. 55, 289-294.
79. Grant, Jr., W.C. (1966). Endocrine induced courtship in three species of European newts. Am. Zool. 6, 585. (Abstract)
80. Guardabassi, A., Campantico, E. and Pattono, P. (1984). Localization of ^{125}I-labelled prolactin in the kidney of normal and subtotally

hypophysectomized Xenopus laevis Daudin adult specimens. A preliminary report. Atti Accad. Sci. Torino, I Cl. Sci. Mat. Fis. Natur. 118, 193-199.

81. Gugten, van der, A.A., Waters, M.J., Murphy, G.S. and Friesen, H.G. (1980). Studies on the irreversible nature of prolactin binding to receptors. Endocrinology 106, 402-411.

82. Hafiez, A.A., Bartke, A. and Lloyd, C.W. (1972). The role of prolactin in the regulation of testis function: The synergistic effects of PRL and luteinizing hormone on the incorporation of [1-^{14}C] acetate into testosterone and cholesterol by testis from hypophysectomized rats in vivo. J. Endocr. 53, 223-230.

83. Hafiez, A.A., Lloyd, C.W. and Bartke, A. (1972). The role of prolactin and luteinizing hormone on the plasma levels of testosterone and androstenedione in hypophysectomized rats. J. Endocr. 52, 327-332.

84. Halliday, T.R. (1976). The libidinous newt. An analysis of variations in the sexual behaviour of the male smooth newt, Triturus vulgaris. Anim. Behav. 24, 398-414.

85. Hengsen, H.T., Rainey, J.M. and Gist, D.H. (1980). Antigonadal effects of prolactin in female Anolis carolinensis. Gen. Comp. Endocrinol. 41, 22-30.

86. Hoar, W.S., Randall, D.J. and Donaldson, E.M. (1983). "Fish Physiology" vol. IXB, Academic Press, New York.

87. Houssay, B.A. (1947). Ovulacion y postura del sapo Bufo arenarum Hensel. V. Transporte de los ovulos por el oviducto y el utero. Rev. Soc. Arg. Biol. 23, 275-287.

88. Howard, K. and Ensor, D.M. (1978). Mechanism of action of prolactin on frog skin. In "Comparative Endocrinology" (P.J. Gaillard, H.H. Boer, eds.), p. 237. Elsevier North Holland, Amsterdam.

89. Johnson, L.F. and Jacob, J.S. (1984). Pituitary activity and reproductive cycle of male Cnemidophorus sexlineatus in West Tennessee. J. Herpetol. 18, 396-405.

90. Katoh, M., Djiane, J., Leblanc, G. and Kelly, P.A. (1984). Characterization of antisera to a partially purified prolactin receptor: effect on prolactin binding in different target tissues. Mol. Cell. Endocrinol. 34, 191-200.

91. Kelly, P.A., Djiane, J., Katoh, M., Ferland, L.H., Houdebine, L.M., Teyssot, B. and Dusanter-Fourt, I. (1984). The interaction of prolactin with its receptors in target tissues and its mechanism of action. Rec. Progr. Horm. Res. 40, 379-439.

92. Kikuyama, S., Yamamoto, K. and Seki, T. (1980). Prolactin and its role in growth, metamorphosis and reproduction in amphibians. Gunma Symp. Endocrinol. 17, 3-13.

93. Kikuyama, S., Nakano, R. and Yasumasu, I. (1975). Synergistic action of prolactin and androgen on the cloacal glands of the newt. Comp. Biochem. Physiol. 51A, 823-826.

94. Klemcke, H.G., Bartke, A., Borer, K.T. and Hogan, M.P. (1984). Regulation of testicular prolactin and luteinizing hormone receptors in golden hamsters. Endocrinology, 114, 594-603.

95. Klicka, J. and Mahmoud, I.J. (1977). The effects of hormones on the reproductive physiology of the painted turtle, Chrysemys picta. Gen. Comp. Endocrinol. 31, 407-413.

96. Licht, P. and Jones, R.E. (1967). Effects of exogenous prolactin on reproduction and growth in adult males of the lizard Anolis carolinensis. Gen. Comp. Endocrinol. 8, 228-244.

97. Lodi, G. and Biciotti, M. (1975-1976). Ricerche preliminari sul trasporto attivo del sodio nel tegumento del tritone crestato in condizioni sperimentali diverse. Atti Acad. Sci. Torino, I Cl. Sci. Mat. Fis. Natur. 110, 141-146.

98. Lodi, G. and Biciotti, M. (1985). Osmoregulatory adaptation of the terrestrial- and aquatic-phase crested newt (Triturus cristatus carnifex Laur.). Renal function. Boll. Zool. 52, 243-245.
99. Lodi, G., Biciotti, M., and Alessio, M.(1984). Effetti della prolattina e dell'ACTH sul metabolismo ionico in Triturus cristatus. Boll. Zool. 51, Suppl., 64.
100. Lodi, G., Biciotti, M. and Sacerdote, M. (1978). Osmoregulatory activity of prolactin in the skin of the crested newt. Gen. Comp. Endocrinol. 36, 7-15.
101. Lodi, G., Biciotti, M. and Viotto, B.(1982). Cutaneous osmoregulation in Triturus cristatus carnifex Laur. (Urodela). Gen. Comp. Endocrinol. 46, 452-457.
102. Loretz, C.A. and Bern, H.A. (1982). Prolactin and osmoregulation in vertebrates. Neuroendocrinology, 35, 292-304.
103. Maderson, P.F.A. and Licht, P. (1967). Epidermal morphology and sloughing frequency in normal and prolactin treated Anolis carolinensis (Iguanidae, Lacertilia). J. Morphol. 123,157-172.
104. Malacarne, G. and Cortassa, R. (1983). Sexual selection in the crested newt. Anim. Behav. 31, 52.
105. Malacarne, G. and Gioacoma, C. (1980). Influence of testosterone on mating behaviour in the male crested newt (Triturus cristatus carnifex Laur.). Gen. Comp. Endocrinol. 47, 139-147.
106. Malacarne, G., Giacoma, C., Vellano, C. and Mazzi, V. (1982). Prolactin and sexual behaviour in the crested newt (Triturus cristatus carnifex Laur.). Gen. Comp. Endocrinol. 47, 139-147.
107. Matt, K.S., Bartke, A., Soares, M.J., Talamantes, F., Hebert, A. and Hogan, M.P. (1984). Does prolactin modify testosterone feedback in the hamster? Suppression of plasma prolactin inhibits photoperiod-induced decreases in testosterone feedback sensitivity. Endocrinology, 115, 2098-2103.
108. Mazzi, V. (1979). Biologia della prolattina. Boll. Zool. 36, 1-60.
109. Mazzi, V. (1970). The hypothalamus as a thermodependent neuroendocrine centre in urodeles. In "The Hypothalamus" (L. Martini, M. Motta and F. Fraschini, eds.), pp. 663-676. Academic Press, New York.
110. Mazzi, V. (1978). Effects on spermatogenesis of permanent lesions to the rostral preoptic area in the crested newt (Triturus cristatus carnifex Laur.). Gen. Comp. Endocrinol. 34,247-250.
111. Mazzi, V. and Biciotti, M. (1976-1977). Possible prolactin-testosterone interaction on the cloacal intumescence of the male crested newt. Atti Accad. Sci. Torino, I Cl. Sci. Mat. Fis. Natur. 111, 75-79.
112. Mazzi, V., Lodi, G. and Guardabassi, A. (1980). Prolactin in the transition from water to land environment: evidence from amphibians. In "Animal Models in Human Reproduction" (M. Serio and L. Martini, eds.), pp. 35-47. Raven Press, New York.
113. Mazzi, V. and Vellano, C. (1968). The counterbalancing effect of follicle-stimulating hormone on the antigonadal activity of prolactin in the male newt, Triturus cristatus carnifex Laur. J. Endocr. 40, 529-530.
114. Mazzi, V., Vellano, C., Peyrot, A. and Lodi, G. (1979-1970). Stimulation of pituitary prolactin release by testosterone in the crested newt. In vivo and in vitro observations. Atti Acad. Sci. Torino, I Cl. Sci. Mat. Fis. Natur. 104, 771-778.
115. Mazzi, V., Vellano, C. and Toscano, C. (1967). Antigonadal effects of prolactin in adult male crested newt (Triturus cristatus carnifex Laur.). Gen. Comp. Endocrinol. 8, 320-324.
116. McNeilly, A.S. (1984). Prolactin and ovarian function. In "Neuroendocrine perspective" (E.E. Müller and R. M. McLeod, eds.), vol. 3, pp. 279-316. Elsevier Science Publ. Amsterdam.

117. Moore, F.L. and Muller, C.H. (1977). Androgens and male mating behaviour in rough-skinned newts, Taricha granulosa. Horm. Behav. 9, 309-320.
118. Moore, F.L. and Zoeller, R.T. (1979). Endocrine control of amphibian sexual behavior. Evidence for a neurohormone-androgen interaction. Horm. Behav. 13,207-213.
119. Moriya, T. (1982). Prolactin induces increase in the specific gravity of salamander, Hynobius retardatus, that raises adaptability to water. J. Exp. Zool. 223, 83-88.
120. N'Diaye, A. and Ozon, R.(1976). Action inibitrice de la prolactine sur les ultrastructures des vesicules seminales de Discoglossus pictus Otth (Amphibien anoure). Gen. Comp. Endocrinol. 28, 434-445.
121. Nicoll, C.S. (1981). Role of prolactin in water and electrolyte balance in vertebrates. In "Prolactin" (R.B. Jaffe, ed.), pp. 127-166. Elsevier, New York.
122. Nicoll, C.S., White, B.A. and Leung, F.C. (1980). Evolution of prolactin, its receptors and its functions. In "Central and Peripheral Regulation of Prolactin Functions" (R.M. McLeod and V. Scapagnini, eds.), pp.11-25. Raven Press, New York.
123. Norris, D.O. and Duvall, D. (1981). Hormone-induced ovulation in Ambystoma tigrinum: influence of prolactin and thyroxine. J. Exp. Zool. 216, 175-180.
123a. Pickford, H.E. and Phillips, W.G. (1959). Prolactin, a factor promoting survival of hypophysectomized killfish in fresh water. Science 130, 454-455.
124. Pierce, B.A., Hoskins, J.B. and Epstein, E. (1984). Acid tolerance in Connecticut wood frogs (Rana sylvatica). J. Herpetol. 18, 159-167.
125. Platt, J.E., (1976). The effects of ergocornine on tail height, spontaneous and T4-induced metamorphosis and thyroidal uptake of radioiodide in neotenic Ambystoma tigrinum. Gen. Comp. Endocrinol. 28, 71-81.
126. Platt, J.E. and LiCause, M.J.(1980). Effects of oxytocin in larval Ambystoma tigrinum: Acceleration of induced metamorphosis and inhibition of the anti-metamorphic action of prolactin. Gen. Comp. Endocrinol. 41, 84-91.
127. Platt, J.E. and Hill, P.L. (1982). Inhibition of the antimetamorphic action of prolactin in larval Ambystoma tigrinum by lysine vasopressin, arginine vasotocin and aldosterone. Gen. Comp. Endocrinol. 48, 356-361.
128. Posner, B.A., Kelly, P.A., Shiu, R.P.C. and Friesen, H.G. (1974). Studies on insulin, growth hormone and prolactin binding: tissue distribution, species variations and characterization. Endocrinology 95, 521-531.
129. Rankin, J.C. and Bolis, L. (1984). Hormonal control of water movement across the gill. In "Fish Physiology" (W. S. Hoar and D. J. Randall, eds.), vol. XB, pp. 177-201. Academic Press, Orlando.
130. Ridley, M. (1978). Paternal care. Anim. Behav. 26, 904-932.
131. Rosa, A.A.M., Djiane, J., Houdebine, L.M. and Kelly, P.A. (1982). Stimulatory effects of prolactin and anti-prolactin receptor serum on prolactin binding sites in rat liver cells in suspension culture. Biochem. Biophys. Res. Comm. 106, 243-249.
132. Rubin, R.T., Poland, R.E. and Tower, B.B. (1976). Prolactin-related testosterone secretion in normal adult men. J. Clin. Endocr. Metab. 42, 112-116.
133. Sacerdote, M. (1958). Cloaca e ghiandole annesse nel maschio di Triturus cristatus carnifex Laur. I. Il ciclo annuo. Arch. Zool. Ital. 43, 105-156.

134. Sampietro, P. and Vercelli, L. (1968). Effetti della prolattina sul tasso ematico del sodio nel tritone crestato normale e ipofisectomizzato. Boll. Zool. 35, 419.
135. Sardet, C., Pisani, M. and Metz, J.(1979). The surface epithelium of teleost fish gills. Cellular and junctional adaptations of chloride cell in relation to salt adaptation. J. Cell. Biol. 80, 96-117.
136. Sasayama, Y. and Oguro, C. (1982). Effects of hypophysectomy and replacement therapy with pituitary homogenates or own prolactin on serum calcium, sodium and magnesium concentrations in bullfrog tadpoles. Gen. Comp. Endocrinol. 46, 75-80.
137. Shani, J., Barkey, R.J. and Amit, T. (1982). Endogenous prolactin maintains its own binding sites in the pigeon crop sac mucosa. J. Receptor Res. 2, 407-417.
138. Shiu, R.P.C. and Friesen, H.G. (1976). Interaction of cell membrane prolactin receptor with its antibody. Biochem. J. 157, 619-626.
139. Shiu, R.P.C. and Friesen, H.G. (1980). Mechanism of action of prolactin in the control of mammary gland function. Ann. Rev. Physiol. 42, 83-96.
140. Singhas, C.A. and Dent, J.N. (1975). Hormonal control of the tail fin and of the nuptial pads in the male red-spotted newt. Gen. Comp. Endocrinol. 26, 382-392.
141. Slijkhuis, H., de Ruiter, A.J.H. Baggerman, B. and Wendelaar Bonga, S.E. (1984). Parental fanning and prolactin cell activity in the male three-spined stickleback _Gasterosteus aculeatus_ L. Gen. Comp. Endocrinol. 54, 297-304.
142. Snart, R.S. and Dalton, T. (1973). Response of toad bladder to prolactin. Comp. Biochem. Physiol. 45A, 307-311.
143. Specker, J.C., King, D.S., Nishioka, R.S. and Bern, H.A. (1984). Chemical and biological characterization of HPLC-purified proteins from organ cultured rostral and proximal pars distalis (RPD and PPD) of tilapia. Gen. Comp. Endocrinol. 53, 434 (abstract).
144. Sundararaj, B.I. and Goswami, S.W. (1965). "Seminal vesicle" response of intact, castrate and hypophysectomized catfish, _Heteropneustes fossilis_ (Bloch) to testosterone propionate, prolactin and growth hormone. Gen. Comp. Endocrinol. 5, 464-474.
144a. Sundararaj, B.I. and Keshvanath, P. (1976). Effects of melatonin and prolactin treatment on the hypophysial-ovarian system in the catfish, _Heteropneustes fossilis_, (Bloch). Gen. Comp. Endocrinol. 29, 84-96.
145. Thibier, C. and Ozon, R. (1983). Regulation of 17β-hydroxysteroid oxydoreductase of the _Xenopus laevis_ follicle by gonadotropin and prolactin. Gen. Comp. Endocrinol. 49, 352-357.
146. Tsim, K.W.T. and Cheng, C.H.K. (1984). Evidence for the involvement of essential sulphydryl group in rat hepatic lactogenic receptor but not in somatogenic receptor. Mol. Cell. Endocrinol. 38, 61-66.
147. Tuchmann-Duplessis, H. (1949). Action de l'hormone gonadotrope et lactogene sur le comportement et les caracteres sexuels secondaires du triton normal et castre. Arch. Anat. Microsc. Exp. 38, 302-317.
148. Vellano, C., Lodi, G., Bani, G., Sacerdote, M. and Mazzi, V. (1970). Analysis of the integumentary effect of prolactin in the hypophysectomized crested newt. Monitore Zool. Ital. (N.S.) 4,115-146.
149. Vellano, C. and Mazzi, V. (1982). Analisi degli effetti esercitati dalla somministrazione di LHRH sull'attività prolattinica nel tritone crestato. Atti Accad. Sci. Torino, I Cl. Sci. Mat. Fis. Natur. 116,] 415-418.
150. Vellano, C., Mazzi, V. and Sacerdote, M. (1970). Tail height, a prolactin-dependent ambisexual character in the newt (_Triturus cristatus carnifex_ Laur.). Gen. Comp. Endocrinol. 14, 535-541.

151. Waters, M.J., Lusins, S. and Friesen, H.G. (1984). Immunological and physicochemical evidence for tissue specific prolactin receptors in the rabbit. Endocrinology, 115, 1-10.

152. Wendelaar Bonga, S.E. (1976). The effect of prolactin on kidney structure of the euryaline teleost, Gasterosteus aculeatus during adaptation to fresh water. Cell Tiss. Res. 166, 319-338.

153. Wendelaar Bonga, S.E., van der Meij J.C.A. and Flik, G. (1984). Prolactin and acid stress in the teleost Oreochromis (formerly Sarotherodon) mossambicus. Gen. Comp. Endocrinol. 55, 323-332.

154. White, B.A. (1981). Occurrence of binding affinity of prolactin receptors in amphibian tissues. Gen. Comp. Endocrinol. 45, 153-161.

155. White, B.A. and Nicoll, C.S. (1979). Prolactin receptors in Rana catesbeiana during development and metamorphosis. Science 204, 851-853.

156. White, B.A., Lebovic, G.S. and Nicoll, C.S. (1981). Prolactin inhibits the induction of its own renal receptors in Rana catesbeiana tadpoles. Gen. Comp. Endocrinol. 43, 30-38.

157. Witorsch, R.J. (1978). Immunohistochemical studies of prolactin binding in sex accessory organs of the male rat. J. Histochem. Cytochem. 26, 565-580.

158. Young, G. and Ball, J.N. (1983). Ultrastructural changes in the adenohypophysis during the ovarian cycle of the viviparous teleost Poecilia latipinna. III The growth hormone, adrenocorticotrophic, and prolactin cells and the pars intermedia. Gen. Comp. Endocrinol. 52, 86-101.

159. Zadunaisky, J.A. (1984). The chloride cell: the active transport of chloride and the paracellular pathways. In "Fish Physiology" (W.S. Hoar and D.J. Randall, eds.), vol. XB, pp. 130-176. Academic Press, Orlando.

160. Zimmer, J.A. and Dent, J.N. (1981). Hormonal regulation of nuptial pads and tail fins in the female red spotted newt. Gen. Comp. Endocrinol. 44, 436-443.

REGULATION OF OVARIAN STEROIDOGENESIS

Giovanni Chieffi and Riccardo Pierantoni

Istituto di Biologia Generale
80138, Napoli, Italy
Universita Degli Studi di Napoli
and Stazione Zoologica
80121, Napoli, Italy

I. INTRODUCTION

The surge of interest in comparative endocrinology over the last thirty years has advanced our knowledge of gonadal steroidogenesis in nonmammalian vertebrates. In particular, the last decade has seen major advances thanks to the isolation and characterization of various compounds and the advent of more sensitive methods for hormone measurement.

The study of gonadal steroidogenesis received further impetus from the progressive discovery of the different physiological roles played by hormonal substances in nonmammalian vertebrates as compared with mammals. In the wake of this finding, comparative endocrinologists began to devise nonmammalian vertebrate models rather than applying mammalian mechanisms to other vertebrate classes.

The present chapter is devoted to the regulation of ovarian steroidogenesis. It is not intended to be exhaustive regarding ovarian morphology and sites of ovarian steroidogenesis, since several excellent reviews are available.[49,82,105] Only the more recent literature will be cited.

II. OVERVIEW OF MAMMALS

With the exception of the monotremes, mammalian eggs lack yolk (some true yolk, however, is formed in the hamster). Therefore, the follicular epithelium of mammals is involved mainly in hormone production. Another peculiarity that mammals share with birds is the blockage of multiplication of the germ cells during development. Hence the number of oocytes is determined before hatching in birds and before birth in most mammals.

During follicular development, Δ^5-3β-hydroxysteroid dehydrogenase (3β-HSD) is concentrated in the thecal cells, which also contain cholesterol and birefringent lipids. However, *in-vitro* biosynthesis carried out on the two follicular components has demonstrated that both thecal tissue and the granulosa produce estrogens, thecal tissue being the major source. Granulosa cells are the site of progesterone production by the corpus luteum. They show a very strong histochemical reaction for 3β-HSD, and a well-developed smooth endoplasmic reticulum. *In-vitro* biosynthesis of progesterone carried out with granulosa cells in tissue culture endorses these results (see Lofts and Bern[82]).

Development of the interstitial tissue is peculiar to mammalian and avian ovaries. It is more or less evident according to the species and it shows all the morphological and histochemical features of steroidogenetic tissues. Interstitial cells contain cholesterol-rich lipids and 3β-HSD activity. *In-vitro* studies with isolated stromal tissue have shown that androgens are the major steroids produced by this tissue (see Lofts and Bern[82]).

The "two cells, two gonadotropins" hypothesis has been postulated to explain the control mechanism of steroidogenesis (see Hillier,[51] Hsueh et al.[53]). In summary it seems ascertained that thecal cells synthesize androgens in response to luteinizing hormone (LH). C_{19}-steroids are converted by aromatization in the granulosa layer into C_{18}-compounds after follicle-stimulating hormone (FSH) stimulation. Some exceptions have been reported; thus, LH also stimulates estrogen production in FSH-primed granulosa cells in rats, and the thecal compartment is reported to be the major site of estrogen formation during some stages of follicular development in horses and humans.

Detailed studies on the hormonal control of granulosa cells are facilitated by simple *in-vitro* cultures which make it possible to study various interactions and mechanisms of action, taking into account also the different populations of granulosa cells (e.g., mural, antral cells, and those surrounding the oocyte). Besides FSH and LH, other factors (i.e., prolactin, epidermal growth factor, fibroblast growth factor, inhibin, plasminogen activator, prostaglandins, proteoglycans, relaxin, oocyte-maturation inhibitor, gonadotropin-releasing substances, platelet-derived growth factor, insulin, adrenergic agents, lipoproteins) may influence the functions of granulosa cells (see Hsueh et al.[53]).

III. OVARIAN MORPHOLOGY

Ovarian differentiation is, according to the more recent theories, "inherent" to the embryo in the case of male digamety; while it is attributed to some unknown female sex inductors in the case of female digamety. This theory implies that the Y chromosome of the heterogametic male is responsible for testicular organogenesis against the "inherent" feminine trend (see Jost[61] as well as Chapter 1). Wachtel et al. (1975) suggested that in mammals and other vertebrate groups with male digamety, the H-Y histocompatibility antigen acts as the testicular inducer, whereas in birds and other vertebrates with female digamety the histocompatibility antigen is produced by the females and therefore induces ovarian differentiation of the gonadal primordium. However, several observations seem to contradict this hypothesis. Recently McLaren et al.[88] reported that certain mice that develop testes and that are unequivocally of the male phenotype lack the H-Y antigen. Hence the question of the inductor(s) of sex differentiation is still far from being settled. Many observations seem to exclude that sex steroids by themselves are involved in gonadal differentiation, whereas, secondary sex characters are irrefutably dependent on the sex steroids secreted by sex-differentiated gonads (see Chieffi[26,27]).

Ovarian development in nonmammalian vertebrates is characterized by the proliferation of cells inward from the surface epithelium. At the onset of oogenesis, each germinal cell is surrounded by a cluster of somatic cells that forms a follicle. Outside the follicle, connective tissue forms a second external sheath. Some eggs undergo atresia; in some nonmammalian vertebrates atretic follicles could, like postovulatory follicles, transform into structures morphologically similar to corpora lutea.

In the following sections attention will focus on the follicular wall and its derivatives; i.e., on the structures involved in ovarian steroidogenesis. Interstitial-like cells, characteristic of both the mammalian and the avian ovary, are not found in other classes of vertebrates. It should be mentioned that one author reported them in the ovary of the gecko, *Hemidactylus flaviviridis* (Dutta, 1944, quoted by Lofts and Bern[82]). It seems likely that interstitial-like cells arose from the thecal cells of atretic follicles as described by Saidapur and Nadkarni[106] in the teleost *Mystus cavasius*.

A. Cyclostomes

The follicular wall shows a great variability among different species of cyclostomes. The follicular epithelium (granulosa) of the growing oocyte of *Petromyzon marinus* is composed of squamous cells, which in the adult lamprey overlap and become multilayered. As the egg grows, the follicular epithelium becomes again single-layered and is confined to one pole of the egg. Theca interna and theca externa surround the follicular epithelium (for references, see Guraya.[49]).

The question of the formation of luteal-like structures in cyclostomes is still unsettled. Busson-Mabillot[17] maintained that in *Lampetra* spp. the occurrence of atresia is responsible exclusively for the resorption of oocytes, and that atresia coincides with the states of pre-vitellogenesis and the beginning of vitellogenesis.

B. Elasmobranchs

The follicular epithelium varies among different species of cartilaginous fishes (see Chieffi,[24]). In some species (*Myliobastis bovina*, *Trygon violacea*, *Torpedo marmorata*, *Torpedo ocellata*, *Chimaera monstrosa*), oocytes with a diameter larger than 0.1 mm are covered by a multilayered, heteromorphic epithelium, which is composed of two cell types: large cells with a reticular nucleus and abundant cytoplasm, and small cells interspersed with large cells. The large cells are sometimes full of globules morphologically and histochemically similar to the yolk platelets. At the end of vitellogenesis, the granulosa becomes flattened and single-layered. The large cells become more rare and finally disappear. In other elasmobranchs (*Scyliorhinus canicula*, *Scyliorhinus stellaris*, *Mustelus laevis*, *Scyliorhinus lichia*, *Acanthias vulgaris*), the follicular epithelium is multilayered, but monomorphic. In all cases the follicular epithelium is surrounded by the theca interna and the theca externa. Between them one can observe a layer of cuboidal or columnar cells, which is particularly evident in follicles of the genera *Scyliorhinus* and *Mustelus*.

The formation of "corpora lutea" from the postovulatory and atretic follicles has been described in numerous representative species of this class of fishes (see Chieffi[24]). The preovulatory corpora lutea (atretic follicles) develop from the ovarian follicles usually during advanced vitellogenesis, after the phagocytosis of the yolk by the granulosa cells. At full development they are characterized by hypertrophy of the granulosa cells, while the theca interna forms the connective stromal tissue.

C. Teleosts

The development and structure of the zona pellucida varies greatly among species of bony fishes depending on their mode of reproduction, whereas the follicular wall is fairly uniform (see Guraya[49]). The follicular epithelium is single-layered, although in some species it might acquire a pseudostratified structure at some stage of oocyte growth. In the final stages of vitellogenesis, the granulosa becomes stretched. The thecal layer is thick and composed of capillary loops and fibroblast-like cells. Intense alkaline phosphatase activity suggests that this layer is involved in the transport of substances across the follicular wall.

The formation of pre- and postovulatory "corpora lutea" has been described in numerous oviparous and viviparous teleostean species. Their development, fate and possible role have been discussed by Pickford and Atz,[96], Ball,[3] Chieffi,[25] and Hoar.[52] These aspects are described in detail in Chapter 9.

D. Amphibians

The follicular epithelium remains monolayered throughout oocyte growth in both Anura and Urodela (see Guraya[49]). The syncytial structure of the granulosa hypothesized by some early authors has been definitely ruled out by ultrastructural studies. The interdigitation observed between granulosa cell processes and oocyte microvilli is strongly indicative of the passage of material from the granulosa into the oocyte.

The theca consists of a thin layer of fibroblasts. Pinocytosis and alkaline phosphatase activity indicate that this layer is involved in the transfer of nutritive material to the oocyte.

The development of postovulatory corpora lutea has been observed in the ovary of both oviparous and viviparous species of Amphibia. It is generally assumed that in oviparous species postovulatory follicles are rapidly absorbed, and that postovulatory corpora lutea are formed only in viviparous species (see Chieffi and Botte[29] for references). However, this antiquated assumption must be revised extensively in lieu of the steroidogenic functions demonstrated for corpora lutea of the oviparous frog, *Rana cyanophlyctis*, and their prolongation by treatment with various gonadotropic preparations.[94a]

The degeneration of ova has been observed with variable frequency in the amphibian ovary.[18] Atresia usually affects the unovulated mature follicles, and it is characterized by hypertrophy of the granulosa and a slight thickening of the theca. It is improbable that these atretic follicles, which rapidly regress, give rise to preovulatory "corpora lutea."

E. Reptiles

The morphology of the follicular epithelium varies greatly among the different groups of reptiles. The granulosa becomes multilayered in the snakes and lizards while it remains monolayered in the chelonians (see Guraya[49]). Three types of cells are observed in the multilayered epithelium: small cells, intermediate cells, and large pyriform cells. The small cells can be subdivided into basal and apical cells. The intermediate cells are derived from small cells and are precursors to the pyriform cells. This morphology is reminiscent of that observed in some elasmobranchs. In the mature follicles, the large pyriform cells degenerate and the follicular epithelium become single-layered and monomorphic.

The follicular epithelium is surrounded by a cellular or glandular theca interna and a fibrous theca externa. The theca interna contains only a few capillaries, while the theca externa is well vascularized.

The formation of postovulatory corpora lutea occurs in many reptilian species (see Chieffi and Botte[29]). Their structure strongly resembles the corpora lutea of mammals. Follicular atresia is not frequent, and its pattern of formation is not unlike that described in other nonmammalian vertebrates.

IV. SITES OF STEROIDOGENESIS

The sites of steroid biosynthesis in the ovary have generally been studied using histochemical and ultrastructural techniques. A biochemical approach to this problem is hindered by the difficulty of isolating homogeneous cell populations from the ovarian follicle. The histochemical localization of cholesterol-positive, sudanophilic material, although useful in the study of the tissue's functional activity, is not sufficient to identify a tissue as a steroid-secreting structure. More specific are the histochemical methods for the detection of enzymes involved in steroid synthesis. Ultrastructural studies have proved to be useful when ovarian follicle cells contain cytoplasmic organelles characteristic of steroidogenic tissues (smooth endoplasmic reticulum, tubular mitochondrial cristae, prominent Golgi zone).

A. Cyclostomes

Activity of 3β-HSD has been reported in the follicular epithelium of

Lampetra fluviatilis by Hardisty and Barnes.[50] This activity reaches a maximum in February and early March when granulosa cells reach their maximal size.[74] No steroidogenic activity was found in the thecal cells.[50,74] However, ultrastructural observations conducted by Busson-Mabillot[16] on *Lampetra planeri* revealed the presence of organelles characteristic of steroidogenic tissues in the thecal cells, but not in the follicular epithelium.

In the postovulatory follicles of *Lampetra* only small quantities of cholesterol are detected by the Schultz test.[29]

B. Elasmobranchs

Investigations on the steroid-secreting cells of the elasmobranch ovary are scarce. Glucose-6-phosphate dehydrogenase (G-6-PDH) activity (Lance, 1968; quoted by Guraya[49]) and 3β-HSD activity[72] have been demonstrated in the granulosa cells. The increased activity of both enzymes with the growth of the oocyte suggests that the follicular epithelium could be the site of estrogen biosynthesis.

In some elasmobranch species, the granulosa cells of postovulatory follicles show a positive histochemical reaction for cholesterol and 3β-HSD,[24,72] while in other species they show no sign of steroid synthesis.[83] 3β-HSD activity has been found in the atretic follicles of *Torpedo marmorata* (ovoviviparous) but not in those of *Scyliorhinus stellaris*.[83]

C. Teleosts

The localization of steroid biosynthesis varies considerably in the species of teleosts studied so far. Lambert[70,71] found a strong reaction for 3β-HSD, and 17β-HSD and G-6-PDH in the granulosa cells of the guppy, *Poecilia reticulata*. In the preovulatory follicles of *Salvelinus leucomaenis* neither granulosa cells nor special thecal cells showed 3β-HSD activity.[62] Bara[5,6] has demonstrated only a weak 3β-HSD activity in the granulosa cells of *Scomber scomber*; the activity is strongest in some thecal cells at beginning of vitellogenesis. 3β-HSD activity has been found in both the granulosa and the thecal cells of *Acanthobrama terraesanctae* and *Sarotherodon niloticus*[130] and *Mystus cavasius*.[106]

Ultrastructural studies conducted by Nicholls and Maple[92] provided evidence for steroid hormone biosynthesis in both the granulosa and the thecal cells of the growing oocytes of the cichlid fishes, *Cichlasoma nigrofasciatum* and *Haplochromis multicolor*. More recently, Kagawa et al.[63] separated the follicular thecal cells from the granulosa cells of the amago salmon, *Oncorhynchus rhodurus*, and assessed their respective roles in estradiol production *in vitro*. The results suggested that thecal cells contribute to estradiol production by synthesizing testosterone, which is transferred to the granulosa layer and aromatized there to estradiol.

While in the majority of oviparous teleosts the postovulatory follicles degenerate and are rapidly absorbed, in some species they transform into well-organized, although in certain cases transitory, glandular corpora lutea (see Saidapur[105]). No glandular structures similar to corpora lutea are formed in the viviparous species that show follicular gestation.[52] Histochemical studies have demonstrated 3β-HSD activity in the granulosa cells of the postovulatory follicles of *Oryzias latipes* (Lambert and van Oordt, 1974, quoted by Saidapur[105]) and *Brachydanio rerio* (Iwasaki, 1973, quoted by Saidapur[105]), in the thecal cells of *Salvelinus leucomaenis*,[62] and in both the granulosa and the thecal cells of *Salmo gairdneri* (Hurk and Peute, 1979, quoted by Saidapur[105]) *Scomber scomber*,[5,6] *Trachurus mediterraneus*,[7] and *Carassius auratus* (Khoo, 1975, and Nagahama et al., 1976, both quoted by Saidapur[105]). In *T. mediterraneus* also 11β-HSD and 20α-HSD activities have been recorded.[7] Activity of 3β-HSD usually parallels the activity of G-6-PDH in *T. mediterraneus*, *Salmo gairdneri* and *Clarias batrachus*. Ultra-structural studies have confirmed the steroid-synthesizing capacities of both the granulosa and thecal cells of *C. auratus* (Lambert, 1978, quoted by Saidapur[105]), *B. rerio* (Iwasaki, 1973, quoted by Saidapur[105]), *S. gairdneri* (Hurk and Peute, 1979, quoted by Saidapur[105]), *C. nigrofasciatum* and *H. multicolor*.[92] In these species, a smooth endoplasmic reticulum and mitocondria with tubular cristae are seen.

Although atretic follicles have been described in a number of teleostean fishes,[18] they do not evolve into glandular structures. In *M. cavasius* the activity of some steroid dehydrogenases (3β-, 17β-, 11β-HSD) and of G-6-PDH has been demonstrated in hypertrophied granulosa cells of early atretic follicles. In later stages only thecal cells remain in the ovarian stroma, and they show a strong enzymatic activity.[106]

The possible role of both pre- and postovulatory corpora lutea in the reproductive physiology of teleostean fishes is still an open question. Because of the lack of consistent experimental work (probably due to the difficulties it entails), it is not possible at present to comment further on this topic (see for discussion Chieffi and Botte,[29] Lofts and Bern,[82] Saidapur,[105] see also Chapter 9).

D. Amphibians

Botte and Cottino[11] detected 3β-HSD activity in both the granulosa and thecal cells of *Rana esculenta* and *Triturus cristatus*. Other authors confirmed these observations in numerous species of anurans and urodeles (for references, see Saidapur,[105] Pancharatna and Saidapur[94a]). In *Xenopus laevis*, 17α-, 17β-, 3α-, and 3β-HSD activities are confined to the granulosa cells.[103] Ultrastructural studies have shown elements of a smooth endoplasmic reticulum in the granulosa cells of *Necturus maculosus* (Kessel and Panje, 1968, quoted by Saidapur[105]), *Salamandra salamandra*,[59] and *Bufo bufo*,[121] while ultra-structural features of steroid-secreting cells have yet to be demonstrated in the thecal cells.

After ovulation, the granulosa cells of the postovulatory follicle show a strong histochemical reaction for 3β-HSD activity and a positive Schultz test for cholesterol (for references, see Saidapur[105]). Weak 3β-HSD activity was demonstrated by Joly[58] in the thecal cells of the postovulatory follicles of *S. salamandra*. Ultrastructural studies have been conducted on the postovulatory follicles of only two species. The granulosa cells of *N. maculosus*, 1968, (Kessel and Panje, 1968, quoted by Saidapur[105]) and *S. salamandra*[59] contain a smooth endoplasmic reticulum, mitochondria with tubular cristae, a well-developed Golgi zone and lipid droplets, all features characteristic of steroidogenic cells.

Neither histochemical (enzymes of steroidogenesis) nor ultrastructural studies have been performed on the atretic follicles of amphibians.

The postovulatory follicles are transitory in oviparous amphibians and generally have been considered to have no functional significance. However, corpora lutea of skipper frogs, *R. cyanophlyctis*, are steroidogenic and last as long as do corpora lutea of mice.[94a] In the case of ovoviviparous and viviparous urodeles, the possible functional role of the postovulatory corpora lutea is still debated. Joly et al.[60] observed that ovariectomy performed in the viviparous *S. salamandra* at the beginning of pregnancy does not interrupt the intrauterine development of the larvae. Xavier[129] has shown that ovariectomy of *Nectophrynoides occidentalis* at the end of gestation does not cause abortion, while ovariectomy performed at the beginning of gestation in females experiencing their first gestation leads to abortion. This finding has been interpreted in terms of low levels of progesterone secreted by the ovaries in females undergoing first gestation. For further details see Chapter 9.

E. Reptiles

Activity of 3 -HSD has been detected in both the granulosa and thecal cells of *Lacerta sicula*.[12] Numerous investigations have confirmed the presence of 3 -HSD activity both in oviparous and in ovoviviparous species (for references, see Guraya[49]). 17 -HSD activity has been found in the granulosa and thecal cells of *Sceloporus cyanogenys* and *Dipsosaurus dorsalis*[19] but not in those of *Lacerta sicula*.[12] In some cases the follicular components show other cytological features of steroidogenesis.

Botte and Delrio[12] demonstrated high 3 -HSD activity after ovulation in both the granulosa and thecal cells of *L. sicula* along with an intense sudanophilia and Schultz reaction for cholesterol; 3 -HSD activity also have been found in the postovulatory follicles of many other species (for references, see Saidapur[105]). Activity of 17 -HSD has been detected in the postovulatory follicles of only a few species (for references, see Saidapur[105]).

In the case of reptiles, the physiological significance of postovulatory corpora lutea is unclear, and ovariectomy has given contradictory results. Clausen[32,33] found that ovariectomy interrupted pregnancy in some viviparous snakes of the genera *Thamnophis*, *Natrix* and *Storeria*, while in the same animals Bragdon,[14,15] found that it did not. Fraenkel et al.[47,48] obtained abortion following ovariectomy of the Brazilian snakes, *Crotalus* and *Bothrops*, and Panigel[95] reported continuation of gestation in *Zootoca vivipara* after ovariectomy.

Follicular atesia is not frequent among reptiles. Unfortunately, nothing is yet known concerning steroidogenesis. Luteogenesis in this class of vertebrates is discussed in more detail in Chapter 9.

V. STEROID IDENTIFICATION AND BIOSYNTHESIS

The first target for early comparative endocrinologists was the identification of steroid hormones, followed by the identification of the biosynthetic pathways (see Ozon,[93,94]; Lance and Callard[73]). Recent studies have demonstrated that some steroids (e.g., 11-ketotestosterone, 17 -20, dihydroxy-4-pregnen-3-one) appear to be peculiar to some vertebrate classes. Since it is difficult to separate the identification of hormones from their biosynthesis, we shall consider them collectively.

A. Cyclostomes

Although it has been demonstrated clearly that the ovary of hagfish can produce steroids, the presence of estrogens is still an open question.

Estrogens have been detected in *Myxine glutinosa* by Lance and Callard,[73] while Kime and Hews[68] failed to detect estrone and estradiol. Moreover, the ovary of *M. glutinosa* contains enzymes that produce 5α-reduced, and 6β- and 7α-hydroxylated steroids.[68]

It has long been known that the ovary of the lamprey, *P. marinus*, contains estradiol, estrone, and progesterone.[13] No sex differences in plasma levels of androstenedione, estrone, or estradiol were found by Katz et al.,[67] who concluded that androstenedione is a stress-related hormone probably originating in the interrenal glands.

B. Elasmobranchs

The presence of estradiol in the ovary of elasmobranchs was reported as long ago as 1958.[128] It also has been established that progesterone, androgens, and estrogens can be synthesized in the ovary.[73,93,94] As far as we are aware, there are no other reports on steroid biosynthesis in the elasmobranch ovary. Indeed, very little is known about reproductive endocrinology in cartilaginous fishes.

C. Teleosts

There is a vast body of literature concerning the reproductive endocrinology of bony fishes (see Lance and Callard[73]). A peculiarity of teleosts is the presence in the ovary of 17α-20β,dihydroxy-4-pregnen-3-one (17α,20βdiOHprog), and its physiological role as the most potent inducer of oocyte maturation. First isolated by Idler et al.[55] from female salmon, *Oncorhynchus nerka*, this steroid and other progestational hormones have since been shown to be synthesized by teleostean ovaries.[116-118] Recently, Nagahama et al.[91] demonstrated that 17α,20βdiOHprog induces oocyte maturation (see also Chapter 6). The only reported exception has been for oocytes of zebrafish (*B. rerio*) in which 17α,20βdiOHprog was less effective than deoxycorticosterone (DOC).[123] The corticosteroids, 11-deoxycortisol (11-DOC) and DOC, are also effective in inducing germinal vesicle breakdown (GVBD) *in vitro*),[115] and the teleostean ovary synthesizes DOC.[34]

Another interesting point to emerge from the literature is the presence of 11-ketotestosterone in the ovary (see Lance and Callard[73]). although it has been claimed that a high plasma concentration of this androgen is typical of the male,[107] 11-ketotestosterone also increases in female *Mugil cephalus* with advancing gonadal development.[1]

Steroid biosynthesis in bony fishes has been studied in ambisexual animals, although not extensively. The presence of sex steroids in the ovotestes of *Serranus scriba*, a synchronous hermaphroditic marine teleost, was discovered by Lupo di Prisco and Chieffi.[84] In *Sparus auratus*, a protandrous fish, the androgen/estrogen ratio is higher in the male phase than in the female phase, and steroid biosynthesis is very low during sex reversal.[35] Recently Reinboth and Becker[104] reported that in *Coris julis*, a protogynous teleost, high 11β-hydroxylation is detectable only during the male phase. Ovarian tissue synthesizes a large amount of 5α-reduced steroids, while 5α-reduction characterizes both primary and secondary males. Both *C. auratus* and *C. julis* have a pattern of steroid metabolism which fits well into present knowledge on the biosynthetic activities of gonochoristic species.

D. Amphibians

Amphibians produce progesterone early in ontogenesis, and the steroid profile correlates well with gonadal differentiation.[30] Adult urodelean amphibians convert testosterone to 11-ketotestosterone in the ovary *in vitro*.[85] There are numerous studies on the synthesis of sex steroids and DOC in amphibians (see Lance and Callard[73]). In particular, the *R. esculenta* ovary produces testosterone and dihydrotestosterone (DHT),[41] and plasma levels of testosterone are similar in both sexes.[43] In the bullfrog, *Rana catesbeiana*, peripheral testosterone is higher in females than in males, but DHT is significantly higher in males.[78] Whether or not these levels are a consequence of peripheral conversion is still a matter for discussion.

E. Reptiles

Several *in-vitro* studies have demonstrated that hormone biosynthesis in reptiles resembles that found in mammals. However, Colombo and coworkers[36,37] have found DOC in a lizard and in a snake, respectively. Further details concerning steroid hormone identification and biosynthesis in reptiles can be found in Ozon[93,94] Lance and Callard,[73] and Delrio et al.[42]

VI. SEASONAL PROFILES

During the first half of the seventies, with the advent of more sensitive methods for hormone measurement, studies on steroid identification underwent a progressive decline, and interest turned to investigations of seasonal hormone profiles and the mechanisms underlying seasonal breeding in all vertebrate classes (see Chieffi[30]).

A. Cyclostomes

The only hagfish known to have seasonal gonadal activity is *Eptatretus burgeri* (see Lance and Callard[73]). To our knowledge, no sex hormone profile has been reported for the Myxiniformes.

In the Petromyzontiformes, circulating steroid levels have been studied by Katz et al.[67] who reported a hormonal stress-response in terms of variations in androstenedione, progesterone, and estradiol levels. Estrone does not show significant changes in response to stress.

B. Elasmobranchs

Dobson (1975, quoted by Dodd[44]), and Sumpter and Dodd (1980, quoted by Jenkins and Dodd[57]) found in *S. canicula* that plasma androgens and estradiol levels are at their lowest in midsummer (when ovarian follicles are atretic), their highest in winter, and are declining during early spring.

C. Teleosts

Hormone profiles, particularly of testosterone, 11-ketotestosterone, and estradiol have been studied widely among teleosts (Lance and Callard[73]). High estradiol levels are found in vitellogenic animals, whereas testosterone is elevated in prespawning females.

In the amago salmon (*O. rhodurus*), a species in which follicles develop synchronously, estrogens decrease prior to final oocyte maturation,[64] while in the goldfish *C. auratus*, in which follicles develop asynchronously, estrogen levels remain high even after ovulation.[65] According to Kagawa et al.,[64] this difference "between the two species can be attributed to the presence of vitellogenic primary and secondary yolk stage oocytes in the ovary of goldfish after the first crop of oocytes are ovulated."

Studies on profiles of estradiol, testosterone, and 17α,20βdiOhprog in rainbow trout,[109] amago salmon,[131] and coho salmon[124] have demonstrated that 17α,20βdiOHprog peaks when oocytes mature. In particular, Van der Kraaf et al.[124] reported that estradiol and testosterone are at low and high concentrations, respectively, during the 17α,20βdiOHprog peak. They suggested that these levels of estradiol and testosterone are necessary to provide a suitable environment for 17α,20βdiOHprog biosynthesis. Moreover, Scott and Sumpter[109] found that gonadotropin (Gn) peaks around ovulation time in two different strains of rainbow trout. These authors investigated Gn profiles using radioimmunoassay to detect ovulation-inducing Gn. Idler and Ng[56] suggested that two different Gns act during vitellogenesis and ovulation, respectively (see also Chapter 10).

Besides an annual cycle, teleosts show circadian variations in plasma steroid levels.[69] In the catfish, *Heteropneustus fossilis*, the amplitude of the circadian pulses of estradiol and testosterone is higher in maturing than in regressed or mature animals.

D. Amphibians

No seasonal variation of progesterone, DHT, testosterone, estrone, or estradiol has been detected in the captive urodele amphibian, *N. maculosus*.[9] In contrast, definite seasonal levels of testosterone, androstenedione, estrone, estradiol, and progesterone have been reported for *Rana esculenta*. According to Follett and Redshaw,[45] the level of estradiol correlates with the vitellogenic period. In addition, androstenedione and progesterone levels are higher in frogs with "ripe" ovaries than in those with "spent" ovaries.[98] Female *R. catesbeiana* are characterized by an elevation in circulating progesterone concentration during the periovulatory gonadotropin surge.[78] Incidentally, progesterone acting synergistically with an insulin-like factor[76] induces GVBD in *X. laevis*.

The above cited studies[78,98] support a postovulatory endocrine quiescence. A peak of plasma estradiol during the postreproductive period has been described in *R. esculenta*.[101] This is the only finding in support of ovarian activity when the gonad is believed to be quiescent[102] and when the morphological parameters used to assess ovarian activity are at their lowest values.[101] In addition studies conducted in our laboratory[43,98] (also Pierantoni et al., unpublished) concerning estradiol profiles in *R. esculenta*, give no evidence of postreproductive increases in estrogen.

E. Reptiles

Sex steroid profiles in reptiles have been reviewed by Lance and Callard.[73] Recently, interest has focused on the reproductive endocrinology of turtles: *Chelydra serpentina*,[17] *Chrysemys picta*,[21] *Chelonia midas*,[81] *Sternotherus odoratus*,[89] and *Lepidochelys olivacea*.[80] Testosterone seems to play a role primarily as a precursor for estradiol synthesis. Estrogen titers correlate well with vitellogenesis, while progesterone appears to have an important role in the ovulatory process. Indeed, Licht et al.[80] demonstrated the presence of a pronounced surge in LH and progesterone (the "ovulatory surge") in *L. olivacea*. Lastly, in oviparous reptiles, progesterone produced by the corpora lutea may cause retention of ova during normal shelling.[38]

VII. GONADOTROPIC CONTROL OF FOLLICULOGENESIS AND STEROIDOGENESIS

It has long been known that gonadal function in vertebrates is under the control of pituitary Gns. However, it has still to be established whether fishes and squamate reptiles are endowed with separate FSH- and LH-like substances (see Lance and Callard;[73] see also Chapter 6 and Chapter 10).

Experimental research on the control of folliculogenesis by Gn is scarce (see Tokarz[122]). The information available is either indirect or based almost exclusively on mammalian hormonal preparations and on non-pituitary materials such as human chorionic gonadotropin (hCG) and pregnant mare serum gonadotropin (PMSG). Even hypophysectomy and the use of anti-hormones has provided only circumstantial information about endogenous endocrine events. The control of ovulation by Gn is better known. Therefore studies are required to relate the levels of endogenous hormones to morphological changes of the follicular layers during the oogenetic cycle. An account of the possible gonadotropic control of "luteogenesis" occurs in Chapter 9.

The control of steroidogenesis by Gns, on the other hand, has been investigated extensively.

A. Cyclostomes

The question of hypothalamus-hypophysial control of the gonads of Myxiniformes is rather confusing (see Lance and Callard;[73] Ball[4]). Gonadotropic control has been demonstrated in lampreys.[75] Although a hypothalamic role has yet to be demonstrated clearly (see Ball[4]), a gonadotropin-releasing hormone (GnRH) has been identified in the lamprey hypothalamus (see Chapter 2).

B. Elasmobranchs

In elasmobranchs it has been shown that the ventral lobe contains most of the Gns that stimulate gonadal steroid production (see Lance and Callard[73]). Using hypothalamic extract and mammalian GnRH, Jenkins and Dodd[57] showed that the pituitary gland increases the plasma concentration of steroids. The ability of animals to respond to treatment was measured in terms of the time course for secretion of androgen and estradiol during two different phases of the sexual cycle. A reduced response during a period of relative sexual dormancy may reflect decreased responsiveness of the pituitary gland to GnRH and/or decreased responsiveness of the ovary to Gn. A cyclic ovarian response to Gn also has been demonstrated in the dogfish.[114] Investigations on gonadotropic control have to rely on pituitary extract since no standard Gns are available for elasmobranchs.

C. Teleosts

Specific standards for both GnRH (see Chapter 2) and Gn are available for bony fishes (see Chapter 10). The existence of the hypothalamus-hypophysial axis emerges clearly from measurements of plasma Gn and steroid hormones after treatment with GnRH (mostly GnRH agonists). Injections of a GnRH agonist (GnRHa) in coho salmon elicits an accelerated rate of GVBD by elevating plasma Gn.[125] This provides indirect evidence that Gn exerts control on the levels of 17α,20βdiOHprog. Van der Kraak et al.[124] measured plasma estradiol and 17α,20βdiOHprog after injections of GnRHa. While the effects of GnRH on testosterone levels were confusing, GnRH treatment clearly increased plasma concentrations of estradiol and 17α,20βdiOHprog. Only fishes that showed the correct balance between 17α,20βdiOHprog and estradiol levels exhibited GVBD. Evidence that Gn exerts control over ovarian activity comes also from studies on the negative control of Gn release by gonadotropin-release-inhibiting factors (GRIF). Dopamine (DA) has GRIF activity and a DA antagonist (pimozide) potentiates the ability of GnRH to stimulate Gn release, GVBD, and ovulation in goldfish.[23] *In-vitro* studies suggest that treatment of the fish ovary with Gn selectively activates 3β- and 20β-HSD.[116] As a consequence, 17α,20βdiOHprog is accumulated in a dose-related manner when challenged with homologous and heterologous Gn preparations.[131] Estrogen and androgen output by ovarian tissue incubated with Gn also has been studied.[8,10,66] Testosterone, produced by specialized thecal cells, is increased by gonadotropic stimulation but is converted into estradiol in the granulosa cells without any Gn action.[63]

In the goldfish *C. auratus*, which has a synchronous-type ovary, hCG treatment induces an increase of estradiol production by primary and secondary stage follicles, and the production of large amounts of testosterone by tertiary yolk-stage follicles.[66] Incubation of tertiary yolk-stage follicles with testosterone does not result in the production of estradiol which suggests that there is a reduction of aromatase activity at this stage.

Incubation with exogenous cyclic adenosine monophosphate (cAMP) or phosphodiesterase inhibitors increases estradiol and 17α,20βdiOHprog levels. This indicates that, as in other vertebrates, cAMP acts as a mediator of Gn action.[8,131] Moreover, treatment with cAMP and phosphodiesterase inhibitors prevents Gn-induced GVBD (Young, 1983, quoted by Young et al.[131]) which suggests that cAMP inhibits a maturation-promoting factor (MPF) within the ooplasm.[108]

D. Amphibians

In amphibians there are two Gns (LH- and FSH-like substances), and GnRH elicits Gn discharge from the pituitary (see Lance and Callard,[73] Daniels and Licht[39] and Chapter 10. Although it is not yet known whether or not FSH and LH have separate functions,[78] it is clear that Gn induces ovarian steroid production. *In vivo*, GnRH and GnRHa elicit a transient increase of Gns followed by a surge of estradiol and androgens.[87] Injections of GnRH in amphibians did not appear to exert an antifertility effect, as observed in mammals, as plasma steroids remained at high concentrations throughout the treatment period. It also should be mentioned that gonadotropic control of steroidogenesis appears to be impaired by capture stress since the level of steroids declines in both male and female *R. catesbeiana* immediately after capture.[78] In male *R. esculenta* the decline is less marked,[99] while in females the effects of captivity on ovarian metabolism are dramatic.[100] It is also interesting to note the enormous increase in androgen levels in response to GnRHa treatment,[87] and the subsequent observation that testosterone and androstenedione are good indicators of the stage of the ovary.[78,98] These findings, together with the demonstrated facilitatory effect of exogenous DHT on Gn release,[86] indicate that enhanced levels of plasma androgens may induce the preovulatory Gn surge in frogs.[86,98]

Incubation of ovarian tissue with several LH preparations results in the production of progesterone and testosterone (see Lance and Callard;[73] Fortune and Tsuang;[46] Pierantoni et al., unpublished), while the same effect is not obtained with FSH (see Lance and Callard[73]). In *R. esculenta* stimulation of progesterone production by ovine-LH (oLH) is greater at 6 h than at 3 h incubation. After 24 h, estradiol production is increased and progesterone production decreased (Pierantoni et al., unpublished data). This suggests that the aromatization is influenced more by progesterone levels than by Gn which influences progesterone production. These results are in accordance with the unpublished observation of Licht and Bona-Gallo (quoted by Licht et al.[78]) that preovulatory follicles produce progesterone but only a small amount of estradiol and with Mulner et al.[90] who found that Gns do not enhance formation of estrogens in *X. laevis*. It is important to mention that Lin and Schuetz[81a] demonstrated in *Rana pipiens* that estradiol inhibits progesterone production.

As far as we are aware, the mechanisms by which Gn induces steroid production have not been investigated fully in amphibians. Thibier-Fouchet et al.[120] suggested that LH treatment of *X. laevis* ovaries stimulates side-chain cleavage of cholesterol. Recently, Thibier and Ozon[119] studied the effect exerted by hCG on 17β-hydroxysteroid oxydoreductase (17β-HSOR) activity, an enzyme which is localized in the follicular envelopes of *X. laevis* and which is involved in the synthesis of testosterone. They found that treatment with Gn inhibits 17β-HSOR by a mechanism that requires Ca^{++} ion rather than cAMP. Moreover they demonstrated that concomitant prolactin treatment prevents the effect of hCG, while prolactin alone does not affect steroidogenesis.

E. Reptiles

Among reptiles, the existence of two Gns similar to mammalian FSH and LH have been reported in chelonians and crocodilians. The responses of turtle and crocodilian ovaries to mammalian FSH and LH preparations have been compared with their responses to homologous FSH and LH preparations (see Lance and Callard[73]). A lack of FSH/LH specificity in terms of steroid production has been observed, and this has been interpreted as a primitive characteristic that depends on hormone-receptor interactions. A comparative analysis of the characteristics of Gn receptors in the various species will probably throw some light on this problem. It is also interesting to note that mammalian GnRH and GnRHa do not elicit plasma steroid responses in turtles, probably due to lack of pituitary sensitivity.[79,80] It should be mentioned that pituitary and subsequent ovarian responses to GnRH have been reported.[20,22]

The types of Gn and variety of ovarian responses in squamates are intriguing. Although it is not clear whether squamates possess one or two, several gonadotropic preparations increase steroid levels in this group (see Lance and Callard[73]). Recent unpublished observations from our laboratory (Pierantoni et al.) indicate a high progesterone output from preovulatory follicles of the lizard *Podarcis sicula sicula* after a 6 h incubation of excised ovaries with oLH. No increase in estrogen output is detectable, which is in accordance with similar observations in turtles reported by Callard and Lance.[20]

VIII. LOCAL CONTROL OF STEROIODOGENESIS

Steroids and other substances (i.e., GnRH, oxytocin and vasopressin) are known to exert an intragonadal regulation (see Hsueh et al.,[53] Lin and Schuetz,[81a] Wathes et al.[127]). According to Hsueh et al.[53] the mammalian ovary is subject to paracrine and autocrine controls. Paracrine control is exerted without traversing the systemic circulation, whereas the autocrine mechanism is described as an "ultra-short loop regulation." The first studies on nonmammalian vertebrates were performed in the male frogs, *R. pipiens* and *R. esculenta*. They were based on *in-vitro* incubation of minced testes treated with GnRH or GnRHa and on long-term treatment *in vivo*.[97,99,110] Because of the short half-life of GnRH, it is conceivable that a local production of testicular GnRH-like substances is responsible for these observations as claimed for rats (see Hsueh and Jones[54]).

Only one paper[54a] has been published on local control of steroidogenesis exerted by nonsteroidal substances in female frogs, *R. pipiens* and *R. catesbeiana*. We recently have incubated follicles from elasmobranchs (*T. ocellata*), bony fishes (*Gobius paganellus*), amphibians (*R. esculenta*), and reptiles (*P. s. sicula*) with combined doses of oLH and a GnRHa (buserelin, Hoechst). After 6 h of incubation estradiol and progesterone output by preovulatory follicles of *T. ocellata* and *G. paganellus* remained essentially unchanged. GnRH and oLH treatment in combination resulted in a small but significant progesterone decrease in media incubated for 6 and 24 h with minced preovulatory follicles of *R. esculenta*. Incubation of preovulatory follicles of *P. s. sicula* with GnRH alone for 6 h resulted in a significant increase in progesterone output, while estradiol levels were unaffected in both species. The lack of effects in *T. ocellata* and *G. paganellus* may be due to an ovarian unresponsiveness caused by the incubation procedure. The results obtained in *P. s. sicula* and *R. esculenta* are in agreement with the biphasic GnRH effect found in mammals, in which the peptide suppresses progesterone production stimulated by FSH or LH, but alone GnRH stimulates steroidogenesis (see Hsueh et al.[53]). Since our preliminary results are in conflict with those of Hubbard and Licht[54a] (who found no effect in either *R. pipiens* or *R. catesbeiana*), it seems appropriate to perform more exhaustive studies on frog species before drawing any conclusions.

As yet there are no studies on the possible effects exerted by oxytocin and related compounds on steroidogenesis. However, an antifertility effect has been observed in intact lizards, *P. s. sicula*, injected daily with arginine vasotocin.[31]

Melatonin exerts a direct control over steroidogenesis by mammalian testes (see Sharpe[113]). Daily injections of melatonin elicit an antigonadal effect in intact females of *Rana ridibunda*;[40] whether or not melatonin or oxytocin directly influences the steroidogenic pathway in nonmammalian vertebrates is still to be investigated.

IX. SUMMARY AND FUTURE RESEARCH DIRECTIONS

Unlike the vertebrate testis whose structure is highly conservative throughout the various classes, the vertebrate ovary varies greatly according to the type of reproduction.

The follicular apparatus of the ovary of nonmammalian vertebrates is basically the same as that described in mammals. Granulosa cells contribute to both yolk production and to the secretion of sex hormones. In some elasmobranch species, the yolk platelets are formed by special granulosa cells. This led some early authors to assume that follicular cells and oogonia have a common origin.[2,111,112]

Postovulatory corpora lutea and atretic follicles (preovulatory corpora lutea) develop in almost all groups of nonmammalian vertebrates. However, because of scarcity of consistent experimental investigations, their physiological role is still uncertain.

Steroidogenesis among nonmammalian vertebrates (except the Myxiniformes) seems to be regulated basically by the same intriguing chain of events that occurs in mammals. Detailed studies regarding the regulatory role of substances other than Gn and research on the mechanisms of action underlying steroid production are hampered by difficulties in isolating cells for cultures and in obtaining protein hormone preparations. Despite these obstacles, some differences in the role played by steroids between mammals and nonmammalian vertebrates have been documented. While in mammals estradiol seems to exert the major regulatory role in the chain of events leading to ovulation (see Hsueh et al.[53]), in the vertebrate classes examined here it appears that progestogens and perhaps androgens play the central role in oocyte maturation and in feedback mechanisms challenging the preovulatory Gn surge. As far as the ovary is concerned, the only function ascertained for estradiol concerns the vitellogenic process.

Studies on the local control of steroidogenesis are *in statu nascenti*, and it has yet to be established whether GnRH-substances play a physiological role. Indeed the presence of specific receptors and the isolation of peptide(s) can furnish a decisive insight into this new reseach area.

X. REFERENCES

1. Azoury, R. and Eckstein, B. (1980). Steroid production in the ovary of the gray mullet (*Mugil cephalus*) during stages of egg ripening. Gen. Comp. Endocrinol. 42, 244-250.
2. Balfour, F.M. (1878). On the structure and development of the vertebrate ovary. Quart. J. Micr. Sc. 18, 383-458.
3. Ball, J.N. (1960). Reproduction in female bony fishes. Symp. Zool. Soc. London 1, 105-135.
4. Ball, J.N. (1981). Hypothalamic control of the pars distalis in fishes, amphibians, and reptiles. Gen. Comp. Endocrinol. 44, 135-170.
5. Bara, G. (1965) Histochemical localization of Δ^5-3β-hydroxysteroid dehydrogenase in the ovaries of teleost fish, *Scomber scomber* L. Gen. Comp. Endocrinol. 5, 284-296.
6. Bara, G. (1965) Glucose-6-phosphate dehydrogenase activity in the ovaries of *Scomber scomber* L. Experientia 21, 638-640.
7. Bara, G. (1974). Localization of steroid hormone production in the ovary of *Trachurus mediterraneus*. *Acta Histochem*. 51, 90-101.
8. Bogomolnaya, A. and Yaron, Z. (1984). Stimulation *in vitro* of estradiol secretion by the ovary of a cichlid fish, *Sarotherodon aureus*. Gen. Comp. Endocrinol. 53, 187-196.
9. Bolaffi, J.R. and Callard, I.P. (1979). Plasma steroid profiles in male and female mudpuppies, *Necturus maculosus* (Rafinesque). Gen. Comp. Endocrinol. 37, 443-450.
10. Bona-Gallo, A. and Licht, P. (1983). Specificity to gonadotropins in the response of *in vitro* estrogen secretion by fish ovaries. Gen. Comp. Endocrinol. 51, 316-319.
11. Botte, V. and Cottino, E. (1964). Richerche istochimiche sulla distribuzione del colesterolo e di alcuni enzimi della steroidogenesi nei follicoli atresici e post-ovulatori de *Rana esculenta* e *Triturus cristatus*. Boll. Zool. 31, 491-500.
12. Botte, V. and Delrio, G. (1965). Ricerche istochimiche sulla distribuzione dei 3- e 17-chetosteroidi e di alcuni enzimi della steroidogenesi nell'ovaio di *Lacerta sicula*. Boll. Zool. 32, 191-197.
13. Botticelli, C.R., Hisaw, F.L., and Roth, W.D. (1963). Estradiol-17β, estrone and progesterone in the ovaries of lamprey (*Petromyzon marinus*). Proc. Soc. Exp. Biol. Med. 114, 255-257.
14. Bragdon, D.E. (1951). The non-essentiality of the corpora lutea for the maintenance of gestation in certain live-bearing snakes. J. Exp. Zool. 118, 419, 435.
15. Bragdon, D.E. (1952). Corpus luteum formation and follicular atresia in the common garter snake, *Thamnophis sirtalis*. J. Morph. 91, 413-445.
16. Busson-Mabillot, S. (1966)., Presence d'une theque interne glandulaire dans le follicule de la lamproie de planer (*L. planeri* Bloch), vertebre cyclostome. C.R.. Acad. Sci. Paris 262, 177-181.
17. Busson-Mabillot, S. (1967). Structure ovarienne de la lamproie adulte (*Lampetra planeri Bloch*). II. Les enveloppes de l'ovocyte: cellules folliculaires et stroma ovarien. J. Microsc. 6, 807-838.
18. Byskov, A.G. (1978). Follicular atresia. In "The Vertebrate Ovary" (R.E. Jones, ed.), pp. 533-562. Plenum Press, New York.
19. Callard, I.P., Doolittle, J., Banks, W.L. Jr. and Chan, S.W.C. (1973). Recent studies on the control of the reptilian reproductive cycle. Gen. Comp. Endocrinol. Suppl. 3, 65-75.
20. Callard, I.P. and Lance, V. (1977). The control of reptilian follicular cycles. In "Reproduction and Evolution" (J. H. Calaby and C.H. Tyndale-Biscoe, eds), pp.199-210, Australian Acad. Sci., Canberra City.

21. Callard, I.P., Lance, V., Shalhanick, A.R. and Barad, D. (1978). The annual ovarian cycle of Chrysemys picta: correlated changes in plasma steroids and parameters of vitellogenesis. Gen. Comp. Endocrinol. 35, 245-257.
22. Callard, I.P. and Scanes, C.G. (1975). Synthesis of radioimmunoassayable LH by suspended pituitary cells derived from turtle Chrysemys picta. In "Proceedings 57th Ann. Meet. Endocrine Soc."New York, abs. 307.
23. Chang, J.P. and Peter, R.E. (1983). Effects of pimozide and del Gly^{10} [DAla^{6}] luteinizing hormone-releasing hormone ethylamide on serum gonadotropin concentrations, germinal vesicle migration and ovulation in female goldfish, Carassius auratus. Gen. Comp. Endocrinol. 52, 30-37.
24. Chieffi, G. (1961). La luteogenesi nei selacei ovovivipari: ricerche istologiche e istochimiche in Torpedo marmorata e Torpedo ocellata. Pubbl. Staz. Zool. Napoli 32, 145-166.
25. Chieffi, G. (1962). Aspetti endocrinologici della riproduzione nei Pesci. Boll. Zool., 29, 150-186.
26. Chieffi, G. (1965). Onset of steroidogenesis in the vertebrate embryonic gonad. In "Organogenesis" (R.L. De Haan and H. Ursprung, eds), pp. 651-669. Holt, Rinehart and Winston, New York.
27. Chieffi, G. (1975). Modern trends in sex differentiation. Pubbl. Staz. Zool. Napoli 39 Suppl., 471-479.
28. Chieffi, G. (1984). Control of reproductive cycles in Metazoa: exogenous and endogenous factors. Boll. Zool. 51, 205-222.
29. Chieffi, G. and Botte, V. (1970). The problem of luteogenesis in nonmammalian vertebrates. Boll. Zool. 37, 85-102.
30. Chieffi, G., Di Grande, F., Gardenghi, G., Pierantoni, R., Stagni, A., Tognatò, G., Vannini, E., Varriale, B. and Zaccanti, F. (1984). Progesterone and testosterone profiles during ontogenesis in the frog Rana dalmatina and in the toad Bufo bufo. Boll. Zool. 51 Suppl., 29.
31. Ciarcia, G., Angelini, F., Picariello, O. and Botte, V. (1983). Arginine-vasotocin and gonadal activity in the lizard Podarcis s. sicula Raf. Experientia 39, 615-616.
32. Clausen, H.J. (1935). The effects of ovariectomy and hypophysectomy on parturition in snakes. Anat. Rec. 64 Suppl., 88.
33. Clausen, H.J. (1940). Effect of ovariectomy and hypophysectomy on gestation in snakes. Endocrinology 27, 700-704.
34. Colombo, L., Bern, H.A., Pieprzyk, J. and Johnson, V.W. (1973). Biosynthesis of 11-deoxycorticosteroids by teleost ovaries and their possible role in oocyte ovulation and maturation. Gen. Comp. Endocrinol. 21, 168-178.
35. Colombo, L., Del Conte, E. and Clemente, P. (1972). Steroid biosynthesis in vitro by the gonad of Sparus auratus L. (Teleostei) at different stages during natural sex reversal. Gen. Comp. Endocrinol. 19, 26-36.
36. Colombo, L. and Yaron, Z. (1976). Steroid 21-hydroxylase activity in the ovary of the snake Storeria dekayi during pregnancy. Gen. Comp. Endocrinol. 28, 403-412.
37. Colombo, L., Yaron, Z., Daniels, E. and Belvedere, P. (1974). Biosynthesis of 11-deoxy-corticosterone by the ovary of the yucca night lizard, Xantusia vigilis. Gen. Comp. Endocrinol. 24, 331-337.
38. Cuellar, H.S. (1979). Disruption of gestation and egg shelling in deluteinized oviparous whiptail lizards Cnemidophorus uniparens (Reptilia: Teiidae). Gen. Comp. Endocrinol. 39, 150-157.
39. Daniels, E. and Licht, P. (1980). Effects of gonadotropin-releasing hormone on the levels of plasma gonadotropins (FSH and LH) in the

bullfrog, Rana catesbeiana. Gen. Comp. Endocrinol. 42, 455-463.
40. Delgado, M.J., Gutierrez, P. and Alonso-Bedate, N. (1983). Effects of daily melatonin injections on the photoperiodic gonadal response of the female frog Rana ridibunda. Comp. Biochem. Physiol. 76A, 389-392.
41. Delrio, G., D'Istria, M., Iela, L. and Chieffi, G. (1979). The possible significance of testosterone in the female green frog Rana esculenta during sexual cycle. Boll. Zool. 46, 1-9.
42. Delrio, G., D'Istria, M., Pierantoni, R. and Fasano, S. (1985). Identification and biosynthesis of androgens in nonmammalian vertebrates. Proc. 9th Intern. Symp. Comp. Endocrinol. p. 104.
43. D'Istria, M., Delrio, G., Botte, V. and Chieffi, G. (1974). Radioimmunoassay of testosterone, 17β-oestradiol and oestrone in the male and female plasma of Rana esculenta during sexual cycle. Steroid Lipids Res. 5, 42-48.
44. Dodd, J.M. (1975). The hormones of sex and reproduction and their effect in fish and lower chordates: twenty years on. Am. Zool. 15, 137-171.
45. Follett, B.K. and Redshaw, M.R. (1974). The physiology of vitellogenesis. In "Physiology of the Amphibia" (B. Lofts ed.), vol. 2, pp. 219-308, Academic press, New York-London.
46. Fortune, J.E. and Tsuang, P.C. (1981). Production of androgen and estradiol-17β by Xenopus ovaries treated with gonadotropins in vitro. Gen. Comp. Endocrinol. 43, 234-242.
47. Fraenkel, L., Martins, T. and Mello, R.F. (1970). Observaciones sobre el embargo de las serpientes viviparas. Arch. Soc. Biol. Montevideo 10, 1-4.
48. Fraenkel, L., Martins, T. and Mallo, R.F. (1940). Studies on the pregnancy of viviparous snakes. Endocrinology 27, 836-838.
49. Guraya, S.S. (1978). Maturation of the follicular wall of nonmammalian vertebrates. In "The Vertebrate Ovary" (R.E. Jones, ed.), pp. 261-329. Plenum Press, New York.
50. Hardisty, M.W. and Barnes, K. (1968). Steroid 3β-ol-dehydrogenase activity in the cyclostome gonad. Nature London 218,880.
51. Hillier, S.G. (1981). Regulation of follicular oestrogen biosynthesis: a survey of current concepts. J. Endocrinol. 89, 3P-18P.
52. Hoar, W.S. (1969). Reproduction. In "Fish Physiology" (W.S. Hoar and J.J. Randall, eds.), vol. III, pp. 1-72. Academic Press, New York.
53. Hsueh, A.J.W., Adashi, E.Y., Jones, P.B.C. and Welsh, T.H. (1984). Hormonal regulation of the differentiation of cultured ovarian granulosa cells. Endocrine Reviews 5, 76-126.
54. Hsueh, A.J.W. and Jones, P.B.C. (1981). Extrapituitary actions of gonadotropin releasing hormone. Endocrine Reviews 2, 437-461.
54b. Hubbard, G.M. and Licht, P. (1985). In vitro study of the direct ovarian effect of gonadotropin-releasing hormone (GnRH) in the frogs, Rana pipiens and Rana catesbeiana. Gen. Comp. Endocrinol. 60, 154-165.
55. Idler, D.R., Fagerlund, U.H.M. and Roland, A.P. (1960). Isolation of pregnene-4-ene-17α,20β-diol-3-one from plasma of pacific salmon (Oncorhynchus nerka). Biochem. Biophys. Res. Commun. 2, 133-137.
56. Idler, D.R. and Ng, T.B. (1979). Studies on two types of gonadotropins from both salmon and carp pituitaries. Gen. Comp. Endocrinol. 38, 421-440.
57. Jenkins, N. and Dodd, J.M. (1980). Effects of synthetic mammalian gonadotrophin releasing hormone and dogfish hypothalamic extracts on levels of androgens and oestradiol in the circulation of the dogfish (Scyliorhinus canicula L.) J. Endocrinol. 86, 171-177.

58. Joly, J. (1965). Mise en evidence histochimique d'une Δ^5-3β-hydroxysteroid dehydrogenase dans l'ovaire de l'urodele Salamandra salamadra (L.) a differents stades du cycle sexual. C.R. Acad. Sci. Paris 261, 1569-1771.
59. Joly, J. and Picheral, B. (1972). Ultrastructure, histochimie et physiologie du follicule pre-ovulatoire et du corps jaune de l'urodele ovovivipare Salamandra salamandra (L.). Gen. Comp. Endocrinol. 18, 235-259.
60. Joly, J., Picheral, B. and Boisseau, C. (1969). Luteinization du follicule ovarien chez une amphibien ovovivipare, Salamandra salamandra L.: etude histochimique et ultrastucturale. Gen. Comp. Endocrinol. 13, 510-511.
61. Jost, A. (1979). Comparative aspects of sexual differentiation. Proc. 9th World Congress of Gynecology and Obstetrics, pp. 10-19, Excerpta Medica, Amsterdam, Oxford, Princeton.
62. Kagawa, H., Takano, K. and Nagahama, Y. (1981). Correlation of plasma estradiol-17β and progesterone levels with ultrastructure and histochemistry of ovarian follicles in the white-spotted char, Salvelinus leucomaenis. Cell Tissue Res. 218, 315-319.
63. Kagawa, H., Young, G., Adachi, S. and Nagahama, Y. (1982). Estradiol-17β production in amago salmon (Oncorhynchus rhodurus) ovarian follicles: role of the thecal and granulosa cells. Gen. Comp. Endocrinol. 47, 440-448.
64. Kagawa, H., Young, G. and Nagahama, Y. (1983). Relationship between seasonal plasma estradiol-17β and testosterone levels and in vitro production by ovarian follicles of amago salmon (Oncorhynchus rhodurus). Biol. Reprod. 29, 301-309.
65. Kagawa, H., Young, G. and Nagahama, Y. (1983). Changes in plasma steroid hormone levels during gonadal maturation in female goldfish Carassius auratus). Gen. Comp. Endocrinol. 54, 139-143.
66. Kagawa, H., Young, G. and Nagahama, Y. (1984). In vitro estradiol-17β and testosterone production by ovarian follicles of the goldfish (Carassius auratus). Gen. Comp. Endocrinol. 54, 139-143.
67. Katz, Y., Dashow, L. and Epple, A. (1982). Circulating steroid hormones of anadromous sea lampreys under various experimental conditions. Gen. Comp. Endocrinol. 48, 261-268.
68. Kime, D.E. and Hews, E.A. (1980). Steroid biosynthesis by the ovary of the hagfish Myxine glutinosa. Gen. Comp. Endocrinol. 42, 71-75.
69. Lamba, V.J., Goswami, S.W. and Sundararaj, B.I. (1983). Circannual and circadian variations in plasma levels of steroids (cortisol, estradiol-17β, estrone, and testosterone) correlated with the annual gonadal cycle in the catfish, Heteropneustes fossilis (Bloch). Gen. Comp. Endocrinol. 50, 205-225.
70. Lambert, J.G.D. (1966). Location of hormone production in the ovary of the guppy Poecilia reticulata. Experientia 22, 476.
71. Lambert, J.G.D. (1970). The ovary of the guppy Poecilia reticulata. The atretic follicle, a corpus atreticum or a corpus luteum preovulationis. Z. Zellforsch. 107, 54-67.
72. Lance, V. and Callard, I.P. (1969). A histochemical study of ovarian function in the ovoviviparous elasmobranch, Squalus acanthias. Gen. Comp. Endocrinol. 13, 255-267.
73. Lance, V. and Callard, I.P. (1978). Hormonal control of ovarian steroidogenesis in nonmammalian vertebrates. In "The Vertebrate Ovary" (R.E. Jones, ed.), pp. 361-407. Plenum Press, New York.
74. Larsen, L.O. (1970). The lamprey egg at ovulation (Lampetra fluviatilis L. Gray). Biol. Reprod. 2,37-47.
75. Larsen, L.O. (1978). Hormonal control of sexual maturation in lampreys. In "Comparative Endocrinology" (P.G. Gaillard and H.H. Boer, eds.), pp. 105-108. Elsevier/North Holland Biomedical Press, Amsterdam.

76. Le Goascogne, C., Hirai, S. and Baulieu, E. E. (1984). Induction of germinal vesicle breakdown in Xenopus laevis oocytes: synergistic action of progesterone and insulin. J. Endocrinol. 101, 7-12.

77. Lewis, J., Mahmoud, I.Y. and Klicka, J. (1979). Seasonal fluctuation in the plasma concentrations of progesterone and estradiol-17β in the female snapping turtle (Chelydra serpentina). J. Endocrinol. 80, 127-131.

78. Licht, P., McCreery, B., Barnes, R. and Pang, R. (1983). Seasonal and stress related changes in plasma gonadotropins sex steroids and corticosterone in the bullfrog Rana catesbeiana. Gen. Comp. Endocrinol. 50, 124-145.

79. Licht, P., Millar, R., King, J.A., McCreery, B.R., Mendonca, M.T., Bona-Gallo, A. and Lofts, B. (1984). Effects of chicken and mammalian gonadotropin releasing hormones (GnRH) on in vivo pituitary gonadotropin release in amphibians and reptiles. Gen. Comp. Endocrinol. 54, 89-96.

80. Licht, P., Owens, D.W., Cliffton, K. and Penaflores, C. (1982). Changes in LH and progesterone associated with the nesting cycle and ovulation in olive ridley sea turtle, Lepidochelys olivacea. Gen. Comp. Endocrinol. 48, 247-253.

81. Licht, P., Rainey, W. and Cliffton, K. (1979). Serum gonadotropin and steroids associated with breeding activities in the green sea turtle Chelonia mydas. II. Mating and nesting in natural populations. Gen. Comp. Endocrinol. 40, 116-122.

81a. Lin, Y.W.P. and Scheutz, A.W. (1985). Intrafollicular action of estrogen in regulating pituitary induced ovarian progesterone synthesis and oocyte maturation in Rana pipiens: temporal relationship and locus of action. Gen. Comp. Endocrinol. 58, 421-435.

82. Lofts, B. and Bern, H.A. (1972). The functional morphology of steroidogenic tissues. In "Steroids in Nonmammalian Vertebrates" (D.R. Idler, ed.), pp. 37-125, Academic Press, New York and London.

83. Lupo di Prisco,C., Botte, V. and Chieffi, G. (1965). Differenze nella capacita di biosintesi degli ormoni steroidi tra follicoli post-ovulatori ed atresici in Torpedo marmorata e Scyliorhinis stellaris. Boll. Zool. 32, 185-191.

84. Lupo di Prisco, C. and Chieffi, G. (1965). Identification of steroid hormones in the gonadal extract of the synchronous hermaphrodite teleost, Serranus scriba. Gen. Comp. Endocrinol. 5, 698.

85. Lupo di Prisco, C., Delrio, G., Chieffi, G., Bellini Cardellini, L. and Polzonetti Magni, A. (1971). Identification and biosynthesis of steroid hormones in the ovary and fat bodies of females Triturus cristatus carnifex. Comp. Biochem. Physiol. 40B, 53-60.

86. McCreery, B.R. and Licht, P. (1984). The role of androgen in the development of sexual differences in pituitary responsiveness to gonadotropin releasing hormone (GnRH) agonist in the bullfrog Rana catesbeiana. Gen. Comp. Endocrinol. 54, 350-359.

87. McCreery, B.R., Licht, P., Barnes, R. Rivier, J. and Vale, W.W. (1982). Actions of agonistic and antagonistic analogs of gonadotropin releasing hormone (GnRH) in the bullfrog Rana catesbeiana. Gen. Comp. Endocrinol. 46, 511-520.

88. McLaren, A., Simpson, E., Tomonari, K., Chandler, P. and Hogg, H. (1984). Male sexual differentiation in mice lacking H-Y antigen. Nature London 312, 552-555.

89. McPherson, R.J., Boots, L.R., McGregor, R. and Marion, R.R. (1982). Plasma steroids associated with seasonal reproductive changes in a multiclutched freshwater turtle, Sternotherus odoratus. Gen. Comp. Endocrinol. 48, 440-451.

90. Mulner, O. Thibier, C. and Ozon, R. (1978). Steroid biosynthesis by ovarian follicles of Xenopus laevis in vitro during oogenesis. Gen. Comp. Endocrinol. 48, 440-451.

91. Nagahama, Y., Hirose, K., Young, G., Adachi, S., Suzuki, K. and Tamaoki, B. (1983). Relative *in vitro* effectiveness of 17α-20β-dihydroxy-4-pregnen-3-one and other pregnene derivatives on germinal vesicle breakdown in oocytes of ayu (*Plecoglossus altivelis*), amago salmon (*Oncorhynchus rhodorus*), rainbow trout (*Salmo gairdneri*), and goldfish (*Carassius auratus*). Gen. Comp. Endocrinol. 51, 15-23.

92. Nicholls, T.J. and Maple, G. (1972). Ultrastructural observations on possible sites of steroid biosynthesis on the ovarian follicular epithelium of two species of cichlid fish, *Cichlasoma nigrofasciatum* and *Haplochromis multicolor*. Z. Zellforsch. Mikrosk. Anat. 128, 317-335.

93. Ozon, R. (1972). Androgens in fishes, amphibians, reptiles and birds. in "Steroids in Nonmammalian Vertebrates" (D.R. Idler, ed.), pp. 329-289, Academic Press, New York and London.

94. Ozon, R. (1972). Estrogens in fishes, amphibians, reptiles and birds. In "Steroids in Nonmammalian Vertebrates" (D.R. Idler, ed.), pp.390-414, Academic Press, New York and London.

94a. Pancharatna, M. and Saidapur, S.K. (1984). The luteotrophic effect of homoplastic gonadotropins pituitary pars distalis homogenate, PMSG, and HCG on the corpora lutea of the hypophysectomized frog, *Rana cyanophlyctis* (SCHN). Gen. Comp. Endocrinol. 53, 375-380.

95. Panigel, M. (1956). Contribution a l'etude de l'ovoviviparite chez les reptiles: gestation et parturition chez le lizard vivipare, *Zootoca vivipara*. Annals Sci. Mat. Zool. 18, 569-668.

96. Pickford, G.E. and Atz, F.W. (1957). "The Physiology of the Pituitary Gland." New York Zool. Soc., New York.

97. Pierantoni, R., Fasano, S., Di Matteo, L., Minucci, S., Varriale, B. and Chieffi, G. (1984). Stimulatory effect of a GnRH agonist (buserelin) in *in vitro* and *in vivo* testosterone production by the frog (*Rana esculenta*) testis. Mol. Cell. Endocrinol. 38, 215-219.

98. Pierantoni, R., Iela, L., Delrio, G. and Rastogi, R.K. (1984). Seasonal plasma sex steroid levels in the female *Rana esculenta*. Gen. Comp. Endocrinol. 53, 126-134.

99. Pierantoni, R., Iela, L., d'Istria, M., Fasano, S., Rastogi, R.K., and Delrio G. (1984). Seasonal testosterone profile and testicular responsiveness to pituitary factors and gonadotrophin releasing hormone during two different phases of the sexual cycle of the frog (*Rana esculenta*). J. Endocrinol. 102, 387-392.

100. Pierantoni, R., Varriale, B., Simeoli, C., Di Matteo, L., Rastogi, R.K., Milone, M. and Chieffi, G. (1983). Fat body and autumn recrudescence of the ovary in *Rana esculenta*. Comp. Bioch. Physiol. 76A, 31-35.

101. Polzonetti Magni, A., Botte, V., Bellini Cardellini, L., Gobetti, A. and Crasto, A. (1984). Plasma sex hormones and post-reproductive period in the green frog *Rana esculenta* complex. Gen. Comp. Endocrinol. 54, 372-377.

102. Rastogi, R.K., Izzo Vitiello, I., Di Meglio, M., Di Matteo, L., Franzese, R., Di Costanzo, M. G., Minucci, S., Iela, L. and Chieffi, G. (1983). Ovarian activity and reproduction in the frog, *Rana esculenta*. J. Zool. Lond. 200, 233-247.

103. Redshaw, M.R. and Nicholls, T.J. (1971). Oestrogen biosynthesis by ovarian tissue of the South African clawed toad, *Xenopus laevis* Daudin. Gen. Comp. Endocrinol. 16, 85-96.

104. Reinboth, R. and Becker, B. (1984). *In vitro* studies on steroid metabolism by gonadal tissues from ambisexual teleosts. I. Conversion of ^{14}C testosterone by males and females of the protogynous wrasse *Coris julis* L. Gen. Comp. Endocrinol. 55, 245-250.

105. Saidapur, S.K. (1982). Structure and function of postovulatory

follicles (corpora lutea) in the ovaries of nonmammalian vertebrates. Intern. Rev. Cytol. 75, 243-285.

106. Saidapur, S.K. and Nadkarni, V.B. (1976). Steroid synthesizing cellular sites in the ovary of catfish Mysus cavasius: a histochemical study. Gen. Comp. Endocrinol. 30, 457-461.

107. Sangalang, G.B. and Freeman, H.C. (1977). A new and precise method for determining testosterone and 11-ketotestosterone in fish plasma by radioimmunoassay. Gen. Comp. Endocrinol. 32, 432-439.

108. Schorderet-Slatkine, S., Schorderet, M. and Baulieu, E.E. (1982). Cyclic AMP-mediated control of meiosis: effects of progesterone, cholera toxin, membrane-active drugs in Xenopus laevis oocytes. Proc. Natl. Acad. Sci. USA 79, 850-854.

109 Scott, A.P. and Sumpter, J.P. (1983). A comparison of the female reproductive cycles of autumn-spawning and winter-spawning strains of rainbow trout (Salmo gairdneri Richardson). Gen. Comp. Endocrinol. 52, 79-85.

110. Segal, S.J. and Adejuwon, C.A. (1979). Direct effect of LHRH on testicular steroidogenesis in Rana pipiens. Biol. Bull. 157, 393.

111. Semper, C. (1975). Das Urogenitalsystem der Plagiostomen und seine Bedeutung fur das der ubrigen Wirbeltiere. Arbeit. a. d. Zool. zoot. Institut. Wurzburg 2, 195-206.

112. Semper, C. (1975). Billdung und Wachstum der Keimdrusen bei der Plagiostomen. Centrabl. med. Wiss. 13, 177-183.

113. Sharpe, R.M. (1982). The hormonal regulation of the Leydig cell. In "Oxford Reviews of Reproductive Biology" (C.A. Finn, ed.), pp. 24-317. Oxford Univ. Press, Oxford.

114. Sumpter, J.P. (1977). Biochemical and physiological studies on the reproductive hormones of the dogfish (Scyliorhinus canicula L.). Ph.D. Thesis, University of Wales.

115. Sundararaj, B.I. and Goswami, S.V. (1977). Hormonal regulation of in vivo and in vitro maturation in the catfish, Heteropneustes fossilis (Bloch). Gen. Comp. Endocrinol. 32,17-28.

116. Suzuki, K., Tamaoki, B. and Hirose, K. (1981). In vitro metabolism of 4-pregnenes in ovaries of a freshwater teleost, the ayu (Plecoglossus altivelis): production of 17α,20β-dihydroxy-4-pregnen-3-one and its 5β-reduced metabolites, and activation of 3β- and 20β-hydroxysteroid dehydrogenases by treatment with a fish gonadotropin. Gen. Comp. Endocrinol. 45, 473-481.

117. Suzuki, K., Tamaoki, B. and Nagahama, Y. (1981). In vitro synthesis of an inducer of germinal vescicle breakdown of fish oocyte, 17α,-20β-dihydroxy-4-pregnen-3-one by ovarian tissue preparation of amago salmon (Oncorhynchus rhodurus). Gen. Comp. Endocrinol. 45, 533-535.

118. Theofan, G. and Goetz, F.W. (1983). The in vitro synthesis of final maturational steroids by ovaries of brook trout (Salvelinus fontinalis) and yellow perch (Perca flavescens). Gen. Comp. Endocrinol. 51, 84-95.

119. Thibier, C. and Ozon, R. (1983). Regulation of 17β-hydroxysteroid oxydoreductase of the Xenopus laevis follicle by gonadotropin and prolactin. Gen. Comp. Endocrinol. 49, 352-357.

120. Thibier-Fouchet, C., Mulner, O. and Ozon, R. (1976). Progesterone biosynthesis and metabolism by ovarian follicles and isolated oocytes of Xenopus laevis. Biol. Reprod. 14, 317-326.

121. Thornton, V.F. and Evennett, P.J. (1973). Changes in the fine structure of the ovarian follicle of the toad (Bufo bufo) prior to induced ovulation. Gen. Comp. Endocrinol. 20, 413-423.

122. Tokarz, R.R. (1978). Oogonial proliferation, oogenesis, and folliculogenesis in nonmammalian vertebrates. In "The Vertebrate Ovary" (R.E. Jones, ed.), pp. 145-179. Plenum Press, New York.

123. Van Ree, G.E., Lok, B. and Bosman, G. (1977). *In vitro* induction of nuclear breakdown in oocytes of the zebrafish *Brachidanio rerio* (Ham. Buch.). Effects of the composition of the medium and of protein and steroid hormones. Proc. K. Ned. Akad. Watensch. C 80, 353-371.
124. Van der Kraak, G., Dye, H.M. and Donaldson, E.M. (1984). Effects of LHRH and des-Gly^{10} [D-Ala^{6}]LHRH-ethylamide on plasma sex steroid profiles in adult female coho salmon. Gen. Comp. Endocrinol. 55, 36-45.
125. Van der Kraak, G., Lin, H. Donaldson, E.M., Dye, H.M. and Hunter, G.A. (1983). Effects of LHRH and des-Gly^{10}[D-Ala^{6}]LHRH-ethylamide on plasma gonadotropin levels and oocyte maturation in adult female coho salmon (*Oncorhynchus kisutch*). Gen. Comp. Endocrinol. 49, 470-476.
126. Wachtel, S.S., Ohno, S., Koo, G.C. and Boyse, E.A. (1975). Possible role of H-Y antigen in primary sex determination. Nature London 257, 235-236.
127. Wathes, D.C. (1984). Possible actions of gonadal oxytocin and vasopressin. J. Reprod. Fert. 71, 315-345.
128. Wotiz, H.H., Botticelli, C., Hisaw, F.L. and Ringler, L. (1958). Identification of estradiol-17β from dogfish ova (*Squalus suckleyi*). J. Biol. Chem. 231, 589-592.
129. Xavier, F. (1970). Analyse du role des corpora lutea dans le mantien de la gestation chez *Nectrophrynoides occidentalis* Angel. C.R. Acad. Sci. Paris 270, 2018-2020.
130. Yaron, Z. (1971). Observations on the granulosa cells of *Acanthobrama terrae-sanctae* and *Tilapia nilotica* (Teleostei). Gen. Comp. Endocrinol. 17, 247-252.
131. Young, G. Ueda, H. and Nagahama, Y. (1983). Estradiol-17β and 17α, 20β-dihydroxy-4-pregnen-3-one production by isolated ovarian follicles of amago salmon (*Oncorhynchus rhodurus*) in response to mammalian pituitary and placental hormones and salmon gonadotropin. Gen. Comp. Endocrinol. 52, 329-335.

ENDOCRINOLOGY OF VITELLOGENESIS

Shuk-mei Ho

Department of Biology
Tufts University
Medford, MA 02155

1. INTRODUCTION

The ovum (female germ cell) is truly the most remarkable of all animal cells: once activated, it can give rise to a complete new individual and it is the only cell, in higher animals, that is capable of doing so. The most distinguished feature of the ovum is its large size, which is required for the storage of reserves for the development of the embryo. During the early part of the developmental program, very rapid rates of cleavage take place in the embryo. This requires a high rate of synthetic activities, which is made possible by storage of large reserves of messenger RNAs, ribosomes, enzymes and other precursors of macromolecular synthesis in the ovum prior to fertilization. The other requirement for development and growth is metabolic energy, the reserve form of which is stored in the oocytes as "yolk." In most vertebrate oocytes, yolk is not synthesized _in situ_ but derived from a precursor protein known as vitellogenin, synthesized in the liver of the animal. A sequence of complex but well-coordinated processes is involved in the production of yolk and its accumulation in vertebrate eggs. This sequence of events, collectively referred to as _vitellogenesis_, includes the following steps: 1) the induction of vitellogenin synthesis in the liver and its subsequent release into blood circulation, 2) the transport of vitellogenin in the blood stream, 3) the uptake of vitellogenin by the growing oocytes and 4) the conversion of vitellogenin into storage forms. Studies of vitellogenesis have recently been reviewed by Tata,[115] Tata & Smith,[6] Wahli,[125] Wallace & Selman,[136] Wiegand,[146] Ho et al.,[64] and Wallace.[131,132]

The scope of these reviews is limited to the cellular and molecular events of hepatic vitellogenesis, and the vitellogenin uptake and yolk formation process. In this article, I will focus on reviewing literature about the involvement of hormones in the regulation of vitellogenesis. As will be evident from my discussion, vitellogenesis is under multihormonal control. Upon activation of the hypothalamo-pituitary axis, gonadotropins are released from the pituitary and stimulate estrogen production in the ovaries. Rising levels of estrogen, in turn, induce vitellogenin production and release by the liver. At the same time, under the stimulation of gonadotropins, ovarian follicles are capable of sequestering circulating vitellogenin at high rate. At various points in the process of vitellogenesis, hormones other than estrogens and gonadotropins are exerting their influences in "fine-tuning" the vitellogenic response.

II. VITELLOGENIN AND VITELLOGENIN GENES

A. Fishes

Vitellogenin has been demonstrated in the plasma of a wide variety of female teleostean and nonteleostean fishes (for reviews, see Wallace,[132a], Wallace & Selman,[136] Wiegand[146]) including *Pseudopleuronectes americanus*,[17] *Salmo gairdneri*,[18,110,121] *Salmo salar*,[70] *Oncorhynchus keta*,[19] *Gobius niger*,[77] *Anguilla japonica*,[56] *Carassius auratus*,[30] *Heteropneustes fossilis*,[87] *Tilapia aurea*,[118] *Gadus morhua*,[99] *Oryzias latipes*,[55] *Fundulus heteroclitus*,[105] *Epinephelus akaara*,[92] *Scyliorhinus canicula*[22,24] *Lampetra fluviatilis*[97] and *Eptatretus stouti*.[158,158] In general, fish vitellogenins are fairly susceptible to proteolytic enzyme degradation;[30,60,137] thus, special precautions against proteolysis are required in the isolation of this protein from fish blood samples.[30,66,105,137] In spite of the technical difficulties in isolation procedures, a number of fish vitellogenins have been isolated and studied biochemically. They generally are less phosphorylated and poorer in serine and calcium content than vitellogenins isolated from other vertebrate groups.[18,30,56] However, the lipid content of these vitellogenins is usually high, averaging over 20% by weight of the protein.[18,22,66] In chum salmon, rainbow trout, the Japanese eel and medaka, vitellogenin exhibits a specific iron-binding capacity which is seldom found in vitellogenins of other vertebrates.[53,55,56] The molecular weights of intact fish vitellogenins range from 300,000 to 600,000 daltons,[18,30,37,54,66,110] whereas the molecular weights of vitellogenin monomers vary between 140,000 to 220,000 daltons.[30,54,66,103,105,120]

Multiple molecular weight forms of the vitellogenin monomer are found in the plasma of the goldfish, *C. auratus*.[30]

Little is known about fish vitellogenin genes or their transcripts except two recent studies on vitellogenin mRNA of the brown bullhead, *Amieurus nebulosus*,[103] and the rainbow trout, *S. gairdneri*.[20,120] Trout vitellogenin mRNA has 7200 nucleotides, a size comparable to other vertebrate vitellogenin mRNA, whereas the bullhead mRNA is significantly smaller.

B. Amphibians

Studies on amphibian vitellogenesis have almost exclusively focused on one frog species, the South African clawed toad, *Xenopus laevis*. Information on other amphibian species is limited. A few scattered studies have been done on the leopard frog, *Rana pipiens*,[59,86] the American bullfrog, *Rana catesbeiana*,[58,109] the green frog, *Rana esculenta*,[50] the toad, *Bufo bufo bufo*,[36] the axolotl, *Ambystoma mexicanum*,[82] and the newt, *Triturus cristatus carnifex*.[107]

Xenopus vitellogenin, purified from plasma of estrogen-treated frogs, is characterized as a phospholipoglycoprotein having a molecular weight of 460,000 daltons.[4,101,129] The circulating form is a dimer molecule composed of two polypeptides, each having a molecular weight of approximately 200,000 daltons. Three vitellogenin polypeptides with distinct molecular weights are found in the plasma of this frog.[149,150] Similarly, multiple forms of vitellogenin polypeptide are also evident in the plasma of the axolotl[82] and the American bullfrog.[58] *Xenopus* vitellogenin is fairly resistant to proteolytic activity; thus, minimal precautions need to be taken in its isolation. In the bloodstream, vitellogenin binds to calcium, and removal of this cation from the protein often leads to the precipitation of the protein.[148]

In *X. laevis*, vitellogenin is encoded by at least four related genes, designated as A1, A2, B1 and B2, which are all actively expressed and under estrogen control.[125,126] There is a 95% coding sequence homology between the two members of both A and B groups, and the two groups exhibit an 80% homology. Wahli et al.[27] have recently shown that the genes A1 and B1 are linked and separated by a 15 kilobase (Kb) region. Four distinct vitellogenin mRNA molecules of 6.3 Kb in length, corresponding to genes A1, A2, B1 and B2, are found in the estrogen-treated *Xenopus* hepatocytes, and each can be translated *in vitro* into vitellogenin polypeptides 200,000 daltons in molecular weight.[40] The isolation of vitellogenins, vitellogenin mRNA and vitellogenin genes gives a new dimension to the study of vitellogenesis in *Xenopus*, since hepatic vitellogenesis can now be investigated at the transcriptional, translational and post-translational levels in this species. Our knowledge about the molecular actions of estrogens and the regulation of gene expression by hormonal factors has been expanded greatly by data generated from these studies.

C. Reptiles

Only one vitellogenin has been purified from a chelonian reptile, *Chrysemys picta*.[46,63] Turtle vitellogenin, like *Xenopus* vitellogenin, can be readily precipitated from female plasma by dimethylformamide, and its monomer has a size of 210,000-220,000 daltons. Using similar procedures, Garstka et al.[47] recently isolated snake vitellogenin from *Thamnophis sirtalis*. However, the biochemical properties of these reptilian vitellogenins have not been studied in detail.

III. ESTROGEN STIMULATION OF VITELLOGENIN SYNTHESIS AND SECRETION

A. Fishes

In fishes, vitellogenesis usually occurs during the prespawning season[74,104,119,121] and is thought to be primarily regulated by ovarian estrogens. High titers of estrogens are detected in circulation either before or during the onset of vitellogenesis in flounder,[37] catfish,[111] rainbow trout,[34,69,104,121] and chum salmon,[119] thereby providing evidence that estrogen is the inducer of vitellogenesis in fishes.

Vitellogenesis can be induced in male and nonvitellogenic female fishes by the administration of estradiol.[18,23,37,69,73,77,87,97,99] When a single dose of estradiol is administered, it induces a steady increase in plasma vitellogenin level in fishes. The effect of estradiol is often long-lasting; for example, in the male flounder, *P. flesus*, vitellogenin accumulated at a constant rate for over 28 days following a single injection of estradiol.[73] In both flounder[39] and rainbow trout[122] the vitellogenic response to estrogen was found to be dose-dependent. A "memory" effect of estrogen (Wallace & Bergink, 1974) was demonstrated in the Indian catfish, *H. fossilis*,[112] and *P. flesus*.[39] Successive injections of the Indian catfish with estrogen after plasma vitellogenin levels had returned to basal levels produced a stronger response than its predecessor.[112] Multiple low-dose injections of estrogen progressively increased the rate of vitellogenesis in flounder.[39,73] Pertinent to these studies is the observation that peak levels of estradiol precede maximal serum vitellogenin levels by 1-3 months in S. gairdneri,[83,143,144] suggesting that once vitellogenin synthesis has been initiated by estrogen, it continues despite the return of circulating estrogen to basal levels.

Although it is generally accepted that estrogen stimulates the liver to synthesize and secrete vitellogenin, direct evidence supporting this notion has never been provided in fishes. Vitellogenin has not been localized cytologically within hepatocytes of fishes, and in-vitro synthesis and release of vitellogenin by fish hepatocytes has not been demonstrated. In the rainbow trout, S. gairdneri, vitellogenin was not detected in the liver after estrogen treatment, but vitellogenin levels in the plasma were elevated.[122] This implies that piscine vitellogenin is not stored in large quantities in the liver and, once synthesized, is released rapidly into the circulation. Indirect evidence in support of the hepatic origin of fish vitellogenin is provided by a number of studies in teleosts and non-teleosts.[39,69,75,78,98,99,112,158] Ultrastructural studies indicate that hepatocytes from female and estrogen-treated fishes are highly synthetic cells, whereas those from males and immature fishes appear quiescent.[93,94,96,105,123,155] Other hepatic changes associated with estrogenization of fishes include liver hypertrophy,[2,83,122] augmented rate of protein synthesis,[30,84,93,94] an increase in protein synthesizing capacity,[38,73] changes in enzyme activities[73,93,145] and an accumulation of hepatic RNA.[35,38,73,84] These cytological and metabolic changes found in the liver of estrogen-treated fishes may be essential for the long-term maintenance or amplification of the vitellogenic response.

At the molecular level, estrogens stimulate vitellogenin gene transcription. This has recently been demonstrated in female catfish[103] and male rainbow trout.[120] Administration of estradiol causes accumulation of vitellogenin mRNA in the liver tissue of these fishes. The mechanism of action of estrogen in vitellogenin induction and its relationship with liver estrogen receptor is reviewed in Chapter 12 of this book.

B. Amphibians

Under natural conditions, female frogs and toads undergo a single cycle of vitellogenesis each year.[36,50,85] In the species studied, vitellogenesis does not start until after spawning in the spring. It continues through the summer months into the fall so that mature eggs are present in the ovaries by September. Ovarian growth and vitellogenesis continue into the fall, and vitellogenesis eventually levels off or undergoes a decline during winter when the ovaries are in a static phase through hibernation. Unfortunately, studies have not been done to correlate changes in plasma vitellogenin levels with those in plasma estrogen levels during the reproductive cycle of these and other amphibian species.

The effects of estrogens on amphibian vitellogenin induction is best studied in *Xenopus*. When male *Xenopus* are exposed to estradiol for the first time, there is a lag period of about 3-12 h before the appearance of vitellogenin in the circulation.[21,32,152] Vitellogenin accumulates in the plasma over the next few days and eventually declines to zero levels. If estrogen is administered a second time after the liver has ceased to secrete the protein, a new round of vitellogenin synthesis occurs. During this secondary response, the lag period is shortened greatly and the synthesis of the protein is more rapid and of a greater magnitude. This "memory" effect[40,132] can be observed weeks, or even months, after the first response has ceased. The cellular and molecular bases of the "memory" effect of estrogen are discussed below.

The direct action of estrogen on hepatic vitellogenin synthesis and the hepatic origin of vitellogenin are best illustrated in amphibians by a series of *in-vitro* experiments. A full vitellogenic response to estrogen added *in vitro* was obtained in organ cultures of *Xenopus* liver explants[51,139] as well as in monolayer hepatocyte cultures of *Xenopus*[140] and of bullfrog.[109] Furthermore, the presence of vitellogenin in cultured hepatocytes was demonstrated by immunofluorescent staining.[140] Intracellularly, vitellogenin or vitellogenin-related proteins were found to associate with rough endoplasmic reticulum, Golgi apparatus, secretory granules and lysosomes of hepatocytes.[58]

The principal action of estrogen on hepatic vitellogenesis is at the level of transcription. In the presence of estrogen, the *Xenopus* vitellogenin genes are transcribed into 30,000 copies of mRNA per hepatocyte during the primary response and 100,000 copies of mRNA per hepatocyte during the secondary response.[40,106,128] Interestingly, during the early stage of the primary response, Group A and Group B vitellogenin genes in male hepatocytes are non-coordinately transcribed. Upon secondary stimulation of the male cells, both groups of vitellogenin genes are expressed coordinately at an enhanced rate. In female hepatocytes, Group A and Group B genes are always expressed coordinately at rates much higher than those found in estrogen-stimulated male cells.[151,154] These findings suggest that estrogen might have induced some long-term or permanent modifications in the vitellogenin genes and/or in the chromosomal organization.[31,42,48] These changes would allow a more efficient transcription of the vitellogenin genes in the female and estrogen-primed male liver cells.

In addition to the aforementioned transcriptional mechanism, the "memory" effect of estrogen can also be explained by several non-transcriptional mechanisms. Estrogen was found to induce higher levels of estrogen receptor in the liver of the male frog[57,142] and suppress the high rate of estrogen metabolism found in the liver of male animals.[117] Either one or both of these mechanisms could explain the higher sensitivity of female *Xenopus* and estrogen-primed males to estrogen stimulation. Cytological changes and enhanced protein synthetic capacity found in estrogen-treated hepatocytes[5,21] might also alter the vitellogenic response to a secondary stimulation.

C. Reptiles

Estrogen is the primary stimulus to vitellogenesis in reptiles, as in amphibians and fishes. Administration of estrogen induces hypertrophy of the liver, increase in hepatic RNA content and changes in hepatic ultrastructure in snakes[10,28,52] and lizards.[12,13] The appearance of a unique, estrogen-induced, plasma protein following estrogen stimulation has been demonstrated electrophoretically in snakes[28] and lizards.[49,100,114,157] Recently, vitellogenesis in a turtle, *Chrysemys picta*, has been studied in some detail with a specific radioimmunoassay developed against *Chrysemys* vitellogenin.[46] Seasonal variations of plasma vitellogenin levels in this species correlate with changes in circulating estradiol levels. Vitellogenin synthesis can be induced in both sexes of *Chrysemys* by estradiol administration.[61] The *in-vivo* vitellogenic response is both estrogen-specific and dose-dependent. A lag period of 6-8 h is observed between hormone administration and the appearance of vitellogenin in plasma, and female animals are shown to be more responsive to estradiol than males. Attempts have been made to induce vitellogenin synthesis *in vitro* in liver explant cultures.[65] Female liver tissues in culture respond readily to estradiol added to the medium whereas male liver tissues are non-responsive to estradiol *in vitro*. These findings are in contrast to those for *Xenopus*[139] and indicate a more complex hormonal requirement for vitellogenin induction in *Chrysemys*.

IV. ESTROGEN SPECIFICITY IN THE INDUCTION OF VITELLOGENIN SYNTHESIS

A. Fishes

The hormonal requirement for vitellogenin induction appears to be less stringent in fishes than in other vertebrates. In a number of studies, steroids other than estrogens were found to be effective in the induction of vitellogenin. In both *C. auratus*[66] and *G. niger*,[77,79] androgens administered in pharmacological doses were able to mimic the action of estrogens. Testosterone, methyltestosterone, ethinyltestosterone and methylandrostendiol were all potent inducers of vitellogenesis in goldfish, while 5α-dihydrotestosterone, a non-aromatizable androgen, was totally inactive. In contrast, 5α-dihydrotestosterone was effective in *G. niger* as a vitellogenesis inducer, and its action can be blocked by tamoxifen, an antiestrogen which has no affinity for androgen receptor. Based on these findings, two probable mechanisms of vitellogenin induction by androgens are suggested: 1) androgens are converted into estrogens which, in turn, stimulate vitellogenesis, and 2) androgens act directly on the induction process as an estrogen-agonist via the estrogen receptor pathway.

In their study on the Pacific hagfish, Yu et al.[159] demonstrated a partial effectiveness of progesterone in vitellogenin induction. However, the mechanism by which progesterone exerts its regulation on hagfish vitellogenesis has not been explored.

In general, estradiol is found to be the most potent estrogen in the induction of vitellogenesis in fishes.[112,121] The potencies of estrone and estriol, relative to that of estradiol, vary according to species. In the Pacific hagfish, estrone was found to be as potent as estradiol,[159] whereas in the rainbow trout, estrone had only 5-12% of the vitellogenic activity of estradiol.[123] In addition to its own vitellogenic activity, estrone was found to potentiate the action of estradiol in vitellogenic induction.[123] Priming of the rainbow trout with estrone increased the fish's vitellogenic response to subsequent estradiol stimulation.[124] Estriol usually is the least active estrogen in vitellogenin induction; however, in a study of *T. aurea*, estriol was found to be the most potent of the three estrogens in increasing plasma calcium and total protein levels.[117]

B. Amphibians

Induction of vitellogenin in *Xenopus* is specific for estrogens; hydrocortisone, cortisol, progesterone and testosterone are all ineffective in inducing the vitellogenin response *in vivo*[102,153] and *in vitro*.[51,139] Among the natural estrogens, estradiol is the most potent estrogen, followed in potencies by estrone and estriol. Interestingly, the weakly estrogenic steroid estradiol-17α is also effective when given in high doses.[153] Tenniswood et al.[117] recently studied the effectiveness of synthetic estrogens at inducing vitellogenin mRNA accumulation in cultured hepatocytes and found that mestranol was as effective as estradiol while 17α-ethinylestradiol and diethylstilbestrol were less potent. Since mestranol is thought to exert its estrogenic action only upon its dimethylation to 17α-ethinylestradiol, it is suggested that the apparent high potency and long-acting effect of mestranol are mainly due to its slow conversion to 17α-ethinylestradiol in hepatocytes. Diethylstilbestrol (DES) has a short-lived response in both male and female cells. Since both estradiol and DES have similar affinities for estrogen receptor in *Xenopus* hepatocytes,[142] a plausible explanation for the ineffectiveness of DES in vitellogenin mRNA induction is a rapid rate of metabolism of this synthetic estrogen in *Xenopus* hepatocytes.

C. Reptiles

Vitellogenin induction in reptiles appears to be estrogen specific. Hormones other than estrogens have never been reported to stimulate hepatic vitellogenesis.

V. INFLUENCE OF NON-ESTROGENIC STEROIDS ON VITELLOGENESIS

A. Fishes

With the exception of the studies discussed in the previous section (IV), non-estrogenic steroids have never been shown to have an effect on vitellogenesis in fishes. A recent report, however, suggests a role of cortisol in vitellogenesis regulation.[111] When administered with estradiol into catfish, cortisol potentiates the effect of estradiol on plasma vitellogenin levels. Pertinent to these findings is the observation that plasma cortisol levels rise simultaneously with estradiol levels during the vitellogenic phase of the catfish reproductive cycle.[74]

B. Amphibians

There are no data pertaining to the possible role of non-estrogenic steroids in amphibian vitellogenesis. However, a recent report[138] seems to indicate a role for glucocorticoids in vitellogenin-reinduction in cultured *Xenopus* hepatocytes (for more detailed discussion see Section VI B).

C. Reptiles

Thus far, few studies have explored the possibility that reptilian vitellogenesis is under multi-hormonal regulation. Studies in *Chrysemys* provide the first evidence that gonadal steroids other than estrogen can inhibit the vitellogenic response.[61] Progesterone and testosterone, given in concert with estradiol to male turtles, suppressed the estrogen-induced vitellogenic response. The inhibitory effect of progesterone was shown to be dose-dependent whereas that of testosterone was inversely related to the dose administered. The site of action and the mechanism by which these steroids suppress vitellogenesis are unclear. However, since both plasma progesterone and testosterone levels rise during the vitellogenic phase of the gonadal cycle,[14] they may act as effective terminators of vitellogenesis at the end of the reproductive season.

VI. INFLUENCE OF NON-STEROIDAL HORMONES ON VITELLOGENESIS

A. Fishes

Thyroid hormones have been shown to influence gonadal development and growth of oocytes in a number of fishes (see Chapter 14 for discussion). However, there is no evidence of a direct effect of thyroid hormones on hepatic vitellogenesis. Information on the role of other non-steroidal hormones in fish vitellogenesis is not available.

B. Amphibians

Using an *in-vitro* hepatocyte culture system, Wangh and coworkers[138,140,141] have investigated the role of thyroid hormones and glucocorticoids in vitellogenin induction in *Xenopus*. Primary hepatocyte cultures were established from purified liver parenchymal cells. when estradiol was added to cultures at the start of the culture period, *in-vitro* vitellogenin synthesis was induced and could be sustained by continuous presence of the hormone.[140] However, if the cultures were first maintained in estrogen-free medium for two weeks before being challenged with estradiol, no vitellogenin induction could be achieved by the addition of physiological doses of the hormone. Addition of thyroxine, together with estradiol, restored the estrogen responsiveness in cultured hepatocytes and allowed re-induction of vitellogenesis in hepatocyte cultures.[141] Addition of dexamethasone in concert with estradiol and thyroxine further enhanced the re-induction response.[140] The relative importance and mode of interaction of thyroid hormones and glucocorticoids in regulating vitellogenesis *in vivo* have not been studied. However, data from these *in-vitro* experiments help to explain the acquisition of competence by tadpole liver to respond to estrogen during metamorphosis[67,72,81] (see Section VII B for more discussion).

C. Reptiles

Callard and co-workers[10,49] first demonstrated the involvement of the pituitary gland for optimal response to estrogen in the iguanid lizard *Dipsosaurus dorsalis*. Following hypophysectomy of female lizards, injection of estradiol did not have a stimulatory effect on plasma phospholipoprotein (an index of vitellogenesis), suggesting dependence on the pituitary gland for induction. Recent studies in *C. picta*[62] also confirm a role by the pituitary gland in vitellogenic regulation. Hypophysectomized turtles showed a reduced response to estrogen in that both the initial rate of vitellogenin synthesis and the final accumulation of vitellogenin in the plasma were lower than those in intact turtles. Administration of mammalian growth hormone to hypophysectomized animals partially restored their vitellogenic response to levels of intact turtles. These findings suggest that a growth-hormone-like factor may act as a synergist to estrogen in reptilian vitellogenesis. Results from a recent *in-vitro* study in *Chrysemys*[64,65] lend further support to this concept.

VII. OTHER NON-HORMONAL FACTORS REGULATING VITELLOGENESIS

A. Fishes

Lamba et al.[74] provided the first evidence that vitellogenin synthesis is tied to a circadian rhythm in catfish. During the month of May to September, plasma vitellogenin levels vary according to the time of the day and peak almost invariably during the dark phase. It is suggested that the circadian variation of plasma vitellogenin levels reflects temporal fluctuations in the rate of vitellogenin synthesis and/or the rate of its deposition in the growing oocytes.

In rainbow trout and other salmonids, photoperiod appears to be the major factor controlling the timing of reproduction (see reviews by Peter and Crim,[95] and Chapter 13). Bromage et al.[6] have shown that the onset of vitellogenesis in female trout can be experimentally manipulated by altering the photoperiod regimen. Furthermore, it has been suggested by Frost[45] that the vitellogenic season for rainbow trout has been significantly altered by artificial selection in fish husbandry within the past hundred years, resulting in distinct autumn- and spring-spawning strains now used in many fish farms.[104] Interestingly, Craik[23] found that the winter sea temperature, rather than winter photoperiod, could stimulate vitellogenin synthesis in the elasmobranch, *S. canicula*

B. Amphibians

Egg production in laboratory-maintained frogs is often unpredictable,[86] and the eggs that are produced do not always support development.[76] It is hypothesized that this situation results from abnormalities in the vitellogenic activity of laboratory-maintained animals. The problem has been resolved partially by Smalley and Nace[103] in their studies on wild-caught and laboratory-maintained *R. pipiens*. Exposure of laboratory-maintained frogs to cold temperature for 4-6 weeks partially corrected the abnormalities in their vitellogenic cycles, resulting in animals with larger numbers of mature eggs. The authors suggested that low estrogen secretion may account for the vitellogenic deficiencies seen in *R. pipiens* kept at constant room temperature.

In *Xenopus*, the appearance of vitellogenin synthesis is a postmetamorphic event, and the tadpole liver does not acquire competence to respond to estrogen until late metamorphosis (stage 57-62).[72] The appearance of estrogen responsiveness in tadpole liver cannot be advanced by estrogen treatment but is obligatorily dependent on thyroid hormones. Thyrostatic tadpoles, in which metamorphosis has been suppressed, failed to respond to effective levels of estrogen even at an age when normally developing animals have become fully responsive to this hormone.[67,81] Similar studies on neotenous amphibians such as *A. mexicanum* would be helpful to ascertain whether acquisition of estrogen responsiveness only occurs with metamorphic transformation. Recently, the ontogeny of the vitellogenic ability in *Xenopus* liver has been studied at the molecular level by Ng and co-workers.[93]

C. Reptiles

Pertinent information is not available in reptiles.

VIII. ENDOCRINE CONTROL OF VITELLOGENIN UPTAKE BY OOCYTES

A. Fishes

Studies with hypophysectomized fishes have demonstrated clearly the indispensable role of gonadotropin in the maintenance of yolky oocytes in teleosts.[3,17] The function of gonadotropin in ovarian maintenance is two-fold: first, it induces ovarian estrogen production for the maintenance of hepatic vitellogenesis,[27,44,156] and second, it acts directly on the ovarian follicles to stimulate pinocytotic uptake of vitellogenin.[1] Ng and Idler[90] injected hypophysectomized female winter flounder with ^{32}P-labelled vitellogenin isolated from estrogen-treated males and demonstrated gonadotropin-sensitive uptake of labels into the yolk of flounder oocytes. Similarly, using an _in-vitro_ system, Campbell and associates[16,19] demonstrated gonadotropin-sensitive and selective uptake of alkylated ^{14}C-labelled vitellogenin by rainbow trout oocytes.

In elasmobranchs and cyclostomes, indication of a gonadotropin-regulated uptake of vitellogenin uptake process in oocytes is not as clear as in teleosts. Hypophysectomy of the elasmobranch _S. canicula_ caused a decrease in the clearance rate of plasma ^{32}P-labelled vitellogenin but did not change the rate of labeled vitellogenin incorporation into yolk granules of its oocytes.[25] Similarly, in the hagfish _E. stouti_ gonadal development and the maintenance of yolky oocytes was not affected by long-term hypophysectomy.[80]

Controversies exist in the literature on the biochemical properties of fish gonadotropins. One concept (see Burzawa-Gerard[9]) maintains that a single gonadotropin, rich in carbohydrates, is responsible for stimulating vitellogenin uptake by fish oocytes as well as for inducing maturation and ovulation. This concept is supported by the findings that a single glycoprotein, adsorbable on Concanavalin A, is isolated from the pituitary of carp,[7] chinook salmon[33] and sturgeon.[9] It is active in stimulating the rate of vitellogenin incorporation by oocytes from a number of teleosts.[8,113] However, Idler and co-workers[68] are able to show two forms of fish gonadotropins in the pituitary extracts from American plaice,[17,90] winter flounder, chum salmon, and carp.[88,89] The carbohydrate-rich gonadotropin induces maturation of oocytes while the carbohydrate-poor gonadotropin (vitellogenic gonadotropin) is responsible for the stimulation of vitellogenin uptake and yolk formation in oocytes. More information on this is provided in Chapter 10 of this book.

B. Amphibians

The uptake of vitellogenin by growing oocytes and its subsequent processing and storage in yolk platelets were shown to be gonadotropin-dependent in both *X. laevis*[43,132,133] and *B. bufo bufo*.[36,71] When intact follicles from *Xenopus* were cultured in vitellogenin-supplemented medium, they responded to human chorionic gonadotropins *in vitro* by increasing their vitellogenin incorporation rates to 35% above control values.[147] However, under similar conditions, oocytes with their follicular investments removed, failed to respond to the hormone.[130] Recent experiments further revealed that naked oocytes incorporated vitellogenin and grew in the absence of any added serum or gonadotropin at a steady rate *in vitro* to sizes that exceeded those normally reached in the ovary. Results from these experiments lead to the conclusion that *Xenopus* oocytes are capable of maintaining a constant rate of vitellogenin incorporation and growth in the absence of external cues like gonadotropins; however, in their follicular envelope, their rate of vitellogenin incorporation is limited or regulated by the follicular cells. During the vitellogenic phase of follicular growth, gonadotropins, by acting on the follicular cells, are able to induce a high rate of vitellogenin incorporation in the oocyte. At the end of the vitellogenic period, a change in follicular architecture renders vitellogenin unavailable to oocytes, thus limiting the size of a full-grown follicle.[131]

C. Reptiles

Both progesterone[11,157] and prolactin[15] have been suggested to be the "antigonadal" hormone in lizards. However, their inhibitory effect on vitellogenin uptake by ovarian follicles has not been established. The endocrine control of vitellogenin uptake in oocytes of other reptiles has not been studied.

IX. SUMMARY OF FUTURE RESEARCH DIRECTIONS

Perhaps the two most important questions to be resolved in the study of the endocrine control of vitellogenesis are: i) localizing the site of hormonal interaction, and ii) understanding the mechanism of action of the various hormones involved in the regulatory process. For example, estrone, androgens, progesterone and growth hormone have all been shown to influence plasma vitellogenin levels, but it is unclear whether they act directly on the liver and as to how they exert their actions. These hormones could change the metabolizing rate of estradiol or modulate the content of estrogen receptor in the liver, thus altering the strength of the stimulus. Alternatively, they might influence the transcriptional efficiency of the vitellogenin gene or modify the nature of the post-transcriptional events such as the translation of the vitellogenin mRNA and the phosphorylation and glycosylation of the vitellogenin polypeptide. o address these questions, it is important to have liver explant or hepatocyte culture systems so that the effect of each regulatory hormone can be studied *in vitro* in a defined manner. Needless to mention, the availability of vitellogenin mRNA, cDNA clones, hybridization assays and other molecular tools is crucial to these studies. Likewise, in the study of vitellogenin incorporation by oocytes, it is important to establish whether gonadotropins are acting directly on the oocyte or indirectly through action on the follicular cells. In *Xenopus*, it would be of interest to reveal the mechanism of gonadotropin-follicular cell interaction. The hormones could exert their action by changing the permeability to vitellogenin in the follicular envelope or by inducing synthesis of a follicular cell product which regulates the pinocytotic activity of the oocyte. In elasmobranchs, cyclostomes and reptiles, the hormonal factors responsible for regulating the uptake process remain to be identified. So far, in the study of vitellogenesis, the emphasis is on the estrogen-caused induction process and little consideration has been given to the de-induction process. It is assumed that once estrogen is removed from the circulation, vitellogenesis will cease. However, it is possible that vitellogenesis is actively terminated by a de-induction process at the end of the reproductive season. The fact that many male fishes, amphibians and reptiles have significant amounts of estrogen in their circulation but do not exhibit vitellogenesis supports this concept. Furthermore, the inability of estrogen to induce vitellogenesis in animals of a certain developmental stage (e.g., tadpoles) or under a certain physiological state indicates the possible existence of this de-induction process. It would be of great interest to verify the existence of the de-induction state and identify the hormonal factors involved in the de-induction of vitellogenesis.

X. REFERENCES

1. Abraham, M., Hilge, V., Lison, S. and Tibika, H. (1984). The cellular envelope of oocytes in teleosts. Cell Tissue Res. 235, 403-410.
2. Aida, K., Hirose, K., Yokote, M. and Hibiya, T. (1973). Physiological studies on gonadal maturation in fishes. II. Histological changes in the liver cells of ayu following gonadal maturation and estrogen administration. Bull. Japan. Soc. Sci. Fish. 38, 1107-1115.

3. Anand, T.C. and Sundararaj, B.I. (1974). Ovarian maintenance in the hypophysectomized catfish, Heteropneustes fossilis (Bloch), with mammalian hypophyseal and placental hormones, and gonadal and adrenocortical steroids. Gen. Comp. Endocrinol. 22, 154-168.
4. Ansari, A.Q., Dolphin, P.J., Lazier, C.B., Munday, K.A. and Akhtar, M. (1971). Chemical composition of an estrogen-induced calcium-binding glycolipophosphoprotein in Xenopus laevis. Biochem. J. 122, 107-113.
5. Berridge, M.V., Farmer, S.R., Green, C.D., Henshaw, E.C. and Tata, J.R. (1976). Characterization of polysomes from Xenopus laevis liver synthesizing vitellogenin and in vitro translation of vitellogenin and albumin messenger RNAs. Eur. J. Biochem. 62, 161-171.
6. Bromage, N.R., Whitehead, C. and Breton, B. (1982). Relationships between serum levels of gonadotropin, estradiol-17β, and vitellogenin in the control of ovarian development in the rainbow trout. I. The effects of alterations in environmental photoperiod. Gen. Comp. Endocrinol. 47, 366-376.
7. Burzawa-Gerard, E. (1971). Purification d'une hormone gonadotrope hypophysaire de poisson teleosteen, la Carpe (Cyprinus carpio L.). Biochemie 53, 545-552.
8. Burzawa-Gerard, E. (1974). Etude biologique et biochimique de l'hormone gonadotrope d'une poisson teleosteen, la Carpe (Cyprinus carpio L.). Mem. Mus. Nat. Hist. Nat. Ser. Zool. 86, 1-77.
9. Burzawa-Gerard, E. (1982). Chemical data on pituitary gonadotropins and their implication to evolution. Can. J. Fish. Aquat. Sci. 39, 80-91.
10. Callard, I.P., Banks, S.H. and Banks, W.L. (1972). Hepatic protein and nucleic acid content in Dipsosaurus dorsalis following hypophysectomy and treatment with estradiol-17β and growth hormone. Comp. Biochem. Physiol. 41B, 503-510.
11. Callard, I.P. and Doolittle, J.P. (1973). The influence of intrahypothalamic implants of progesterone on ovarian growth and function in the ovoviviparous iguanid lizard, Sceloporus cyanogenys. Comp. Biochem. Physiol. 44A, 625.
12. Callard, I.P., Doolittle, J.P., Banks, W. L. and Chan, S. W. (1972). Recent studies on the control of the reptilian ovarian cycle. Gen. Comp. Endocrinol. Suppl. 3, 65-75.
13. Callard, I.P. and Klotz, K.L. (1973). Sensitivity of parameters of estrogen action in the iguanid lizard Dipsosaurus dorsalis. Gen. Comp. Endocrinol. 21, 314-321.
14. Callard, I.P., Lance, V., Salhanik, A.R. and Barad, D. (1978). The annual ovarian cycle of Chrysemys picta: correlated changes in plasma steroids and parameters of vitellogenesis. Gen. Comp. Endocrinol. 35, 245-257.
15. Callard, I.P. and Zeigler, H. (1970). Inhibitory effects of prolactin upon gonadotrophin-stimulated ovarian growth in the iguanid lizard, Dipsosaurus dorsalis. J. Endocrinol. 47, 131-132.
16. Campbell, C.M. (1978). In vitro stimulation of vitellogenin incorporation into trout oocyte by salmon pituitary extracts. Ann. Biol. Anim. Biochem. Biophys. 18, 1013-1018.
17. Campbell, C.M. and Idler, D.R. (1976). Hormonal control of vitellogenesis in hypophysectomized winter flounder (Pseudopleuronectes americanus Walbaum). Gen. Comp. Endocrinol. 28, 143-150.
18. Campbell, C.M. and Idler, D.R. (1980). Characterization of an estradiol-induced protein from rainbow trout as vitellogenin by the cross reactivity to ovarian yolk fractions. Biol. Reprod. 22, 605-617.
19. Campbell, C.M. and Jalabert, B. (1979). Selective protein incorporation by vitellogenic Salmo gairdneri oocytes in vitro. Ann. Biol. Anim. Biochem. Biophys. 19, 429-437.

20. Chen, T.T., Agellon, L.B., Reid, P.C. and Howard, D.A. (1982). Estrogen control of gene expression in rainbow trout. Proc. Can. Fed. Biol. Sci. 25, 45. Abst. 169.
21. Clemens, M.J., Lofthouse, R. and Tata, J.R. (1975). Sequential changes in the protein synthetic activity of male Xenopus laevis liver following induction of egg yolk proteins by estradiol-17β. Biochem. J. 250, 2213-2218.
22. Craik, J.C.A. (1978). Plasma levels of vitellogenin in the elasmobranch Scyliorhinus canicula L. (lesser spotted dogfish). Comp. Biochem. Physiol. 60B, 9-18.
23. Craik, J.C.A. (1978). Kinetic studies of vitellogenin metabolism in the elasmobranch Scyliorhinus canicula L. Comp. Biochem. Physiol. 61A, 355-361.
24. Craik, J.C.A. (1978). The effects of oestrogen treatment on certain plasma constituents associated with vitellogenesis in the elasmobranch Scyliorhinus canicula L. Gen. Comp. Endocrinol. 35, 455-464.
25. Craik, J.C.A. (1978). Effects of hypophysectomy on vitellogenesis in the elasmobranch Scyliorhinus canicula L. Gen. Comp. Endocrinol. 36,63-67.
26. Crim, L.W. (1982). Environmental modulation of annual and daily rhythms associated with reproduction in teleost fishes. Can. J. Fish. Aq. Sci. 39, 17-21.
27. Crim, L.W. and Idler, D.R. (1978). Plasma gonadotropin, estradiol, and vitellogenin and gonad phosvitin levels in relation to the seasonal reproductive cycle of female brown trout. Ann. Biol. Anim. Biochem. Biophys. 18, 1001-1005.
28. Dessauer, H.C. and Fox, W. (1959). Changes in ovarian follicle composition with plasma levels of snakes during estrus. Am. J. Physiol. 197, 360-366.
29. de Vlaming, V.L., Vodicnik, M.J., Baner, G., Murphy, T. and Evans, D. (1977). Estradiol-17β effects on lipid and carbohydrate metabolism and on the induction of a yolk precursor in goldfish, Carassius auratus. Life Sci. 20, 1945-1952.
30. de Vlaming, V.L., Wiley, H.S., Delahunty, G. and Wallace, R.A. (1980). Goldfish (Carassius auratus) vitellogenin: induction, isolation, properties and relationship to yolk proteins. Comp. Biochem. Physiol. 67B, 613-623.
31. Dimitriadis, G.J. and Tata, J.R. (1982). Differential sensitization to deoxyribonuclease I of Xenopus vitellogenin and albumin genes during primary and secondary induction of vitellogenesis by estradiol. Biochem. J. 202, 491-497.
32. Dolphin, P.J., Ansari, A.Q., Lazier, C.B., Munday, K.A. and Akhtar, M. (1971). Studies on the induction and biosynthesis of vitellogenin, an estrogen-induced glycolipophosphoprotein. Biochem. J. 124, 751-758.
33. Donaldson, E.M. Yamazaki, F., Dye, H.M. and Philleo, W.W. (1972). Preparation of gonadotropin from salmon (Oncorhynchus tshawytscha) pituitary glands. Gen. Comp. Endocrinol. 18, 469-481.
34. Elliott, J.A.K., Bromage, N.R. and Whitehead, C. (1979). Effects of estradiol-17β on serum calcium and vitellogenin levels in rainbow trout. J. Endocrinol. 83, 54P-55P.
35. Emmersen, B.K. and Emmersen, J. (1976). Protein, RNA and DNA metabolism in relation to ovarian vitellogenic growth in the flounder Platichthys flesus (L.). Comp. Biochem. Physiol. 55B, 315-321.
36. Emmersen, B.K. and Kjaer, K. (1974). Seasonal and hormonally induced changes in the serum level of the precursor protein vitellogenin in relation to ovarian vitellogenic growth in the toad Bufo bufo bufo L.). Gen. Comp. Endocrinol. 22, 261-267.

37. Emmersen, B.K. and Petersen, I.M. (1976). Natural occurrence, and experimental induction by estradiol-17β, of a lipophosphoprotein (vitellogenin) in flounder (Platichyths flesus L.). Comp. Biochem. Physiol. 54B, 443-446.
38. Emmersen, J. and Korsgaard, B. (1982). Measurements of amino acid incorporation capacity of liver cell-free systems through the breeding season and following estradiol treatment of flounder (Platichthys flesus L.). Mol. Physiol. 2, 181-192.
39. Emmersen, J., Korsgaard, B. and Petersen, I. (1979). Dose-response kinetics of serum vitellogenin, liver DNA, RNA, protein and lipid after induction by estradiol-17β in male flounders (Platichthys flesus L.). Comp. Biochem. Physiol. 63B, 1-6.
40. Farmer, S.R., Henshaw, E.C., Berridge, M.V. and Tata, J.R. (1978). Translation of Xenopus vitellogenin mRNA during primary and secondary induction. Nature 273, 401-403.
41. Felber, B.K., Maurhofer, S., Jaggi, R.B., Wyler, T., Wahli, W. Ryffel, G.U. and Weber, R. (1980). Isolation and translation in vitro of four related vitellogenin mRNAs of estrogen-stimulated Xenopus laevis. Eur. J. Biochem. 105, 17-24.
42. Folger, K., Anderson, J.N., Hayward, M.A. and Shapiro, D.J. (1983). Nuclease sensitivity and DNA methylation in estrogen regulation of Xenopus laevis vitellogenin gene expression. J. Biol. Chem. 258, 8908-8914.
43. Follett, B.K. and Redshaw, M.R. (1974). The physiology of vitellogenesis. In "The Physiology of the Amphibia" (B. Lofts, ed.), pp. 219-308. Academic Press, New York.
44. Fostier, A., Breton, B. and Jalabert, B. (1979). Stimulation hypophysaire de la secretion oestradiol-17β chez la carpe commune, Cyprinus carpio L. Ann. Endocrinol. 40, 83-84.
45. Frost, W.E. (1974). A survey of the rainbow trout (Salmo gairdneri) in Britain and Ireland. Salmon and Trout Association, London.
46. Gapp, D.A., Ho, S.M. and Callard, I.P. (1979). Plasma levels of vitellogenin in Chrysemys picta during the annual gonadal cycle: Measurement by specific radioimmunoassay. Endocrinology 103, 784-790.
47. Garstka, W.R., Camazine, B. and Crews, D. (1982). Plasma levels of vitellogenin in Chrysemys picta during the annual gonadal cycle: Measurement by specific radioimmunoassay. Endocrinology 103, 784-790.
48. Gerber-Huber, S., Felber, B.K., Weber,R. and Ryffel, G.U. (1981). Estrogen induces tissue specific changes in the chromatin conformation of the vitellogenin gene in Xenopus. Nucl. Acids Res. 9, 2475-2494.
49. Gerstle, J.F. and Callard, I.P. (1972). Reproduction and estrogen-induced vitellogenesis in Dipsosaurus dorsalis. Comp. Biochem. Physiol. 42A, 791-801.
50. Giorgi, F., Gobbetti, A. and Polzonetti-Magni, A. (1982). Variations in the vitellogenin titre during reproductive cycle of Rana esculenta L. Comp. Biochem. Physiol. 72B, 501-506.
51. Green, C. D. and Tata, J.R. (1976). Direct induction of estradiol of vitellogenin synthesis in organ cultures of male Xenopus laevis liver. Cell 7, 131-139.
52. Hahn, W.E., Church, R.B. and Gorbman, A. (1969). Synthesis of new RNA species prior to estradiol-induced vitellogenesis. Endocrinology 84, 738-745.
53. Hara, A. (1976). Iron-binding activity of female-specific serum proteins of rainbow trout (Salmo gairdneri) and chum salmon (Oncorhynchus keta). Biochim. Biophys. Acta. 427, 549-557.

54. Hara, A. and Hirai, H. (1978). Comparative studies on immunochemical properties of female-specific serum protein and egg yolk proteins in rainbow trout (Salmo gairdneri). Comp. Biochem. Physiol. 59B, 339-343.
55. Hara, A., Takano, K. and Hirai, H. (1983). Immunochemical identification of a female-specific serum protein, vitellogenin, in the medaka, Oryzias latipes (Teleosts). Comp. Biochem. Physiol. 76, 135-141.
56. Hara, A., Yamauchi, K. and Hirai, H. (1980). Studies on female-specific serum protein (vitellogenin) and egg yolk protein in Japanese eel (Anguilla japonica). Comp. Biochem. Physiol. 65B, 315-320.
57. Hayward, M.A., Mitchell, R.A. and Shapiro, D.J. (1980). Induction of estrogen receptor and reversal of the nuclear-cytoplasmic receptor ratio during vitellogenin synthesis and withdrawal in Xenopus laevis. J. Biol. Chem. 225, 11308-113.2
58. Herbener, G.H., Bendayan, M. and Feldhoff, R.C. (1984). The intracellular pathway of vitellogenin secretion in the frog hepatocyte as revealed by protein A-gold immunocytochemistry. J. Histochem. Cytochem. 32, 697-704.
59. Herbener, G.H., Feldhoff, R.C. and Fonda, M.L. (1983). A correlated morphometric and biochemical study of estrogen-induced vitellogenesis in male Rana pipiens. J. Ultrastruct. Res. 83, 28-42.
60. Hickey, E.D. and Wallace, R.A. (1974). A study of the vitellogenic protein in the serum of estrogen-treated Ictalurus nebulosus. Biol. Bull. 147, 481.
61. Ho, S.M., Danko, D. and Callard, I.P. (1981). Effect of exogenous estradiol-17β on plasma vitellogenin levels in male and female Chrysemys and its modulation by testosterone and progesterone. Gen. Comp. Endocrinol. 43, 413-421.
62. Ho, S.M., Kleis, S., McPherson, R., Heisermann, G.J. and Callard, I.P. (1982). Regulation of vitellogenesis in reptiles. In "Reproductive Biology of Reptiles." Herpetologica 38, 40-50.
63. Ho, S.M., L'Italien, J. and Callard, I.P. (1980). Studies on reptilian yolk: Chrysemys vitellogenin and phosvitin. Comp. Biochem. Physiol. 65B, 139-144.
64. Ho, S.M., Taylor, S. and Callard, I.P. (1982). Effect of hypophysectomy and growth hormone on estrogen-induced vitellogenesis in the freshwater turtle, Chrysemys picta. Gen. Comp. Endocrinol. 48, 254-260.
65. Ho, S.M., Wangh, L.J. and Callard, I.P. (1985). Sexual differences in the in vitro induction of vitellogenesis in the turtle: Role of the pituitary and growth hormone. Comp. Biochem. Physiol. 81B, 467-472.
66. Hori, S.H., Kodama, T. and Tanahashi, K. (1979). Induction of vitellogenin synthesis in goldfish by massive doses of androgens. Gen. Comp. Endocrinol. 37, 306-320.
67. Huber, S., Ryffel, G.U. and Weber, R. (1979). Thyroid hormone induces competence for oestrogen-dependent vitellogenin synthesis in developing Xenopus laevis liver. Nature 278, 65-67.
68. Idler, D.R. (1982). Some perspectives on fish gonadotropins. In "Reproductive Physiology of Fish." (C.J.J. Richter and H.J.T. Goos, eds.), pp. 4-13. Pudoc, Wageningen, Netherlands.
69. Idler, D.R. and Campbell, C.M. (1980). Gonadotropin stimulation of estrogen and yolk precursor synthesis in juvenile rainbow trout. Gen. Comp. Endocrinol. 41, 384-391.
70. Idler, D.R., Hwang, S.J. and Crim, L.W. (1979). Quantification of vitellogenin in Atlantic salmon (Salmo salar) plasma by radioimmunoassay. J. Fish. Res. Bd. Can. 36, 574-578.

71. Kjaer, K. and Jorgensen, C.B. (1971). Effects of mammalian gonadotropins on the ovary in the toad Bufo bufo bufo (L.). Gen. Comp. Endocrinol. 17, 424.

72. Knowland, J. (1978). Induction of vitellogenin synthesis in Xenopus laevis tadpoles. Differentiation 12, 47-51.

73. Korsgaard, B., Emmersen, J. and Petersen, I. (1983). Estradiol-induced hepatic protein synthesis and transaminase activity in the male flounder, Platichthys flesus L. Gen. Comp. Endocrinol. 50, 11-17.

74. Lamba, V.J., Goswami, S.V. and Sundararaj, B.I. (1983). Circannual and circadian variations in plasma levels of steroids (cortisol, estradiol-17β, estrone, and testosterone) correlated with the annual gonadal cycle in the catfish, Heteropneustes fossilis (Bloch). Gen. Comp. Endocrinol. 50, 205-225.

75. Larsen, L.O. (1978). Subtotal hepatectomy in intact or hypophysectomized river lampreys (Lampetra fluviatilis L.): Effects on regeneration, blood glucose regulation and vitellogenesis. Gen. Comp. Endocrinol. 35, 197-204.

76. Lehman, G.C. (1977). Environmental influence on ovulation and embryonic development in Rana pipiens. J. Exp. Zool. 199, 51-56.

77. Le Menn, F. (1979). Induction de vitellogenine par l'oestradiol et par des androgenes chez un teleosteen marin: Gobius niger L. C.R. Acad. Sci. Paris Ser. D. 289, 413-416.

78. Le Menn, F. and Lamy, M.M. (1977). Proteins vitellines seriques et hepatiques chez un poisson teleosteen, Gobius niger L., soumis a des traitements hormonaux. Bordeaux Medical 10, 213-216.

79. Le Menn, F., Rochefort, H. and Garcia, M. (1980). Effect of androgen mediated by the estrogen receptor of fish liver: vitellogenin accumulation. Steroids 35, 315-328.

80. Matty, A.J., Tsureki, K., Dickhoff, W.W. and Gorbman, A. (1976). Thyroid and gonadal function in hypophysectomized hagfish, Eptatretus stouti. Gen. Comp. Endocrinol. 30, 500-516.

81. May, F.E.B. and Knowland, J. (1980). The role of thyroxine in the transition of vitellogenin synthesis from noninducibility to inducibility during metamorphosis in Xenopus laevis. Devel. Biol. 77, 419-430.

82. May, F.E.B., Westley, B.R. and Knowland, J. (1981). Vitellogenin synthesis and characterization of the liver estrogen receptor in the neotenous salamander Ambystoma mexicanum. Devel. Biol. 82, 350-357.

83. McBride, J.R. and van Overbeeke, A.P. (1971). Effects of androgens, estrogens, and cortisol on the skin, stomach, liver, pancreas and kidney in gonadectomized adult sockeye salmon (Oncorhynchus nerka). J. Fish. Res. Bd. Can. 28, 485-490.

84. Medda, A.K., Dasmahapatra, A.K. and Ray, A.K. (1980). Effect of estrogen and testosterone on the protein and nucleic acid contents of liver, muscle, and gonad and plasma protein content of male and female (vitellogenic and nonvitellogenic) singi fish, Heteropneustes fossilis Bloch. Gen. Comp. Endocrinol. 42, 427-436.

85. Mizell, S. (1964). Seasonal differences in spermatogenesis and oogenesis in Rana pipiens. Nature 202, 875-876.

86. Mohanty-Hejmadi, P., Safriel, O.J. and Nace, G.W. (1978). Serum proteins and vitellogenesis in the frog Rana pipiens. Indian J. Exp. Biol. 16, 531-537.

87. Nath, P. and Sundararaj, B.I. (1981). Isolation and identification of female-specific serum lipophosphoprotein (vitellogenin) in the catfish, Heteropneustes fossilis (Bloch). Gen. Comp. Endocrinol. 43,184-190.

88. Ng, T.B. and Idler, D.R. (1978). "Big" and "little" forms of plaice vitellogenic and maturational hormones. Gen. Comp. Endocrinol. 34, 408-420.

89. Ng, T.B. and Idler, D.R. (1978). A vitellogenic hormone with a large and a small form from salmon pituitaries. Gen. Comp. Endocrinol. 35, 189-195.
90. Ng, T.B. and Idler, D.R. (1979). Studies on two types of gonadotropins from both American plaice and winter flounder pituitaries. Gen. Comp. Endocrinol. 38, 410-420.
91. Ng, T.B. and Idler, D.R. (1980). Gonadotropic regulation of androgen production in flounder and salmonids. Gen. Comp. Endocrinol. 38,410-420.
92. Ng, T.B., Woo, N.Y.S., Tam, P.P.L. and Au, C.Y.W. (1984). Changes in metabolism and hepatic ultrastructure induced by estradiol and testosterone in immature female Epinephelus akaara (Teleostei, Serranidae). Cell Tissue Res. 236, 651-659.
93. Ng, W.C., Wolffe, A.P. and Tata, J.R. (1984). Unequal activation by estrogen of individual Xenopus vitellogenin genes during development. Devel. Biol. 102, 238-247.
94. Olivereau, M. and Olivereau, J. (1979). Effect of estradiol-17β on the cytology of the liver, gonads and pituitary, and on plasma electrolytes in the female freshwater eel. Cell Tissue Res. 199, 431-454.
95. Peter, R.E. and Crim, L.W. (1979). Reproductive endocrinology of fishes: gonadal cycles and gonadotropin in teleosts. Annu. Rev. Physiol. 41, 323-335.
96. Peute, J., van der Gaag, A. and Lambert, J.G.D. (1978). Ultrastructure and lipid content of the liver of the zebrafish, Brachydanio rerio, related to vitellogenin synthesis. Cell Tissue Res. 186, 297-308.
97. Pickering, A.D. (1976). Effects of gonadectomy, oestradiol and testosterone on the migrating river lamprey, Lampetra fluviatilis (L.). Gen. Comp. Endocrinol. 28, 473-480.
98. Plack, P.A. and Fraser, N.W. (1971). Incorporation of L-[14C] leucine into egg proteins by liver slices from cod. Biochem. J. 121, 857-862.
99. Plack, P.A., Pritchard, D.J. and Fraser, N.W. (1971). Egg proteins in cod serum: natural occurrence and induction by injections of oestradiol-3-benzoate. Biochem. J. 121, 847-856.
100. Rao, C.P.A. and David, G.F. (1967). The effect of certain steroids on the serum protein concentrations of the lizard, Uromastix hardwicki Gray. Gen. Comp. Endocrinol. 9, 227-233.
101. Redshaw, M.R. and Follett, B.K. (1971). The crystalline yolk-platelet proteins and their soluble plasma precursor in an amphibian, Xenopus laevis. Biochem. J. 124, 759-766.
102. Redshaw, M.R., Follett, B.K. and Nicholls, T.J. (1969). Comparative effects of oestrogens and other steroid hormones on serum lipids and proteins in Xenopus laevis Daudin. J. Endocrinol. 43, 47-53.
103. Roach, A.H. and Davies, P.L. (1980). Catfish vitellogenin and its messenger RNA are smaller than their chicken and Xenopus counterparts. Biochim. Biophys. Acta. 610, 400-412.
104. Scott, A.P. and Sumpter, J.P. (1983). A comparison of the female reproductive cycles of autumn-spawning and winter-spawning strains of rainbow trout (Salmo gairdneri Richardson). Gen. Comp. Endocrinol. 52, 79-85.
105. Selman, K. and Wallace, R.A. (1983). Oogenesis in Fundulus heteroclitus, III. Vitellogenesis. J. Exp. Zool. 226, 441-457.
106. Shapiro, D.J., Baker, H.J. and Stitt, D.T. (1976). In vitro translation and estradiol-17β induction of Xenopus laevis vitellogenin messenger RNA. J. Biol. Chem. 251, 3105-3111.
107. Signorini, G., Bucci-Innocenti, S., Ragghianti, M. and Giorgi, F.

(1981). Vitellogenin synthesis inducted by estradiol-17β in the newt *Triturus cristatus carnifex* (Laurenti). Comp. Biochem. Physiol. 69B, 121-126.

108. Smalley, K.N. and Nace, G.W. (1983). Vitellogenic cycles in laboratory-maintained females of the leopard frog, *Rana pipiens*. J. Exp. Zool. 226, 211-219.

109. Stanchfield, J.E. and Yager, J.D., Jr. (1978). An estrogen-responsive primary amphibian liver cell culture system. Exp. Cell. Res. 116, 239-252.

110. Sumpter, J.P. (1985). The purification, radioimmunoassay, and plasma levels of vitellogenin from the rainbow trout *Salmo gairdneri*. *In* "Current Trends in Comparative Endocrinology" (B. Lofts and W.N. Holmes, eds.), Vol. 1, 355-357. Hong Kong University Press, Hong Kong.

111. Sundararaj, B. I., Goswami, S.V. and Lamba, V.J. (1982). Role of testosterone, estradiol-17β, and cortisol during vitellogenin synthesis in the catfish, *Heteropneustes fossilis* (Block). Gen. Comp. Endocrinol. 48, 390-397.

112. Sundararaj, B.I. and Nath, P. (1981). Steroid-induced synthesis of vitellogenin in the catfish, *Heteropneustes fossilis* (Bloch). Gen. Comp. Endocrinol. 43, 201-210.

113. Sundararaj, B.I., Nath, P. and Burzawa-Gerard, E. (1982). Synthesis of vitellogenin and its uptake by the ovary in the catfish, *Heteropneustes fossilis* (Block) in response to carp gonadotropin and its subunits. Gen. Comp. Endocrinol. 46, 93-98

114. Suzuki, H.K. and Prosser, R.L. (1968). The effects of estradiol valerate upon the serum and bone of the lizard, *Sceloporus cyanogenys*. Proc. Soc. Exp. Biol. Med. 127, 4-7.

115. Tata, J.R. (1976). The expression of the vitellogenin gene. Cell 9, 1-14.

116. Tata, J.R. and Smith, D.F. (1979). Vitellogenesis: A versatile model for hormonal regulation of gene expression. Rec. Prog. Horm. Res. 35, 47-95.

117. Tenniswood, M.P.R., Searle, P.F., Wolffe, A.P. and Tata, J. R. (1983). Rapid estrogen metabolism and vitellogenin gene expression in *Xenopus* hepatocyte cultures. Mol. Cell. Endocrinol. 30, 329-345.

118. Terkatin-Shimony, A. and Yaron, Z. (1978). Estrogens and estrogenic effects in *Tilapia aurea* (*Cichlidae*, *Teleostei*). Ann. Biol. Anim. Biochem. Biophys. 18, 1007-1012.

119. Ueda, H., Hiroi, O., Hara, A., Yamauchi, K. and Nagahama, Y. (1984). Changes in serum concentrations of steroid hormones, thyroxine, and vitellogenin during spawning migration of the chum salmon, *Oncorhynchus keta*. Gen. Comp. Endocrinol. 53, 203-211.

120. Valotaire, Y., Tenniswood, M., LeGuellec, C. and Tata, J.R. (1984). The preparation and characterization of vitellogenin messenger RNA from rainbow trout (*Salmo gairdneri*). Biochem. J. 217, 73-77.

121. van Bohemen, C.G. and Lambert, J.G.D. (1981). Estrogen synthesis in relation to estrone, estradiol, and vitellogenin plasma levels during the reproductive cycle of the female rainbow trout, *Salmo gairdneri*. Gen. Comp. Endocrinol. 45, 105-114.

122. van Bohemen, C.G., Lambert, J.G.D., Goos, H.J.T. and van Oordt, P.G.W.J. (1982). Estrone and estradiol participation during exogenous vitellogenesis in the female rainbow trout, *Salmo gairdneri*. Gen. Comp. Endocrinol. 46, 81-92.

123. van Bohemen, C.G., Lambert, J.G.D. and Peute, J. (1981). Annual changes in plasma and liver in relation to vitellogenins in the female rainbow trout, *Salmo gairdneri*. Gen. Comp. Endocrinol. 46, 136-139.

124. van Bohemen, C.G., Lambert, J.G.D. and van Oordt, P.G.W.J. (1982). Vitellogenin induction by estradiol in estrone-primed rainbow trout, Salmo gairdneri. Gen. Comp. Endocrinol. 46, 136-138.
125. Wahli, W., Dawid, I.B., Ryffel, G.U. and Wever, R. (1981). Vitellogenesis and the vitellogenin gene family. Science 212, 298-304.
126. Wahli, W., Dawid, I.B., Wyler, T., Jaggi, R.B., Weber, R. and Ryffel, G.U. (1979). Vitellogenin in Xenopus laevis is encoded in a small family of genes. Cell 16, 535-549.
127. Wahli, W., Germond, J.E., Heggeler, B. and May, F.E.B. (1982). Vitellogenin genes A1 and B1 are linked in the Xenopus laevis genome. Proc. Nat. Acad. Sci. 79, 6832-6836.
128. Wahli, W., Wyler, T., Weber, R. and Ryffel, G.U. (1976). Size, complexity and abundance of a specific poly(A)-containing RNA of liver from male Xenopus induced to vitellogenin synthesis by estrogen. Eur. J. Biochem. 66, 457-465.
129. Wallace, R.A. (1970). Studies on amphibian yolk IX. Xenopus vitellogenin. Biochim. Biophys. Acta. 215, 176-183.
130. Wallace, R.A. (1970). Some factors involved in protein incorporation and transformation by isolated amphibian oocytes. J. Cell. Biol. 47, 219a.
131. Wallace, R.A. (1983). Interactions between somatic cells and the growing oocyte of Xenopus laevis. In "Current Problems in Germ Cells" (A. McLaren and C.C. Wylie, eds.), pp. 285-306. Cambridge University Press, Cambridge.
132. Wallace, R.A. and Bergink, E.W. (1974). Amphibian vitellogenin: Properties, hormonal regulation of hepatic synthesis and ovarian uptake, and conversion to yolk proteins. Amer. Zool. 14, 1159-1175.
132a. Wallace, R.A. (1985). Vitellogenesis and oocyte growth in nonmammalian vertebrates. In "Developmental Biology" (L. Browder, ed.), Vol.1, pp. 127-177. Pergamon Press, New York.
133. Wallace, R.A. and Dumont, J.N. (1968). The induced synthesis and transformation of yolk proteins and their accumulation by the oocyte in Xenopus laevis. J. Cell Physiol. 72, 73-89.
134. Wallace, R.A., Misulovin, Z. and Etkin, L.D., (1981). Full-grown oocytes from Xenopus laevis resume growth when placed in culture. Proc. Nat. Acad. Sci. 78, 3078-3082.
135. Wallace, R.A., Misulovin, Z. and Wiley, H.S., (1980). Growth of anuran oocytes in serum-supplemented medium. Reprod. Nutr. Develop. 20, 699-708.
136. Wallace, R.A. and Selman, K. (1981). Cellular and dynamic aspects of oocyte growth in teleosts. Am. Zool. 21, 325-343.
137. Wallace, R.A. and Selman, K. (1982). A new procedure for the isolation of intact vitellogenin from teleosts. In "Reproductive Physiology of Fish" (C.J.J. Richter and H.J.T. Goos, eds.), p. 161. Pudoc, Wageningen, Netherlands.
138. Wangh, L.J. (1982). Glucocorticoids act together with estrogens and thyroid hormones in regulating the synthesis and secretion of Xenopus vitellogenin, serum albumin, and fibrinogen. Devel. Biol. 89, 294-298.
139. Wangh, L.J. and Knowland, J. (1975). Synthesis of vitellogenin in cultures of male and female frog liver regulated by estradiol treatment in vitro. Proc. Nat. Acad. Sci. 72, 3172-3175.
140. Wangh, L.J., Osborne, J.A., Hentschel, C.C. and Tilly, R. (1979) Parenchymal cells purified from Xenopus liver and maintained in primary culture synthesize vitellogenin in response to estradiol-17β and serum albumin in response to dexamethasone. Devel. Biol. 70, 479-499.

141. Wangh, L.J. and Schneider, W. (1982). Thyroid hormones are corequisites for estradiol-17β *in vitro* induction of *Xenopus* vitellogenin synthesis and secretion. Devel. Biol. 89, 287-293.

142. Westley, B. and Knowland, J. (1979). Estrogen causes a rapid, large and prolonged rise in the level of nuclear estrogen receptors in *Xenopus laevis* liver. Biochem. Biophys. Res. Comm. 88, 1167-1172.

143. Whitehead, C., Bromage, N.R. and Breton, B. (1983). Changes in serum levels of gonadotropin, estradiol-17β and vitellogenin during the first and subsequent reproductive cycles of female rainbow trout. Aquaculture 34, 317-326.

144. Whitehead, C., Bromage, N.R. and Forster, J.R.M. (1978). Seasonal changes in reproductive function of the rainbow trout (*Salmo gairdneri*). J. Fish. Biol. 12, 601-608.

145. Whiting, S.J. and Wiggs, A.J. (1978). Effect of sexual maturation and estradiol-17β on liver glycogen and tyrosine aminotransferase activity of brook trout, *Salvelinus fontinalis* Mitchill. Comp. Biochem. Physiol. 60B, 463-465.

146. Wiegand, M.D. (1982). Vitellogenesis in fishes. *In* "Reproductive Physiology of Fish" (C.J.J. Richter and H.J.T. Goos, eds.), pp. 136-146. Pudoc, Wageningen, Netherlands.

147. Wiley, H.S. and Dumont, J.N. (1978). Stimulation of vitellogenin uptake in stage IV *Xenopus* oocytes by treatment with chorionic gonadotropin *in vitro*. Biol. Reprod. 17, 762-771.

148. Wiley, H. S., Opresko, L. and Wallace, R.A. (1979). New methods for the purification of vertebrate vitellogenin. Analyt. Biochem. 97, 145-152.

149. Wiley, H.S. and Wallace, R.A. (1978). Three different molecular weight forms of the vitellogenin peptide from *Xenopus laevis*. Biochem. Biophys. Res. Comm. 85, 153-159.

150. Wiley, H.S. and Wallace, R.A. (1981). The structure of vitellogenin. Multiple vitellogenins in *Xenopus laevis* give rise to multiple forms of the yolk proteins. J. Biol. Chem. 256, 8626-8634.

151. Williams, J.L. and Tata, J.R. (1983). Simultaneous analysis of conformation and transcription of A and B groups of vitellogenin genes in male and female *Xenopus* during primary and secondary activation of estrogen. Nucl. Acids Res. 11, 1151-1166.

152. Wittliff, J.L. and Kenney, F.T. (1972). Regulation of yolk protein synthesis in amphibian liver. I. Induction of lipovitellin synthesis by estrogen. Biochim. Biophys. Acta. 269, 485-492.

153. Witliff, J.L. and Zelson, P.R. (1974). Estrogen specificity of the induction of lipovitellin synthesis and evidence for a specific estrogen-binding component. Endocrinol. Res. Commun. 1, 117-124.

154. Wolffe, A.P. and Tata, J.R. (1983). Coordinate and non-coordinate estrogen-induced expression of A and B groups of vitellogenin genes in male and female *Xenopus* hepatocytes in culture. Eur. J. Biochem. 130, 365-372.

155. Yamamoto, M. and Egami, N. (1974). Sexual differences and age changes in the fine structure of hepatocytes in the medaka, *Oryzias latipes*. J. Fac. Sci. Univer. Tokyo, Ser. IV, 13, 199-210.

156. Yaron, Z. and Barton, C. (1980). Stimulation of estradiol-17β output from isolated ovarian fragments of the plaice (*Pleuronectes platessa*) by homologous pituitary extract. Gen. Comp. Endocrinol. 42, 151-154.

157. Yaron, Z. and Widzer, L. (1978). The control of vitellogenesis by ovarian hormones in the lizard, *Xantusia vigilis*. Comp. Biochem. Physiol. 60, 279-284.

158. Yu, J.Y.L., Dickhoff, W.W., Inui, Y. and Gorbman, A. (1980). Sexual patterns of protein metabolism in liver and plasma of hagfish, *Eptatretus stouti* with special reference to vitellogenesis. Comp. Biochem. Physiol. 65B, 111-117.
159. Yu, J.Y.L., Dickhoff, W. W., Swanson, P. and Gorbman, A. (1981). Vitellogenesis and its hormonal regulation in the Pacific hagfish, *Eptatretus stouti* L. Gen. Comp. Endocrinol. 43, 492-502.

ENDOCRINE CONTROL OF OOCYTE MATURATION

Yoshitaka Nagahama

Laboratory of Reproductive Biology
National Institute for Basic Biology
Okazaki, 444 Japan

I. INTRODUCTION

Embryogenesis in vertebrates is initiated by fusion of male and female gametes. In most vertebrates, however, full-grown postvitellogenic oocytes in the ovary are still in prophase I of meiosis and cannot be fertilized; for the oocytes to be fertilizable, they must complete the first meiotic division. Under appropriate hormonal stimulation, full-grown oocytes resume their first meiotic division which involves breakdown of the germinal vesicle (GVBD), chromosome condensation, assembly of the first meiotic spindle, and extrusion of the first polar body (Fig. 1). In most cases, however, meiosis is again arrested at the metaphase II stage. Shortly thereafter, mature fertilizable oocytes are ovulated. The meiotic process leading to extrusion of the second polar body is resumed again at the time of fertilization, immediately after sperm penetration. The time period between the resumption of meiosis and the second meiotic metaphase has been referred to as the period of oocyte maturation (meiotic maturation). Thus, the process of oocyte maturation is a prerequisite for successful fertilization.

The investigation of hormonal regulation of oocyte maturation in nonmammalian vertebrates has been restricted almost exclusively to teleosts and amphibians. The ovaries of these animals, in general, contain large numbers of oocytes that are relatively easy to maintain *in vitro*. Thus, the development of an *in-vitro* system, using GVBD as a biological indicator of hormone action, was a great stimulus to work in this field. Accumulating evidence suggests that oocyte maturation in these organisms is regulated by a series of interdependent hormonal

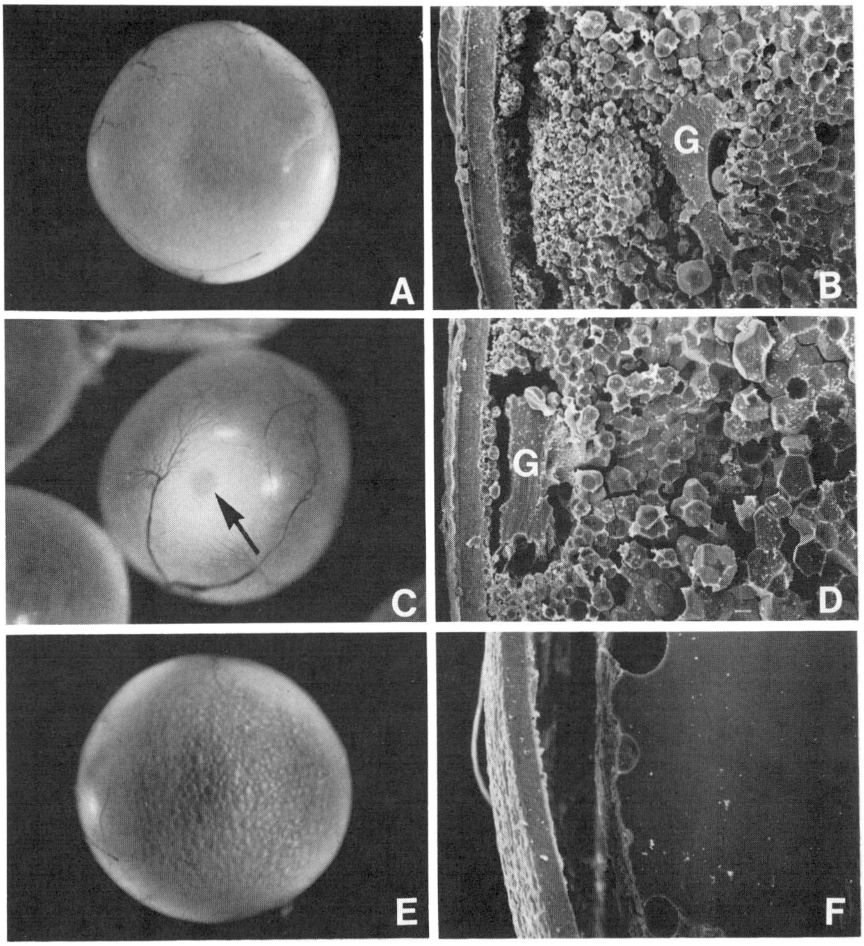

Fig. 1. Morphology of meiotic maturation of salmonid oocytes. B, D, and F: scanning electron micrographs of the fractured surface of the oocyte. A and B: initial controls. Germinal vesicle (G) is located half way between the center and the oocyte periphery. C and D: 24 hr after the onset of in-vitro incubation with salmon gonadotropin. Germinal vesicle (arrow and G) can be observed under the dissecting microscope. E and F: 72 hr after the onset of in-vitro incubation with salmon gonadotropin. Note the disappearance of germinal vesicle and translucent ooplasm. Reproduced from Bull. Jap. Soc. Sci. Fish. 46 (9), p. 1102, with permission.

actions. Three major mediators have been described; gonadotropin (Gn), maturation-inducing substance (MIS), and maturation-promoting factor (MPF) (for reviews, see Wasserman and Smith,[147] Masui and Clarke,[83] Schuetz,[118] Kanatani and Nagahama,[62] Maller and Krebs[80]). While a great deal now is known about various aspects of oocyte maturation in mammals, very little is known about the hormonal control of its induction (see Channing et al.,[15] Tsafriri[139]). Nevertheless, some progress recently has been made to account for the mechanisms of maintenance of meiotic arrest and gonadotropin-induced resumption of meiosis in mammalian oocytes. This chapter attempts to summarize our current understanding of the hormonal control of oocyte maturation in ectothermic vertebrates.

II. PRIMARY HORMONES INVOLVED IN TRIGGERING OOCYTE MATURATION

A. Overview of Mammals

Most oocytes of sexually mature mammals are arrested at the diplotene (dictyate) stage of the first meiotic division. Oocyte maturation *in vivo* is the result of the endogenous preovulatory surge of luteinizing hormone (LH) or the administration of exogenous Gns (for review see Tsafriri[139]). *In vitro*, follicle-enclosed oocytes remain arrested if the follicles are removed from the female before the preovulatory LH surge and cultured in a hormone-free medium. Addition of LH to the medium causes the oocytes within these follicles to initiate maturation (see Lindner et al.,[75] Tsafriri,[139] Masui and Clarke[83]). In this respect, oocyte maturation is under the positive control of Gns. However, when full-grown oocytes or oocyte-cumulus cell complexes are removed from antral follicles and cultured, they undergo meiotic maturation spontaneously without the addition of Gns or other stimulating agents.[32,105] This meiotic maturation, therefore, is referred to as "spontaneous maturation." This observation led to the generally held conclusion that some component of the antral follicle inhibits oocyte maturation until that inhibition is overcome by the action of Gns (see Section V,A).

B. Fishes

The number of Gns present in the teleost pituitary gland has been debated and is currently unresolved. A glycoprotein-rich Gn with a molecular weight of 25,000-40,000 has been purified in several teleosts (see Idler and Ng;[48] see also Chapter 10). This type of Gn, termed maturational gonadotropin, has been reported to stimulate almost all ovarian activities including yolk formation, final oocyte maturation, and steroidogenesis. There is also some evidence that fish produce a second Gn with a low carbohydrate content. This second type of Gn, termed vitellogenic gonadotropin, has been shown to stimulate the uptake of vitellogenin into the oocyte of certain teleosts (see Idler and Ng[48]). In this review, the term fish Gn is taken to refer to the glycoprotein-rich Gn.

Detailed preovulatory changes in blood levels of Gn have been studied in several teleostean species. As a result, it has become apparent that the profiles of preovulatory changes in Gn differ according to species. In salmonids, for example, increased Gn levels are evident during the 1-2 week period before ovulation, with the highest levels seen at the time of final oocyte maturation and ovulation.[39] In rainbow trout, Salmo gairdneri, it has been reported that plasma Gn continues to rise for several weeks after ovulation.[39,58] On the other hand, a distinct surge of serum Gn prior to oocyte maturation and ovulation has been reported in the goldfish Carassius auratus (Stacey et al.,[132] Kobayashi, personal communication). This ovulatory surge of Gn lasts about 12 h, during which Gn levels increase about tenfold above basal levels. These differences certainly reflect changes in reproductive strategy among different teleostean species. Recently, it has been shown that in the coho salmon Oncorhynchus kisutch,[143] the ayu Plecoglossus altivelis[47] and goldfish,[103] the induction of GVBD using mammalian gonadotropin-releasing hormones (GnRH) was related to the elevation of plasma Gn levels. Evidence for the hypothalamic control of a preovulatory surge of Gn has been extensively investigated in goldfish (see Peter[102]).

C. Amphibians

Two chemically distinct Gns, which appear homologous to mammalian LH and follicle-stimulating hormone (FSH), have been purified from pituitaries of several amphibian species (see Licht et al.,[69] Licht;[66] see also Chapter 10). Biochemical, immunological, and biological data all indicate a marked structural homology between the two Gns of amphibians and the FSH and LH of mammals. Both in-vivo and in-vitro experiments, utilizing these homologous hormones in female frogs, have established that LH plays a major role in the ovulatory (maturational) process (see Licht et al.[69]). Continuous infusion of mammalian GnRH into Rana catesbeiana resulted in a significant elevation ("surge") in plasma LH, concomitant with the occurrence of ovulation.[85] Plasma FSH also increased following GnRH infusion, but to much lower levels than LH.

D. Reptiles

Two separate Gns have been purified from some reptilian pituitaries (see Licht et al.,[69] Licht;[67] see also Chapter 10). Even though they appear homologous to mammalian LH and FSH, their roles in the physiology of reproduction are poorly understood. Using a homologous RIA, Licht and his colleagues[70,71] have demonstrated the presence of a pronounced preovulatory surge in serum LH concentrations in the green sea turtle *Chelonia mydas*. Later, they confirmed a similar LH surge in the olive ridley sea turtle *Lepidochelys olivacea*.[68] Thus, available data on temporal changes in plasma Gn levels appear to indicate that LH plays a major role in the process of ovulation (oocyte maturation) in reptiles. However, these findings are not entirely consistent with results of physiological studies utilizing exogenous hormones in reptiles (for review, Licht et al.[69]).

III. MATURATION-INDUCING SUBSTANCES

Although it was known for some time that both Gn and steroid hormones can stimulate oocyte maturation and ovulation in anuran oocytes *in vivo* and *in vitro*, it was not until the late 1960's that Gns were found to induce the follicles to secrete a hormone (MIS), which in turn acts on the oocytes to initiate maturation. These findings have facilitated investigations of the hormonal regulation of final oocyte maturation in a variety of vertebrates.[82,115,116,128]

A. Overview of Mammals

As discussed earlier (Section II,A), LH triggers oocyte maturation in the preovulatory follicles of mammals both *in vivo* and *in vitro*. LH appears to act indirectly on the oocyte via the surrounding somatic cells. It is likely that cyclic adenosine monophosphate (cAMP) is involved in the early steps leading to meiosis, since rapid increases in follicular cAMP levels have been demonstrated following LH stimulation. Nevertheless, the exact role of cAMP in stimulating oocyte maturation in mammals still remains unclear. It is known that Gn-induced maturation in follicle-enclosed mammalian oocytes is accompanied by sequential changes in follicular steroidogenesis.[1] Thus, it is possible that steroids are involved in the physiological regulation of the resumption of meiosis. It has been proposed, however, that steroids are not the major follicular mediator of Gn-induced meiotic maturation since addition of steroid inhibitors, such as cyanoketone and aminoglutethimide, to culture medium did not prevent LH-induced maturation in follicle-enclosed rat oocytes.[72] Nevertheless, it is possible that a specific intra-follicular steroid environment is required for the completion of the full maturation process, since alterations in the normal profile of steroids secreted during maturation induce both biochemical and developmental abnormalities in the oocyte.[100]

B. Fishes

It has been shown that follicle-enclosed, full-grown, postvitellogenic oocytes of some teleosts undergo GVBD *in vitro* when they are incubated with Gn. However, denuded oocytes are incapable of responding to Gn (see Goetz[45]). Cyanoketone, a specific inhibitor of 3β-hydroxy-Δ^5-steroid dehydrogenase (HSD), completely abolished the maturational effects of Gn and pregnenolone, but not of Δ^4 steroids such as progesterone or 17α,20β-dihydroxy-4-pregnen-3-one (17α,20β-diOH prog).[153] Thus, the action of Gn in inducing oocyte maturation appears to be dependent on the synthesis of a second steroidal mediator of meiotic maturation.

In teleosts, a variety of C_{21}-steroids have been shown to be potent initiators of GVBD *in vitro*. These include progesterone, 17α-hydroxyprogesterone, 17α,20β-diOHprog, cortisol and deoxycorticosterone. Recent *in-vitro* studies have shown that 17α,20β-diOHprog is the most effective agent for inducing GVBD in salmonids as well as in nonsalmonid fishes[42,95] (see also Goetz[45] for references). Testosterone as well as other C_{19}-steroids have been shown to induce GVBD only at high concentrations. Estradiol and other C_{18}-steroids are generally not effective at inducing oocyte maturation in teleostean oocytes. In the catfish *Heteropneustes fossilis* the terminal hormones that act on the oocytes to induce maturation appear to be corticosteroids.[134]

Theophan and Goetz[137] have identified several steroidal inducers of GVBD in two species of teleosts, the brook trout, *Salvelinus fontinalis*, and the yellow perch, *Perca flavescens*, by incubating ovarian tissues with radioactively labelled steroid precursors and testing the metabolites produced in a homologous *in-vitro* assay based on GVBD. It was reported that 17α,20β-diOHprog was a major metabolite of cell-free ovarian homogenates of Gn-primed ayu. In homogenates from nonprimed fish a range of steroids were identified, including a number of 5-reduced hydroxysteroids.[136] More recently, the MIS of a salmonid, the amago salmon, *Oncorhynchus rhodurus*, has been identified from media in which immature but full-grown folliculated oocytes had been incubated with Gn.[94] In this study, reversed-phase high performance liquid chromatography (HPLC) was employed to fractionate medium in which oocytes had been induced to mature with Gn. Of the fractions prepared by HPLC, only one exhibited maturation-inducing activity as assessed by an *in-vitro* GVBD assay. This fraction had a retention time that coincided exactly with authentic 17α,20β-diOHprog standard. The purity and final characterization of the residue of this fraction were further confirmed by a comparison with authentic 17α,20β-diOHprog using thin layer chromatography and mass spectroscopy.

Temporal changes in blood concentrations of steroid hormones during the final oocyte maturation and ovulation period have been followed in several species of salmonids (see Goetz;[45] Fostier et al.,[40] Nagahama et al.[98]). Levels of plasma progestogens such as 17α-hydroxyprogesterone and 17α,20β-diOHprog increase in conjunction with the preovulatory increase in plasma Gn, reaching peak levels at or after the period of oocyte maturation.[13,39,123,124,148,151] Certainly, the elevation in these steroids coincides with oocyte maturation. It is also evident that there is a significant decrease in estradiol levels and that the level of testosterone remains high during the preovulatory period. Thus, the preovulatory period in salmonids involves a shift in steroidogenesis from estradiol to 17α,20β-diOHprog[124,150] (Fig. 2). A similar predominance of progestogens in plasma also was observed in females which underwent oocyte maturation following the injection of mammalian GnRH.[47,103,142] The decrease in plasma estradiol levels immediately prior to oocyte maturation may be due, in part, to a decrease in aromatase activity in the granulosa cells.[153] At present, no definite physiological function has been assigned to the observed high plasma levels of testosterone during the preovulatory period.

Similar preovulatory steroid changes have been observed in some nonsalmonid species. The presence of 17α,20β-diOHprog in plasma during oocyte maturation and ovulation was reported in the white sucker, *Catostomus commersoni*,[122] and goldfish.[61,133]

The ovarian production of progestogens, especially 17α,20β-diOHprog, has been confirmed in several teleosts by *in-vitro* incubation studies.[41,151,152,155] In rainbow trout, the levels of 17α,20β-diOHprog increased significantly in a dose-dependent manner from 24 to 72 h of incubation in Gn-containing medium, whereas only very low levels were measured in Gn-free incubates.[41] In amago salmon, high levels of 17α,20β-diOHprog were present in media in which follicle-enclosed oocytes had been induced to mature with Gn. In contrast, in media in which exogenous cyanoketone inhibited Gn-induced maturation, little or no 17α,20β-diOHprog was found.[152]

Ovarian follicles of teleosts, like those of other vertebrates, are composed of two major layers, an outer thecal layer and an inner granulosa layer, which are separated by a substantial basal lamina. The thecal layer consists of fibroblasts, collagen fibers and capillaries, and in

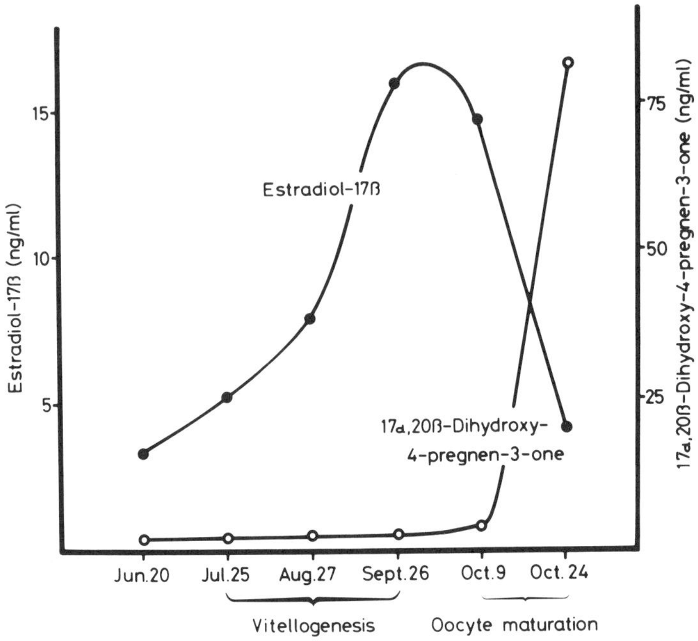

Fig. 2. Changes in plasma estradiol-17β and 17α,20β-dihydroxy-4-pregnen-3-one levels of female amago salmon during sexual maturation.

some species, large glandular cells known as special thecal cells (steroid-producing cells), while the granulosa layer consists of a single layer of uniform cells (see Nagahama[92]). Development of a simple dissection technique to separate the ovarian follicle of the medaka, *Oryzias latipes*,[50] and salmonids[60] into two layers has made it possible to investigate the role of the thecal and granulosa layers in Gn-induced oocyte maturation. In medaka, removal of all follicular constituents, but not removal of the thecal layer alone, prevented Gn-induced maturation of oocytes. Denuded oocytes underwent maturation in response to Gn when co-cultured with isolated granulosa cells.[50] These results suggest that medaka granulosa cells alone are capable of producing MIS in response to Gn. This situation appears to differ from that in salmonids, in which both thecal and granulosa layers are necessary for the follicular production of MIS in response to Gn (see details below). In this regard, it is noteworthy that the medaka thecal layer has been reported to lack steroid-producing cells,[53] whereas special thecal cells with steroidogenic features are present in the thecal layer of salmonids (see Nagahama[92]).

The identification of the MIS of amago salmon as 17α,20β-diOHprog made it possible to investigate the role of each follicular layer in the synthesis of MIS. Using all the *in-vitro* data, a two-cell type model has been proposed for the follicular production of MIS; this is summarized in Fig. 3.[93,154] In this model, 17α-hydroxyprogesterone produced by the thecal layer in response to Gn is converted to 17α,20β-diOHprog in the granulosa layer, where Gn acts to enhance the activity of 20β-HSD, the key enzyme involved in the conversion of 17α-hydroxyprogesterone to 17α,20β-diOHprog. The first step of the stimulating effect of Gn in the thecal layer appears to be receptor-mediated activation of adenylate cyclase and formation of cAMP, the major site of action probably occurring at the steroidogenic step between cholesterol and pregnenolone. Furthermore, Gn-induced 17α-hydroxyprogesterone production by the thecal layer is abolished by cycloheximide, a translational inhibitor, but not by actinomycin D, a transcriptional inhibitor, suggesting that stimulating effects of Gn in the thecal layer require the synthesis of a new protein (Nagahama, unpublished). However, the extent to which thecal layers contribute to the production of 17α,20β-diOHprog is still not clear, since in some cases thecal layer preparations produced this steroid in response to Gn. Although these granulosa layer preparations were completely free of attached thecal layers, it is extremely difficult to obtain pure thecal layer preparations from postvitellogenic follicles.

The granulosa cells of salmonids provide a unique system in which to study Gn regulation of a steroidogenic enzyme, in this case 20β-HSD, since precursor steroids cannot be produced by these cells. This allows the indirect quantification of 20β-HSD activity by the measurement of 17α,20β-diOHprog production when these cells are incubated with exogenous 17α-hydroxyprogesterone.[97] Granulosa cells obtained from full-grown follicles responded to Gn by enhancing 20β-HSD activity when exogenous 17α-hydroxyprogesterone was added to the incubation medium. The action of Gn on 20β-HSD enhancement was mimicked by forskolin, an adenylate cyclase activator, by dibutyryl cAMP (dbcAMP), but not dbcGMP, and by two phosphodiesterase inhibitors, theophylline and 3-isobutyl-1-methylxanthine (IBMX)[96] (Fig. 4). Furthermore, cAMP concentrations in granulosa cells were markedly elevated immediately after the addition of Gn or forskolin into the incubation medium (Kanamori and Nagahama, unpublished). These findings are consistent with the view that cAMP is the second messenger of Gn action.

Recent *in-vitro* experiments utilizing both cycloheximide and actinomycin D have revealed that these inhibitors completely block Gn stimulation of the induction of 20β-HSD activation in granulosa cells.[97] Similarly, dbcAMP-induced 20β-HSD activation also was prevented by these

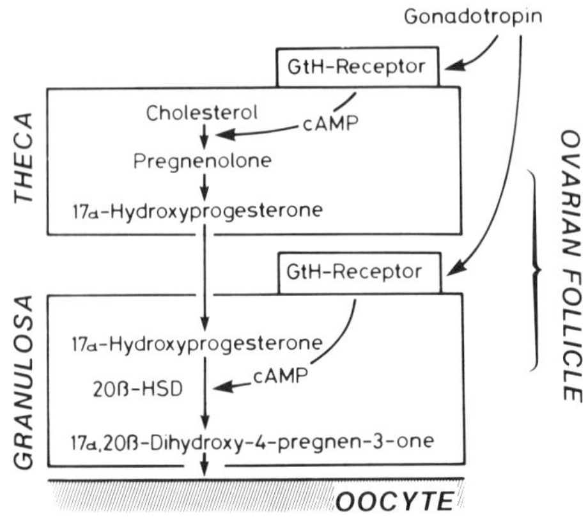

Fig. 3. Two-cell type model for the production of 17α,20β-dihydroxy-4-pregnen-3-one in the ovarian follicle of salmonids (see the text for details). GtH = gonadotropin. Reproduced from Gunma Symposium on Endocrinology 21 (1984), p. 176, with permission.

inhibitors. These findings suggest that Gn acts directly on granulosa clls to enhance the *de novo* synthesis of 20β-HSD by a mechanism involving new RNA synthesis.[93] Jalabert[57] reported that, in rainbow trout, both translational and transcriptional inhibitors blocked Gn- or pituitary-induced GVBD *in vitro*, whereas only cycloheximide affected 17α,20β-diOHprog-induced final maturation. It is noteworthy that the capacity of amago salmon follicles to respond to Gn by synthesizing and secreting this steroid is acquired immediately prior to the natural maturation period.[151] Taken together, these results suggest that the major action of Gn in inducing oocyte maturation in salmonids is to enhance the activity of 20β-HSD in the granulosa cells required for the production of MIS (17α,20β-diOHprog), a step probably mediated by a genetic signal.

Although detailed studies on the action of MIS on the oocyte to induce GVBD have not been conducted in fish oocytes, some data are available. It has been shown that denuded oocytes are still responsive to steroid stimulation, indicating the direct action of steroids on the induction of GVBD (see Goetz[45]). 17α,20β-DiOHprog was found to be ineffective in inducing oocyte maturation when microinjected into full-grown immature oocytes of goldfish but was effective when applied externally (Nagahama and Kishimoto, unpublished). These data suggest that the site of action of steroidal inducers is at the oocyte surface. The only other data available are the results of experiments utilizing translational and transcriptional inhibitors. Dettlaff and Skoblina[26] have

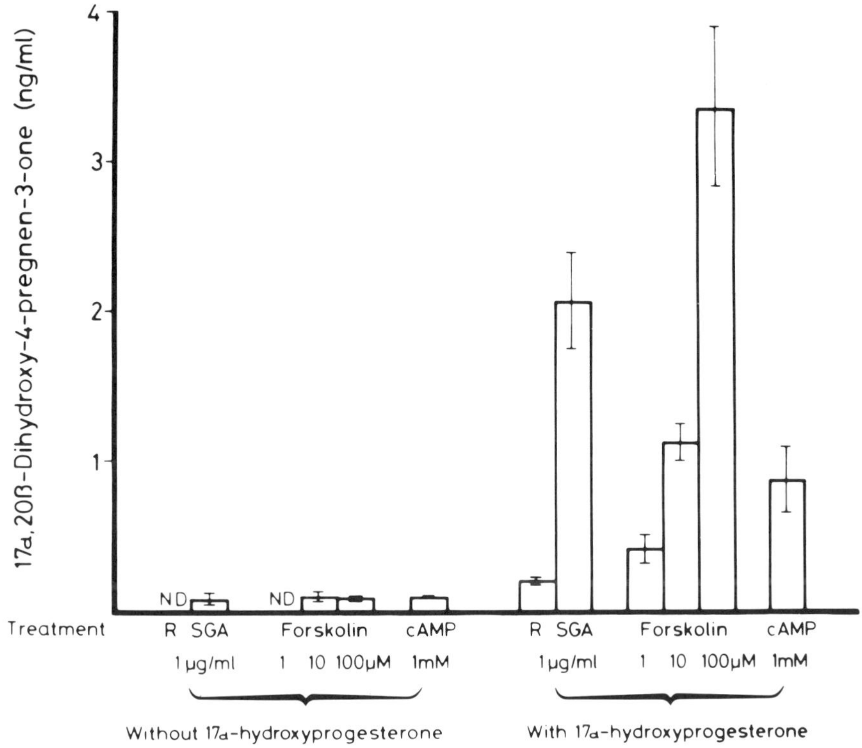

Fig. 4. Effect of chum salmon gonadotropin (SGA), forskolin, and dibutyryl cAMP (CAMP) on 17α,20β-dihydroxy-4-pregnen-3-one production by amago salmon granulosa cells with or without 17α-hydroxyprogesterone (100 ng/ml).

shown that puromycin, a translational inhibitor, prevented progesterone-induced maturation in oocytes of the sturgeon, *Acipenser stellatus*, but that actinomycin D did not. This was confirmed by subsequent experiments with several species of teleosts (see Goetz[45]).

C. Amphibians

The mechanism of action of Gns in inducing oocyte maturation in amphibians appears to be analogous to that in teleosts. Defolliculated amphibian oocytes are incapable of responding to Gn, whereas their response to steroids such as progesterone is unimpaired. When cyanoketone was added to Gn-treated *Rana pipiens* follicles incubated *in vitro*, oocyte maturation was inhibited;[129] this inhibition could be overcome by addition of exogenous progesterone. These results strongly suggest that the conversion of pregnenolone to progesterone is a necessary step in Gn-induced meiotic maturation of frog oocytes.

The *in-vitro* effects of a wide range of steroids have been demonstrated in several amphibian species. Certain androgenic, progestational, and adrenocortical steroids, but not estrogenic steroids, induce maturation of follicle-enclosed or denuded oocytes.[55,115,128] In most anuran species studied, progesterone has been found to be one of the most potent inducers of maturation (see Schuetz,[117] Masui and Clarke[83]). It also has been demonstrated that progesterone is an effective inducer of oocyte maturation *in vitro* in a urodele, *Notophthalmus viridescens*.[104] Morrill and Block,[88] who examined the structure-function relationship of various steroids relative to their capacity for inducing oocyte maturation and ovulation in *Rana pipiens*, reported that the spatial arrangement of substituents on the upper, β-surface of the steroid molecule is of critical importance in the induction of nuclear breakdown. The 3,20-dione,21-ol configuration was found to be the most active in inducing oocyte maturation.

It has been reported that progesterone and other steroids are incorporated extensively into isolated oocytes and undergo rapid metabolic conversion, with the major metabolites generally being hydroxylated progestational steroids and 5α- or 5β-reduced derivatives.[106,110,111,138] Since all of the metabolites identified were less effective than progesterone in the induction of GVBD, it has been hypothesized that these conversions by amphibian oocytes represent an inactivation process and would play a role in controlling the biological activity of intracellular progesterone.

While the specific steroid (MIS) produced by amphibian ovarian follicle cells to induce oocyte maturation in response to Gn has not been identified, the results described above strongly suggest that it is progesterone or a similar steroid. This hypothesis has been reinforced by the results of measurements of blood or ovarian concentrations of progesterone. *In-vivo* treatment with Gn preparations significantly stimulated ovarian production of progesterone in *Xenopus laevis*.[38] Recent infusion studies by McCreery and Licht[85] have provided direct evidence for the important contribution of LH and progesterone to the induction of ovulation (oocyte maturation). They found that continuous infusion of GnRH into *Rana catesbeiana* caused depression of serum testosterone and estradiol levels followed by surges in LH and progesterone; ovulation was observed coincident with the surges. It also was reported that superfusion of bullfrog pituitary homogenate into *Rana pipiens* caused a significant increase in progesterone production; this progesterone increase was observed only in fragments that underwent maturation (Hubbard and Licht, personal communication). Fortune[37] reported that the capacity of *Xenopus* ovarian follicles to secrete progesterone in response to Gns increased dramatically with follicular development, with a maximal stimulation in full-grown large follicles *in vitro* (Stages V and VI) which were capable of undergoing meiotic maturation. This finding is in accord with the work of Mulner et al.,[91] who showed that Stage-VI follicles converted much more radioactive pregnenolone to progesterone than did Stage-IV follicles. It also was shown that addition of frog pituitary homogenate to *Rana pipiens* follicles *in vitro* stimulated endogenous production of progesterone;[74] peak levels of progesterone occurred 6-8 h after frog pituitary homogenate stimulation and preceded the occurrence of GVBD by 4-6 h.[74,120] Finally, in *Rana pipiens*, medium in which isolated follicle cells had previously been incubated with frog pituitary homogenate was able to induce maturation in denuded oocytes; subsequently it was found that this medium contained a large amount of progesterone. Although it remains to be determined whether isolated follicular cells secrete maturational steroids in addition to progesterone, these results provide very strong support for the hypothesis that the MIS in amphibians is progesterone.[119]

The ovarian follicle layer of amphibians is composed of three discrete cell tissue layers: surface epithelium, thecal layer, and follicle or granulosa cells.[31] However, involvement of each cellular component in Gn-stimulated oocyte maturation and steroid production is not fully understood. Microdissection studies by Schuetz and Lessman[120] have shown that, in *Rana pipiens*, removal of the surface epithelium alone or together with the thecal layer reduces frog pituitary homogenate (FPH)-induced maturation, whereas removal of the entire follicle wall completely blocks the effect of FPH. In contrast, progesterone-induced oocyte maturation was not affected by removal of any or all follicular layers. Follicles from which the surface epithelium and the thecal layer were removed produced progesterone in response to FPH, although the amount of progesterone produced by these follicles was much smaller than that produced by intact follicles. These results, together with the fact that isolated follicular cells of *Rana* are capable of secreting progesterone in response to FPH,[119] indicate that the major site of Gn action on progesterone production is the follicular (granulosa) cell. However, the extent to which the surface epithelium and thecal layers contribute to the production of progesterone is still not clear.

The site of progesterone action in inducing oocyte maturation in amphibians is considered to be at or near the oocyte surface (see Maller and Krebs,[80] Maller[76]). It has been demonstrated that progesterone is ineffective in inducing oocyte maturation when microinjected into oocytes of R. pipiens,[84,126,127] and X. laevis,[29,55] but that steroid applied externally is effective. A surface effect of steroids also has been postulated as a result of studies involving the use of hormone linked to agarose beads or to a polymer.[4,44,49] Furthermore, the animal pole of Rana oocytes was found to be much more sensitive to progesterone than the vegetal pole.[18] Recently, two reports have described specific progesterone receptors in X. laevis oocytes. Sadler and Maller[109], using photoaffinity labeling with the synthetic progestogen, 17,21-dimethyl-19-norpregn-4,9-diene-3,20-dione (R5020), identified a 110,000-dalton protein as a specific steroid receptor on the oocyte surface. Using a similar approach with the same species, Blondeau and Baulieu[7] reported a 30,000-dalton peptide as a progesterone receptor. This receptor was not present within the melanosome-enriched, microsomal or cytosol fractions. Thus, maturational steroids appear to act initially at the cell surface, with subsequent cytoplasmic events regulated at a post-transcriptional level during amphibian oocyte maturation; this mode of action of the steroid hormone contrasts to the generally accepted transcriptional mechanism in somatic target tissues (see Gorski and Gannon,[46] Schrader[114]).

Tso et al.,[141] however, reported that microinjection of progesterone dissolved in paraffin oil induces the reinitiation of oocyte maturation in the Xenopus oocyte. This result led them to suggest that progesterone acts initially at an intracellular site to trigger maturation of Xenopus oocytes. In support of this hypothesis, specific progesterone receptors were demonstrated to be present within melanosomes of Xenopus[6,55] and Pleurodeles waltl[101] oocytes. The physiological significance of this melanosome receptor is doubtful, especially in light of the fact that oocytes from the albino mutant of Xenopus, completely devoid of melanosomes as well as premelanosomes, are capable of undergoing meiotic maturation.[8] In addition, it was demonstrated that the receptor-like properties of melanosomes are due to eumelanin itself rather than to an associated protein.[19]

Besides steroid hormones, some polypeptides also have been reported to initiate maturation in Xenopus oocytes; these include melittin, a simple peptide isolated from bee venom,[23] insulin,[33] and insulin-like growth factor.[34] El-Etr et al.[33] further found that modified insulin analogs were inactive and that proinsulin was less effective than insulin.

In addition, microinjection of insulin had no effect. From these results they suggested that the maturational action of insulin is mediated through an insulin receptor. In fact, Maller and Koontz[78] have demonstrated a remarkable number of high-affinity, specific, insulin-binding sites on *Xenopus* oocytes. However, based on their analysis of binding data and results using an anti-receptor antibody, these authors suggested that this insulin action is not mediated by the high-affinity binding component.

D. Reptiles

Details of the cytology of the terminal phase of oogenesis, including two meiotic divisions, have been investigated in the triploid parthenogenetic lizard *Cnemidophorus uniparens*.[20] Although there have been no reports describing the effects of hormones on oocyte maturation in reptiles, studies involving measurements of circulating steroid hormones during various reproductive cycles suggest the involvement of progesterone in the process of oocyte maturation. A pronounced surge of serum progesterone levels occurred during the nesting cycle (the time of follicular maturation and ovulation) of both green sea turtle and sea turtle[68,70,71] (see also Licht[67]). This rise was accompanied by a surge of LH; it is possible that the LH surge is responsible for the observed progesterone peak. A similar increase in circulating levels of progesterone around the time of ovulation has also been reported for the painted turtle *Chrysemys picta*,[11] the stink pot turtle *Sternotherus odoratus*,[86] the snapping turtle *Chelydra serpentina*,[65] and the Chinese cobra *Naja naja*.[9] Certainly, further comparative studies on the relationship between circulating levels of various hormones and the reproductive physiology of the animal, coupled with *in-vitro* investigations examining the effects of hormones on the induction of oocyte maturation, are required to ascertain whether progesterone is the major MIS of reptiles.

IV. MATURATION-PROMOTING FACTOR

A. Amphibians

As discussed earlier (Section III, B and C), the microinjection of maturation-inducing steroids into full-grown oocytes fails to induce maturation. These results suggest that maturational steroids act indirectly on germinal vesicles to induce resumption of meiosis. It was therefore anticipated that some cytoplasmic factor which directly brings about oocyte maturation is newly produced under the influence of MIS. Dettlaff et al.[25] first reported that nuclear sap taken from the germinal vesicle of Bufo viridis oocytes at the time of GVBD could by itself induce maturation when microinjected into immature oocytes. Further, Masui and Markert[84] demonstrated that the cytoplasm taken from progesterone-treated oocytes induced GVBD and all subsequent maturational events when injected into untreated immature oocytes. This cytoplasmic factor has been designated maturation-promoting factor (MPF). It was found that enucleation affected neither the time of appearance nor the effectiveness of MPF in inducing maturation in recipient oocytes.[107,112]

MPF appears shortly before GVBD and attains its maximal activity at the time of GVBD. The activity remains at a high level until the oocytes are activated, at which time it begins to decrease rapidly.[84] Masui and Markert[84] further demonstrated that repeated serial transfers of cytoplasm from progesterone-treated donor oocytes to untreated recipient oocytes continued to induce maturation in secondary or tertiary recipients, despite the fact that the initial donor oocyte was the only one exposed to progesterone. Based on this observation, these authors hypothesized that MPF is produced by autocatalytic amplification. This was later confirmed in oocytes of X. laevis.[30,107]

In Xenopus oocytes, it appears that MPF can induce GVBD when injected into first-recipient oocytes pretreated with cycloheximide, indicating that the action of MPF to induce GVBD does not depend on protein synthesis.[30,144] This is in contrast with the initial production of MPF triggered by progesterone, which involves synthesis of a new protein called the "initiator."[144]

The extraction of MPF from maturing oocytes is difficult since MPF activity is so unstable that it is easily lost when oocytes are homogenized. Despite this, some studies revealed that Rana pipiens MPF was Mg^{2+}-dependent and Ca^{2+}-sensitive, and was associated with protein but not with RNA.[145] The crude preparation exhibited three different peaks on linear sucrose gradients, corresponding to S values of approximately 32S, 15S, and 4S. More recently, Wu and Gerhart[149] partially purified (20- and 30-fold) and characterized X. laevis MPF by ammonium sulfate precipitation and chromatography on pentyl-agarose and arginine-agarose. The final material was a protein with a molecular weight of approximately 100,000 daltons, and injection of 50 ng of this proteinaceous material induced maturation in 50% of the recipient oocytes. Subsequently it was found that small phosphorylated molecules, which may act to inhibit phosphoprotein phosphatases, stabilized MPF activity. Furthermore, ATP enhanced MPF activity at least two-fold. Taken together, these results suggest that Xenopus MPF may be a phosphoprotein.

It has been reported that amphibian MPF is not species-specific and in fact not even order-specific;[107] mature cytoplasm from *Rana* or *Xenopus* (anuran) oocytes induces maturation when injected into *Ambystoma* (urodele) oocytes and *vice versa*.

B. Others

Although the investigation of MPF in vertebrates has been restricted almost exclusively to amphibians, some evidence is also available from mammals and fishes. The appearance of a cytoplasmic MPF in maturing mouse oocytes was first suggested by oocyte fusion experiments.[2,3] A similar oocyte fusion experiment by Clarke and Masui[17] confirmed the presence of MPF in mouse oocytes at metaphase II. MPF also has been demonstrated in sturgeon.[24] Furthermore, MPF activity was detected in the cytoplasm of goldfish oocytes matured by treatment with 17α,20β-diOHprog *in vitro* (Nagahama and Kishimoto, unpublished).

As discussed in Section IV, A, amphibian MPF appears to lack species specificity. In fact, recent studies indicate that MPF is similar even among vertebrates and invertebrates. MPF obtained from mature oocytes of mouse, *Xenopus*, *Bufo bufo* or goldfish was found to induce maturation of immature starfish, while starfish MPF was effective in inducing maturation of *Xenopus* oocytes[63,64] (and Kishimoto, unpublished). Cytoplasmic extracts of meiotically mature mouse oocytes induced GVBD in recipient immature *Xenopus* oocytes.[130] In addition, MPF activity also has been detected in the cytoplasm from fertilized eggs and early embryos of *Xenopus*[146] and starfish.[63] This activity changes in parallel with the mitotic cell cycle; in the embryos, the peak MPF activity occurs immediately before mitosis.

It also has been shown that MPF activity is present in mammalian mitotic somatic cells and that this factor is not species specific. An extract of mammalian cultured cells such as HeLa cells and Chinese hamster ovarian (CHO) cells synchronized in M-phase of the cell cycle could induce maturation when injected into *Xenopus* oocytes.[99,135] Similarly, cytoplasmic extracts of CHO and V-79 cells of Chinese hamster origin in M-phase were effective in inducing GVBD in starfish oocytes, whereas extracts of CHO cells in S-phase had little activity.[63] Finally, Miake-Lye et al.[87] found that the MPF purified from cytoplasm of unfertilized *Xenopus* mature oocytes could induce the breakdown of embryonic nuclear envelopes during the early cleavage cycle; this is the first demonstration that MPF purified from unfertilized mature oocytes can cause an effect in somatic cells. From these results, it appears that MPF is ubiquitous both in mitotic and meiotic cells. Thus, MPF is not merely maturation-promoting factor, but can be referred to as a more general factor responsible for the initiation of nuclear membrane breakdown and subsequent cell division of both mitosis and meiosis.

V. MATURATION-INHIBITING SUBSTANCES

A. Overview of Mammals

As discussed in Section II, A, isolated mammalian oocytes, unlike those of nonmammalian species, undergo spontaneous maturation *in vitro*. This observation led to the suggestion that some component of the antral follicle inhibits oocyte maturation. Much attention has been focused on three potential physiological inhibitors of meiotic maturation: these are a factor(s) in follicular fluid, cAMP, and steroid hormones. However, discussion of a physiological role for steroid hormones in the maintenance of meiotic arrest will be omitted, since the literature in this area has been inconsistent.

It has been demonstrated that follicular fluid of several mammalian species contains a meiotic-arresting substance, termed "oocyte maturation inhibitor" (OMI). Porcine OMI has been purified partially (about 15,000 fold) as a polypeptide with a molecular weight of less than 2,000 daltons,[140] and apparently originates from granulosa cells. OMI activity appears not to be species-specific. Human follicular fluid from mature Graafian follicles was found to be inhibitory to progesterone-induced meiotic maturation of *Xenopus* oocytes.[12] At present, the exact mechanism(s) by which OMI exerts its action on oocyte maturation remains obscure (see Channing et al.[15]). It has been shown, though, that the inhibitory action of OMI is dependent upon an intact cumulus oophorus, since GVBD was suppressed in cumulus cell-enclosed oocytes but not in denuded oocytes. Furthermore, it has been reported recently that follicular fluid from mature, fertilizable human follicles contains significantly less OMI activity than similar fluid from immature oocytes.[14]

The evidence supporting a role for cAMP in the control of maturation of mammalian oocytes comes mainly from the finding that treatment of both cumulus-enclosed and denuded mammalian oocytes with cAMP analogs or phosphodiesterase inhibitors blocks spontaneous maturation.[16,21] In mouse oocytes, a decrease in cAMP levels occurs prior to GVBD.[121] Dekel and Beers[21] reported that Gns reversed the inhibitory action of dbcAMP and IBMX on the maturation of cumulus cell-enclosed rat oocytes. In contrast, the Gn failed to overcome the inhibitory effect of either dbcAMP or IBMX on the maturation of denuded rat oocytes.[22] Based on these results, it has been hypothesized that meiotic arrest is maintained by the passage of cAMP from the cumulus cells into the oocyte through gap junctions.[22,43] It also was proposed that Gns trigger resumption of meiosis by the physical disruption of intercellular communication between cumulus cells and oocytes.

In contrast to this hypothesis, it has been demonstrated recently that treatment of cumulus cell-oocyte complexes with agents which elevated cAMP levels in the complexes but not in denuded oocytes, resulted in a significant inhibition of GVBD. Furthermore, a suboptimal concentration of dbcAMP had a greater inhibitory effect on cumulus cell-enclosed oocytes than on denuded oocytes.[36] These results, taken together, suggest that cumulus cells play a significant role in suppressing oocyte maturation by a cAMP-dependent process, and that this process may involve the generation and/or activation of a maturation-inhibiting factor other than cAMP itself[28] (see also Eppig and Downs[35]). It was shown that this inhibitory factor is present both in porcine follicular fluid and serum. Recently, Downs et al.[27] have identified hypoxanthine as the principal inhibitory component of porcine follicular fluid. It is important to establish if similar inhibitory compounds exist in follicular fluid or serum from other species.

B. Fishes

It was found that exogenous dbcAMP and agents, such as IBMX and theophylline, that elevate intracellular cAMP levels all blocked both Gn (human chorionic gonadotropin, hCG) and 17α,20β-diOHprog-induced meiotic maturation of follicle-enclosed goldfish oocytes (Kagawa and Nagahama, unpublished). Similar results were obtained with these compounds in medaka oocytes. In addition, forskolin was found to block 17α,20β-diOH-prog-induced GVBD in both follicle-enclosed and denuded oocytes (Iwamatsu, personal communication). Taken together, these results suggest that maintenance of elevated intracellular levels of cAMP is responsible for prophase arrest in these species.

Iwamatsu and Takama[54] have demonstrated the presence of a factor in rabbit serum which inhibits Gn- and steroid-induced GVBD in follicle-enclosed medaka oocytes. This inhibitory activity was not affected by treatment with proteolytic enzymes or heat treatment, suggesting that this factor is not a peptide. Activity was dialyzable, with a molecular weight of less than 3,500 daltons, and could be removed by charcoal extraction. The presence of a similar activity was confirmed in sera from chicken and lizard as well as in several mammalian species.[51] Later, it was demonstrated that this factor exhibits maturational activity; when follicle-enclosed medaka oocytes were exposed to these sera for 9-12 h and then incubated in serum-free medium for an additional 12 h, oocytes underwent GVBD.[52] It is interesting to note that the nature of this serum factor described by Iwamatsu closely resembles that of an inhibitor of oocyte maturation present in porcine follicular fluid that recently was identified as hypoxanthine[27] (see Section V, A).

Certain steroid hormones also have been reported to block hormone-induced oocyte maturation *in vitro*. In rainbow trout, estradiol was shown to prevent Gn-induced GVBD, but not 17α,20β-diOHprog-induced GVBD.[56] Later, this inhibitory effect of estradiol was found to be due to the blockage of 17α,20β-diOHprog production in the follicular layer.[59]

C. Amphibians

Cyclic AMP may be an important intracellular regulator of meiotic maturation in amphibian oocytes. Exogenous dbcAMP and agents that elevate oocyte cAMP levels have been reported to inhibit progesterone-induced GVBD.[89,113] These observations suggest that the prophase block is maintained by an elevated level of oocyte cAMP. Consistent with this hypothesis, it has been reported that soon after exposure to progesterone the level of oocyte cAMP decreases rapidly,[10,77,89,113,131] due at least in part to inhibition of adenylate cyclase by progesterone.[108] Thus, a decrease in cAMP may be a necessary step in the release of the prophase block of the oocyte.

Maller and Krebs,[79] using microinjection techniques, have shown that the regulatory subunit of cAMP-dependent protein kinase or the heat-stable inhibitor of the catalytic subunit of cAMP-dependent protein kinase mimics progesterone action by inducing maturation of *X. laevis* oocytes, whereas the catalytic subunit inhibits progesterone-induced maturation. These results suggest that decreased cAMP levels in the oocytes immediately following hormone stimulation may result in the suppression of cAMP-dependent protein kinase, thus enhancing the dephosphorylation of a phosphoprotein. It is interesting to note that a burst of non-cAMP-dependent protein phosphorylation occurs in progesterone-treated oocytes shortly before GVBD or immediately following MPF injection.[81] Although the precise relationship between these biochemical events and GVBD is still unclear, these processes may be an essential step in the formation of MPF. Several models to account for the mechanisms by which progesterone induces meiotic maturation in amphibian oocytes have been proposed (see reviews, Smith,[125] Schuetz,[118] Masui and Clarke,[83] Maller and Krebs,[80] Morrill et al.,[90] Baulieu and Schorderet-Slatkine[5]).

Estradiol, under *in-vitro* conditions, also has been shown to block maturation of amphibian oocytes. Prolonged pretreatment with estradiol inhibited pituitary homogenate-induced maturation in follicle-enclosed oocytes of *R. pipiens*.[73] Later, this inhibitory effect of estrogen was shown to be due to blockage of progesterone production in follicular cells.[74] Estradiol also prevents progesterone-induced GVBD in both follicle-enclosed and denuded oocytes. Thus, it is possible that estrogen also acts directly on the oocyte to block the biological action of progesterone.[73]

VI. SUMMARY AND FUTURE RESEARCH DIRECTION

It appears that a preovulatory surge of blood levels of Gns, in most cases LH, occurs in nearly all the vertebrates and is responsible for triggering oocyte maturation. In teleosts and amphibians, however, maturational action of Gn is not direct, but is mediated by the follicular production of MIS (Fig. 5). Accumulating evidence suggests that progesterone is the natural MIS of amphibians. It is now well-established in salmonids that 17α,20β-diOHprog is the major MIS. A two-cell type model has been proposed for the follicular production of 17α,20β-diOHprog. It was further demonstrated that the major action of Gns in inducing oocyte maturation in salmonids is to stimulate the activity of 20β-HSD in the granulosa cells due to the *de novo* synthesis of this enzyme through a mechanism involving RNA synthesis. Thus, the induction of 20β-HSD by Gn in salmonid granulosa cells is a good example of a differentiated function expressed by the cells in response to Gn. Isolated salmonid granulosa cells will provide a useful model to understand the basic mechanisms underlying Gn-induced steroidogenesis and also the genetic control of granulosa cell differentiation.

One of the interesting features of study of oocyte maturation is that maturational steroids act at the surface of the oocyte. Progesterone receptors recently have been demonstrated at the surface of *Xenopus* oocytes. This mode of action of steroid hormones contrasts with the generally accepted mechanism in somatic target cells. However, very few data are available for other species. Certainly, the oocyte surface receptor of 17α,20β-diOHprog merits further study. Unfortunately, at present almost nothing is known about the hormonal control of oocyte maturation in cyclostomes and elasmobranchs, and in reptiles only a little is known. Although some data are available to account for the mechanisms

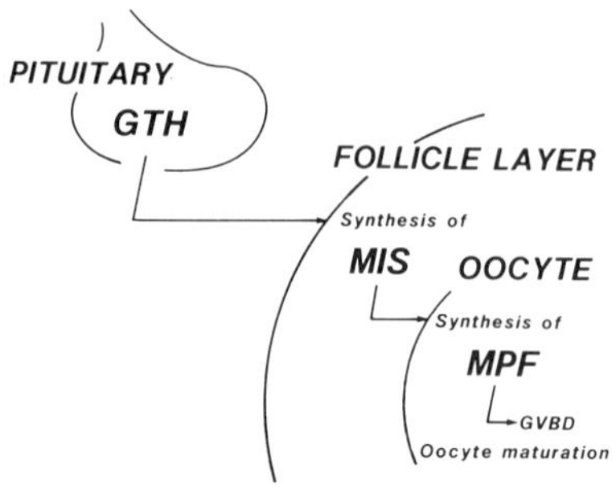

Fig. 5. Hormonal control of oocyte maturation in teleosts and amphibians. Three major mediators, gonadotropin (GTH), maturation-inducing substance (MIS) and maturation-promoting factor (MPF) are involved. They function sequentially at the level of the follicle layer, the oocyte surface and the oocyte cytoplasm.

of maintenance of meiotic arrest in mammalian oocytes, very little is known about the action of Gn in inducing meiotic resumption. Thus, further comparative studies are needed for a complete understanding of the hormonal control of oocyte maturation in vertebrates.

One of the important questions in the study of oocyte maturation concerns the mechanism controlling the process of GVBD. It has been shown that the early steps following MIS action involve increases in free calcium levels and decreases in oocyte cAMP levels, followed by the formation of the major cytoplasmic mediator of MIS, MPF. However, the chemical nature of MPF is not yet known. Recent studies suggest that MPF is not merely a maturation-promoting factor common to maturing oocytes of both invertebrates and vertebrates, but plays a more general role in initiating nuclear breakdown and subsequent cell division both in meiosis and mitosis. Thus, the determination of the chemical nature and the biochemical action of MPF is considered to be most important, not only for an understanding of the mechanisms of meiotic maturation but also for understanding the initiation of mitosis in general. Since the data presented for MPF have come almost exclusively from studies with amphibian oocytes, research along this line should be conducted in more species.

VII. ACKNOWLEDGMENTS

Thanks is extended to Dr. A. W. Schuetz for supplying copies of manuscripts in press, and Drs. P. Licht, T. Iwamatsu, G. M. Hubbard and M. Kobayashi for permission to include some of their unpublished observations. Thanks is also extended to Dr. R. E. Jones for helpful comments during preparation of this chapter. The research of Y. Nagahama was supported by Grant-in-Aid for Special Project Research (Project No. 58119008 and 60129035) and Scientific Research from the Ministry of Education, Sciences, and Culture of Japan.

VIII. REFERENCES

1. Ainsworth, L., Tsang, B.K., Downey, B.R., Marcus, G.T. and Armstrong, D.T. (1980). Interrelationships between follicular fluid steroid levels, gonadotropic stimuli and oocyte maturation during preovulatory development of porcine follicles. Biol. Reprod. 23, 621-627.
2. Balakier, H. (1978). Induction of maturation in small oocytes from sexually immature mice by fusion with meiotic or mitotic cells. Exp. Cell Res. 112, 137-141.
3. Balakier, H. and Czolowska, R. (1977). Cytoplasmic control of nuclear maturation in mouse oocytes. Exp. Cell Res. 110, 466-469.
4. Baulieu, E.-E., Godeau, F., Schorderet, M. and Schorderet-Slatkine, S. (1978). Steroid-induced meiotic division in _Xenopus laevis_ oocytes: surface and calcium. Nature 275, 593-598.
5. Baulieu, E.-E. and Schorderet-Slatkine, S. (1983). Steroid and peptide control mechanisms in membranes of _Xenopus laevis_ oocytes resuming meiotic division. Molecular biology of egg maturation. (Ciba Foundation symposium 98), pp. 137-158. Pitman Books, London.
6. Bellé, R., Schorderet-Slatkine, S., Drury, K.C. and Ozon, R. (1975). _In vitro_ progesterone binding to _Xenopus laevis_ oocytes. Gen. Comp. Endocrinol. 25,339-345.
7. Blondeau, J.-P. and Baulieu, E.-E. (1984). Progesterone receptor characterized by photoaffinity labelling in the plasma membrane of _Xenopus laevis_ oocytes. Biochem. J. 219, 785-792.
8. Bluemink, J.G. and Hoperskaya, O.A. (1975). Ultrastructural evidence for the absence of premelanosomes in eggs of the albino mutant (a^P) of _Xenopus laevis_. Wilhelm Roux Arch. Entwichlungsmech. Org. 177, 75-79.
9. Bona-Gallo, A., Licht, P., MacKenzie, D.S. and Lofts, B. (1980). Annual cycles in levels of pituitary and plasma gonadotropin, gonadal steroids, and thyroid activity in the Chinese cobra (_Naja naja_). Gen. Comp. Endocrinol. 42, 477-493.
10. Bravo, R., Otero, C., Allende, C.C., and Allende, J.E. (1978). Amphibian oocyte maturation and protein synthesis: Related inhibition by cyclic AMP, theophylline, and papaverine. Proc. Natl. Acad. Sci. USA 75, 1242-1246.
11. Callard, I.P., Lance, V., Salhanick, A.R. and Barad, D. (1978). The annual ovarian cycle of _Chrysemys picta_: correlated changes in plasma steroids and parameters of vitellogenesis. Gen. Comp. Endocrinol. 35,245-257.
12. Cameron, I.L., Jum, J.B., Nations, C., Asch, R.H. and Silverman, A.Y. (1983). Assay for characterization of human follicular oocyte maturation inhibitor using Xenopus oocytes. Biol. Reprod. 28, 817-822.
13. Campbell, C.M., Fostier, A., Jalabert, B. and Truscott, B. (1980). Identification and quantification of steroids in the serum of rainbow trout during spermiation and oocyte maturation. J. Endocr. 85, 371-378.

14. Channing, C. P., Liu, C.-Q., Jones, G.S. and Jones, H. (1983). Decline of follicular oocyte maturation inhibitor coincident with maturation and achievement of fertilizability of oocytes recovered at midcycle of gonadotropin-treated women. Proc. Natl. Acad. Sci. USA 80, 4184-4188.
15. Channing, C.P., Schaerf, F.W., Anderson, L.D. and Tsafriri, A. (1980). Ovarian follicular and luteal physiology. Int. Rev. Physiol. 22, 117-201.
16. Cho, W.K., Stern, S. and Biggers, J.D. (1974). Inhibitory effect of dibutyryl cAMP on mouse maturation in vitro. J. Exp. Zool. 187, 383-386.
17. Clarke, H.J. and Masui, Y. (1985). Inhibition by dibutyryl cyclic AMP of the transition to metaphase of mouse oocyte nuclei and its reversal by cell fusion to metaphase oocytes. Dev. Biol. 108, 32-37.
18. Cloud, J.G. and Schuetz, A.W. (1977). Interaction of progesterone with all or isolated portions of the amphibian (*Rana pipiens*) oocyte surface. Dev. Biol. 60, 359-370.
19. Coffman, G.K., Keem, K. and Smith, L.D. (1979). The progesterone receptor-like properties of *Xenopus laevis* oocyte melanosomes are probably due to eumelanin. J. Exp. Zool. 207, 375-382.
20. Cuellar, O. (1971). Reproduction and the mechanism of meiotic restitution in the parthenogenetic lizard *Cnemidophorus uniparens*. J. Morph. 133, 139-166.
21. Dekel, N. and Beers, W.H. (1978). Rat oocyte maturation *in vitro*: Relief of cyclic AMP inhibition by gonadotropins. Proc. Natl. Acad. Sci. USA 75, 4369-4373.
22. Dekel, N. and Beers, W.H. (1980). Development of the rat oocyte *in vitro*: inhibition and induction of maturation in the presence or absence of the cumulus oophorus. Dev. Biol. 75, 247-254.
23. Deshpande, A.T. and Koide, S.S. (1982). In vitro induction of germinal vesicle breakdown in *Xenopus laevis* oocytes by melittin. Differentiation 21, 127-132.
24. Dettlaff, T.A., Felgengauer, P.E. and Chulitzkaya, E.V. (1977). Effect of cycloheximide on disintegration of the germinal vesicle membrane and changes of the cortical layer in the oocytes of *Xenopus laevis* and *Acipenser stellatus* during their maturation under the influence of active cytoplasm. Ontogenesis 8, 478-486.
25. Dettlaff, T.A., Nikitina, L.A. and Stroeva, O.G. (1964). The role of the germinal vesicle in oocyte maturation in anurans as revealed by the removal and transplantation of nuclei. J. Embryol. Exp. Morph. 12, 851-873.
26. Dettlaff, T.A. and Skoblina, M.N. (1969). The role of germinal vesicle in the process of oocyte maturation in *Anura* and *Acipenseridae*. Ann. Embryol. Morphog. Suppl. 1, 133-151.
27. Downs, S.M., Coleman, D.L., Ward-Bailey, P.F. and Eppig, J.J. (1985). Hypoxanthine is the principal inhibitor of murine oocyte maturation in a low molecular weight fraction of porcine follicular fluid. Proc. Natl. Acad. Sci. USA 82, 454-458.
28. Downs, S.M. and Eppig, J.J. (1985). A follicular fluid component prevents gonadotropin reversal of cyclic adenosine monophosphate-dependent meiotic arrest in murine oocytes. Gamete Res. 11, 83-97.
29. Drury, K.C. and Schorderet-Slatkine, S. (1975). Les effets du cortisol et de la progesterone directement injectés à l'intérieur de l'ovocyte de *Xenopus laevis*. C. R. Acad. Sci. Paris 280, 1273-1275.
30. Drury, K.C. and Schorderet-Slatkine, S. (1975). Effects of cycloheximide on the "autocatalytic" nature of the maturation promoting factor (MPF) in oocytes of *Xenopus laevis*. Cell 4, 269-274.

31. Dumont, J.N. and Brummett, A.R. (1978). Oogenesis in Xenopus laevis (Daudin). V. Relationships between developing oocytes and their investing follicular tissues. J. Morph. 155, 73-98.
32. Edwards, R.G. (1965). Maturation in vitro of mouse, sheep, cow, pig, rhesus monkey and human ovarian oocytes. Nature 208, 349-351.
33. El-Etr, M., Schorderet-Slatkine, S. and Baulieu, E.-E. (1979). Meiotic maturation in Xenopus laevis oocytes initiated by insulin. Science 205, 1397-1399.
34. El-Etr, M., Schorderet-Slatkine, S. and Baulieu, E.-E. (1980). The role of zinc and follicle cells in insulin-initiated meiotic maturation of Xenopus laevis oocytes. Science 210, 928-930.
35. Eppig, J.J. and Downs, S.M. (1984). Chemical signals that regulate mammalian oocyte maturation. Biol. Reprod. 30, 1-11.
36. Eppig, J.J., Freter, R.R., Ward-Bailey, P.F. and Schultz, R.M. (1983). Inhibition of oocyte maturation in the mouse: participation of cAMP, steroid hormones, and a putative maturation-inhibitory factor. Dev. Biol. 100, 39-49.
37. Fortune, J.E. (1983). Steroid production by Xenopus ovarian follicles at different developmental stages. Dev. Biol. 99, 502-509.
38. Fortune, J.E., Concannon, P.W. and Hansel, W. (1975). Ovarian progesterone levels during in vitro oocyte maturation and ovulation in Xenopus laevis. Biol. Reprod. 13, 561-567.
39. Fostier, A., Breton, B., Jalabert, B. and Marcuzzi, O. (1981). Evolution des niveaux plasmatiques de la gonadotropine glycoprotéique et de la 17α-hydroxy-20βdihydroprogesterone au cours de la maturation et de l'ovulation chez la Triute arc-en-ciel, Salmo gairdnerii. C. R. Acad. Sci. Paris. Ser. III 293, 817-820.
40. Fostier, A., Jalabert, B., Billard, B., Breton, B. and Zohar, Y. (1983). The gonadal steroids. In "Fish Physiology" (W.S. Hoar, D.J.Randall and E.M. Donaldson, eds.). Vol. IXA, pp. 277-372. Academic Press, New York.
41. Fostier, A., Jalabert, B., Campbell, C., Terqui, M. and Breton, B. (1981). Cinétique de libération in vitro de 17αhydroxy-20β-dihydroprogesterone par des follicules de Truite arc-en-ciel, Salmo gairdnerii. C. R. Acad. Sci. Paris Ser. III 292, 777-780.
42. Fostier, A., Jalabert, B. and Terqui, M. (1973). Action prédominante d'un dérivé hydroxylé de la progestérone sur la maturation in vitro des ovocytes de la Truite arc-en-ciel, Salmo gairdnerii. C. R. Acad. Sci. Paris Ser. D 277, 421-424.
43. Gilula, N.B., Epstein, M.L. and Beers, W.H. (1978). Cell-to-cell communication and ovulation. A study of the cumulus-oocyte complex. J. Cell Biol. 78, 58-75.
44. Godeau, J.F., Schorderet-Slatkine, S., Hubert, P. and Baulieu, E.-E. (1978). Induction of maturation in Xenopus laevis oocytes by a steroid linked to a polymer. Proc. Natl. Acad. Sci. USA 75, 2353-2357.
45. Goetz, F.W. (1983). Hormonal control of oocyte final maturation and ovulation in fishes. In "Fish Physiology" (W.S. Hoar, D.J. Randall and E.M. Donaldson, eds.), Vol. IXB, pp. 117-170. Academic Press, New York.
46. Gorski, J. and Gannon, F. (1976). Current models of steroid hormone action: a critique. Ann. Rev. Physiol. 38, 425-450.
47. Hirose, K., Nagahama, Y., Adachi, S. and Wakabayashi, K. (1983). Changes in serum concentrations of gonadotropin, 17α-hydroxyprogesterone and 17α,20β-dihydroxy-4-pregnen-3-one during synthetic LH-RH-induced final oocyte maturation and ovulation in the ayu Plecoglossus altivelis. Bull. Jap. Soc. Sci. Fish. 49, 1165-1169.

48. Idler, D.R. and Ng, T.B. (1983). Teleost gonadotropins: isolation, biochemistry, and function. In "Fish Physiology" (W.S. Hoar, D.J. Randall and E.M. Donaldson, eds.), Vol. IXA, pp 187-221. Academic Press, New York.
49. Ishikawa, K., Hanaoka, Y., Kondo, Y. and Imai, K. (1977). Primary action of steroid hormone at the surface of amphibian oocyte in the induction of germinal vesicle breakdown. Mol. Cell. Endocrinol. 9, 91-100.
50. Iwamatsu, T. (1980). Studies on oocyte maturation of the medaka, *Oryzias latipes*. VIII. Role of follicular constituents in gonadotropin- and steroid-induced maturation of oocytes in vitro. J. Exp. Zool. 211, 231-239.
51. Iwamatsu, T. (1981). Inhibitory effect of sera from various animals on *in vitro* maturation of the oocyte. Annot. Zool. Japon. 54, 157-163.
52. Iwamatsu, T. (1983). Evidence for the presence of a low molecular weight factor in sera from various animals that induces *in vitro* maturation of fish oocytes. Develop. Growth Differ. 25, 211-216.
53. Iwamatsu, T. and Ohta, T. (1981). On a relationship between oocyte and follicle cells around the time of ovulation in the medaka, *Oryzias latipes*. Annot. Zool. Japon. 54, 17-29.
54. Iwamatsu, T. and Takama, K. (1980). Inhibitory effect of rabbit serum factor on *in vitro* oocyte maturation of the teleost fish, *Oryzias latipes*. Develop. Growth Differ. 22,229-235.
55. Jacobelli, S., Hanocq, J., Baltus, E. and Brachet, J. (1974). Hormone-induced maturation of *Xenopus laevis* oocytes: effects of different steroids and study of the properties of a progesterone receptor. Differentiation 2, 129-135.
56. Jalabert, B. (1975). Modulation par différents stéroides non maturants de l'efficacite de la 17α-hydroxy-20β-dihydroprogestérone ou d'un extrait gonadotrope sur la maturation intrafolliculaire *in vitro* des ovocytes de la Truite arc-en-ciel *Salmo gairdneri*. C. R. Acad. Sci. Paris Ser. D. 281, 811-814.
57. Jalabert, B. (1976). *In vitro* oocyte maturation and ovulation in rainbow trout (*Salmo gairdneri*), northern pike (*Esox lucius*), and goldfish (*Carassius auratus*). J. Fish. Res. Board Can. 33, 974-988.
58. Jalabert, B. and Breton, B. (1980). Evolution de la gonadotropine plasmatique t-GtH après l'ovulation chez la Truite arc-en-ciel (*Salmo gairdneri* R.) et influence de la rétention des ovules. C. R. Acad. Sci. Paris Ser. D 290, 799-801.
59. Jalabert, B. and Fostier, A. (1984). The modulatory effect *in vitro* of estradiol-17α, testosterone or cortisol on the output of 17α-hydroxy-20β-dihydroprogesterone by rainbow trout (*Salmo gairdneri*) ovarian follicles stimulated by the maturational gonadotropin s-GtH. Reprod. Nutr. Develop. 24, 127-136.
60. Kagawa, H., Young, G., Adachi, S. and Nagahama, Y. (1982). Estradiol-17β production in amago salmon (*Oncorhynchus rhodurus*) ovarian follicles: role of the thecal and granulosa cells. Gen. Comp. Endocrinol. 47, 440-448.
61. Kagawa, H., Young, G. and Nagahama, Y. (1983). Changes in plasma steroid hormone levels during gonadal maturation in female goldfish *Carassius auratus*. Bull. Jap. Soc. Sci. Fish. 49, 1783-1787.
62. Kanatani, H. and Nagahama, Y. (1980). Mediators of oocyte maturation. Biomed. Res. 1, 273-291.
63. Kime, D.E. and Dolben, I.P. (1985). Hormonal changes during induced ovulation of the carp, *Cyprinus carpio*. Gen. Comp. Endocrinol. 58, 137-149.
63a Kishimoto, T., Kuriyama, R., Kondo, H. and Kanatani, H. (1982) Generality of the action of various maturation-promoting factors. Exp. Cell Res. 137, 121-126.

64. Kishimoto, T., Yamazaki, K., Kato, Y., Koide, S.S. and Kanatani, H. (1984). Induction of starfish oocyte maturation by maturation-promoting factor of mouse and surf clam oocytes. J. Exp. Zool. 231, 293-295.
65. Lewis, J., Mahmoud, I.Y. and Klicka, J. (1979). Seasonal fluctuations in the plasma concentrations of progesterone and oestradiol-17β in the female snapping turtle (*Chelydra serpentina*). J. Endocr. 80, 127-131.
66. Licht, P. (1979). Reproductive endocrinology of reptiles and amphibians: gonadotropins. Ann. Rev. Physiol. 41, 337-351.
67. Licht, P. (1982). Endocrine patterns in the reproductive cycle of turtles. Herpetologica 38, 51-61.
68. Licht, P., Owens, D.W., Cliffton, K. and Penaflores, C. (1982). Changes in LH and progesterone associated with the nesting cycle and ovulation in the olive ridley sea turtle, *Lepidochelys olivacea*. Gen. Comp. Endocrinol. 48, 247-253.
69. Licht, P., Papkoff, H., Farmer, S.W., Muller, C.H., Tsui, H.W. and Crews, D. (1977). Evolution of gonadotropin structure and function. Rec. Prog. Hor. Res. 33, 169-248.
70. Licht, P., Rainey, W. and Cliffton, K. (1980). Serum gonadotropin and steroids associated with breeding activities in the green sea turtle, *Chelonia mydas*. II. Mating and nesting in natural populations. Gen. Comp. Endocrinol. 40, 116-122.
71. Licht, P., Wood, J., Owens, D.W. and Wood, F. (1979). Serum gonadotropins and steroids associated with breeding activities in the green sea turtle *Chelonia mydas*. I. Captive animals. Gen. Comp. Endocrinol. 39, 274-289.
72. Lieberman, M.E., Tsafriri, A., Bauminger, S., Collins, W.P., Ahren, K. and Lindner, H.R. (1976). Oocyte meiosis in cultured rat follicles during inhibition of steroidogenesis. Acta Endocrinol. 83, 151-157.
73. Lin, Y.-W.P. and Schuetz, A.W. (1983). In vitro estrogen modulation of pituitary and progesterone-induced oocyte maturation in *Rana pipiens*. J. Exp. Zool. 226, 281-291.
74. Lin, Y.-W.P. and Schuetz, A.W. (1985). Intrafollicular action of estrogen in regulating pituitary-induced ovarian progesterone synthesis and oocyte maturation in *Rana pipiens*: Temporal relationship and locus of action. Gen. Comp. Endocrinol. 58, 421-435.
75. Lindner, H.R., Tsafriri, A., Lieberman, M.E., Zor, U., Koch, Y., Bauminger, S. and Barnea, A. (1974). Gonadotropin action on cultured Graafian follicles: induction of maturation division of the mammalian oocyte and differentiation of the luteal cell. Rec. Prog. Hor. Res. 30, 79-138.
76. Maller, J. (1983). Interaction of steroids with the cyclic nucleotides system in amphibian oocytes. Adv. Cyclic Nucleotide Res. 15, 295-336.
77. Maller, J.L., Butcher, F.R. and Krebs, E.G. (1979). Early effect of progesterone on levels of cyclic adenosine 3':5'-monophosphate in *Xenopus* oocytes. J. Biol. Chem. 254, 579-582.
78. Maller, J.L. and Koontz, J.W. (1981). A study of the induction of cell division in amphibian oocytes by insulin. Dev. Biol. 85, 309-316.
79. Maller, J.L. and Krebs, E.G. (1977). Progesterone-stimulated meiotic cell division in *Xenopus* oocytes. Induction by regulatory subunit and inhibition by catalytic subunit of adenosine 3':5'-monophosphate-dependent protein kinase. J. Biol. Chem. 252, 1712-1718.
80. Maller, J.L. and Krebs, E.G. (1980). Regulation of oocyte maturation. Curr. Top. Cell Reg. 16, 271-311.

81. Maller, J., Wu, M. and Gerhart, J.C. (1977). Changes in protein phosphorylation accompanying maturation of *Xenopus laevis* oocytes. Dev. Biol. 58, 295-312.
82. Masui, Y. (1967). Relative roles of the pituitary, follicle cells, and progesterone in the induction of oocyte maturation in *Rana pipiens*. J. Exp. Zool. 166, 365-376.
83. Masui, Y. and Clarke, H.J. (1979). Oocyte maturation. Int. Rev. Cytol. 57, 185-282.
84. Masui, Y. and Markert, C.L. (1971). Cytoplasmic control of nuclear behavior during meiotic maturation of frog oocytes. J. Exp. Zool. 177, 129-146.
85. McCreery, B.R. and Licht, P. (1983). Induced ovulation and changes in pituitary responsiveness to continuous infusion of gonadotropin-releasing hormone during the ovarian cycle in the bullfrog, *Rana catesbeiana*. Biol. Reprod. 29, 863-871.
86. McPherson, R.J., Boots, L.R., MacGregor, III R. and Marion, K.R. (1982). Plasma steroids associated with seasonal reproductive changes in a multiclutched freshwater turtle, *Sternotherus odoratus*. Gen. Comp. Endocrinol. 48, 440-451.
87. Miake-Lye, R., Newport, J. and Kirschner, M. (1983). Maturation-promoting factor induces nuclear envelope breakdown in cyclohexamide-arrested embryos of *Xenopus laevis*. J. Cell Biol. 97, 81-91.
88. Morrill, G.A. and Bloch, E. (1977). Structure-function relationships of various steroids relative to induction of nuclear breakdown and ovulation in isolated amphibian oocytes. J. Steroid Biochem. 8, 133-139.
89. Morrill, G.A., Schatz, F., Kostellow, A.B. and Poupko, J.M. (1977). Changes in cyclic AMP levels in the amphibian ovarian follicle following progesterone induction of meiotic maturation. Differentiation 8, 97-104.
90. Morrill, G.A., Ziegler, D. and Kostellow, A.B. (1981). The role of Ca^{+2} and cyclic nucleotides in progesterone initiation of the meiotic divisions in amphibian oocytes. Life Sci. 29, 1821-1835.
91. Mulner, O., Thibier, C. and Ozon, R. (1978). Steroid biosynthesis by ovarian follicles of *Xenopus laevis in vitro* during oogenesis. Gen. Comp. Endocrinol. 34, 287-295.
92. Nagahama, Y. (1983). The functional morphology of teleost gonads. *In* "Fish Physiology" (W.S. Hoar, D.J. Randall and E.M. Donaldson, eds), Vol. IXA, pp. 223-275. Academic Press, New York.
93. Nagahama, Y. (1984). Mechanism of gonadotropin control of steroidogenesis in the teleost ovarian follicle. Gunma Symp. Endocrinol. 21, 167-182.
94. Nagahama, Y. and Adachi, S. (1985). Identification of maturation-inducing steroid in a teleost, the amago salmon (*Oncorhynchus rhodurus*). Dev. Biol. 109, 428-435.
95. Nagahama, Y., Hirose, K., Young, G., Adachi, S., Suzuki, K. and Tamaoki, B. (1983). Relative *in vitro* effectiveness of 17α,20β-dihydroxy-4-pregnen-3-one and other pregnene derivatives on germinal vesicle breakdown in oocytes of ayu (*Plecoglossus altivelis*), amago salmon (*Oncorhynchus rhodurus*), rainbow trout (*Salmo gairdneri*), and goldfish (*Carassius auratus*). Gen. Comp. Endocrinol. 51, 15-23.
96. Nagahama, Y., Kagawa, H., Adachi, S. and Young, G. (1985). Stimulation of 17α,20β-dihydroxy-4-pregnen-3-one production in the granulosa cells of amago salmon, *Oncorhynchus rhodurus* by cyclic nucleotides. J. Exp. Zool. 236, 371-375.
97. Nagahama, Y., Young, G. and Adachi, S. (1985). Effect of actinomycin D and cycloheximide on gonadotropin-induced 17α,20β-dihydroxy-4-preg-

nen-3-one production by intact follicles and granulosa cells of the amago salmon, Oncorhynchus rhodurus. Develop. Growth Differ. 27, 213-221.

98. Nagahama, Y., Young, G., Ueda, H., Kagawa, H. and Adachi, S. (1985). Endocrine control of final gamete maturation in salmonids. In "Salmonid Reproduction" (R.N. Iwamoto and S. Sower, eds.), pp. 8-19. Washington Sea Grant Program, University of Washington, Seattle.

99. Nelkin, B., Nichols, C. and Vogelstein, B. (1980). Protein factor(s) from mitotic CHO cells induce meiotic maturation in Xenopus laevis oocytes. FEBS Lett. 109, 233-238.

100. Osborn, J.C. and Moor, R.M. (1983). The role of steroid signals in the maturation of mammalian oocytes. J. Steroid Biochem. 19, 133-137.

101. Ozon, R. and Bellé, R. (1973). Progesterone receptor associated with the "melanosome" fraction of Pleurodeles waltlii (Urodela Amphibia). Biochim. Biophys. Acta 320, 588-593.

102. Peter, R.E. (1983). The brain and neurohormones in teleost reproduction. In "Fish Physiology" (W.S. Hoar, D.J. Randall and E.M. Donaldson, eds.), Vol. IXA, pp. 97-135. Academic Press, New York.

103. Peter, R.E., Sokolowska, M., Truscott, B., Walsh, J. and Idler, D.R. (1984). Secretion of progestogens during induced ovulation in goldfish. Can. J. Zool. 62, 1946-1949.

104. Pilone, G.B. and Humphries, A.A. (1975). Progesterone-induced in vitro maturation in oocytes of Notophthalmus viridescens (Amphibia Urodela) and some observations on cytological aspects of maturation. J. Embryol. Exp. Morph. 34, 451-466.

105. Pincus, G. and Enzmann, E.V. (1935). The comparative behavior of mammalian eggs in vivo and in vitro. I. The activation of ovarian eggs. J. Exp. Med. 62, 665-675.

106. Reynhout, J.K. and Smith, L.D. (1973). Evidence for steroid metabolism during the in vitro induction of maturation in oocytes of Rana pipiens. Dev. Biol. 30, 392-402.

107. Reynhout, J.K. and Smith, L.D. (1974). Studies on the appearance and nature of a maturation-inducing factor in the cytoplasm of amphibian oocytes exposed to progesterone. Dev. Biol. 38, 394-400.

108. Sadler, S.E. and Maller, J.L. (1981). Progesterone inhibits adenylate cyclase in Xenopus oocytes. J. Biol. Chem. 256, 6368-6373.

109. Sadler, S.E. and Maller, J.L. (1982). Identification of a steroid receptor on the surface of Xenopus oocytes by photoaffinity labeling. J. Biol. Chem. 257, 355-361.

110. Sanyal, M.K. and Sibre, E.R. (1973). Breakdown of germinal vesicle of frog oocytes with 5α-reduced products of progesterone in vitro. Proc. Soc. Exp. Biol. Med. 144, 483-486.

111. Schatz, F. and Morrill, G.A. (1975). 5α-Reductive pathway of progesterone metabolism in the amphibian ovarian cytosol. Biol. Reprod. 13, 408-414.

112. Schorderet-Slatkine, S. and Drury, K.C. (1973). Progesterone induced maturation in oocytes of Xenopus laevis. Appearance of a 'maturation promoting factor' in enucleated oocytes. Cell Differ. 2, 247-254.

113. Schorderet-Slatkine, S., Schorderet, M. and Baulieu, E.-E. (1982). Cyclic AMP-mediated control of meiosis: Effects of progesterone, cholera toxin, and membrane-active drugs in Xenopus laevis oocytes. Proc. Natl. Acad. Sci. USA 79, 850-854.

114. Schrader, W.T. (1984). New model for steroid hormone receptors? Nature 308, 17-18.

115. Schuetz, A.W. (1967). Effect of steroids on germinal vesicle of oocytes of the frog (*Rana pipiens*) *in vitro.* Proc. Soc. Exp. Biol. Med. 124, 1307-1310.
116. Schuetz, A.W. (1967). Action of hormones on germinal vesicle breakdown in frog (*Rana pipiens*) oocytes. J. Exp. Zool. 166, 347-354.
117. Schuetz, A.W. (1974). Role of hormones in oocyte maturation. Biol. Reprod. 10, 150-178.
118. Schuetz, A.W. (1979). Control mechanisms in oocyte growth and maturation. J. Steroid Biochem. 11, 695-699.
119. Schuetz, A.W. and Glad, R. (1985). In vitro production of meiosis inducing substance (MIS) by isolated amphibian (*Rana pipiens*) follicle cells. Develop. Growth Differ. 27, 201-211.
120. Schuetz, A.W. and Lessman, C. (1982). Evidence for follicle wall involvement in ovulation and progesterone production by frog (*Rana pipiens*) follicles in vitro. Differentiation 22, 79-84.
121. Schultz, R.M., Montgomery, R.R. and Belanoff, J.R. (1983). Regulation of mouse oocyte meiotic maturation: implication of a decrease in oocyte cAMP and protein dephosphorylation in commitment to resume meiosis. Dev. Biol. 97, 264-273.
122. Scott, A.P., MacKenzie, D.S. and Stacey, N.E. (1984). Endocrine changes during natural spawning in the white sucker, *Catostomus commersoni*. II. Steroid hormones. Gen. Comp. Endocrinol. 56, 349-359.
123. Scott, A.P., Sheldrick, E.L. and Flint, A.P.E. (1982). Measurement of 17α,20β-dihydroxy-4-pregnen-3-one in plasma of trout (*Salmo gairdneri* Richardson): seasonal changes and response to salmon pituitary extract. Gen. Comp. Endocrinol. 46, 444-451.
124. Scott, A.P., Sumpter, J.P. and Hardiman, P.A. (1983). Hormone changes during ovulation in the rainbow trout (*Salmo gairdneri*). Gen. Comp. Endocrinol. 49, 128-134.
125. Smith, L.D. (1975). Molecular events during oocyte maturation. *In* "Biochemistry of Animal Development" (R. Weber, ed.) Vol. III, pp. 1-46. Academic Press, San Francisco.
126. Smith, L.D. and Ecker, R.E. (1969). Role of the oocyte nucleus in physiological maturation in *Rana pipiens*. Dev. Biol. 19, 281-309.
127. Smith, L.D. and Ecker, R.E. (1971). The interaction of steroids with *Rana pipiens* oocytes in the induction of maturation. Dev. Biol. 25, 232-247.
128. Smith, L.D., Ecker, R.E. and Subtelny, S. (1968). *In vitro* induction of physiological maturation in *Rana pipiens* oocytes removed from their ovarian follicles. Dev. Biol. 17, 627-643.
129. Snyder, B.W. and Schuetz, A.W. (1973). *In vitro* evidence of steroidogenesis in the amphibian (*Rana pipiens*) ovarian follicle and its relationship to meiotic maturation and ovulation. J. Exp. Zool. 183, 333-342.
130. Sorensen, R.A., Cyert, M.S. and Pedersen, R.A. (1985). Active maturation-promoting factor is present in mature mouse oocytes. J. Cell Biol. 100, 1637-1640.
131. Speaker, M.G. and Butcher, F.R. (1977). Cyclic nucleotide fluctuations during steroid induced meiotic maturation of frog oocytes. Nature 267, 848-850.
132. Stacey, N.E., Cook, A.F. and Peter, R.E. (1979). Ovulatory surge of gonadotropin in the goldfish, *Carassius auratus*. Gen. Comp. Endocrinol. 37, 246-249.
133. Stacey, N.E., Peter, R.E., Cook, A.F., Truscott, B., Walsh, J.M. and Idler, D.R. (1983). Changes in plasma concentrations of gonadotropin, 17α-estradiol, testosterone, and 17α-hydroxy-20β-dihydro-

progesterone during spontaneous and brain lesion induced ovulation in goldfish. Can. J. Zool. 61, 2646-2652.

134. Sundararaj, B.I., Goswami, S.V. and Lamba, V. (1979). Some aspects of oocyte maturation in catfish. J. Steroid Biochem. 11, 701-707.

135. Sunkara, P.S., Wright, D.A. and Rao, P.N. (1979). Mitotic factors from mammalian cells induce germinal vesicle breakdown and chromosome condensation in amphibian oocytes. Proc. Natl. Acad. Sci. USA 76, 2799-2802.

136. Suzuki, K., Tamaoki, B. and Hirose, K. (1981). In vitro metabolism of 4-pregnenes in ovaries of a freshwater teleost, the ayu (Plecoglossus altivelis): production of 17α,20β-dihydroxy-4-pregnen-3-one and its 5α-reduced metabolites, and activation of 3β- and 20β-hydroxysteroid dehydrogenases by treatment with a fish gonadotropin. Gen. Comp. Endocrinol. 45,473-481.

137. Theofan, G. and Goetz, F.W. (1983). The in vitro synthesis of final maturational steroids by ovaries of brook trout (Salvelinus fontinalis) and yellow perch (Perca flavescens). Gen. Comp. Endocrinol. 51, 84-95.

138. Thibier-Fouchet, C., Mulner, O. and Ozon, R. (1976). Progesterone biosynthesis and metabolism by ovarian follicles and isolated oocytes of Xenopus laevis. Biol. Reprod. 14, 317-326.

139. Tsafriri, A. (1978). Oocyte maturation in mammals. In "The Vertebrate Ovary" (R.E. Jones, ed.), pp. 409-442. Plenum Press, New York.

140. Tsafriri, A., Pomerantz, S.H. and Channing, C.P. (1976). Inhibition of oocyte maturation by porcine follicular fluid: partial characterization of the inhibitor. Biol. Reprod. 14, 511-516.

141. Tso, J., Thibier, C., Mulner, O. and Ozon, R. (1982). Micro-injected progesterone reinitiates meiotic maturation of Xenopus laevis oocytes. Proc. Natl. Acad. Sci. USA 79, 5552-5556.

142. Van Der Kraak, G., Dye, H.M. and Donaldson, E.M. (1984). Effects of LH-RH and des-Gly^{10}[D-Ala^{6}]LH-RH-ethylamide on plasma sex steroid profiles in adult female coho salmon (Oncorhynchus kisutch). Gen. Comp. Endocrinol. 55, 36-45.

143. Van Der Kraak, G., Lin, H.R., Donaldson, E.M., Dye, H.M. and Hunter, G.A. (1983). Effects of LH-RH and des-Gly^{10}[D-Ala^{6}]LH-RH-ethylamide on plasma gonadotropin levels and oocyte maturation in adult female coho salmon (Oncorhynchus kisutch). Gen. Comp. Endocrinol. 49, 470-476.

144. Wasserman, W.J. and Masui, Y. (1975). Effects of cycloheximide on a cytoplasmic factor initiating meiotic maturation in Xenopus oocytes. Exp. Cell Res. 91, 381-388.

145. Wasserman, W.J. and Masui, Y. (1976). A cytoplasmic factor promoting oocyte maturation; Its extraction and preliminary characterization. Science 191, 1266-1268.

146. Wasserman, W.J. and Smith, L.D. (1978). The cyclic behavior of a cytoplasmic factor controlling nuclear membrane breakdown. J. Cell Biol. 78, R15-R22.

147. Wasserman, W.J. and Smith, L.D. (1978). Oocyte maturation in nonmammalian vertebrates. In "The Vertebrate Ovary" (R.E. Jones, ed.), pp. 443-468. Plenum Press, New York.

148. Wright, R.S. and Hunt, S.H.V. (1982). A radioimmunoassay for 17α,20β-dihydroxy-4-pregnen-3-one: its use in measuring changes in serum levels at ovulation in Atlantic salmon (Salmo salar), coho salmon (Oncorhynchus kisutch), and rainbow trout (Salmo gairdneri). Gen. Comp. Endocrinol. 47, 475-482.

149. Wu, M. and Gerhart, J.C. (1980). Partial purification and characterization of the maturation-promoting factor from eggs of Xenopus laevis. Dev. Biol. 79, 465-477.

150. Yamauchi, K., Kagawa, H., Ban, M., Kasahara, N. and Nagahama, Y. (1984). Changes in plasma estradiol-17β and 17α,20β-dihydroxy-4-pregnen-3-one levels during final oocyte maturation of the masu salmon, *Oncorhynchus masou*. Bull. Jap. Soc. Sci. Fish. 50, 2137.

151. Young, G., Crim, L.W., Kagawa, H., Kambegawa, A. and Nagahama, Y. (1983). Plasma 17α,20β-dihydroxy-4-pregnen-3-one levels during sexual maturation of amago salmon (*Oncorhynchus rhodurus*): correlation with plasma gonadotropin and *in vitro* production by ovarian follicles. Gen. Comp. Endocrinol. 51, 96-105.

152. Young, G., Kagawa, H. and Nagahama, Y. (1982). Oocyte maturation in the amago salmon (*Oncorhynchus rhodurus*): *In vitro* effects of salmon gonadotropin, steroids, and cyanoketone (an inhibitor of 3β-hydroxy-Δ^5-steroid dehydrogenase). J. Exp. Zool. 224, 265-275.

153. Young, G., Kagawa, H. and Nagahama, Y. (1983). Evidence for a decrease in aromatase activity in the ovarian granulosa cells of amago salmon (*Oncorhynchus rhodurus*) associated with final oocyte maturation. Biol. Reprod. 29, 310-315.

154. Young, G., Kagawa, H. and Nagahama, Y. (1984). Role of the thecal and granulosa cells in the production of maturation-inducing steroid (MIS) by ovarian follicles of salmonid fishes. Gen. Comp. Endocrinol. 53, 455.

155. Young, G., Ueda, H. and Nagahama, Y. (1983). Estradiol-17β and 17α,20β-dihydroxy-4-pregnen-3-one production by isolated ovarian follicles of amago salmon (*Oncorhynchus rhodurus*) in response to mammalian pituitary and placental hormones and salmon gonadotropin. Gen. Comp. Endocrinol. 52, 329-335.

OVULATION: INSIGHTS ABOUT THE MECHANISMS BASED ON A COMPARATIVE APPROACH

Richard E. Jones

Laboratory of Comparative Reproduction
Department of Environmental, Population, and Organismic Biology
University of Colorado
Boulder, Colorado 80309-0334

I. INTRODUCTION

The ovary of all vertebrates contains oocytes in various stages of growth and maturation. Each oocyte is housed in a follicle (from *folliculus*, L., dim. of *follis*, "a leather bag"). In most sexually reproducing animals, the oocyte leaves the follicle and migrates to an extraovarian site, where fertilization occurs. The process by which the oocyte escapes the confines of the follicular wall is termed "ovulation." In reality, however, the process of ovulation encompasses several changes in function of the oocyte and follicular wall prior to and during oocyte expulsion. These changes include separation of the oocyte from its surrounding follicular cells ("follicular separation"), rupture of the follicular wall, and expulsion of the oocyte. For the purposes of this discussion, all of these phenomena of the ovulatory process will be included under the rubric "ovulation." Final meiotic maturation of the oocyte (germinal vesicle breakdown, or GVBD) also is concomitant with ovulation, but this topic is discussed by Y. Nagahama in Chapter 7.

Several recent reviews have emphasized the action of gonadotropic hormones in causing ovulation *in vivo* and *in vitro* in fishes, amphibians and reptiles; these include Licht et al.,[116] Goetz,[75] Donaldson and Hunter,[50] Dodd and Sumpter,[49] Licht,[113] and Lofts.[124] I shall discuss the role of gonadotropins (luteinizing hormone, LH; follicle-stimulating hormone, FSH; human chorionic gonadotropin, hCG) in ovulation only in context with the other points being covered about the mechanisms of ovulation. That is, I initially emphasize what is known about the mechanisms of ovulation and the biophysics of the ovulatory process (i.e., the changes in the follicular wall that precede and accompany ovulation). Then, I discuss the role of gonadotropins, steroid hormones, enzymes, prostaglandins and follicular contraction in causing the described changes. Since the great majority of information on this topic exists for mammals and birds, I briefly review what is known about the mechanisms of ovulation in these endothermic groups before embarking on a discussion of similar mechanisms in ectothermic vertebrates. Finally, I discuss how comparative knowledge of the ovulatory process can add new insight about the mechanisms of ovulation, and I propose a modified theory based on this comparative view.

II. GENERAL DESCRIPTION OF THE PROCESS OF OVULATION

A. Mammals and Birds

In most eutherian mammals, large, graafian follicles that will soon ovulate appear as fluid-filled, blister-like bulges on the ovarian surface. The portion of the graafian follicle protruding from the ovarian surface is termed the "apex."[222] The base of the follicle, representing a majority of the follicular surface, is embedded in ovarian stroma; this internal portion contains the oocyte.

External changes in the preovulatory follicle after the LH surge are studied most easily in an induced ovulator such as the laboratory rabbit. Numerous investigators have described the appearance of the apex of the preovulatory rabbit follicle after copulation or after injection of hCG.[19-24,90,101,106,123,125,131,132,222,232] After the LH surge, the earliest change in the rabbit follicle visible externally is a rapid increase in its size (swelling) accompanied by marked dilatation of vessels on the apical follicular surface. About 2 h before ovulation, a small region of the apex (the "stigma"; see below) appears pale in contrast to the surrounding dilated blood vessels. Then the follicular wall in this pale area bulges to form a small, primary "ovulation cone," and the vessels on the apex blanch. Immediately before rupture, a tiny secondary cone(s) may form,[123] small amounts of fluid may leak from the surface of the cone, and internal hemorrhage in the cone wall may be visible. Then, a rupture develops in the cone(s), and the oocyte-cumulus complex oozes through while assuming a dumbbell configuration.[191]

Descriptions of the external appearance of the preovulatory and ovulating follicles of other eutherian mammals basically agree with the above descriptions for the rabbit; see studies of the bat *Myotis* lucifugus,[238] mouse,[143] rat,[26,164,165] sheep,[128] hamster,[145,213] horse,[89] monkeys,[218,6,97,172] and human.[51]

At this point, the terms "stigma," "macula pellucida," and "ovulation cone" should be discussed. Greenwald[81] equated the stigma with the macula pellucida, and described this eventual site of rupture as ". . . a bleb-like protrusion usually at the apex of the follicle . . .". Markee and Hinsey[125] stated that, in the rabbit, formation of the stigma is preceded by the development of an area of dilated vessels surrounding a round, relatively avascular area. Taken together, then, these authors equate the terms stigma, macula pellucida, and ovulation cone, all being the cone-like protrusion at the future rupture site. Others (e.g.,[237]) refer to the avascular area from which the cone develops as the macula pellucida, and to the ovulation cone itself as the stigma. Asdell,[10] on the other hand, preferred to use the term stigma for the avascular region, and ovulation cone for the protrusion developing from it.

In this review, I will use the following definitions: stigma (Gr. "mark"), defined medically as any physical mark or peculiarity which aids in the identification or diagnosis of a condition (Dorland's Medical Dictionary, 25th ed.). The follicular stigma, then, is the pale area developing on the vascular apical surface of the follicle, and this is equivalent to macula pellucida (L., "a translucent region distinguished by color from its surroundings"). The ovulation cone, then, is the nipple-like protrusion forming from all or part of the stigma. The present definitions agree in essence with those used by Van Blerkom and Motta.[222]

For comparative purposes, a description of the appearance of the avian preovulatory follicle is pertinent here. Most of what we know comes from studies of the chicken follicle.[140,141,169,233] These studies describe the stigma of the hen, which Asdell[10] equates with the stigma of the mammalian follicle as being an avascular band in the preovulatory follicular wall running in the anterior-posterior axis of the ovary and surrounded by dilated blood vessels. Nalbandov and James[141] noted that, although it appears avascular, the hen stigma does contain several small blood vessels entering it from parallel, larger vessels running along its length. Immediately before ovulation, the blood vessels in the follicular wall and stigma blur and obliterate, the stigma widens and bulges, and a pinpoint hole develops at one end of the stigma band. Then, the stigma tears and the egg exits through the rupture while assuming a dumbbell shape. Bartelmez' description of the stigma of the domestic pigeon follicle agrees with the above description of the hen follicle.[12]

B. Fishes

The process of ovulation has been described in a few cyclostome species. In the lamprey *Petromyzon marinus*, fluid accumulates between the apical follicle cells and oocyte several days before ovulation; the fluid is the result of holocrine secretion of the follicle cells at the apex. Also, the follicle cells behind the oocyte change shape (from cuboidal to columnar). Thus, the long axis of the follicle shortens, the follicle constricts at one end, and an apical rupture appears in the follicle wall.[240] A similar process occurs in the preovulatory follicle of *Lampetra planeri*.[39]

Surprisingly, there are only a few descriptions of the external appearance of the ovulating follicle of teleostean fishes. Robinson and Rugh[176] describe the stigma of the medaka (*Oryzias latipes*) as a mound-like protrusion in the center of a vascular plexus. Goetz[75] states that the preovulatory follicle of teleosts exhibits follicular separation followed by apical rupture. He also states (pg. 148) that "The fact that the oocyte is constricted as it is expelled may indicate that a very specific area of the follicle is weakened at ovulation and involved in hole formation."

C. Amphibians

Rugh[183] provided a description of ovulation in *Rana pipiens* that still is useful today. He described the stigma as an avascular region, surrounded by blood vessels, on the follicular wall. A pinpoint hole then develops in the stigma, and the oocyte escapes slowly, assuming an hour-glass shape since the rupture is smaller than the oocyte. Rugh's observations of *R. pipiens* follicles have been confirmed by Kraus[104] and Anderson and Yatvin,[6] and the process of ovulation appears similar in *Hyla regilla*[107] and *Xenopus laevis*.[52] Smith[200] described ovulation in a urodele, the hellbender (*Cryptobranchus alleghenlensis*), as an initial appearance of a small hole in the follicle wall followed by exit of the egg through the hole, the egg assuming an hour-glass shape.

D. Reptiles

As far as I know, there are no detailed, published descriptions of the external appearance of ovulation in a reptile. Our preliminary observations of ovulation in the lizard *Anolis carolinensis*, however, revealed a well-formed, circular, relatively avascular stigma surrounded by dilated blood vessels. A few small vessels radiate from the surrounding rosette of vessels toward the center of the stigma, like spokes on a wheel. The few times that we have observed *in vivo* ovulation of an *A. carolinensis*

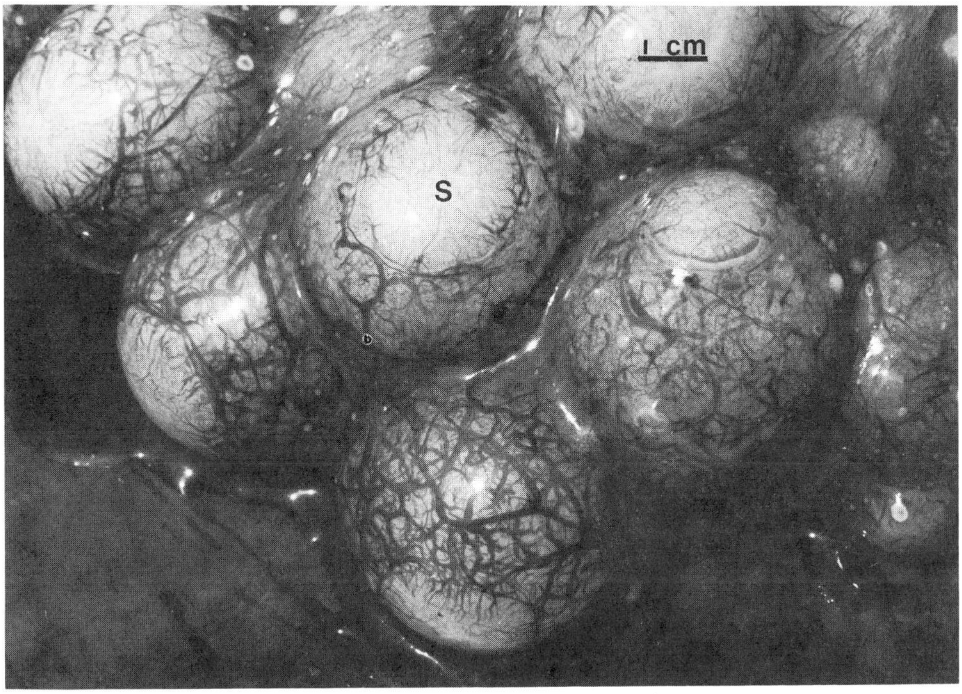

Fig. 1. Preovulatory follicles of the American alligator, Alligator mississippiensis. Note that each follicle exhibits a pale, circular stigma (S) surrounded by the dilated vessels of the remainder of the follicular wall. Note also that the stigma contains fewer and smaller vessels, in a different arrangement than the surrounding region. Photograph courtesy of Dr. Valentine Lance.

follicle, the oocyte escapes in an hour-glass configuration, as previously described briefly.[109,114] We have observed similar stigmata on preovulatory follicles of several other lizards and the turtle, Chrysemys picta. The stigma of alligator preovulatory follicles (Fig. 1) also appears similar to that of Anolis.

III. CELLULAR CHANGES IN THE FOLLICULAR WALL PRECEDING ITS RUPTURE

A. Mammals and Birds

In the few hours after the LH surge and before ovulation, several changes occur in the follicular wall of eutherian mammals. These changes have been carefully described in the rabbit,[19-24,43,56,65a,131,132] cat,[46] guinea pig,[173a] hamster,[125a,145] mouse,[40,50a] rat,[164] and human[150] and have been summarized in several reviews.[11,62,64,120,222] The general impression (e.g., see Bjersing and Cajander[21,23,24]) is that, although these changes occur throughout the follicle wall, they are most pronounced at the apex. An initial response to the LH surge is follicular hyperemia; the thecal blood vessels dilate and become engorged, and the permeability of capillaries increases. The result is edema of the thecal layers and migration of leucocytes and macrophages, cells that are characteristic of an inflammatory response, into the follicular wall.[64] Then, there is a gradual alteration of the intercellular ground substance and dissociation and degeneration of fibrillar and cellular components of the tunica albuginea, theca externa, and theca interna of the follicular wall; collagenolysis is apparent. Also, the cell-cell connections of the surface epithelium as well as the membrana granulosa dissociate. The result of this degradation is a marked thinning of the apical wall as well as a great increase in wall distensibility and a decrease in its tensile strength.[32,57,58,120,178-180] The increase in distensibility, along with leakage of fluid, may explain why intrafollicular fluid pressure does not increase near the time of ovulation.[179] As the primary ovulation cone forms from the stigmal surface, some of the layers of the follicular wall (the surface epithelium, theca interna, and membrana granulosa) may disappear from the cone wall.

At about the time of degradation of the follicular wall, there are concomitant changes in the cumulus-oocyte complex. The oocyte microvilli and cytoplasmic process of the cells of the corona radiata, which interdigitate within the zona pellucida and are a possible source of intercellular communication,[196] begin to be withdrawn.[55] There is also pycnosis and dissociation of cumulus cells, and the cumulus stalk detaches, leaving the cumulus-oocyte complex floating free in the follicular fluid.[48,222]

The cellular changes in the stigma of the hen preovulatory follicle have been well-studied, and these changes appear very similar to those seen in mammals.[241-243]

B. Fishes, Amphibians, and Reptiles

Some type of preovulatory degeneration of the follicular wall at the site of future rupture has been described for three cyclostomes,[39,108,240] the teleostean fish *O. latipes*,[166] and the anuran *R. pipiens*.[6] As far as I know, the preovulatory changes in the stigma of a reptilian follicle have not been described; although Laughran et al.[109] observed general changes in the follicular wall of a lizard, *A. carolinensis*, the origin of the tissue (stigma or nonstigma) was not clear.

In ovaries of ectothermic vertebrates, as in mammals, there are microvillar connections, or even intercellular bridges, between oocytes and follicular cells. In fish, these connections (gap junctions) become separated immediately before ovulation.[75] A similar event occurs in the preovulatory follicles of at least some amphibians.[34,107,194] In squamate reptiles, the membrana granulosa of previtellogenic follicles contains large pyriform cells having intercellular connections with the oocyte.[82] The pyriform cells, however, disappear as soon as the oocyte begins to incorporate yolk, a long time before ovulation will occur.[109]

IV. GONADOTROPINS, STEROID HORMONES, AND OVULATION

A. Mammals and Birds

It is very clear that a surge of LH causes ovulation in mammals, although there is some evidence that FSH synergizes with LH in the ovulatory process;[64,81,105] a recent study demonstrates that FSH alone induces ovulation in the rat.[197] Administration of LH or hCG is highly efficient in inducing ovulation *in vivo*.[81] Also, LH or hCG can induce ovulation *in vitro* in a system using perfusion of the whole rabbit ovary.[85] In order for gonadotropin to induce ovulation of a single follicle cultured *in vitro*, however, the follicle must be co-cultured with extrafollicular (interstitial) tissue, at least in the rabbit.[217,218] The substance from the interstitial tissue that permits LH-induced ovulation *in vitro* may be a prostaglandin[218] or a progestogen.[144] Although steroid biosynthesis is a prerequisite for LH-induced ovulation,[121,210] progesterone alone is not an ovulation-inducing hormone in mammals.[121] In the rabbit, LH, cAMP, or progesterone increases distensibility of follicular strips, and the fact that cyanoketone blocks the LH effect on distensibility argues that this effect is progesterone-mediated.[180,181]

Injection of LH can induce ovulation in the hen.[70,140] In addition, the large, preovulatory follicle of the hen, when excised from the ovary, will ovulate *in vitro* if removed within 2 h of the expected ovulation.[142,148] Also, LH can induce premature ovulation of the hen follicle *in vitro*.[133]

B. Fishes

In natural ovarian cycles, there is a surge (e.g., in goldfish) or a gradual increase (e.g., in salmonids) of plasma gonadotropin concentration preceding spontaneous ovulation in several teleostean species.[49,75] Piscine or mammalian gonadotropins, when administered *in vivo* or *in vitro*, can induce ovulation of teleostean follicles.[50,75] In contrast to amphibians, however, the steroid hormones that induce GVBD in fishes (see Chapter 6) *in vitro* do not induce ovulation *in vitro* in most species.[75] However, steroids readily induce *in-vitro* ovulation of yellow perch (*Perca flavescens*) follicles, and gonadotropin priming is not required.[80] In the Indian catfish, (*Heteropneustes fossilis*), gonadotropin stimulates ovulation by causing secretion of 11-deoxygenated and 11-oxygenated corticosteroids from the interrenal tissue.[209] The general suspicion, therefore, is that the ovulatory process is under complex control in teleosts; follicular separation may require a gonadal or interrenal steroid, but oocyte expulsion may be more directly dependent on gonadotropin stimulation.[75]

C. Amphibians

Since Rugh's studies of *R. pipiens*,[183] it has been standard procedure in many laboratories to induce amphibian ovulation with hCG *in vivo* as well as *in vitro*. Recent examples are the studies using *X. laevis*[115] and *Ambystoma tigrinum*.[115,146] Ovulation of amphibian follicles also is induced *in vitro* by the progestational steroid hormones that induce GVBD,[146,193] and LH-induced ovulation from anuran ovaries requires preovulatory progesterone biosynthesis.[201] Hormone-induced, *in-vitro* ovulation of *A. tigrinum* follicles is enhanced by previous *in-vivo* treatment with prolactin and inhibited by *in-vivo* thyroxine administration.[146]

D. Reptiles

Gonadotropin administration *in vivo* has been shown to induce ovulation in lizards.[114,117] Inasmuch as squamate reptiles appear to secrete only one gonadotropin with FSH-like properties,[116] it is interesting that turtle FSH is more effective than turtle LH in causing rupture of the follicular wall of a lizard,[114] a process stimulated by LH-like hormones in other vertebrates. There are no reports of the induction of ovulation *in vitro* in reptiles, and nothing is known of the role of steroid hormones in reptilian ovulation.

V. PROSTAGLANDINS AND OVULATION

A. Mammals and Birds

The mammalian follicle secretes $PGF_{2\alpha}$ and PGE_2; synthesis occurs in both the thecal and granulosa layers.[4] Furthermore, there is good evidence that prostaglandins of the "2" series are involved in LH-induced ovulation in mammals.[16,64] For example, ovulation from the perfused rabbit ovary can be induced by $PGF_{2\alpha}$. Furthermore, indomethacin (which inhibits prostaglandin synthesis) blocks hCG-induced ovulation from the perfused rabbit ovary, and $PGF_{2\alpha}$ restores ovulation from indomethacin-blocked ovaries.[85,86,91,185] Indomethacin also inhibits mammalian ovulation *in vivo*[7-9,61,135,224] by preventing the preovulatory thinning of the apical wall.[50a,65a] Martin and Talbot blocked ovulation by placing an inhibitor of PGF synthesis, flufenamate, into the ovarian bursa of hamsters.[127] Injection of $PGF_{2\alpha}$ directly into the rabbit follicle caused ovulation, and injection of an antiserum to $PGF_{2\alpha}$ blocked LH-induced ovulation; an antiserum to PGE_2, however, was less effective. Intrafollicular injection of indomethacin also completely blocked LH-induced ovulation.[8,9] It should be noted that the positive influence of $PGF_{2\alpha}$ on ovulation probably does not involve stimulation of steroidogenesis.[91]

A role for prostaglandins in ovulation is supported by correlative studies of ovarian and plasma prostaglandin levels in relation to the ovulatory event. Administration of LH *in vitro* increases prostaglandin secretion from ovaries of rats,[13] cows,198 and rabbits.[102] Most studies have shown an increase in both $PGF_{2\alpha}$ and PGE_2 in follicular fluid and plasma before spontaneous ovulation and after gonadotropin treatment; since there is a greater increase of PGE, the E/F ratio increases; immediately before ovulation, however, a decline in PGE secretion lowers the E/F ratio again.[3,9,33,111,112,134,135,239] Indomethacin inhibits the LH-induced rise in plasma prostaglandin.[9,239] The mechanism of action of prostaglandins as ovulatory substances will be discussed later in relation to their myotonic and proinflammatory effects.

The largest follicle of the hen ovary exhibits a great increase in $PGF_{2\alpha}$ secretion, and a lesser increase in PGE secretion, before ovulation.[47,87,199] Indomethacin, however, fails to block ovulation in the hen *in vivo*.[47]

B. Fishes

Prostaglandins may play a critical role in piscine ovulation. As in mammals, indomethacin has been shown to block spontaneous or hCG-induced ovulation *in vivo*.[204,205] Furthermore, the ovulatory effect of prostaglandins *in vivo* is at least in part due to a direct action on the ovary. Jalabert and Szollosi found that $PGF_{2\alpha}$ (but not PGE_2) induced ovulation of *Salmo gairdneri* follicles *in vitro* following meiotic maturation and follicular separation *in vivo*.[95] Similarly, goldfish follicles ovulate *in vitro* in response to PGE_1, PGE_2, PGF_1 or $PGF_{2\alpha}$ (with $PGF_{2\alpha}$ being most effective) after hCG treatment *in vivo*.[99] Furthermore, as pH of the culture medium increases, so does spontaneous *in-vitro* ovulation of goldfish follicles. Since this pH effect is blocked by indomethacin and restored by $PGF_{2\alpha}$, it may be mediated by stimulation of $PGF_{2\alpha}$ secretion.[78,79] A noteworthy finding, however, is that indomethacin did not inhibit spontaneous, *in-vitro* ovulation in the brook trout even though PGE and PGF production were reduced.[17] Goetz and Theofan demonstrated that yellow perch (*P. flavescens*) oocytes ovulate *in vitro* after treatment with 17α,20β-dihydroxy-4-pregnen-3-one (17α,20β-DHP); indomethacin inhibited steroid-induced ovulation, and either PGE_2, PGE_1, or $PGF_{2\alpha}$ restored it (with PGE_2 being the most effective[80]). Carp follicles ovulate *in vitro* in response to $PGF_{2\alpha}$ over a temperature range of 5-25°C.[54]

A biological role for prostaglandins in fish ovulation is further supported by studies of prostaglandins in ovarian tissue and plasma. For example, Bouffard found that hCG injection increased PGF, and decreased PGE, levels in the plasma of postovulatory goldfish.[29] Similarly, Ogata and co-workers in their study of the pond loach (*Misgurnus anguillicaudatus*), found that hCG increased plasma PGF and that indomethacin blocked this gonadotropic effect.[147] In brook trout, carp pituitary extract increased plasma and ovarian PGF, but not until after ovulation had occurred.[42] Nevertheless, there is a rise in plasma and ovarian PGF at the time of spontaneous ovulation of brook trout, and this rise is blocked by indomethacin.[76]

C. Amphibians and Reptiles

Prostaglandin $F_{2\alpha}$ induces *in-vitro* ovulation of *R. pipiens* eggs, but only if the surface epithelium is present.[195] To my knowledge, there are no published studies of the effects of prostaglandins or their inhibitors on *in-vivo* or *in-vitro* ovulation in reptiles.

VI. ENZYMES AND OVULATION

A. Mammals and Birds

An increasingly convincing array of evidence suggests that the preovuatory weakening and rupture of the wall of the mammalian follicle is caused by enzymatic action. In fact, in a comprehensive review of ovulation, Espey proposed a central role for enzymes (as part of an LH-induced, inflammatory-like response) in causing follicular rupture.[64]

Ever since Schochet[186] proposed a role for enzymes in degeneration of the mammalian follicular wall, there has been a continuing series of studies of the enzymes present in the follicular wall near the time of ovulation. The concentrations of proteolytic enzymes (including collagenase) have been reported to increase in the follicular wall and fluid near the time of ovulation in the rabbit,[65,68] rat,[43a,67,119] hamster,[92,93] and human.[73] There is an increase in collagen breakdown and a resultant decrease in collagen content of follicles after the LH surge.[50a,125a,130a,175] That <u>de novo</u> RNA and protein synthesis (and therefore perhaps enzyme synthesis) is required for ovulation is indicated by blockage of ovulation with transcriptional and translational inhibitors.[171,208]

Proteolytic enzymes, when applied to the follicular wall, reduce wall tensile strength[58] and induce normal-appearing degenerative changes in the wall.[59,60,179] Espey and Lipner injected collagenase and trypsin into the follicular fluid of the rabbit; a normal stigma was formed, and the oocyte was expelled.[67] A collagenase inhibitor blocked <u>in-vitro</u> ovulation of the hamster follicle as did several other proteinase inhibitors.[92,93]

Recent work also suggests a role for serum proteases in ovulation.[15,174,175] Application of plasmin to the follicular wall decreases wall tensile strength.[14] Furthermore, release of plasminogen activator into the follicular fluid increases as the follicle nears ovulation.[207] When inhibitors of serum proteases are injected into the bursa of the proestrus rat ovary, ovulation is inhibited in a dosage-dependent manner.[174,175] Protease inhibitors also block ovulation in the hamster.[92,166a] In summary,[64,175] FSH (or LH) stimulates release of plasminogen activator from granulosa cells, which in turn cleaves serum or follicular fluid plasminogen into plasmin. Relaxin produced by the granulosa cells may also increase production of plasminogen activator.[221] Plasmin then activates collagenase, which in turn nicks the collagen helices and allows degradation of collagen in the follicular wall by a cascade of nonspecific proteases. Plasmin also activates proteoglycanase, and the active enzyme degrades proteoglycans, the matrix gel of connective tissue. It should be noted, however, that lysosomal enzymes released from surface epithelial cells also may play a role in activation of procollagenase.[25]

What is the cellular source of enzymes involved in ovulation? The source of the plasminogen activator is the granulosa cells.[208] Other enzymes in the cascade leading to ovulation could be produced from lysosome-like, acid phosphatase-containing dense bodies in the surface epithelial cells[21,25,41] --but see a rebuttal of this idea by Rawson and Espey.[173] Multivesicular bodies protruding from the plasmalemma of proliferating fibroblasts in the theca externa and tunica albuginea also could be a source of proteolytic enzymes,[59,60] Motta and co-workers proposed, however, that these bodies are products of cytolysis,[131] and Downs and Longo[50a] found no such bodies in mouse follicles. Finally, invading cells associated with the inflammatory response could produce enzymes.[64]

There is also good evidence for a role of enzymes in stigma formation and follicular rupture in birds. Morphological studies, as well as assays for enzyme activity, indicate an increase in enzyme production by the hen follicular wall before ovulation.[71,147a,242] Application of crude collagenase or pronase to the hen follicular wall causes oocyte release in 2 to 3 hours; there is no effect, however, of hyaluronidase or lysosomal enzymes.[138] Similarly, Yoshimura and Fujii produced normal-appearing stigmata and ovulations in the hen after applying collagenase or trypsin locally to the follicular wall.[241] It should be mentioned, however, that Tojo and co-workers found no increase in follicular collagenase or neutral or acid protease activity prior to ovulation in the hen.[220] In fact, Nakamura and Nakajo found a decrease in overall proteolytic activity in the Japanese quail follicle 30 min before ovulation.[139] It must be kept in mind, however, that the amount of an enzyme present at any given time in a tissue is related to rates of synthesis, storage, and degradation, and thus may not reflect the level of activity *in vivo*.[64] Also, enzyme assays may not be sensitive enough to detect the small changes that occur during activation. Finally, gonadotropin-induced changes in enzymatic activity may differ in stigmal versus nonstigmal regions;[147a] i.e., different parts of the follicular wall may yield different enzymatic activities.

B. Fishes

There is scant information on the role of enzymes in piscine ovulation, although degenerative changes of the wall of cyclostome and teleostean follicles suggest enzymatic action. Oshiro and Hibiya, using frozen sections, found proteolytic enzymes in the wall of preovulatory follicles of the pond loach.[158] Szollosi and colleagues detected chymotrypsin, peptidase, phosphatase, esterase, and glycosidase (but not collagenase) in the coelomic fluid of rainbow trout 24 h before ovulation.[212] Exposure of follicles of rainbow trout to a trypsin inhibitor blocked $PGF_{2\alpha}$-induced ovulation.[95] There are two major phases of *de novo* RNA synthesis in follicles of the yellow perch treated with $17\alpha,20\beta$-DHP *in vitro*; the early phase may be related to enzyme synthesis. Actinomycin D (a transcriptional inhibitor) completely blocked steroid-induced *in-vitro* ovulation of perch follicles if given during the early phase (0-6 h).[219] Therefore, RNA synthesis at that time may be required for the synthesis of enzymes involved in ovulation.

C. Amphibians

The earliest studies of the role of enzymes in ovulation utilized amphibian follicles. Hartog[88] detected proteolytic enzymes in the frog egg, and Rugh[183] suggested that enzymatic digestion of the frog follicular wall allows rupture of the stigma; Rugh, however, failed to induce ovulation by applying either pepsin or trypsin to the follicular wall, and only a solution of pepsin plus HCl caused rupture. Anderson and Yatvin suggested a role for collagenase in ovulation of *R. pipiens* eggs based on morphological evidence.[6] Inhibitors of transcription[6] and translation[130,192,234] block ovulation of anuran follicles suggesting that *de novo* RNA and protein synthesis is required before ovulation and that enzymes may be synthesized at that time.

D. Reptiles

The role of enzymes in reptilian ovulation has not been studied. Laughran and co-workers, however, report that the theca externa of the preovulatory follicle of the lizard *A. carolinensis* contains fibroblasts possessing multivesicular bodies,[110] the type Espey[59,60] suggested plays a role in enzymatic digestion of the mammalian follicle.

VII. FOLLICULAR CONTRACTION AND OVULATION

A. Evidence for a Role of Follicular Contraction in Ovulation

1. *Mammals and Birds*. In three reviews of the control of ovulation in mammals, Espey casts doubt on any role for follicular smooth muscle contraction in ovulation.[62-64] Recent evidence, however, strongly supports a role for follicular smooth muscle contraction in ovulation in mammals as well as in other vertebrates and invertebrates.[191]

In eutherians, several cytological characteristics are diagnostic for smooth muscle cells.[74,203] These include the presence of sarcoplasmic reticulum, thin (actin-containing), thick (myosin-containing) and intermediate filaments (desmin) which run parallel to the long axis of the uncontracted cell, dense bodies in the cytoplasm and dense attachment plaques adjacent to the plasmalemma that anchor the myofilaments, caveolae (micropinocytotic vesicles) on the cell surface, hemidesmosomes along the inner plasmalemma that bind the smooth muscle cell to the basal lamina, adhering zonuli (desmosomes) attaching apposing cells, and nuclear irregularities (indicating contraction). Cells with most or all of these characteristics have been identified clearly in the theca externa of follicles from a wide variety of mammals, including the gerbil,[129] cat,[35,36,151,228] rabbit,[23,35,56,104,151] guinea pig,[35,228] hamster,[126,129] monkey,[161] and human.[151,230] The number of these thecal smooth muscle cells (or myofibroblasts) increases as the follicle enlarges.[36,118,155]

Several studies have shown that the muscle cells of the theca externa are more abundant at the base of the follicle, and are sparse or absent from the apex; clearly, the stigma itself has no smooth muscle cells.[35,36,56] Martin and Talbot demonstrated that the smooth muscle cells at the base of the hamster follicle radiate meridionally, from the base towards the apex.[126] It should also be mentioned that extrafollicular regions of the mammalian ovary (hilus, medulla) also contain smooth muscle cells.[36,63,161]

Referring to mammals, Espey[63] stated that "ovarian contractility is not necessary for ovulation to take place." He based this statement on three lines of reasoning: (1) there is no increase, and instead usually a decrease, in intrafollicular pressure before ovulation;[28,32,66,178-180,190] (2) the cells identified as smooth muscle cells in the theca externa could be fibroblasts, and (3) to that date, only O'Shea and Phillips had demonstrated rhythmic contraction of ovarian follicular strips,[157] and these strips could have included stromal smooth muscle. Concerning point (1), it is not surprising that there is a decrease in intrafollicular pressure, due to an increase in wall distensibility and fluid leakage, but this decrease is not incompatible with presence of follicular contraction before and during ovulation. Concerning (2), I have already referred to strong evidence of myofibroblasts or true smooth muscle cells in the follicular wall of mammals. About point (3), it is understandable that only tonic contraction of follicular strips had been demonstrated to that time, since most of these studies used strips from the apex of follicles, and there are no smooth muscle cells in the apex. In contrast, a study using strips including the base of the sheep follicle demonstrated rhythmic contractions.[157]

Schroeder and Talbot recently reviewed evidence that strongly suggests a role for follicular contraction in ovulation in a wide variety of animals.[191] In the rabbit, for example, follicles in ovaries autotransplanted to the anterior chamber of the eye contract and change shape after hCG treatment or cervical stimulation.[122] Similarly, smooth muscle cells at the base of the hamster follicle contract before ovulation (after the LH surge), producing a V-shaped constriction.[126] Furthermore, several smooth muscle inhibitors block ovulation when placed in the hamster ovarian bursa; all of the 17 inhibitors which were effective are known to block calcium influx in other systems.[127] Preovulatory hamster follicles will not ovulate *in vitro* unless smooth muscle cell contraction had begun *in vivo*, about 12.0-12.5 h after an hCG injection.[215] Recording of intrafollicular pressure in hamster follicles *in vitro* just before ovulation detected a slow, steady fall in most follicles, with no evidence of rhythmic fluctuations.[190] Such recordings, however, would be a poor indicator of rhythmic contraction because of the distensibility and permeability of the follicular wall, or it may be the contractions are tonic at first or slow to develop. To my knowledge, there are no published direct recordings of the contraction of whole mammalian follicle walls *in vitro*.

Previous authors have suggested various roles for follicular contraction, some involving ovulation and some not.[62,63,149,151,161,191,214] These possible roles include regulation of blood flow and capillary filtration rate in thecal vessels, dissociation and degeneration of apical tissue, breaking of the cumulus-oocyte complex, apical rupture, expulsion of the oocyte, postovulatory collapse and repair, and formation of the corpus luteum. Although it appears that follicular contraction may support expulsion of the oocyte during normal ovulation, artificially poked hamster follicles expel the oocyte without contracting.[214]

In the chicken ovary, muscle cells are found not only in the theca externa of follicles but also in the loose connective tissue outside the theca, where muscle fibers radiate meridionally from an abundance at the stalk of the follicle; the stigma itself is devoid of muscle cells.[72,84,104,167]

Follicular contraction appears to play an important role in avian ovulation. Phillips and Warren observed that the blood vessels in the follicular wall of the chicken blurred or obliterated just before ovulation, and then filled again after ovulation.[169] Since manually-induced tension on the follicle also caused a similar blurring, they suggested that contraction of the follicular wall causes rupture. They also noted that the turgidity of the hen follicle changed periodically, again suggesting preovulatory, rhythmic contractions. Meridional strips taken from the proximal, extrafollicular smooth muscle layer of the hen follicle exhibit strong, rhythmic contractions *in vitro*, whereas extrafollicular strips taken across the equator of the follicle exhibit only mild, irregular contractions. Both meridional and equatorial strips taken from the theca externa, however, contracted rhythmically.[244] Soliman and Walker also demonstrated rhythmic contraction of strips of the hen follicular wall (theca externa plus extrafollicular tissue).[202]

2. *Fishes*. Microfilament-bearing cells resembling smooth muscle cells have been described in the theca of the rainbow trout.[95,211,212] Similar cells are present in the theca of large follicles of the medaka[166] and goldfish.[137] Busson-Mabillot suggested, on histological evidence, that the follicle cells of the lamprey *L. planeri* contract before ovulation.[39]

Similarly, contraction and change in shape of basal follicular cells, which acquire myofibrils, may facilitate ovulation in the lamprey P. marinus.[240]

Evidence for a role of follicular contraction in piscine ovulation comes from studies using cytochalasin B, which inhibits actin depolymerization and thus inhibits the microfilament contractile system.[236] This drug partially inhibits rainbow trout ovulation,[95] and completely blocks in-vitro ovulation of medaka follicles.[189] Also, inhibitors of calcium influx block rainbow trout ovulation.[95] When brook trout ovaries, previously primed with hCG in vivo, are placed in vitro, cAMP inhibits theophylline-induced ovulation.[79] This anti-ovulatory influence may have been on follicular contraction, since cAMP inhibits smooth muscle contraction by interfering with phosphorylation of myosin, thus relaxing smooth muscle.[1,182]

In Chapter 18 of this volume, L. Guillette discusses the possible role of follicular contraction in "parturition" of teleostean species exhibiting intrafollicular gestation; that is, embryonic development within the ovarian follicles. Inasmuch as expulsion of the embryo from the follicle requires an ovulatory-like rupture of the wall, the factors (including smooth muscle contraction) normally operative in ovulation of other teleostean species may have been delayed and now function in parturition in these viviparous forms.

3. Amphibians. The initial studies of the possible role of smooth muscle contraction in ovulation were performed using anuran follicles. These studies described smooth muscle in the theca (cyst wall; inner ovarian epithelium) of Rana, and suggested that contraction of these cells causes ovulation.[2,31,104,168,18] The theca of H. regilla is a sheet of smooth muscle cells,[107] and this smooth muscle layer also has been identified in the newt Taricha granulosa[107] as well as Ascaphus, Ensatina, and Plethodon.[188] Dumont and Brumett, however, described smooth muscle cells only in the outer ovarian (surface) epithelium of X. laevis follicles.[52] Removal of the outer ovarian epithelium of R. pipiens follicles prevented follicular contraction and ovulation.[195] Finally, Cytochalasin B blocked in-vitro ovulation of H. regilla follicles.[187] Schroeder and Talbot[191] conclude that contraction of the theca of the amphibian follicle forces the oocyte to emerge through the enzyme-induced apical rupture. As with mammalian and avian follicles, the site of future rupture (stigma) does not contain smooth muscle cells, at least in R. pipiens.[6]

4. Reptiles. In their study of the follicular wall of the lizard A. carolinensis, Laughran and co-workers provided evidence for the presence

of myofibroblasts, with most of the characteristics of smooth muscle cells, in the surface epithelium and theca externa of preovulatory vitellogenic follicles.[109,110] The surface epithelial cells contain microfilaments that apparently do not contain actin. The myofibroblasts of the theca externa, however, contain microfilaments and intermediate filaments, and these cells do contain actin.[110] The spontaneous, rhythmic contractions observed when whole follicular walls of this lizard are placed *in vitro* are probably the result of the contractile activity of these thecal myofibroblasts.[98] To date, however, there is no evidence for a role of follicular contraction in reptilian ovulation.

B. Adrenergic and Cholinergic Effects on Follicular Contraction

1. *Mammals and Birds*. The thecal layer of mammalian and avian follicles contains a rich autonomic innervation. Adrenergic nerve endings are found near blood vessels and smooth muscle cells in the theca of mammals[37,161,163,206,251] and birds.[37,44] Cholinergic fibers also are present around smooth muscle cells in the theca externa of mammals.[37,162,206,229] It is likely that the autonomic fibers play a role in regulation of the microcirculation of the follicular wall, but they also may influence follicular contraction and therefore perhaps ovulation. But, since ovulation can occur *in vitro* in response to gonadotropins, nerves may only influence, but not be required for, ovulation.[37]

Alpha-adrenoreceptor agonists increase contractility of mammalian follicular strips *in vitro*.[157,162,225,226,228,230] Generally, these agonists induce an increase in tone of the strips, with alpha agonists being more effective than beta agonists. Beta-adrenoreceptor agonists, on the other hand, decrease contractility of strips from mammalian follicles.[157,162,226]

Similar effects of alpha agonists have been described on contractility of strips from the hen follicle.[202,244] LH-induced ovulation of hen follicles *in vitro* is inhibited by phenoxybenzamine, but not propanolol.[133] Therefore, LH is ineffective in inducing ovulation if alpha (but not beta) adrenoreceptors are blocked. Alpha inhibitors also block chicken ovulation *in vivo*.[69,100] Finally, beta agonists decrease contractility of hen follicular strips.[202]

Cholinergic nerve endings, as mentioned, are present in the theca of mammalian follicles.[37,229] Furthermore, acetylcholine increases rhythmic contraction of strips from sheep follicles *in vitro*[157] and increases the tone of apical strips from cow[229] and human[230] follicles. It should also be mentioned that serotonin stimulates follicular contraction, but not ovulation, of whole hamster follicles *in vitro*.[216]

2. Teleostean Fishes. Epinephrine causes *in-vitro* ovulation of follicles of the rainbow trout[94] as well as yellow perch.[53,204] Epinephrine has a direct effect on the follicle by acting through alpha, but not beta, adrenoreceptors.[94] The stimulatory action of alpha agonists may be mediated by an increase in prostaglandin synthesis.[204] It is not known, however, if teleostean follicles possess autonomic innervation.

3. Amphibians and Reptiles. There is no information on the innervation of reptilian or amphibian follicles, and the effects of alpha and beta andrenoreceptor agonists as well as cholinergic stimulation on reptilian and amphibian follicular contraction and ovulation remain unexplored.

C. Prostaglandins and Follicular Contraction

The aforementioned effects of prostaglandins on ovulation may be explained by the influence of these chemicals on contraction of the follicular wall. Hamada and colleagues found that $PGF_{2\alpha}$ stimulates contraction of follicles in the perfused rabbit ovary.[85] Furthermore, $PGF_{2\alpha}$ increases rhythmic contactility of strips from sheep follicles[157] and causes tonic contraction of apical strips from human follicles.[149,230] In their study of hamster follicles *in vitro*, Martin and Talbot found that $PGF_{2\alpha}$ stimulated tonic basal contraction and that flufenamate blocked contraction and ovulation.[127]

Prostaglandin E_2 decreases contraction of follicles in the perfused rabbit ovary.[85] PGE_1 also decreases rhythmic contraction of strips from sheep follicles *in vitro*[157] and blocks smooth muscle contraction and ovulation in the hamster.[127] PGE_1 also decreases the tone of apical strips of human follicles *in vitro*.[149]

Indomethacin blocks ovulation in rats by causing intraovarian release of oocytes. That is, instead of ovulating through the apical surface, oocytes are either entrapped within luteinized follicles or are "ovulated" through ruptures in the basal or lateral follicular wall, the oocytes ending up in extrafollicular tissue.[160] Thus, prostaglandins may be involved in follicular rupture at the normal location (the apical stigma), whereas dissolution and rupture of the follicular wall can occur in other areas of the follicular wall if prostaglandin synthesis is disrupted. Failure of apical rupture after indomethacin treatment may relate to the absence of contraction of muscle fibers oriented in a direction critical to application of tension on the normal stigma region.

In rainbow trout, $PGF_{2\alpha}$-induced ovulation may be due to an effect on follicular contraction, since indomethacin blocks contraction and ovulation.[95] Furthermore, $PGF_{2\alpha}$-induced *in-vitro* ovulation of *R. pipiens* oocytes is accompanied by follicular contraction.[195] Other than these observations there is no information on the role of prostaglandins in contraction of follicles from ectothermic vertebrates.

D. Oxytocic Hormones and Follicular Contraction

Although oxytocin has been shown to influence contraction of the whole mammalian ovary or strips thereof,[63,235] there are only a few studies of the influence of neurohypophysial nonapeptide hormones on follicular contraction. Such studies, by the way, become more relevant now that it is revealed that the mammalian ovary secretes oxytocin and arginine vasopressin.[184,235] O'Shea and Phillips found that oxytocin increased the tone of sheep follicular strips, but did not change the rhythmic contraction pattern.[157] Hoshimura and co-workers reported that oxytocin increased the frequency and amplitude of rhythmic contraction of the extrafollicular layer of smooth muscle of hen follicles but decreased contractility of the thecal muscle layer.[244] Arginine vasotocin, the oxytocic hormone of nonmammalian vertebrates, did not influence spontaneous *in-vitro* contraction of the follicular wall of a lizard.[98]

E. Follicular Wall Tension and Follicular Contraction

As the mammalian follicle enlarges, the tension on the wall increases even though the wall becomes more distensible.[178] For example, the tension (stress) on the follicular wall of the rabbit increases 2.5 times as ovulation nears.[120] Although (as discussed above) there usually is a reduction in intrafollicular pressure in the mammalian follicle approaching ovulation, the pressure within the yolky oocyte of lower vertebrate follicles may actually increase near ovulation, resulting in even more tension on the wall. For example, the intrafollicular pressure in *X. laevis* follicles increases 1-2 mm Hg during ovulation.[188] In the plaice *Limanda yokohamae*,[159] water absorption swells the oocyte, perhaps leading to an increase in tension on the wall and rupture of the follicle.

Stretch-induced increase in tension on the follicular wall may influence follicular contraction, since stretch increases the intensity of subsequent smooth muscle contraction.[45,103] Increasing the tension on the wall of a preovulatory follicle from the lizard *A. carolinensis* greatly increases the amplitude of spontaneous, rhythmic contractions *in vitro*.[98] Also, when Espey and Lipner[66] increased intrafollicular pressure in the rabbit follicle and then allowed the wall to relax, contraction waves appeared on the follicular surface. Prostaglandins are released when smooth muscle is stretched,[170] and Martin and Talbot[127] hypothesized that stretch of the hamster follicular wall during its final growth spurt may cause $PGF_{2\alpha}$ release, which then would increase the intensity of follicular contraction. This hypothesis, however, has not been tested directly.

If an increase in tension on the follicular wall somehow assists follicular rupture, the stigma region must be at a critical low tensile strength for rupture to occur. That is, an increase in wall tension will not cause rupture unless the stigma is weakened. For example, Guttmacher and Guttmacher injected fluid into the pig follicle and did not cause premature ovulation.[83] Artificially increasing intrafollicular pressure in the hen follicle also did not cause rupture.[69]

VIII. SUMMARY AND FUTURE RESEARCH DIRECTIONS

Present theory on the mechanism of ovulation of mammalian follicles can be summarized as follows:[64] The surge of LH in thecal blood plasma increases cAMP and then steroid (progestogen) production from the granulosa and/or interstitial cells. The local increase in progestogen concentration then stimulates the release of mediators of the inflammatory response (histamine, serotonin, leukotrienes) which in turn cause hyperemia and an inflammatory-like response throughout the follicular wall. Typical cells of the inflammatory response migrate into the extravascular spaces of the follicular wall. Prostaglandins E_2 and $F_{2\alpha}$ are released by the thecal and granulosa cells as well as perhaps cells involved in the inflammatory response. $PGF_{2\alpha}$ stimulates fibroblast proliferation, and these fibroblasts in turn release precollagenase. PGE_2 enhances the follicular hyperemia and edema initially induced by histamine. Serum proteinases (e.g., plasminogen activator; plasmin) are also released, and these activate procollagenase. The resultant collagenase degrades the follicular wall, along with a cascade of other proteolytic enzymes characteristic of the inflammatory response. A preovulatory increase in $PGF_{2\alpha}$, and a decrease in PGE_2, secretion may cause the vascular stasis seen immediately before ovulation, and this may contribute to stigma necrolysis.[135,136] Weakening of the stigma allows the tension on the follicular wall to tear the ovulation cone. Thus, follicular rupture and ovulation occur. Espey[64] played down the role of smooth muscle contraction in ovulation. However, a recent comparative review of the role of follicular contraction in ovulation gives strong support to the theory that smooth muscle contraction plays an important role in ovulation throughout the animal kingdom.[191] Follicular contraction could function in follicular vasoregulation, follicle-oocyte separation, apical rupture, oocyte expulsion, postovulatory follicle collapse, and luteal formation. Limited data on nonmammalian vertebrates suggest that fundamental mechanisms, common to follicles of all vertebrates, are operative during ovulation.

Although many details of the ovulatory process need to be studied, especially in lower vertebrates, the above theory about ovulation seems to be supported by the evidence. It does not deal adequately, however, with one major question: i.e., how is maximal degeneration and follicular rupture localized to one small region on the follicular surface (e.g., see Bjersing and Cajander[19-24])? To propose a possible answer to this question, I first must backtrack to the stigma--when it forms, and how it forms.

First, when does the stigma form? Many authors, in reporting on the process of ovulation in mammals, describe the avascular stigma as forming shortly before ovulation; that is, several hours after the peak of the LH surge.[19,26,27,30,125,128,223,232,238] For example, in his classic review, Brambell states that "Immediately before rupture a non-vascular area, the macula pellucida, appears at the apex of the protruding follicle, where rupture will occur. . . ."[30] Although Johansson and co-workers report that the monkey stigma is present by 2 days before ovulation,[97] Betteridge and colleagues criticize their observations,[18] and Jewett and Dukelow report stigma formation in the monkey follicle 8-10 h before ovulation.[96]

Second, what factors cause the stigma to form? The general feeling in the present literature on the mammalian follicle is that the stigma region, before the LH surge, has a blood supply no different from that of the rest of the follicular wall. The stigma then becomes avascular (actually, ischemic is more accurate) because the blood vessels are compressed[19,27,237] or retracted.[56] The reduction in blood flow to the stigma region is accentuated by a general vascular stasis that occurs throughout the follicle soon before rupture.[38,135] The factors suggested to cause this vascular stasis in mammalian and avian follicles include $PGF_{2\alpha}$,[135,136] desensitization of the adenylate cyclase system response of follicular cells to LH,[64] smooth muscle contraction,[22,169] and mechanical factors (see below). Whatever factors cause this periovulatory vascular stasis, however, they must have their greatest effect on the small stigma region.

Although the apex of the mammalian follicle bulges from the ovarian surface, the avascular stigma is only a small portion of the total apical region.[222] How, then, does the stigma form only in one small region of the apex? The presently accepted theory is that the stigma forms purely because of mechanical reasons.[19,26,56,61,104] This mechanical theory, discussed by Rodbard,[177] is as follows: the enzyme-induced degenerative changes, although occurring to some extent throughout the follicle, are most marked in the apex because of mechanical factors. That is, because the apex protrudes above the ovarian surface and thus is not surrounded by stromal connective tissue, it degenerates to a greater degree. It follows then that the stigma forms at the weakest point on the already weakened apex, also for mechanical reasons.

The mechanical theory about the mechanism of stigma formation is compatible with a theory proposed earlier by Nalbandov[140] based mainly on observations of the chicken follicle; that is, that ischemia of the stigma region is the primary cause of follicular rupture by causing necrolysis and withdrawal of tropic hormones (e.g., FSH). Thus, ovulation may be both a phenomenon of FSH withdrawal and LH stimulation, an idea based on the following information from the hen:[152-154] LH will induce ovulation of the largest follicle in the size hierarchy in intact hens, but will induce additional ovulation of smaller follicles (multiple ovulation) if LH is injected after hypophysectomy. The longer the interval after hypophysectomy, the greater the number of ovulations induced by LH. Thus, ischemia and/or withdrawal of FSH from the follicle sensitizes smaller follicles to the ovulatory effects of LH. It follows then, that if the stigma suffers relatively greater ischemia than the rest of the follicle, it is sensitized to the ovulatory effects of LH more than other regions of the follicle. In support of Nalbandov[140], Bjersing and Cajander, in their study of the rabbit preovulatory follicle, proposed that stigma formation and rupture is caused by ischemia and necrolysis.[19]

Thus, a marriage of Nalbandov's and Espey's theories is possible. That is, the LH surge induces an inflammatory response and related enzyme release, which weakens the follicular wall, especially at the stigma. The vascular stasis immediately before ovulation, as well as the relative ischemia of the stigma, causes the final necrosis leading to rupture. Although this combines the theories of Nalbandov,[140] Espey,[64] and Rodbard[177] into a logical whole, there may be a major flaw in this synthesis which, if true, would change present views on the mechanism of ovulation and direct new research emphasizing the comparative approach.

As an initial approach to a new view of the mechanisms of ovulation, let me propose why the ideas about stigma formation discussed above may need to be modified. As mentioned, an early and pronounced effect of the LH surge is hyperemia and an increase in permeability of thecal capillaries.[64]

Biophysical principles would suggest that the thecal vessels with greatest external pressure (those at the base of the follicle) would be more compressed than those in the "free" apex. Bjersing and Cajander, however, in their study of hCG-induced follicular changes in rabbits, found that the apical vessels compressed while the basal vessels dilated.[24] Then, an avascular stigma formed within the already ischemic apical region. This observation seems to contradict predictions of the mechanical theory.

Another argument against the mechanical theory of stigma formation comes from comparative studies. Biophysical principles suggest a stigma region that is circular in shape if purely mechanical factors are involved. Although this is true for many vertebrate stigmata, the avian stigma assumes a band shape across the anterior-posterior axis of the follicular wall, which suggests that other factors besides the mechanical may cause stigma formation.

Another major argument against the mechanical theory is that the prevoulatory follicle of several vertebrate types hangs free (often suspended by a stalk) in the pelvic cavity or within a large intraovarian cavity. Clearly, there are no differential external presures involved in local stigma formation in follicles of these animals. Assuming that stigma formation is an evolutionarily stable process, this comparative view casts further suspicion on the mechanical theory of stigma formation.

A comparative survey instead offers the possibility that the stigma is a preformed, hypovascular area of the follicular wall in small, immature follicles. That is, the ischemic stigma seen by observers of mammalian follicles a few hours before ovulation actually was present much earlier, and simply _becomes visible_ right before ovulation because of concurrent dilatation of the surrounding apical blood vessels. It would be difficult to test this idea using mammalian follicles, because the small follicles are surrounded by stroma and it would be difficult to locate a small, hypovasular region in the theca of these small follicles. Since the incompletely developed follicles of lower vertebrates are often more visible, it is perhaps here that a test of the preformed stigma idea would be feasible. Indeed, there is some evidence in support of this theory. Rugh, in his classic study of frog ovulation, states that the avascular stigma is formed very early in follicular maturation.[183] Similarly, we (R. Jones et al., unpublished) have seen a stigma on the surface of very small follicles of the lizard _Anolis equestris_. Bartelmez, in his study of pigeon follicles, states that the avascular stigma forms when the small follicle reaches the ovarian surface (i.e., long before it enlarges and becomes suspended on a stalk).[12] Phillips and Warren[169] describe a "good stigma" at least 24 h before ovulation, and photographs of the hen follicle in Nalbandov and James[141] show stigmata on small vitellogenic follicles. Also, one is reminded that Nalbandov[140] described ovulation of immature chicken follicles, complete with a stigma, after hypophysectomy and LH treatment.

The aforementioned demonstration that indomethacin treatment of rats causes "ovulation" through nonapical regions of the follicular wall[7,160] may, at first glance, argue that the stigma is not preformed in mammals but can form anywhere on the follicular wall. However, these studies may simply show that prostaglandins normally are involved in rupture of an enzymatically weakened, preformed stigma; once prostaglandin synthesis and apical rupture are inhibited; however, another part of the follicular wall will eventually be thin enough under the action of enzymes to rupture, thus releasing the entrapped oocyte.

If the stigma is a preformed area on the follicular surface and does not appear immediately before ovulation, aforementioned theories about the mechanism of ovulation must be modified. Thus, I propose the following hypothesis: The stigma (macula pellucida) is present as a special, hypovascular region on the surface of small, growing follicles. It forms either as the theca is formed or when the growing follicle first reaches the ovarian surface. This preformed, hypovascular region receives less blood flow (and therefore less of both gonadotropic hormones and growth-supporting materials) than does the rest of the follicular wall. The LH surge induces hyperemia and an inflammatory-like response in the wall of the entire follicle. Release of proteolytic enzymes weakens the entire follicular wall, but the degeneration is more pronounced at the stigma because of its previous thinness and weakness (due to ischemia). The enzymes could reach the stigma via the few small vessels in the region, by a simple diffusion from the surrounding follicular wall or, in mammals, via follicular fluid. Contraction of meridionally-arranged smooth muscle cells in the follicular wall causes the preovulatory vascular stasis; since the stigma has few vessels to begin with, the ischemia and resultant necrotic changes are greatest in that region. The muscle contraction also maintains the tension on the stigma region, thus combating the increase in distensibility of the follicular wall. When the decreasing tenacity of the stigma approaches the tension on the stigma wall, the ovulation cone bulges. When the tenacity becomes less than the tension on the cone wall, rupture occurs. Smooth muscle contraction then assists (or is necessary for) oocyte expulsion and postovulatory follicular collapse.

The major difference between this theory and the theories of Espey, Nalbandov, and Rodbard are (1) the stigma is a preformed, hypovascular area of the follicular wall, and (2) LH acts directly on all areas of the follicular wall <u>except</u> the stigma to cause rupture. That is, because of the poor blood flow to the stigma, it is more likely that LH acts on the extrastigmal regions of the follicular wall, with the resultant changes in those regions then influencing stigma weakening and rupture. The theory presented above does support the role of an inflammatory response and enzymes,[64] supports the ischemia theory of Nalbandov, and emphasizes the role of smooth muscle contraction presented by Nalbandov[140] and Schroeder & Talbot[191] but denied by Espey.[62-64] It does not support that the stigma forms soon before ovulation for purely mechanical reasons.[177] The existence of a preformed, hypovascular stigma on the follicular surface of maturing follicles may help explain why LH-induced ovulation <u>in vitro</u> is possible. A localized response (degeneration and rupture) to LH is difficult to explain <u>in vitro</u>, inasmuch as the entire follicular surface is exposed to an equal concentration of ovulating hormone.

Much more information is needed about the microanatomical changes in the follicular wall preceding ovulation. Is a periovulatory, inflammatory-like response ubiquitous among the vertebrates? How localized are the vasular and degenerative changes around the follicular surface? What is the orientation of smooth muscle fibers in relation to the point of rupture? if the vascular changes accompanying the inflammatory response are required for ovulation,[64] how is it that ovulation can be induced <u>in vitro</u>, where presumably no inflammation can occur? Does LH bind to the stigma region, and does this same hormone bind to vascular and thecal smooth muscle as well as to thecal and granulosa endocrine cells? Is there an evolutionary pattern to the steroid that mediates LH-induced ovulation? Are prostaglandins mediators of LH, or just permissive to LH action? What are the relative roles of $PGF_{2\alpha}$, PGE_1, and PGE_2 in the ovulatory process? Since there are sensory fibers in the theca of some mammals[35] and even specialized thecal sensory receptors in the avian theca,[44] perhaps a role of sensory feedback from the follicle in controlling ovulation should be investigated. The large size of the ovarian follicles of reptiles and birds offers an excellent opportunity to study ovulation, because the follicular wall can be manipulated and assayed more readily, the microarchitecture of the follicle can be mapped more easily, and the pattern of electrical and muscular activity across the follicular wall can more readily be observed.

IX. ACKNOWLEDGMENTS

I am grateful to several people for their critique of this manuscript; these include Drs. Fred Goetz, Paul Schroeder and Prudence Talbot as well as my graduate students (Harriet Austin, Kristin Lopez, Matt Rand, and Cliff Summers). The author's present research on ovulation is supported by NSF grant PCM-8417083.

X. REFERENCES

1. Adelstein, R.S., and Klee, C.B. (1980). Smooth muscle myosin light chain kinase. In "Calcium and Cell Function" (W.Y. Cheung, ed.), Vol. 1, pp. 167-182. Academic Press, New York.
2. Aeby, C. (1859). Uber glatten Muskelfasern in Ovarium und Mesovarium non Wirbelthieren. Arch. Anat. Physiol. Wiss. Med. 5, 675-676.
3. Ainsworth, L., Baker, R.D., and Armstrong, D.T. (1975). Pre-ovulatory changes in follicular fluid prostaglandin F levels in swine. Prostaglandins 9, 915-925.
4. Ainsworth, L., Tsang, B.K., Marcus, G.J., and Downey, B.R. (1984). Prostaglandin production by dispersed granulosa and theca interna cells from porcine preovulatory follicles. Biol. Reprod. 31, 115-12.
5. Amsterdam, A., Lindner, H.R., and Groschel-Stewart, U. (1977). Localization of actin and myosin in the rat oocyte and follicular wall by immunofluorescence. Anat. Rec. 187, 311-327.
6. Anderson, J.W., and Yatvin, M.B. (1970). Metabolic and ultrastructural changes in the frog ovarian follicle in response to pituitary stimulation. J. Cell Biol. 46, 491-505.
7. Armstrong, D.C., and Grinwich, D.L. (1972). Blockade of spontaneous and LH induced ovulation in rats by indomethacin, on inhibitor of prostaglandin synthesis. Prostaglndins 1, 21-28.
8. Armstrong, D.T., Grinwich, D.L., Moon, Y.S., and Zamecnik, J. (1974). Inhibition of ovulation in rabbits by intrafollicular injection of indomethacin and prostaglandin F antiserum. Life Sci. 14, 129-140.
9. Armstrong, D.T., Moon, Y.S., and Zamecnik, J. (1974). Evidence for a role of ovarian prostaglandins in ovulation. In "Gonadotropins and Gonadal Function" (N.R. Mougdal, ed.), pp. 343-356. Academic Press, New York.
10. Asdell, S.A. (1962). The mechanism of ovulation. In: "The Ovary" (S. Zuckerman, ed.), pp. 435-449. Academic Press, New York.
11. Balboni, G.C. (1983). Structural changes: ovulation and luteal phase. In "The Ovary: comprehensive Endocrinology" (G.B. Serra, ed.), pp. 123-142. Raven Press, New York.
12. Bartelmez, G.W. (1912). The bilaterality of the pigeon's egg. I. A study in egg organization from the first growth period of the oocyte to the beginning of cleavage. J. Morphol. 23, 269-329.
13. Bauminger, S., and Lindner, H.R. (1975). Periovulatory changes in ovarian prostaglandin formation and their hormonal control in the rat. Prostaglandins 9, 737-751.
14. Beers, W.H. (1975). Follicular plasminogen and plasminogen activator and the effect of plasmin on follicular wall. Cell 6, 379-386.
15. Beers, W.H., Strickland, S., and Reich, E. (1975). Ovarian plasminogen activator: relationship to ovulation and hormonal regulation. Cell 6, 387-394.
16. Behrman, H.R. (1979). Prostaglandins in hypothalamo-pituitary and ovarian function. Ann. Rev. Physiol. 41, 685-700.

17. Berndtson, A.K., and Goetz, F.W. (1986). Investigations on the control of *in vitro* spontaneous brook trout (*Salvelinus fontinalis*) ovulation. Gen. Comp. Endocrinol. 61, 134-141.
18. Betteridge, K.J., Kelley, W.A., and Marston, J.H. (1970). Morphology of the rhesus monkey ovary near the time of ovulation. J. Reprod. Fertil. 22 453-459.
19. Bjersing, L., and Cajander, S. (1974). Ovulation and the mechanism of follicle rupture. I. Light microscopic changes in rabbit ovarian follicles prior to induced ovulation. Cell Tiss. Res. 149, 287-300.
20. Bjersing, L., and Cajander, S. (1974). Ovulation and the mechanism of follicle rupture. II. Scanning electron microscopy of rabbit germinal epithelium prior to induced ovulation. Cell Tiss. Res. 149, 301-312.
21. Bjersing, L., and Cajander, S. (1974). Ovulation and the mechanism of follicle rupture. III. Transmission electron microscopy of rabbit germinal epithelium prior to induced ovulation. Cell Tiss. Res. 149, 313-327.
22. Bjersing, L., and Cajander, S. (1974). Ovulation and the mechanism of follicle rupture. IV. Ultrastructure of membrana granulosa of rabbit graafian follicles prior to induced ovulation. Cell Tiss. Res. 153, 1-14.
23. Bjersing, L., and Cajander, S. (1974). Ovulation and the mechanism of follicle rupture. V. Ultrastructure of tunica albuginea and theca externa of graafian follicles prior to induced ovulation. Cell Tiss. Res. 153, 15-30.
24. Bjersing, L., and Cajander, S. (1974). Ovulation and the mechanism of follicle rupture. VI. Ultrastructure of theca interna and the inner vascular network surrounding rabbit graafian follicles prior to induced ovulation. Cell Tiss. Res. 153, 31-44.
25. Bjersing, L., and Cajander, S. (1975). Ovulation and the role of the ovarian surface epithelium. Experientia 31, 605-608.
26. Blandau, R. J. (1955). Ovulation in the living albino rat. Fertil. Steril. 6, 391-404.
27. Blandau, R.J. (1970). Growth of the ovarian follicle and ovulation. *In* "Progress in Gynecology" (S. Sturgis and M. Taymor, eds.), Vol. V, pp. 58-76. Grune and Stratton, New York.
28. Blandau, R.J., and Rumery, R.E. (1963). Measurements of intrafollicular pressure in ovulatory and preovulatory follicles of the rat. Fertil. Steril. 14, 330-341.
29. Bouffard, R.E. (1979). The role of prostaglandins during sexual maturation, ovulation and spermiation in the goldfish, *Carassius auratus*. M.Sc. Thesis, Univ. British Columbia, Vancouver, B.C., Canada.
30. Brambell, F.W.R. (1956). Ovarian changes. *In* "Marshall's Physiology of Reproduction" (A.S. Parker, ed.), Vol. 1, pp. 397-542. 3rd ed. Longmans Green, London and New York.
31. Brandt, A. (1877). Fragementarische Bemerkungen Uber de Ovarium des Frosches. Z., Wiss. Zool. 28, 575-586.
32. Bronson, R., Bryant, G., Balk, M., and Emanuele, N. (1979). Intrafollicular pressure within preovulatory follicles of the pig. Fertil. Steril. 31, 205-213.
33. Brown, C.G., and Poyser, A.L. (1984). Studies on ovarian prostaglandin production in relation to ovulation in the rat. J. Reprod. Fertil. 72, 407-414.
34. Browne, C.L., Wiley, H.S., and Dumont, J.N. (1979). Oocyte-follicle cell gap junctions in *Xenopus laevis* and the effects of gonadotropin on their permeability. Science 203, 182-183.
35. Burden, H.W. (1972). Ultrastructural observation on ovarian perifollicular smooth muscle in the cat, guinea pig, and rabbit. Amer. J. Anat. 133, 125-142.

36. Burden, H.W. (1973). The distribution of smooth muscle in the cat ovary with a note on its adrenergic innervation. J. Morphol. 140, 467-476.
37. Burden, H.W. (1978). Ovarian innervation. In "The Vertebrate Ovary--Comparative Biology and Evolution." (R.E. Jones, ed.), pp. 615-638. Plenum Publ. Corp., New York.
38. Burr, J.H., and Davies, J.I. (1951). The vascular system of the rabbit ovary and its relationship to ovulation. Anat. Rec. 111, 273-292.
39. Busson-Mabillot, S. (1967). Structure ovarienne chez la lamproie de planer adulte, Lampetra planeri (Bloch). Arch. Zool. Exp. Gen. 108, 413-446.
40. Byskov, A.G.S. (1969). Ultrastructural studies on the preovulatory follicle in the mouse ovary. Zeit. Zell. Mikrosk. Anat. 100, 285-299.
41. Cajander, S., and Bjersin, L. (1975). Fine structural demonstration of acid phosphatase in rabbit germinal epithelum prior to induced ovulation. Cell Tiss. Res. 164, 279-289.
42. Cetta, F., and Goetz, F.W. (1982). Ovarian and plasma prostaglandin E and F levels in brook trout (Salvelinus fontinalis) during ovulation. Biol. Reprod. 27, 1216-1221.
43. Cherney, D.D., Didio, L.J.A., and Motta, P. (1975). The development of rabbit ovarian follicles following copulation. Fertil. Steril. 26, 257-270.
43a. Curry, T.E., Jr., Dean, D.D., Woessner, J.F., Jr., and LeMaire, W.J. (1985). The extraction of a tissue collagenase associated with ovulation in the rat. Biol. Reprod. 33, 981-991.
44. Dahl, E. (1970). Studies on the fine structure of ovarian interstitial tissue. 3. The innervation of the thecal gland of the domestic fowl. Z. Zellforsch. 109, 212-226.
45. Daniel, E.E., and Sarna, S. (1978). The generation and conduction of activity in smooth muscle. Ann. Rev. Pharmacol. Toxicol. 18, 145-166.
46. Dawson, A.B., and Friedgood, H.B. (1940). The time and sequence of preovulatory changes in the cat ovary after mating or mechanical stimulation of the cervix uteri. Anat. Rec. 76, 411-429.
47. Day, S.C., and Nalbandov, A.V. (1977). Pressure of prostaglandin F (PGF) in hen follicles and its physiological role in ovulation and oviposition. Biol. Reprod. 16, 486-494.
48. Dekel, N., Kraicer, P.F., Phillips, D.M., Sanchez, R.S., and Segal, S.J. (1978). Cellular associations in the rat oocyte-cumulus complex: morphology and ovulatory changes. Gamete Res. 1, 47-58.
49. Dodd, J.M., and Sumpter, J.P. (1984). Fishes. In: "Marshall's Physiology of Reproduction," Fourth ed. (G.E. Lamming, ed.)., Vol. 1, pp. 1-126. Churchill Livingston, New York.
50. Donaldson, E.M., and Hunter, G.A. (1983). Induced final maturation, ovulation and spermation in cultured fish. In: "Fish Physiology" (W.S. Hoar, D.J. Randall, and E.M. Donaldson, eds.). Vol. IX, Part B, pp. 351-404.
50a. Downs, S.M. and Longo, F.J. (1983). An ultrastructural study of preovulatory apical development in mouse ovarian follicles: effects of indomethacin. Anat. Rec. 205, 159-160.
51. Doyle, J.B. (1954). Ovulation and the effects of selective uterotubal denervation; direct observation by culdotomy. Fertil. Steril. 5, 105-130.
52. Dumont, N.J., and Brumett, A.R. (1978). Oogenesis in Xenopus laevis (Daudin). V. Relationships between developing oocytes and their investing follicular tissues. J. Morphol. 155, 73-98.
53. Epler, P. (1981). Effects of steroid and gonadotropic hormones on the maturation of carp ovaries. Part V. Ovulation, fertilization, and

embryonal development of carp oocytes *in vitro*. Pol. Arch. Hydrobiol. 28, 119-125.

54. Epler, P., Bieniarz, K., and Marosz, E. (1985). Effect of temperature, 17α-hydroxy-20β-dihydroprogesterone and prostaglandin $F_{2\alpha}$ on carp oocyte maturation and ovulation *in vitro*. Gen. Comp. Endocrinol. 58, 192-201.
55. Eppig, J.J., and Ward-Bailey, P.F. (1982). The mechanism of cumulus cell-oocyte uncoupling: evidence for the participation of both cumulus cells and oocytes. Gamete Res. 6, 145-154.
56. Espey, L.L. (1967). Ultrastructure of the apex of the rabbit Graafian follicle during the ovulation process. Endocrinology 81, 267-276.
57. Espey, L.L. (1967). Tenacity of porcine graafian follicle as it approaches ovulation. Amer. J. Physiol. 212, 1397-1401.
58. Espey, L.L. (1970). Effect of various substances on tensile strength of sow ovarian follicles. Amer. J. Physiol. 219, 230-233.
59. Espey, L.L. (1971). Multivesicular structures in proliferating fibroblasts of rabbit ovarian follicles during ovulation. J. Cell Biol. 48, 437-442.
60. Espey, L.L. (1971). Decomposition of connective tissue in rabbit ovarian follicles by multivesicular structures of thecal fibroblasts. Endocrinology 88, 437-444.
61. Espey, L.L. (1974). Ovarian proteolytic enzymes and ovulation. Biol. Reprod. 10, 216-233.
62. Espey, L.L. (1978). Ovulation. In: "The Vertebrate Ovary-Comparative Biology and Evolution." (R.E. Jones, ed.), pp. 503-532. Plenum Publ. Corp., New York.
63. Espey, L.L. (1978). Ovarian contractility and its relationship to ovulation: A review. Biol. Reprod. 19, 540-551.
64. Espey, L. (1980). Ovulation as an inflammatory reaction--A hypothesis. Biol. Reprod. 22, 73-106.
65. Espey, L.L., and Coons, P.J. (1976). Factors which influence ovulatory degradation of rabbit ovarian follicles. Biol. Reprod. 14, 233-245.

65a. Espey, L.L., Coons, P., Marsh, J., and LeMaire, W.J. (1981). Effect of indomethacin on preovulatory changes in the ultrastructure of rabbit Graafian follicles. Endocrinology 108, 1040-1048.

66. Espey, L., and Lipner, H. (1963). Measurements of intrafollicular pressures in the rabbit ovary. Amer. J. Physiol. 205, 1067-1072.
67. Espey, L.L., and Lipner, H. (1965). Enzyme-induced rupture of rabbit Graafian follicles. Amer. J. Physiol. 208, 208-213.
68. Espey, L.L., and Rondell, P. (1968). Collagenolytic activity in the rabbit and sow graafian follicle during ovulation. Amer. J. Physiol. 214, 326-320.
69. Ferrando, G., and Nalbandov, A.V. (1969). Direct effect on the ovary of the adrenergic blocking drug dibenzyline. Endocrinology 85, 38-42.
70. Follett, B.K. (1984). Birds. *In*: "Marshall's Physiology of Reproduction," Fourth Ed. (G.E. Lamming, ed.), Vol. 1, pp. 283-350. Churchill Livingstone, New York.
71. Fujii, M., Tojo, H., and Koga, K. (1981). Detection and properties of collagenase in ovarian follicle wall of domestic fowl (*Gallus domesticus*). Int. J. Biochem. 13, 1043-1046.
72. Fujii, S., and Yoshimura, Y. (1979). Morphological observations on the stigma of the follicular wall concerning the mechanism of ovulation in hens. J. Fac. Appl. Biol. Sci., Hiroshima Univ. 18, 185-196.
73. Fukumoto, M., Yajima, Y., Okamura, H., and Midorikawa, O. (1981). Collagenolytic enzyme activity in human ovary: an ovulatory enzyme system. Fertil. Steril. 36, 746-750.
74. Gabella, G. (1973). Fine structure of smooth muscle. Phil. Trans. Roy. Soc. Lond. 265B, 7-16.

75. Goetz, F.W. (1983). Hormonal control of oocyte final maturation and ovulation in fishes. In "Fish Physiology" (W.S. Hoar, D.J. Randall, and E.M. Donaldson, eds.), Vol. IXB, pp.117-170. Academic Press, New York.
76. Goetz, F.W., and Cetta, F. (1983). Ovarian and plasma PGE and PGF levels in naturally ovulating brook trout (Salvelinus fontinalis) and the effects of indomethacin on prostaglandin levels. Prostaglandins 26, 387-395.
77. Goetz, F.W., and Nagahama, Y. (1985). The effects of pH on in vitro ovulation of goldfish (Carassius auratus) oocytes. J. Exp. Zool. 235, 81-86.
78. Goetz, F.W., and Smith, D.C. (1980). In vitro effects of theophylline and dibutyryl adenosine 3'5'-cyclic monophosphoric acid on spontaneous and prostaglandin $PGF_{2\alpha}$-induced ovulation of brook trout (Salvelinus botinalis) oocytes. Biol. Reprod. 22, Suppl. 1, 114A
79. Goetz, F.W., Smith, D.C., and Krickl, S.P. (1982). The effects of prostaglandins, phosphodiesterase inhibitors, and cyclic AMP on ovulation of brook trout (Salvelinus fontinalis) oocytes. Gen. Comp. Endocrinol. 48, 154-160.
80. Goetz, F.W., and Theofan, G. (1979). In vitro stimulation of germinal vesicle breakdown and ovulation of yellow perch (Perca flavescens) oocytes. Effects of 17α-hydroxy-20β-dihydroprogesterone and prostaglandins. Gen. Comp. Endocrinol. 37, 273-285.
81. Greenwald, G.S. (1974). Role of follicle-stimulating hormone and luteinizing hormone in follicular development and ovulation. In "Handbook of Physiology," Sec. 7, "Endocrinology," Vol. IV. (R.O. Greep and E.B. Astwood, eds.), pp. 293-323. Amer. Physiol. Soc., Washington, D.C.
82. Guraya, S.S. (1978). Maturation of the follicular wall of nonmammalian vertebrates. In "The Vertebrate Ovary-Comparative Biology and Evolution" (R.E. Jones, ed.), pp. 216-330. Plenum Publ. Corp., New York.
83. Guttmacher, M.S., and Guttmacher, A.F. (1921). Morphological and physiological studies of the musculature of the mature Graafian follicle of the sow. Johns Hopkins Hosp. Bull. 32, 394-399.
84. Guzsal, E. (1966). Histological studies on the mature and post-ovulation ovarian follicle of the fowl. Acta Vet. Acad. Sci. Hung. 16, 37-44.
85. Hamada, Y., Bronson, R.A., Wright, K.H., and Wallach, E.E. (1977). Ovulation in the perfused rabbit ovary: the influence of prostaglandins and prostaglandin inhibitors. Biol. Reprod. 17, 58-63.
86. Hamada, Y., Wright, K.H., and Wallach, E.E. (1978). In vitro reversal of indomethacin-blocked ovulation by prostaglandin $F_{2\alpha}$. Fertil. Steril. 30, 702-706.
87. Hammond, R. W., Olson, D.M., Frenkel, R.B., Biellier, H.V., and Hertelendy, F. (1980). Prostaglandins and steroid hormones in plasma and ovarian follicles during the ovulation cycle of the domestic hen (Gallus domesticus). Gen. Comp. Endocrinol. 42, 195-202.
88. Hartog, M. (1904). Note on embryonic ferments. J. Physiol. 31, p. xlvii.
89. Heinze, H. (1975). Pelviscopy in the mare. J. Reprod. Fertil, Suppl. 23, 319-321.
90. Hill, R.T., Allen, E., and Kramer, T.C. (1935). Cinemicrographic studies of rabbit ovulation. Anat. Rec. 63, 239-245.
91. Holmes, P.V., Janson, P.O., Sogn, J., Kallfelt, B., LeMaire, W.J., Ahren, K.B., Cajander, S., and Bjersing, L. (1983). Effects of $PGF_{2\alpha}$ and indomethacin on ovulation and steroid production in the isolated perfused rabbit ovary. Acta Endocrinol. 104, 233-239.
92. Ichikawa, S., Morioka, H., Ohta, M., Oda, K., and Murao, S. (1983). Effect of various proteinase inhibitors on ovulation of explanted

hamster ovaries. J. Reprod. Fertil. 68, 407-412.

93. Ichikawa, S., Ohta, M., Morioka, H., and Murao, S. (1983). Blockage of ovulation in the explanted hamster ovary by a collagenease inhibitor. J. Reprod. Fertil. 68, 17-19.

94. Jalabert, B. (1976). *In vitro* oocyte maturation and ovulation in rainbow trout (*Salmo gairdneri*), northern pike (*Esox lucius*) and goldfish *Carassius auratus*). J. Fish. Res. Bd. Canad. 33, 974-988.

95. Jalabert, B., and Szollosi, D. (1975). *In vitro* ovulation of trout oocytes: Effects of prostaglandins on smooth muscle-like cells of the theca. Prostaglandins 9, 765-778.

96. Jewett, D.A., and Dukelow, W.R. (1972). Serial observations of follicular morphology near ovulation in *Macaca fascicularis*. J. Reprod. Fertil. 31, 287-290.

97. Johansson, E.D.B., Neill, J.D., and Knobil, E. (1968). Periovulatory progesterone concentration in the peripheral plasma of the Rhesus monkey with a methodological note of the detection of ovulation. Endocrinology 82, 143-148.

98. Jones, R.E., Austin, H.A., and Summers, C.H. (1984). Spontaneous rhythmic contractions of the ovarian follicular wall of a lizard (*Anolis carolinensis*). Gen. Comp. Endocrinol. 56, 252-257.

99. Kagawa, H., and Nagahama, Y. (1981). *In vitro* effects of prostaglandins on ovulation in goldfish *Crassius auratus*. Bull. Jpn. Soc. Sci. Fish. 47, 1119-1121.

100. Kao, L.W.L., and Nalbandov, A.V. (1972). The effect of antiadrenergic drugs on ovulation in hens. Endocrinology 90, 1343-1349.

101. Kelly, G.L. (1931). Direct observation of rupture of the graafian follicle in a mammal. J. Fla. Med. Assoc., 17, 422-423.

102. Koos, R.D., Clark, M.R., Janson, P.O., Ahren, K.E.B., and LeMaire, W.J. (1983). Prostaglandin levels in preovulatory follicles from rabbit ovaries perfused *in vitro*. Prostaglandins 25, 715-724.

103. Kosterlitz, H.W., and Watt, A.J. (1975). Stimulation by stretch. *In* "Methods in Pharmacology" (E.E. Daniel and D.M. Paton, eds.), Vol. 3,pp. 347-358. Plenum Publ. Corp., New York.

104. Kraus, S.D. (1947). Observations on the mechanism of ovulation in the frog, hen, and rabbit. West. J. Surg. Obstet. Gynecol. 55, 424-437.

105. Labhsetwar, A.P. (1972). Further evidence for synergism between LH and FSH in the induction of ovulation in rats; lack of effects of prolactin. J. Reprod. Fertil. 29, 435-438.

106. Lambertsen, C.J., Greenbaum, D.F., Wright, K.H., and Wallach, E.E. (1976). *In vitro* studies of ovulation in the perfused rabbit ovary. Fertil. Steril. 27, 178-187.

107. Larsen, J.H., Schroeder, P.C., and Waldo, A.E. (1977). Structure and function of the amphibian follicular epithelium during ovulation. Cell Tiss. Res. 181, 505-518.

108. Larsen, L.O. (1970). The lamprey egg at ovulation (*Lampetra fluviatilis*). Biol. Reprod.. 2, 37-47.

109. Laughran, L.J., Larsen, J.H., Jr., and Schroeder, P.C. (1981). Ultrastructure of developing ovarian follicles and ovulation in the lizard *Anolis carolinensis* (Reptilia). Zoomorphology 98, 191-208.

110. Laughran, L.J., Larsen, J.H., Jr., and Schroeder, P.C. (1981). Microfilaments interacting with heavy meromyosin and deoxyribonuclease I in cells of the ovarian follicle of a lizard. Cell Tiss. Res. 218, 537-543.

111. LeMaire, W.J., Leidner, R., and Marsh, J.M. (1975). Pre- and postovulatory changes in the concentration of prostaglandins in rat Graafian follicles. Prostaglandins 9, 221-229.

112. LeMaire, W.J., Yang, N.S.T., Behrman, H.R., and Marsh, J.M. (1973). Preovulatory changes in the concentration of prostaglandins in rabbit Graafian follicles. Prostaglandins 3, 367-376.

113. Licht, P. (1984). Reptiles. In: "Marshall's Physiology of Reproduction," Fourth ed. (G.E. Lamming, ed.), Vol. 1, pp. 206-282. Churchill Livingston, New York.
114. Licht, P., and Crews, D.P. (1975). Stimulation of ovarian and oviductal growth and ovulation in female lizards by reptilian (turtle) gonadotropins. Gen. Comp. Endocrinol. 25, 466-471.
115. Licht, P., Farmer, S.W., and Papkoff, H. (1975). The nature of the pituitary gonadotropins and their role in ovulation in a urodele amphibian (Ambystoma tigrinum). Life Sci. 17, 1049-1054.
116. Licht, P., Papkoff, H., Farmer, S., Muller, C., Tsui, H.W., and Crews, D. (1977). Evolution of gonadotropin structure and function. Rec. Prog. Horm. Res. 33, 169-248.
117. Licht, P., and Tsui, H.W. (1975). Evidence for the intrinsic activity of ovine FSH on spermatogenesis, ovarian growth, steroidogenesis and ovulation in lizards. Biol. Reprod. 12, 346-350.
118. Lindner, H.R., Amsterdam, A., Saloman, V., Tsafriri, A., Nimrod, A., Lamprecht, S.A., Zor, U., and Koch, Y. (1977). Intraovarian factors in ovulation: determinants of follicular response to gonadotropins. J. Reprod. Fertil. 51, 215-235.
119. Lipner, H. (1965). Induction of Graafian follicle proteolytic enzyme activity by HCG in rat Graafian follicle. Amer. Zool. 5, 167 (abstract).
120. Lipner, H. (1973). Mechanism of mammalian ovulation. In "Handbook of Physiology," Part 1, "Endocrinology," Vol. II (R.O. Gree, ed.), pp. 409-437. Amer. Physiol. Soc. Wash., D.C.
121. Lipner, H., and Greep, R.O. (1971). Inhibition of steroidogenesis at various sites in the biosynthetic pathway in relation to induced ovulation. Endocrinology 88, 602-607.
122. Lipner, H., and Maxwell, B. (1960). Hypothesis concerning the role of follicular contractions in ovulation. Science 131, 1737-1738.
123. Lofman, C.O., Janson, P.O., Kallfelt, B.J., Ahren, K., and LeMaire, W.J. (1982). The study of ovulation in the isolated perfused rabbit ovary. II. Photographic and cinematographic observations. Biol. Reprod. 26, 467-474.
124. Lofts, B. (1984). Amphibians. In: "Marshall's Physiology of Reproduction," Fourth ed. (G.E. Lamming, ed.), Vol. 1, pp. 127-205. Churchill Livingston, New York.
125. Markee, J.E., and Hinsey, J.C. (1936). Observations on ovulation in the rabbit. Anat. Rec. 64, 309-319.
125a. Martin, G.G. and Miller-Walker, C. (1983). Visualization of the three-dimensional distribution of collagen fibrils over preovulatory follicles in the hamster. J. Exp. Zool. 225, 311-319.
126. Martin, G.C., and Talbott, P. (1981). The role of follicular smooth muscle cells in hamster ovulation. J. Exp. Zool. 216, 468-482.
127. Martin, G.G., and Talbot, P. (1981). Drugs that block smooth muscle contraction inhibit in vivo ovulation in hamsters. J. Exp. Zool. 216, 483-491.
128. McKenzie, F.F., and Terrill, C.F. (1937). Estrus, ovulation, and related phenomena in the ewe. Res. Bull. Mo. Agric. Exp. Sta., no. 264.
129. McReynolds, H., Siraki, C., Branson, P., and Pollock, R., Jr. (1973). Smooth muscle-like cells in ovaries of the hamster and gerbil. Z. Zellforsch. 140, 1-8.
130. Merriam, R.W. (1972). On the mechanism of action in gonadotropic stimulation of oocyte maturation in Xenopus laevis. J. Exp. Zool. 180, 421-426.
130a. Morales, T.I., Woessner, J.F., Jr., Marsh, J.M. and LeMaire, W.J. (1983). Collagen, collagenase and collagenolytic activity in rat Graafian follicles during follicular growth and ovulation. Biochem. Biophys. Acta 756, 119-122.

131. Motta, P., Cherney, D.D., and Didio, L.J.A. (1971). Scanning and transmission electron microscopy of the ovarian surface in mammals with special reference to ovulation. J. Submicroscop. Cytol. 3, 85-100.
132. Motta, P., and Van Blerkom, J. (1975). A scanning electron microscopic study of the luteo-follicular complex. II. Events leading to ovulation. Amer. J. Anat. 143, 241-263.
133. Mougdal, R.P., Razdan, M.N. (1981). Induction of ovulation in in vitro by LH and catecholamines in hens is mediated by α-adrenergic receptors. Nature Lond., 293, 738-739.
134. Murdoch, W.J., Dailey, R.A., and Inskeep, E.K. (1981). Preovulatory changes in prostaglandin E_2 and $F_{2\alpha}$ in ovine follicles. J. Anim. Sci. 53, 192-205.
135. Murdoch, W.J., and Myers, D.A. (1983). Effect of treatment of estrous ewes with indomethacin on the distribution of ovarian blood flow to the periovulatory follicle. Biol. Reprod. 29, 1229-1232.
136. Murdoch, W.J., Nix, K.J., and Dunn, T.G. (1983). Dynamics of ovarian blood supply to periovulatory follicles of the ewe. Biol. Reprod. 28, 1001-1006.
137. Nagahama, Y., Chan, K., and Hoar, W.S. (1976). Histochemistry and ultrastructure of pre- and post-ovulatory follicles in the ovary of the goldfish, Carassius auratus. Can. J. Zool. 54, 128-139.
138. Nakajo, S., Zakaria, A.H., and Imai, K. (1973). Effect of local administration of proteolytic enzymes on the rupture of the ovarian follicle in the domestic fowl, Gallus domesticus. J. Reprod. Fertil. 34, 235-240.
139. Nakamura, A., and Nakajo, S. (1980). Proteolytic activity in the ovarian follicle wall of the Japanese quail, Coturnix corturnix japonica. Jap. J. Fertil. 25, 98-102.
140. Nalbandov, A.V. (1961). Mechanisms controlling ovulation of avian and mammalian follicles. In "Control of Ovulation" (C.A. Villee, ed.), pp. 122-132. Pergamon Press, New York.
141. Nalbandov, A.V., and James, M.F. (1949). The blood-vascular system of the chicken ovary. Amer. J. Anat. 85, 347-377.
142. Neher, B.H., Olsen, M.W., and Fraps, R.M. (1950). Ovulation of excised ovum in the hen. Poultry Sci. 29, 554-557.
143. Nilsson, O., and Munshi, S.F. (1973). Scanning electron microscopy of mouse follicles at ovulation. J. Submicroscop. Cytol. 5, 1-6.
144. Norman, R.L., and Greenwald, G.S. (1971). Effect of phenobarbital, hypophysectomy and x-irradiation on preovulatory progesterone levels in the cyclic hamster. Endocrinology 89, 598-605.
145. Norman, R.L., and Greenwald, G.S. (1972). follicular histology and physiological correlates in the preovulatory hamster. Anat. Rec. 173, 95-108.
146. Norris, D.O., and Duvall, D. (1981). Hormone-induced ovulation in Ambystoma tigrinum: Influence of prolactin and thyroxine. J. Exp. Zool. 216, 175-180.
147. Ogata, H., Normua, T., and Hata, M. (1979). Prostaglandin $F_{2\alpha}$ changes induced by ovulatory stimuli in the pond loach, Misgurnus anguillicaudatus. Bull. Jpn. Soc. Sci. Fish. 45, 929-931.
147a. Ogawa, K. and Goto, K. (1984). Effect of gonadotropins, sex steroids and adenohypophysectomy on the activities of proteolytic enzymes in the ovarian follicle wall of the domestic fowl (Gallus domesticus). J. Reprod. Fertil. 71, 545-549.
148. Ogawa, K., and Nishiyama, H. (1979). Studies on the mechanism of ovulation in the fowl. Mem. Foc. Agric. Kagoshima Univ. 7, 1-13.
149. Okamura, H., Okazaki, T., and Nakajima, A. (1974). Effects of neurotransmitters and prostaglandins on human ovarian contractility. Obstet. Gynecol. 44: 720-726.

150. Okamura, H., Takenaka, A., Yajima, Y., and Nishimura, T. (1980). Ovulatory changes in the wall at the apex of the human Graafian follicle. J. Reprod. Fertil. 58, 153-155.

151. Okamura, H., Virutamasen, P., Wright, K.H., and Wallach, E.E. (1972). Ovarian smooth muscle in the human being, rabbit, and cat. Amer. J. Obstet. Gynecol. 44, 720-726.

152. Opel, H., and Nalbandov, A.V. (1958). A study of hormonal growth and ovulation of follicles in hypophysectomized hens. Poultry Sci. 37, 1230-1231.

153. Opel, H., and Nalbandov, A.V. (1961). Follicular growth and ovulation in hypophysectomized hens. Endocrinology 69, 1016-1028.

154. Opel, H., and Nalbandov, A.V. (1961). Ovulability of ovarian follicles in the hypophysectomized hen. Endocrinology 69, 1029-1035.

155. O'Shea,J.D. (1970). An ultrastructural study of smooth muscle-like cells in the theca externa of ovarian follicles in the rat. Anat. Rec. 167, 127-140.

156. O'Shea, J.D. (1971). Smooth muscle-like cells in the theca externa of ovarian follicles in the sheep. J. Reprod. Fertil. 24, 283-285.

157. O'Shea, J.D., and Phillips, R.E. (1974). Contractility *in vitro* of ovarian follicles from sheep, and the effects of drugs. Biol. Reprod. 10, 370-379.

158. Oshiro, T., and Hibiya, T. (1975). Presence of ovulation-inducing enzymes in the ovarian follicles of loach. Bull. Jpn. Soc. Sci. Fish. 41, 115.

159. Oshiro, T., and Hibiya, T. (1981). Water absorption of oocytes in the plaice *Limanda yokohamae* during meiotic maturation and its role in rupturing follicles. (in Japanese). Bull. Jap. Soc. Fish. 47, 835-842.

160. Osman, P., and Dullaart, J. (1976). Intraovarian release of eggs in the rat and monkey. J. Reprod. Fertil. 47, 101-103.

161. Osvaldo-Decima, L. (1970). Smooth muscle in the ovary of the rat and monkey. J. Ultrastr. Res. 29, 218-227.

162. Owman, Ch., Sjoberg, N.-O., Svennson, K.-G., and Walles, B. (1975). Autonomic nerves mediating contractility in the human Graafian follicle. J. Reprod. Fertil. 45, 553-566.

163. Owman, Ch., Sjoberg, N.-O., Wallach, E., Walles, B., and Wright, K. (1979). Neuromuscular mechanisms of ovulation. *In* "Human Ovulation" (E.S.E. Hafez, ed.), pp. 57-100. Elsevier/North Holland Biomedical, New York.

164. Parr, E.L. (1974). Histological examination of the rat ovarian follicle wall prior to ovulation. Biol. Reprod. 11, 483-503.

165. Parr, E.L. (1975). Rupture of ovarian follicles at ovulation. J. Reprod. Fertil., Suppl. 22, 165-207.

166. Pendergrass, P., and Schroeder, P.C. (1976). The ultrastructure of the thecal cell of the teleost, *Oryzias latipes*, during ovulation *in vitro*., J. Reprod. Fertil. 47, 229-233.

166a. Penkala, J.E. and Talbot, P. (1984). Inhibitors of protease activity block *in vivo* ovulation in the golden hamster. J. Cell Biol. 99, 391a.

167. Perry, M.M., Gilbert, A.B., and Evans, A.J. (1978). Electron microscope observations on the ovarian follicle of the domestic fowl during the rapid growth phase. J. Anat. 125, 481-497.

168. Pfluger, E. (1859). Ueber die Bewegungen der Ovarien. Arch. Anat. Physiol. Wirs. Med. 30-32.

169. Phillips, R.E., and Warren, C.D. (1937). Observations concerning the mechanisms of ovulation in the fowl. J. Exp. Zool. 76, 117-136.

170. Piper, P. and Vane, J. (1971). The release of prostaglandins from lung and other tissues. Ann. N.Y. Acad. Sci. 180, 363-385.

171. Pool, W.R., and Lipner, H. (1966). Inhibition of ovulation by antibiotics. Endocrinology 79, 858-864.

172. Rawson, J.M.R., and Dukelow, W.R. (1973). Observation of ovulation in *Macada fascicularis*. J. Reprod. Fertil. 34,187-190.
173. Rawson, J.M.R., and Espey, L.L. (1977). Concentration of electron dense granules in the rabbit ovarian surface epithelium during ovulation. Biol. Reprod. 17, 561-566.
173a. Reed, M., Burton, F.A., and Diest, P.A. (1979). Ovulation in the guinea pig. I. The ruptured follicle. J. Anat. 128, 1195-1206.
174. Reich, R., Miskin, R., and Tsafriri, A. (1985). Follicular plasminogen activator: Involvement in ovulation. Endocrinology 116, 516-521.
175. Reich, R., Tsafriri, A., and Mechanic, G.L. (1985). The involvement of collagenolysis in ovulation in the rat. Endocrinology 116, 522-527.
176. Robinson, E.J., and Rugh, R. (1943). The reproductive processes of the fish, *Oryzias latipes*. Biol. Bull. Woods Hole 84, 115-125.
177. Rodbard, D. (1968). Mechanics of ovulation. J. Clin. Endocrinol. 28, 849-861.
178. Rondell, P. (1964). Follicular pressure and distensibility in ovulation. Amer. J. Physiol. 207, 590-594.
179. Rondell, P. (1970). Biophysical aspects of ovulation. Biol. Reprod., Suppl. 2, 64-89.
180. Rondell, P. (1970). Follicular processes in ovulation. Fed. Proc. 29, 1875-1879.
181. Rondell, P. (1974). Role of steroid synthesis in the process of ovulation. Biol. Reprod. 10, 199-215.
182. Ruegg, R.P., Kriger,F.L., and Davey, K.G. (1981). Ovarian ecdysone elicits release of a myotropic ovulation hormone in *Rhodnius*. Int. J. Invert. Reprod. 3, 357-361.
183. Rugh, R. (1935). Ovulation in the frog. II. Follicular rupture to fertilization. J. Exp. Zool. 71, 163-193.
184. Schaeffer, J.M., Liu, J., Hsueh, A.J.W., and Yen, S.C.Y. (1984). Presence of oxytocin and arginine vasopressin in human ovary, oviduct, and follicular fluid. J. Clin. Endocrinol. Metab. 59, 970-973.
185. Schlaff, S., Kobayashi, Y., Wright, K.H., Santulli, R., and Wallach, E.E. (1983). Prostaglandin $F_{2\alpha}$, an ovulatory intermediate in the *in vitro* perfused rabbit ovary model. Prostaglandins 26, 111-121.
186. Schochet, S.S. (1916). A suggestion as to the process of ovulation and ovarian cyst formation. Anat. Rec. 10, 447-457.
187. Schroeder, P.C. (1973). The inhibition of amphibian ovulation *in vitro* by cytochalasin B. Experientia 29, 617-618.
188. Schroeder, P.C., and Larsen, J.H., Jr. (1985). Mechanism of ovulation in lower vertebrates. *In*: "Current Trends in Comparative Endocrinology" (B. Lofts and W.N. Holmes, eds.), Vol. 2, pp. 383-384. Hong Kong Univ. Press, Hong Kong.
189. Schroeder, P.C., and Pendergrass, P. (1976). The inhibition of *in vitro* ovulation from follicles of the teleost *Oryzias latipes* by cytochalasin B. J. Reprod. Fertil. 48, 327-330.
190. Schroeder, P.C., and Talbot, P. (1982). Intrafollicular pressure decreases in hamster preovulatory follicles during smooth muscle cell contraction *in vitro*. J. Exp. Zool. 224, 417-426.
191. Schroeder, P.C., and Talbot, P. (1985). Ovulation in the animal kingdom: A review with an emphasis on the role of contractile processes. Gamete Res. 11, 191-221.
192. Schuetz, A.W. (1967). Action of hormones on germinal vesicle breakdown in frog (*Rana pipiens*) oocytes. J. Exp. Zool. 166, 347-354.
193. Schuetz, A.W. (1971). *In vitro* induction of ovulation and oocyte maturation in *Rana pipiens* ovarian follicles: Effects of steroidal and nonsteroidal hormones. J. Exp. Zool. 178, 377-386.

194. Schuetz, A.W. (1972). Induction of structural alterations in the preovulatory amphibian ovarian follicle by hormones. Biol. Reprod. 6, 67-77.
195. Schuetz, A.W., and Lessman, A.L. (1981). Role of the surface epithelium of the ovarian follicle in ovulation of frog (Rana pipiens) oocytes. J. Cell Biol. 91, 187a.
196. Schultz, R.M. (1985). Roles of cell-to-cell communication in development. Biol. Reprod. 32, 27-42.
197. Shaykh, M., LeMaire, W.J., Papkoff, H., Curry, T.E., Jr., Sogn, J.H., and Koos, R.D. (1985). Ovulation in rat ovaries perfused in vitro with follicle-stimulating hormone. Biol. Reprod. 33, 629-636.
198. Shemesh, M., and Hansel, W. (1975). Stimulation of prostaglandin synthesis in bovine ovarian tissues by arachidonic acid and luteinizing hormone. Biol. Reprod. 13, 448-452.
199. Shimada, K., Olson, D.M., and Etches, R.J. (1984). Follicular and uterine prostaglandin levels in relation to uterine contraction and the first ovulation of a sequence in the hen. Biol. Reprod. 31, 76-82.
200. Smith, B.G. (1916). The process of ovulation in Amphibia. Rep. Mich. Acad. Sci. 18, 102-105.
201. Snyder, B.S., and Schuetz, A.W. (1973). In vitro evidence of steroidogenesis in the amphibian (Rana pipiens) ovarian follicle and its relationship to meiotic maturation and ovulation. J. Exp. Zool. 183, 333-341.
202. Soliman, K., and Walker, C. (1976). In vitro motility of the avian ovarian follicles. Gen. Comp. Endocrinol. 30, 372-378.
203. Somlyo, A.P., and Somlyo, A.V. (1976). Ultrastructural aspects of activation and contraction of vascular smooth muscle. Fed. Proc. 35, 1288-1293.
204. Stacey, N.E., and Goetz, F.W. (1982). Role of prostaglandins in fish reproduction. Can. J. Fish. Aquat. Sci. 39, 92-98.
205. Stacey, N.W., and Pandey, S. (1975). Effects of indomethacin and prostaglandins on ovulation of goldfish. Prostaglandins 9, 597-607.
206. Stefenson, A., Owman, Ch., Sjoberg, N.-O, Sporrong, B., and Walles, B. (1981). Comparative studies of the autonomic innervation of the mammalian ovary, with particular regard to the follicular system. Cell Tiss. Res. 215,47-62.
207. Strickland, S., and Beers, W.H. (1976). Studies on the role of plasminogen activation in ovulation. J. Biol. Chem. 251, 5694-5702.
208. Strickland, S., and Beers, W.H. (1979). Studies of the enzymatic basis and hormonal control of ovulation. In "Ovarian Follicular Development and function" (A.R. Midgley and W.A. Sadler, eds.), p. 143. Raven Press, New York.
209. Sundararaj, B.I., and Goswami, S. (1979). Role of interrenal in luteinizing hormone-induced ovulation and spawning in catfish Heteropneustes fossilis (Bloch). Gen. Comp. Endocrinol., Suppl. 2, 374-384.
210. Swanson, R.J., and Lipner, H. (1977). Mechanism of ovulation: effect of intrafollicular progesterone antiserum in rabbits. Fed. Proc. 36, 390.
211. Szollosi, D., and Jalabert, B. (1974). La theque de follicule ovarien de la Truite. J. Microscop. (Paris) 20, 92 (abstract).
212. Szollosi, D., Jalabert, B., and Breton, B. (1978). Postovulatory changes in the theca folliculi of the trout. Ann. Biol. Anim. Biochem. Biophys. 18, 883-891.
213. Talbot, P. (1983). Videotape analysis of hamster ovulation in vitro. J. Exp. Zool. 225, 141-148.
214. Talbot, P. (1983). Intrafollicular pressure promotes ovulation of the antrum during hamster ovulation in vitro. J. Exp. Zool. 226, 129-135.

215. Talbot, P., and Chacon, R.S. (1982). In vitro ovulation of hamster oocytes depends on contraction of follicular smooth muscle cells. J. Exp. Zool. 224, 409-415.
216. Talbot, P., and Schroeder, P.C. (1982). 5-hydroxytryptamine causes contraction of smooth muscle cells in preovulatory hamster follicles. J. Exp. Zool. 224, 427-436.
217. Testart, J., Thebault, A., and Lefevre, B. In-vitro ovulation of rabbit ovarian follicles isolated after the endogenous gonadotropin surge. J. Reprod. Fertil. 68, r13-418.
218. Thebault, A., Lefevre, B., and Testart, J. (1983). Role of the extra-follicular compartment in the ovulation of isolated rabbit ovarian follicles. J. Reprod. Fertil. 68, 419-424.
219. Theofan, G., and Goetz, F.W. (1981). The in vitro effects of transcriptional and translational protein synthesis inhibitors on final maturation and ovulation of yellow perch (Perca flavescens) oocytes. Comp. Biochem. Physiol. 69A, 557-561.
220. Tojo, H., Fujii, M., and Ogawa, K. (1982). Proteolytic enzymes and gonadal hormones of the ovarian follicle wall during ovulation in the domestic fowl (Gallus domesticus). J. Reprod. Fertil. 64, 73-77.
221. Too, C.K.L., Bryant-Greenwood, G.D., and Greenwood, F.C. (1984). Relaxin increases the release of plasminogen activator, collagenase, and proteoglycanase from rat granulosa cells in vitro. Endocrinology 115, 1043-1050.
222. Van Blerkom, J., and Motta, P. (1979). "The Cellular Basis of Mammalian Reproduction." Urgan and Schwarzenberg, Baltimore. 252 pp.
223. Velardo, J.T. (1958). The anatomy and endocrine physiology of the female reproductive system. In "The Endocrinology of Reproduction" (J.T. Velardo, ed.), pp. 101-212. Oxford Univ. Press, New York.
224. Wallach, E.E., Bronson, R., Hamada, Y., Wright, K.H., and Stevens, V.C. (1975). Effectiveness of prostaglandin $F_{2\alpha}$ in restoration of HMG-HCG induced ovulation in indomethacin-treated rhesus monkeys. Prostaglandins 10, 129-138.
225. Walles, B., Edvinsson, L., Falck, B., Nybell, B., Owman, Ch., Sjoberg, N.-O., and Svensson, K.-G. (1974). Modification of ovarian and follicular contractility by amines: A mechanism involved in ovulation? Eur. J. Obstet. Gynecol., Suppl. 4, S103-S107.
226. Walles, C., Edvinsson, L., Falck, B., Owman, Ch., Sjoberg, N.-O., and Svennson, K.-G. (1975). Mechanical response in the wall of ovarian follicles mediated by adrenergic receptors. J. Pharm. Exper. Ther. 193, 460-473.
227. Walles, B., Edvinsson, L., Falck, B., Owman, Ch., Sjoberg, N.-O., and Svennson, K.-G. (1975). Evidence for a neuromuscular mechanism involved in contractility of the ovarian follicular wall; fluorescence and electron microscopy and effects of tyramine on follicle strips. Biol. Reprod. 12,239-248.
228. Walles, B., Edvinsson, L., Nybell, G., Owman, Ch., Sjoberg, N.-O. (1974). Amine-induced influence on spontaneous ovarian contractility in the guinea pig and the cat. Fertil. Steril. 25, 602-611.
229. Walles, B., Edvinsson, L., Falck, B., Owman, Ch., Sjoberg, N.-O., and Sporrong, B. (1976). Cholinergic nerves and receptors mediating contraction of the Graafian follicle. Biol. Reprod. 15, 565-572.
230. Walles, B., Falck, B., Owman, Ch., Sjoberg, N.-O., and Svennson, K.-G. (1977). Characterization of autonomic receptors in the smooth musculature of human Graafian follicles. Biol. Reprod. 17, 423-431.
231. Walles, B., Groschel-Stewart, U., Owman, Ch., Sjoberg, N.-O., and Unsicker, K. (1978). Fluorescence histochemical demonstration of a relationship between adrenergic nerves and cells containing actin and myosin in the rat ovary, with special regard to the follicle wall. J. Reprod. Fertil. 52, 175-178.

232. Walton, A., and Hammond, J. (1928). Observations on ovulation in the rabbit. J. Exp. Biol. 6, 190-204.
233. Warren, D.C., and Scott, H.M. (1934). Ovulation in the domestic hen. Science 80, 461-462.
234. Wasserman, W.J., and Masui, Y. (1974). A study on gonadotropin action in the induction of oocyte maturation in _Xenopus laevis_. Biol. Reprod. 11, 133-144.
235. Wathes, D.C. (1984). Possible actions of gonadal oxytocin and vasopressin. J. Reprod. Fertil. 71, 315-345.
236. Weichung, R.R. (1978). _In vitro_ interactions of cytochalasins with contractile proteins. _In_ "Cytochalasins - Biochemical and Cell Biological Aspects" (S. W. Tanenbaum, ed.), pp. 431-444. Elsevier-North-Holland Biomedical, New York.
237. Weir, B.J., and Rowlands, J.W. (1977). Ovulation and atresia. _In_ "The Ovary," sec. ed. (L. Zuckerman and B.J. Weir, eds.), Vol. 1, pp. 265-302. Academic Press, New York.
238. Wimsatt, W.A. (1944). Growth of the ovarian follicle and ovulation in _Myotis lucifugus lucifugus_. Amer. J. Anat. 74, 129-173.
239. Yang, N.S.T., Marsh, J.M., and LeMaire, W.J. (1973). Prostaglandin changes induced by ovulatory stimuli in rabbit Graafian follicles. The effect of indomethacin. Prostaglandins 4, 395-404.
240. Yorke, M.A., and McMillan, D.B. (1980). Structural aspects of ovulation in the lamprey, _Petromyzon marinus_. Biol. Reprod. 27, 897-912.
241. Yoshimura, Y., and Fujii, S. (1981). Comparison of decomposition of the stigma of hen's ovarian follicle in spontaneous and proteolytic enzyme-induced ovulation. Jap. Poultry Sci. 18, 60-65.
242. Yoshimura, Y., and Koga, O. (1982). Ultrastructural changes of the stigma of the follicle during the process of ovulation in the hen. Cell. Tiss. Res. 224, 349-359.
243. Yoshimura, Y., and Koga, O. (1983). Ultrastructural changes in follicular stigma during the ovulation process in the hen. _In_ "Avian Endocrinology: Environmental and Ecological Perspectives" (S. Mikami et al., eds.), pp. 107-115. Japan Sci. Soc. Press, Tokyo/Springer Verlag, Berlin.
244. Yoshimura, Y., Tanaka, K., and Koga, O. (1983). Studies on the contractility of follicular wall with special reference to the mechanism of ovulation in hens. Brit. Poult. Sci. 24, 213-218.

FUNCTIONAL MORPHOLOGY AND REGULATION OF THE CORPUS LUTEUM

Françoise Xavier

Ecole Normale Superieure, CNRS U.A. 686
Biochimie et Physiologie du Developpement
46 rue d'Ulm, F-75230 Paris Cedex 05, France

I. INTRODUCTION

Corpora lutea are ephemeral endocrine glands occurring in all vertebrate groups. These structures develop in the ovaries after ovulation, from the membranes of postovulatory follicles. The processes of luteogenesis, luteal maintenance and luteolysis have been studied extensively in mammals. Luteinization of the cells lining the follicular cavity is accompanied by a striking increase in progesterone secretion. Progesterone is the hormone required for the establishment of pregnancy, and the main function of the mammalian corpus luteum is to synthesize and secrete this steroid. Autonomy of progesterone secretion is short-lived in mammals, and the ability of the corpus luteum to maintain progesterone secretion depends on the action of permissive and/or stimulatory factors. Regression of progesterone secretion results from the luteolytic action of other factors. The regulation of the corpus luteum activity in mammals is complex and not completely understood, especially as the mammalian corpus luteum may secrete hormones other than steroids (polypeptide hormones, prostaglandins), the roles of which remain to some extent to be determined.

Compared with mammals, little attention has been paid to the study of corpus luteum function in nonmammalian vertebrates and, unfortunately, there is often much confusion and controversy in the literature about the formation and the endocrine nature of the nonmammalian corpus luteum. In this chapter, after an overview of mammals, I will focus on the current knowledge of corpus luteum physiology in lower vertebrates. This requires analysis of both functional morphology of the corpus luteum and regulatory control of luteal activity in fishes, amphibians and reptiles. Such a functional approach leaves the matter unsettled because of the limited amount of information available and the large number of questions that remain without conclusive answers. However, it is hoped that this comparative review will contribute to a better understanding of corpus luteum physiology in lower vertebrates and give substance for evolutionary considerations.

II. FORMATION AND MORPHOLOGY OF THE CORPUS LUTEUM

A. Overview of Mammals

The corpus luteum results from the morphological and secretory transformation of the steroidogenic layers of the follicle following the gonadotropic surge and rupture of the follicle. After ovulation, granulosa and thecal cells invade the follicular cavity, and this process lasts between two to three days. The use of monoclonal antibodies raised against granulosa and thecal cells demonstrated that, at the time of the corpus luteum formation, the large luteal cells originate from the transformed granulosa cells whereas the small luteal cells originate from the theca interna.[2] Progressively, small luteal cells hypertrophied and became large luteal cells through the process of ploidization.[2,48] It is noteworthy that progressively both monoclonal antibodies did not recognize most luteal cells, indicating an important change in the structural proteins of the luteal cells especially from mid-gestation.[2] Large and small luteal cells are observed in the corpus luteum of all mammals thus far studied, but the ratio of cell types varies for each species (see review by Levasseur and Thibault[122]). Luteal cells, which are recognizable by an abundant smooth endoplasmic reticulum (SER) and mitochondria with a complex system of internal tubular cristae, represent about half of the corpus luteum cells. The other corpus luteum cells belong to the vascular and connective tissues.[173] Like all endocrine glands, corpora lutea have a considerable microvascular network. The blood vessels originate from the vascular system forming part of the thecal layer before ovulation. It appears that the corpus luteum cells can have an angiogenic effect (neovascularization of transplanted corpora lutea) and the existence of a corpus luteum angiogenic factor, responsible for proliferation of capillary blood vessels and neovascularization of the corpus luteum, has been postulated.[66] Lastly, adrenergic nerves enter the corpus luteum along with blood vessels. Completion of the corpus luteum, with a definitive vascular and nervous network, usually takes place between 5 to 7 days.

B. Fishes

In most of the elasmobranchs so far studied, as well as some teleosts, the postovulatory follicle gives rise to a gland-like structure, or corpus luteum, in both oviparous and viviparous species (see review by Lofts and Bern,[138] Browning,[25] Guraya,[74] Saidapur,[176] Nagahama;[152] see Chapter 18 for definitions of parity types). Corpora lutea are not formed in the viviparous fishes showing follicular gestation.[25,79,109] The corpora lutea are short-lived in oviparous species but persist longer in live-bearing forms. In teleosts, the very short life span of the corpus luteum may explain the difference of findings with regard to the fate of the ovulated follicle; the absence of corpora lutea reported in some species may have been due to sampling the ovaries too late after ovulation.

The corpora lutea vary greatly both in structure (cavitory or compact corpora lutea) as well as in cellular architecture. This variability is not associated with either phylogenetic affinities or reproductive mode of the species considered. In all piscine species developing corpora lutea, the cells of the folded granulosa hypertrophy immediately after ovulation and transform into luteal cells characterized by a well-developed SER and a prominent Golgi apparatus.[88,97,154,155,159,202] Hypertrophy of thecal cells is observed in some species only. Granulosa cells may or may not fill up the postfollicular cavity; in either case, mitotic figures are rarely reported during the formation of the corpus luteum. The blood vessels being located only in the thecal layer, the vascularization of the granulosa luteal cell mass is dependent on the behavior of the thecal tissue which may or may not invade the follicular cavity along with the granulosa cells after ovulation.

Two basic types of corpora lutea can be distinguished in fishes. In the first type, only the granulosa cells hypertrophy and transform into luteal cells. The thecal tissue either forms a distinct sheath surrounding the central granulosa luteal cells as in some oviparous teleosts (_Notopterus notopterus_,[187] _Merluccius merluccius_,[39] _Oryzias latipes_,[96] _Perca fluviatilis_,[116]) and ovoviviparous elasmobranchs (_Torpedo marmorata_,[35] _Squalus acanthias_,[114]) or invades the follicular cavity with the hypertrophied granulosa cells and supplies the connective tissue (oviparous bony fish, _Tor_ (_Barbus_) _tor_[170]). In the second type, both thecal and granulosa cells hypertrophy after ovulation. In some species, both hypertrophied granulosa and thecal cells form the luteal cell mass as in oviparous and viviparous elasmobranchs (_Rhinobatus granulatus_,[119] _Scyliorhinus stellaris_,[36,37] several species of the genus _Raja_[38]) and an oviparous bony fish (_Clarias batrachus_,[117]); in others, the hypertrophied thecal cells remain separated from the granulosa luteal cells, as in the viviparous elasmobranch *Mustelus canis*[190] and a large number of oviparous teleosts (*Scomber scomber*,[13] *Cichlosoma nigrofasciatum* and *Haplochromis multicolor*,[159] *Cyprinus carpio*,[18,75] *Brachydanio rerio*,[110] *Carassius auratus*,[153] *Oncorhynchus kisutch* and *Oncorhynchus gorbuscha*,[154] *Salmo gairdneri*,[88] *Salvelinus leucomaenis*,[97]). In teleosts, "special thecal cells" (STC), which have most of the ultrastructural features of the typical steroid-producing cells such as an extensive SER and large mitochondria with tubular cristae, are organized in small clusters frequently near blood capillaries. In elasmobranchs, the thecal cells form prominent islands which tend to fuse into short and long strips, whereas the originally thick collagenous layer is reduced. A postovulatory reduction of the collagen fiber framework has been reported in various species. In the trout, Szollosi and associates[189] have shown that the thecal smooth muscle-like cells became phagocytotically active after ovulation and incorporate and break down collagen fibers. Phagocytosis of collagen appeared to be a generally occurring process in the postovulatory ovarian connective tissue compartment.

In summary, most oviparous and viviparous cartilaginous and bony fishes develop corpora lutea. The granulosa layer is the main source of the luteal cells; in some species, the thecal cells hypertrophy as well and either form the luteal cell mass with the granulosa cells or remain located in the thecal tissue.

C. Amphibians

The development of corpora lutea has been observed in the ovary of both oviparous and viviparous species of caecilian, anuran and urodelean amphibians. The postovulatory luteal bodies degenerate rapidly in oviparous species but are more persistent in ovoviviparous and viviparous species (reviews by Chieffi and Botte,[38] Lofts,[137] Guraya,[74], Wake,[201] Saidapur[176]).

In oviparous species ovulation is immediately followed by hypertrophy of the granulosa cells and then of the thecal cells. In some species, the granulosa becomes multilayered due to the shrinkage of the empty follicle. Granulosa cells fill the cavity leading to the formation of a fully-developed corpus luteum having enlarged thecal cells on the outside and a granulosa cell mass in the center (*Rana esculenta* and *Triturus cristatus*,[21] *Rana catesbeiana*,[72] *Rana cyanophlyctis*,[177] *Rana verrucosa*[176]). In others, as in the oviparous marsupial frog, *Gastrotheca riobambae*,[167] the granulosa cells never fill the follicular cavity.

The corpora lutea of ovoviviparous and viviparous amphibian species develop in the same way as in oviparous forms. There is thickening of the

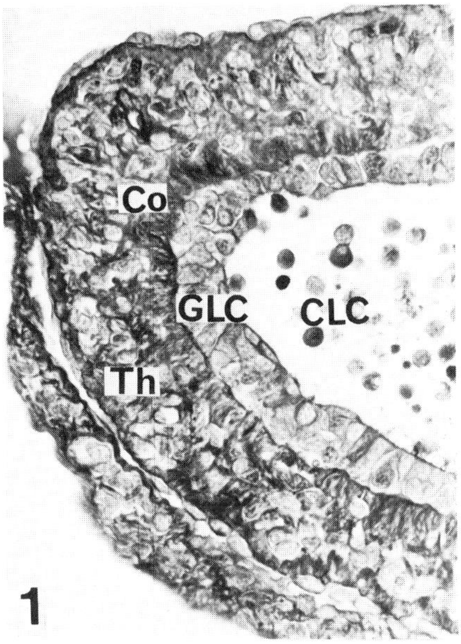

Fig. 1. Portion of Salamandra salamandra terrestris corpus luteum showing theca (Th) with numerous collagenous fibers (Co) and granulosa luteal cells (GLC). Some luteal cells are observed in the corpus luteum cavity (CLC) (x 300). Photomicrograph courtesy of Pr. J. Joly, Rennes University, France.

theca and hypertrophy of the granulosa cells which are characterized by the presence of abundant SER and mitochondria with tubular cristae.[92] The granulosa cells form either cavitory (Salamandra salamandra terrestris,[90] (Fig. 1) or compact corpora lutea (S. s. fastuosa,[90] Nectophrynoides occidentalis,[111,112] N. tornieri[113] (Fig. 2)). In all above-mentioned species, the granulosa cells never proliferate after ovulation,[98,112] and the capillary blood vessels remain localized in the thecal layers and are never observed in granulosa cell mass. By contrast, in the viviparous apodan Typhlonectes compressicaudus, Exbrayat and Collenot[45] reported that the granulosa cells proliferate and form the luteal cell mass which is invaded thereafter by the thecal cells accompanied by capillary blood vessels (Fig. 3).

To summarize, amphibians develop either cavitory or fully-formed corpora lutea. Granulosa cells appear to be solely responsible for formation of luteal cell mass, and no invasion of fibroblasts and capillaries occurs except in the only apodan studied.

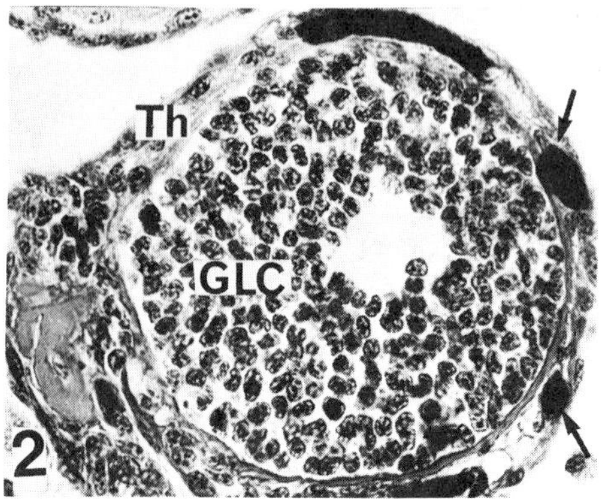

Fig. 2. Early *Nectophrynoides occidentalis* corpus luteum. Central cavity is not completely filled by the granulosa luteal cells (GLC). Capillary blood vessels (arrows) are located in the thecal layer only (Th) (x 300).

D. Reptiles

The majority of reptiles, including all crocodilians and chelonians, are oviparous, but some lizards and snakes are viviparous. All of them, however, develop true corpora lutea whose life span is correlated closely with the retention of eggs or young in the female genital tract (reviews by Weekes,[206] Browning,[25] Fox,[51] Saidapur;[176] see also Chapter 18).

After ovulation, the granulosa cells of collapsed follicles proliferate and hypertrophy, fill the follicular cavity, and transform into luteal cells characterized by the presence of an abundant SER and close association of lipid droplets with mitochondria and SER.[42] As in other lower vertebrates, there is variation among species concerning the participation of theca interna and vascularization in development of the reptilian corpus luteum. This varibility in structure cannot be associated either with phylogenetic relationships or with mode of reproduction. The theca interna may be implicated in providing theca luteal cells and/or the connective tissue of the luteal cell mass.

Concerning the behavior of the theca interna as connective tissue of the corpus luteum, three different types of corpora lutea can be distinguished among reptiles. In the first type, the granulosa cells appear to be solely responsible for formation of luteal tissue. The thecal layers remain distinct from the luteal cell mass in which no invasion of fibroblasts and capillaries occurs; such corpora lutea are observed in several viviparous lizards, such as *Sceloporus jarrovi*,[64] *Sphenomorphus quoyii*, *Leiolopisma quadridigitatum* and *Egernia whitii*.[206] In the second type, fibroblasts from theca interna invade the adjacent luteal cells and tend to produce

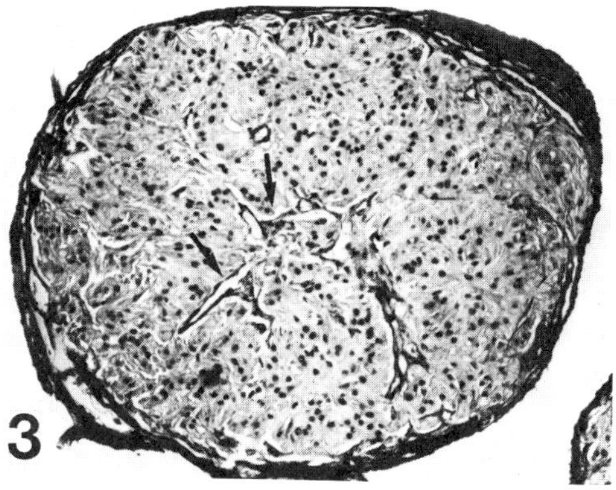

Fig. 3. Typhlonectes compressicaudus corpus luteum. Capillary blood vessels (arrows) are observed in the luteal cell mass (x 70). Photomicrograph courtesy of J. M. Exbrayat, Lyon Catholic University, France.

superficial septa, but no capillaries penetrate (e.g., the viviparous lizards Xantusia vigilis,[148] Hoplodactylus maculatus,[22] Chalcides ocellatus[11] and in the viviparous water-snake Cerebus rhyncops[180]). In the third type, fibroblasts of the theca interna along with blood vessels invade the luteal cell mass either moderately as in the oviparous snakes Naja naja[115] and Lycodon aulicus[73] and the viviparous snake Natrix rhombifera,[16] or extensively as in oviparous turtles (Chelydra serpentina),[42] and lizards (Calotes versicolor,[198] Uromastix hardwicki[6]) and viviparous lizards and snakes (Leiolopisma weekesae and Leiolopisma entrecasteauxi,[206] Thamnophis elegans,[78] Vipera aspis[52]). In all species mentioned above, the penetration of fibroblasts is confined to septa; however, in the viviparous lizard Lacerta vivipara and the viviparous sea-snake Hydrophis cyanocinctus, the fibroblasts penetrate between individual luteal cells.[162,180] Furthermore, in some species, the theca interna cells immediately adjacent to the basement membrane, as well as the granulosa cells, hypertrophy. The hypertrophied theca interna cells either remain separated from the granulosa luteal cells by a distinct membrana propria as reported in Naja naja[115] or combine with the granulosa luteal cells to form the luteal cell mass as in U. hardwicki[6] and in L. vivipara in which theca luteal cells are organized in superficial small clusters (Xavier and Yvorra, unpublished data, Fig. 4).

In summary, contrary to the other lower vertebrates, all oviparous and viviparous reptiles so far studied exhibit fully-developed corpora lutea. In most species, luteal cells originate from granulosa only, and the luteal cell mass possesses a more or less well-defined vascular system. In some species, both theca interna and granulosa cells transform into luteal cells; in these species, the corpus luteum is well-vascularized and closely

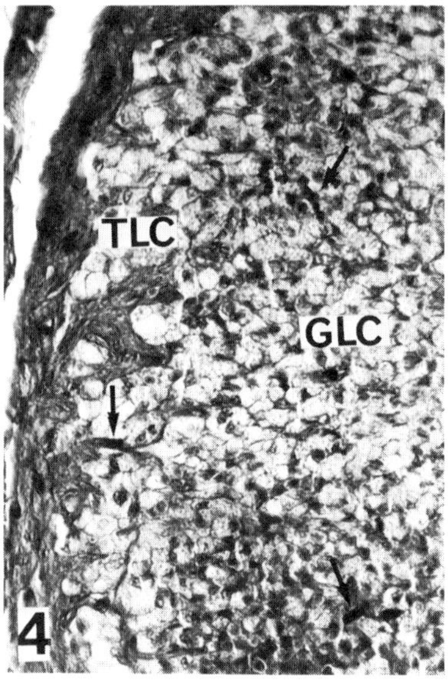

Fig. 4. Portion of *Lacerta vivipara* corpus luteum. Hypertrophied theca luteal cells (TLC) constitute with the granulosa luteal cells (GLC) the luteal cell mass in which capillary blood vessels are observed (arrows) (x 350).

resembles those of mammals. The variability in reptilian luteal structure cannot be associated with the mode of reproduction.

III. HORMONE PRODUCTION BY THE CORPUS LUTEUM

A. Overview of Mammals

In all mammalian species, the main function of the corpus luteum is to synthesize and secrete progesterone which is required for the establishment of pregnancy. Progesterone is secreted at a higher basal rate by large cells than by small cells (sow,[119] cow,[108,195] sheep[48,172]). Progesterone release by the corpus luteum may involve an active secretory process. The hormone, possibly bound to protein, is packaged into granules and secreted by exocytosis.[56,164,169,181,209] The increase in progesterone production generally is accompanied by a reduction, it is not a complete loss, in aromatase activity.[77,194] However, the primate corpus luteum synthesizes and secretes estrogens as well as progesterone,[26,171,185] and a high aromatase activity has been demonstrated in the rat corpus luteum.[57,100]

The corpus luteum in mammals secretes not only steroid hormones but also, and sometimes in large quantities, oxytocin, vasopressin, relaxin and prostaglandins. Oxytocin has been immunologically identified in the cyclic corpus luteum of numerous mammals.[49,101,168,203,205] The oxytocin gene is highly transcribed in the corpus luteum which can produce 250 times more oxytocin than the hypothalamus (cow[89]). It now seems likely that, at least in some species, it is the corpus luteum rather than the neurohypophysis which supplies much of the circulating oxytocin during the estrous cycle (cow, sheep; see Wathes[204]). Oxytocin, concomitantly with neurophysin I,[183] is secreted, as well as progesterone, in a pulsatile manner.[203] In cyclic mammals, the luteal secretion patterns of oxytocin and progesterone can be either in parallel (monkey,[102] ewe[50]) or in opposition (pig[168]). Luteal and circulating oxytocin levels fall at the same time whether or not the animals are pregnant.[49,204] The corpus luteum also synthesizes and secretes vasopressin but in smaller amounts than oxytocin. Relaxin is present at low levels in the cyclic corpus luteum (e.g., ewe[33]). High concentrations of relaxin are produced by the corpus luteum during the second half of gestation[186,207] or after hysterectomy when this treatment prolongs corpus luteum activity.[3] This suggests that embryo or placenta might not be required for the maintenance of elevated relaxin levels (see Levasseur and Thibault[122]). As well as progesterone, relaxin is sequestered into secretory granules of the luteal cells.[4,5,15] Lastly, the mammalian corpus luteum has the capacity to produce prostaglandins (PGs) of series E (PGE) and series F (PGF). The production of PGE is higher than that of PGF during the early corpus luteum development, and the balance of PGE/PGF decreases as the age of the corpus luteum progresses (cow,[149] monkey[9]). Luteal $PGF_{2\alpha}$ concentrations increase in all mammals at a time when luteolysis would be commencing.

B. Fishes

There is a paucity of information about the endocrine activity of the elasmobranch corpus luteum. Data are limited to some histochemical studies (hydroxysteroid dehydrogenases: HSD) and biochemical *in-vitro* experiments for the evaluation of the capacity to form steroid hormones (see review by Guraya[74]). A positive 3β-HSD activity has been reported in the corpus luteum of oviparous *S. stellaris*, several oviparous species of *Raja*, and the ovoviviparous *S. acanthias*[114,140] while enzyme activity could not be localized in the postovulatory follicle of the ovoviviparous *T. marmorata*;[140] in the latter species, only the atretic follicles showed 3β-HSD activity. In *S. acanthias*, the 3β-HSD activity decreased as gestation proceeded. Tested for biosynthetic capacity, the corpus luteum from *S. stellaris* yielded progesterone *in vitro* and no estrogens, whereas those of *T. marmorata* did not produce progesterone but could synthesize estrogens.[139]

More data are available on the secretory activity of the corpora lutea in teleosts. According to the structure of the corpus luteum (see II B),

3β-HSD activity was reported in the granulosa cells only, as in O. latipes,[96] or in both granulosa and special thecal cells (STC). High activity of STC has been demonstrated by histochemical and ultrastructural features in younger corpora lutea from a large number of species (S. scomber,[13] C. nigrofasciatum and H. multicolor,[159] C. auratus,[103,153] O. kisutch,[154] S. gairdneri,[88] S. leucomaenis[97]), but the maintenance of this activity varies from species to species. STC gave often more 3β-HSD-positive tests than the granulosa cells. However, a negative observation based on a single technique should be treated with caution. For example, in S. leucomaenis, although granulosa cells of the corpus luteum did not show 3β-HSD activity, they contained some ultrastructural features generally observed in steroid-producing cells.[97]

In-vitro studies have shown that corpora lutea from teleosts are able to synthesize progesterone, 17α-hydroxyprogesterone (17α-OH-P) and 17α,20β-dihydroxy-4-pregnen-3-one (17α,20β-diOHprog) and to a lesser extent androgens but no estrogens. In S. leucomaenis, the high plasma progesterone concentrations, observed for several days after ovulation, coincided with high 3β-HSD activity in the STC.[97] In Oncorhynchus rhodurus, the enhanced 17α,-20β-diOHprog production after ovulation was associated with the hypertrophy of the granulosa cells.[217] These data suggest that the STC and the granulosa cells are the major sites, respectively, of progesterone and 17α,20β-diOHprog synthesis by the corpora lutea. This is consistent with the two cell-type model proposed for the production of steroid hormones by the teleostean ovarian follicle.[152] The thecal cells synthesize progesterone, 17α-OH-P and testosterone; 17α-OH-P and testosterone are transferred to the granulosa and converted, respectively, into 17α,20β-diOHprog and estradiol. The very low levels of plasma estradiol observed after ovulation seem not to be due to the lack of aromatizable substrate (as in the rat during luteinization), since the thecal cells continue to produce androgens, but are due to the loss of aromatase activity in the granulosa cells after the completion of vitellogenesis.[218]

The production by the piscine corpus luteum of hormones other than steroids is not known. As concerns prostaglandins, it would be interesting to determine whether or not the corpus luteum is involved in the increasing ovarian PGF production observed in some species of teleosts which probably is dependent on the presence of ovulated oocytes within the ovarian lumen.[62,63,188]

In summary, the corpora lutea of fishes are able to produce progestogens (progesterone, 17α,20β-diOHprog) and, to a lesser extent, androgens but generally no estrogens. As concerns the much debated question regarding the cellular site of steroid hormone synthesis, it must be pointed out that, in teleosts, thecal cells are responsible for the production of progesterone and androgens whereas the granulosa cells secrete 17α,20β-diOHprog.

C. Amphibians

Presence of steroid-synthesizing enzymatic activities and, in some species, biochemical *in-vitro* studies indicate that the corpus luteum in amphibians secretes steroids. The presence of an intense 3β-HSD activity has been demonstrated histochemically in the granulosa cells of the postovulatory follicle in different species of oviparous and ovoviviparous anurans and urodeles (reviews by Guraya,[74] Saidapur[176]) and recently in a viviparous apodan (*T. compressicaudus*[45]). In oviparous anurans, 3β-HSD activity was maximal on the day of spawning and decreased rapidly thereafter.[21,72,178] Very low 3β-HSD activity has been shown in the thecal cells of postovulatory follicles in the oviparous *R. esculenta* and *T. cristatus*[21] and in the ovoviviparous *S. s. terrestris*.[92] No activity has been observed in the thecal cells of the postovulatory follicle in anuran and apodan viviparous species (*N. occidentalis*,[215] *T. compressicaudus*.[45]

In the ovoviviparous salamander, *S. s. terrestris*, *in-vitro* incubations of corpora lutea with labeled precursors have shown that the enzymes necessary for converting pregnenolone to progesterone and progesterone to 17α-OH-P and 20α- and 20β-hydroxyprogesterone are present in these structures; little if any 17α-OH-P converted to C_{19} steroids.[92] An increase in progesterone content per ovary was observed during the first stage of the gestation, when the corpora lutea are still well-developed, but the plasma progesterone levels remained low (Table 1; Garnier and Joly[53] and personal communication).

In the viviparous toad *N. occidentalis*, the ovarian tissue *in vitro* converted labeled pregnenolone exclusively to progesterone when tested during gestation. The amount of progesterone formed *in vitro* was maximal during the underground period of the pregnant female, lasting six months; it declined after emergence and was minimal at the time of parturition, three months later; similar activity was observed in the ovary of pseudopregnant female *N. occidentalis*.[214] By contrast, the ovarian steroidogenic activity in the ovoviviparous toad *N. tornieri*, analyzed by *in-vitro* incubation with labeled precursors, was weak during the first month of gestation (gestation lasts from 62 to 78 days) and then increased progressively until the next ovulation (the sexual cycle lasts from 90 to 150 days according to the season). This increase of ovarian capacity to produce steroids corresponded to oocyte growth and vitellogenesis.[213] These data suggest that, in

Table 1. Plasma Progesterone and Progesterone Content in the Ovary of the Ovoviparous Urodele, Salamandra salamandra terrestris, During the Annual Cycle. (From Garnier and Joly, personal communication.)

	Non-Pregnant			Pregnant		
	Jan	Mar	May	Jun-Jul	Oct	Dec-Jan
Plasma progesterone (ng/ml)	0.8 ± 0.2	0.71 ± 0.08	0.70 ± 0.21	1.00 ± 0.23	0.70 ± 0.09	0.59 ± 0.07
Ovarian progesterone (pg/mg tissue)	6.9 ± 0.2	5.8 1 1.4	7.0 ± 2.6	65.0 ± 8.2	19.0 ± 3.5	11.0 ± 1.9
(pg/ovary)	3200 ± 348	3000 ± 901	5500 ± 1800	8100 ± 1100	4200 ± 534	5000 ± 1090

the ovoviparous Nectophyrnoides, the corpora lutea are poorly steroidogenic, contrary to what is observed in the viviparous form.

There is a distinct correlation between the decrease of the corpora lutea and the growth of oocytes, indicative of the involvement of the corpus luteum in the inhibition (or slowing down) of follicular development during gestation. This correlation has been reported in Salamandra atra[200] and in pregnant and pseudopregnant N. occidentalis.[112,211] (Pseudopregnancy occurs in females isolated from males. The females ovulate and eggs are retained in utero for 7-8 months, a time comparable to the normal nine-month gestation.) During the pseudogestation of female N. occidentalis, the regression of corpora lutea occurred earlier than during gestation and correlatively oocyte growth was precocious. In consequence, the ovarian cycle lasted only 9 months in pseudopregnant females instead of 12 months as in normal females. It is noteworthy that in S. s. terrestris, in which corpora lutea regress earlier than in S. s. fastuosa, follicular growth and vitellogenesis are observed during gestation (annual cycle), whereas these occur only after parturition in S. s. fastuosa (biennial cycle).[90] It could be of importance to analyze, during gestation, the steroid production by the corpora lutea of S. s. fastuosa and compare it to that of S. s. terrestris.

D. Reptiles

Histochemical and biochemical in-vitro studies suggest that corpora lutea of reptiles function as endocrine glands by secreting progesterone. A strong 3β-HSD activity has been demonstrated in the corpora lutea of all oviparous and viviparous reptiles so far studied (reviews by Callard et al.,[27] Guraya,[74] Saidapur[176]) and in both theca interna layer and luteal cell

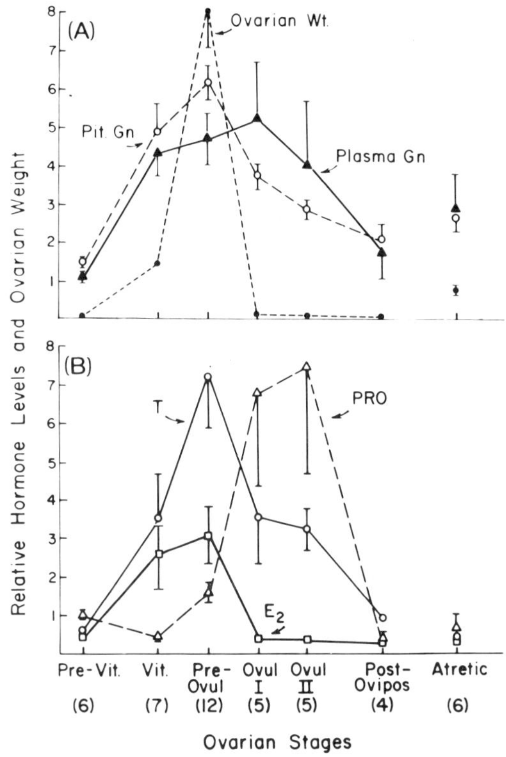

Fig. 5. Relative ovarian weight (ovarian wt) and hormonal levels (means ± SEM) in relation to the stage of the ovarian cycle in the cobra Naja naja. (Pre-vit.: previtellogenic; Vit.: vitellogenic; Pre-Ovul.: preovulatory; Ovul.I: ovulated, unshelled eggs; Ovul.II: ovulated, shelled eggs; Post-ovip.: postoviposition). Sample sizes in parentheses below each group. Units for each variable are equivalent to following: ovarian weight (10g); pituitary Gn (pituitary gonadotropin, 15ng/ml); plasma Gn (1.5ng/ml); PRO (progesterone 1.5ng/ml); Testo and E_2 (testosterone and estradiol 4 pg/ml); From Bona Gallo et al. (1980), Gen. Comp. Endocrinol. 42, 477-493, with permission of Academic Press.

mass during early gestation in some viviparous species.[28,151] Moreover, *in-vitro* incubation studies have shown that the reptilian corpora lutea are able to convert pregnenolone to progesterone as reported in an oviparous turtle[105] and viviparous snakes and lizards.[31,212] Colombo and Yaron[40] have mentioned in the snake, *Storeria dekayi*, *in-vitro* production of 11-deoxy-corticosterone (DOC) as well as progesterone, but other workers failed to detect DOC.[32,212]

Numerous reports on plasma progesterone levels in oviparous and viviparous reptiles strengthen the finding that corpora lutea secrete progesterone. In turtles and in oviparous lizards and snakes progesterone increases only during or after ovulation (*Chelydra serpentina*;[123] *Chelonia* mydas;[135,136] Sternotherus odoratus;[145] Naja naja;[20] Fig. 5; Agama atra[196] review by Licht[126]). An exception is Chrysemys picta in which progesterone peaks shortly before ovulation.[30] Sequential blood sampling in the sea turtle, C. mydas, revealed a pronounced surge in progesterone of about 24-hr. duration; plasma progesterone then rapidly returned to baseline levels, indicating that corpora lutea were relatively inactive in the ovigerous stage.[135,136] In C. picta, the organization of the corpus luteum after ovulation is accompanied by declining, although still high, progesterone titers. The persistence of elevated progesterone levels in the ovigerous cobra N. naja[20] and in the lizard U. hardwicki[7] suggests that, unlike turtles, the corpora lutea of the squamates remain active up to near the time of oviposition. In viviparous lizards and snakes, plasma progesterone, which begins to increase during the vitellogenic phase, peaks either just after ovulation (Chamaeleo p. pumilis[199]) or at mid- or two-thirds gestation (S. cyanogenys,[28] Nerodea sipedon,[34] T. elegans[78]) or even in late pregnancy as in S. jarrovi[71] and L. vivipara in which plasma progesterone concentrations are tremendously high compared to those observed in other reptiles (Table 2). It should be noted that, in L. vivipara, progesterone

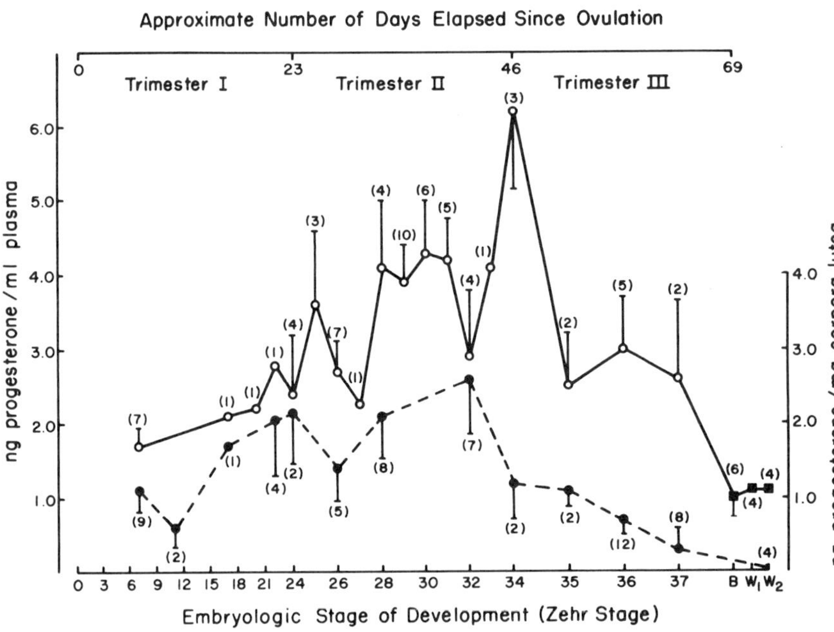

Fig. 6. Luteal progesterone concentrations in pooled corpora lutea from individual snakes Thamnophis elegans (lower curve) and peripheral plasma levels of progesterone (upper curve) in snakes with intact corpora lutea (means $\pm$ SEM). Sample sizes in parentheses. b: mean birth, W_1: means week one, W_2: means week two. From Highfill and Mead (1975), Gen. Comp. Endocrinol. 27, 389-400, with permission of Academic Press.

Table 2. Plasma Progesterone Levels (ng/ml) Before and After Ovulation and During Pregnancy in Some Oviparous and Viviparous Reptiles. (See also Table 1, Chapter 18.)

Oviparous Species	Ref.	Preovulatory Phase	Postovulatory Phase (eggs in utero)
Chrysemys picta	29	5.00 ± 1.02	0.46 ± 0.10
Naja naja	20	∿ 2.00	∿ 11.00
Uromastix hardwicki	7	1.66 ± 0.30	13.41 ± 1.43

Viviparous Species	Ref.	Preovulatory Phase	Early Pregnancy	Mid-Pregnancy	Late Pregnancy	Post-Partum
Nerodia s. pictiventris	29	3.93 ± 0.83	4.95 ± 1.41	6.94 ± 0.78	2.81 ± 0.44	-
Thamnophis elegans	76, 144	3.30 ± 2.30	1.70 ± 0.30	6.20 ± 1.00	∿ 3.00	1.00 ± 0.20
Sceloporus cyanogenys	29	0.90 ± 0.38	3.30 ± 0.48	-	3.50 ± 0.34	1.60 ± 0.22
Bradypodion (Chamaeleo) *p. pumilus*	197	0.95 ± 0.71	4.95 ± 3.90	2.30 ± 0.34	-	-
Lacerta vivipara	210	57.90 ± 4.81	132.65 ± 16.79	216.22 ± 17.92	347.20 ± 17.38	3.71 ± 0.56

is bound strongly to a specific protein in plasma.[212] A positive correlation exists between plasma progesterone concentrations and histological activity of corpora lutea in all species referred to above except in *S. jarrovi*. In this species, plasma progesterone peak occurs after the regression of corpora lutea suggesting another source of progesterone which remains to be determined.[71] Recently, Guillette and Fox[68] provided indirect evidence that the adrenal glands are a source of progesterone in *Anolis carolinensis*. Changes in luteal progesterone content were not investigated in these species. In fact, whereas progesterone levels have been evaluated in the plasma of a number of oviparous and viviparous reptiles, the changes in luteal progesterone concentrations have been determined in one species only, the garter snake *T. elegans*.[78] In this species, the changes in luteal progesterone content closely parallel the plasma progesterone levels (Fig. 6). Luteal and plasma progesterone concentrations increase during the first trimester of pregnancy when the theca interna appears glandular and the luteal cell mass is invaded by blood vessels. Highest concentrations in corpora lutea and in plasma are observed during the second trimester of pregnancy; at this time, the thecal layers begin compaction and theca interna loses its glandular appearance and probably contributes little to progesterone synthesis.[78] In *L. vivipara*, the non-luteal ovarian tissue produces negligible amounts of progesterone *in vitro* compared to the corpus luteum (Fig. 7A) which secretes increasing amounts of this steroid during gestation (Fig. 7B; Xavier and Yvorra, unpublished data) as reflected in the plasma progesterone pattern[212] (Fig. 8).

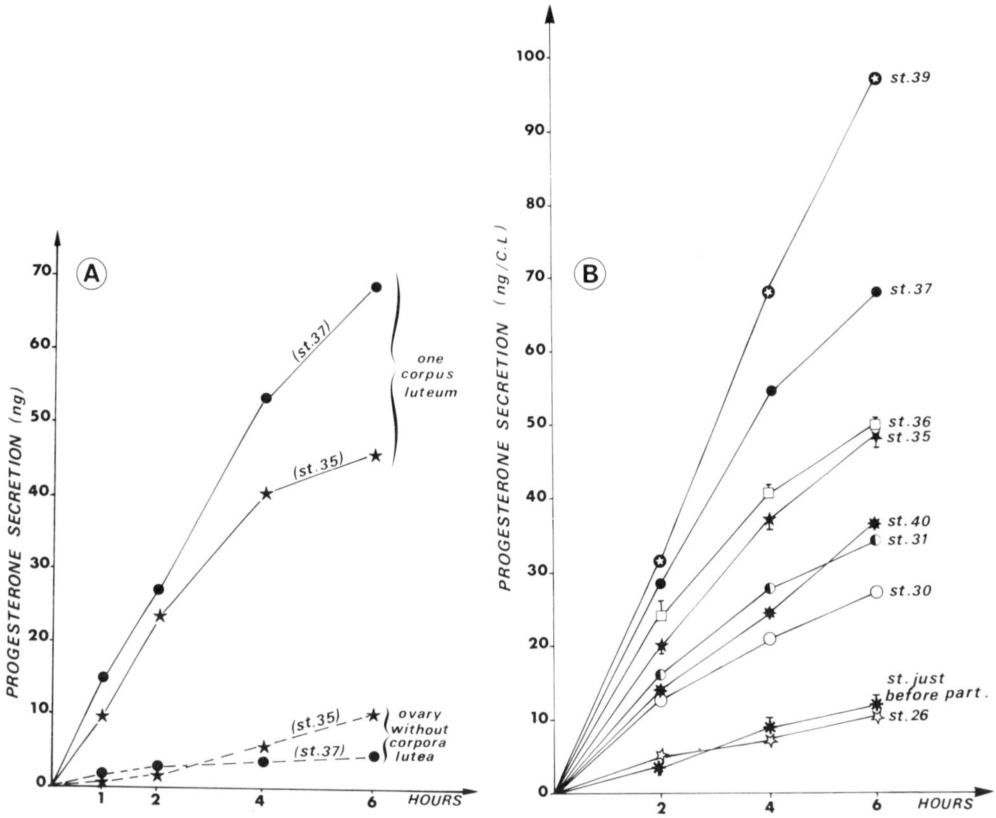

Fig. 7. Progesterone secretion _in vitro_ by: (A) one corpus luteum (C.L.) and the remainder ovary of _Lacerta vivipara_ pregnant females, (B) one corpus luteum from the ovary of _L. vivipara_ pregnant females at various stages of the embryonic development (st.). From Xavier and Yvorra, unpublished data.

The increasing luteal progesterone secretion during pregnancy corresponds toan increase of progesterone production by the corpus luteum. Preliminary short-term kinetics of progesterone secretion by dispersed luteal cells in an open perifusion system _in vitro_ tend to show that the rate of progesterone secretion is ten times higher in mid- than in early pregnancy (Xavier, Burzawa-Gerard and Yvorra, unpublished data). Lastly, the decrease in plasma progesterone following removal of corpora lutea confirmed that the corpora lutea are the main source of plasma progesterone (Xavier and Yvorra, unpublished data). It is noteworthy that luteal progesterone production markedly increases at a time in which theca luteal cells, organized as superficial small clusters, together with granulosa luteal cells constitute the luteal cell mass (see II, D) and show an intense 3β-HSD activity.[151] These findings

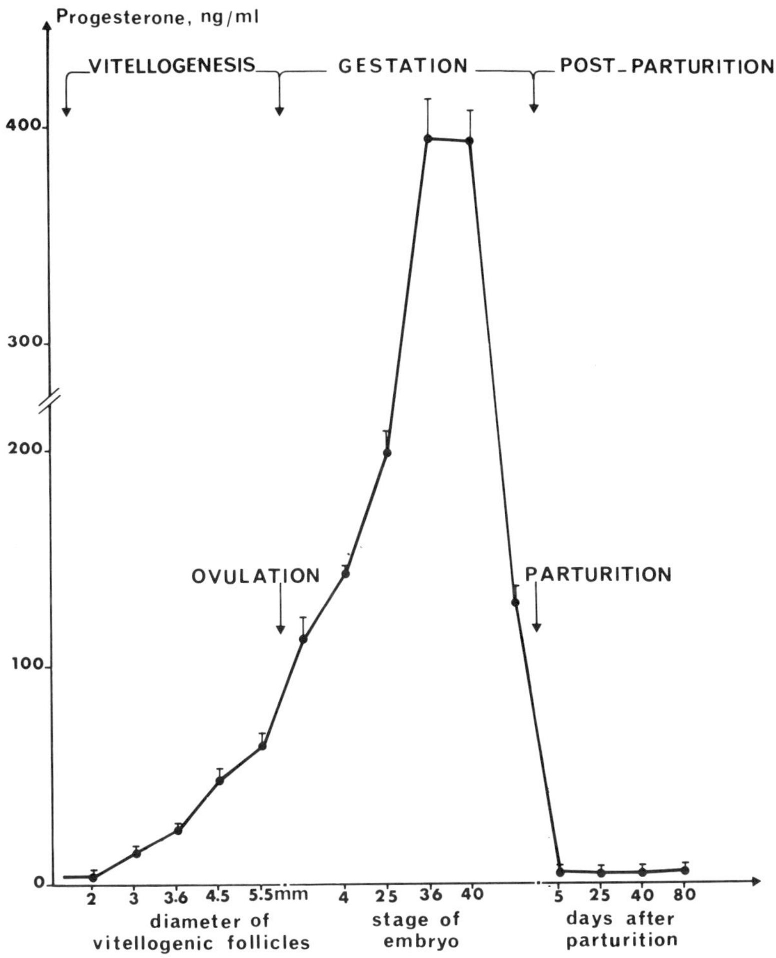

Fig. 8. Peripheral progesterone levels associated with reproduction in *Lacerta vivipara* females. Points represent means for 2-5 females; vertical lines indicate SE. From Xavier (1982), Herpetologica, 38, 62-70, with permission.

suggest that theca luteal cells can contribute to the luteal progesterone production during the gestation in *L. vivipara*.

The corpus luteum secretes mainly progesterone in most reptiles. However, in some oviparous squamate reptiles, corpora lutea also may synthesize either estradiol as in *A. carolinensis*[93,94] or androgens and estradiol as reported in *U. hardwicki*. In the latter species, luteal concentrations of both hormones are significantly higher than follicular concentrations in preovulatory as well as in gravid animals; plasma levels of progesterone, estradiol and testosterone peak during the middle of the gravid period.[7] In the cobra *N. naja*, testosterone does not show an abrupt decline as does estradiol after ovulation; i.e., the corpus luteum may secrete androgen.[20] In viviparous lizards, circulating estrogen levels could be uniformly low during pregnancy as reported in *T. elegans*.[146] In *L. vivipara*, estradiol release by the ovary *in vitro* is depressed just before ovulation whereas an increasing output of progesterone and androgens is observed.[55] *In-vitro* incubation of ovaries in the presence of labeled androstenedione or testosterone, performed at various times of the reproductive cycle, confirm these results and even more show that the corpus luteum, as well as the remainder of the ovary, are not able to aromatize labeled androgens (Gavaud and Xavier, unpublished data).

In conclusion, regarding patterns of progesterone production by the reptilian corpus luteum, it is possible to distinguish two basic types as reported by Callard and Lance.[29] The first type is exhibited by turtles in which a preovulatory or an ovulatory progesterone surge is observed; in turtles, the corpora lutea, in comparison to preovulatory follicles, are relatively poorly steroidogenic in the ovigerous phase. The second type is observed in squamate reptiles in which plasma progesterone peaks either shortly after ovulation (as in oviparous species) or at mid- or even late pregnancy as in viviparous species. The corpus luteum secretes mainly progesterone; in some species, however, it is able to produce androgens and estrogens as well. Invasion of blood vessels and theca interna cells into the luteal cell mass seems to contribute markedly to a higher production of progesterone by the corpus luteum. As observed in amphibians, the maintenance in the ovary of active corpora lutea inhibits or retards the development of the ovarian follicles[25,27,28,67,107] (also Yvorra and Xavier, unpublished data).

IV. REGULATION OF CORPUS LUTEUM ACTIVITY (LUTEOTROPIC FACTORS)

A. Overview of Mammals

Once the corpus luteum is formed, it spontaneously secretes steroid hormones and then becomes dependent on luteotropic factors. These factors consist of two or more hormones which may originate from the pituitary, the ovary and, during pregnancy, the conceptus (see reviews by Yoshinaga,[216] Peters and McNatty,[166] Rothchild,[175] Levasseur and Thibault[122]).

Experiments using hypophysectomy, hormonal treatment and drug administration have shown that luteinizing hormone (LH) and prolactin (PRL) are generally mandatory for the maintenance of luteal function.[19,104,147,157,197] Their actions and interrelationships in the luteal cells are best understood in the rat. LH stimulates luteal progesterone production by increasing the intracellular level of cyclic adenosine monophosphate (cAMP) and the uptake of lipoprotein which is an important source of cholesterol for luteal steroid hormone synthesis.[76] PRL enhances the number of LH receptors,[60,82] stimulates cholesterol turnover and increases progesterone secretion by inhibiting reduction of progesterone to the biologically inactive 20α-dihydroprogesterone.[43,208] It should be noted that, during the earlier stage of the luteal phase, rat corpora lutea have a greater responsivness to catecholamines than to LH; this could indicate a specific role for catecholamines in corpus luteum function during a period when the LH response is poor.[160] Increase of luteal production by β-adrenergic agonists has been shown in various species of mammals.[41,87,95]

Estradiol can have a luteotropic action in some species and even can be mandatory for the maintenance of corpus luteum function (see Nalbandov[158]). Thus, in the rabbit, estradiol alone supports luteal function in the hypophysectomized animal.[17] In this species, estradiol acts directly on the luteal tissue, via estrogen receptors at the nuclear level,[80,81,83] by increasing the surface areas of SER and inner mitochondrial membranes, tightly coupled to progesterone production;[24] estradiol also could stimulate the formation of receptors (see Levasseur and Thibault[122]). In other species, action of estradiol might depend upon PRL (rat;[59]pig[49]), but the function of PRL in the regulation of corpus luteum activity is not clear and requires further investigations.

Lastly, prostaglandins of the E series and prostacyclin (PGI_2) can be considered as luteotropic and angiogenic factors. They have been found to stimulate adenyl cyclase synthesis and progesterone secretion in luteal tissue[142,149] and promote luteal blood flow (see Levasseur and Thibault[122]).

In mammals, the corpus luteum is necessary for the establishment of gestation. In species in which the corpus luteum of pregnancy functions for a longer period during a pregnancy than during a non-fertile cycle, luteal maintenance depends on signals produced by the conceptus which ensure that the corpus luteum is exposed to more luteotropin than luteolysin. In primates, the embryonic luteotropin (chorionic gonadotropin) directly stimulates the luteal secretion of progesterone and estradiol[86,171] and diminishes the capacity of the corpus luteum to synthesize luteolytic $PGF_{2\alpha}$.[12]

Foetoplacental luteotropins acting on the activity of the corpus luteum also have been identified in non-primate species.[54,58] For example, in the sheep, trophoblastin[143] or trophoprotein I[61] act indirectly on the corpus luteum activity through the blocking of the uterine luteolytic $PGF_{2\alpha}$ pulses (see V. A) and via the stimulation of the luteotropic PGE synthesis.

In vitro studies, using purified luteal cells, have shown that small and large cells differ in their abilities to respond to luteotropic factors. Small luteal cells appear to be much better targets for LH than do the large ones (sow,[119] cow,[195] sheep[48,174]), and it has been suggested that LH might play a role in regulating the differentiation of small luteal cells into large luteal cells (see II. A, and Hosyer and Niswender[85]). Lack of response to LH by large luteal cells seems to be due to a lack of receptors for this hormone. In contrast, receptors for PGs (particularly for PGE_2 which stimulates progesterone secretion) have been found in large luteal cells only (ewe[47]). Progesterone secretion in large luteal cells is not subject to the same regulation as in small luteal cells; contrary to small cells, large cells are cyclic AMP-independent.[85] Furthermore, secretion of progesterone by the large cells could be stimulated by the small ones.[120] Additional work on regulation of the corpus luteum shall have to take into consideration the action of the luteotropic factors not only on the whole corpus luteum but also on each population of luteal cells.

B. Fishes

There is very little information concerning the existence and importance of luteotropic factors in fishes. In viviparous elasmobranchs, the corpus luteum could be formed and maintained independently of the pituitary (see Dodd[44]); however, this suggestion is supported by only one study on hypophysectomized pregnant sharks, M. canis, and it is difficult to draw any definite conclusion on the participation of gonadotropin in the control of luteal activity of elasmobranchs.

A voluminous literature on teleosts has accumulated in the last few years concerning the hormonal control of ovarian activity, principally during oocyte maturational and ovulatory processes, but there is as yet little information on the hormonal control of luteal activity. The major advances have been made in salmonids in which gonadotropin (Gn) seems responsible for progestogen production; isolated corpora lutea of amago salmon, O. rhodurus, soon after ovulation (1-2 days) produce a large amount of progesterone and 17α,20β-diOHprog in response to salmon Gn whereas older corpora lutea (one week) respond only slightly.[155,156,218] Salmon Gn also stimulates testosterone production by the corpora lutea, but not estradiol production.[99,218]

As suggested by Manning and Kime[141] in the common carp, *Cyprinus carpio*, the "switch off" of aromatase activity could be actuated by an ovarian rather than a pituitary factor but the mechanism by which the switch operates remains unclear. The relationships between 17α,20β-diOHprog and prostaglandins merit attention. Young and associates[217] have speculated that 17α,20β-diOHprog could influence, either directly or indirectly, the synthesis of prostaglandins.

C. Amphibians

The hormonal factors responsible for regulating corpus luteum activity remain to be identified in amphibians. Information is scanty and confined to very few species. In the oviparous ranid *R. cyanophlyctis*, the life span of the corpora lutea, which is normally one day, is extended several days by exogenous homoplastic pituitary extract or human chorionic gonadotropin (hCG).[161] Concerning pituitary-corpus luteum relationships in viviparous species, it has been suggested by cytological criteria that both PRL and LH are required for the luteal maintenance (*S. Salamandra*,[90] *N. occidentalis*[219]). In the viviparous toad *N. occidentalis*, the adenohypophysis contains numerous and active prolactin cells and an increasing number of gonadotropic cells during the first period of gestation, which is characterized by the presence of active corpora lutea in the ovaries (see III. C); the prolactin cells decrease in number thereafter as the corpora lutea are regressing.[219] In this species, hypophysectomy in early pregnancy provoked a premature regression of the corpora lutea and atresia of young oocytes.[210] It must be pointed out that identical results were obtained after hysterectomy or removal of embryos on the same date.[210] This suggests a role of the pregnant uterus or embryo on luteal activity, presumably due to its action at the hypothalamus-pituitary level.

D. Reptiles

There is some evidence that reptilian corpora lutea are able to produce steroids and constitute generally the main source of progesterone production during the oviducal retention of eggs; however, little is known about the regulatory mechanisms of corpus luteum secretory activity. Most of the available information on pituitary control of luteal production of progesterone has been obtained from turtles. Turtles, as well as crocodilians, possess two Gns which are biochemically homologous to FSH and LH of mammals and other tetrapods.[124,131,134] In *Chelydra serpentina* progesterone release *in vitro* by corpora lutea may be stimulated by homologous pituitary homogenates,[106] by both homologous FSH and LH or by ovine FSH but not by ovine LH.[129] The biological activity ascribed to a specific Gn is dependent on the state of the corpus luteum at the time of testing. In the young corpus luteum, just after ovulation, turtle LH was equipotent with ovine FSH and turtle FSH was weakly active; in the older, fully formed corpus luteum, turtle LH was less potent than ovine FSH and only slightly more potent than turtle FSH.[129] In another turtle, *Chrysemys picta*, results of *in-vitro* studies of progesterone biosynthesis from labeled exogenous precursors are not consistent with those mentioned above, since ovine (or avian) LH has been demonstrated to be more active in stimulating conversion of labeled cholesterol or pregnenolone to progesterone than is ovine FSH; furthermore, stimulatory effects of ovine LH appear to be reduced by mammalian PRL.[32]

Characteristics of FSH-binding sites in the turtle ovary supports the conclusion that FSH may have broader effects on steroidogenesis.[124] FSH-binding sites in turtle ovaries are specific for mammalian FSH but not for various nonmammalian Gns.[128,132] Owing to the fact that progesterone production by dispersed luteal cells *in vitro* is stimulated by dibutyryl cAMP in the range found effective in mammals, Callard and Lance[29] have suggested that Gns could act in a similar manner in reptiles and mammals. Receptor studies and biological tests indicate a significant divergence in gonadotropic function between reptiles and mammals. Both FSH and LH appear to share a common receptor site and to elicit comparable physiological responses.[125] Measurements of endogenous plasma LH levels in *Chelonia mydas* have shown that, along with progesterone, LH displays a pronounced "ovulatory surge" that is responsible for ovulation and luteinization.[125,135,136] Periovulatory surges in LH and progesterone also have been also observed in *Lepidochelys olivacea*[133] and *C. picta*.[29] Unfortunately, the lack of available data on plasma FSH levels precludes any evaluation of the involvement of this Gn in ovulation and luteinization processes.

Concerning the extent of the pituitary involvement in the control of the luteal ativity in squamate reptiles, researchers have encountered difficulties developing suitable assays for the evaluation of the gonadotropic activities. Indeed, squamate reptiles may have a single Gn which is distinct from either FSH or LH of mammals and other tetrapods including turtles and crocodiles.[127,130] So, data from physiological studies using heterologous FSH and LH must be interpreted with caution.[127] A radioimmunoassay system for the Gn of the oviparous snake, the cobra *N. naja*, has been developed. In this species, highest levels of plasma Gn are observed after ovulation when plasma concentrations of progesterone reach maximal values; circulating Gn, progesterone and testosterone levels remained elevated until oviposition

(Fig. 5; also Bona Gallo et al.[20]). These results suggest that the maintenance of luteal activity in the gravid cobra is supported by Gn; persistent high levels of Gn binding by the luteal tissue reinforce this finding.[20] In viviparous squamate reptiles, there is no direct information on the changes of the circulating Gn concentrations. In order to obtain some information on the extent of pituitary involvement in the maintenance of the luteal activity, hypophysectomy and injection of heterologous pituitary hormones have been performed in some live-bearing species. Apparently, the presence of the pituitary is not essential for the development and/or the maintenance of the corpus luteum in *Thamnophis sirtalis*[23] and *L. vivipara*.[163] Callard and associates[27,28] demonstrated that the corpus luteum of hypophysectomized blue spiny lizards (*S. cyanogenys*) is secretory, although it did not produce normal levels of progesterone; treatment for 8 days following hypophysectomy with various mammalian pituitary hormones [either FSH alone, FSH + growth hormone (GH) or PRL + GH] restored progesterone level in the circulation, whereas injection of PRL alone depressed plasma progesterone still further. However, it is unfortunate that the authors failed to provide sufficient information on the time during the 12-week pregnancy at which hypophysectomy was performed as well as, before and after treatment, the structural aspects and the secretory activity of the corpora lutea and the remainder of the ovary. Furthermore, the validity of these studies using mammalian pituitary hormones is debatable as reported above. In the viviparous lizard, *L. vivipara*. Panigel[163] has reported that hypophysectomy did not affect formation and development of corpora lutea (morphological observations). Adenohypophysectomy performed in mid-pregnant *L. vivipara* failed to reduce plasma progesterone levels which dropped at the same time as in intact animals (Table 3; Xavier and Yvorra, unpublished data). Rates of progesterone release *in vitro* by corpora lutea from adenohypophysectomized lizards are similar to those from intact animals (Xavier and Yvorra, unpublished data). Furthermore, hysterectomy or removal of embryos and embryonic membranes performed in early pregnancy does not influence either the life span of the corpus luteum or luteal progesterone production, suggesting that the feto-placental unit does not have any role in corpus luteum activity of *L. vivipara* (Table 3; Xavier and Yvorra, unpublished data). Similar results have been reported in the viviparous snake, *T. elegans*.[78]

In conclusion, homologous radioimmunoassays have shown that LH is associated with the ovulatory and luteinization processes in turtles. In oviparous squamate reptiles, Gn moreover could support the luteal activity during the short period of oviducal retention of eggs. Apparently, in the viviparous squamate reptiles, the presence of the pituitary is not essential

Table 3. Mean (±SE) Plasma Progesterone Levels in Female *Lacerta vivipara* Subjected to Removal of Eggs, Hysterectomy and Adenohypophysectomy and in Control Females. Xavier and Yvorra, unpublished data.

Treatment	Duration of Treatment (number of days)	Plasma Progesterone Levels (ng/ml)	
		Treated females	Control females
Removal of eggs	12 [a]	274.0 ± 23.0 (3)	250.0 ± 50.0 (3)
	20 [a]	330.0 ± 49.0 (5)	294.0 ± 24.0 (5)
	12 [b]	131.0 ± 13.0 (3)	121.0 ± 34.0 (3)
Hysterectomy	12 [b]	214.5 ± 44.0 (5)	180.5 ± 60.0 (4)
Adenohypophysectomy	16 [b]	163.5 ± 33.5 (3)	201.0 ± 17.0 (3)
	21 [c]	43.0 ± 17.0 (3)	56.5 ± 3.5 (4)

(n): sample size; (a): from early- to mid-gestation; (b): from mid- to late gestation; (c): from mid-gestation to near term.

for the maintenance of luteal secretory activity. However, it is likely that Gn is involved in ovulation and luteinization and, as in oviparous forms, supports the activity of the corpus luteum during a short time after ovulation. Authors have failed to investigate this time period; however, it should be noted that, in turtles and lizards, unlike in some mammals, PRL reduces the stimulatory effect of Gn; however, the use of mammalian hormones may place some suspicion on these results. During pregnancy, the fetoplacental unit could not effect luteal maintenance. These studies are limited to only two species, and additional experimental investigations are needed.

V. REGRESSION OF THE CORPUS LUTEUM (LUTEOLYTIC FACTORS)

A. Overview of Mammals

Luteal regression occurs in two stages: the rapid decline of the functional capacity of the tissue (decrease of progesterone production, loss of Gn receptors) followed by the degeneration of the tissue itself. The involution of the luteal cells, generally characterized by a marked distention of the SER and an accumulation of large lipid-filled vacuoles, is associated with a reduction in blood flow to luteal tissue (see Peters and McNatty[166]).

Autophagy and heterophagy (macrophages) seem to be involved in the regression of the corpus luteum which is gradually invaded by connective tissue. The increase of intercellular collagen may eventually result in a corpus albicans; however, in several species (e.g., rodents), the corpus luteum disappears without leading to any scar.

In some mammals (e.g., species with epitheliochorial placentae: artiodactyls) the uterus plays an important role in cyclic luteal regression. In these species, hysterectomy prolongs the life and the activity of the corpus luteum, and $PGF_{2\alpha}$ has been identified as the uterine luteolytic factor (see reviews by Golberg and Ramwell,[65] Horton and Poyser,[84] Thibault and Levasseur,[191] Peters and McNatty,[166] Levasseur,[121]). $PGF_{2\alpha}$ initiates functional luteolysis by acting directly on luteal cells (loss of Gn receptors; Thomas et al.,[192] Behrman et al.[14] Fitz et al,[46] Agudo et al.[1]) and indirectly by decreasing luteal blood flow; $PGF_{2\alpha}$ also provokes structural luteolysis by causing release of lysosomal enzymes (see Thibault and Levasseur[191]). The action of uterine $PGF_{2\alpha}$ on luteal tissue is allowed by an intimate relationship between uterine vein and ovarian artery which avoids the passage of $PGF_{2\alpha}$ through the general circulation and its metabolization by the lungs. Estradiol interferes in the control mechanisms of the luteal regression process (see review by Levasseur and Thibault[122]: (1) by stimulating the first $PGF_{2\alpha}$ pulse from the progesterone-primed uterus, and (2) by increasing the number of uterine oxytocin receptors and consequently the sensitivity of the uterus to luteal oxytocin which controls the successive uterine $PGF_{2\alpha}$ pulses (see IV. A). Prolonged luteal function during gestation is due to the action of the fetoplacental unit which blocks the uterine pulsatile secretion of $PGF_{2\alpha}$, probably by acting on the uterine oxytocin receptors.[144]

In other mammals (species with hemochorial or endotheliochorial placentation), hysterectomy either slows down the gradual drop in plasma progesterone levels (rodents except the guinea-pig) or has no effect on the life span and progesterone secretion from corpora lutea (primates and carnivora except the dog). In primates, the luteolytic factors, $PGF_{2\alpha}$ and estradiol, originate from the corpus luteum itself: the primate corpus luteum has a self-destruct mechanism, triggered by an increase within the corpus luteum of estrogen concentration. Estradiol would promote luteolysis either directly by inhibiting the luteal 3β-HSD, Δ^{4-5} isomerase activity,[118] or indirectly by stimulating the luteal secretion of $PGF_{2\alpha}$[8] which has an intraluteal action and can induce luteolysis.[9,165] Oxytocin also may induce luteolysis in primates,[10] but its role is still poorly understood.[102,122] In pregnant primates, secretion of hCG by the fetus prevents luteolysis by diminishing the capacity of the corpus luteum to produce $PGF_{2\alpha}$.[12] The luteolytic factors that provoke regression of the cyclic corpus luteum in primates are similar to those of ruminants or even rodents. The main difference is the source of the luteolytic factors. Concerning most of the carnivora, the existence of luteolytic factors is questionable; the life span of the corpus luteum might reflect the life span of the luteal cells themselves (see Levasseur and Thibault.[122]).

B. Fishes

Little is known concerning luteal regression in fishes. The duration of the functional luteal phase in teleosts is limited to a few days as in *S. gairdneri*,[88] *S. leucomaenis*[98] and *O. rhodurus*,[155] or even some hours (*C. auratus*,[153] *O. latipes*[96]). Involution of the luteal cells (both granulosa cells and STC) is characterized by an increase of lysosome-like bodies and a disappearance of cell organelles; SER appears less prominent and many of the mitochondria have shrunken tubular cristae and their matrices are less dense. Finally, numerous autophagocytotic bodies containing organelles in degradation are observed.[96,97] In general, corpora lutea persist through gestation in live-bearing elasmobranchs but may begin a protracted regression early during gestation as in the ovoviviparous spiny dogfish, *S. acanthias* (see Browning[25]). Studies on the control mechanisms of the luteal regression in fishes are lacking.

C. Amphibians

As in fishes, corpora lutea have a very short life in oviparous species and generally disappear within two weeks.[137,176,182] They persist longer in ovoviviparous and viviparous species, and luteal regression can occur at two-thirds gestation (*N. occidentalis*,[112,210] *S. atra*[200]) or even towards term (*T. compressicaudus*[45]). The first signs of the corpus luteum regression are a decrease in size and pycnosis of the nucleus in some luteal cells. Ultrastructurally, SER becomes less abundant and mitochondria are scarce and small.[92,193] These degenerative changes in organelles are accompanied by an accumulation of lipid droplets.[72,92] The activity of enzymes involved in steroid hormone synthesis gradually diminish.[90,214] The control mechanisms of the luteal regression in amphibians have not been studied.

D. Reptiles

The first sign of structural degeneration of the corpus luteum is an increase of connective tissue inside the luteal cell mass. The degeneration of the luteal cells may first begin at the center of the corpus luteum, whereas some luteal cells at the periphery still have the appearance of healthy cells in *L. vivipara*.[163,206] The nuclei of the luteal cells become pycnotic. Ultrastructurally, as reported by Cyrus and associates[42] in *C. serpentina*, there is a general disruption of the endoplasmic reticulum and mitochondria; the tubular endoplasmic reticulum becomes extremely vesiculated, the Golgi bodies disappear, but there is no significant change in the number of the lipid droplets. During the final stage of regression, the luteal cells disintegrate and are replaced by connective tissue which becomes part of the ovarian tissue.

Reptilian corpora lutea persist for several days to several months depending on the species and the mode of reproduction; the life span of the corpora lutea has been correlated to some extent with the retention of eggs *in utero*. In some oviparous species, corpora lutea begin to regress before oviposition (most turtles and some squamate reptiles; e.g., *P. scripta*,[150] *C. serpentina*,[42] *A. muricatus*,[206] *S. a. aenus*;[67] see review by Fox[51]). In others, structural regression of the corpora lutea sets in immediately after oviposition (*C. versicolor*,[198] *U. hardwicki*,[6] *N. naja*[115]). In viviparous species corpora lutea may begin to regress either during gestation or around the time of parturition (see reviews by Fox,[51] Saidapur[176]), but the life span of the corpus luteum is usually longer than in the oviparous reptiles. For example, in *S. aeneus* structural luteolysis occurs when embryos are stage 25-28 (just prior to oviposition) in the oviparous subspecies *S. a aenus*, and when embryos are stage 30-31 in the viviparous subspecies *S. a. bicanthalis*.[67]

As in mammals, the luteal regression is characterized by a rapid decline in the capacity of the tissue to produce progesterone and by a slower disappearance of the tissue itself. In turtles, after a preovulatory or ovulatory surge, plasma progesterone rapidly declines to basal levels. Callard and associates[30] have suggested that progesterone in *C. picta* may depress Gn secretion and thus induce the postovulatory phase of regression of progesterone secretion accompanied by follicular atresia. In squamate reptiles, high levels of plasma progesterone are maintained either until oviposition in most oviparous species or for a longer time in viviparous species, but plasma progesterone levels always fall before oviposition or parturition.[93] An exception, however, is the lizard *A. carolinensis*. This species alternates ovulation, so a single corpus luteum is always present during the breeding season and progesterone secretion is continuous, even during oviposition.[94]

Rothchild[175] has postulated that progesterone could be the primary stimulus of its own secretion, and when this steroid reaches a critical level it could reduce the rate of further increase in progesterone and switch on an increase in PG production. The self-stimulating quality of progesterone secretion and PG production and their mutually opposing effects could lead to regression of progesterone secretion and eventual disappearance of the corpora lutea.[175] Guillette and associates[70] have shown that exogenous $PGF_{2\alpha}$ induced luteolysis in the oviparous lizard *Anolis carolinensis*: 24 h after treatment, luteal cells exhibit an increase in lysosome area and number and a decrease in lipid vesicles correlated with a decrease in plasma progesterone concentration. The ability of the reptilian reproductive system to synthesize PGs has been determined recently: *in-vitro* studies indicate that ovarian and oviducal tissues from reproductively quiescent lizards (*S. jarrovi*) are capable of synthesizing PGE_2 and $PGF_{2\alpha}$; similar results have been obtained for turtles.[69] Changes in PG synthesis during the reproductive period remain to be demonstrated. In mammals, the main source of $PGF_{2\alpha}$ is the uterus, except in primates in which the corpus luteum might produce its own PGs (see V. A). The interference of the uterus in the luteal production of progesterone has been studied in only two species (see IV, D). In the lizard *L. vivipara*, progesterone secretion by the fully formed corpus luteum did not appear to be affected by hysterectomy, removal of eggs or adenohypophysectomy (see Table 3). These data suggest that the life span of the corpus luteum is self-determined in *L. vivipara*. Important questions remain to be solved in regard to the mechanism of regression of the corpus luteum, and the source and the nature of the luteolytic factors. The above observation, that exogenous $PGF_{2\alpha}$ may induce luteolysis in an oviparous reptile, suggests that the regression of the reptilian corpus luteum may be related, as in most mammals, to the luteolytic action of this prostaglandin. Further comparative studies are needed to ascertain whether prostaglandin-induced luteolysis is a common feature in reptiles.

VI. SUMMARY AND FUTURE RESEARCH DIRECTIONS

Although our knowledge of corpus luteum physiology in lower vertebrates has substantially increased in the last few years, a number of questions are yet unsettled, and this concluding section is an attempt to highlight some of the current fundamental issues.

Corpora lutea develop in both oviparous and viviparous lower vertebrates. They vary greatly in degree of development (cavitory or compact corpora lutea) and in structure (origin of luteal cells, vascularization), but this variability cannot be associated with either phylogenetic affinities or reproductive mode of the species considered. In some lower vertebrates, only granulosa cells hypertrophy had present the ultrastructural and histochemical characteristics of steroid-producing cells; in others, a hypertrophy of thecal cells also is observed. Hypertrophied granulosa and thecal cells either remain separated (e.g., fishes, squamate reptiles) or constitute together the luteal cell mass (e.g., elasmobranchs, apodans, squamate reptiles). The role of the theca luteal cells remains to be determined. It is now well-established that corpora lutea in lower vertebrates are steroidogenic and secrete mainly progesterone as in mammals. Do theca luteal cells contribute to the progesterone production? Is there any interaction between granulosa and theca luteal cells in the progesterone secretion process, as observed in some mammals? Analytical studies *in vitro*, using purified luteal cells in an open perifusion system, should help to resolve these questions. The thecal layer may provide not only luteal cells but also the supporting connective tissue and the microvascular network of the luteal cell mass. Thus, corpora lutea in lower vertebrates may possess a more or less well-defined vascular system depending on the behavior of the thecal layer. Development of the luteal vascularization plays an important role in corpus luteum activity, especially by influencing the quantities of tropic hormones that reach individual cells. In mammals, blood flow to the luteal tissue is correlated highly with progesterone secretion by this gland. Unfortunately, little attention has been given to this phenomenon in lower vertebrates, and investigations on luteal vasodynamics might help to clarify some important aspects of corpus luteum function.

The life span of corpora lutea in lower vertebrates depends on the species and the mode of reproduction. Corpora lutea are short-lived in oviparous species but persist longer in live-bearing species; this is particularly relevant for squamate reptiles in which the life span of corpora lutea generally is related closely to the retention of eggs *in utero*. Our knowledge of what controls the functioning of the corpus luteum in lower vertebrates is limited. The involvement of the pituitary in this control still is debatable. Gonadotropin seems to be responsible for the luteinization process and to support the corpus luteum activity for some time after ovulation (e.g., teleosts, turtles, oviparous squamates). In some amphibians, prolactin might be required with LH for luteal maintenance as observed in some mammals, but this action could depend on signals produced by the conceptus (e.g., the viviparous anuran *N. occidentalis*). On the contrary, the presence of the pituitary might not be essential for the maintenance of the corpus luteum in viviparous elasmobranchs and squamate reptiles; however, this finding is only founded on effects of hypophysectomy and heterologous hormone treatment, and it is difficult to draw any definite conclusion on the involvement of hypophysial hormones in the control of the luteal activity in these vertebrate groups. Moreover, in squamate reptiles the fetoplacental unit could not have any effect on the corpus luteum so far as the progesterone secretion was considered. Thus, activity of the corpus luteum in lower vertebrates could be dependent on hypophysial support either continuously or only for a limited period of time after which it becomes autonomous.

According to Rothchild,[175] the three characteristics common to all vertebrate corpora lutea are: (1) ability to synthesize and secrete progesterone which could be the primary stimulus of its own secretion, (2) responsiveness to the luteolytic action of $PGF_{2\alpha}$, (3) ability to synthesize prostaglandins. However, production of prostaglandins by the corpus luteum has not been demonstrated in lower vertebrates and important questions remain to be solved in regard to the source and nature of the luteolytic factors as well as the mechanisms causing corpus luteum regression. Although it has been shown that exogenous $PGF_{2\alpha}$ may induce functional luteolysis in an oviparous lizard, further comparative studies are needed to ascertain whether prostaglandin-induced luteolysis is a common phenomenon in lower vertebrates. Lastly, it is now well-established for several species of mammals that the corpus luteum is able to secrete, sometimes in large quantities, hormones other than steroids and prostaglandins, such as oxytocin, vasopressin and relaxin. In regard to the role of the corpus luteum in the timing of oviposition or parturition (see Chapter 18), it should be of great interest to determine whether or not the corpus luteum in lower vertebrates is also capable of secreting such hormones.

VII. ACKNOWLEDGMENTS

I am grateful to Pr. C. Thibault (INRA, Jouy-en-Josas) and Dr. E. Burzawa (M.N.H.H., Paris) for providing constructive criticisms of various drafts of the manuscript and Pr. D. Norris and Pr. R. Jones for reviewing the manuscript. I am also grateful to Pr. J. Joly and Dr. D. Garnier from the University of Rennes, Dr. J. Exbryat from the Catholic University of Lyon, and A. Yvorra from my laboratory for making unpublished data and micrographs available for this manuscript. I wish to thank also Pr. P. Licht and Pr. R. Mead for permission to reproduce diagrams. I thank my colleagues Dr. R. Lafont and Dr. J. Gavaud for reading the manuscript. Special acknowledgment is due to my brother-in-law who though not scientifically oriented patiently pored over the text to improve it into proper English form. Thanks are also due to Mrs. M. Th. Guilbot for technical assistance and Mrs. M. Modol for typing the manuscript. Unpublished work from my laboratory was supported by grants from CNRS (U.A. 686) and Ecole Normale Superieure.

VIII. REFERENCES

1. Agudo, L., Sp. Zahler, W.S. and Smith, M.F. (1984). Effect of prostaglandin $F_{2\alpha}$ on adenylate cyclase and phosphodiesterase activity of ovine corpora lutea. J. Anim. Sci. 58, 955-962.
2. Alila, H.W. and Hansel, W. (1984). Origin of different cell types in the bovine corpus luteum as characterized by specifical monoclonal antibodies. Biol. Reprod. 31, 1015-1025.
3. Anderson, L.L., Adair, V., Stromer, M.H. and Donald, W.G. (1983). Relaxin production and release after hysterectomy in the pig. Endocrinology 113, 677-686.
4. Anderson, M.B. and Sherwood, O.D. (1984). Ultrastructural localization of relaxin immunoreactivity in corpora lutea of pregnant rats. Endocrinology 114, 1124-1127.
5. Anderson, M.B., Vaupel, M.R. and Sherwood, O.D. (1984). Pregnant mouse corpora lutea: immunocytochemical localization of relaxin ultrastructure. Biol. Reprod. 31, 391-397.
6. Arslan, M., Jalali, S. and Qazi, M.H. (1976). Reproductive cycle of the female spiny-tailed lizard, Uromastix hardwicki Gray. Biologia 22, 53-60.
7. Arslan, M., Zaidi, P., Lobo, J., Zaidi, A.A. and Qazi, M.H. (1978). Steroid levels in preovulatory and gravid lizards (Uromastix hardwicki). Gen. Comp. Endocrinol. 34, 300-303.
8. Auletta, F.J., Agins, H. and Scommegna, A. (1978). Prostaglandin $F_{2\alpha}$ mediation of the inhibitory effect of estrogen on the corpus luteum of the rhesus monkey. Endocrinology 104, 1183-1187.
9. Auletta, F.J., Kamps, D.L., Portes, S., Bisset, J. and Gibson, M. (1984). An intra corpus luteum site for the luteolytic action of prostaglandin $F_{2\alpha}$ in the rhesus monkey. Prostaglandins 27, 285-298.
10. Auletta, F.J., Paradis, D.K., Wesley, M. and Duby, R.T. (1984). Oxytocin is luteolytic in the rhesus monkey (Macaca mulatta). J. Reprod. Fert. 72, 401-406.
11. Badir, N. (1968). Structure and function of corpus luteum during gestation in the viviparous lizard Chalcides ocellatus. Anat. Anz. 122, 1-10.
12. Balmaceda, J.P., Valenzuela, G., Eddy, C.A. and Asch, R.H. (1981). Effects of hCG on prostaglandin synthesis and function of corpus luteum. Obstetr. Gynecol. U.S.A. 57, 505-508.
13. Bara, G. (1965). Histochemical localization of Δ^5-3β-hydroxysteroid dehydrogenase in the ovaries of a teleost fish Scomber scomber L. Gen. Comp. Endocrinol. 5, 284-296.
14. Behrman, H.R., Luborsky-Moore, J.L., Pang, C.Y., Wright, K., and Dorflinger, L.J. (1979). Mechanisms of $PGF_{2\alpha}$ action in functional luteolysis. Adv. Exp. Med. Biol. 112, 557-571.
15. Belt, W.D., Anderson, L.L., Cavozos, L.F. and Melampy, R.M. (1971). Cytoplasmic granules and relaxin levels in porcine corpora lutea. Endocrinology 89, 1-10.
16. Betz, T.W. (1963). The ovarian histology of the diamond-backed water snake, Natrix rhombifera during the reproductive cycle. J. Morph. 113, 245-260.
17. Bill, C.H. and Keyes, P.L. (1983). 17β-estradiol maintains normal function of corpora lutea throughout pseudopregnancy in hypophysectomized rabbits. Biol. Reprod. 28, 608-617.
18. Blane-Livni, N. (1971). Ovarian histochemistry of the fishes Cyprinus carpio, Mugil carpito and Tilapia aurea (Teleostei). Histochem. J. 3, 405-415.

19. Boehm, N., Plas-Roser, S. and Aron, Cl. (1984). Does corpus luteum function autonomously during estrous cycle in the rat? A possible involvement of LH and prolactin. J. Steroid Biochem. 20, 663-670.
20. Bona-Gallo, A., Licht, P., MacKenzie, D.S. and Lofts, B. (1980). Annual cycles in levels of pituitary and plasma gonadotropin, gonadal steroids and thyroid activity, in the Chinese cobra (Naja naja). Gen. Comp. Endocrinol. 42, 477-493.
21. Botte, V. and Cottino, E. (1964). Ricerche istochimiche sulla distribuzion del colesterolo et di alcuni enzimi della steroidogenesi nei follicoli ovarici e post-ovulatori di Rana esculenta e Triturus cristatus. Boll. Zool. 31, 491-500.
22. Boyd, M.M.M. (1940). The structure of the ovary and the formation of the corpus luteum in Hoplodactylus maculatus Gray. Quart. J. Micr. Sci. 82, 337-375.
23. Bragdon, D.E. (1952). Corpus luteum formation and follicular atresia in the common garter snake Thamnophis sirtalis. J. Morph. 91, 413-445.
24. Bronstein, R.D., Holt, J.A. and Zirkin, B.R. (1984). Estradiol-induced changes in progesterone secretion by rabbit corpora lutea are associated with quantitative ultrastructural changes in luteal cells. Endocrinology 115, 342-349.
25. Browning, H.C. (1973). The evolutionary history of the corpus luteum. Biol. Reprod. 8, 128-157.
26. Butler, W.R., Hotchkiss, J. and Knobil, E. (1975). Functional luteolysis in the Rhesus monkey: ovarian estrogen and progesterone during the luteal phase of the menstrual cycle. Endocrinology 96, 1509-1512.
27. Callard, I.P., Chan, S.W.C. and Potts, M.A. (1972). The control of reptilian gonad. Amer. Zool. 12, 273-287.
28. Callard, I.P., Doolittle, J., Banks, W.L. Jr. and Chan, S.W.C. (1972). Recent studies on the control of the reptilian ovarian cycle. Gen. Comp. Endocrinol. Suppl. 3, 65-75.
29. Callard, I.P. and Lance, V. (1977). The control of reptilian follicular cycles. In "Reproduction and Evolution" (J.H. Calaby and C.H. Tyndale-Biscoe, eds), pp. 199-210. Australian Acad. Sci., Canberra City.
30. Callard, I.P., Lance, V., Salhanick, A.R. and Barad, D. (1978). The annual ovarian cycle of Chrysemys picta: correlated changes in plasma steroids and parameters of vitellogenesis. Gen. Comp. Endocrinol. 35,
31. Callard, I.P. and Leathem, J.H. (1965). In vitro steroid synthesis by the ovaries of elasmobranchs and snakes. Arch. Anat. Micros. Morphol. Exp. 54, 35-48.
32. Chamley, W.A., Stelmasiak, T., and Bryant, G.D. (1975). Plasma relaxin immunoactivity during the oestrous cycle of the ewe. J. Reprod. Fert. 45, 455-461.
33. Callard, I.P., McChesney, I., Scanes, C. and Callard, G.V. (1976). The influence of mammalian and avian gonadotropins on in vitro ovarian steroid synthesis in the turtle. (Chrysemys picta). Gen. Comp. Endocrinol. 28, 2-9.
34. Chan, S.W.C., Ziegel, S. and Callard, I.P. (1973). Plasma progesterone in snakes. Comp. Biochem. Physiol. 44A, 631-637.
35. Chieffi, G. (1961). La luteogenesi nei selaci ovovivipari. Ricerche istologiche e istochimiche in Torpedo marmorata e Torpedo ocellata. Publ. Staz. Zool. Napoli 32, 145-166.
36. Chieffi, G. (1962). Endocrine aspects of reproduction in elasmobranch fishes. Gen. Comp. Endocrinol. Suppl. 1, 275-285.
37. Chieffi, G. and Botte, V. (1961). La luteogenesi in Scylliorhinus stellaris selacio oviparo. Boll. Zool. 28, 203-209.
38. Chieffi, G. and Botte, V. (1970). The problem of "luteogenesis" in non-mammalian vertebrates. Boll. Zool. 37, 85-102.

39. Christiansen, H.E. (1971). La reproduccion de la merluzza en el mar argentino (merlucciidae, Merluccius merluccius hubbsi). Bol. Inst. Biol. Mar. 20, 5-41.
40. Colombo, L. and Yaron, Z. (1976). Steroid 21-hydroxylase activity in the ovary of the snake Storeria dekayi during pregnancy. Gen. Comp. Endocrinol. 28, 403-412.
41. Condon, W.A. and Black, D.L. (1976). Catecholamine induced stimulation of progesterone by bovine corpus luteum in vitro. Biol. Reprod. 15, 573-578.
42. Cyrus, R.V., Mahmoud, I.Y. and Klicka, J. (1978). Fine structure of the corpus luteum of the snapping turtle, Chelydra serpentina. Copeia 1978, 622-627.
43. De la Llosa-Hermier, M.P., Leboulleux, P., Evrard, M. and Hermier, C. (1979). In vitro effect of prolactin, prostaglandin $F_{2}\alpha$ and cycloheximidine on 20β-dihydroprogesterone synthesis in pseudopregnant rat ovaries. J. Steroid Biochem. 10, 689-693.
44. Dodd, J.M. (1983). Reproduction in cartilaginous fishes (Chondrichthyes). In "Fish Physiology" Vol. IXA (W.S. Hoar, D.J. Randall and E.M. Donaldson, eds), pp. 31-95. Academic Press, New York.
45. Exbrayat, J.M. and Collenot, G. (1983). Quelques aspects de l'evolution de l'ovaire de Typhlonectes compressicaudus (Dumeril et Bibron, 1841), Batracien Apode vivipare. Etude quantitative et histochimique des corps jaunes. Reprod. Nutr. Develop. 23, 889-898.
46. Fitz, T.A., Fleeger, J.L., Smith, M.F. and Harms, P.G. (1980). Human chorionic gonadotropin (hCG) binding and adenylate cyclase (AC) activity in developing and regressing bovine corpora lutea (CL). Biol. Reprod. 22, 61a.
47. Fitz, T.A., Hoyer, P.B. and Niswender, G.D. (1984). Interactions of prostaglandins with subpopulations of ovine luteal cells. Stimulatory effects of prostaglandins E_1, E_2 and I_2. Prostaglandins 28, 119-126.
48. Fitz, T.A., Mayan, M.H., Sawyer, H.R. and Niswender, G.D. (1982). Characterization of two steroidogenic cell types in the ovine corpus luteum. Biol. Reprod. 27, 703-711.
49. Flint, A.P.F., Burton, R.D., Gadsby, J.E., Heap, R.B. and Sheldrick, E.L. (1983). Embryonic steroid synthesis and luteal oxytocin production: controlling mechanisms for the maternal recognition of pregnancy. J. Steroid Biochem. 19, 973-978.
50. Flint, A.P.F. and Sheldrick, E.L. (1983). Evidence for a systemic role of ovarian oxytocin in luteal regression in sheep. J. Reprod. Fert. 67, 215-225.
51. Fox, H. (1977). The urogenital system of reptiles. In "Biology of the Reptilia" Vol. 6 (C. Gans and T.S. Parsons, eds), pp. 1-157. Academic Press, New York.
52. Gabe, M. and Saint Girons, H. (1962). Donnees histophysiologiques sur l'elaboration d'hormones sexuelles au cours du cycle reproducteur chez Vipera aspis (L). Acta Anat. 50, 22-51.
53. Garnier, D.H. and Joly, J. (1985). Steroides sexuels et cycles saisonniers chez la femelle de Salamandra salamandra. "Endocrine Regulations as Adaptative Mechanisms to the Environment." Chize, France, 1-5 July.
54. Gadsby, J.E. and Keyes, M. (1984). Control of corpus luteum function in the pregnant rabbit; role of the placenta ("Placental luteotropin") in regulating responsiveness of corpora lutea to estrogen. Biol. Reprod. 31, 16-24.
55. Gavaud, J. (1985). Role de l'hibernation dans le determinisme de la vitellogenese de Lacerta vivipara J. (Reptile Squamate): etudes morphologiques, endocrines and biochimiques. These de Doctorat es-Sciences and Publ. Lab. Zool. E.N.S. 27, 276 pages.

56. Gemmel, R.T., Stacy, B.D. and Thornburn, G.D. (1974). Ultrastructural study of secretory granules in the corpus luteum of the sheep during the estrous cycle. Biol. Reprod. 11, 447-462.

57. Gibori, G., Antczak, E. and Rothchild, I. (1977). The role of estrogen in the regulation of luteal progesterone secretion in the rat after day 12 of pregnancy. Endocrinology 100, 1483-1495.

58. Gibori, G. Kalison, G., Basuray, R., Rao, M.C. and Hunzicker-Dunn, M. (1984). Endocrine role of the decidual tissue: decidual luteotropin regulation of luteal adenylyl cyclase activity, luteinizing hormone receptors and steroidogenesis. Endocrinology 115, 1157-1163.

59. Gibori, G. and Keyes, P. (1980). Luteotropic role of estrogen in early pregnancy in the rat. Endocrinology 106, 1584-1588.

60. Gibori, G. and Richards, J.S. (1978). Dissociation of two distinct luteotrophic effects of prolactin: regulation of luteinizing hormone-receptor content and progesterone secretion during pregnancy. Endocrinology 102, 767-774.

61. Godkin, J.C., Bazer, F.W., Moffatt, J., Sessions, F. and Roberts R.M. (1982). Purification and properties of a major low molecular weight protein released by the trophoblast of sheep blastocysts at day 13-21. J. Reprod. Fert. 65, 141-150.

62. Goetz, F.W. (1983). Hormonal control of oocyte final maturation and ovulation in fishes. In "Fish Physiology" Vol. IXB (W.S. Hoar, D.J. Randall and E.M. Donaldson, eds.) pp. 117-170. Academic Press, New York.

63. Goetz, F.W. and Cetta, F. (1983). Ovarian and plasma PGE and PGF levels in naturally ovulating brook trout (*Salvenilus fontinalis*) and the effects of indomethacin on prostaglandin levels. Prostaglandins 26, 387-395.

64. Goldberg, S.R. (1970). Seasonal ovarian histology of the ovoviviparous iguanid lizard *Sceloporus jarrovi* Cope. J. Morph. 132, 265-276.

65. Goldberg, V.J. and Ramwell, P.W. (1975). Role of prostaglandins in reproduction. Physiol. Rev. 55, 325-351.

66. Gospodarowicz, D. and Thakral, K.K. (1978). Production of a corpus luteum angiogenic factor responsible for proliferation of capillaries and neovascularization of the corpus luteum. Proc. Natl. Acad. Sci. U.S.A. 75, 847-851.

67. Guillette, L.J., Jr. (1982). The evolution of viviparity and placentation in the high elevation, Mexican lizard *Sceloporus aeneus*. Herpetologica 38, 94-103.

68. Guillette, L.J., Jr., and Fox, S.L. (1985). Effect of deluteinization on plasma progesterone concentration and gestation in the lizard, *Anolis carolinensis*. Comp. Biochem. Physiol. 80A, 303-306.

69. Guillette, L.J., Jr., and Herman, C.A. (1985). Synthesis of series two prostaglandins by the reptilian reproductive tract. 10th Intern. Symp. Comp. Endocrinol., Colorado U.S.A., 21-26 July.

70. Guillette, L.J., Jr., Lavia, L.A., Walker, N.J. and Roberts, D.K. (1984). Luteolysis induced by prostaglandin F_2 in the lizard, *Anolis carolinensis*. Gen. Comp. Endocrinol. 56, 271-277.719.

71. Guillette, L.J., Jr., Spielvogel, S. and Moore, F.L. (1981). Luteal development, placentation and plasma progesterone concentration in the viviparous lizard *Sceloporus jarrovi*. Gen. Comp. Endocrinol. 43, 20-29.

72. Guraya, S.S. (1968). Histochemical study of granulosa (follicular) cells in the preovulatory and postovulatory follicles of amphibian ovary. Gen. Comp. Endocrinol. 10, 138-146.

73. Guraya, S.S. (1973). Morphological and histochemical observations on the corpora lutea of the snake (*Lycodon aulicus*) ovary. Acta Morph. Acad. Sci. Hung. 21, 239-247.

74. Guraya, S.S. (1976). Recent advances in the morphology, histochemistry and biochemistry of steroid-synthesizing cellular sites in the non-mammalian vertebrate ovary. Int. Rev. Cytol. 44, 365-409.
75. Guraya, S.S. and Kaur, S. (1979). Morphology of the postovulatory follicle (corpus luteum) of teleost (Cyprinus carpio) ovary. Zool. Beitr. 25, 381-390.
76. Gwynne, J.T. and Strauss, J.F. (1982). The role of lipoproteins in steroidogenesis and cholesterol metabolism in steroidogenic glands. Endocr. Rev. 2, 299-329.
77. Henderson, K.M. and Moon, Y.S. (1979). Luteinization of bovine granulosa cells and corpus luteum formation associated with loss of androgen-aromatizing ability. J. Reprod. Fert. 56, 89-97.
78. Highfill, D.R. and Mead, R.A. (1975). Sources and levels of progesterone during pregnancy in the garter snake, Thamnophis elegans. Gen. Comp. Endocrinol. 27, 389-400.
79. Hoar, W.S. (1969). Comparative physiology: hormones and reproduction in fishes. In "Fish Physiology" Vol. 3 (W.S. Hoar and D-J. Randall, eds.) pp. 51-70. Academic Press, New York.
80. Holt, J.A., Dinerstein, J. and Lorincz, M.A. (1982). Site of estrogen action in rabbit ovarian tissue. Amer. J. Physiol. 243, E188-E195.
81. Holt, J.A., Lorincz, M.A. and King, W.J. (1983). Monoclonal antibody recognition of estrogen receptor in rabbit corpus luteum. Amer. J. Physiol. 244, E494.
82. Holt, J.A., Richards, J.S., Migley, A.R., Jr., and Reichert, I.E., Jr. (1976). Effect of prolactin on LH receptors in rat luteal cells. Endocrinology 98, 1005-1013.
83. Holt, J.A., Schreiber, J.R. and Zirkin, B.R. (1983). Estradiol induced changes in rabbit luteal cell progestin production and cholesterol and cholesterol ester content. Biochem. Biophys. Res. Comm. 113, 1026-1033.
84. Horton, E.W. and Poyser, N.L. (1976). Uterine luteolytic hormone: a physiological role for prostaglandin $F_{2\alpha}$. Physiol. Rev. 56, 595-651.
85. Hoyer, P.B. and Niswender, G.D. (1985). The regulation of steroidogenesis is different in the two types of ovine luteal cells. Can. J. Physiol. Pharmacol. 63, 240-248.
86. Hunter, M.G. and Baker, T.G. (1981). Effect of HCG, cAMP and FSH on steroidogenesis by human corpora lutea in vitro. J. Reprod. Fert. 63, 285-288.
87. Hunzicker-Dunn, M. (1982). Epinephrine-sensitive adenylyl cyclase activity in rabbit ovarian tissues. Endocrinology 110, 233-240.
88. Hurk, R. van den and Peute, J. (1979). Cyclic changes in the ovary of the rainbow trout, Salmo gairdneri, with special reference to sites of steroidogenesis. Cell Tissue Res. 199, 289-306.
89. Ivell, R. and Richter, D. (1984). The gene for the hypothalamic peptide hormone oxytocin is highly expressed in the bovine corpus luteum: biosynthesis, structure and sequence analysis. Embo J. 3, 2351-2354.
90. Joly, J. (1966). Ecologie et cycle sexuels de Salamandra salamandra (L.). These de Doctorat-es-Sciences, CNRS AO 1095, 269 pages.
91. Joly, J. (1970). Etude autoradiographique de l'incorporation de thymidine tritiee dans le follicule ovarien mur et dans le corps jaune de Salamandra salamandra (L.), Amphibian ovovivipare. C. R. Soc. Biol. 164, 2102-2106.
92. Joly, J. and Picheral, B. (1972). Ultrastructure, histochimie et physiologie du follicule pre-ovulatoire et du corps jaune de l'Urodele ovovivipare Salamandra salamandra (L.). Gen. Comp. Endocrinol. 18, 235-259.
93. Jones, R.E. and Guillette, L.J., Jr. (1982). Hormonal control of oviposition and parturition in lizards. Herpetologica 38, 80-93.

94. Jones, R.E., Guillette, L.J., Jr., Summers, C.H., Tokarz, R.R. and Crews, D. (1983). The relationship among ovarian condition, steroid hormones, and estrous behavior in Anolis carolinensis. J. Exp. Zool. 227, 145-154.
95. Jordan, A.W., Caffrey, J.L. and Niswender, G.D. (1978). Catecholamine-induced stimulation of progesterone and adenosine 3',5'-monophosphate production by dispersed ovine luteal cells. Endocrinology 103, 385-392.
96. Kagawa, H. and Takano, K. (1979). Ultrastructure and histochemistry of granulosa cells of pre- and post-ovulatory follicles in the ovary of the medaka, Oryzias latipes. Bull. Fac. Fish., Hokkaido Univ. 30, 191-204.
97. Kagawa, H., Takano, K. and Nagahama, Y. (1981). Correlation of plasma estradiol-17β and progesterone levels with ultrastructure and histochemistry of ovarian follicles in the white-spotted char, Salvelinus leucomaenis. Cell. Tissue Res. 218, 315-329.
98. Kagawa, H., Young, G. and Nagahama, Y. (1982). Estradiol-17β production in isolated amago salmon (Oncorhynchus rhodurus) ovarian follicles and its stimulation by gonadotropins. Gen. Comp. Endocrinol. 47, 361-365.
99. Kagawa, H., Young, G. and Nagahama, Y. (1983). Relationship between seasonal plasma estradiol-17β and testosterone levels and in vitro production by ovarian follicles of amago salmon (Oncorhyncus rhodurus). Biol. Reprod. 29, 301-309.
100. Keyes, P.L., Yuh, K.C.M. and Miller, J.B. (1979). Estrogen action in the corpus luteum. In "Ovarian Follicular and Corpus Luteum Function." (C.P. Channing, J.A. Marsh, W.A. Sadler, eds.), pp. 447-465. Raven Press, New York.
101. Khan-Dawood, F.S. and Dawood, M.Y. (1983). Human ovaries contain immunoreactive oxytocin. J. Clin. Endocrinol. Metab. 57, 1129-1132.
102. Khan-Dawood, F.S., Marut, E.L. and Dawood, M.Y. (1984). Oxytocin in the corpus luteum of the Cynomolgus monkey (Macaca fascicularis). Endocrinology 115, 570-574.
103. Khoo, K.H. (1975). The corpus luteum of goldfish (Carassius auratus L.) and its functions. Can. J. Zool. 53, 1306-1323.
104. Kim, I. and Greenwald, G.S. (1984). Hormonal requirements for maintenance of follicular and luteal function in the hypophysectomized cyclic hamster. Biol. Reprod. 30, 1063-1072.
105. Klicka, J. and Mahmoud, I.Y. (1972). Conversion of pregnenolone-$4^{14}C$ to progesterone-$4^{14}C$ by turtle corpus luteum. Gen Comp. Endocrinol. 19, 367-369.
106. Klicka, J. and Mahmoud, I.Y. (1973). Conversion of cholesterol to progesterone by turtle corpus luteum. Steroids 21, 483-495.
107. Kicka, J. and Mahmoud, I.Y. (1977). The effects of hormones on the reproductive physiology of the painted turtle, Chrysemys picta. Gen. Comp. Endocrinol. 31, 407-413.
108. Koos, R.D. and Hansel, W. (1981). The large and small cells of the bovine corpus luteum. Ultrastructural and function differences. In "Dynamics of Ovarian Function" (N.B. Schwartz and M. Hunzicker-Dunn, eds.) pp. 197-203. Raven Press, New York.
109. Lambert, J.G.D. (1966). Location of hormone production in the ovary of the guppy, Poecilia reticulata. Experientia 22, 476-477.
110. Lambert, J.G.D. and van Oordt, P.G.W.J. (1974). Ovarian hormones in teleosts. Fortschr. Zool. 22,340-349.
111. Lamotte, M., Rey, P. and Vilter, V. (1956). Evolution ovarienne au cours de la gravidite chez un Batracien vivipare, Nectophrynoides occidentalis. C. R. Soc. Biol. 150, 393-396.
112. Lamotte, M., Rey, P. and Vogeli, M. (1964). Recherches sur l'ovaire de Nectophrynoides occidentalis, Batracien Anoure vivipare. Arch. Anat. microsc. Morphol. exp. 53,179-224.

113. Lamotte, M. and Xavier, F. (1972). Les Amphibiens Anoures a developpement direct d'Afrique. Observations sur la biologie de Nectophrynoides tornieri (Roux). Bull. Soc. Zool. Fr. 97, 413-418.
114. Lance, V. and Callard, I.P. (1969). A histochemical study of ovarian function in the ovoviviparous elasmobranch, Squalus acanthias. Gen. Comp. Endocrinol. 13, 255-267.
115. Lance, V. and Lofts, B. (1978). Studies on the annual reproductive cycle of the cobra, Naja naja. IV. Ovarian histology. J. Morph. 157, 161-180.
116. Lang, I. (1981). Electron microscopic and histochemical study of the postovulatory follicles of Perca fluviatilis L. (Teleostei). Gen. Comp. Endocrinol. 45, 219-233.
117. Lehri, G.K. (1968). Cyclical changes in the ovary of the catfish, Clarias batrachus (Linn.). Acta Anz. 69, 105-124.
118. Lemaire, W.J., Williams, M.T., Thibier, M., Clark, M.R., El-Hassan, N.D., Roth, M.S. and Marsh, J.M. (1980). Some aspects of the control of the human corpus luteum. In "Endocrine Physiopathology of the Ovary" (R.I. Tozzini, G. Reeves and R.L. Pineda, eds.) pp.179-190. Elsevier/North-Holland Biomedical Press.
119. Lemon, M. and Loir, M. (1977). Steroid release in vitro by two luteal cell types in the corpus luteum of the pregnant sow. J. Endocrinol. 72, 351-359.
120. Lemon, M. and Mauleon, P. (1982). Interaction between two luteal cell types from the corpus luteum of the sow in progesterone synthesis in vitro. J. Reprod. Fertil. 64, 315-323.
121. Levasseur, M.C. (1983). Utero-ovarian relationships in placental mammals: role of uterus and embryo in the regulation of progesterone secretion by the corpus luteum. A review. Reprod. Nutr. Develop. 23, 793-816.
122. Levasseur, M.C. and Thibault, C. (1985). La cellule luteale: reflexions sur ses origines et son devenir. Actualites gynecologiques 16, 95-110.
123. Lewis, J. Mahmoud, I.Y. and Klicka, J. (1979). Seasonal fluctuations of plasma progesterone and oestradiol-17β in the female snapping turtle, Chelydra serpentina. J. Endocrinol. 80,127-131
124. Licht, P. (1979). Reproductive endocrinology of reptiles and amphibians: gonadotropins. Annu. Rev. Physiol. 41, 337-351.
125. Licht, P. (1980). Evolutionary and functional aspects of pituitary gonadotropins in the green turtle, Chelonia mydas. Amer. Zool. 20, 565-574.
126. Licht, P. (1982). Endocrine patterns in the reproductive cycle of turtles. Herpetologica 38, 51-61.
127. Licht, P. (1983). Evolutionary divergence in the structure and function of pituitary gonadotropins of tetrapod vertebrates. Amer. Zool. 23, 673-683.
128. Licht, P., Bona-Gallo, A. and Daniels, E.L. (1977). In vitro binding of radioiodinated sea turtle (Chelonia mydas) follicle-stimulating hormone to reptilian gonadal tissues. Gen. Comp. Endocrinol. 33, 226-230.
129. Licht, P. and Crews, D. (1976). Gonadotropin stimulation of in vitro progesterone production in reptilian and amphibian ovaries. Gen. Comp. Endocrinol. 29, 141-151.
130. Licht, P., Farmer, S.W., Bona-Gallo, A. and Papkoff, H. (1979). Pituitary gonadotropins in snakes. Gen. Comp. Endocrinol. 39, 34-52.
131. Licht, P., Farmer, S.W. and Papkoff, H. (1976). Further studies on the chemical nature of reptilian gonadotropins: FSH and LH in the American alligator and green sea turtle. Biol. Reprod. 14, 222-232.

132. Licht, P. and Midgley, A.R., Jr. (1977). Autoradiographic localization of binding sites for human follicle-stimulating hormone in reptilian testes and ovaries. Biol. Reprod. 16, 117-121.
133. Licht, P., Owens, D.W., Cliffton, K. and Penaflores, C. (1982). Changes in LH and progesterone associated with the nesting and ovulation in the olive ridley turtle, Lepidochelys olivacea. Gen. Comp. Endocrinol. 48, 247-253.
134. Licht, P. and Papkoff, H. (1974). Separation of two distinct gonadotropins from the pituitary gland of the snapping turtle (Chelydra serpentina). Gen. Comp. Endocrinol. 22, 218-237.
135. Licht, P., Rainey, W. and Cliffton, K. (1980). Serum gonadotropin and steroids associated with breeding activities in the green sea turtle Chelonia mydas. II. Mating and nesting in natural populations. Gen. Comp. Endocrinol. 40, 116-122.
136. Licht, P., Wood, D., Owens, W. and Food, F. (1979). Serum gonadotropin and steroids associated with breeding activities in the green sea turtle Chelonia mydas. I. Captive animals. Gen. Comp. Endocrinol. 39, 276-289.
137. Lofts, B. (1974). Reproduction. In "Physiology of the Amphibia" Vol. II (Lofts, B. ed.) pp. 107-218. Academic Press, New York.
138. Lofts, B. and Bern, H.A. (1972). Functional morphology of steroidogenic tissues. In "Steroids in Nonmammalian Vertebrates" (D.R. Idler, ed.) pp. 37-125. Academic Press, New York.
139. Lupo di Prisco, C. (1968). Biosintesi di ormoni steroidi nell'ovario di due specie di selacio: Scyliorhinus stellaris (oviparo) e Torpedo marmorata (ovoviviparo). Riv. Biol. 61, 113-146.
140. Lupo di Prisco, C., Botte, V. and Chieffi, C. (1965). Differenze nella capacita di biosintesi degli ormoni steroidi tra folliculi post-ovulatori ed atresici in Torpedo marmorata e Scyliorhinus stellaris. Boll. Zool. 32, 185-191.
141. Manning, N.J. and Kime, D.E. (1984). Temperature regulation of ovarian steroid production in the common carp, Cyprinus carpio L., in vivo and in vitro. Gen. Comp. Endocrinol. 56, 376-388.
142. Marsh, J.M. and Lemaire, W.J. (1974). Cyclic AMP accumulation and steroidogenesis in the human corpus luteum: effect of gonadotrophins and prostaglandins. J. Clin. Endocrinol. Metab. 38, 99-106.
143. Martal, J., Lacroix, M.C., Loudes, C., Saunier, M. and Winterberger-Torres, S. (1979). Trophoblastin, an antiluteolytic protein present in early pregnancy in sheep. J. Reprod. Fert. 56, 63-73.
144. McCracken, J.A., Schram, W. and Okulicz, W.C. (1984). Hormone receptor control of pulsatile secretion of $PGF_{2\alpha}$ from the ovine uterus during luteolysis and its abrogation in early pregnancy. Anim. Reprod. Sc. 7, 31-55.
145. McPherson, R.J., Boots, L.R., MacGregor, R. and Marion, K.R. (1982). Plasma steroids associated with seasonal reproductive changes in a multiclutched freshwater turtle, Sternotherus odoratus. Gen. Comp. Endocrinol. 48, 440-451.
146. Mead, R.A., Eroschenko, V.P. and Highfill, D.R. (1981). Effects of progesterone and estrogen on the histology of the oviduct of the garter snake, Thamnophis elegans. Gen. Comp. Endocrinol. 45, 345-354.
147. Mednick, D.L., Barkley, M.S. and Geschwind, I.I. (1980). Regulation of progesterone secretion by LH and prolactin during the first half of pregnancy in the mouse. J. Reprod. Fert. 60, 201-207.
148. Miller, M.R. (1948). The seasonal histological changes occurring in the ovary, corpus luteum and testis of the viviparous lizard Xantusia vigilis. Univ. Calif. Publ. Zool. 47, 197-223.
149. Milvae, R.A. and Hansel, W. (1983). Prostacyclin, prostaglandin $F_{2\alpha}$ and progesterone production by bovine luteal cells during the estrous cycle. Biol. Reprod. 29, 1063-1068.

150. Moll, E.O. and Legler, J.M. (1971). The life story of a neotropical slider turtle, Pseudemys scripta (Schoepff) in Panama. Bull. Los Angeles Mus. Nat. Hist. Sci. 11, 1-102.
151. Morat, M. (1969). Contribution a l'etude de l'activite Δ^5-3β-hydroxysteroide deshydrogenasique chez quelques Reptiles du Massif Central. Ann. Stat. Biol. Besse-en-Chandesse 4, 1-74.
152. Nagahama, Y. (1983). The functional morphology of teleost gonads. In "Fish Physiology" Vol. IXA (W.S. Hoar, D.J. Randall and E.M. Donaldson, eds.) pp. 223-275. Academic Press, New York.
153. Nagahama, Y., Chan, K., and Hoar, W.S. (1976). Histochemistry and ultrastructure of pre- and postovulatory follicles in the ovary of the goldfish, Carassius auratus. Can. J. Zool. 54, 1128-1139.
154. Nagahama, Y., Clarke, W.C. and Hoar, W.S. (1978). Ultrastructure of putative steroid-producing cells in the gonads of coho (Oncorhynchus kisutch) and pink salmon (Oncorhynchus gorbuscha). Can. J. Zool. 56, 2508-2519.
155. Nagahama, Y. and Kagawa, H. (1982). In vitro steroid production in the postovulatory follicles of the amago salmon, Oncorhynchus rhodurus, in response to salmon gonadotropin. J. Exp. Zool. 219, 105-109.
156. Nagahama, Y.H., Kagawa, H. and Young, G. (1982). Cellular sources of sex steroids in teleost gonads. Can. J. Fish. Aquat. Sci. 39, 56-64.
157. Nalbandov, A.V. (1970). Comparative aspects of corpus luteum function. Biol. Reprod. 2, 7-13.
158. Nalbandov, A.V. (1973). Control of luteal function in mammals. In "Handbook of Physiology" Section 7, Vol. 2, part. 1 (R.O. Greep, ed.) pp. 153-168. Amer. Physiol. Soc., Washington.
159. Nicholls, T.J. and Maple, G. (1972). Ultrastructural observations on possible sites of steroid biosynthesis in the ovarian follicular epithelium of two species of cichlid fish, Cichlosoma nigrofasciatum and Haplochromis multicolor. Z. Zellforsch. 128, 317-335.
160. Norjavaara, E., Rosberg, S., Gåfvels, M. and Selstam, C. (1984) β-adrenergic receptor concentration in corpora lutea of different ages obtained from pregnant mare serum gonadotropin-treated rats. Endocrinology 114, 2154-2159.
161. Pancharatna, M. and Saidapur, S.K. (1984). The luteotrophic effect of homoplastic pituitary pars distalis homogenates, PMSG and hCG on the corpora lutea of the hypophysectomized frog, Rana cyanophlyctus (SCHN). Gen. Comp. Endocrinol. 53, 375-380.
162. Panigel, M. (1951). L'evolution du corps jaune au cours de la gestation chez le Lezard Zootoca vivipara W. (Lacerta vivipara J.) Ann. Endocrinol. 12,206-212.
163. Panigel, M. (1956). Contribution a l'etude de l'ovoviviparite chez les Reptiles: gestation et parturition chez le Lezard vivipare, Zootoca vivipara. Ann. Sci. Nat. 18, 569-668.
164. Parry, D.M., Willcox, D.L. and Thornburn, G.D. (1980). Ultrastructural and cytochemical study of the bovine corpus luteum. J. Reprod. Fert. 60, 349-357.
165. Patwardhan, V.V. and Lanthier, A. (1984). Effect of prostaglandin $F_{2\alpha}$ on the hCG-stimulated progesterone production by human corpora lutea. Prostaglandins 27, 465-473.
166. Peters, H. and McNatty, K.P. (1980). Corpus luteum function. In "The Ovary" pp. 85-97. Granada Publishing Ltd., Great Britain.
167. del Pino, E.M. and Sanchez, G. (1977). Ovarian structure of the marsupial frog Gastrotheca riobambae (Fowler). J. Morphol. 153, 153-162.
168. Pitzel, L., Welp, K., Holtz, W. and Koenig, A.M. (1984). Neurohypophyseal hormones in the corpus luteum of the pig. Neuroendocrinol. Lett. 6, 1-5.

169. Quirk, S.J., Willcox, D.L., Parry, D.M. and Thornburn, G.D. (1979). Subcellular location of progesterone in the bovine corpus luteum: a biochemical, morphological and cytochemical investigation. Biol. Reprod. 20, 1133-1145.

170. Rai, B.P. (1966). The corpora atretica and the so-called corpora lutea in the ovary of *Tor (Barbus) tor* (Ham.). Anat. Anz. 119, 459-465.

171. Richardson, M.C. and Mason, G.M. (1981). Stimulation by human chorionic gonadotrophin of oestradiol production by dispersed cells from human corpus luteum: comparison with progesterone production; utilization of exogenous testosterone. J. Endocrinol. 91, 197-203.

172. Rodgers, R.J. and O'Shea, J.D. (1982). Purification, morphology and progesterone production and content of three cell types isolated from the corpus luteum of the sheep. Aust. J. Biol. Sci. 35, 441-455.

173. Rodgers, R.J., O'Shea, J.D. and Bruce, N.W. (1984). Morphometric analysis of cellular composition of the ovine corpus luteum. J. Anat. 138, 757-769.

174. Rodgers, R.J., O'Shea, J.D. and Findlay, J.K. (1983). Progesterone production *in vitro* by small and large ovine luteal cells. J. Reprod. Fert. 69, 113-124.

175. Rothchild, I. (1981). The regulation of the mammalian corpus luteum. Rec. Progr. Horm. Res. 37, 183-298.

176. Saidapur, S.K. (1982). Structure and function of postovulatory follicles (corpora lutea) in the ovaries of non-mammalian vertebrates. Int. Rev. Cytol. 75, 243-285.

177. Saidapur, S.K. and Nadkarni, V.B. (1972). Studies on postovulatory follicles in the skipper frog *Rana cyanophlyctis* (Schn.). Karnatak Univ. J. (Science) 17, 93-95.

178. Saidapur, S.K. and Nadkarni, V.B. (1974). Steroid-synthesizing cellular sites in amphibian ovary. A histochemical study. Gen. Comp. Endocrinol. 22, 459-462.

179. Samuel, M. (1943). Studies on the corpus luteum in *Rhinobatus granulatus* Cuv. Proc. Indian Acad. Sci. Sect. B 18, 133-157.

180. Samuel, M. (1951). Histogenesis and development of the corpus luteum in *Cerberus rhyncops* (Schneider). J. Zool. Soc. India 3, 291-310.

181. Sawyer, H.R., Abel, J.H., Jr., McClellan, M.C., Schmitz, M. and Niswender, G.D. (1979). Secretory granules and progesterone secretion by ovine corpora lutea *in vitro*. Endocrinology 104, 476-486.

182. Saxena, P.K., Rastogi, R.K. and Chieffi, G. (1977). Hydroxysteroid dehydrogenase activity in the ovary of *Rana esculenta* L. during the annual sexual cycle. Monitore Zool. Ital. (N.S.) 11, 21-32.

183. Schams, D., Schallenberger, E. and Legros, J.J. (1985). Evidence for the secretion of immunoreactive neurophysin I in addition to oxytocin from the ovary in cattle. J. Reprod. Fert. 73, 165-171.

184. Schams, D., Walters, D.L., Schallenberger, E., Bullerman, B. and Karg, H. (1983). Ovarian oxytocin in the cow. Acta Endocrinol. 102, suppl. 253, 147.

185. Schoomaker, J.N., Bergman, K.S., Steiner, R.A. and Karsch, F.J. (1982). Estradiol-induced luteal regression in the Rhesus monkey: evidence for an extrovarian site of action. Endocrinology 110, 1708-1715.

186. Sherwood, O.D., Martin, P.A., Chang, C.C. and Dzink, P.J. (1977). Plasma relaxin levels in pigs with corpora lutea induced during late pregnancy. Biol. Reprod. 17, 97-100.

187. Shrivastava, S.S. (1969). Formation of the corpora atretica in *Notopterus notopterus* (Pallas). Acta. Zool. 50, 77-89.

188. Stacey, N.E. and Goetz, F.W. (1982). Role of prostaglandins in fish reproduction. Can. J. Fish Aquat. Sci. 39, 92-98.

189. Szollosi, D., Jalabert, B. and Breton, B. (1978). Post-ovulatory changes in the theca folliculi of the trout. Ann. Biol. Anim. Biochem. Biophys. 18, 383-391.
190. Te Winkel, L.E. (1972). Histological and histochemical studies of postovulatory and preovulatory atretic follicles in Mustelus canis. J. Morphol. 136, 433-458.
191. Thibault, C. and Levasseur, M.C. (1979). Le corps jaune. In "La fonction ovarienne chez les Mammiferes." (INRA, ed.) pp. 57-80. Masson, Paris.
192. Thomas, J.P., Dorflinger, L.J. and Behrman, H.R. (1978). Mechanism of the rapid antigonadotropic action of prostaglandins in cultured luteal cells. Proc. Natl. Acad. Sci. USA 75, 1344-1348.
193. Thornton, V.F. and Evennett, P.J. (1973). Changes in the fine structure of the ovarian follicle of the toad (Bufo bufo) prior to induced ovulation. Gen. Comp. Endocrinol. 20, 413-423.
194. Ursely, J., Darbon, J.M. and Leymarie, P.(1981). Aromatization of testosterone in large and small bovine luteal cells. Conflicting results between radioimmunoassay and cocrystallisation data. Steroids 38, 271-279.
195. Ursely, J. and Leymarie, P. (1979). Varying response to luteinizing hormone of two luteal cell types isolated from bovine corpus luteum. J. Endocrinol. 83, 303-310.
196. van Wyk, J.H. (1984). Physiological changes during the ovarian cycle of the female rock lizard, Agama atra (Sauria: agamidae). S. Afr. J. Zool. 19, 254-260.
197. Vande Wiele, R.L., Bogumil, J., Dyrenfurth, I., Ferin, M., Jewelewicz, R., Warren, M., Rizkallah, T. and Mikhail, G. (1970). Mechanisms regulating the menstrual cycle in women. Rec. Progr. Horm. Res. 26, 63-103.
198. Varma, S.K. (1970). Morphology of ovarian changes in the garden lizard, Calotes versicolor. J. Morph. 131, 195-210
199. Veith, W.J. (1974). Reproductive biology of Chamaeleo pumilus pumilus with special references to the role of the corpus luteum and progesterone. Zool. Afr. 9, 161-183.
200. Vilter, V. and Vilter, A. (1964). Sur l'evolution des corps jaunes ovariens chez Salamandra atra Laur. des Alpes vaudoises. C. R. Soc. Biol. 158, 457-461.
201. Wake, M.H. (1977). The reproductive biology of caecilians. An evolutionary perspective. In "The Reproductive Biology of Amphibians" (D.H. Taylor and S.I. Guttman, eds.), pp. 73-102. Plenum Press, New York.
202. Wallace, R.A. and Selman, K. (1980). Oogenesis in Fundulus heteroclitus. II. The transition from vitellogenesis into maturation. Gen. Comp. Endocrinol. 42, 345-354.
203. Walters, D.L., Schams, D. and Schallenberger, E. (1984). Pulsatile secretion of gonadotrophins, ovarian steroids and ovarian oxytocin during the luteal phase of the oestrous cycle in the cow. J. Reprod. Fert. 71, 479-491.
204. Wathes, D.C. (1984). Possible actions of gonadal oxytocin and vasopressin. J. Reprod. Fert. 71, 315-345.
205. Wathes, D.C. and Swann, R.W. (1982). Is oxytocin an ovarian hormone? Nature 297, 225-227.
206. Weekes, H.C. (1934). The corpus luteum in certain oviparous and viviparous reptiles. Proc. Linn. Soc. N.S.W. 59, 380-391.
207. Weiss, G., O'Byrne, E.M., Hochman, J.A., Goldsmith, L.J., Rifkin, I. and Steinetz, B.G. (1977). Secretion of progesterone and relaxin by the human corpus luteum at mid-pregnancy and at term. Obstet. Gynaecol. 50, 679-681.

208. Wiest, W.G., Kidwell, W.R. and Balogh, K. (1968). Progesterone catabolism in the rat ovary: a regulatory mechanism for progestational potency during pregnancy. Endocrinology 82, 844-859.
209. Willcox, D.L., Parry, D.M. and Morris, J.F. (1983). Storage of progesterone in secretory granules. Spec. Sci. Techn. 6, 3-9.
210. Xavier, F. (1971). Recherches sur l'endocrinologie sexuelle de la femelle de Nectophyrnoides occidentalis Angel (Amphibien Anoure vivipare). These de Doctorat es-Sciences, CNRS n A.O 6385, 223
211. Xavier, F. (1974). La pseudogestation chez Nectophrynoides occidentalis Angel. Gen. Comp. Endocrinol. 22, 98-116.
212. Xavier, F. (1982). Progesterone in the viviparous lizard Lacerta vivipara: ovarian biosynthesis, plasma levels, and binding to transcortin-type protein during the sexual cycle. Herpetologica 38, 62-70.
213. Xavier, F. (1986). La reproduction des Nectophrynoides. In "Traite de Zoologie" XIV, I B. Amphibiens. (P.P. Grasse and M. Delsol, eds.), pp. 497-513. Masson, Paris.
214. Xavier, F. and Ozon, R. (1971). Recherches sur l'activite endocrine de l'ovaire de Nectophyrnoides occidentalis Angel (Amphibien Anoure vivipare). II. Synthese in vitro des steroides. Gen. Comp. Endocrinol. 16, 30-40.
215. Xavier, F., Zuber-Vogeli, M. and LeQuang-Trong, Y. (1970). Recherches sur l'activite endocrine de l'ovaire de Nectophyrnoides occidentalis Angel (Amphibien Anoure vivipare). I. Etude histochimique. Gen. Comp. Endocrinol. 15, 425-431.
216. Yoshinaga, K. (1978). Cyclic hormone secretion by the mammalian ovary. In "The Vertebrate Ovary. Comparative Biology and Evolution" (R.E. Jones, ed.) pp. 691-729. Plenum Press, New York.
217. Young, G., Crim, L.W., Kagawa, H., Kambegawa, A. and Nagahama, Y. (1983). Plasma 17α,20β-dihydro-4-pregnen-3-one levels during sexual maturation of amago salmon (Oncorhynchus rhodurus): correlation with plasma gonadotropin and in vitro production by ovarian follicles. Gen. Comp. Endocrinol. 51, 96-105.
218. Young, G., Kagawa, H. and Nagahama, Y. (1983). Evidence for a decrease in aromatase activity in the ovarian granulosa cells of amago salmon (Oncorhynchus rhodurus) associated with final oocyte maturation. Biol. Reprod. 29, 310-315.
219. Zuber-Vogeli, M. (1983). L'hypophyse de Nectophrynoides occidentalis Angel (Amphibien Anoure vivipare). Arch. Biol. (Bruxelles) 94, 109-154.

TESTICULAR FUNCTION

Brian Lofts

Department of Zoology
University of Hong Kong
Hong Kong

I. INTRODUCTION

In mammals, the gametogenetic and steroidogenic tissues are not only structurally distinct, but are also separated physiologically by cellular barriers. The gametogenetic component consists of a mass of convoluted seminiferous tubules embedded in a loose connective tissue stroma, the interstitium, and the whole mass is encapsulated in a tough fibrous outer coat, the tunica albuginea. The interstitial tissue contains blood vessels, lymphatic vessels, and also the steroidogenic Leydig cells, which pack the interstices between the tubules.

Each seminiferous tubule is lined by a permanent germinal epithelium composed of a basement membrane and various layers of germ cells closely arranged in specific associations around the large Sertoli cells. Electron microscopy has shown that the Sertoli cells are attached to the basement membrane and extend into the lumen, enveloping the developing germ cells interspersed between them, so that in cross section the latter appear sometimes to occupy crypts in the Sertoli cell cytoplasm. Tight-junctional complexes bind adjacent Sertoli cells together to form a blood-testis barrier which divides the tubule into a basal compartment containing the spermatogonia and a much larger adluminal compartment containing the primary spermatocytes and more advanced germinal stages (Fig. 1). Thus, the meiotically dividing germinal stages are bathed in a fluid of quite different composition from that surrounding the mitotically active spermatogonia in the basic compartment which are bathed in fluid similar in composition to the interstitial fluid. The passage of the developing germ cells from one

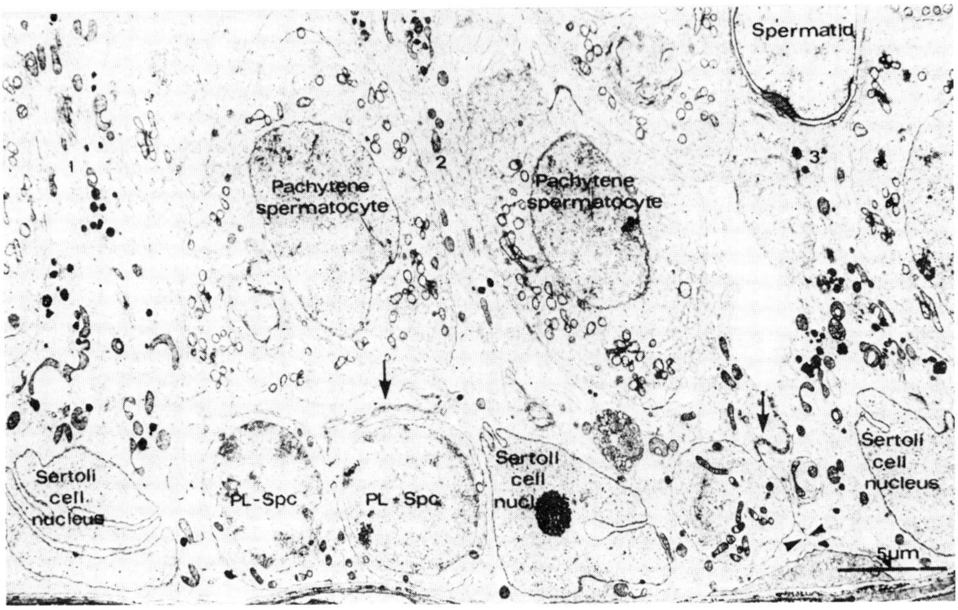

Fig. 1. Electron micrograph of the germinal epithelium of rat seminiferous tubule. Note the way the Sertoli cell cytoplasm surrounds the developing germ cells, and the junctional complexes (arrows) joining adjacent Sertoli cells.

compartment to another is allowed by the transient breakdown of the junctional complexes, facilitated by a plasminogen activator.

The commonly held concept of mammalian testicular regulation implicates the secretion of two highly specific adenophypophysial hormones, namely, FSH (follicle-stimulating hormone) which stimulates spermatogenesis, and LH (luteinizing hormone), which stimulates steroidogenesis in the Leydig cells. The main steroid secreted in mammals is testosterone which, as well as maintaining the accessory sexual structures, also has a spermatokinetic effect on the germinal epithelium. Thus, exogenous testosterone can maintain spermatogenesis in hypophysectomized rats; LH injections can also stimulate spermatogenesis in such animals, presumably by hyperstimulating testosterone secretion by the Leydig cells. This has led to the belief that testosterone is essential for the maintenance of normal mammalian spermatogenesis and passes from the interstitial compartment through the blood-testis barrier into the adluminal compartment to stimulate meiosis in the primary spermatocytes. FSH is necessary for stimulating spermatogonial mitotic activity and Sertoli cell activity. Whereas LH has been shown to bind specifically to the Leydig cells, FSH binds specifically to the Sertoli cells and stimulates, among other things, their production of androgen-binding protein which transports androgens in the luminal fluid to the epididymis. Whether it is testosterone itself that stimulates spermatogenesis, or whether it is a metabolite, is still a matter of conjecture. Some investigators believe that rather than testosterone, the Sertoli cells may be involved in first converting it into some other product, such as 5α-dihydrotestosterone (DHT), which acts as the stimulant.

Our knowledge of the functioning of the mammalian testis is largely based on investigations using laboratory rodents and domestic animals; until recently, comparatively little information has been available on seasonally breeding wild species. Yet, the great majority of extant vertebrate species, both mammalian and nonmammalian, are seasonal breeders with well-defined testicular cycles interspersed with periods of sexual quiescence. A study of nonmammalian forms, representing as they do an earlier evolutionary stage in the development of vertebrate reproductive structures and strategies, provides us, therefore, with a much broader understanding of the fundamental regulatory mechanisms controlling testicular function. During the past decade, our knowledge of the factors regulating reproductive cycles in ectothermic vertebrates has increased greatly. The present chapter summarizes our understanding of testicular function in these nonmammalian groups, and emphasizes where they diverge from the mammalian pattern briefly outlined above.

II. TESTICULAR STRUCTURE

A. Fishes

The cyclostomes have a single, medially placed testis that discharges spermatozoa into the body cavity, from where they pass to the exterior via abdominal pores. Gnathostome fishes, however, have paired testes and discharge their genital products into gonaducts. The piscine testicular structure differs from the mammalian pattern in that the tubular components lack a permanent germinal epithelium and are thin, fibrous-walled compartments that, with the exception of the cyclostomes, eventually link up with the efferent duct system. For this reason, they are commonly referred to as "lobules". The lobules accommodate the developing germ cells which are organized into coordinated clones of cells ensheathed in fibrous capsules known as 'germinal cysts.'

In lampreys, breeding is closely followed by death, and consequently, the testis produces only one generation of spermatozoa. There is rapid testicular growth prior to spawning, and the single gonad comes to fill most of the body cavity. Testes examined at different times during the anadromous migration to the freshwater breeding grounds show a uniform distribution of germinal cysts encased in lobules, all in a similar spermatogenetic stage of development. What little is known about the hagfish indicates that the basic organization is similar, and although a broader spectrum of germ cell stages have been noted at any one time in the testicular lobules, unequivocal evidence of more than one breeding cycle per lifetime has not yet been established.

Gnathostome fishes have multiple breeding cycles and, in many species, the testes show a distinct zonation. In elasmobranchs, for example, a meso-ventral germinal (ampullogenic) ridge runs the length of the gonad, and propagates germinal cysts which mature as they move away in a dorso-lateral direction. They are enclosed within large, spherical, lobular chambers called 'ampullae' (Fig. 2A). In transverse section the testis shows a distinct concentric zonation of successively more advanced germinal stages radiating out from the ampullogenic ridge. Such an organization also has been noted in some holocephalan species.

Zonal testes also are found in some teleosts. In the cyprinodont, *Fundulus heteroclitus*, for example, a subcapsular germinal zone surrounds the main lobular mass. The lobules have a radial distribution, converging on a centrally placed efferent duct system, and the germinal cysts mature as they move towards the duct (Fig. 2B). A similar organization also is seen in atheriniform species. In many other teleosts the lobules are much more loosely organized, and there is no apparent zonation or focalized propagating zone.

The interlobular (interstitial) stromal tissue of cyclostomes contains clusters of cells which, morphologically and ultrastructurally, resemble mammalian Leydig cells. Similar cells also have been reported in several squaliform and raiiform species of elasmobranchs. Most of the teleostean species so far studied also show an interlobular distribution of Leydig cells, but in some species Leydig cells appear to be absent and are replaced by another type of glandular cell for which Marshall and Lofts[132] coined the name 'lobule boundary cell'. In the pike, *Esox lucius*, these cells become seasonally prominent in the lobule walls at the time when the germinal cysts are becoming sexually mature.[123] In a more recent electron microscopic study in the same species, Grier and Linton[66] confirmed the absence of any identifiable Leydig cells in the interlobular stroma, but concluded that the lobule boundary cells were homologous to Sertoli cells rather than Leydig cells, a belief that also is shared by some others.[13,146]

In some teleosts, Leydig-type cells are concentrated in areas quite separate from the seminiferous elements. In *Poecilia latipinna*, for example, these cells are concentrated around the efferent ducts,[71a] whereas

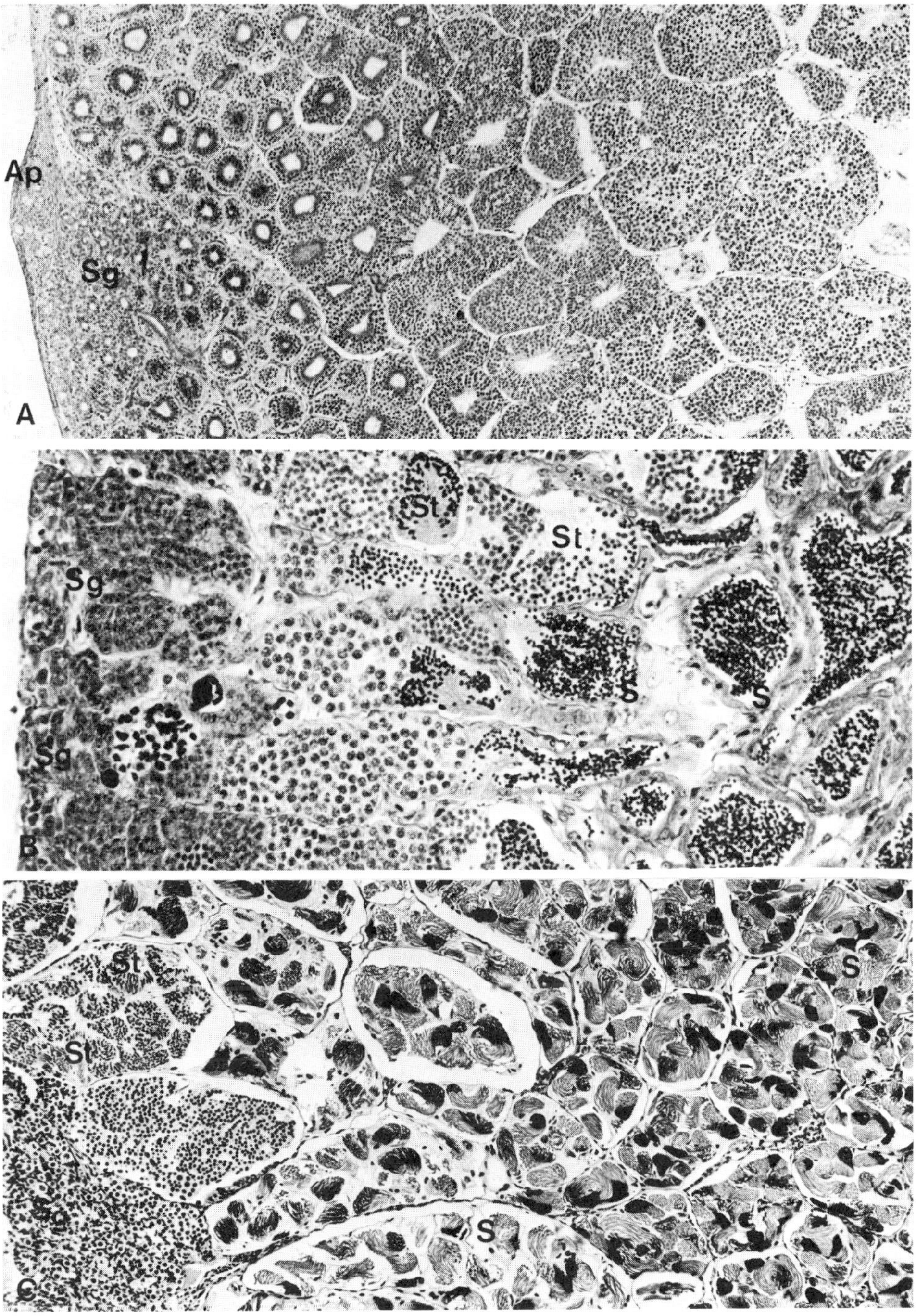

Fig. 2. Zonal testis. A. Scyliorhinus canicula. B. Fundulus heteroclitus. C. Trituroides hongkongenesis. Ap ampullogenic ridge, Sg spermatogonia, St spermatids, S spermatozoa.

in *Gobius paganellus*[171a] and *Gobius jozo*,[29a] a glandular mass, quite distinct from the seminiferous region, lies along the mesorchium. In various species of *Blennius* too, the Leydig cells are concentrated in a distinct accessory gland.

B. Amphibians

Little is known about the testicular structure of apodan amphibians, but the testicular organization of the Anura and Urodela (Caudata) has been extensively studied. The anuran testis, as in mammals, consists of a mass of convoluted seminiferous tubules with a permanent germinal epithelium and well-developed interstitial Leydig cells. The urodelean testis, on the other hand, is quite different and consists of piscine-like lobules which display a zonal distribution of germinal cysts (Fig. 2C). A further distinction is that whereas anurans have compact ovoid testes, many (but not all) urodeles have segmented testes, each consisting of a variable number of lobes joined in a linear chain by narrow bridges of tissue (Fig. 3A). The number of lobes increases with age and are generated from residual embryonic primordial cells in the tissue cords linking the lobes. Each lobe has the structural organization of a miniature testis, with zones of germinal cysts showing a cephalocaudal axis of increasing maturity (Fig. 3B).

Unlike anurans, many urodeles lack conspicuous secretory cells in the interlobular stroma, but display lobule boundary cells similar to those described in teleosts. After discharge of their seminal content, the evacuated germinal cysts and lobules become compressed into a distinct nodule of tissue which frequently is referred to as the 'glandular tissue.'

C. Reptiles

The reptilian testis is similar to the mammalian pattern; convoluted seminiferous tubules are embedded in an interstitial stroma which, in most species, shows a mammalian-type Leydig cell distribution. In several *Cnemidophorus* species, however, the Leydig cells are concentrated into a circumtesticular layer beneath the tunica albuginea, and fewer cells are scattered throughout the interstitium.[38] Within the seminiferous tubules, there is a permanent germinal epithelium, and there is no overall zonal distribution of the germinal stages. The individual germ cells do, however, show a graded radial axis of maturation within each tubule, so that the spermatozoa are the most centrally located, and the spermatogonia the most peripherally located, germ cells.

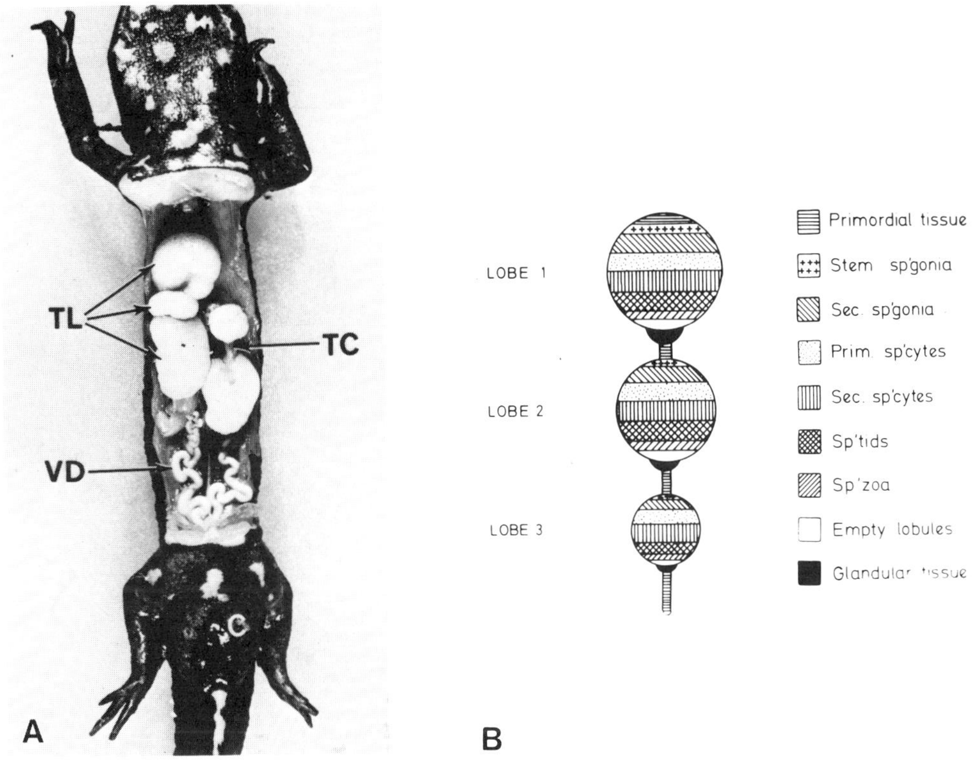

Fig. 3. A. Dissection of *Trituroides hongkongenesis* showing the segmented testis consisting of three testicular lobes (TL) joined by cords of tissue (TC). VD, vas deferens. B. Diagram illustrating the cephalo-caudal zonation.

III. THE SPERMATOGENETIC CYCLE

A. Cystic and Non-cystic Spermatogenesis

In fishes and amphibians spermatogenesis is cystic, and each stem (primary) spermatogonium is surrounded by one or more flattened follicular cells which form the ensheathing membranous wall of the germinal cyst (Fig. 4A). Such primary spermatogonial cysts are located in the basal region or all anuran seminiferous tubules, but in urodeles and elasmobranchs they are restricted to the cephalic or meso-ventral (ampullogenic) testicular zone, respectively. The seasonal recrudescence of mitotic activity in these cells repopulates the seminiferous elements with germinal cysts. In bony fish a residual population of intralobular stem spermatogonia has sometimes been reported to be responsible for this replenishment, but in many teleosts they are seasonally derived from primordial germ cells of interlobular origin which migrate through the lobule wall to establish the next generation of stem spermatogonia (e.g., *Couesius plumbeus*,[3] *E. lucius*,[123] *Culaea inconstans*[160]). In *Salmo gairdneri*, such a seasonal replenishment by

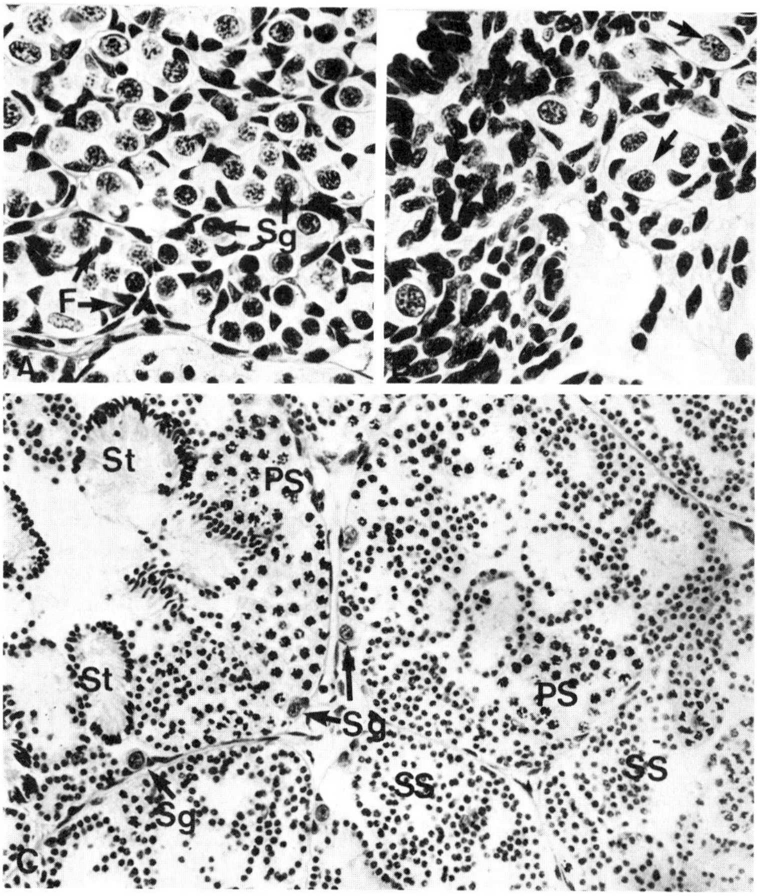

Fig. 4. A. Spermatogonial cysts of Trituroides hongkongenesis showing the primary spermatogonia (Sg) surrounded by follicle cells (F). B. Primordial germ cells (arrows) in the testicular cord of T. hongkongenesis. C. Seminiferous tubules of Rana temporaria containing germinal cysts in different stages of development. Sg stem spermatogonium, PS primary spermatocysts, SS secondary spermatocytes, St spermatids.

primordial germ cells of interlobular origin has been shown by electron microscopy,[72] and Ruby and McMillan[160] have provided evidence of such a mechanism in the testis of C. inconstans by means of [^{3}H]-thymidine labelling. Migratory primordial germ cells also occur in the testicular cords of the segmental-type testis of urodeles (Fig. 4B), and Takahashi and Hanaoka[179] have demonstrated the propagation of such cells from cord tissue of Cynops (Triturus) pyrrhogaster, in tissue culture. Similar cells also have been reported in this tissue in Notophthalmus (Triturus) viridiscens, Triturus alpestris, and Trituroides hongkongenesis.

Stem spermatogonia are the largest germ cells and often have irregularly shaped nuclei. Each divides repeatedly to produce a cluster of daughter cells with rounded nuclei, which are of similar size and appearance, and all are at an identical stage of development. Within this clone of secondary spermatogonia each cell is interconnected by means of cytoplasmic bridges[12,64,65] and these remain throughout the subsequent meiotic division, persisting, in the fish Oryzias latipes, right up to the time of spermatid formation. Within any particular cyst, the cells all develop at a similar rate and divide synchronously, and Grier[65] has suggested that the intercellular bridges help maintain this synchrony within the individual cysts. In this way a spermatogonial cyst, once the germinal content enters meiotic prophase, transforms into a primary spermatocytic cyst; then, with the subsequent synchronous division of its germ cells, the cyst transforms into a secondary spermatocytic cyst and then into a spermatid cyst. The latter eventually ruptures to liberate spermatozoa.

At the height of the spermatogenetic cycle, all the seminiferous elements in anurans and non-zonal teleosts contain a broad spectrum of maturing germinal cysts (Fig. 4C), but in zonal testes each zone contains only one type of germinal stage (Fig. 2).

During their gradual transformation into a greatly enlarged post-meiotic cyst containing numerous spermatids, the follicular cells change from the flattened cells of the juvenile cyst into a more glandular form, and take on the appearance and ultrastructural characteristics of Sertoli cells. Then, as the spermatids mature into spermatozoa (spermiogenesis), the sperm heads embed into their adluminal surface. An ultrastructural investigation of the spermatozoal cysts of Aphanius dispar (a teleost), using horseradish peroxidase, has demonstrated that tight junctional complexes join these Sertoli-like cells, forming an effective blood-testis barrier during spermiogenesis.[1] A similar blood-testis barrier has also been reported in Poecilia reticulata.[131] Recently, the presence of tight junctions linking adjacent Sertoli cells has been confirmed in the testes of the common carp, Cyprinus carpio, and immunohistochemical studies have indicated that a blood-testis barrier forms around the maturing spermatozoa in the lumina but is not present in the cysts surrounding the germinal stages up to spermatid formation.[155a]

There is a paucity of information concerning spermiogenesis in hagfish, but in lampreys and teleosts the eventual rupturing of the sperm-filled cysts liberates the spermatozoa into the lobule lumina which become filled with an amorphous mass of spermatozoa. This differs from the elasmobranch testis, where, during spermiogenesis, the spermatozoa become organized into discrete bundles embedded in the Sertoli cells which remain attached to the

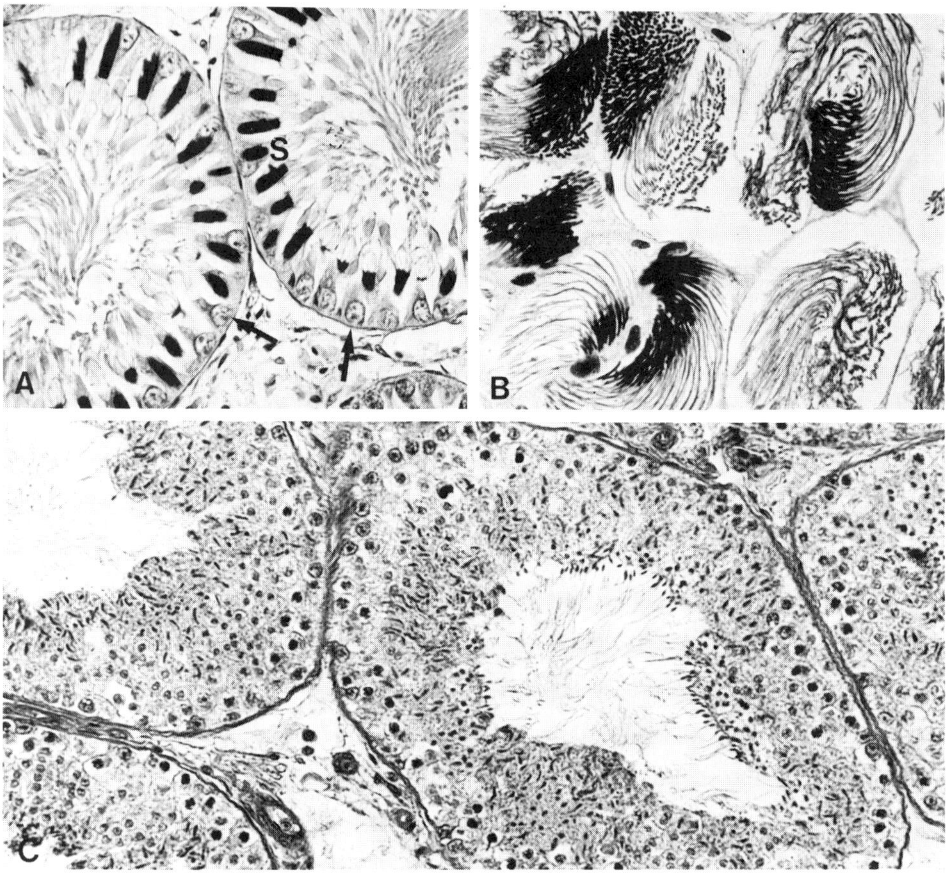

Fig. 5. A. Spermatozoal ampullae of *Scyliorhinus canicula* showing bundles of sperm heads (*S*) embedded in the Sertoli cells (*arrows*). B. Detached sperm bundles in the lobule lumen of *Trituroides hongkongenesis*. C. Seminiferous tubules of *Naja naja* during the breeding season.

ampullar wall (Fig. 5A). The same pattern is also seen in the anuran seminiferous tubules, but in urodeles, once the mature germinal cyst ruptures, the Sertoli cell-sperm bundle units lie free within the lobule lumina (Fig. 5B). Eventually, the spermatozoa are released and pass out through the efferent ducts. In anurans, the Sertoli cells become detached from the tubular wall and accumulate in the lumen after evacuation of spermatozoa (spermiation).

In reptiles, the spermatogenetic mechanism is similar to that of birds and mammals. The stem spermatogonia adjacent to the basement membrane are not associated with ensheathing follicular cells, and the secondary spermatogonia produced as a result of their mitotic division remain individual and are not interlinked by cytoplasmic bridges into a coordinated clone within a common germinal cyst. Thus, spermatogenesis is noncystic, and each secondary spermatogonium individually progresses through the various spermatogenetic stages to the formation of spermatozoa (Fig. 5C). After the spermatozoa become released into the tubular lumen, the Sertoli cells remain attached to the tubule wall and are not replaced by a new generation every spermatogenetic cycle as is the case in cystic spermatogenesis.

B. Spermatogenetic Cyclicity

1. Fishes. Little is known about spermatogenetic cyclicity in chondrichthyans, and in the few species where testicular histology has been followed throughout an annual cycle, a complete spectrum of spermatogenetic stages has been noted in all seasons. But even though, at first sight, this might indicate a continuous cycle, there are seasonal variations in testicular weight, and marked changes occur in the relative areas of the different spermatogenetic zones. In *Mustelus griseus* and *M. manazo*, for example, in May the testis contains predominantly spermatozoal ampullae and other germinal zones are greatly reduced, whereas in September, following the extrusion of ripe spermatozoa during the copulatory phase, the testis is almost completely devoid of spermatozoal ampullae, but the zone of spermatocytic ampullae is greatly enlarged.[182] An indication of spermatogenetic discontinuity also is evident in the testis of *Squalus acanthias* where, in May, a zone of degenerating ampullae develops at the rim of the ampullogenic region between the spermatogonia zone and primary spermatocytic ampullae.[167] By July, spermatogonial proliferation has resumed, and as new germinal cysts mature and move away from the ampullogenic region, the degenerative strip is gradually displaced outwards towards the area of spermatozoal discharge. When the degenerate zone reaches this region, spermatozoa no longer become released into the efferent ducts. This interruption of sperm discharge remains until the new generation of germinal cysts, moving behind the degenerative zone eventually reaches this region (Fig. 6). There is, therefore, an effective break of several months in the release of spermatozoa. Dodd et al.[42] have noted that removal of the ventral pituitary lobe in *Scyliorhinus canicula* causes a similar zonal degeneration.

The spermatogenetic cycle of teleosts is commonly discontinuous, and incorporates a postnuptial resurgence of high mitotic activity in stem spermatogonia that rapidly restocks the spent lobules with germinal cysts. Spermatogonial proliferation then ceases, but the cysts already formed continue to mature and eventually discharge ripe spermatozoa into the

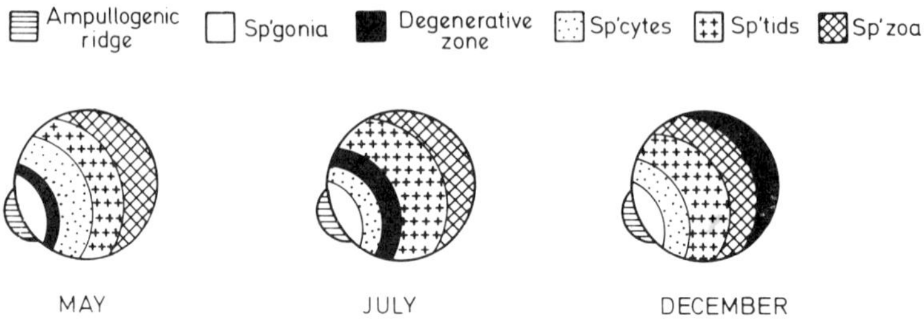

Fig. 6. Diagram of the testicular cycle in Squalus acanthias showing the seasonal development of a degenerative zone which gradually becomes displaced outwards towards the region of ampullar discharge.

lobular lumina. In most teleostean species of northern latitudes the annual cycle is of the 'postnuptial' pattern; fully developed spermatozoa are formed before the onset of winter. These spermatozoa are stored for several months, either in the lobules or sometimes in the efferent ducts, before being discharged in the following spring. At the time of spawning, therefore, the lobules are almost exclusively filled with masses of spermatozoa. In some species, however, the cycle becomes halted at an intermediate stage through the winter, and the final maturation into spermatozoa resumes during the following spring. In such a cycle there is no protracted period of sperm storage, and spermatozoa are only produced shortly before spawning. Such a cycle is classified as 'prenuptial.' In such species a post-spawning period of hypothalamic refractoriness to environmental stimulation imposes a spermatogenetically quiescent break before gonadal recrudescence begins.[8,49,85] Then, once this has finished the degree of spermatogenetic advancement reached before the onset of winter often depends on the environmental temperature.

2. Amphibians. In many tropical and subtropical anurans spermatogonial activity is continuous and germinal cysts are propagated throughout the year. However, in some subtropical species with a geographical range extending into northern latitudes (e.g., Rana esculenta, Discoglossus pictus), the cycle is interrupted in the northern populations during the colder winter months, and germinal cysts do not develop further than primary spermatocytes. Under experimental conditions, however, they can always be stimulated into full spermatogenesis by increasing the environmental temperature, and are therefore classified as 'potentially continuous' breeders to distinguish them from the truly 'discontinuous' cycles shown by many temperate anurans that have a postnuptial testicular refractory phase incorporated into their annual cycle. In Rana temporaria, for example,

after spawning a refractory phase of three months delays the reestablishment of spermatogenesis. Artificially exposing such refractory frogs to elevated environmental temperature has no effect on spermatogenesis, but once the refractory phase is over spermatogenesis rapidly resumes and germinal cyst production is completed before the onset of winter.[118]

As in elasmobranchs, the zonal testes of urodeles show seasonal variations in the areas occupied by any particular germinal zone. Generally, in spring-breeding forms the lobules of winter specimens mainly contain spermatozoal cysts, so that morphologically the gonad is distinguished by a large caudal concentration of stored spermatozoa and a small cephalic region of resting spermatogonia and spermatocytic cysts. Usually, there is a relatively brief spawning period, followed by a postnuptial phase of intense spermatogenetic activity after evacuation of the ripe spermatozoa from the caudal region. This results in the enlargement of the cephalic area as the succeeding generation of germinal cysts proliferate and commence their development. The timing of these various sequences differs from species to species, and the duration of sperm storage also may show considerable variations. In *Cryptobranchus alleganiensis* and *Necturus maculosus*, for example, spermatogenesis is prenuptial; spermatozoa are produced only shortly before spawning as a result of a very rapid phase of spermiogenesis which transforms nearly all the immature germinal cysts into spermatozoal cysts shortly before breeding.[120] This differs from the semitropical newt *T. hongkongenesis*, where a much more protracted and slower spermiogenic phase results in the release of spermatozoa over a much longer period.[184] In some urodeles spermatozoa are transferred from the testicular caudal zone to become stored in the efferent ducts some weeks before breeding.

3. *Reptiles*. The majority of temperate-zone reptiles have discontinuous spermatogenetic cycles, and even in lower latitudes many species show pronounced cyclicity. Northern species generally breed shortly after emergence from hibernation in the spring, and both pre- and postnuptial patterns are found. Prenuptial spermatogenesis occurs in most lizards, and spermatozoa are produced just before the breeding season without any significant period of sperm storage. There is little information about members of the Crocodilia, but the one species that has been studied in any detail, *Alligator mississippiensis*, also shows a prenuptial cycle.[93] A postnuptial pattern occurs in most turtles, but the situation in snakes is more variable, and while most temperate zone species and a few tropical snakes (e.g. *Elaphe taeniura*, *Pytas korrus*) exhibit a postnuptial pattern,

some tropical species (e.g., *Naja naja*) and some vipers of northern latitudes (*Vipera berus*, *V. ursinis*, *V. ammodytes*) have prenuptial cycles.[98]

Following the copulatory phase the seminiferous tubules regress and, although a few species may show spermatogenetic recrudescence almost immediately, most enter a quiescent phase which may last from a few weeks to several months before the restoration of mitotic activity in stem spermatogonia. In some species there is evidence of a postnuptial period characterized by refractoriness to environmental stimuli,[5,10,121] and in *N. naja* this phase lasts for about three months. The degree of spermatogenetic advancement reached before winter differs between species. In some colubrid snakes and turtles, following an intense period of prenuptial spermatogenesis and spermiogenesis, spermatozoa become transferred from the tubules to the accessory sexual ducts for a protracted period of storage. In *Trionyx sinensis*, for example, spermatozoa are stored in the epididymal canals from about October until their discharge in the following April.[126]

IV. STEROID CYCLES

A. Steroidogenesis

As in mammals, studies of *in-vitro* conversions of exogenous labelled precursors, or isolation and identification of endogenous steroids, have indicated testosterone to be a major testicular androgen in ectothermic vertebrates. There are, however, indications that differences from the mammalian pattern exist in some groups.

1. Fishes. In cyclostomes, testosterone has been found in the blood of *Petromyzon marinus*[84,192] and *Eptatretus stouti*[84,133,192] and in *in-vitro* incubates of testicular tissue from *P. marinus*[22] and *Myxine glutinosa*.[89] The biosynthetic pathway leading to androgen formation is not known, but in *Myxine* there are 6β- and 7β-hydroxylase enzymes that have not been detected in the gonads of other nonmammals.[40] In the ammocoete larvae of *P. marinus*, at least eight steroids have been identified by radioimmunoassay in the plasma; namely, progesterone, corticosterone, cortisol, androstenedione, testosterone, DHT, estrone and estradiol.[36] The presence of so many steroids has led these investigators to speculate that early chordates also may have contained such an array, from which the most suitable ones may have been selected during the subsequent evolution of any particular group.

A few elasmobranch species have been studied, and testosterone has been identified in the blood of male *Raja radiata* and *Raja ocellata*.[79] Testosterone, androstenedione, progesterone and estradiol have been identified in testicular extracts of Scyliorhinus stellaris,[28] and Simpson et al.[168] have identified testosterone as the major product of S. acanthias testicular tissue when incubated in vitro with labelled progesterone. In S. canicula, in-vitro incubation of testis with labelled progesterone or pregnenolone yields androstenedione and testosterone.[86]

In teleosts, a deviation from the mammalian pattern seems to exist since 11-ketotestosterone, a steroid that is not produced by the mammalian testis, has been identified in the blood of Oncorhynchus nerka,[74] Salmo salar[75] and Cyprinus carpio[71] and has been isolated from the testicular incubates of a wide variety of teleosts.[56,77] In some teleosts it has been shown to have an androgenic potency greater than that of testosterone.[6,78] Plasma levels of this steroid are correlated with the spermatogenetic cycle, rising slowly during spermatozoa production and increasing sharply prior to spermiation. It decreases significantly in the spent fish.[54,55,75,165,166] In Carassius auratus, testes incubated with testosterone form testosterone and 11-ketotestosterone glucuronides at 26-31°C, but at lower temperatures only form 11-ketotestosterone.[40] An analogous observation also has been made with S. gairdneri testis[87] and has led to the suggestion that one function of testicular glucuronyl transferase may be to confine the secretion of free androgens to the environmentally preferred temperature for reproduction. The biosynthetic pathway of this steroid is, however, still a matter of controversy.

In addition to 11-ketotestosterone, testosterone also appears to be a major testicular product and has been isolated from the testes of a variety of species. Testosterone also has been identified in the blood, and plasma levels have been reported to show seasonal fluctuations that parallel the seasonal testicular cycle; e.g., in Salvelinus fontinalis,[161] Salmo trutta,[89a] S. gairdneri,[165] and Pleuronectes platessa.[196] In a number of these species 11-ketotestosterone is quantitatively predominant to testosterone only during spermiation.

2. Amphibians. Although in-vitro incubations have demonstrated the ability of amphibian testicular tissue to synthesize testosterone from precursors such as pregnenolone and progesterone, anuran amphibians are apparently unique among the tetrapods in also producing large amounts of DHT, and in D. pictus,[155] R. temporaria,[154] Rana catesbeiana,[139] R. esculenta[88] and Nectophrynoides occidentalis,[58] DHT has been identified as the major metabolite. The situation in urodeles is different, in that DHT is of minor importance, and in Pleurodeles waltl,[153] Triturus cristatus carnifex,[128] Taricha granulosa,[137] Pseudoeurycea smithi[139] and Ambystoma tigrinum[139,142,147]

testosterone is the major steroid produced. 11-Ketotestosterone has been identified in testicular incubates of _T. c. carnifex_.[128]

The data on steroid levels in amphibian blood confirm the data from testicular incubates. Thus, testosterone is the major component, and DHT a minor one, in the plasma of urodeles, and in _Cynops pyrrhogaster_, where the plasma levels have been analyzed throughout the year, DHT levels are always lower than testosterone except during the spermiating period in May where they rise sharply and, for the only time, exceed those of testosterone.[81] In _A. tigrinum_, however, plasma testosterone levels always exceed DHT levels even during spermiation.[147] In anurans, on the other hand, in addition to testosterone, DHT has been reported to be a major plasma component in _Rana pipiens_,[191] _Rana sylvatica_ and _R. catesbeiana_.[139]

3. _Reptiles_. Steroidogenesis in the Reptilia is similar to that observed in mammals, and is closer to the urodelean pattern than to the anuran one. Testosterone and androstenedione are the two major steroids that have been isolated from the testicular tissue of _L. sicula_[129] and from testicular incubates of _N. naja_[60] and _Nerodea sipedon pictiventris_.[23] Testosterone also has been identified in the plasma of a number of species. However, a recent report on plasma androgens and _in-vitro_ testicular secretions of _Tiliqua rugosa_ identified epitestosterone (17α-hydroxy-4-androsten-3-one) as the major circulating androgen.[18a] It is not known how widespread the secretion of epitestosterone might be among reptiles.

B. Interstitial Leydig Cells

The testes are the main source of circulating androgens, and in all groups castration results in a decrease in plasma androgen levels with a consequent regression of the secondary sexual characteristics. Furthermore, histochemical, radioimmunological and ultrastructural techniques have identified the interlobular/interstitial Leydig cells as the main loci for androgen synthesis. In mammals, a well-developed smooth endoplasmic reticulum (SER), rounded mitochondria with tubular or vesiculate cristae (rather than stacked lamellar style cristae), and a positive reaction to tests for Δ^5-3β-hydroxysteroid dehydrogenase (3βHSD), a key enzyme in the biosynthesis of testosterone, are all features that are characteristic of a steroidogenic tissue.

1. _Fishes_. Cells having the ultrastructural criteria of steroidogenic tissue have been identified in the interlobular tissue of cyclostomes (_E. stouti_,[188] _P. marinus_[193]). In _Lampetra fluviatilis_ the Leydig cells are

concentrated into compact acinar clusters of heavily lipoidal and cholesterol-positive cells[68] which first become conspicuous at the time when the secondary sexual structures develop.[94] A positive 3βHSD response has been reported in these cells in both L. fluviatilis[67] and P. marinus.[193]

Whereas lampreys have large, easily distinguishable Leydig cells, in elasmobranchs they are much smaller and less observable, and have sometimes been said to be absent. However, such cells have been identified in several squaliform and raiiform species and have been described as being lipoidal, cholesterol-positive and 3βHSD-positive.[27,39]

In teleosts, Leydig cells have been identified in the interlobular stroma of a wide variety of species and have been shown to be 3βHSD-positive and to have the ultrastuctural characteristics of steroid-producing cells (e.g., Lebistes reticulatus,[53] S. gairdneri,[72], Anguilla japonica[173]). A seasonal appearance and prenuptial depletion of cholesterol-rich cytoplasmic lipids is evident in some teleosts, and may be indicative of seasonal fluctuations in androgen secretion. Thus, in Monopterus albis, the prenuptial nest-building activity, formation of a foam mass, and guarding territory, are all associated with a high androgen titer, and these events occur concomitantly with the rapid depletion of lipids from the Leydig cells.[26] In Cymatogaster aggregata, too, an accumulation of cytoplasmic lipid droplets occurs in the Leydig cells as the breeding season approaches, and Wiebe[194] has demonstrated that the 3βHSD activity closely parallels this lipid cycle. A similar build-up of 3βHSD activity (also 3βHSD and glucose-6-phosphate dehydrogenase) occurs in S. gairdneri,[72] and coincides with the increase in plasma androgen levels.[163] In some teleosts these histochemical manifestations of Leydig cell seasonal cyclicity are less apparent, and lipid droplets may be observed only in the Leydig cells during periods of high sexual activity (e.g., F. heteroclitus[125]). In A. japonica, administration of hCG stimulates steroid production, and this is paralleled by the development of Leydig cell SER and steroid-type mitochondria.[173]

2. Amphibians. Anuran Leydig cells have similar histochemical and ultrastructural cycles that closely parallel the seasonal development of the secondary sexual structures. In temperate-zone frogs, the winter testis possesses Leydig cells that are largely lipid free, but as the breeding season approaches, numerous small lipid droplets gradually accumulate in the cell cytoplasm. This increasing lipophilia is accompanied by an increase in nuclear size and a build-up of SER and steroid-type mitochondria.[20] The cells become strongly 3βHSD-positive during this phase, and thumbpads show evidence of accelerated growth and glandular development corresponding to

this increasing activity. After spawning, the Leydig cells rapidly become densely lipoidal and seem to be filled with an amorphous mass of strongly cholesterol-positive lipid globules. Lofts[116,118] has interpreted this sudden accumulation as being indicative of a sharp decline in androgen synthesis with a consequent build-up of precursor material in the cytoplasm. This is supported by a recent study on plasma steroids in R. esculenta which reports a significant decline in plasma androgen levels at this time.[157] There is a loss of the 3βHSD reaction and SER during this heavily lipoidal phase, and this is paralleled by a regression of the thumbpads and atrophy of their glandular structures. Hypophysectomy produces a similar effect.[115] In R. temporaria this condition is succeeded by the gradual weakening of the lipid and cholesterol reactions as the cells regress and become indistinguishable from ordinary stromal connective tissue cells, but in R. esculenta and B. bufo, they remain in a densely lipoidal state for 2-3 months, before the cholesterol-rich material becomes depleted and thumbpad development resumes. A recrudescence of 3βHSD activity also occurs at this time.[120]

3. Reptiles. In reptiles, interstitially located 3βHSD has been reported in a number of species. Ultrastructural studies have confirmed the presence of SER and steroid-type mitochondria in the Leydig cells of L. vivipara, L. viridis, Anguis fragilis and Vipera aspis.[45,46] They also have been noted in the circumtesticular layer of Leydig cells of Cnemidophorus l. lemniscatus which appear identical with the few cells scattered throughout the more conventional intertubular locations.[38] In the cobra (N. naja) the winter testis contains Leydig cells filled with numerous discrete lipid droplets. These cells are cholesterol-positive but 3βHSD-negative. Lipid depletion occurs rapidly in spring, and by the breeding season these cytoplasmic lipids have all but disappeared, and the cells have become strongly 3βHSD-positive.[119] There is an abundance of SER and steroid-type mitochondria in the Leydig cells at this time. After the breeding season there is a rapid metamorphosis into a non-secretory state; Leydig cells become filled with a dense mass of cholesterol-rich lipid globules, they lose their 3βHSD response, SER becomes sparse, and mitochondria with lamellar-type cristae reappear. Plasma testosterone levels closely parallel these events, showing a rapid prenuptial rise as interstitial tissue lipid depletion begins and an equally rapid decline coincident with the lipophilic postnuptial phase.[17] A similar prenuptial peak in plasma testosterone levels also has been recorded in T. rugosa,[19] C. picta,[24] U. hardwicki,[7] L. vivipara[30] and A. granulatus.[59] The histochemical and ultrastructural changes displayed by chelonian Leydig cells are similar to those of squamate reptiles.[119]

There are clear differences in the temporal relationships between seminiferous tubule activity and interstitial cell steroidogenic activity in prenuptial and postnuptial cycles. In prenuptial types such as N. naja, Leydig cell secretion and plasma androgen levels closely parallel the annual spermatogenetic cycle, culminating in a peak coincident with the peak of spermatogenesis, whereas in a postnuptial species such as T. sinensis, there is an uncoupling of the spring plasma testosterone peak from the spermatogenetic cycle. At the time of peak Leydig cell secretory activity, as judged by histochemical and ultrastructural criteria, the seminiferous tubules are inactive. Later, when recrudescence of germ cell propagation begins, the interstitial tissue is inactive, heavily lipoidal and 3βHSD-negative. A similar reverse relationship has been observed in a number of other temperate turtles.

C. Lobule Boundary Cells

1. Teleostean Fishes. The annual lobule boundary cell cycle in E. lucius is essentially similar to that of the Leydig cell cycle. After the spawning period the spent lobules are bounded by a wall composed of unmodified fibroblasts, which gradually transform into glandular elements, become filled with cholesterol-positive lipids and eventually become depleted during the prespawning phase.[123] It was this similarity to the established interstitial cell cycle that led Lofts and Marshall[123] to suggest that, in a gonad apparently lacking any normal Leydig cells, these cells were the probable source of testicular androgens. O'Halloran and Idler[148] have recorded a positive 3βHSD and 3αHSD reaction in these cells in the testis of S. salar, and, on the basis of their histochemical studies, have concluded that they are homologous with mammalian Leydig cells. Evidence of a physiological homology between these cells and normal interstitial cells has been provided by Ahsan.[3] Injections of LH into hypophysectomized C. plumbeus have been reported to stimulate the lobule boundary cells and produce an intensification of the cholesterol reaction which cannot be produced by injections of FSH.

Some of the more recent reviews on the teleostean gonad[44,143] tend to discount this hypothesis, and point to recent electron microscopic observations indicating that some teleosts have both interstitial and lobule boundary cells. Furthermore, a positive 3βHSD response has been recorded in both locations in Oreochromis (Tilapia) mossambicus.[197] The basis for this confusion probably lies in the loose use of the term 'lobule boundary cell' sometimes to define the glandular cells derived from the follicular cells of the germinal cysts which, as already mentioned, metamorphose into glandular elements during maturation. A clear distinction must be made between these cells and the cells of the lobule wall. Teleosts which have been shown ultrastructurally to have normal interstitial cells do not possess lobule boundary cells but do have Sertoli cells which are separated from the interlobular stroma by a basement membrane. These cells also become lipoidal and, in some teleosts, possess ultrastructural features that have been interpreted as suggesting phagocytosis of germ cell debris (e.g., Billard,[11] Billard et al.,[13] Gresik et al.,[63] Grier[64]). More extensive studies need to be made, however, before this controversy can be resolved.

2. Amphibians. The situation in urodeles is much clearer, and ultrastructural studies have demonstrated both lobule boundary cells and Sertoli cells in the same testis (Fig. 7), with a distinct collagen layer and a thin basement membrane separating the former from the latter within the lobule.[29,80,83,184] Although morphologically lobule boundary cells appear to be very different from the anuran Leydig cells, they show a very similar sequence of histochemical and ultrastructural changes. In the spermatogonial and primary spermatocytic zones lobule boundary cells are flattened, with little cytoplasm, and are devoid of lipids, cholesterol or 3βHSD. Ultrastructurally, they contain only rough ER, and mitochondria are small and few in number.[118] As maturation progresses, the lobule boundary cells transform into a glandular tissue; they increase in size, the cytoplasm begins to accumulate small lipid droplets, and they start reacting to 3βHSD tests. In *N. maculosus*, Pudney et al.[158] have given a detailed description of the ultrastructural changes associated with the transformation of fibroblast-like cells into glandular steroidogenic elements surrounding the spermiating lobules. By the time that spermiation begins, these cells have become strongly positive for 3βHSD and contain the ultrastructural features of steroid-secreting cells, with SER occupying a large area of the cytoplasm.[80,158,184] The evacuated lobules become compressed into the glandular tissue, and positive 3βHSD responses have been recorded in this tissue from several urodelean species. The seasonal fluctuations in the circulating androgen levels closely parallel the lobule boundary cell cycle, and in those species where seasonal patterns of plasma androgen levels have been studied, an autumnal increase coincides with the period of hypertrophy of the lobule boundary cells (*T. granulosa*,[171] *C. pyrrhogaster*,[181], *A. tigrinum*[147]).

D. Sertoli Cells

1. Fishes. In elasmobranchs, the Sertoli cells are well-defined, particularly in mature ampullae where the radiating Sertoli cell-sperm bundle

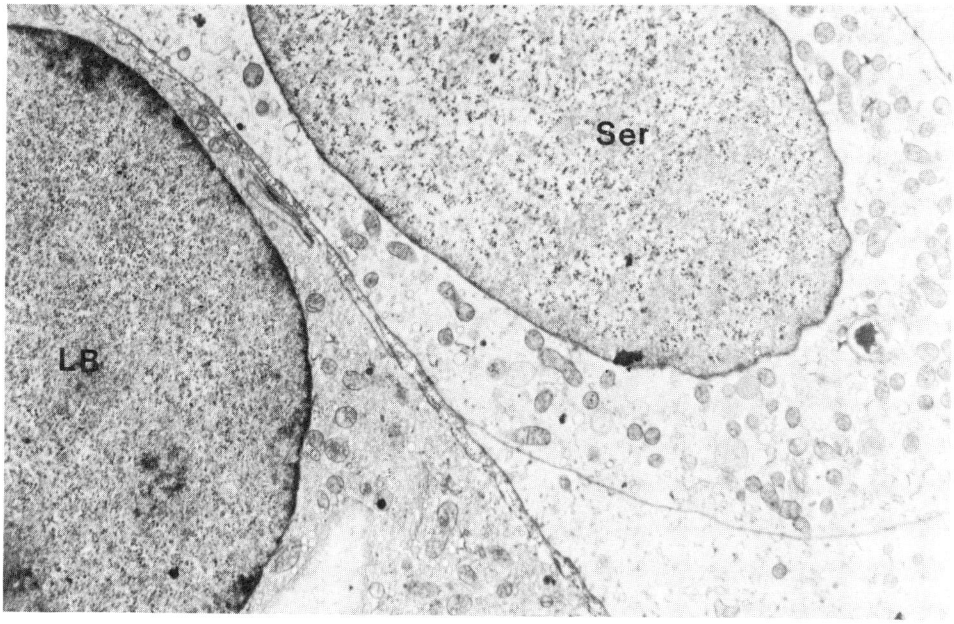

Fig. 7. Electron micrograph of a lobule boundary cell (LB) and Sertoli cell (Ser) in the testis of Batrachoseps attenuatus. Note the basement membrane and collagen layer separating the two. The mitochondria in the lobule boundary cell have steroid type tubular cristae, whereas those in the Sertoli cell have lamellar cristae.

units are similar to those seen in anurans. Several investigators have suggested they may be the main source of androgenic steroids, and 3βHSD has been recorded in these cells. They are probably responsible for the considerable concentrations of steroids reported in the semen of Squalus by Simpson and co-workers.[168] The elasmobranch Sertoli cells become densely lipoidal and cholesterol-positive after the release of their sperm bundles and eventually become resorbed in a cycle very similar to the urodelean pattern (see IV. D. 3 below).

In the sea perch C. aggregata, Wiebe[194] has described a seasonal transition from small Sertoli cells lying adjacent to the basement membrane in winter, to a large columnar layer of lipoidal cells surrounding the mass of spermatozoa in the sexually mature gonad in summer. The Sertoli cells exhibit strong 3βHSD activity. A similar positive response has been recorded in the Sertoli cells of F. heteroclitus[9] and in the Sertoli cells of post-spermatogonial cysts of S. gairdneri in November.[72] Billard et al.[13] failed to find clear ultrastructural evidence for a steroidogenic role in the Sertoli cells of the latter and five other teleostean species, concluding that the data were more indicative of a phagocytotic role; but SER has been noted in Sertoli cells of S. gairdneri by van den Hurk et al.[72] Abraham et al.[1] have described numerous rounded mitochondria with tubular cristae and abundant SER in the Sertoli cell cytoplasm of *Aphanius dispar*.

2. *Amphibians*. There is some evidence that metamorphosis of the follicular cells into Sertoli cell homologues in cystic spermatogenesis (particularly in anurans) is accompanied by the development of some of the histochemical and ultrastructural steroidogenic features noted in reptilian Sertoli cells (see IV. D. 4 below). Thus, once the fibroblast-like cells ensheathing the juvenile germinal cysts in *R. temporaria* have hypertrophied and expanded into the glandular elements surrounding the meiotic and postmeiotic germinal clusters, their nuclei become more rounded, mitochondria become more numerous and a well-developed SER is seen in the cytoplasm.[20] Small lipid granules[127] and 3βHSD activity[150] become demonstrable in the Sertoli cells at this time. Following the release of sperm bundles, Sertoli cells become densely lipoidal, detach from the tubule wall, and aggregate in the tubular lumen. Ultrastructurally, they no longer show the characteristics of steroidogenic cells, and histochemically, they react strongly to tests for cholesterol and acid phosphatase but are negative for 3βHSD. In *R. temporaria*, where the spermatogenetic cycle is a discontinuous one, at the time of spermiation the whole Sertoli cell population becomes released more or less synchronously and is compressed into a lipoidal, central luminal mass. These cholesterol-rich tubular lipids disappear with the resurgence of the succeeding spermatogenetic cycle. In a continuous or potentially continuous species such as *R. esculenta*, where the production of germinal cysts is never interrupted, the seminiferous tubules contain a spectrum of Sertoli cells in different phases of their cycle, and loose, spent lipoidal cells may be seen in the lumina throughout the year. In urodeles, too, the Sertoli cells remain in the discharged lobules after the release of their sperm bundles, and similarly becomes densely lipoidal and cholesterol-rich. They form part of the glandular tissue, and, as in frogs, eventually become resorbed concomitant with the upsurge in germinal cyst propagation. Evidence of a secretory phase in urodele Sertoli cells is sparse, but in the newt, *T. hongkongenesis*, Tso[183] has recorded the presence of SER and round mitochondria with a dense matrix and tubular cristae in follicular (Sertoli) cells during the period when spermatids are forming within the cyst lumen. Tso and Lofts,[184] however, failed to obtain any 3βHSD reaction at any stage of the cycle, and the same negative result has also been noted in the Sertoli cells of the red-bellied newt, *C. p. pyrrhogaster*.[80] In the glandular tissue, it is the cells derived from the lobule boundary cells, and not the

spent Sertoli cells, that show the ultrastructural and histochemical characteristics of steroid secretory activity.[80,118,158]

3. Reptiles. Over the past decade or so, data have accumulated indicating that Sertoli cells may be loci of intratubular steroid secretion. The reptilian Sertoli cell is permanently attached to the basal lamina of the tubule wall and demonstrates a well-defined histochemical and ultrastructural cyclicity which closely correlates with the seasonal spermatogenetic cycle. This involves a rapid genesis of large quantities of cholesterol-rich lipid material within the Sertoli cells following the seasonal cessation of spermatogenesis, then its rapid depletion coincident with spermatogenetic recrudescence. As is the case with the interstitial lipid cycle, 3βHSD activity develops at the time of the Sertoli cytoplasmic lipid depletion (Fig. 8A), and ultrastructural observations indicate that these cells have masses of SER and mitochondria with the characteristic tubular cristae during this phase (Fig. 8B).[47,117,119]

Once the spermatogenetic cycle is over, and Sertoli cells become densely lipoidal, these enzymatic and ultrastructural manifestations of steroidogenic activity disappear until the succeeding spermatogenetic wave resumes some months later.[124] Gouder and Nadkarni[61,62] similarly have reported the presence of both 3β- and 17αHSD in the Sertoli cells of a wide variety of Indian reptilian species during the reproductive phase. Lofts[117] suggested that these cyclic criteria indicate a seasonal pattern of steroid hormone production by the reptilian Sertoli cell, and this hypothesis has gained support by the demonstration that isolated seminiferous tubules of *Naja*[122] and *Trionyx*[119] have the ability to synthesize testosterone from labelled progesterone *in vitro*. Furthermore, the production of tubular androgen varies seasonally and shows maximum synthesis at times when there is also a high level of spermatogenetic activity. In a prenuptial species like *Naja*, tubules show a high synthesizing ability in spring and early autumn, but low activity during the period of testicular regression in June, whereas a postnuptial-type species like *Trionyx* shows a high level in late summer and early autumn only.

Measurements of testicular androgen content support these observations. Thus, Arslan et al.[7] have reported two peaks of testicular androgen in the lizard *U. hardwicki*, with a smaller peak developing in the autumn and corresponding to the onset of spermatogenesis. Courty and Dufaure[30] also have observed two seasonal peaks in testicular androgen content in *L. vivipara*. In this species the later minor peak develops in June, immediately

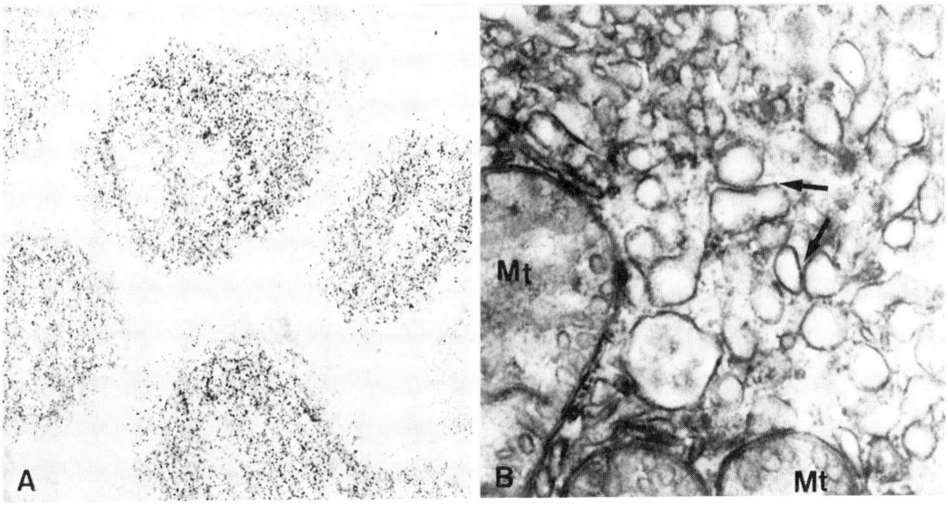

Fig. 8. Testis of *Trionyx sinensis* at the peak of the spermatogenetic cycle. A. A strong 3βHSD reaction is seen in the seminiferous tubules but the Leydig cells are inactive. B. Electron micrograph of the Sertoli cell cytoplasm showing steroid type mitochondria (*Mt*) with tubular cristae, and extensive vesiculate smooth endoplasmic reticulum (*arrows*).

preceding the recovery of spermatogonial mitotic activity which begins towards the end of the same month. In both these species, the secondary peak in testicular androgens was not accompanied by an elevation of plasma testosterone levels which show only a single peak coincident with the primary spring peak of testicular androgens. The secondary rise in testicular testosterone level may reflect intratubular synthesis by Sertoli cells at times of high spermatogenetic activity, whereas the primary testicular androgen peak and accompanying plasma testosterone peak are more likely expressions of the interstitial cell cycle which is related to the seasonal regulation of secondary sexual structure.[24,117] It is of interest to note that in laboratory rats, a recent report has recorded that a complete destruction of the interstitial tissue *in vivo*, produced as a result of a single intraperitoneal injection of ethane-1,2-dimethanesulphonate, does not cause any significant disruption of the spermatogenetic activity in the seminiferous tubules.[85a]

A recent investigation of testicular androgens in *Naja*[17] has revealed only a single peak which coincides with the plasma androgen cycle. No obvious secondary peak occurs at the time of spermatogenetic recovery, although *in-vitro* studies show an elevation of synthetic ability at this time, in either whole testicular tissue incubates[180] or seminiferous tubule incubates.[119]

V. PITUITARY-TESTIS INTERRELATIONSHIPS

A. Fishes

In fishes, as in mammals, testicular steroidogenesis and spermatogenesis are regulated by the secretory activity of the pituitary gland; hypophysectomy invariably leads to a decline in plasma androgen levels with a consequent regression of the secondary sexual structures and a cessation of spermatogenetic propagation. Only in cyclostomes does spermatogenetic activity continue in hypophysectomized animals,[50,95,133] and even in these primitive vertebrates there is evidence of some pituitary influence, since removal of the gland early in the spermatogenetic buildup in *L. fluviatilis* has a delaying effect and also prevents the appearance of androgen-dependent secondary sexual characters.[50,94]

In elasmobranchs, although some gonadotropic activity occurs throughout the pituitary, the main concentration of gonadotropes is in the ventral lobe. Ventral lobectomy inhibits spermatogonial proliferation and causes a reduction in plasma testosterone levels, whereas injections of ventral lobe extracts into such hypophysectomized specimens stimulates an increase in plasma androgen levels and prevents suppression of spermatogonial development. Purification and partial characterization of a gonadotropic hormone (Gn) has been carried out on ventral lobe material from *S. canicula*, and antibodies raised against this Gn cause a significant drop in plasma androgens when injected into intact fish.[174-176] Unlike the mammalian pituitary, where two chemically and biologically distinct Gns have been identified, only a single Gn has so far been isolated from the elasmobranch pituitary,[176] and one putative gonadotrope, which reacts specifically with fluorescein-labelled antibody to ovine or avian LH, has been identified in the ventral lobe.[41,136]

There is a lack of unanimity of opinion as to whether the teleostean gonad is controlled by the secretion of a single Gn, or whether two Gns are involved. The earlier histological studies, combined with the use of purified mammalian Gns injected into hypophysectomized or seasonally regressed fish, generally tended to show a positive response only to LH. More recently, however, Bona-Gallo and Licht[16] have shown that whereas *in-vitro* secretion of testosterone in *Gillichthys mirabilis* similarly is stimulated selectively by purified preparations of tetrapod LH but not FSH, the testes of *Cichlasoma citrinellum*, *O. mossambicus*, and *S. gairdneri* lack such specificity and respond significantly to FSH as well. These results point to interspecific variability in gonadal sensitivity to Gns of tetrapod origin, and demonstrate that any generalizations concerning a single or dual hormonal control of piscine testicular function based on such data are speculative.

A dicotomy of opinion also is evident in the literature concerning teleostean pituitary cell-types.[149,151] The gonadotropic basophils in teleosts are largely concentrated in the ventral region of the proximal pars distalis, and while in a number of species only a single gonadotropic cell-type has been identified, in others two gonadotropes have been identified. Slijkhuis[170] has similarly reported two gonadotropes on the basis of ultrastructural criteria, in the pituitary of Gasterosteus aculeatus, and has correlated the secretory activity of one with the spermatogenetic cycle and the other with Leydig cell activity. Such a conclusion also has been supported by data on the ultrastructural changes in pituitary cell-types in the loach, Misgurnus anguillicaudatus, after gonadectomy and testosterone injections.[189]

Isolation, purification and preliminary characterization of Gns have been carried out on pituitary glands of several teleosts, and in most of these only one type of Gn has been isolated with any certainty. Salmon Gn and carp Gn have been shown to stimulate spermatogenesis in H. fossilis[177] and A. anguilla,[96] respectively, and both tilapia Gn[73] and flounder Gn[145] can restore spermatogenesis in hypophysectomized specimens. All these preparations as well as Gn preparations from Hippoglossoides platessoides[144] and Muraenesox cinereus,[144] also can stimulate teleostean testicular steroidogenesis, both in vivo[114,144] and in vitro.[70,96]

Recently, Idler and coworkers[76] have isolated two putatively distinct Gns from the pituitary glands of O. keta, C. carpio, H. platessoides and P. americanus (see also Chapter 6). The relevance of these hormones to testicular regulation is unclear since only one of the isolated hormones, the so-called 'maturational gonadotropin,' has any biological action on testicular activity, whereas the 'vitellogenic gonadotropin' is ineffective and only has an influence on the female gonad. The maturational Gn is thus similar to the other teleostean Gn preparations in that it has a stimulatory effect on spermatogenesis in both intact[190] and hypophysectomized fish.[145] It also stimulates 11-ketotestosterone and testosterone production in hypophysectomized flounder[144,145] and induces androgen synthesis in intact juvenile rainbow trout.[144] Conversely, treatment of intact S. salar sebago with an antiserum to chum salmon (O. keta) maturational Gn, causes a dramatic decrease in plasma levels of 11-ketotestosterone.[144] Studies utilizing homogenates of mature rainbow trout testis have indicated that hCG binds specifically to such preparations, and whereas the vitellogenic Gn has no inhibitory effect on this binding, the more LH-like maturation of Gn successfully competes for the testicular receptor sites and inhibits hCG binding.[162,164]

It seems possible that the primary influence of this piscine Gn (maturational hormone) may be on testicular steroidogenic activity, and the consequent rise in endogenous androgen levels, in turn, stimulates spermatogenesis. There is much evidence that exogenous androgens stimulate teleostean spermatogenesis, and it had already been noted that plasma androgen levels seasonally increase concomitant with spermatogenetic development. In addition, high levels of plasma androgens are associated with precocious gonadal development in Atlantic salmon.[43] Crim and Evans[32] have reported that intraperitoneally administered testosterone has no effect on spermatogenesis in juvenile *S. gairdneri*, but such lack of stimulation can be attributed to the fact that the immature gonads contain only spermatogonia. Earlier studies have indicated that intraperitoneal injections of methyl testosterone can significantly stimulate post-spermatogonial stages but have no effect on spermatogonia,[125] a result which also has been reported after stimulating endogenous androgen levels with injections of LH.[156] Salmon Gn binds specifically to Gn receptors in the testis of *Glossogobius olivaceus*. Although mammalian FSH has no effect on this reaction, LH partially inhibits the binding, thus indicating that piscine testicular receptors have an affinity for LH-type Gns.[4]

As in mammals, there is some evidence that testicular androgens exert a negative feedback effect on the pituitary gland of adult fish. Thus, castration of *S. gairdneri*, leads to a significant increase in plasma Gn levels[14] which can be suppressed by androgen injections. Unilateral castration in this species, as in higher vertebrates, leads to a compensatory hypertrophy of the remaining testis.[159] Receptor sites for testosterone have been located by autoradiography in the hypothalamic region and pituitary gland in *Lepomis cyanellus*[138] and *Macropodus opercularis*.[37]

In immature fish androgenic steroids appear to exert a positive feedback effect, and peritoneal testosterone implants into juvenile *S. gairdneri* increase the pituitary Gn content.[32,34] A similar increase also has been recorded in immature *A. anguilla* pituitaries[48] and is in good agreement with histological studies which have shown enlargement of pituitary gonadotropes in juvenile, gonadectomized *O. nerka* after androgen injection.[152] In all these studies circulating Gn levels were reported to have remained low in steroid-treated fish, but a more recent study in *S. gairdneri* has shown significant increases in plasma Gn levels after testosterone administration.[33] In landlocked salmon, castration causes a fall in pituitary Gn, again indicating a positive-feedback mechanism.[31]

B. Amphibians

There is unequivocal evidence that in the Amphibia, as in all other tetrapod vertebrates, testicular steroidogenesis and spermatogenesis are dependent upon the secretory activity of the pars distalis. There is much ambiguity in the early literature, however, whether these two functional roles are regulated by the secretion of a single Gn (as appears to be indicated in fishes) or whether a mammalian-type dual hormonal system is involved. Much of this confusion has been derived from contradictory data about the specificity of the testicular response to mammalian FSH and LH preparations, and to differences of opinion about the correlation of gonadal function with specific adenohypophysial cell-types. During the past decade, however, two chemically and biologically distinct Gns, that appear to be chemically homologous to the FSH and LH of other tetrapods, have been isolated from the amphibians pituitary gland.[52,107] Furthermore, in *R. catesbeiana*, both circulating FSH and LH have been shown to have different clearance rates, both in intact and gonadectomized bullfrogs.[134] Early histometric studies using partially purified mammalian Gns, and more recent studies using highly purified amphibian Gns, have shown that whereas FSH stimulates spermatogenesis, LH selectively stimulates androgen production in *R. pipiens*, *R. temporaria* and *R. catesbeiana*. A similarly high specificity also has been demonstrated in three urodelean amphibians, *N. maculosus*,[15] *A. tigrinum* and *Bolitoglossa* sp.[139,142] Interestingly, in *N. maculosus*, plasma estrogen, as well as testosterone, levels rise significantly within 24 h of an injection of Gn[15,25] which correlates well with the presence of high aromatase activity reported in the testis of this species, particularly in the Leydig cells.[21] Estrogen receptors have been characterized in these cells and exhibit seasonal changes which parallel the spermatogenetic cycle, showing their highest activity at the time when the *Necturus* testis is nearing the end of the annual spermatogenetic wave.[130]

Such a high specificity of testicular Gn action is not a universal characteristic of all amphibian species, however, and there are considerable interspecific variations in response to FSH and LH with regard to androgen production. For example, while Muller[140] has recorded that a highly purified homologous LH preparation is 300-1000 times more potent than FSH in stimulating androgen secretion from *R. catesbeiana* testicular tissue *in vitro*, and selectively stimulates increased plasma androgen concentrations *in vivo*,[141] *in-vitro* androgen secretion by testes of *B. marinus* and *P. smithi* can be stimulated by both amphibian FSH and LH, and also by ovine FSH and LH.[142] In these two latter species, and also in *Xenopus laevis* and *Taricha torosa*, LH and FSH are, more or less, equipotent, but in *T. granulosa* FSH is more potent than LH. In *X. laevis*, this result is supported by studies which have shown an elevation of testicular 3βHSD activity in adults with either mammalian LH or FSH,[195] and also an elevation of plasma androgens in larval and juvenile forms treated with either Gn.[172] Takada and Ishii,[178] using purified preparations of bullfrog Gns, have reported that both FSH and LH bind to the same receptors in *Xenopus* testis, but there is a higher affinity for LH. The same is true also in *H. retardatus*.[92] In both *Hynobius* and *Xenopus*, and also in *C. pyrrhogaster*,[90] purified mammalian FSH binds *in vitro* to the same interstitial receptors[2] and may thus explain why mammalian FSH mimics LH activity in these amphibians.

There is good evidence of a feedback relationship between gonadal androgen secretion and the secretion of Gns by the amphibian pars distalis. There are numerous cytological studies describing the hypertrophy and degranulation of pituitary gonadotropes following castration, and their prevention by exogenously administered androgens. Furthermore, compensatory hypertrophy of residual testicular material following unilateral castration has been recorded in a number of species. The use of recently developed, highly specific radioimmunoassays for FSH and LH in *R. catesbeiana*[35] has endorsed these earlier observations, and has shown that castration is followed by a gradual rise in plasma LH and FSH levels that are detectable after 2 wks and continue to rise for many months. Silastic implants of either testosterone or DHT, on the other hand, reduce the levels of plasma Gns in castrated animals, thereby confirming the negative-feedback effects of circulating androgens.[102] The close correlation between plasma Gn levels and plasma testosterone indicated by these results is, however, less clear when continuous monitoring of plasma levels is carried out over a protracted period on natural populations. A five-year study on *R. catesbeiana* has shown that Gns and steroids become elevated simultaneously, leading to the conclusion that other factors (e.g., temperature) also must be involved in the testicular response to circulating LH. There is some indication that DHT may have both negative-feedback effect at the hypothalamic level and a positive one at the pituitary level.[135]

C. Reptiles

As in the Amphibia, controversy over whether the reptilian testis is regulated by a single or dual pituitary Gn system is evident in much of the literature. The histophysiology of the reptilian pituitary gland has been studied less extensively than that of the other vertebrate groups, and attempts to identify gonadotropes have produced conflicting results. In chelonians, however, there is reasonable evidence that a dual hormonal control of testicular function may be operating, and in *T. sinensis* two Gn cell-types have been identified, and their secretory activity is correlated closely with the activity of the seminiferous tubules and interstitial Leydig cells, respectively.[198] This attribution of an FSH-like secretory function to one cell-type and an LH-like function to another agrees with the conclusions of Herlant and Grignon[69] about the pituitary of *Testudo mauritanica*, and is supported by the isolation of distinct FSH- and LH-like fractions from the pituitary of the turtles *C. serpentina*[108] and *C. mydas*.[101] In the latter species, a temporal dissociation between peaks in circulating levels of FSH and LH supports the suggestion of independent mechanisms regulating the secretion of two Gns.[111,113] FSH-like and LH-like factors also have been isolated from the pituitary of the alligator, *A. mississipiensis*.[101]

As in some amphibian species, even though there is evidence of two Gns in chelonians, their testicular response to FSH and LH, whether of mammalian or reptilian origin, is equivocal. *In-vivo* experiments using injections of purified ovine FSH and LH have indicated that the turtle testis is more responsive to FSH than LH, both with regard to spermatogenesis and steroidogenesis,[24] but injections of turtle Gns (*C. mydas*) show more or less equal potency in stimulating testicular growth and plasma androgen levels in *A. carolinensis*.[97] A similar demonstration of equal potency with regard to stimulation of plasma androgen levels also has been reported for *C. serpentina* FSH and LH when injected into hypophysectomized *Kinosternum subrubrum*.[86] *In-vitro* experiments support the results obtained from *in-vivo* studies. Thus, whereas ovine FSH stimulates testosterone production in minced *Chelydra* testis, ovine LH is relatively inactive. However, reptilian FSH and LH (*Chelonia* and *Chelydra*) are both stimulatory, with LH consistently being slightly more potent than FSH.[97] This apparent lack of hormonal specificity also is supported by autoradiographic evidence showing that turtle LH is as potent as turtle FSH in competing for binding with ^{125}I-turtle FSH.[97]

It would, however, be premature to conclude that chelonian testicular Gn receptors are nonspecific on the basis of such results. Attention has already been drawn to data suggesting a steroidogenic ability in Sertoli cells as well as the interstitial Leydig cells, and this is supported in *T. sinensis* where isolated tubules have been shown to synthesize androgens from labelled precursors.[119] It is possible, therefore, that *in-vitro* androgen secretion in response to FSH or LH by minced *Chelydra* testis may be produced by these hormones stimulating the tubules and interstitial tissue, respectively. When radioiodinated human FSH is used, the hormone binds to both the seminiferous tubules and interstitial tissue in *Chrysemys*[104] and, consistent with the *in-vitro* and *in-vivo* data, shows a high specificity for this hormone.[105] In this, the turtles differ from snakes and lizards, where mammalian LH and reptilian LH preparations show relatively high activity in competition for ^{125}I-hFSH binding sites.[97,105] Recently, an estrogen receptor has been reported in the testis of _C. picta_, though the precise role that estrogens play in male reproduction is not known.[130]

In squamate reptiles, experiments involving highly purified mammalian FSH and LH preparations have shown that the testicular tissue is nonspecific in its response, and both have the ability to stimulate steroidogenesis and spermatogenesis to varying degrees.[98] Although Eyeson[51] has reported that administration of mammalian FSH and LH to hypophysectomized _Agama agama_ selectively stimulates spermatogenesis and interstitial cell activity respectively, in many other lizards and snakes either hormone has the ability to stimulate both functions. Studies using FSH and LH antisera in combination with the two hormones in hypophysectomized _Anolis_ have confirmed that a steroidogenic activity is intrinsic to either hormone.[106,112] The _in-vitro_ data support _in-vivo_ observations, and minced testicular preparations from several species have been shown to respond to both FSH and LH obtained from all four tetrapod groups.[186] Again, neutralization studies with relevant antisera support these results.[109] In the cobra _Naja_, the biosynthetic production of androgens from labelled precursors by testicular tissue, is similarly stimulated by both mammalian FSH and LH, but isolated seminiferous tubules are stimulated selectively by FSH alone.[117,122]

Unlike chelonians, there is no substantive evidence of a dual hormonal regulation of testicular function in squamate reptiles, and the investigations of Licht and coworkers suggest that a major divergence from the tetrapod pattern may have occurred in this group.[98] Separation and purification procedures on pituitary glands from representative species of five genera of snakes (_Crotalus atrox_, _P. korrus_, _P. molossus_, _Elaphe_ sp., _Bungarus_ sp., _N. naja_) have only revealed a single Gn which appears to be unique among tetrapod Gns in terms of biological cross-reactivity, biochemical composition, and immuno-chemical properties.[99] Snake Gn tends to be relatively inactive in radioreceptor assays and bioassays employing nonreptilian tetrapods, and also is unable to stimulate significantly _in-vitro_ testosterone production in turtle testes.[187] This Gn is, however, able to stimulate _in-vivo_ testicular growth and androgen production in other squamate species, such as the lizard _A. carolinensis_,[109] and appears to be more potent than mammalian FSH in stimulating _in-vitro_ androgen production in snake testes.[99,187] Antiserum raised against _Naja_ Gn blocks the biological activity of other snake Gns, and also inhibits the stimulatory effects of the FSH fractions of turtle, alligator and bird.[99]

In a prenuptial species such as *Naja*, where the seasonal patterns of spermatogenesis and androgen secretion parallel one another, the seasonal pattern of secretion of one Gn stimulating both activities could account for the annual build-up of these two testicular functions, and the data on the seasonal pattern of plasma and pituitary Gn levels in this species support such an interpretation.[17] On the other hand, in postnuptial species such as *Ptyas* and *Elaphe*, where the peaks of interstitial cell activity and spermatogenesis are separated temporally, the question arises as to how the same Gn can stimulate the one activity without influencing the other and, at another time of year, produce the reverse effect. A differential sensitivity of the two testicular compartments to the same Gn at different times of the annual cycle could account for such a pattern; a seasonal variation in testicular responsiveness to mammalian FSH and LH has been noted in *A. carolinensis*[110] and *N. naja*.[117] Seasonal variations also have been noted in the ability of isolated seminiferous tubules to synthesize testosterone from labelled precursors *in vitro* in *N. naja* and *T. sinensis*.[119] It is possible that seasonal fluctuations in environmental temperature differentially might affect the Gn sensitivities of the interstitial and spermatogenetic tissues, and there is some evidence of such a differential response to exogenous FSH in lizards experimentally maintained at different temperatures. What little information is available[98] indicates that Leydig cell function is less inhibited than spermatogenesis at lower temperatures, and there also is evidence that fluctuations in temperature may affect the binding efficiency of Gn receptors in the testis.[82,91] There is, however, a need for much research to be done on this area before firm conclusions can be drawn on squamate testicular control, particularly in those with postnuptial spermatogenetic cycles.

VI. REFERENCES

1. Abraham, M., Rahamim, E., Tibika, H., Golenser, E. and Kieselstein, M. (1980). The blood-testis barrier in *Aphanius dispar* (Teleostei). Cell Tiss. Res. 211, 207-214.
2. Adachi, T., Pandey, A.K. and Ishii, S. (1979). Follicle-stimulating hormone receptors in the testis of the frog, *Xenopus laevis*. Gen. Comp. Endocrinol. 37, 177-185.
3. Ahsan, S.N. (1966). Cyclical changes in the testicular activity of the lake chub, *Couesius plumbeus* (Agassiz). Can. J. Zool. 44, 149-159.
4. Aida, K. and Ishii, S. (1985). *In vitro* binding of salmon gonadotropin to the testis of teleost fish. *In* "Current Trends in Comparative Endocrinology" (B. Lofts and W.N. Holmes, eds.) Vol. II, pp. 745-746. Hong Kong University Press, Hong Kong.
5. Angelini, F., Botte, V. and Boschi, R. (1980). The regulation of the reproductive cycle in *Podarcis s. sicula* Raf. (Reptilia: Lacertidae): Influence of the refractory period on the springtime testicular activity. Mon. Zool. Ital. 14,27-39.

6. Arai, R. (1967). Androgenic effects of 11-ketotestosterone on some sexual characteristics in the teleost Oryzias latipes. Annot. Zool. Jpn. 40, 1-5.
7. Arslan, M., Lobo, J., Zaidi, A.A., Jalali, S. and Qazi, M.H. (1978). Annual androgen rhythm in the spring-tailed lizard, Uromastix hardwicki. Gen. Comp. Endocrinol. 36, 16-22.
8. Baggerman, B. (1972). Photoperiodic responses in the stickleback and their control by a daily rhythm of photosensitivity. Gen. Comp. Endocrinol. Suppl. 3, 466-476.
9. Bara, G. (1969). Histochemical demonstration of 3β-, 3α-, 11β- and 17β-hydroxysteroid dehydrogenases in the testis of Fundulus heteroclitus. Gen Comp. Endocrinol. 5, 284-296.
10. Bartholomew, G.A. (1953). The modification by temperature of the photoperiodic control of gonadal development in the lizard Xantusia vigilis. Copeia 1953, 45-50.
11. Billard, R. (1970). La spermatogènese de Poecilia reticulata. III. Ultrastructure des cellules de Sertoli. Ann. Biol. Anim. Biochim. Biophys. 10, 37-50.
12. Billard, R. and Flechon, J.E. (1969). Spermatogonies et spermatocytes flagélles chez Poecilia reticulata (Téléostéen cyprinodontiformes). Ann. Biol. Anim., Biochim. Biophys. 9, 281-286.
13. Billard, R., Jalabert, B. and Breton, B. (1972). Les cellules de Sertoli des poissons téléostéens. I. Etude ultrastructurale. Ann. Biol. Anim. Biochim. Biophys. 12, 19-32.
14. Billard, R., Richard, M., and Breton, B. (1977). Stimulation of gonadotrophin secretion after castration in rainbow trout. Gen. Comp. Endocrinol. 33, 163-165.
15. Bolaffi, J.L. and Callard, I.P. (1981). In vivo regulation of steroidogenesis by ovine gonadotropins in male and female mudpuppies, Necturus maculosus (Rafinesque). Gen. Comp. Endocrinol. 44, 108-116.
16. Bona-Gallo, A. and Licht, P. (1981). Gonadotropin specificity of in vitro testosterone secretion by fish testes. Gen. Comp. Endocrinol. 43, 467-478.
17. Bona-Gallo, A., Licht, P., McKenzie, D.S. and Lofts, B. (1980). Annual cycles in pituitary and plasma gonadotropin, gonadal steroids and thyroid activity in the chinese cobra (Naja naja). Gen. Comp. Endocrinol. 42, 477-493.
18. Bourne, A.R., Taylor, J.L. and Watson, T.G. (1985). Identification of epitestosterone in the plasma and testis of the lizard Tiliqua (Trachydosaurus) rugosa. Gen. Comp. Endocrinol. 58, 394-401.
19. Bourne, A.R. and Seamark, R.F. (1975). Seasonal changes in 17β-hydroxysteroids in the plasma of a male lizard (Tiliqua rugosa). Comp. Biochem. Physiol. 50, 535-536.
20. Brokelmann, J. (1964). Uber die Stutz - und Zwischenzellen des Froschodens wahrend des spermatogenetischen Zyklus. Z. Zelforsch. Mikrosk. Anat. 64, 429-461.
21. Callard, G.V., Canick, J.A. and Pudney, J. (1980). Estrogen synthesis in Leydig cells: Structural-functional correlations in Necturus testis. Biol. Reprod. 23, 461-479.
22. Callard, G.V., Petro, Z. and Ryan, K.J. (1980). Aromatization and 5α-reduction in brain and nonneural tissues of a cyclostome Petromyzon marinus. Gen. Comp. Endocrinol. 42, 155-159.
23. Callard, I.P. (1967). Testicular steroid synthesis in the snake, Natrix sipedon pictiventris. J. Endocrinol. 37, 105-106.
24. Callard, I.P., Callard, G.V., Lance, V. and Eccles, S. (1976). Seasonal changes in testicular structure and function and the effects of gonadotrophins in the freshwater turtle, Chrysemys picta. Gen. Comp. Endocrinol. 30, 347-356.

25. Callard, I.P., Callard, G.V., Lance, V., Bolaffi, J.L. and Rosset, J.S. (1978). Testicular regulation in nonmammalian vertebrates. Biol.-Reprod. 18, 16-43.
26. Chan, S.T.H. (1968). The endocrinology of sex-reversal in the ricefield eel, Monopterus albis (Zuiew). Thesis. University of Hong Kong.
26a. Chieffi, G. and Botte, V. (1964). Osservazioni sul significato funzionale della ghiandola annessa del testicolo dei Blennidii. Biol. Zool. 31, 471-477.
27. Chieffi, G. and Botte, V. (1964). Osservazioni dul significato sul tessuto interstiziale del testicolo dei Selacei. biol. Zool. 28, 211-217.
28. Chieffi, G. and Lupo di Prisco, C. (1961). Identification of estradiol-17β, testosterone and its precursors from Scyliorhinus stellaris testes. Nature (London) 190, 169-170.
29. Christensen, A.K. (1965). Testicular fine structure in the Californian slender salamander during the postnuptial regression of the testis. Amer. Zool. 5, 678-679.
29a. Colombo, L. and Burighel, P. (1974). Fine structure of the testicular gland of the black goby, Gobius jozo. Cell Tiss. Res. 154, 34-49.
30. Courty, Y. and Dufaure, J.P. (1979). Levels of testosterone in the plasma and testis of the viviparous lizard (Lacerta vivipara Jacquin) during the annual cycle. Gen. Comp. Endocrinol. 39, 336-342.
31. Crim, L.W. (1985). Feedback regulation of gonadotropic hormone during the annual reproductive cycle of the male landlocked Atlantic salmon. In "Current Trends in Comparative Endocrinology" (B. Lofts and W.N. Holmes, eds) Vol. I, pp. 213-216. Hong Kong University Press, Hong Kong.
32. Crim, L.W. and Evans, D.M. (1979). Stimulation of pituitary gonadotropin by testosterone in juvenile rainbow trout (Salmo gairdneri). Gen. Comp. Endocrinol. 37, 192-196.
33. Crim, L.W. and Evans, D.M. (1983). Influence of testosterone and/or luteinizing hormone releasing hormone analogue on precocious sexual development in the juvenile rainbow trout. Biol. Reprod. 29, 137-142.
34. Crim, L.W., Peter, R.E. and Billard, R. (1981). Onset of gonadotropic hormone accumulation in the immature trout pituitary gland in response to estrogen or aromatizable androgen steroid hormones. Gen. Comp. Endocrinol. 44, 374-381.
35. Daniels, E.L., Licht, P., Farmer, S.W. and Papkoff, H. (1977). Immunochemical studies on the pituitary gonadotropins (FSH and LH) from the bullfrog, Rana catesbeiana. Gen. Comp. Endocrinol. 32, 146-147.
36. Dashaw, L., Katz, Y., Trachtman, M.S. and Epple, A. (1984). Plasma steroids in the ammocoete of Petromyzon marinus. Gen. Comp. Endocrinol. 55, 361-366.
37. Davis, R.E., Morrell, J.I. and Pfaff, D.W. (1977). Autoradiographic localization of sex steroid-concentrating cells in the brain of the teleost Macropodus opercularis (Osteichthyes: Belontiidae). Gen. Comp. Endocrinol. 33, 496-505.
38. Del Conte, E. (1978). The Leydig cells of a tropical lizard in periods of different activity; Ultrastructure and apparent secretion toward intercellular canaliculi. Biol. Cellulaire 31, 177-180.
39. Della Corte, F., Botte, V. and Chieffi, G. (1961). Ricerca istochimica della steroide-3β-olodeidrogenasi nel testicolo di Torpedo marmorata Risso e di Scyliorhinus stellaris (L). Atti Soc. Peloritana Sci. Fis. Mat. Nat. 7, 393-397.
40. Delrio, G., d'Istria, M., Pierantoni, R. and Fasano, S. (1985). Identification and biosynthesis of androgens in non-mammalian vertebrates. In "Comparative Trends in Comparative Endocrinology" (B. Lofts and W.N. Holmes, eds.), pp. 209-211. Hong Kong University Press, Hong Kong.

41. Dodd, J.M. and Dodd, M.H.I. (1985). The ventral lobe and rachendachypophyse (buccal lobe) of the pituitary gland of cartilagenous fishes. In "Current Trends in Comparative Endocrinology" (B. Lofts and W.N. Holmes, eds.) Vol. I, pp. 127-130. Hong Kong University Press, Hong Kong.
42. Dodd, J.M., Evennett, P.J. and Goddard, C.K. (1960). Reproductive endocrinology in cyclostomes and elasmobranchs. Symp. Zool. Soc. Lond. 1, 77-103.
43. Dodd, J.M., Stuart-Kregor, P.A.C., Sumpter, J.P., Crim, L.W. and Peter, R.E. (1978). Premature sexual maturation in the Atlantic salmon (Salmo salar L.). In "Comparative Endocrinology" (P.J. Gaillard and H.H. Boer, eds.), pp. 101-104. Elsevier/North-Holland Biomedical Press, Amsterdam.
44. Dodd, J.M. and Sumpter, J.P. (1984). Fishes. In "Marshall's Physiology of Reproduction" (G.E. Lamming, ed.) Vol. I, pp. 1-26. Churchill Livingstone, London.
45. Dufaure, J.P. (1969). Ultrastructural features of steroid secreting cells in reptiles. Gen. Comp. Endocrinol. 13, 502.
46. Dufaure, J.P. (1970). L'ultrastructure du testicule de lézard vivipare (Reptile, Lacertilien). I. Les cellules interstitielles. Z. Zellforsch. Mikrosk. Anat. 103, 33-45.
47. Dufaure, J.P. (1971). L'ulstrastructure du testicule de lézard vivipare (Reptile, Lacertilien). II. Les cellules de Sertoli, étude de glycogène. Z. Zellforsch. Mikrosk. Anat. 115, 565-578.
48. Dufour, S., Delerue-Le Belle, N. and Fontaine, Y.A. (1983). Effects of steroid hormones on pituitary immunoreactive gonadotropin in European freshwater eel, Anguilla anguilla L. Gen. Comp. Endocrinol. 52, 190-197.
49. Egami, N. and Hosokawa, K. (1973). Responses of gonads to environmental changes in the fish, Oryzias latipes. In "Responses of Fish to Environmental Changes" (W. Chavin, ed.), pp. 279-301. Thomas, Springfield.
50. Evennett, P.J. and Dodd, J.M. (1963). Endocrinology of reproduction in the river lamprey. Nature (London) 197, 715-716.
51. Eyeson, K.N. (1971). The role of the pituitary gland in testicular function in the lizard, Agama agama. Gen. Comp. Endocrinol. 16, 342-355.
52. Farmer, S.W., Licht, P., Papkoff, H. and Daniels, E. (1977). Purification of gonadotropins in the leopard frog (Rana pipiens). Gen. Comp. Endocrinol. 32, 158, 162.
53. Follénius, E. and Porte, A. (1960). Cytologie fine des cellules interstitielle du testicule du poisson Lebistes reticulatus R. Experientia 16, 190-192.
54. Fostier, A., Billard, R. and Breton, B. (1984). Plasma 11-oxotestosterone and gonadotrophin in relation to the arrest of spermiation in rainbow trout (Salmo gairdneri). Gen. Comp. Endocrinol. 54, 378-381.
55. Fostier, A., Billard, R., Breton, B., Legendre, M. and Marlot, S. (1982). Plasma 11-oxytestosterone and gonadotropin during the beginning of spermiation in rainbow trout Salmo gairdneri. Gen. Comp. Endocrinol. 46, 428-434.
56. Fostier, A., Jalabert, B., Billard, R., Breton, B. and Zohar, Y. (1983). The gonadal steroids. In "Fish Physiology" (W.S. Hoar, D.J. Randall and E.M. Donaldson, ed.) Vol. IX, pp. 277-372. Academic Press, New York and London.
57. Fratini, L. (1953). Osservazioni sulla spermatogenesi di Scyliorhinus canicula. Publ. Stn. Zool. Napoli 24, 201-206.
58. Gavaud, J. (1975). Étude expérimentale du rôle des facteurs externes sur la spermatogenese et la stéroidogènese des males de Nectophrynoides occidentalis Angel. J. Physiol. (Paris) 70, 549-559.
59. Gorman, G.C., Licht, P. and McCollum, F. (1981). Annual reproductive patterns in three species of marine snakes from the central Philippines. J. Herpetol. 15, 335-354.

60. Gottfriend, H., Huang, D.P., Lofts, B., Phillips, J.G. and Tam, W.H. (1967). In vitro production of steroids by the adrenal and testicular tissues of the cobra (Naja naja). Gen. Comp. Endocrinol. 8, 18-31.
61. Gouder, B.Y.M. and Nadkarni, V.B. (1974). A histochemical study of the steroid synthesizing sites in the testes of the reptiles, Chamaeleon calcaratus Boulenger, Python molurus Gray and Crocodilus porosus Schneider. Indian J. Exp. Biol. 12, 326-329.
62. Gouder, B.Y.M. and Nadkarni, V.B. (1976). Histochemical study of the steroidogenic sites in the testes of the lizards, Hemidactylus flaviviridis and Mabuya carinata. Curr. Sci. 45, 102-103.
63. Gresik, E.W., Quirk, J.G. and Hamilton, J.B. (1973). A fine structural and histochemical study of the Leydig cell in the testis of the teleost, Oryzias latipes (Cyrinidontiformes). Gen. Comp. Endocrinol. 20, 86-98.
64. Grier, H.J. (1975). Aspects of germinal cyst and sperm development in Poecilia latipinna (Teleosti: Poeciliidae). J. Morph. 146, 229-250.
65. Grier, H.J. (1976). Sperm development in the teleost Oryzias latipes. Cell Tiss. Res. 168, 419-431.
66. Grier, H.J. and Linton, J.R. (1977). Ultrastructural identification of the Sertoli cell in the testis of the northern pike, Esox lucius. Amer.-J. Anat. 149, 283-288.
67. Hardisty, M.W. and Barnes, K. (1968). Steroid 3β-ol-dehydrogenase activity in the cyclostome gonad. Nature (London) 218, 880.
68. Hardisty, M.W., Rothwell, B. and Steele, I. (1967). The interstitial tissue of the testis of the river lamprey, Lampetra fluviatilis. J. Zool. 152, 9-18.
69. Herlant, M. and Grignon, G. (1961). Les modifications hypophysaires chez la Tortue terrestre (Testudo mauritanica) au cours du cycle génitol. C.R. Acad. Sci. 252, 2303-2305.
70. Huang, F.L., Huang, C.J., Lin, S.H., Lo, T.B. and Papkoff, H. (1981). Isolation and characterization of gonadotropin isohormones from the pituitary gland of pike eel (Muraenesox cinereus). Int. J. Pept. Protein Res. 18, 69-78.
71. Huang, F.L., Liu, T.C. and Lo, T.B. (1985). Biosynthesis of 11-ketotestosterone by carp testis in vitro. In "Current Trends in Comparative Endocrinology" (B. Lofts and W.N. Holmes, eds.), pp. 233-234. Hong Kong University Press, Hong Kong.
71a. Hurk, R. van den (1973). The localization of steroidogenesis in the testis of oviparous and viviparous teleosts. Proc. kon. ned. Akad. Wet. Ser.C. 76, 270-279.
72. Hurk, R. van den, Peute, J. and Vermeij, J.A.J. (1978). Morphological and enzyme cytochemical aspects of the testis and vas deferens of the rainbow trout, Salmo gairdneri. Cell Tiss. Res. 186, 309-325.
73. Hyder, M., Shah, A.V. and Hartree, A.S. (1979). Methallibure studies on Tilapia. III. Effects of tilapian partially purified pituitary gonadotropic fractions on the testes of methallibure-treated Sarotherodon spirulus (= Tilapia nigra). Gen. Comp. Endocrinol. 39, 475-480.
74. Idler, D.R., Fagerlund, V.H.M. and Ronald, A.P. (1960). Isolation of pregn-4-ene-17α,20β-diol-3-one from the plasma of Pacific salmon (Oncorhynchus nerka). Biochem. Biophys. Res. Commun. 2, 133-137.
75. Idler, D.R. Horne, D.A. and Sangalang, G.B. (1971). Identification and quantification of the major androgens in testicular and peripheral plasma of Atlantic salmon (Salmo salar) during sexual maturation. Gen. Comp. Endocrinol. 16, 257-267.
76. Idler, D.R. and Ng, T.B. (1983). Teleost gonadotropins: Isolation, biochemistry, and function. In "Fish Physiology" (W.S. Hoar, D.J. Randall and E.M. Donaldson, eds.) Vol. IX,pp. 187-221. Academic Press, New York and London.

77. Idler, D.R., Reinboth, R., Walsh, J.M. and Truscott, B. (1976). A comparison of 11-hydroxytestosterone and 11-ketotestosterone in blood of ambisexual and gonochoristic teleosts. Gen. Comp. Endocrinol. 30, 517-521.
78. Idler, D.R., Schmidt, P.J. and Biely, J. (1961). The androgenic activity of 11-ketotestosterone, a steroid in salmon plasma. Can. J. Biochem. Physiol. 39, 317-320.
79. Idler, D.R. and Truscott, B. (1966). Identification and quantification of testosterone in peripheral plasma of skate. Gen. Comp. Endocrinol. 7, 375-383.
80. Imai, K. and Tanaka, S. (1978). Histochemical and electronic microscopic observations on the steroid hormone-secreting cells in the testis of the Japanese red-bellied newt Cynops pyrrhogaster pyrrhogaster. Devel. Growth Different. 20, 151-167.
81. Imai, K.,, Tanaka, S. and Takikawa, H. (1985). Annual cycle of gonadotropin and testicular steroid hormones in the Japanese red-bellied newt. In "Current Trends in Comparative Endocrinology" (B. Lofts and W.N. Holmes, eds.), pp.247-250. Hong Kong University Press, Hong Kong.
82. Ishii, S. and Kubokawa, K. (1985). Adaptation of vertebrate gonadotropin receptors to environmental temperature. In "Current Trends in Comparative Endocrinology" (B. Lofts and W.N. Holmes, eds.) Vol. II, pp. 751-754. Hong Kong University Press, Hong Kong.
83. Joly, J. (1971). Les cycles sexuels de Salamandra (L.). I. Cycle sexual des males. Ann. Sci. Nat. Zool. (Paris) 13, 451-504.
84. Katz, Y., Dashaw, L. and Epple, A. (1982). Circulating steroid hormones of anadromous sea lampreys under various experimental conditions. Gen.-Comp. Endocrinol. 48, 261-268.
85. Kaya, C.M. (1973). Effects of temperature and photoperiod on seasonal regression of gonads of green sunfish, Lepomis cyanellus. Copeia 1973, 369-373.
85a. Kerr, J.B., Donachie, K. and Rommerts, F.F.G. (1985). Selective destruction and regeneration of rat Leydig cells in vivo. A new method for the study of seminiferous tubular-interstitial tissue interaction. Cell Tiss. Res. 242, 145-156.
86. Kime, D.E. (1978). Steroid biosynthesis by the testis of the dogfish Scyliorhinus caniculus. Gen. Comp. Endocrinol. 34, 6-17.
87. Kime, D.E. (1980). Comparative aspects of testicular androgen biosynthesis in nonmammalian vertebrates. In "Steroids and their Mechanism of Action in Nonmammalian Vertebrates" (G. Delrio and J. Brachet, eds.), pp.17-31. Raven Press, New York.
88. Kime, D.E. and Hews, E.A. (1978). Androgen biosynthesis in vitro by testes from Amphibia. Gen. Comp. Endocrinol. 35, 280-288.
89. Kime, D.E., Hews, E.A. and Rafter, J. (1980). Steroid biosynthesis by the testes of the hagfish, Myxine glutinosa. Gen. Comp. Endocrinol. 41, 8-12.
89a. Kime, D.E. and Manning, N.J. (1982). Seasonal patterns of free and conjugated androgens in the brown trout Salmo trutta. Gen. Comp. Endocrinol. 48, 222-231.
90. Kubokawa, K. and Ishii, S. (1980). Follicle-stimulating hormone (FSH) receptors in the testis of the newt, Cynops pyrrhogaster, and comparison of temperature dependency of the receptors with those of the other vertebrates. Gen. Comp. Endocrinol. 40, 425-433.
91. Kubokawa, K. and Ishii, S. (1984). Adaptation of testicular follicle-stimulating hormone receptors to ambient temperatures in vertebrates: Equilibrium analysis. Gen. Comp. Endocrinol. 54, 277-284.
92. Kubokawa, K., Moriya, T. and Ishii, S. (1985). FSH receptors in the testis of the salamander, Hynobius retardatus. In "Current Trends in Comparative Endocrinology" (B. Lofts and W.N. Holmes, eds), Vol. II, pp. 755-758. Hong Kong University Press, Hong Kong.

93. Lance, V. (1985). Reproduction in the American alligator, Alligator mississippiensis. In "Current Trends in Comparative Endocrinology" (B. Lofts and W.N. Holmes, eds.), pp. 253-256. Hong Kong University Press, Hong Kong.

94. Larsen, L.O. (1973). Development in adult, freshwater river lampreys and its hormonal control. Starvation, sexual maturation and natural death. Thesis, University of Copenhagen.

95. Larsen, L.O. and Rothwell, B. (1972). Adenohypophysis. In "The Biology of Lampreys" (M.W. Hardisty and I.C. Potter, eds.) Vol. II, pp. 1-67. Academic Press, New York and London

96. Leloup-Hatey, J., Oudinet, J.P. and Lopez, E. (1985). Testicular steroidogenesis during gonadotropin-induced spermatogenesis in male European eel (Anguilla anguilla L.). In "Current Trends in Comparative Endocrinology" (B. Lofts and W.N. Holmes, eds.) Vol I, pp. 229-232. Hong Kong University Press, Hong Kong.

97. Licht, P. (1977). Evolution in the roles of gonadotropins in the regulation of the tetrapod testis. In "Reproduction and Evolution" (J.H. Calaby and C.H. Tyndale-Biscoe, eds.), pp. 101-110. Australian Academy of Science, Canberra.

98. Licht, P. (1984). Reptiles. In "Marshall's Physiology of Reproduction" (G.E. Lamming, ed.) Vol. I, pp. 206-282. Churchill Livingstone, London.

99. Licht, P., Farmer, S.W., Bona-Gallo, A. and Papkoff, H. (1979). Pituitary gonadotropins in snakes. Gen. Comp. Endocrinol. 39, 34-52.

100. Licht, P., Farmer, S.W. and Papkoff, H. (1975). The nature of the pituitary gonadotropins and their role in ovulation in a urodele amphibian (Ambystoma tigrinum). Life Sci. 17, 1049-1054.

101. Licht, P., Farmer, S.W. and Papkoff, H. (1976). Further studies on the chemical nature of reptilian gonadotropins: FSH and LH in the American alligator and green sea turtle. Biol. Reprod. 14, 222-232.

102. Licht, P. and McCreery, B.R. (1985). Regulation of secretion and physiological actions of gonadotropins in amphibians. In "Current Trends in Comparative Endocrinology" (B. Lofts and W.N. Holmes, eds.) Vol. I, pp. 191-194. Hong Kong University Press, Hong Kong.

103. Licht, P., McCreery, B.R., Barnes, R., and Pang, R. (1983). Seasonal and stress related changes in plasma gonadotropins, sex steroids, and corticosterone in the bullfrog, Rana catesbeiana. Gen. Comp. Endocrinol. 50, 124-145.

104. Licht, P. and Midgley, A.R. Jr. (1976). Competition for the in vitro binding of radioiodinated human follicle-stimulating hormone in reptilian, avian and mammalian gonads by nonmammalian gonadotropins. Gen. Comp. Endocrinol. 30, 364-371.

105. Licht, P. and Midgley, A.R. Jr. (1975). In vitro binding of radioiodinated human follicle-stimulating hormone to reptilian and avian gonads: Radioligand studies with mammalian hormones. Biol. Reprod. 15, 195-205.

106. Licht, P. and Papkoff, H. (1973). Evidence for an intrinsic gonadotropic activity of ovine LH in the lizard. Gen. Comp. Endocrinol. 20, 172-176.

107. Licht, P. and Papkoff, H. (1974). Separation of two distinct gonadotropins from the pituitary gland of the bullfrog, Rana catesbeiana. Endocrinology 94, 1587-1594.

108. Licht, P. and Pakoff, H. (1974). Separation of two distinct gonadotropins from the pituitary gland of the snapping turtle (Chelydra serpentina). Gen. Comp. Endocrinol. 22, 218-237.

109. Licht, P., Pakkoff, H., Farmer, S.W., Muller, C.H., Tsui, H.W. and Crews, D. (1977). Evolution in gonadotrophin structure and function. Rec. Prog. Horm. Res. 33, 169-248.

110. Licht, P. and Pearson, A.K. (1969). Effects of mammalian gonadotropin (FSH and LH) on the testes of the lizard Anolis carolinensis. Gen. Comp. Endocrinol. 13, 367-381.

111. Licht, P., Rainey, W. and Cliffton, K. (1980). Serum gonadotropin and steroids associated with the breeding activities in the green sea turtle *Chelonia mydas*. II. Mating and nesting in natural populations. Gen. Comp. Endocrinol. 40, 116-122.
112. Licht, P. and Tsui, H.W. (1975). Evidence for intrinsic activity of ovine FSH on spermatogenesis, ovarian growth, steroidogenesis and ovulation in lizards. Biol. Reprod. 12, 346-350.
113. Licht, P., Wood, J., Owens, D.W. and Wood, F. (1979). Serum gonadotropin and steroids associated with breeding activities in the green turtle *Chelonia mydas*. I. Captive animals. Gen. Comp. Endocrinol. 39, 274-289.
114. Lo, T.B., Huang, F.L., Chang, Y.S., Huang, C.J. and Chang, G.D. (1985). Gonadotropins of the pike eel (*Muraenesox cinereus*). *In* "Current Trends in Comparative Endocrinology" (B. Lofts and W.N. Holmes, eds.) Vol. I, pp. 181-185. Hong Kong University Press, Hong Kong.
115. Lofts, B. (1961). The effects of follicle-stimulating hormone and luteinizing hormone on the testis of hypophysectomized frogs (*Rana temporaria*). Gen. Comp. Endocrinol. 1, 179-189.
116. Lofts, B. (1968). Patterns of testicular activity. *In* "Perspectives in Endocrinology: Hormones in the Lives of Lower Vertebrates" (E.J.W. Barrington and C.B. Jorgensen, eds.), pp. 239-404. Academic Press, New York and London.
117. Lofts, B. (1972). The Sertoli Cell. Gen. Comp. Endocrinol. Suppl. 3, 636-648.
118. Lofts, B. (1974). Reproduction. *In* "Physiology of the Amphibia" (B. Lofts, ed.) Vol. II, pp. 107-218. Academic Press, New York and London.
119. Lofts, B. (1977). Patterns of spermatogenesis and steroidogenesis in male reptiles. *In* "Reproduction and Evolution" (J.H. Calaby and C.H. Tyndale-Biscoe, eds.), pp. 127-136. Australian Academy of Science, Canberra.
120. Lofts,B. (1984). Amphibians. *In* "Marshall's Physiology of Reproduction" (G.E. Lamming, ed.) Vol. I, pp. 127-205. Churchill Livingstone, London.
121. Lofts, B. (1985). Environmental control of reptilian reproduction. *In* "The Endocrine System and the Environment" (B.K. Follett and S. Ishii, eds.), pp. 93-103. Japan Scientific Press, Tokyo.
122. Lofts, B. and Choy, L.Y.L. (1971). Steroid synthesis by the seminiferous tubules of the snake, *Naja naja*. Gen. Comp. Endocrinol. 17, 588-591.
123. Lofts, B. and Marshall, A.J. (1957). Cyclical changes in the distribution of the testis lipids of a teleost fish, *Esox lucius*. Quart. J. Microsc. Sci. 98, 79-88.
124. Lofts, B., Phillips, J.G. and Tam, W.H. (1966). Seasonal changes in the testis of the cobra, *Naja naja* (Linn). Gen. Comp. Endocrinol. 6, 466-475.
125. Lofts, B., Pickford, G.E. and Atz, J.W. (1966). effects of methyl testosterone on the testes of a hypophysectomized cyprinodont fish, Fundulus heteroclitus. Gen. Comp. Endocrinol. 6, 74-88.
126. Lofts, B. and Tsui, H.W. (1977). Histological and histochemical changes in the gonads and epididymides of the male soft-shelled turtle, Trionyx sinensis. J. Zool. 181, 57-68.
127. Lofts, B., Wellen, J.J. and Benraad, Th.J. (1972). Seasonal changes in endocrine organs of the male common frog, *Rana temporaria*. III. The gonads and cholesterol cycles. Gen. Comp. Endocrinol. 18, 344-363.
128. Lupo di Prisco, C., Basile, C., Delrio, G. and Chieffi, G. (1972). *In vitro* metabolism of cholesterol-4-^{14}C in testes and fat bodies of *Treiturus cristatus carnifex*. Comp. Biochem. Physiol. 41, 245-249.
129. Lupo di Prisco, C., Chieffi, G. and Delrio G. (1967). Identification of steroid hormones from *Lacerta sicula* testes. Experientia 23, 73-74.

130. Mak, P., Callard, I.P. and Callard, G.V. (1983). Characterization of an estrogen receptor in the testis of the urodele amphibian Necturus maculosus. Biol. Reprod. 28, 261-270.

131. Marcaillou, C. and Szöllosi, A. (1980). The "blood-testis" barrier in a nematode and a fish: a generalizable concept. J. Ultrastruc. Res. 70, 128-136.

132. Marshall, A.J. and Lofts, B. (1956). The Leydig cell homologue in certain teleost fishes. Nature (London) 177, 704-705.

133. Matti, A.J., Tsuneki, K., Dickhoff, W.W. and Gorbman, A. (1976). Thyroid and gonadal function in hypophysectomized hagfish, Eptatretus stouti. Gen. Comp. Endocrinol. 30, 500-516.

134. McCreery, B.R. and Licht, P. (1983). The effects of gonadectomy on polymorphism in stored and circulating gonadotropins in the bullfrog, Rana catesbeiana: I. Clearance profiles. Biol. Reprod. 29, 637-645.

135. McCreery, B.R. and Licht, P. (1984). Effects of gonadectomy and sex steroids on pituitary gonadotrophin release and response to gonadotrophin-releasing hormone (GnRH) agonist in the bullfrog, Rana catesbeiana. Gen. Comp. Endocrinol. 54, 282-396.

136. Mellinger, J. and Dubois, M.P. (1973). Confirmation, par l'immuno-fluorescence, de la fonction corticotrope du lobe rostral et de la fonction gonadotrope du lobe ventral de l'hypophyse d'un poisson cartilagineux, la torpille marbree (Torpedo marmorata). C.R. Acad. Sci. Paris, Ser. D. 276, 1879-1881.

137. Moore, F.L., Muller, C.H. and Specker, J.L. (1979). Origin and regulation of plasma dihydrotestosterone and testosterone in the rough-skinned newt, Taricha granulosa. Gen. Comp. Endocrinol. 38, 451-456.

138. Morrel, J.I., Kelly, D.B. and Pfaff, D.W. (1975). Sex steroid binding in the brains of vertebrates. In "Brain-Endocrine Interactions" (K.M. Kniggie, D.E. Scott, H. Kobayashi and S. Ishii, eds.), pp. 230-256. Karger, Basel.

139. Muller, C.H. (1976). Steroidogenesis and spermatogensis in the male bullfrog, Rana catesbeiana: Regulation by purified bullfrog gonadotropins. Ph.D. Dissertation. University of California, Berkeley.

140. Muller, C.H. (1977). In vitro stimulation of 5α-dihydrotestosterone and testosterone secretion from bullfrog testes by nonmammalian and mammalian gonadotropins. Gen. Comp. Endocrinol. 33, 109-121.

141. Muller, C.H. (1977). Plasma 5α-dihydrogestosterone and testosterone in the bullfrog, Rana catesbeiana: stimulation by bullfrog LH. Gen. Comp. Endocrinol. 33, 122-132.

142. Muller, C.H. and Licht, P. (1980). Gonadotropin specificity of androgen secretion by amphibian testes. Gen. Comp. Endocrinol. 42, 365-377.

143. Nagahama, Y. (1983). The functional morphology of teleost gonads. In "Fish Physiology" (W.S. Hoar, D.J. Randall and E.M. Donaldson, eds.) Vol. IX, pp. 223-275. Academic Press, New York and London.

144. Ng, T.B. and Idler, D.R. (1980). Gonadotropic regulation of androgen production in flounder and salmonids. Gen. Comp. Endocrinol. 42, 25-38

145. Ng, T.B., Idler, D.R. and Burton, M.P. (1980). Effects of teleost gonadotropins and their antibodies on gonadal histology in winter flounder. Gen. Comp. Endocrinol. 42, 355-364.

146. Nicholls, T.J. and Graham, G.P. (1972). The ultrastructure of lobule boundary cells and Leydig cell homologs in the testis of a cichlid fish, Cichlasoma nigrofasciatum. Gen. Comp. Endocrinol. 19, 133-146.

147. Norris, D.O., Norman, M.F., Pancak, M.K. and Duvall, D. (1985). Seasonal variations in spermatogenesis, testicular weights, vasa deferentia and androgen levels in neotenic male tiger salamanders, Ambystoma tigrinum. Gen. Comp. Endocrinol. 60, 51-57.

148. O'Halloran, M.J. and Idler, D.R. (1970). Identification and distribution of the Leydig cell homolog in the testis of sexually mature Atlantic salmon (Salmo salar). Gen. Comp. Endocrinol. 15, 361-364.

149. Olivereau, M. and Nagahama, Y. (1983). Immunocytochemistry of

gonadotropic cells in the pituitary of some teleost species. Gen. Comp. Endocrinol. 50, 252-260.

150. Oordt, P.G.W.J. van and Brands, F. (1970). The sertoli cells in the testis of the common frog, Rana temporaria. J. Endocrinol. 48, 1.

151. Oordt, P.G.W.J. van and Peute, J. (1983). The cellular origin of pituitary gonadotropins in teleosts. In "Fish Physiology" (W.S. Hoar, D.J. Randall and E.M. Donaldson, eds.) Vol. IX, pp. 137-186. Academic Press, New York and London.

152. Overbeeke, A.P. van and McBride, J.R. (1971). Histological effects of 11-ketotestosterone, 17α-methyltestosterone, estradiol, estradiol cyprionate, and cortisol on the interrenal tissue, thyroid gland, and pituitary gland of gonadectomized sockeye salmon (Oncorhynchus nerka). J. Fish. Res. Board Can. 28, 477-484.

153. Ozon, R. (1967). Synthèse in vitro des hormones steroides dans le testicule et l'ovaire de l'amphibien urodèle Pleurodeles waltlii. Micah. Gen. Comp. Endocrinol. 8, 214-227.

154. Ozon, R., Breur, H. and Lisboa, B.P. (1964). Étude du metabolisme des hormones steroides chez les vertèbres inférieurs. III. Metabolism in vitro de la testosterone par l'ovaire et le testicule de la grenouille Rana temporaria. Gen. Comp. Endocrinol. 4, 577-583.

155. Ozon, R. and Stocker, C. (1974). Formation in vitro de 5-hydrotestosterone par le testicule de Discoglossus pictus. Gen. Comp. Endocrinol. 23, 224-236.

155a. Parmentier, H.K., Boogaart, van den J.G.M. and Timmermans, L.P.M. (1985). Physiological compartmentation in gonadal tissue of the common carp (Cyprinus carpio L.). Cell Tiss. Res. 242, 75-81.

156. Pickford, G.E., Lofts, B., Bara, G. and Atz, J.W. (1972). Testis stimulation in hypophysectomized male killifish, Fundulus heteroclitus, treated with mammalian growth hormone and/or luteinizing hormone. Biol. Reprod. 7, 370-386.

157. Polzonetti-Magni, A., Botte, V., Bellini-Cardellini, L., Gobbetti, A. and Crasto, A. (1984). Plasma sex hormones and post-reproductive period in the green frog, Rana esculenta Complex. Gen. Comp. Endocrinol. 54, 373-377.

158. Pudney, J., Canick, J.A., Mak, P. and Callard, G.V. (1983). The differentiation of Leydig cells, steroidogenesis, and the spermatogenetic wave in the testis of Necturus maculosus. Gen. Comp. Endocrinol. 50, 43-66.

159. Robertson, O.H. (1958). Accelerated development of testis after unilateral gonadectomy, with observations on normal testis of rainbow trout. Bull. Fish. Wildlife Serv. 58, 9-30.

160. Ruby, S.M. and McMillan, D.B. (1975). The interstitial origin of germinal cells in the testis of the stickleback. J. Morph. 145, 295-318.

161. Sangalang, G.B. and Freeman, H.C. (1974). Effects of sublethal cadmium on maturation and testosterone and 11-ketotestosterone production in vivo in brook trout. Biol. Reprod. 11, 429-435.

162. Schlaghecke, R. (1983). Binding of ^{125}I-hCG to rainbow trout (Salmo gairdneri) testis in vitro. Gen. Comp. Endocrinol. 49, 261-269.

163. Schreck, G.B., Flickinger, S.A. and Hopwood, M.L. (1972). Plasma androgen levels in intact and castrated rainbow trout, Salmo gairdneri. Proc. Soc. Symp. Biol. Med. 140, 1009-1011.

164. Schulz, R. and Schlaghecke, R. (1984). A receptor preparation from gestes of Salmo gairdneri (Richardson): In vitro binding studies with ^{125}-labelled hCG and trout pituitary proteins. Gen. Comp. Endocrinol. 54, 132-138.

165. Scott, A.P., Bye, V.J., Baynes, S.M. and Springate, J.R.C. (1980). Seasonal variations in plasma concentrations of 11-ketotestosterone and testosterone in male rainbow trout, Salmo gairdneri. J. Fish. Biol. 17, 495-505.

166. Scott, A.P., MacKenzie, D.S. and Stacey, N.E. (1984). Endocrine changes during natural spawning in the white sucker, Catostomus commersoni. II. Steroid hormones. Gen. Comp. Endocrinol. 56, 349-359.

167. Simpson, T.H. and Wardle, C.S. (1967). A seasonal cycle in the testis of the spurdog, Squalus acanthias and the site of 3β-hydroxysteroid dehydrogenase activity. J. Mar. Biol. Assn. U.K. 47, 699-708.

168. Simpson, T.H., Wright, R.S. and Hunt, S.V. (1964). Steroid biosynthesis in the testis of dogfish (Squalus acanthias). J. Endocrinol. 31, 29-38.

169. Singh, D.P. (1977). Annual sexual rhythm in relation to environmental factors in a tropical pond turtle, Lissemys punctata granosa. Herpetelogica 33, 190-194.

170. Slijkhuis, H. (1978). Ultrastuctural evidence for two types of gonadotropic cells in the pituitary gland of the male three-spined stickleback, Gasterosteus aculeatus. Gen. Comp. Endocrinol. 36, 639-641.

171. specker, J.L. and Moore, F.L. (1980). Annual cycle of plasma androgens and testicular composition in the rough-skinned newt, Taricha granulosa. Gen. Comp. Endocrinol. 42, 297-303.

171a. Stanley, H., Chieffi, G. and Botte, V. (1964). Histological and histochemical observations on the testis of Gobius paganellus. Z. Zellforsch. Mikrosk. Ant. 65, 350-362.

172. Stehle, B.E. (1978). Stimulation of androgen production in male immature Xenopus laevis by mammalian gonadotropins. In "Comparative Endocrinology" (P.J. Gaillard and H.H. Boer, eds.), p. 86. Elsevier/ North-Holland Biomedical Press, New York.

173. Sugimoto, Y. and Takahashi, H. (1979). Ultrastructural changes of testicular interstitial cells of silver Japanese eels, Anguilla japonica, treated with human chorionic gonadotropin. Bull. Fac. Fish., Hokkaido Univ. 30, 23-33.

174. Sumpter, J.P., Follett, B.K., Jenkins, N. and Dodd, J.M. (1978). Studies on the purification and properties of gonadotrophin from ventral lobes of the pituitary gland of the dogfish (Scyliorhinus canicula L.). Gen. Comp. Endocrinol. 36, 264-274.

175. Sumpter, J.P., Jenkins, N. and Dodd, J.M. (1978). Studies on the purification of gonadotrophin from dogfish (Scyliorhinus canicula L.) pituitary glands. Ann. Biol. Anim. Biochim. Biophys. 18, 787-791.

176. Sumpter, J.P., Jenkins, N. and Dodd, J.M. (1978). Gonadotrophic hormone in the pituitary gland of the dogfish (Scyliorhinus canicula L.): Distribution and physiological significance. Gen. Comp. Endocrinol. 36, 275-285.

177. Sundararaj, B.L., Nayyar, S.K., Anand, T.C. and Donaldson, E.M. (1971). Effects of salmon pituitary gonadotropin, ovine luteinizing hormone and testosterone on the testes and seminal vesicles of hypophysectomized catfish, Heteropneustes fossilis (Bloch). Gen. Comp. Endocrinol. 71, 73-83.

178. Takada, K. and Ishii, S. (1985). Mammalian FSH acts as LH in the toad, Xenopus laevis. In "Current Trends in Comparative Endocrinology" (B. Lofts and W.N. Holmes, eds.) Vol. I, pp. 189-190, Hong Kong University Press, Hong Kong.

179. Takahashi, H. and Hanaoka, K.I. (1969). Cultivation in vitro of the testis cord of the newt, Triturus pyrrhogaster. Develop. Growth Differentiat. 11, 186-202.

180. Tam, W.H., Phillips, J.G. and Lofts, B. (1967). Seasonal variation in histology and in vitro steroid production by the cobra (Naja naja Linn.) testis and adrenal gland. In "Proc. 3rd Asia Ocean. Endocr. Cong." (A. Litonjua, ed.), pp. 309-314. Manila.

181. Tanaka, S. and Takikawa, H. (1983). Seasonal changes in plasma testosterone and 5α-dihydrotestosterone levels in the adult male newt, Cynops pyrrhogaster pyrrhogaster. Endocrinol. Japon. 30, 1-6.

182. Teshima, K. (1981). Studies on the reproduction of Japanese smooth dogfish, Mustelus manazo and M. griseus. J. Shimonoseki Univ. Fish. 29, 113-199.
183. tso, E.C.F. (1971). Seasonal changes in the ultrastructure and histochemistry of the newt (Trituroides hongkongenesis, Myer and Leviton) testis. M. Phil. Dissertation. University of Hong Kong.
184. Tso, E.C.F. and Lofts, B. (1977). Seasonal changes in the newt, Trituroides hongkongenesis testis. I. A histological and histochemical study. Acata Zool. (Stockholm) 58, 1-8.
185. Tso, E.C.F. and Lofts, B. (1977). Seasonal changes in the newt, Trituroides hongkongenesis testis. II. An ultrastructural study of the lobule boundary cell. Acta Zool. (Stockholm) 58, 9-15.
186. Tsui, H.W. (1976). Gonadotropin control of androgen production in reptilian and avian testes. Ph.D. Dissertation. University of California.
187. Tsui, H.W. and Licht, P. (1977). Gonadotropin regulation of in vitro androgen production by reptilian testes. Gen. Comp. Endocrinol. 31, 422-434.
188. Tsuneki, K. and Gorbman, A. (1977). Ultrastructure of the testicular interstitial tissue of the hagfish Eptatretus stouti. Acta Zool. (Stockholm) 58, 17-25.
189. Ueda, H. and Takahashi, H. (1980). Responses of two different types of pituitary gonadotrophs of the loach, Misgurnus anguillicaudatus, to gonadectomy and to exogenous sex steroids. Gen. Comp. Endocrinol. 40, 463-472.
190. Upadhyay, S.N. (1977). Morphologie des gonades immatures et etude expérimentale des l'induction de la gamétogénèse chez la Truite arc-en-ciel juvenile (Salmo gairdneri R.). Ph.D. Dissertation. University of Paris.
191. Wada, M., Wingfield, J.C. and Gorbman, A. (1976). Correlation between blood level of androgens and sexual behaviour in male leopard frogs, Rana pipiens. Gen. Comp. Endocrinol. 29, 72-77.
192. Weisbart, M., Dickhoff, W.W., Gorbman A. and Idler, D. (1980). The presence of steroids in the sera of the Pacific hagfish Eptatretus stouti and the sea lamprey Petromyzon marinus. Gen. Comp. Endocrinol. 41, 506-519.
193. Weisbart, M., Youson, J.H. and Wiebe, J.P. (1978). Biochemical, histochemical and ultrastructural analyses of presumed steroid-producing tissues in the sexually mature sea lamprey Petromyzon marinus L. Gen. Comp. Endocrinol. 34, 26-37.
194. Wiebe, J.P. (1969). Steroid dehydrogenases and steroids in gonads of seaperch Cymatogaster aggregata Gibbons. Gen. Comp. Endocrinol. 12, 256-266.
195. Wiebe, J.P. (1970). The mechanism of action of gonadotropic hormones in amphibians: The stimulation of Δ^5-3β-hydroxysteroid dehydrogenase activity in testes of Xenopus laevis Daudin. J. Endocrinol. 47, 439-450.
196. Wingfield, J.C. and Grimm, A.S. (1977). Seasonal changes in plasma cortisol, testosterone and oestradiol-17β in the plaice. Pleuronectes platessa L. Gen. Comp. Endocrinol. 31, 1-11.
197. Yaron, Z. (1966). Demonstration of 3β-hydroxyteroid dehydrogenase in the testis of Tilapia mossambica. J. Endocrinol. 34, 127-128.
198. Yip, D.Y. and Lofts, B. (1976). Adenohypophysial cell-types in the pituitary gland of the soft-shelled turtle, Trionyx sinensis. I. Seasonal cycles. Cell Tiss. Res. 170, 523-537.

REGULATION OF MALE GONADUCTS AND SEX ACCESSORY STRUCTURES

David O. Norris

Laboratory of Comparative Endocrinology
Department of Environmental, Population and Organismic Biology
University of Colorado
Boulder, CO 80309-0334

I. INTRODUCTION

In this chapter I shall review briefly the development of the male genital tract (including the gonaducts and associated glands as well as derivatives of the urogenital sinus), secondary sexual characters and the hormonal factors influencing development, structure and function. Emphasis is placed on fishes, amphibians and reptiles, but occasional observations concerning birds and mammals are included.

The gonads are often termed primary sexual characters. The gonaducts and associated structures (derived from wolffian ducts, mullerian ducts and urogenital sinus) are termed secondary sexual characters by some authors and sex accessory structures by others. The latter terminology is employed here. Other structures that depend upon gonadal steroids for their development and/or function are termed secondary sexual characters. Only the glandular secondary sexual characters are emphasized in this chapter. A listing of sex accessory and secondary sexual characters of nonmammalian vertebrates is provided in Table 1.

II. WOLFFIAN DUCTS, MESONEPHRIC KIDNEYS AND THEIR DERIVATIVES

Wolffian (archinephric, mesonephric) ducts and frequently mullerian (paramesonephric) ducts develop in fishes, amphibians, reptiles, birds and mammals regardless of sex. Each wolffian duct develops as a kidney duct that initially drains one of the pronephric kidneys. (The pronephros is the most anterior portion of the embryonic kidney.) Later, this same duct is coopted by the developing mesonephros, while the pronephros degenerates.

Table 1. Male Sex Accessory Structures and Secondary Sexual Characters in Fishes, Amphibians and Reptiles. (Based on Noble, 1931; Duellman and Trueb, 1986; van Tienhoven, 1983; Dodd and Sumpter, 1983; Evans, 1984; Young, 1981)

Class	Sex Accessory Structures	Secondary Sexual Characters
Agnatha	No duct system	Urogenital swellings and modified dorsal fin in lampreys
Chondrichthyes	Wolffian duct gives rise to vas deferens which stores sperm Persistent remnant of mullerian ducts Leydig gland Claspers	
Osteichthyes	Vas deferens usually present	Gonopodium from anal fin, especially in viviparous species
	Mullerian remnant persists in sturgeon (Chondrostei)	Nuptial tubercles on head (Cyprinids) Nuptial colorations
Amphibia	Vas deferens for sperm storage	Dorsal crest (newts) and tail height (many urodeles)
	Sexual kidney region	Thumb pads (anurans)
	Phallodeum (apodans)	Nuptial pads (urodeles)
	Seminal vesicles in some anurans	Variety of skin glands; hedonic glands; coloration patterns
	Intromittent organ in *Ascaphus* (anuran)	Vocalizations (anurans) Clasping musculature (newts and anurans)
	Cloacal gland complex (urodeles)	Teeth (desmognathid urodeles) several anurans)
	Persistent mullerian ducts common	"Hair" of *Trichobatrachus robustus*
Reptilia	Vas deferens with epididymis	Femoral glands (lizards)
	Cloacal glands (snakes, alligators)	Mental glands (chelonians)
	Sexual kidney (metanephric)	Skin glands (crocodilians)
	Hemipenis (squamates) or penis (chelonians, crocodilians)	Nuccal crest (some lizards) Dewlap (some lizards) Foreclaws (chelonians) Coloration patterns (squamates)

Castration blocks formation of the wolffian ducts, implying that a testicular factor is necessary for their development.[39]

In males the wolffian duct becomes the sperm duct or vas deferens (ductus deferens). Each vas deferens becomes attached to a testis by way of mesonephric tubules. These modified tubules persist even in amniotes after the mesonephric kidney degenerates and is replaced by the metanephric kidney, and they become the rete testis and vasa efferentia. In fishes and amphibians these ducts continue to function for urine conduction as well as for transportation of spermatozoa from the testes to the cloaca. The vasa deferentia of amniotes, however, carry only spermatozoa since the metanephric kidneys have their own separate ducts, the ureters.

In addition to forming vasa deferentia, the wolffian ducts give rise to other structures. The anterior portion of the wolffian duct in amniotes differentiates into an epididymis that is interposed between the vasa efferentia and the vas deferens. The androgen-dependent epididymis functions for storage and maturation of spermatozoa. In mammals it is secretory and usually exhibits seasonal changes related to breeding. Forbes[39] and van Tienhoven[112] contend that the epididymis is not secretory in birds and reptiles, but a secretory epididymis has been described in one lizard.[43]

The posterior-most region of the wolffian duct gives rise to androgen-dependent secretory structures often called seminal vesicles (vesicular glands). Androgenic stimulation of the seminal vesicles of rats[16] and squirrels[89] is enhanced by treatment with prolactin (PRL).

A. Fishes

Cyclostomes lack gonaducts, and spermatozoa are released from the testes directly into the coelomic cavity. From there the spermatozoa enter the urinary sinus through small pores and escape via the genital papilla (see Dodd and Sumpter,[31] Forbes,[39] and Hoar[51]).

Vasa deferentia are well-developed in chondrichthyan and bony fishes, although they are secondarily absent in some groups (e.g., the salmonids). They appear to develop from the wolffian ducts and are homologous to the gonaducts of tetrapods (see Dodd and Sumpter,[31] Hoar[51]). The posterior end of the vas deferens is expanded in many jawed fishes, and this ampulla may serve the functions of both epididymis and seminal vesicles. According to Mathews,[69] the vasa deferentia of the basking shark are about two meters in length, and the expanded ampullar region of each tube stores as much as 27 liters of seminal fluid. In elasmobranchs, seminal fluid is secreted by the Leydig gland which develops from the cranial portion of the mesonephros (see Dodd,[30] Dodd and Sumpter,[31] van Tienhoven[113]).

Some teleosts have seminal vesicle-like structures, but it is not known if these glands are homologous to the seminal vesicles of other vertebrates (see van Tienhoven[113]). An early study of the mudsucker *Gillichthyes mirabilis* reported that the seminal vesicles did not respond to androgens (see Table 1). A later study, however, found that not only did testosterone stimulate growth and secretory activity in seminal vesicles of this species, but ovine prolactin (oPRL) enhanced the effect when given with testosterone.[29] Similarly, oPRL synergized with testosterone to stimulate growth of the seminal vesicles in the Asian catfish, *Heteropneustes fossilis*.[108]

For fishes in general (see van Tienhoven[113]), spermatozoa must pass through the kidney in order to reach the vas deferens. In elasmobranchs this connection is limited to the cranial portion of the kidney, but in most bony fishes connections between the testis and the vas deferens occur along the entire length of each kidney. In the South American lungfishes, however, only the posterior portion of the kidney is used, and the kidney is bypassed completely in the primitive bichir (Polypterus).

The anterior portion of the kidney of elasmobranchs apparently functions as an epididymis with respect to final maturation of spermatozoa (see van Tienhoven[113]). This so-called "sexual kidney" probably is not involved with sperm storage.

B. Amphibians

Appearance of embryonic primordia for the wolffian ducts precedes gonadal differentiation and may be independent of gonadal hormones (see Adkins-Regan[3]). Later differentiation of the wolffian ducts is stimulated by androgens in larval and postmetamorphic amphibians (Table 2). In *Rana japonica*, treatment with estradiol enhanced the stimulatory action of testosterone on the wolffian ducts.[111] As a rule, each wolffian duct becomes a vas deferens and does not give rise to a distinct epididymis or to seminal vesicles.

The vasa efferentia do not connect the testes of most amphibians directly to the kidneys but usually form a common longitudinal duct (Bidder's duct) which then reforms into small tubules that laterally join to the kidney tubules. The plethodontid salamanders *Gyrinophilus danielsi*,[106] *Eurycea longicauda* and *Eurycea lucifuga*[119] have no longitudinal duct, and the testis communicates with the vas deferens through long transverse tubules. There are no longitudinal ducts in some anurans (e.g., *Discoglossus*, *Alytes*), and spermatozoa are conducted around the kidney and directly into the vasa deferentia (see van Tienhoven[113]).

Most amphibians store spermatozoa in the expanded posterior ampullae of the vasa deferentia. Several anuran species (e.g., *Discoglossus pictus*[74,75] and some species of the genus *Rana*[52]) have a true seminal vesicle in which spermatozoa are stored. Treatment with testosterone or 5α-dihydrotestosterone (DHT) can stimulate formation of secretory granules in the epithelium of seminal vesicles of castrated *D. pictus*.[75] Curiously, oPRL was shown later to antagonize the action of testosterone on the seminal vesicles of this species.[74]

Table 2. Effect of Pituitary and Steroid Hormones on Wolffian Ducts and Derivatives.

Class and Species	Hormone[a]	Effect[b]	Reference
CHONDRICHTHYES			
Scyliorhinus sp., female	T	(+) growth of vas deferens	39
OSTEICHTHYES			
Gillichthyes mirabilis	T	(0) growth of seminal vesicles	117
	T	(+) growth of seminal vesicles	29
	T + PRL	(+) growth of and secretion by seminal vesicles	29
Heteroptneustes fossilis	T	(+) growth of and secretion by seminal vesicles	107
	T + PRL	(+) growth of and secretion by seminal vesicles	107
	T + GH	(+) growth of and secretion by seminal vesicles	107
AMPHIBIA			
Ambystoma punctatum	T	(+) development of wolffian ducts	14
Ambystoma tigrinum	T	(+) development of wolffian ducts	37
Ambystoma maculatum	T	(+) development of wolffian ducts	37
Ambystoma opacum	T	(+) development of wolffian ducts	38
Discoglossus pictus	T, DHT	(+) secretion by seminal vesicles	75
Rana japonica	T, T+E	(+) hypertrophy of wolffian ducts	111
REPTILIA			
Anolis carolinensis	T	(+) growth of vas deferens	77
Lizard embryos (species?)	E	(0) development of wolffian duct	c
Lacerta vivipara	T	(+) secretion by epididymis	43, 44
Cnemidophorus uniparens	T, DHT	(0) development of wolffian duct	10

[a] T=testosterone, E=estrogen; PRL=prolactin; DHT=dihydrotestosterone

[b] (+)=stimulation; (-)=inhibition; (0)=no effect.

[c] Danchakoff, 1938, cited by Forbes[39]

The most caudal portion of the urodelan vas deferens is enlarged for storage of spermatozoa (see van Tienhoven[113]) and appears to be secretory in the mudpuppy Necturus maculosus,[17] E. lucifuga and E. longicauda[119] and in the tiger salamander Ambystoma tigrinum (Norris, unpublished). In contrast, cells of the posterior collecting ducts that drain the caudal portion of the kidney in urodeles are filled with secretory granules during the breeding season. These collecting ducts are stimulated by androgen treatment (Table 3; see also Forbes[39]), and become secretory during the breeding season in Notophthalmus viridescens,[1] Taricha torosa,[71] and A. tigrinum (Norris, unpublished). The role of these secretory portions of the kidney in

Table 3. Effects of Pituitary and Steroid Hormones on Secretory State of Kidney Collecting Ducts (Amphibia) and Renal Sexual Segments (Reptilia)

Class and Species	Hormone[a]	Effect[b]	Reference
AMPHIBIA			
Ambystoma punctatum	T	(+)	14
Ambystoma maculatum	T	(+)	37
Ambystoma opacum	T	(+)	38
Ambystoma tigrinum	T	(+)	37, c
REPTILIA			
Crocodylus palustris, immature	T	(0)	86
Anolis carolinensis	GH	(+)	62
A. carolinensis	T, T+GH	(+)	42
A. carolinensis	PRL	(0)	62
A. carolinensis	T	(+)	77
Eumeces fasciatus	And	(+)	d
Hemidactylus flaviviridis, hypophysectomized	T, FSH	(+)	90
Uromastix acanthoclinus	A	(+)	e
Cnemidophorus uniparens, treated pre- and post-hatching	T	(+)	9
Thamnophis radix, intact or castrates both sexes, immatures	T	(+)	f

[a]T=testosterone, A=androsterone, And=androgens, PRL=prolactin, GH=growth hormone, FSH=follicle-stimulating hormone.

[b](+)=stimulation, (-)=inhibition, (0)=no effect.

[c]Norris, unpublished.

[d]Reynold, 1943, cited by Forbes[39]

[e]Kehl, 1944, cited by Forbes[39]

[f]Waters, 1940, cited by Forbes[39]

reproduction is not clear. They may be homologous to the sexual segment of the reptilian kidney discussed below.[50]

In most anurans, the vas deferentia fuse just prior to entering the cloaca,[9] and they conduct both urine and spermatozoa to the cloaca.[113] The vasa deferentia are always separate in urodeles and in the more advanced families (Ambystomatidae,[6] Plethodontidae,[106] Salamandridae[5]) transport only spermatozoa. Multiple ducts drain the secretory kidney and either join the wolffian duct where it enters the cloaca or enter separately. Thus, the derivative of the wolffian duct could be termed a vas deferens in most urodeles. In more primitive families (Crytobranchidae[88] and Sirenidae[118]), however, the wolffian ducts transport both spermatozoa and urine.

C. Reptiles

Degeneration of the mesonephros and the appearance of the metanephric kidney in reptiles separates the genital tract from the urinary tract (see van Tienhoven[113]). Vasa efferentia connect each testis to an epididymis that develops from the anterior end of the wolffian duct. The epididymis of at least one lizard species is a secretory structure,[43] and its secretions may be responsible for activation of spermatozoa.[27] The vas deferens carries spermatozoa to the cloaca directly whereas urine produced in the metanephric kidney reaches the cloaca via the ureter. Seminal vesicles have not been described in reptiles.

Testosterone stimulates growth and secretion in vasa deferentia of some lizards (see Table 2), but both testosterone and DHT were ineffective in an all female (parthenogenetic) species, *Cnemidophorus uniparens*.[10] The effect of hormones on wolffian duct derivatives has not been examined in other reptilian groups.

Squamate reptiles have a renal sexual segment in each metanephric kidney that appears to develop from the ureteric bud, a derivative of the lower portion of the wolffian duct (see Forbes,[39] Fox,[40] Hildebrand[50]). Prasad and Reddy[82] reviewed the secretory nature of the sexual segment and concluded it was homologous to the seminal vesicles of mammals. The probable origin of the sexual segment from the wolffian duct would support this conclusion. Fox[40] disputes this relationship but provides no evidence contrary to the view of Prasad and Reddy. If the reptilian sexual segment is derived from wolffian duct, it might be more closely related (homologous) to the sexual kidney of amphibians than its final location in the metanephric kidney would suggest.

Secretory activity in the sexual segment is testosterone-dependent (see Table 3), and in the lizard, *Anolis carolinensis*, ovine growth hormone (oGH) but not oPRL enhances this action of testosterone, although oGH alone had no effect.[42] The specific functions of these sexual segments are not known, and they deserve closer scrutiny. Possible functions include providing nutrition required for motility of spermatozoa (see Cuellar et al.[24]), contributing to formation of copulatory plugs in some snakes (see Devine[28]), blocking urine flow during mating (see Fox[40]) and producing pheromones involved in courtship (see van Tienhoven[113]).

III. MULLERIAN DUCT DEVELOPMENT IN MALES

In mammals the mullerian ducts differentiate from coelomic epithelium covering the mesonephric ridges and grow caudally along the medial side of the wolffian ducts (see Jirasek[53]). They become the paired oviducts, uteri and a portion of the vagina. In males the mullerian ducts degenerate early during embryonic development. Mammalian testes secrete a mullerian-regressing factor (mullerian-inhibiting substance) which appears to be a glycoprotein (see Donahoe et al.[32]).

A. Fishes

In male elasmobranchs and sturgeon (Chondrostei), the mullerian ducts do not degenerate but persist in a rudimentary condition in adults (see van Tienhoven[113]). Mullerian ducts apparently do not develop in male or female teleosts, but when "oviducts" are present in some of the viviparous species, these structures may be homologous to a portion of the mammalian female genital tract. Wake[115] describes the origin of the oviduct in viviparous teleostean fishes as a continuation of the muscular layer covering the ovary and states it is not homologous to oviducts of other vertebrate groups. However, in mammals it appears that the outermost layer of longitudinal muscle of the uterus may be derived from peritoneum and not from mullerian duct (see Kawarabayashi and Marshall,[56] Bengtsson[7]). Consequently, this muscular layer may be homologous to the muscular layer of the teleostean oviduct which also may be derived from peritoneum.

B. Amphibians

Mullerian ducts persist in male urodeles[5,6,88,106,118] and most anurans[9] as they do in some fishes. In cases of sexual reversal of bufonid males to females, the rudimentary mullerian ducts can differentiate into completely functional oviducts (see Zaccanti and Tognato[124]). This phenomenon is accomplished through the activation of a rudimentary mass of ovarian-like tissue, the Bidder's organ, which can be found at the anterior end of each testis. The Bidder's organ can develop into a functional ovary and apparently stimulates the rudimentary mullerian ducts to develop.

In apodan amphibians, adult males retain mullerian ducts that function as "mullerian glands." These glands secrete seminal fluid necessary for the direct transfer of spermatozoa to the females. It has been suggested[115] that they are homologous to the prostate gland of mammals (see section IV).

C. Reptiles

Persistence of mullerian ducts in male reptiles is much reduced as compared to the amphibians (see Forbes,[39] Fox[40]). In oviparous lizards and snakes the mullerian ducts degenerate prior to hatching. Only in alligators, some turtles, and certain snakes do the ducts persist until after hatching.

Presumably, reptilian testes produce a mullerian-regressing factor similar to birds and mammals, but no direct evidence is available (see Adkins-Regan[3]).

IV. DERIVATIVES OF THE UROGENITAL SINUS

In male mammals the major sex accessory structure to develop from the anterio-ventral portion of the urogenital sinus is the prostate gland (see Jirasek[53]). Much of the stromal component of the gland develops from mesenchyme, with the epithelium being contributed by the urogenital sinus. According to Nabaitz,[73] however, a portion of the prostate epithelium is derived from the posterior ends of the mullerian ducts prior to their degeneration. This observation supports a dual origin for the prostate gland.

The bulbo-urethral glands (Cowper's glands) are smaller glands which also are derived from the urogenital sinus.[53] They too may have homologies to cloacal glands in amphibians, reptiles and birds. At least five additional glands derived from the urogenital sinus have been described, but each is limited to only a small number of species (see van Tienhoven[113]).

The prostate is the major secretory gland in the male genital tract and is possibly homologous to the cloacal glands of other tetrapods. Functional development of the mammalian prostate is dependent upon its ability to convert testosterone to DHT. Prolactin enhances the actions of androgens on the prostate of castrate and hypophysectomized Sprague-Dawley rats[16] and does so in the Indian palm squirrel as well.[89]

A. Fishes

Fishes do not have structures that are known to be homologous to the mammalian prostate gland. The Leydig gland of elasmobranchs performs a similar function but is clearly not homologous to the prostate (see section II. A). The "seminal vesicle-like" structures of certain teleosts (see van Tienhoven[113]) have not been studied sufficiently to relate their possible homologies to any mammalian genital structures.

B. Amphibians

Cloacal glands have been described in a number of plethodontid, ambystomatid and salamandrid urodeles.[25,57,98-101,120] Four types of cloacal glands have been identified largely by their anatomical positions: pelvic glands, Kingsbury's glands, ventral glands, and vent glands.[100] All types except for the vent glands have been subdivided into two or three groups depending on the species. The criteria used for subdividing the glands combines anatomical features such as where the tubules terminate and release their secretions and what type of reactions occur in response to various histochemical staining techniques (e.g., see Williams et al.[120]).

All cloacal glands of male Eurycea lucifuga and Eurycea longicauda show a distinct seasonal cycle in secretory activity, becoming active prior to breeding and regressing thereafter.[120] In male tiger salamanders there is a distinct seasonal cycle in the mass of the cloacal complex that correlates positively with circulating levels of testosterone and DHT.[79] This externally visible seasonal cycle in the cloacal complex of the tiger salamander is due mainly to changes in the ventral glands (Norris, unpublished). Similar observations have been reported for the red-bellied newt, Cynops pyrrhogaster. Based on histochemical grounds, the ventral glands are believed to secrete the stalk of the spermatophore (see section V. B.). Treatment of immature tiger salamander larvae with testosterone induces formation of ventral glands but not of any other cloacal gland types (Norris, unpublished). Kikuyama et al.[57] found that testosterone stimulated quiescent ventral glands for C. pyrrhogaster, and that this effect was enhanced by treatment with PRL. Receptors for both testosterone and DHT were demonstrated in the cloacal glands.[58] In immature tiger salamander larvae testosterone-induced differentiation of the ventral glands also was enhanced by PRL (Norris, unpublished; see Table 4).

Seasonal glandular development in the cloacal epithelium has been reported for some anurans,[76,84] but the significance of these observations is not clear. They may represent glands that produce pheromones used in mating[66] or that produce secretions related to oviposition and/or subsequent development of eggs.

C. Reptiles

Cloacal glands in reptiles seem to be involved in chemical communication.[23,66] Paired scent glands in some snakes exhibit distinct sexual differences[80] suggesting a definite reproductive function in males. Male crocodilians also have well-developed cloacal scent glands that are most active during the breeding season. Behavioral studies implicate secretions from these glands in courtship activities (see Madison[66]). Gekkonid lizards have bilateral cloacal sacs that may be homologous to the scent glands of snakes and crocodilians, but their functions are not known. The cloacal gland of the lacertid lizard Lacerta sicula regresses in males treated with the antiandrogen, cyproterone acetate,[87] indicating dependence of this gland on testicular androgens. In males of four species of Sceloporus there is a

Table 4. Effects of Steroid and Pituitary Hormones on Cl
and Function.

Class and Species	Hormone[a]	Effect[b]	
AGNATHA			
Lamprey	T, E_3	(+) cloacal distention	
AMPHIBIA			
Ambystoma punctatum	T	induce male-type cloaca	14
Ambystoma tigrinum	T	induce male-type cloaca	78, 37, c
Ambystoma tigrinum	T+PRL	enhances induction of ventral glands	c
Cynops pyrrhogaster	T, T+PRL	(+) cloacal gland secretion	57
Triturus cristatus	T	induces male-type cloaca	92, 93
Notophthalmus viridescens, castrate	T, DHEA	maintains normal secretory state	8
REPTILIA			
Anolis carolinensis	T	(+) cloacal keratinization	77
Lizard embryos	E	(+) cloacal epithelium	d

[a]E=estrogens; E_3=estrone, T=testosterone, DHEA=dihydroepiandrosterone, PRL=prolactin

[b](+)=stimulation

[c]Norris, unpublished

[d]Dantchakoff, 1938, cited by Forbes[39]

sacular subcutaneous gland located immediately anterior to the ventral lip of the cloaca. These proctodeal glands are active during the breeding season and for some time afterwards.[13] Their possible homology to cloacal glands of other reptiles has not been determined.

Chemical communication may be influenced by androgens in more than one way. Androgens and estrogens can stimulate the cloacal epithelium of lizards and thus influence secretion (Table 4). Testosterone-treated male skinks (Eumeces laticeps) are more interested in cloacal secretions from conspecific females implying that androgens alter the threshold of sensitivity to pheromones from females.[23] Cloacal glands of terrestrial male turtles may secrete a chemical substance detectable by the females that could be important in courtship (Weaver, 1967, cited by Madison[66]).

D. Birds

The cloacal glands of male Japanese quail, Coturnix coturnix japonicus, are stimulated by long photoperiods and regress following castration. Regression also occurs if photo-stimulated birds are placed under short

...ses. These glands appear to be similar to mammalian glands derived ...m the urogenital sinus in that testosterone must be converted first to either androstenedione or DHT to stimulate their secretion.[68]

V. INTROMITTANT ORGANS

In mammals intromittant organs develop from tissues associated with the urogenital sinus in the vicinity of the urogenital orifice.[50,105,115] The penis is the mammalian intromittant organ used to transfer spermatozoa into the genital tract of the female. The glans penis develops from the genital tubercle, and the shaft of the penis results from the genital folds. The scrotum, when present, develops from the lateral genital swellings. Development of the penis is dependent upon testicular androgens, but, unlike derivatives of the wolffian ducts, these structures require conversion of testosterone into DHT.

A. Fishes

Agnathan fishes are distinguished in being the only class of vertebrates that lacks intromittant organs. All fertilization is external in these fishes.

Chondrichthyan fishes (elasmobranchs and holocephalans) have evolved complicated copulatory structures known as claspers. These intromittant organs are modified pelvic fins and are found only in males. Claspers are not only used to hold the female, but one clasper may be inserted into the cloaca of the female during transfer of spermatozoa along the grooved surface of the clasper (see Dodd[30]). Holocephalans have an additional clasper anterior to the dorsal fin, but it is not used directly for transfer of spermatozoa.[122]

As one might expect, many kinds of copulatory organs have evolved among the viviparous teleosts that require internal fertilization. In some of these teleosts the genital papilla extends beyond the cloacal orifice and is modified for use as a penis-like organ. Elongation of the genital papilla of *Sarotherodon macrocephala* is stimulated by androgens (see Liley and Stacey[63]). In other species the anal fin is modified into a gonopodium with a groove along which spermatozoa can travel into the female's genital opening (see Dodd and Sumpter,[31] van Tienhoven[113]). Formation of the gonopodium is dependent upon androgens,[20] and 11-ketotestosterone is more effective than testosterone.[87] The gonopodium is in fact a secondary sexual character and is not homologous to other intromittant organs discussed here.

B. Amphibians

Fertilization is considered to be external in most anurans. Copulatory organs consequently are uncommon, occurring only in the "tailed frog," *Ascaphus truei*, which has a permanently everted cloaca that serves as an intromittant organ.[76] Apparently spermatozoa can be transferred in some species such as the viviparous frogs of the genus *Nectophrynoides* when the male places his cloacal opening against that of the female.[114]

Among the apodan amphibians internal fertilization is the rule rather than the exception. Transfer of spermatozoa is aided by utilization of an elaborate copulatory organ, the phallodeum which is a modification of the cloacal wall.[114]

External fertilization is employed only by the most primitive of living urodelean amphibians (see van Tienhoven[113]). Most urodeles accomplish internal fertilization without the aid of any anatomical copulatory organs. Instead, transfer of spermatozoa is accomplished by production of an intermediary structure, the spermatophore, and subsequent complicated courtship procedures (see Arnold,[4] Halliday,[48,49] Madison,[66] Salthe and Mecham.[94] A coagulated mass (cap) of spermatozoa is placed on top of the stalk to produce the completed spermatophore. Several of the cloacal glands appear to participate in secretion of the cap.[102] Then the male induces the female to walk over the spermatophore and pick up the mass of spermatozoa with her cloacal lips. In one species the seemingly impatient male grasps the female and drops her onto the spermatophore (see Halliday[49]). The spermatozoa are stored within the female's cloaca until the time of ovulation and oviposition. Successful transfer of spermatozoa seems to involve both precise behavioral patterns and pheromones.

C. Reptiles

Van Tienhoven[113] recently summarized the literature concerned with reptilian copulatory organs. All reptiles exhibit internal fertilization regardless of whether they are viviparous or oviparious. Male lizards and snakes possess two cloacal extensions each of which is termed a hemipenis. During copulation the male can insert either hemipenis into the female's cloaca. Spermatozoa travel along a longitudinal groove on the hemipenis and enter the female's genital tract. Chelonians and crocodilians have a single penis modified from the cloacal floor. The rhynchocephalian *Sphenodon* has no penis and uses cloacal apposition for transfer of spermatozoa.[123]

Internal fertilization and storage of spermatozoa in the genital tract of the female provides a source of spermatozoa when males are not present.

Females of some species can carry viable spermatozoa for a year or more (see van Tienhoven[113]).

D. Birds

As in reptiles, fertilization in birds is internal (see van Tienhoven,[112,113]). Most birds lack true intromittant organs, and the cloacal openings of male and female birds are briefly held in contact while spermatozoa are transferred. Some birds such as ducks and rheas have prominent intromittant organs which are everted so that sperm may be inserted directly into the cloaca of the female. These copulatory organs develop from the cloacal lining and are therefore of urogenital sinus origin.

VI. SECONDARY SEXUAL CHARACTERS

Secondary sexual characters in male vertebrates are androgen-dependent structures that may not participate directly in gamete production and transfer but which may be essential for successful courtship and mating. A listing of secondary sexual characters of nonmammalian vertebrates is provided in Table 1. In mammals the male sex can often be identified by features such as differences in size, development of body musculature, patterns of hair growth, presence or absence of horns or antlers, and numbers and secretory activities of skin glands. The emphasis of this discussion will focus on the skin and associated glands.

In male mammals the sebaceous glands of the skin are stimulated by androgens but inhibited by estrogens (see Ebling[34]). Androgen-dependent, specialized sebaceous and sudoriparous (sweat) glands have been described in males of many species. They occur, for example, behind the horns of goats, behind the ears of shrews, on the face of bats, in the temporal region of elephants and in the inguinal region of rabbits. Stimulation of these glands requires the conversion of testosterone to DHT. This conversion appears to be dependent upon the presence of pituitary PRL.[34]

A. Fishes

Among the cyclostomes, the hagfishes exhibit no morphologically discernible secondary sexual characters. However, in male lampreys there is an enlargement of fins and a swelling of the cloaca. Both events are attributable to increased vascularity (Dodd and Sumpter,[31] Forbes,[39]).

A number of integumentary secondary sexual characters have been identified in bony fishes. Colors and pigmentation patterns are important as recognition cues for use in courtship and mating. Nuptial coloration patterns among teleosts are stimulated by androgens; e.g., in sticklebacks, certain minnows, tilapia, poeciliids and salmon (see Liley and Stacey[63]).

Other skin modifications have been described. For example, males of a number of minnow species develop nuptial tubercles on the head, and these tubercles can be induced to form by treatment with androgens.[85,103] Treatment of the fathead minnow, *Pimephales promelas*, with either mammalian PRL or human chorionic gonadotropin (hCG) does not cause formation of nuptial tubercles although pituitary extracts prepared from *P. promelas* are effective in intact males.[103]

The skin of teleosts contains numerous mucus-secreting glands. These glands in the discus fish, *Symphysodon aequifasciatus* are influenced by PRL to secrete mucus which is fed to the young (see Liley and Stacey[63]). The skin may be a source of pheromones as well, but most studies implicate the gonads as the source of pheromones in fishes.[63]

B. Amphibians

Amphibians are characterized as having a moist, glandular skin. Mucus-secreting glands are found throughout the epithelial layer of the skin; their secretions serve primarily to moisten the skin. Granular (poison) glands are found concentrated on the dorsal surface, especially in the head region and along the tail (urodeles). Most of these granular glands secrete alkaloid toxins that are probably used as defensive chemicals. (See Noble[76] and Duellman and Trueb[33] for discussions of amphibian integumentary glands.)

Some skin glands in male urodeles are modified for secreting pheromones used to influence the behavior of females during courtship and mating.[4,64,66] These glands are termed generally as hedonic (pleasure-giving) glands, although some may be named specifically for the location in which they occur (e.g., the chin or mental gland of plethodontid salamanders).[4] Nasolabial hedonic glands may be involved in pheromone production (e.g., Rogoff[91]), but Madison[66] concludes that the evidence for this is not strong. Nasolabial and mental glands in the dwarf salamander, *Eurycea quadridigitatta*, exhibit a seasonal cycle in which maximal secretory activity coincides with breeding.[95] Furthermore, treatment of females with testosterone causes induction of mental glands and development of labial cirri as a result of stimulation of the nasolabial glands.[97]

Anuran abdominal skin glands and dermal glands of the thumb pads are thought to produce adhesive substances that aid the male while clasping the female during mating[22,71] (see also Madison[66]). Thumb pads and associated glands exhibit distinct seasonal changes correlated positively with breeding (see Lofts et al.,[65] Ting and Boring[110]). In *Bufo melanostictus* growth of the thumb-pad glands is stimulated by testosterone, whereas estradiol causes regression.[55] In contrast, development of thumb pads following testosterone treatment is enhanced by estradiol in *R. japonica*.[111] Appearance of thumb pads in the salamander *Pleurodeles waltl* coincides with maximal blood androgen levels.[41] Some male urodeles develop nuptial pads on the forelimbs during the breeding season (see Lofts[64]) that are used as aids in grasping females during courtship.[48] Their development also is induced by testosterone, and this induction is enhanced by PRL (see Chapter 4).

Pigmentation patterns have not been studied extensively in amphibians with respect to endocrine regulation by gonadal hormones. In the red-bellied newt, *C. pyrrhogaster*, nuptial coloration appears during the breeding season, and presumably androgens are responsible for their appearance.[108] In most cases of sexual color dimorphism in amphibians the pattern is determined by the genetic sex, and there is no evidence of any hormonal regulation (see Lofts[64]).

In male newts an increase in tail height occurs during the breeding season. This change is caused by testosterone and is enhanced by PRL (for references, see Chapter 4).

C. Reptiles

Reptilians have a number of integumentary glands that can be considered secondary sexual characters (see Cole,[21] Madison[66]). These glands appear to be involved in courtship and might be influenced by androgens. Lizards have glands in the ventral abdominal region anterior to the cloaca and have femoral glands along the midventral surface of each thigh. The femoral glands of males are better developed than those of females and exhibit their greatest secretory activity during the breeding season. Treatment of male *L. sicula* with an antiandrogen (cyproterone acetate) caused regression of the femoral pores (Del Corte, unpublished, cited by Rastogi and Chieffi[87]). Femoral glands are absent in females of some species; e.g., *Hemidactylus flaviviridis*.[67]

Inguinal and axillary glands called Rathke's glands occur in terrestrial and semiterrestrial chelonians. Mental (chin) hedonic glands occur in some of these same turtles. Male crocodilians have glands in the throat region that secrete musk similar to their cloacal glands (see section IV. C.). As in lizards, these glands may be involved in reproductive behaviors (see Madison[66]).

VII. THE "PARADOXICAL" ACTION OF ANDROGENS

The paradoxical effect occurs when an androgen produces a feminizing effect or an estrogen causes masculinization. Numerous examples of the paradoxical action of androgens have been documented in elasmobranchs, amphibians, reptiles and birds (Table 5). The most marked examples involve the apparent stimulation (hypertrophy) of mullerian ducts and their derivatives in androgen-treated males or females. Although other sex accessory structures or secondary sexual characters may respond normally in the appropriate direction following treatment with androgens or estrogens, oviducts of many nonmammalian vertebrates respond similarly to androgens and estrogens. Paradoxical effects generally result from treatment with large amounts of steroids. However, in at least one instance a relatively small dose of testosterone (0.5 μg/g body weight) produced a paradoxical effect in larval tiger salamanders (Table 6).

VIII. EFFECTS OF ARGININE VASOTOCIN ON GENITAL DUCTS

Arginine vasotocin (AVT) has long been known to stimulate contractions of oviducts in nonmammals (see LaPointe[61]). Even though androgens or estrogens can cause hypertrophy of oviductal tissue, progesterone treatment of immature tiger salamanders is necessary for development of the contractile response in the oviduct to AVT.[47]

A parallel relationship between male gonaducts and AVT exists in urodeles. The vasa deferentia of mature rough-skinned newts, *Taricha granulosa*, also contract in the presence of AVT,[125] but the role of steroid hormones in this response has not been examined. There are no data on effects of AVT on male gonaducts in other nonmammals.

IX. SUMMARY AND FUTURE RESEARCH DIRECTIONS

It should be obvious from the preceding survey that there is much to be learned about male ducts, sex accessory glands and secondary sexual characters in male nonmammalian vertebrates. Males have not received nearly the attention that has been directed toward females. Although more work is needed in every group, more relative emphasis should be placed on nonteleostean bony fishes, apodan amphibians and reptiles.

A few generalizations stand out. There obviously existed early in the evolution of jawed vertebrates an intimate association between the kidneys and the gonads. This relationship persists through mammals. There has been a distinct separation of urinary and genital tracts which became complete in the amniotes. Most of the components of the male genital tract (i.e., rete

Table 5. Paradoxical Effects of Steroids on Sex Accessory Structures and Secondary Sexual Characters in Vertebrates

Class and Species	Hormone[a]	Effect[b]	Reference
CHONDRICHTHYES: (Elasmobranchii)	T	(+) oviductal hypertrophy; feminized wolffian ducts	18
OSTEICHTHYES: (Teleostei) *Pimpehales notatus*	E_2	(+) nuptial tubercle formation on head	85
AMPHIBIA: (Caudata)			
Cynops pyrrhogaster	T	(+) oviductal secretion	54
Notophthalmus viridescens	T	(+) oviductal weight	2
Ambystoma tigrinum	T	(+) oviductal hypertrophy	47, 78
Ambystoma opacum	T	(+) oviductal development	38
Ambystoma mexicanum	T	(+) oviductal hypertrophy	72
AMPHIBIA: (Anura)			
Acris gryllus	T	(+) oviductal weight	46
Rana pipiens	T	(+) growth of vestigial oviducts in castrate male	121
Bufo arenarum	T	(+) enhanced (+)-effect of progesterone on oviductal secretion	c
Rana japonica	E	(+) enhanced (+)-effect of testosterone on wolffian duct hypertrophy	111
REPTILIA: (Chelonia)			
Malaclemmys centrata	T	(+) oviductal growth in male or female	d
Pseudemys elegans	E	(+) foreclaw growth (a male secondary sexual character)	39
REPTILIA: (Squamata)			
Lacerta vivipara	T	(+) uterine gland development	81
Lacerta sicula	T	(+) avidin secretion by oviduct	12
	T	(+) oviductal weight	11
Hemidactylus flaviviridis	T	(+) oviductal weight	83
Cnemidophorus inornatus	T	(+) oviductal weight	19

Class and Species	Hormone[a]	Effect[b]	Reference
REPTILIA: (Squamata) (Cont'd)			
Cnemidophorus neomexicanus	T	(+) oviductal secretion	19
Cnemidophorus uniparens	T	(+) oviductal hypertrophy	10
Anolis carolinensis	T	(+) oviductal weight (+) oviductal hypertrophy	36 77
Uromastix acanthoclinus	A	(+) oviductal growth and glandular development	e
Sceloporus occidentalis	T	(+) oviduct	45
Thamnophis sirtalis	T	(+) oviductal development	f
REPTILIA: (Crocodilia)			
Alligator mississippiensis	T	(+) oviductal growth	39
Crocodylus niloticus	T	(+) oviductal development	g
AVES:			
Domestic chicken embryos	T	(+) mullerian duct hypertrophy	14
MAMMALS: (Metatheria)			
Oppossum	T	(+) mullerian ducts	14
MAMMALS: (Eutheria)			
Rat	E_2	(+) wolffian ducts	105

[a]T=testosterone, A=androsterone, E=estrogen, E_2=estradiol

[b](+)=stimulation or increase

[c]Penhos and Naller, 1956, cited by Forbes[39]

[d]Risley, 1946, cited by Forbes[39]

[e]Kehl, 1944, cited by Forbes[39]

[f]Hartley, 1945, cited by Adkins-Regan[3]

[g]Godet, 1961, cited by Adkins-Regan[3]

ct of Steroid Pretreatments (0.5 μg/g) on Mean (± SEM) ıctal Weight (g) in Larval Tiger Salamanders. (These data ummarized in Guillette et al., 1985.)

	Saline	Testosterone	Estradiol	Testosterone + Estradiol
...ature females (54.8 ± 1.07g)	0.07 ± 0.035^{a}	0.21 ± 0.024^{a}	0.23 ± 0.006^{a}	0.542 ± 0.095^{b}
Immature males (46.9 ± 1.36 g)	0.02 ± 0.001^{a}	0.04 ± 0.009^{a}	0.05 ± 0^{a}	0.13 ± 0.12^{b}
Mature females (63.6 ± 1.48 g)	1.37 ± 0.074^{a}	1.67 ± 0.195^{a}	$0.863 \pm 0.136^{a,b}$	2.59 ± 0.363^{b}

One-way ANOVA: Immature females: $F(3,8) = 10.008$, $p<0.01$

One-way ANOVA: Immature males $F(3,11) = 22.216$, 0.001

One-way ANOVA: Mature females: $F(3,12) = 4.182$, $p<0.05$

a-b Values having different superscripts within a row are significantly different. (Duncan's multiple range test, $p<0.05$ per comparison).

testis, vasa efferentia, epididymis, vas deferens and seminal vesicles) are derived from the primitive kidney and its ducts. Mullerian ducts (when present) are retained in male fishes and amphibians and may even become functional structures. In amniotes the mullerian ducts degenerate early in life. The remainder of the genital system consists of secretory glands (especially the prostate in mammals and the various cloacal glands of other vertebrates) derived from the urogenital sinus. Integumentary glands are present in most vertebrates, and their secretions typically are related to reproductive behaviors.

A few general statements can be made about endocrine involvement in development and functioning of the male system. Testosterone and/or DHT are necessary for development and/or functioning of many of the male gonaducts and related glands. In those instances where DHT appears to be stimulatory, no one has demonstrated whether or not conversion of testosterone is possible in these tissues, whether there are multiple androgen receptor sites for different androgens or whether receptors are not discriminating among the various androgens. Prolactin and/or GH appear to play a synergistic role in almost every case where this role has been investigated. Thyroid hormones (see Chapter 14) and corticosteroids (see Chapters 5, 15) have not been examined thoroughly for possible synergistic or antagonistic actions on accessory structures or secondary sexual characters. Responsiveness of male gonaducts to neurohypophysial hormones or prostaglandins is virtually unknown. Testes of amniotes may produce a mullerian-regressing substance responsible for degeneration of mullerian ducts. There is no reason to postulate such a substance for fishes and amphibians in which the mullerian ducts usually persist. In many amphibians, however, mullerian ducts eventually degenerate in males.

Some specific suggestions for future research are listed in the following sections.

A. Wolffian Duct, Mesonephric Kidneys and Their Derivatives

Homologues of wolffian duct derivatives are not clear. More surveys are needed with respect to the condition of the vas deferens, epididymis and seminal vesicles on a seasonal basis as well as to the influence of steroids, PRL and possibly other hormones on their seasonal development. Embryological studies of fish and amphibian species that possess putative "seminal vesicles" are necessary to establish their homologies. Potential secretory functions of vasa deferentia during sperm storage should be examined as well as potential contributions from "sexual kidneys" in both amphibians and reptiles.

B. Mullerian Duct Development in Males

Research that has been concentrated on identification of mullerian-regressing substance in birds and mammals needs to be extended to reptiles and comparisons made to fishes and amphibians where the mullerian ducts may persist. Crocodilians and turtles, in which mullerian ducts persist for some time after hatching (see Forbes[39], Fox[40]), may prove to be especially useful species for these studies.

C. Derivatives of the Urogenital Sinus

A considerable amount of embryological work is needed to establish possible homologies between mammalian glands derived from the urogenital sinus and the cloacal glands of nonmammals. Endocrinological studies are needed concerning development and functions of the various cloacal glands.

D. Secondary Sexual Characters

Most endocrinological studies of secondary sexual characters have employed testosterone, and it would be useful to know more about the fate of testosterone and the nature of androgen receptors in these various target tissues. Limited information is available only for receptors of the anuran thumb pads.[26,87]

Which if any of these glands produce pheromones? Homologies are not known. The possible influence of other hormones on development and functioning of these glands is needed.

E. The "Paradoxical" Action of Androgens

There is no explanation for paradoxical effects in most cases. The most likely explanation for feminization by androgens in oviductal tissue might be their conversion by an aromatase enzymatic system into estrogens (see Chapter 12) so that the resulting action is a female-specific response. Another possibility is that the receptor cannot distinguish between an estrogen and testosterone. Receptors for at least one teleostean tissue cannot distinguish among estrogens, testosterone, and DHT (see Chapter 6). Studies employing the use of DHT, which is considered to be essentially nonaromatizable, might help to clarify this situation. Greater understanding of the relationship between dose of administered hormone and normal circulating levels is needed as well as an explanation for why the animal's own hormones do not cause these paradoxical effects.

X. ACKNOWLEDGMENTS

I wish to thank Harriet Austin, Richard E. Jones, Kristen Lopez and Martha K. Pancak for their contributions to preparation of this chapter. Some of the unpublished data reported here were supported in part by NSF grant PCM 78-22335.

XI. REFERENCES

1. Adams, A.E. (1940). Sexual conditions in Triturus viridescens. III. The reproductive cycle of the adult aquatic form of both sexes. Amer. J. Anat. 66, 235-275.
2. Adams, A.E., Gay, H. and Terzian, A. (1941). The effect of injections of testosterone propionate in adult female newts. Anat. Rec. 79, suppl. 2, 67.
3. Adkins-Regan, E. (1981). Early organizational effects of hormones: an evolutionary perspective. In "Neuroendocrinology of Reproduction" (N.T. Adler, ed.), 159-228. Plenum Press, N.Y.
4. Arnold, S.J. (1977). The evolution of courtship behavior in new world salamanders with some comments on old world salamanders. In "The Reproductive Biology of Amphibians" (D.H. Taylor and S. Guttman, eds.), pp. 141-183. Plenum Publishing Corp., New York.
5. Baker, C.L. (1965). The male urogenital system of the Salamandridae, J. Tenn. Acad. Sci. 40, 1-5.
6. Baker, C.L. and Taylor, W.W., Jr. (1964). The urogenital system of the male Ambystoma. J. Tenn. Acad. Sci. 39, 1-10.
7. Bengtsson, B. (1982). Factors of importance for regulation of uterine contractile activity. Acta Obstet. Gynecol. Scand. Suppl. 108, 13-16.

8. Benson, D.G., Jr. (1965). Chemical and morphological changes in the cloacal glands of the newt, Triturus viridescens, as induced by hormonal alterations. Ph.D. Dissertation, University of Virginia.
9. Bhaduri, J.L. (1953). A study of the urinogenital system of Salientia. Proc. Zool. Soc. Bengal 6, 1-111.
10. Billy, A.J. and Crews, D. (1986). The effects of sex steroid treatments on sexual differentiation in a unisexual lizard, Cnemidophorusuniparens (Teiidae). J. Morphol. 187, 129-142.
11. Botte, V. (1974). The hormonal control of the oviduct in a lizard Lacerta sicula Raf. I. The effects of ovariectomy and steroid replacement. Monit. zool. Ital. 8, 47-54.
12. Botte, V., Segal, S.J. and Koide, S.S. (1974). Induction of avidin synthesis in the oviduct of the lizard Lacerta sicula, by sex hormones. Gen. Comp. Endocrinol. 23, 357-359.
13. Burkholder, G.L. and Tanner, W.W. (1974). A new gland in Sceloporus graciosus males (Sauria: iguanidae). Herpetologica 30, 368-371.
14. Burns, B.K. (1939). The effects of crystalline sex hormones on sex differentition in Amblystoma. II. Testosterone propionate. Anat. Rec. 73, 73-93.
15. Burns, B.K. (1961). Role of hormones in the differentiation of sex. In "Sex and Internal Secretion" (W.C. Young, ed.), Vol. 1, pp. 76-158. Williams and Wilkins, Baltimore.
16. Chase, M.D., Geschwind, I.I. and Bern, H.A. (1957). Synergistic role of prolactin in response of male rat sex accessories to androgen. Proc. Soc. Exp. Biol. Med. 94, 680-683.
17. Chase, S.W. (1923). The mesonephros and urogenital ducts of Necturus maculosus Rafinesque. J. Morphol. 37, 547-531.
18. Chieffi, G. (1967). The reproductive system of elasmobranchs: developmental and endocrinological aspects. In "Sharks, Skates and Rays," (P.W. Gilbert, R.F. Mathewson and D.P. Ralls, eds.), pp. 553-580. The Johns Hopkins Press, Baltimore.
19. Christiansen, J.L. (1973). Natural and artificially induced oviductal and ovarian growth in two species of Cnemidophorus (Sauria: Teiidae). Herpetologica 29, 195-294.
20. Cohen, H. (1946). Effects of sex hormones on the development of the platyfish, Platypoecilius maculatus. Zoologica 31, 121-128.
21. Cole, C. G. (1966). Femoral glands in lizards: a review. Herpetologica 22, 199-206.
22. Conaway, C.H. and Metter, D.E. (1967). Skin glands associated with breeding in Microhyla cardinensis. Copeia 1967, 672-673.
23. Cooper, W.E., Jr. and Vitt, L.J. (1984). Conspecific odor detection by the male broad-headed skink, Eumeces latipes: effects of sex and site of odor source and of male reproductive condition. J. Exp. Zool. 230, 199-209.
24. Cuellar, H.S., Roth, J.J., Fawcett, J.D. and Jones, R.E. (1972). Evidence for sperm sustenance by secretions of the renal sexual segment of male lizards, Anolis carolinensis. Herpetologica 28, 53-57.
25. Dawson, A.B. (1922). The cloaca and cloacal glands of the male Necturus. J. Morphol. 36, 447-465.
26. Delrio, G. and D'Istria, M. (1973). Androgen receptor in the thumb pads of Rana esculenta. Experientia 19, 1412-1413.
27. Depeiges, A. and Dufaure, J.-P. (1982). Isolation and characterization of a major soluble protein secreted by the epididymis of a nonmammalian vertebrate: immunohistochemical evidence of its binding to spermatozoa. Ann. N.Y. Acad. Sci. 383, 442-443.
28. Devine, M.C. (1975). Copulatory plugs: enforced chastity. Science 187, 844-845.
29. de Vlaming, V.L. and Sundararaj, B.I. (1972). Endocrine influences on seminal vesicles in the estaurine gobiid fish, Gillichthyes mirabilis. Biol. Bull. 142, 243-259.

30. Dodd, J.M. (1983). Reproduction in cartilaginous fishes (Chondrichthyes). In "Fish Physiology" (W.S. Hoar, D.R. Randall and E.M. Donaldson, eds.), Vol. 10A, pp. 31-85. Academic Press, New York.

31. Dodd, J.M. and Sumpter, J.P. (1984). Fishes. In "Marshall's Physiology of Reproduction, Vol. 1, Reproductive Cycles of Vertebrates," (G.E. Lamming, ed.), pp. 1-126. Plenum Publishing Corp., New York.

32. Donahoe, P.K., Budzik, G.P., Irelstad, R., Mudgett-Hutner, M., Fuller, A., Jr., Hutson, J.M., Ikawa, H., Hayashi, A., and MacLaughlin, D. (1982). Mullerian-inhibiting substance: an update. Rec. Prog. Horm. Res. 38, 279-230.

33. Duellman, W.E. and Trueb, L. (1986). "Biology of Amphibians." McGraw-Hill Book Company, New York.

34. Ebling, F.J. (1972). The response of the cutaneous glands to steroids. Gen. Comp. Endocrinol., Suppl. 3, 228-237.

35. Evans, L.T. (1948). The effects of gonadotropic and androgenic hormones upon dorso-nuchal crest of the lizard. Anat. Rec. 100, 657-658. (Abstract)

36. Fawcett, J.D. (1975). Effects of season, ovariectomy and hormone replacement therapy on the oviduct of _Anolis carolinensis_ (Reptilia: Iguanidae). Ph.D. Dissertation, University of Colorado

37. Foote, C.L. (1940). Responses of gonads and gonoducts of _Ambystoma_ larvae to treatment with sex hormones. Proc. Soc. Exp. Biol. Med. 43, 519-523.

38. Foote, C.L. (1941). Modifications of sex development in the marbled salamander by administration of synthetic sex hormones. J. Exper. Zool. 86, 291-319.

39. Forbes, T.R. (1961). Endocrinology of reproduction in cold-blooded vertebrates. In "Sex and Internal Secretions" (W.C. Young, ed.), Vol. 2, pp. 1035-1087. Williams and Wilkins, Baltimore.

40. Fox, H. (1977). The urogenital system of reptiles. In "Biology of the Reptilia" (C. Gans and T.S. Parson, eds.), Vol. 6, pp. 1-157. Academic Press, New York.

41. Garnier, D.H. (1985). Androgen and estrogen levels in the plasma of _Pleurodeles waltl_ Michah. during the annual cycle. 1. Male cycle. Gen. Comp. Endocrinol. 58, 376-385.

42. Gerrard, A.M. (1974). Interaction of growth hormone with testosterone in stimulating the renal sexual segment of _Anolis carolinensis_. Gen. Comp. Endocrinol. 22, 281-288.

43. Gigon, A. and Dufaure,J.-P. (1975). Mise en evidence per incorporation de leucine radioactive d'une elaboration proteique dansl'epididyme de lezard. Etude au cours du cycle sexuel et apres injection de testosterone chez l'animal entier et castre. C.R. Acad. Sci. 280, 2241-2244.

44. Gigon-Depeiges, A. and Dufaure, J.-P. (1977). Secretory activity of the lizard epididymis and its control by testosterone. Gen. Comp. Endocrinol. 33, 473-479.

45. Gorbman, A. (1939). Action of mammalian sex hormones in the lizard, _Sceloporus occidentalis_. Proc. Soc. Exp. Biol. Med. 42, 811-813.

46. Greenberg, B. (1942). Some effects of testosterone on the sexual pigmentation and other sex characters of the cricket frog (_Acris gryllus_). J. Exp. Zool. 91, 435-466.

47. Guillette, L., Norris, D.O. and Norman, M.F. (1985). Response of amphibian (_Ambystoma tigrinum_) oviduct to arginine vasotocin and acetylcholine _in-vitro_: influence of steroid hormone pretreatment _in-vitro_. Comp. Biochem. Physiol. 80C, 151-154.

48. Halliday, T.R. (1977). The courtship of European newts: an evolutionary perspective. In "The Reproductive Biology of Amphibians" (D.H. Taylor and S.I. Guttman, eds.), pp. 185-232.

49. Halliday, T.R. (1982). "Sexual Strategy." University of Chicago Press, Chicago.

50. Hildebrand, M. (1982). "Analysis of Vertebrate Structure." John Wiley & Sons, New York.
51. Hoar, W.S. (1969). Reproduction. In "Fish Physiology" (W.S. Hoar and D.J. Randall, eds.), Vol. 3, pp. 1-72. Academic Press, New York.
52. Iwasawa, H. and Michibata, M. (1972). Comparative morphology of sperm storage portion of Wolffian duct in Japanese anurans. Annot. Zool. Jpn. 45, 218-233.
53. Jirasek, J.E. (1980). Normal development of the male accessory glands. In "Male Accessory Sex Glands," (E. Spring-Mills and E.S.E. Hafez, eds.), pp. 3-16. Elsevier/North Holland Biomedical Press, Amsterdam.
54. Kambara, S. (1964). Activity of alkaline phosphatase in the oviduct epithelium of the newt, Triturus pyrrhogaster, injected with estriol, diethylstilbestrol, estrone, testosterone or desoxycorticosterone acetate, single or in combination with progesterone. Proc. Jpn. Acad. 40, 536-540.
55. Kanamadi, R.D. and Saidapur, S.K. (1982). Effect of estradiol-17β, estradiol-17β + homoplastic pars distalis extract, estradiol-17β + testosterone & testosterone on spermatogenesis, Leydig cells & thumb pads in the toad Bufo melanostictus (Schn.). Ind. J. Exper. Biol. 20, 209-215.
56. Kawarabayashi, T. and Marshall, B.M. (1981). Factors influencing circular muscle activity in the pregnant rat uterus. Biol. Reprod. 24, 373-379.
57. Kikuyama, S., Nakano, R. and Yasumasu, I. (1975). Synergistic action of prolactin and androgen on the cloacal glands of the newt. Comp. Biochem. Physiol. 51A, 823-826.
58. Kikuyama, S., Yonezawa, Y. and Noumura, T. (1976). Cytoplasmic receptors for sex steroids in oviduct and cloacal glands of the newt. Zool. Mag. (Tokyo) 85, 463.
59. Kingsbury, B. F. (1895). The spermatheca and methods of fertilization in some American newts and salamanders. Amer. Micro. Soc. 17, 261-395.
60. Knowles, F.G.W. (1939). The influence of anterior-pituitary and testicular hormones on the sexual maturation of lampreys. J. Exper. Biol. 16, 535-547.
61. LaPointe, J. (1977). Comparative physiology of neurohypophysial hormone action on the vertebrate oviduct-uterus. Amer. Zool. 17, 763-773.
62. Licht, P. and Jones, R.E. (1967). Effects of exogenous prolactin on reproduction and growth in adult males of the lizard Anolis carolinensis. Gen. Comp. Endocrinol. 8, 228-244.
63. Liley, N.R. and Stacey, N.E. (1983). Hormones, pheromones, and reproductive behavior in fish. In "Fish Physiology" (W.S. Hoar, D.J. Randall and E.M. Donaldson, eds.), Vol. 10B, pp. 1-64. Academic Press, New York.
64. Lofts, B. (1984). Amphibians. In "Marshall's Physiology of Reproduction, Vol. 1, Reproductive Cycles of Vertebrates" (G.E. Lamming, ed.), pp. 127-205. Churchill-Livingston, Edinburgh.
65. Lofts, B., Wellen, J.J. and Benraad, Th.J. (1972). Seasonal changes in endocrine organs of the male common frog Rana temporaria. III. The gonads and cholesterol cycles. Gen. Comp. Endocrinol. 18, 344-363.
66. Madison, D.M. (1977). Chemical communication in amphibians and reptiles. In "chemical Signals in Vertebrates" (D. Muller-Schwartz and M.M. Mozell, eds.), pp. 135-168. Plenum Publishing corp., New York.
67. Mahenda, B.C. (1953). Contributions to the bionomics, anatomy, reproduction and development of the Indian house-gecko Hemidactylus flaviviridis Ruppel. Part V. The urogenital system. Proc. Ind. Acad. Sci. 38, 215-230.
68. Massa, R. (1980). The role of androgens in male birds' reproduction. In "Steroids and Their Mechanisms of Action in Nonmammalian Vertebrates" (G. Delrio and J. Brachet, eds.), pp. 147-160. Raven Press, New York.

69. Mathews, L.H. (1950). Reproduction in the basking shark, Cetorhinus maximus (Gunner). Phil. Trans. R. Soc. London, Ser. B 234, 247-316.
70. Metter, D. and Conaway, C. (1969). The influence of hormones on the development of breeding glands in Microhyla. Copeia 1969, 621-622.
71. Miller, M.R. and Robbins, M.E. (1954). The reproductive cycle in Taricha torosa (Triturus torosus). J. Exper. Zool. 125, 415-445.
72. Mintz, B. (1947). Effects of testosterone propionate on sex development in female Ambystoma larvae. Physiol. Zool. 20, 355-373.
73. Narbaitz, R. (1974). Embryology, anatomy and histology of the male sex accessory glands. In "Male Accessory Sex Organs" (D. Brandes, ed.), pp. 3-15. Academic Press, New York.
74. N'Diaye, A. and Ozon, R. (1976). Action inhibitrice de la prolactine sur les ultrastructures des vesicules seminales de Discoglossus pictus Otth. (Amphibien anoure). Gen. Comp. Endocrinol. 28, 434-445.
75. N'Diaye, A., Sandoz, D., Boisvieux-Ulrich, E. and Ozon, R. (1974). Action des androgenes chez l'amphibien anoure Discoglossus pictus (Otth.) III. Effets de la castration et action des hormones androgenes sur les ultrastructures de la vesicule seminale. Gen. Comp. Endocrinol. 24, 286-304.
76. Noble, G.K. (1931). "Biology of the Amphibia." McGraw-Hill, New York.
77. Noble, G.K. and Greenberg, G. (1941). Effects of seasons, castration and crystalline sex hormones upon the urogenital system and sexual behavior of the lizard (Anolis carolinensis). J. Exper. Zool. 88, 451-479.
78. Norris, D.O. and Moore, F.L. (1975). Antagonism of testosterone-induced cloacal development of estradiol-17β in immature larval tiger salamanders (Ambystoma tigrinum). Herpetologica 31, 255-263.
79. Norris, D.O., Norman, M.F., Pancak, M.K. and Duvall, D. (1985). Seasonal variations in spermatogenesis, testicular weights, vasa deferentia, and androgen levels in neotenic male tiger salamanders, Ambystoma tigrinum. Gen. Comp. Endocrinol. 60, 51-57.
80. Oldak, P. (1976). Comparison of the scent gland secretion lipids for twenty-five species of snakes: implications for biochemical systematics. Copeia 1976, 320-326.
81. Panigel, M. (1956). Contribution a l'etude de l'ovoviviparite chez les reptiles; gestation et parturition chez le lezard vivipare, Zootoca vivipara. Ann. Sc. Nat. Zool. 18, 569-668.
82. Prasad, M.R.N. and Reddy, P.R.W. (1972). Physiology of the sexual segment of the kidney in reptiles. Gen. Comp. Endocrinol. Suppl. 3, 649-662.
83. Prasad, M.R.N. and Sanyal, M.K. (1969). Effect of sex hormones on the sexual segment of the kidney and other accessory reproductive organs of the Indian house lizard Hemidactylus flaviviridis Ruppell. Gen. Comp. Endocrinol. 12, 110-118.
84. Rabb, G. B. and Rabb, M.S. (1963). Additional observation on breeding behavior of the Surinam toad, Pipa pipa. Copeia 1963, 636-642.
85. Ramaswami, L.S. and Hasler, A.D. (1955). Hormones and secondary sex characters in the minnow Hyborhynchus. Physiol. Zool. 28, 62-68.
86. Ramaswami, L.S. and Jacob, B. (1965). Effect of testosterone propionate on the urogenital organs of immature crocodile Crocodilus palustris Lesson. Experientia 21, 206-207.
87. Rastogi, R.K. and Chieffi, G. (1975). The effects of antiandrogens and antiestrogens in nonmammalian vertebrates. Gen. Comp. Endocrinol. 26,-79-91.
88. Ratcliff, M.A., Jr. (1965). The male urogenital system in Cryptobranchus. J. Tenn. Acad. Sci. 40, 52-57.
89. Reddi, A.H. (1969). Role of prolactin in the growth and secretory activity of the prostate and other accessory glands of mammals. Gen. Comp. Endocrinol. Suppl. 2, 81-86.

90. Reddy, P.R.K. and Prasad, M.R.N. (1970). Hormonal control of the maintenance of spermatogenesis and sexual segment in the Indian house lizard Hemidactylus flaviviridis Ruppell. Gen. Comp. Endocrinol. 14, 15-24.
91. Rogoff, J.L. (1927). The hedonic glands of Triturus viridescens: a structural and functional study. Anat. Rec. 34, 132-133. (Abstract)
92. Sacerdote, M. (1960). Il determinismo endocrinolgico della regione cloacle in Triturus cristatus carnifex Laur. I. Maxulinizzazione nella femina castrata trattata con propionato de testosterone. Boll. Zool. 27, 35-48.
93. Sacerdote, M. and Clerici, P.(1965). Nouvi data sulle mucine epitheliali della regione pelvica e di quella cloacle del tritone crestata. Boll. Zool. 39, 299-305.
94. Salthe, S.N. and Mecham, J.S. (1974). Reproduction and courtship. In "Physiology of the Amphibia" (B. Lofts, ed.), vol. 2, pp. 309-521. Academic Press, New York.
95. Sever, D.M. (1975). Morphological and seasonal variation of the mental hedonic glands of the dwarf salamander, Eurycea quadridigitatta (Holbrook). Herpetologica 31, 241-251.
96. Sever, D.M. (1976). Morphology of the mental hedonic glands clusters of plethodontid salamanders (Amphibia, Urodela, Plethodontidae). J. Herpetol. 10, 227-239.
97. Sever, D.M. (1976). Induction of secondary sexual characters in Eurycea quadridigitta. Copeia 1976, 830-833.
98. Sever, D.M. (1978). Male cloacal glands of Plethodon cinereus and Plethodon dorsalis (Amphibia: Plethodontidae). Herpetologica 34, 1-20.
99. Sever, D.M. (1980). Cloacal anatomy of male brook salamanders (Eurycea). Herpetologica 36, 51-60.
100. Sever, D.M. (1981). Cloacal anatomy of male salamanders in the fami lies Ambystomatidae, Salamandridae and Plethodontidae. Herpetologica 37, 142-155.
101. Sever, D.M. (1983). Cloacal anatomy of male salamanders in the plethodontid subfamily Desmognathidae. Herpetologica 39, 16-27.
102. Sever, D.M. and Houck, L.D. (1985). Spermatophore formation in Desmognathus ochrophaeus (Amphibia: Plethodontidae). Copeia 1985, 394-402.
103. Smith, R.J.F. (1974). Effects of 17α-methyltestosterone on the dorsal pad and tubercles of fathead minnows (Pimephales promelas). Can. J. Zool. 52, 1031-1038.
104. Steinberger, E. (1979). Genetics, anatomy, fetal endocrinology. In "Endocrinology" (L.J. DeGroot et al., eds.), Vol. 3, pp. 1309-1316.
105. Stinnakre, M.-G. (1972). Etude de l'action "paradoxale" des oestrogenes sur la canaux de Wolff des foetus femelles de rat, a l'aide d'un anti-androgene, l'acetate de "cyproteron." C.R. Acad. Sci. 275, 101-103.
106. Strickland, P. (1966). The male urogenital system of Gyrinophilus danielsi dunni. J. Tenn. Acad. Sci. 41, 26-31.
107. Sundararaj, B.I. and Goswami, S.V. (1965). "Seminal vesicle" response of intact, castrate and hypophysectomized catfish, Heteropneustes fossilis (Bloch) to testosterone proprionate, prolactin and growth hormone. Gen. Comp. Endocrinol. 5,464-474.
108. Tanaka, S. and Iwasawa, H. (1979. Annual change in testicular structure and sexual character of the Japanese red-bellied newt Cynops pyrrhogaster. Zool. Mag. 88, 295-305.
109. Tanaka, S. and Takikaawa, H. (1983). Seasonal changes in plasma testosterone and 5α-dihydrotestosterone levels in the adult male newt, Cynops pyrrhogaster pyrrhogaster. Endocrinol. Japan. 30, 1-6.

110. Ting, H.-P. and Boring, A.M. (1939). The seasonal cycle in the reproductive organs of the Chinese toad Bufo bufo and the pond frog Rana nigromaculata. Peking Nat. Hist. Bull. 14, 49-80.

111. Tojio, Y. and Iwasawa, H. (1977). Effects of sex hormones on the development of gonads and sexual characters in young frogs of *Rana japonica*. Zool. Mag. 86, 20-28.

112. van Tienhoven, A. (1961). Endocrinology of reproduction in birds. *In* "Sex and Internal Secretions" (W.C. Young, ed.), Vol. 2, 1088-1169. Williams and Wilkins, Baltimore.

113. van Tienhoven, A. (1983). "Reproductive Physiology of Vertebrates." second edition. Cornell University Press, Ithaca, New York.

114. Wake, M.H. (1977). The reproductive biology of caecilians: an evolutionary perspective. *In* "The Reproductive Biology of Amphibians" (D.H. Taylor and S.I. Guttman, eds.), pp. 73-102. Plenum Publishing Corp., New York.

115. Wake, M.H. (1979). The comparative anatomy of the urogenital system. *In* "Hyman's Comparative Vertebrate Anatomy" (M.H. Wake, ed.), pp. 555-614. University of Chicago Press, Chicago.

116. Wake, M.H. (1981). Structure and function of the male mullerian duct in caecilians, with comments on its evolutionary significance. J. Herpetol. 15, 17-22.

117. Weisel, G.F. (1949). The seminal vesicles and testes of *Gillichthyes*, a marine teleost. Copeia 1949, 101-110.

118. Willett, J.A. (1965). The male urogenital system in the Sirenidae. J. Tenn. Acad. Sci. 40, 9-17.

119. Williams, A.A., Brandon, R.A. and Martan, J. (1984). Male genital ducts in the salamander *Eurycea lucifuga* and *Eumeces longicauda*. Herpetologica 40, 322-330.

120. Williams, A.A., Martan, J. and Brandon, R.A. (1985). Male cloacal gland complex of *Eurycea lucifuga* and *Eurycea longicauda* (Amphibia: Plethodontidae). Herpetologica 41, 272-281.

121. Wolf, O.M. (1939). The effect of testosterone propionate injections into castrate male frogs, *Rana pipiens*. Anat. Rec. Suppl. 75, 55.

122. Wourms, J.P. (1977). Reproduction and development in chondrichthyan fishes. Amer. Zool. 17, 379-410.

123. Young, J.Z. (1981). "The Life of Vertebrates." Oxford University Press, New York.

124. Zaccanti, F. and Tognato, G. (1976). Effects of different doses of diethylstilbestrol dipropionate on the Bidder's organ of intact and orchiectomized adult males of *Bufo bufo*. Monit. Zool. Ital. 10, 105-117.

125. Zoeller, R.T., Lais, L.T. and Moore, F.L. (1983). Contractions of amphibian wolffian duct in response to acetylcholine, norepinephrine and arginine vasotocin. J. Exper. Zool. 226, 53-57.

SEX STEROID RECEPTORS AND NON-RECEPTOR BINDING PROTEINS

Ian P. Callard and Gloria V. Callard

Department of Biology
Boston University
2 Cummington Street
Boston, MA 02115

I. MECHANISMS OF STEROID HORMONE ACTION: AN OVERVIEW

Steroid hormones are primary regulators in many systems. Estrogens, androgens and progestogens regulate reproduction and attendant functions; corticosteroids control protein, water and ion balance; and the secosteroids are responsible for calcium homeostasis. Reviews of interest to comparative endocrinologists have described the physiological aspects of steroid actions in each of these subject areas.[30,51,62,82,100a] Although this paper is intended primarily to review sex steroid receptors in subavian vertebrates, reference is made to birds and mammals for purposes of orientation.

A. Genomic Mechanisms

Present knowledge of the molecular events which underlie steroid hormone actions is based primarily on studies of the rat uterus (see Jensen et al.[51]) and chick oviduct.[100a] These show that target cells are regulated at the level of gene transcription, hence the term "genomic" mechanism of action. According to the Jensen-Jacobson model, circulating steroids accumulate in target cells by binding to specific, high affinity receptors. Ligand binding at physiological temperatures somehow alters the properties of the receptor, resulting in an increased affinity for target cell nuclei, a process designated "activation" in this paper. Consequent interaction of the hormone-receptor complex with nuclear constituents modulates gene transcription, causing changes in the rate of synthesis of specific mRNA's and proteins. Tissue-specific responses follow, and these vary from export of a product (e.g., yolk protein secretion in liver) to triggering of a neuronal pathway (e.g., lordosis).

This model is by no means worked out in all details. Indeed, some areas remain highly controversial, and new ideas continue to evolve. For example, within the recent past it has been shown in several laboratories that receptors present in the soluble fraction of tissue homogenates (cytosol) are in reality artifacts of preparation.[55,118] Because ligand-free (empty, unoccupied or non-activated) receptors have a lower affinity for chromatin than those complexed with hormone (filled, occupied or activated receptors), they are selectively extracted from nuclei during homogenization and appear mainly in cytosolic subfractions after centrifugation. By contrast, occupied receptors adhere more tenaciously to nuclear constituents and are therefore measured primarily in nuclear subfractions. It is this dual distribution of receptors which originally led to the premise that empty receptors were extranuclear and translocated to the nucleus after hormone binding. When cells are enucleated prior to homogenization, however, virtually all receptors are associated with nuclei.[118] This also has been verified using receptor antibodies and immunocytochemical methods[55] and by homogenizing whole tissue in non-aqueous solvents which avoid receptor solubilization.[100] At present, the two-step translocation theory virtually has been abandoned for all classes of steroid receptors.

It is noteworthy that in certain systems receptors appear exclusively or predominantly in nuclear subfractions, even when subjected to the same procedures which cause redistribution in classical mammalian and avian systems. These include the estrogen receptor from livers of the adult hen[7] and of the amphibian *Xenopus laevis*;[43] the estrogen receptors of dogfish testis;[17] the vitamin D receptor of chick intestine;[42] and the ecdysterone receptor of *Drosophila melanogaster* imaginal disc.[128] Recently, a large number of exceptions to the general rule have been documented.[115] Since they went against accepted dogma, early reports indicating that cytosolic receptors were absent in these systems were dismissed as being due to technical problems. This explanation has now been ruled out in the dogfish testis, for example, by showing that addition of cytosol to nuclear extracts did not diminish the quantity of measurable receptors.[17] Also, both occupied and unoccupied forms were simultaneously present in nuclei, thus negating an essential translocation step.

Why receptors in certain species or target tissues are readily extracted from nuclei whereas others are not is an intriguing question. Because it is well known that salt-containing buffers can disrupt the relatively weak bonds between receptors and nuclei, we reasoned that estrogen receptors with strong nucleophilic properties evolved in dogfish as an adaptation to high body fluid osmolarity.[17] This was supported experimentally by showing that higher ionic strengths were required for extracting dogfish receptors from DNA-cellulose or homologous nuclei than were reported previously for avian or mammalian receptors. By contrast, a freshwater elasmobranch (*Potamotrygon*) with body fluid osmolarity in the mammalian range had receptors with extraction properties more closely resembling mammals. To our knowledge, no other systems have been examined with this in mind.

A second area of controversy addresses the nature of the interaction between the hormone-receptor complex and the nucleus. There are at least two schools of thought, one which proposes the existence of "acceptors" comprised of nucleoacidic proteins plus DNA[108] and a second theory which favors receptor binding directly to specific gene sequences.[126] Evidence supporting the acceptor hypothesis shows that these binding sites have a high affinity and limited capacity and are undetectable in non-target tissues. Recent data from *Squalus acanthias* using the same methodology has contributed to this theory by showing that binding sites are species- and tissue-specific.[94] Thus, estrogen-receptor complex from *S. acanthias* testis will bind to testicular chromatin of the same species but not to rabbit uterine chromatin. Moreover, *S. acanthias* oviduct receptor will not interact with testicular chromatin from the same species. As more data accumulate, it may become evident that the acceptor and DNA theories are not mutually exclusive; rather, both types of binding may be involved. Many important questions remain. Exactly where in the nucleus do the empty or unoccupied receptors reside and by what mechanism are they retained? What is the role of the nuclear matrix? What happens to the steroid-receptor complex following gene activation? Is there degradation or recycling? How is receptor synthesis regulated, and where does it occur? As discussed elsewhere in this chapter, some of these important questions may be profitably studied through the use of nonmammalian vertebrates.

B. Nongenomic Mechanisms

It is important to recognize that steroids probably have more than one mechanism of action, although these alternatives have not received the wide recognition of the Jensen-Jacobson model. There are numerous reports of short latency responses (msec range) which cannot be explained by an action at the transcriptional level.[29,31,51a] Also, in the frog oocyte progesterone induces maturation by a mechanism not requiring steroid entry into the cell.[6] Alterations in several membrane-linked events have been described in this system.[9] A direct action of steroids on membranes is supported by studies describing specific, high-affinity binding sites on plasma membranes and in microsomal subfractions.[84,125] Szego[110] has proposed an alternative mechanism by which steroids associate with lysosomal structures as they undergo their centripetal migration from the cell surface to the nucleus and thereby affect nontranscriptional events.

II. RECEPTORS

A. General Characteristics

Steroid receptors involved in conventional genomic actions are recognized on the basis of the following criteria: high affinity for ligand ($K_d = 10^{-9}$ to 10^{-10} M); limited capacity ($B_{max} = 10^{-12}$ moles/g tissue or equivalent protein); ligand and tissue specificity. In those systems which have been studied adequately, separate classes of receptor have been found for each major category of gonadal steroid, namely estrogen, androgen and progesterone receptors. These exhibit little or no cross-reactivity with steroids of other classes, although even here there are exceptions (see C1, below). Also, because more than one class of receptor may be present in a single target tissue, specificity studies conducted with crude tissue extracts may sometimes give equivocal results. In general, recognized target tissues have receptors, whereas non-target tissues do not; however, sensitive binding assays sometimes reveal the presence of receptors prior to the elucidation of a regulatory function. This has obvious importance for investigators working with unstudied animal species, because it may be difficult to select an appropriate "non-target" tissue control.

It is accepted generally that receptors are proteins, although their primary structure is not known, and there appear to be multiple receptor forms which may be transition states of the same molecule.[51] For example, among estrogen receptors 8S, 5S and 4S forms are all recognized. Aggregation of receptor molecules to form dimers may occur under certain circumstances, whereas the S-value may become smaller after dissociation into monomeric units. For estrogen receptors in rat uterus a change in sedimentation properties from 4S to 5S is considered to be an index of activation, presumably a conformational change which accompanies increased receptor affinity for nuclear components.[121]

In addition to the primary receptor characteristics described above, two other criteria have importance because they address the function of the receptor. One of these is the stability of the ligand-receptor complex as measured by its dissociation rate. Classical mammalian and avian receptors have a half-time of dissociation ($t_{1/2}$) at 4°C measured in hours, whereas non-receptor steroid-binding proteins dissociate much more rapidly

($t_{1/2}$ = minutes). This is in keeping with their roles as a steroid concentrating mechanism in the case of receptors and as a reservoir and transfer system in the case of binding proteins. Despite lower body temperatures in ectotherms, dissociation kinetics are not dissimilar from those in endotherms (see Salhanick et al.,[98] Ho and Callard[45]). As discussed above, recent evidence indicates that steroid receptors are primarily or exclusively nuclear, whereas cytosolic receptors are artifacts of tissue preparation. Hence, the ultimate proof that a steroid-binding component is a receptor is its identification in cell nuclei or nuclear extracts. Among the subavian species reviewed here, nuclear receptors have been studied in only a few instances (see Tables 1 to 4).

A useful method by which cytosolic receptors may be distinguished from non-receptor binding proteins in a crude tissue extract is by DNA-cellulose affinity chromatography (see Alberts and Herrick,[1] Fox and Pardee[39]). Although adherence to DNA in itself is not proof of receptor function, a steroid-binding protein which has been exposed to ligand at elevated temperatures (activation) but is not retained on DNA-cellulose is unlikely to be a true receptor. This methodology has been especially useful in identifying receptors in fetal and neonatal mammals[37] and in all subavian classes in which high levels of tissue- and serum-binding proteins interfere in binding assays.[17,66,68,85,98] A second method of approach is to demonstrate shifts in nuclear *versus* cytosolic receptor levels following steroid hormone treatment. In this instance, receptor occupancy *in vivo* causes receptor activation and increased nuclear adherence; hence an increase in the number of receptors recovered from nuclear subfractions. This latter technique has been applied only infrequently to nonmammals: the turtle testis,[66] snake oviduct,[56,57] and salmon liver.[60]

B. Phylogenetic Distribution

Although the majority of steroid-receptor studies has been carried out on mammalian and avian systems, when other species have been examined, steroid receptors of the classical type have been identified. Reports in subavian vertebrates are listed in Tables 1 to 4 and further discussed below. As in vertebrates, steroids are present throughout the animal and plant kingdoms, suggesting they may be universal bioregulators.[99] The studies of Watanabe and Watanabe[117] and Watanabe et al.[116] first demonstrated the possible involvement of a steroid-binding protein in the transport of testosterone into the cells of the bacterium *Pseudomonas testosteroni*.[117] This protein was located in the periplasmic space and had high affinity for ligand (K_d 6.7 x 10^{-9} M). Duchman and Trager[27] demonstrated a progesterone-binding protein in Streptomyces hydrogenans with some characteristics of a steroid hormone receptor. Recently Feldman et al.[35] described an estrogen-binding protein in cytosolic extracts of the yeast Saccharomyces and suggested it may be a primitive receptor. The fungus Achlya ambisexualis may be the most primitive eukaryote known to respond to steroids. In the presence of antheridiol, a steroidal pheromone secreted by female strains of A. ambisexualis, male strains are stimulated to produce antheridial branches which eventually differentiate into male sex organs.[3] Subsequently, Sutherland and Horgen[109] showed that antheridiol treatment of cultures resulted in an increase in chromatin template activity and transcription due to an increased number of RNA-polymerase binding and initiation sites. Recently, an antheridiol-binding protein resembling a classical steroid receptor was identified in cytosols of the same species.[91] Moreover, when preparations of chromatin were incubated with cytosol plus antheridiol, transcription was stimulated although neither of these substances alone was effective.[47] In insects, ecdysone is a recognized hormone with nuclear receptors.[128] Thus we can conclude that the elements of the classical genomic mechanism of steroid hormone action predate vertebrates and possibly originated in prokaryotes (3000×10^6 years before present).

Despite a resemblance in physicochemical properties, it is unlikely that any one receptor class is identical from one species to another. For example, antibodies prepared to the estrogen receptor of the calf uterus cross-reacted with the estrogen receptor of chick oviduct[51] but not with turtle oviduct receptor eluted from DNA-cellulose columns (Jensen, personal communication). However, the same estrogen receptor antibody reacted with a variety of estrogen target tissues in other mammalian species, indicating some common immunoreactive determinants.

C. Major Classes of Steroid Receptors

1. Androgen Receptors. Androgen receptors have been studied relatively little compared to estrogen receptors (Table 1), possibly because of low tissue levels and the greater instability of the molecule itself. One aspect of androgen receptor function of special interest to comparative endocrinologists is whether testosterone, dihydrotestosterone and 11-ketotestosterone act via the same or separate receptor mechanisms. Each of these steroids is a biologically active androgen in one vertebrate group or another. To our knowledge, separate receptors for testosterone and dihydrotestosterone have not been identified in mammals, nor are there marked differences in the affinity constants of these steroids. This leaves open to conjecture why

testosterone is first converted to dihydrotestosterone in the male accessory ducts and other targets. Similarly, in teleosts, although a classical testosterone receptor is present, it reacts poorly with 11-ketotestosterone, at least in brain.[85] Since the latter is a potent androgen in peripheral targets, the possible existence of a separate 11-ketotestosterone receptor merits investigation.

2. Estrogen Receptors. Estrogen receptors of reptiles, amphibians and fishes (Tables 2 and 3) are of high affinity (K_d 10^{-9} to 10^{-10} M) and low capacity (< 500 fmoles/mg protein or tissue equivalent) (Table 1). Receptor levels vary from tissue to tissue (e.g., oviduct/uterus > pituitary > liver > brain), but whether this reflects receptor number per cell or number of receptor-positive cells per weight of tissue is unknown. Sedimentation coefficients of estrogen receptors in sucrose gradients vary somewhat, but in the presence of high salt concentration (0.4-0.5 M) values range from 3.5 to 5.5S, and in low salt (0.05 M) from 4 to 8S depending on tissue and species. In *Xenopus* liver the molecular weight of the nuclear estrogen receptor is approximately 40,000 daltons,[119] whereas in chicken oviduct the cytosolic estrogen receptor was reported to be 60,000.[59] Although estrogen receptors in general exhibit a high degree of specificity for estradiol binding, there is significant interaction with 5α-androstan-3β,17β-diol in several mammalian species[65,111] as well as in the elasmobranch *S. acanthias*.[17] This suggests that the conformation of the steroid-binding site has been highly conserved through evolution. Due to an unusually high affinity of the estrogen receptor from *S. acanthias* for nuclei, this model has been used for investigating the nature of the interaction between the steroid-receptor complex and native chromatin (for further discussion, see I A and I D2).

3. Progesterone Receptors. The literature describing progesterone receptors in lower vertebrates is scanty (Table 4), and this may be due in part to a persistent technical problem, namely, interference from the more abundant corticoid receptors when [^{3}H]progesterone is used as a ligand. The new high affinity, synthetic ligand R5020 may overcome this problem. Using both ligands, progesterone receptors have been identified recently in the oviduct of turtles[45] and snakes.[56,57] A progesterone receptor also has been described by Baulieu and colleagues[8] in the plasmalemma of amphibian oocytes; however, this surface receptor is clearly unlike those involved in classical genomic steroid actions (see also Section I B above).

Table 1. Putative Androgen Receptors in Teleostean Fishes and Amphibians

Vertebrate Class	Genus	Tissue	Location	K_d (nM)	S-Value	Specificity	Concentration	Reference
Teleosts	<u>Carassius</u>	Brain	cytosol, nucleus	1.0		T=DHT 11KT>E_1>E_2>P	100-500 fmol/mg protein	85
	<u>Gobius</u>	Liver	cytosol, nucleus	3.2 6.7	3.4	T=DHT=11KT=CA>P>E_2>F	178 fmol/mg protein 23 fmol/mg protein	61
Amphibia	<u>Rana</u>	Thumb pad	cytoplasm	0.3				24

Abbreviations used in Tables 1-4:

5ADIOL = 5αandrostane-3β17β-diol
B = Corticosterone
CA = Cyproterone acetate
DES = Diethylstilbestrol
DHP = 3β,5α-Dihydroprogesterone
DHT = Dihydrotestosterone
DEX = Dexamethasone
DOC = Desoxycorticosterone
E_1 = Estrone
E_2 = Estradiol
E_3 = Estriol
F = Cortisol
17OHP = 17α-hydroxyprogesterone
11KT = 11-Ketotestosterone
MOXE = Moxestrol
P = Progesterone
PREG = Pregnenolone
R5020 = Promegestone
T = Testosterone

Table 2. Putative Estradiol Receptors in Fishes

Vertebrate Class	Genus	Tissue	Location	K_d (nM)	S-Value	Specificity[a]	Concentration	Reference
Agnatha	*Eptatretus*	Liver	nucleus	0.4		$E_1=E_3>E_3=DES$	0.13 pmol/mg DNA	114
Elasmobranchs	*Scyliorhinus*	Brain	cytosol	0.9		$E_2>DES>E_1>E_3>T=DHT$	8.26 fmol/mg protein	50
	Squalus	Testis	nucleus	1.5	4.5	$E_2>E_1$=MOXE=3 5ADIOL>E_3	250 fmol/g tissue	17
Teleosts	*Gobius*	Liver	cytosol nucleus				51.0 fmol/mg protein 26 fmol/mg protein	61
	Pseudopleuronectes	Liver	cytosol	0.6	8.0	E_2=DES E_1>T>P>B=DHT	29 fmol/mg protein	104
	Cynoscion	Liver	cytosol			DES>$E_2>E_1=E_3$>IIKT[2]	10-100 fmol/mg protein	106
	Salmo	Liver	cytosol nucleus	3.0 5.5	 3.6	E_2>DES>DHT=P>F	0.6 pmol/g tissue 0.15-6.0 pmol/g tissue	 60

[a]See Table 1 for explanation of abbreviations.

Table 3. Putative Estradiol Receptors in Amphibians and Reptiles

Vertebrate Class	Genus	Tissue	Location	K_d (nM)	S-Value	Specificity[a]	Concentration	Reference
Amphibia	*Necturus*	Testis	cytosol nucleus	0.3 0.5	4.7-5	E_2>DES>E_1>E_3>T>DHT~P	2.5-fmol/mg protein 0.2-1.2 fmol/μg DNA	67
	Xenopus	Liver	cytosol nucleus	0.5	 3.5	E_2~DES>T E_2~DES>E_1,E_3>T_1,DEX	100 sites/cell 100 sites/nucleus	119
	Ambystoma	Liver	nucleus	0.3	3.7	E_2>T=P_4=DEX	1300 sites/nucleus	75
	Triturus	Skin	cytosol	2.2		E_2~E_1>P>F		25
Reptilia	*Chrysemys*	Oviduct	cytosol nucleus	1.0	4-5	E_2>E_1>E_3~DES	10-20 fmol/mg protein	98
		Brain	cytosol nucleus	0.7		E_2>DES>E_1>E_3	0.8-6 fmol/mg protein	67
		Epidydi-mis	cytosol	0.8	4-5	E_2~DES>E_1>E_3>DHT=T=P	20 fmol/mg protein	28
		Testis	cytosol nucleus	0.7 0.4	4-5	E_2~DES>E_1>E_3>P=DHT	1-4 fmol/mg protein 0.4-0.6 fmol/μg DNA	68
		Liver nucleus	cytosol	2.0		E_2~DES E_1>T>P E_2~DES>E_1>E_3>DHT>T>P		93
	Nerodia	Liver	cytosol nucleus	1.3 5.7	6.0 3.5	E_2~DES>E_1>E_3>DHT=P=T	18-500 fmol/g tissue	93

[a]See Table 1 for explanation of abbreviations.

Table 4. Putative Progesterone (R5020) Receptors in Amphibians and Reptiles

Vertebrate Class	Genus	Tissue	Location	K_d (nM)	S-Value	Specificity[a]	Concentration	Reference
Reptilia	*Chrysemys*	Oviduct	cytosol	1.5		R5020=P>5 DHP> DHT>E_2>B	4.6 fmol/mg protein	45
			nucleus	0.5		R5020>P>DHP>DHT>E_2>B		
	Nerodia	Oviduct	cytosol	1.5		R5020=P>E_2>T=DHT>B	16-1000 fmol/g tissue	56,57
			nucleus	1.7		P>DHT>17OHP>T>B >E_2=E_1	19-100 fmol/g tissue	
	Anolis	Brain	cytosol	0.1			3-7 fmol/mg protein	112
Amphibia	*Rana*	Oviduct	cytosol	0.21		P>DOC=PREG>E_2	0.12 pmol/mg protein	107

[a]See Table 1 for explanation of abbreviations.

D. Receptors in Different Physiological States and Their Regulation

A question of obvious importance to physiologists is whether the quantity and quality of receptors are determinants of target organ responsiveness or sensitivity. If receptor abundance can determine whether or not a response will be elicited with a given level of circulating hormone, then it becomes important to know how receptor levels are regulated. One approach to this question is to investigate receptor properties in relation to natural and experimentally induced changes in physiological status. Seasonally breeding species, in which changes in the reproductive hormones and their target organs are part of the normal life cycle, would seem to be ideal for studies of this type although they have been explored relatively little to date. Described below are the few studies with data pertinent to this topic.

1. The Liver and Vitellogenesis.

Synthesis of yolk protein by the liver is a cyclic process absolutely dependent on circulating estrogen and the presence of estrogen receptors (see Chapter 6). Although seasonal changes in total receptor number have not been documented in any species, seasonal shifts in cytosolic versus nuclear estrogen receptors reflect circulating hormone levels and the vitellogenic cycle in the turtle *Chrysemys picta*.[93] Data for the water snake *Nerodia sipedon* are similar.[92]

Using male *Xenopus* liver *in vitro*, Westley and Knowland[119] have shown that the estradiol level necessary to saturate the nuclear receptor is similar to the concentration needed to stimulate vitellogenesis (2×10^{-9} M). The abundance of estrogen receptors in the liver of male *Xenopus* is extremely low, and even when increased tenfold by prolonged estrogen treatment, the level is 5 to 10 times below that found in other steroid target tissues.[14] Ontogenetic studies with *X. laevis* tadpoles have demonstrated that nuclear estrogen-receptor levels increase from Stage 54 to Stage 66 larvae; however, sufficient receptor is present to allow the onset of vitellogenesis during metamorphic climax (Stage 62).[24] An interesting observation from this same laboratory is that the appearance of estrogen receptors in hepatic nuclei precedes the time at which the liver becomes responsive to endogenous estrogen.[74] We can infer from this that additional, undefined components are critical for a mature response.

In the axolotl *Ambystoma mexicanum*, May and co-workers[75] have correlated vitellogenin synthesis with the presence and abundance of a high affinity nuclear estrogen-binding component ($K_d = 10^{-9}$ M). They calculated a value of 1300 sites per nucleus in the male axolotl, or 10 times the levels in male *Xenopus* hepatocytes; however, this species difference disappears if a 12-fold difference in the size of the nuclei are taken into account.

Unlike in Xenopus, the quantity of this nuclear receptor was not increased by a single estrogen injection in either male or female axolotls; however, in view of the recently revised concept of steroid hormone action, the techniques used probably failed to reveal shifts in occupied and unoccupied forms.

In immature male or female Atlantic salmon (Salmo salar), specific, high-affinity estradiol-binding sites have been demonstrated in liver cytosol K_d = 2 to 4 nM, 60 fmol/mg/tissue).[60] Levels of nuclear estradiol-binding sites in immature salmon (17 fmol/mg/tissue) were increased to 700 fmol/mg protein after estradiol treatment. Studies from our laboratory have demonstrated a specific estradiol-binding moiety which binds to DNA cellulose in liver cytosol of the winter flounder (Riley and Callard, unpublished).

A single report by Turner et al.[114] suggests that an estrogen receptor is associated with hepatic nuclei in a cyclostome, the pacific hagfish (Eptatretus stouti). Moreover, high affinity binding sites (0.13 pmol/mg DNA) were more numerous in vitellogenic adults than in non-vitellogenic adults or immatures (0.01-0.03 pmol/mg DNA). Although these workers did not indicate whether an effort was made to measure receptors in cytosol, because hagfish have exceptionally high body fluid osmolarity (1000 mOsm), their receptors may be exclusively nuclear as occurs in S. acanthias.

2, Gonadal Functions. Although the gonads are the primary source of sex steroids, they are also targets of sex steroid hormone action. In the mammalian testis, estrogen and androgen receptors have been identified in both Leydig and Sertoli cells,[13,81] and high-affinity binding sites for both steroids are reported to occur in round germ cells and spermatozoa.[113,123] This suggests that steroids may affect gametogenesis directly as well as having indirect effects via control of somatic cell functions. Recently, in an effort to identify the exact spermatogenetic stages that are steroid-dependent, we analyzed receptors in the testis of the dogfish (S. acanthias) and mudpuppy (Necturus maculosus). In contrast to the testis of mammals in which 3 to 4 germ cell generations are simultaneously present in a single tubular cross-section, these lower vertebrates have a testis in which different spermatogenetic stages are topographically segregated and readily separable under a dissecting microscope.[18,89] In S. acanthias, the zone of the testis composed of stem cells, spermatogonia and undeveloped Sertoli cells had the highest concentration of estrogen receptors, and the number decreased as germ cells matured (Zone I > Zone II / Zone III).[18] This result would be predicted if the estrogen receptors were localized in Sertoli cells since germ cell proliferation would dilute somatic elements during the course of maturation. On the other hand, it is possible that estrogen receptors are present only in immature germ cells. Which of these possibilities is correct cannot be determined from the present data. It is important, nonetheless, that the fraction of total receptors occupied with endogenous ligand was also highest in spermatogonial stages, good evidence that estrogen actions predominate early in spermatogenesis.

In contrast to *S. acanthias*, in which Leydig cells are undeveloped at all stages,[88] estrogen-binding activity in mudpuppy (*Necturus maculosus*) testes is localized mainly in regions composed of differentiating or fully differentiated Leydig cells.[89] This result is consistent with mammalian studies indicating that estrogens have inhibitory effects on androgen production by Leydig cells[78] and suggests that estrogen may be involved in regulating the demise of the Leydig cells in *N. maculosus* at the end of its cycle of differentiation.[66] It is important to note, however, that estrogen production and estrogen-binding activity are highest in regions immediately adjacent to the site of initiation of the new spermatogenetic wave and in this respect may have a role analogous to that in *S. acanthias*. In the turtle, *C. picta* the exact localization of estrogen receptors within the testis has not been determined.[68] However, unpublished data from this laboratory indicate that the estrogen receptor concentration in whole testis is highest in May, when Leydig cells are beginning to atrophy and just prior to the start of next season's spermatogenetic cycle.[64]

Although androgen and estrogen receptors have been identified in mammalian ovary and have an important role in regulating follicular maturation,[32,90] classical steroid receptors have not been explored in the ovary of submammalian species despite the obvious advantages of using clearcut follicular hierarchies.

3. Female Reproductive Tract. Progesterone receptors have been studied extensively in the reproductive tract of mammals and birds where they are involved in development and secretory functions. Recent evidence indicates that progesterone receptors have a key role in regulating the duration of estrogen receptor occupancy in cell nuclei.[33]

Progesterone receptors are present in cytosolic and nuclear extracts prepared from the oviduct of both oviparous and viviparous reptiles. In the turtle, progesterone receptors in cytosol varied in concentration from 563 to 6125 fmoles/g tissue during the year and were greatest in preovulatory animals.[45] Paradoxically, nuclear receptors did not vary over the same time period. In a viviparous water snake (*N. sipedon*), progesterone receptors in both nuclear and cytosolic extracts were highest during follicular growth when plasma progesterone levels were also high.[57] After ovulation and through mid-gestation, when plasma estradiol levels fall and progesterone values rise, cytosolic and nuclear progesterone receptors declined in parallel. This correlation suggests that progesterone receptors in reptiles may be stimulated by estrogen as has been reported in mammals.[33]

4. Neuroendocrine Tissues. There is good evidence that sex-steroid receptors mediate the actions of hormones on brain centers controlling reproduction and sexual behavior via conventional genomic mechanisms. In all vertebrate groups, injection with radiolabeled estradiol, testosterone or DHT leads to labeling of cell nuclei in the preoptic area, hypothalamus and other limbic areas (for review, Callard[16]). Moreover, in animals exhibiting sex-dimorphic behaviors (e.g., singing in birds, clasping in frogs), there are corresponding sex differences in steroid-binding cells in discrete neuronal cell clusters.[2,34,51] Biochemical studies in *S. acanthias*,[50] *C. picta*[67] and the goldfish *Carassius auratus* (Pasmanik and Callard, unpublished) have confirmed autoradiographic localization of estrogen receptors in the same species.[50,53,54] Despite exceedingly high levels of aromatase in goldfish brain, estrogen receptors are low in concentration, thus making characterization difficult (Pasmanik and Callard, unpublished). On the other hand, there are high levels of androgen receptors with properties resembling those of mammals.[85]

5. Other Tissues. Delrio et al.[23] attempted to show that cytosolic androgen receptors in the thumb pad of *Rana esculenta* were dependent on the testis. Thus, castration or seasonal testicular inactivity reduced receptors, whereas androgen treatment or seasonal testicular activity increased receptor levels in cytosol. Presumably, cytosolic levels reflect receptor reserves. However, no nuclear receptors were measured; hence, inferences about occupied versus unoccupied receptor sites cannot be made.

III. NON-RECEPTOR STEROID BINDING PROTEINS (SBPs)

A. General Characteristics

In addition to steroid receptors, numerous non-receptor steroid-binding proteins (SBPs) are detectable in both serum and tissue extracts. Depending on their source and ligand-specificity, they have been variously termed progesterone-binding globulin, sex-hormone-binding globulin, estrogen-binding protein, testosterone/estradiol-binding globulin, androgen-binding protein and α-fetoprotein.

The SBPs are distinguishable from receptors by several criteria. In general they have a higher binding capacity (B_{max} = 10^{-6} to 10^{-9} M), lower affinity (K_d = 10^{-7} to 10^{-9} M) and are not as strict in their ligand-specificity, often binding two and sometimes three classes of sex steroids equally well. There are exceptions to these generalizations, however. For example, the SBP in _S. acanthias_ testis has a K_d for estradiol in the same range as that of the estrogen receptor (10^{-9} M) although it binds testosterone and progesterone with equal avidity and is present in amounts 10 to 1000 times greater than estrogen receptors. Similarly, cytosol obtained from the liver of the winter flounder (_Pseudopleuronectes americanus_) has a binding component that saturates in the same range as estrogen receptor (K_d = 10^{-10} M); however, estradiol, testosterone and progesterone are all effective competitors.[104] By contrast, the liver of both _Gallus domesticus_[26] and _X. laevis_[44] have medium affinity SBPs (K_d = 10^{-8} M) with strict specificity for estrogen. Note that all of these species have, in addition, classical estrogen receptors associated with nuclei (Tables 2-3).

In addition to the moderate affinity, limited capacity steroid-binding proteins described above, serum albumin is a relatively non-specific, low-affinity binder. Nonetheless, because of its high concentrations, it should not be overlooked as a significant component in the total binding spectrum.

How then does one differentiate among the various binding components? Although no one property will provide the key, rarely, if ever, are significant levels of non-receptor proteins found in nuclei or nuclear extracts of hormone-exposed animals. Additionally, SBPs generally do not adhere to DNA-cellulose even following exposure to hormone at elevated temperatures (activation). Thus, we and others have used affinity chromatography successfully to measure true receptors in the presence of high concentrations of non-receptor SBPs.[98] A second distinguishing feature of SBPs is the relative instability of the protein-ligand interaction ($t_{1/2}$ = minutes). Less reliable criteria are sedimentation properties of these binding proteins in dense sucrose or glycerol and their electrophoretic mobilities, since these properties vary considerably even when different receptors are compared.

Of the binding activities listed in Tables 1 to 5, only a few have been identified rigorously as receptors; thus it remains open whether some are in fact tissue or serum SBPs.

B. Phylogenetic and Tissue Distribution

Serum and tissue SBPs are widely distributed through the vertebrates from fish to mammals (see Table 5). This in itself signifies a vital function. Sera in many mammalian species are characterized by two major steroid-binding systems: (1) a corticoid-binding globulin (CBG) or transcortin type, binding primarily adrenal steroids and (2) sex-hormone-binding proteins (SHBP) which bind estrogen and androgen preferentially. Progesterone may interact with one or both of these binding components. A review of the literature suggests some phylogenetic trends in the differentiation of steroid-binding systems in blood (see Table 5 for details and references; also previous reviews by Salhanick and Callard,[96] Wingfield[122]). In the most primitive vertebrates (Cyclostomata) two moderate affinity SBPs have been described in serum.[10] One binds progesterone preferentially, whereas the other binds estradiol. The specificity is low, however, as both globulins bind testosterone and corticosterone to some extent. Elasmobranchs have a protein that binds C_{18}-, C_{19}- and C_{21}-steroids without any clear differentiation of a separate CBG.[40,41,48,49] In teleosts separate transcortin-type and sex-hormone binding proteins first appear[37,41] (see also Pasmanik and Callard[86]). This separation of steroid transport functions is well-defined in the amphibians, which have both a SHBP and a CBG.[71,72] Available reptilian data indicate the presence of separate CBGs and SHBPs.[73,97] In both amphibians and reptiles, however, SHBPs interact somewhat less effectively with progesterone and corticosterone. The data of Sernia[102] for several Australian reptiles appears to be somewhat at odds with this phyletic pattern, since a single steroid-binding moiety for C_{18}-, C_{19}- and C_{21}-steroids was detected. It is possible that seasonal variations in the quantity of CBG might preclude its detection under some circumstances. Also, with crude tissue and serum samples it is difficult to distinguish a single, relatively non-specific binding component from a mixture of ligand-specific binding components.

A second trend seen through phylogeny is progessively decreased concentrations of binding proteins in circulation.[41] This correlates well with decreases in plasma estrogen and testosterone levels (10^{-7} M versus 10^{-9} M when fish and mammals are compared). Nonetheless, the capacity of the SBPs is almost always sufficient under normal conditions to bind most of the circulating steroid, leaving only a small fraction in the free form.[101]

We wish to emphasize here that SBPs within cells are not always identical to those in the circulation. There are numerous reports of novel SBPs having characteristics that differ from both classical receptors and conventional serum SBPs. An example of one such protein is the estrone-binding protein identified in the cytosol of chicken liver.[26] Recently, a novel SBP was characterized in chicken oviduct.[95] It is noteworthy that no binding activity occurs in the serum of this species.

Table 5. Sex-Steroid-Binding Proteins in the Blood of Fishes, Amphibians and Reptiles

Animal and Class	Binding[a] System	Ligands[b] Bound	K_d (nM)	Capacity M	Reference
A. AGNATHA					
1. *Petromyzon marinus*	SHBP α_1-globulin	E_2	-	-	10
		P	-	-	
	β-globulin	P	-		
		E_2			
B. CHONDRICHTHYES					
1. *Raja radiata*	SHBP	T	12.5-20	1-2	41
		E_2	-		49
		P	-		
2. *Raja eglanteria*	SHBP	E_2	10.0	-	96
		P	10.0	-	
3. *Squalus acanthias*	SHBP	E_2	37.0	0.18	46
		T	25.0	0.11	
		P	24.0	0.13	
4. *Scyliorhinus canicula*	SHBP	E_2	2.9	2.3	70
		T	13.0	1.6	
		P	2.5	1.7	
C. OSTEICHTHYES					
1. *Salmo gairdneri*	SHBP	T	20.8	8.0	37
		P	25.0	10.0	
		E_2	28.5	11.6	
2. *Salmo salar*	SBP	T	3.3	2.4	41
		E_2	6.3	2.0	
	TeBG	T	8.6	1.0	
		11-KT	90.9	2.5	
3. *Salmo salar*	SBP	E, DHT	13.0	0.02	60
4. *Gadus morhua*	TeBG	T	52.6	2.5	41
5. *Carassius auratus*	TeBG	T	2.8	0.27	86
		E	2.1	0.29	
D. AMPHIBIA					
1. *Pleurodeles waltl*	SBP	E_2	0.9	0.58	72
		T	1.9	1.70	

Animal and Class	Binding System	Ligands bound	K_d (nM)	Capacity	Reference
2. *Salamandra salamandra*	SBP	E_2	1.0	0.64	72
		T	1.1	1.20	
	SHBP	B	6.6	0.48	
	("CBG"-like)	P	--		
3. *Taricha granulosa*	SBP (TeBG)	T	20	.03	79
		DHT			
4. *Ambystoma tigrinum*	TeBG	T	0.5	4.0	15
		E	0.9		
5. *Rana temporaria*	SBP	T	0.8	0.35	72
		E_2	2.0	0.33	
	SHBG	B	1.2	0.13	
	("CBG"-like)	P	--	--	
		T	--	--	
6. *Discoglossus pictus*	SBP	T	1.3	1.0	72
		E_2	2.5	1.4	
	SHBP ("CBG"-like)	B	4.5	0.64	
7. *Xenopus laevis*	SBP (SHBP)	E_2	200	0.30	105
		T		--	
		DHT			
E. REPTILIA					
1. *Chrysemys picta*	SHBP	E_2	4.1	0.35	97
		T			
2. *Lacerta vivipara*	SBP	E_2	0.5	0.30	73
		T			
		DHT			
3. *Nerodia sipedon*	SHBP	P	58.2	2.0	93a
		E_2	38.3	0.43	
		B			
		T			

[a] SHBP = sex-hormone-binding protein; SBP = steroid-binding proteins; TeBG = testosterone/estradiol-binding globulin, CBG = corticoid-binding globulin

[b] E_2 = estradiol; P = progesterone; T = testosterone; B = corticosterone; 11-KT = 11-ketotestosterone; DHT = dihydrotestosterone

One of the best known examples of a tissue-specific SBP is the androgen-binding protein (ABP) found in the testes of several mammals.[5] This binding protein is synthesized in Sertoli cells and, although a fraction is secreted into the circulation, it is found in highest concentrations in the seminiferous tubules and excurrent ducts. Presumably, ABP and consequent high levels of testosterone have a role in sperm maintenance or in delivery of hormone to sites of action outside the testis.[19] Where careful comparisons have been made, testicular ABP is distinctly different from the testosterone/estradiol-binding globulin (TeBG) in serum.[20] An SHBP also has been identified in testis of *S. acanthias* (Mak and Callard, unpublished). Based on physicochemical properties, this SHBP differs from that in serum[46] and from the testicular estrogen receptor in the same species.[17] That it is synthesized within the testis, possibly by Sertoli cells, is suggested by its intra-testicular distribution. Greatest concentrations are measured in regions with highly developed Sertoli cells and mature spermatozoa, whereas levels are correspondingly lower in tissues from less mature testicular regions. This topographical distribution of SHBP also correlates with the distribution of key enzymes for androgen biosynthesis, suggesting that synthesis of the protein may somehow be related to high rates of androgen production. The chicken liver and the dogfish testis, two systems in which estrogen receptors are resistant to redistribution artifacts, may provide attractive models in which to test hypotheses about the role of SBPs in steroid uptake and cytoplasmic transport.

C. Functions

There is good evidence that binding of hormone to plasma proteins prevents the small molecular weight steroids from being excreted directly or after metabolism in the liver, kidney or other tissues.[83,103] A clear correlation between plasma steroid concentrations and SBP concentrations through phylogeny further supports the idea that binding has an important role in maintaining a reservoir of steroid within the vascular system. Even when different species within a single vertebrate group are compared, it can be seen that binding capacity of serum for a given steroid is related to circulating hormone levels. To illustrate, in the lizard *Lacerta vivipara*,[124] progesterone binding is at least an order of magnitude greater than that in the turtle, *C. picta*[97] and may be a mechanism by which exceedingly high levels of progesterone are attained during pregnancy (400 ng/ml).

Because steroids are lipophilic and thought to pass freely through the cell membrane, it is generally assumed that only unbound (free) steroid is biologically active.[83] Binding proteins may therefore regulate steroid actions by limiting the fraction available for entry into target cells. In the trout, for example, Fostier and Breton[37] found that addition of plasma significantly increased the ED_{50} of 17α-hydroxy,20β-dihydroprogesterone in the *in vitro* oocyte maturation assay. In another system, the dasyurid marsupial *Antechinus stuarti*, decreases in CBG which occur in males following mating are thought to be responsible for the symptoms of corticoid excess (i.e., suppression of the immune and inflammatory systems and, ultimately, death).[12] The presence of binding proteins outside the cell may actually facilitate steroid secretion in steroidogenic tissues, presumably by maintaining a concentration gradient.[34]

Contrary to prevailing views, however, Siiteri and co-workers[103] recently advanced the thesis that SBPs somehow assist in the transmembrane transport of steroids. Once inside the cell there are a number of alternatives: the steroid (1) remains associated with the serum steroid-binding moiety; (2) is transferred to a second, tissue-specific SBP; or (3) is specifically bound to receptors in the nuclear compartment. Presumably, there is a dynamic equilibrium involving all binding components, and at any one point in time the fraction of active hormone associated with one or the other is determined by the concentration and physicochemical properties of each component of the system. Intuitively, this is a satisfying concept, as it would explain why free steroids don't become "trapped" in lipophilic cell membranes but does not rule out the possibility that some steroid enters the cell in its free form. Also, taking into account new evidence showing that steroid receptors are exclusively nuclear, Siiteri's theory provides a reasonable mechanism whereby steroids can be concentrated in the extranuclear compartment prior to interaction with the genome. Several lines of experimental evidence are supportive of this concept. Thus, the intracellular localization of plasma-like components by immunocytochemical techniques[7,11] and their presence in homogenates of enzyme-dispersed and cultured cells[22,58] suggest that steroid target cells either selectively take up such proteins from the blood or synthesize them *in situ*.

The subject of the foregoing discussion is by no means generally accepted, nor are all existent data supportive. For example, α-fetoprotein (AFP), a non-receptor estrogen-binding protein, has been observed intracellularly in greater amounts in brain regions outside of known target areas,[7] suggesting an inactivating rather than a transport and delivery function. This is consistent with physiological and behavioral studies showing that fetal and neonatal rodents are insensitive to the masculinizing effects of secreted and injected estrogen when serum AFP levels are high.[76]

As originally suggested by Dower and Ryan,[26] the estrone-binding component in chicken liver may serve as an estrogen "sink," preventing the rapid metabolism which is characteristic of liver. Because the chicken lacks a plasma SBP, a mechanism of this kind may be required to maintain a pool of estrogen in close proximity to receptor sites. On the other hand, since liver SBPs have been identified in species in which serum SBPs are also present (e.g., *Xenopus*, *Ambystoma*), the two binding components may operate in tandem.

Finally, consideration must be given to the possibility that certain molecules identified as putative receptors or SBPs are, in fact, steroidogenic enzymes or other proteins which accept steroids as ligands. Recently, a subclass of estrogen-binding sites in rat uterus was shown to correlate temporally and quantitatively with increased peroxidase activity ascribed to invading eosinophils.[63] The implication of these findings have, in turn, been rebutted.[69]

D. Natural Changes and Regulation of SBPs

Despite the probable importance of SBPs in regulating hormone access to target cells, relatively little is known of natural changes or control mechanisms outside of mammals. Nonetheless, there is scattered evidence to indicate that regulation may be achieved through the steroids themselves.

In the human, SBP concentrations increase after the administration of estrogens and during pregnancy.[4,80,87,120] Estrogens also appear to play a role in regulating the concentration of SBP in some amphibians,[72] as ovariectomy caused a decrease in binding activity of plasma which could be restored by estradiol treatment. Unlike the human, no change in SBP occurred during pregnancy in the ovoviviparous salamander, *Salamandra salamandra*.[72] In the elasmobranch, *S. acanthias*, no differences in steroid-binding capacity were seen in animals from different stages of pregnancy.[46] In the viviparous lizard *L. vivipara*, a two-fold increase in progesterone binding occurred during the later stages of pregnancy. This was associated with a two-fold increase in the dissociation constant,[124] suggesting that new protein had been synthesized or, alternatively, that some conformational change at the binding site had occurred. No sex differences were observed in B_{max} or K_d of the steroid-binding protein in the plasma of dogfish sharks, *S. acanthias*[46] and *Scyliorhinus canicula*,[70] or in goldfish, *C. auratus*.[86] However, in the urodele *Pleurodeles waltl*, Martin and Ozon[72] found that males had a lower concentration of SBP than females. Neofemales, which are genetically male, have levels similar to females, indicating that SBP levels in this species may be hormone-dependent.

thyroid hormone has been implicated in the induction or increased synthesis of SBP in several studies.[120] Also, ontogenetic changes are well known. Marked decreases in α-fetoprotein are observed in rodents through early development such that little or no capacity for binding estrogen remains in mature animals. This pattern in rodents is consistent with the postulated functions of this estrogen-binding protein in protecting the fetus from overexposure to maternal estrogens.[76] In adult S. acanthias the binding capacity for estradiol in late embryos was half that of adults.[46]

IV. SUMMARY AND FUTURE DIRECTIONS

Throughout this review we have attempted to highlight unanswered questions and to indicate when nonmammalian species might be especially useful for addressing problems of general relevance (e.g., see Section IA). Clearly, there is a large family of steroid-binding molecules which have different properties and functions. These include (a) non-specific, low-affinity binders of high capacity (e.g., albumins); (b) somewhat more specific, moderate-affinity binders present in lesser quantities (e.g., CBG, SHBP); (c) highly specific, high-affinity, low-capacity receptors which interact with chromatin. From an evolutionary perspective, we might ask how the different steroid-binding proteins are related. It is tempting to suggest an evolutionary sequence in which steroid-binding proteins became progessively more specialized and less abundant as the regulation of cellular functions in multicellular organisns became more precise. However, since proteins with characteristics of classical steroid receptors as well as SBPs have been described in unicellular organisms, it is not possible to rationalize such an evolutionary sequence. Rather, it would appear that multicellular organisms have conserved primitive mechanisms for the purpose of gene regulation but also have retained and experimented with a variety of other proteinaceous carriers of steroids. Perhaps it is in the study of the latter group of steroid carriers that a phylogenetic approach and unconventional animal models may yield dividends.

V. ACKNOWLEDGMENTS

The original research from the authors' laboratories was carried out in part at the Mount Desert Island Biological Laboratory, Salsbury Cove, Maine. We gratefully acknowledge the invaluable contributions of Gary Heiserman, Shuk-mei Ho, Tom Koob, Paul Mak, Metsada Pasmanik, Deborah Riley, Susan Kleis-San Francisco and Adele Salhanick. This project was supported by NSF PCM 81-04144 to I.P.C. and NSF PCM 82-08248 and HD 16715 to G.V.C.

VI. REFERENCES

1. Alberts, B. and Herrick, G. (1970). DNA-cellulose chromatography. In "Methods in Enzymology," Vol. XXI "Nucleic Acids" (L. Grossman and K. Moldare, eds.), pp. 198-217. Academic Press, New York.
2. Arnold, A. and Saltiel, A. (1979). Sexual differences in pattern of hormone communication in the brain of a songbird. Science 205, 702-705.
3. Banksdale, A.W. (1969). Sexual hormones of Achyla and other fungi. Science, 166, 831-836.
4. Bardin, C.W. and Mahoudeau, J.A. (1970). Dynamics of androgen metabolism in women with hirsutism. Ann. Clin. Res. 2, 251-262.
5. Bardin, C.W., Musto, N., Gunsalus, G., Kotite, N., Chey, S.-L., Laurea, F. and Bechei, R. (1981). Extracellular androgen binding proteins. Ann. Rev. Physiol. 43, 189-239.
6. Baulieu, E.E., Goleaan, F., Schorderet, M., and Schorderet-Slatkine, S. (1978). Steroid-induced meiotic division in Xenopus laevis oocytes: Surface and calcium. Nature (London) 275, 593-598.
7. Benno, R.H. and Williams, T.H. (1978). Evidence for intracellular localization of alpha-feto-protein in the developing rat brain. Brain Res. 142, 182-186.
8. Blondeau, J.-P and Beaulieu, E.E. (1984). Progesterone receptor characterized by photoaffinity labelling in the plasma membrane of Xenopus laevis oocytes. Biochem. J. 219, 785-792.
9. Blondeau, J.-P. and E.E. Beaulieu. (1985). Progesterone-inhibited phosphorylation of an unique M_r 48,000 protein in the plasma membrane of Xenopus laevis oocytes. J. Biol. Chem. 260, 3617-3625.
10. Boffa, G.A., Martin, B., Winchenne, J.J. and Ozon, R. (1972). Steroid-protein interaction in human, amphibian and cyclostomic sera. Analytical ion exchange chromatography. Biochimie 54, 1137-1145.
11. Bordin, S., and Petra, P.H. (1980). Immunocytochemical localization of the sex steroid-binding protein of plasma in tissues of the adult monkey Macaca nemestrina. Proc. Natl. Acad. Sci. USA 77, 5678-5682.
12. Bradley, A.J., McDonald, I.R. and Lee, A.K. (1980). Stress and mortality in a small marsupial (Antechenus stuarti Macleay). Gen. Comp. Endocrinol. 40, 188-200.
13. Brinkmann, A.O., Mulder, E., Lamers-Stahlhofen, G.J.M., Mechielsen, M.J. and van der Molen, W.J. (1972). An estradiol receptor in rat testis interstitial tissues. FEBS Lett. 26, 301-305.
14. Bullen, R.E. and O'Malley, B.W. (1976). The biology and mechanism of steroid hormone receptor interaction with the eukaryotic nucleus. Biochem. Pharmacol. 25, 1-12.
15. Burns, J.M. and Rose, F.L. (1980). Testosterone, estrogen binding protein in sexually mature larvae of Ambystoma tigrinum. Gen. Comp. Endocrinol. 41, 314-320.
16. Callard, G.V. (1983). Androgen and estrogen actions in the vertebrate brain. Amer. Zool. 23, 607-620.
17. Callard, G.V. and Mak, P. (1985). Exclusive nuclear location of estrogen receptors in Squalus testis. Proc. Natl. Acad. Sci. (USA) 82, 1336-1340.
18. Callard, G.V., Pudney, J.A., Mak, P. and Canick, J. (1985). Stage-dependent changes in steroidogenic enzymes and estrogen receptors during spermatogenesis in the testis of the dogfish, Squalus acanthias. Endocrinology 117, 1328-1335.
19. Cavreau, S., Drosdowsky, M.A. and Courot, M. (1984). Androgen-binding proteins in sheep epididymis: Age-related effects on androgen receptor and testosterone concentrations. Correlation with histological studies. J. Endocr. 103, 281-296.

20. Cheng, C.Y., Musto, N.A., Gunsalus, G.L., Frich, J. and Bardin, C.W. (1985). There are two forms of androgen binding protein in human testes. J. Biol. Chem. 260, 5631-5640.
21. Corvol, P. and Bardin, C.W. (1973). Species distribution of testosterone-binding globulin. Biol. Reprod. 8, 277-282.
22. DeKloet, E.R. and McEwen, B.S. (1976). A putative glucocorticoid receptor and a transcortin-like macromolecule in pituitary cytosol. Biochim. Biophys. Acta 421, 115-123.
23. Delrio, G., Citarella, F. and D'Istria, M. (1980). Androgen receptor in the thumb pad of *Rana esculenta*: Dynamic aspects. Gen. Comp. Endocrinol. 26, 281-283.
24. D'Istria, M., Citarella, Iela, L. and Delrio, G. (1979). Characterization of a cytoplasmic androgen receptor in the male secondary sexual character of Green Frog (*Rana esculenta*). J. Steroid Biochem. 10, 53-59.
25. D'Istria, M., Delrio, G. and Chieffi, G. (1975). Receptors for sex hormones in the skin of the amphibia. Gen. Comp. Endocrinol. 26, 281-283.
26. Dower, W.J. and Ryan, K.J. (1976). A cytoplasmic estrone-specific binding protein in hen liver. Fed. Proc. 35, 1366.
27. Duchman, H., and L. Trager. (1979). Progesterone binding protein from *Streptomyces hydrogenans*. J. Steroid Biochem. 10, 277-283.
28. Dufaure, J.-P, Mak, P., and Callard, I.P. (1983). Estradiol binding activity in epididymal cytosol of the turtle, *Chrysemys picta*. Gen. Comp. Endocrinol. 51, 61-65.
29. Dufy, B., Vincent, J.-D, Fleury, H., duPasquier, P., Gowrdji, D. and Tixier-Vidal, A. (1979). Dopamine interaction of action potentials in a prolactin secretory cell line is modulated by oestrogen. Nature (London) 282, 855-857.
30. Edelman, I.S. and Fimignani, G.M. (1968). On the biochemical mechanism of action of aldosterone. Rec. Prog. Horm. Res. 24:1-44.
31. Edwards, P.M., Jacquemyns, C.R. and Rousseau, G.G. (1981). Melanosome aggregation by corticosteroids: evidence for a novel type of steroid action. J. Steroid Biochem. 15, 17-23.
32. Erickson, G.F., Magoffin, D.A., Dyer, C.A. and Hafeditz, C. (1985). The ovarian androgen producing cells: a review of structure/function relationships. Endocr. Rev. 6, 371-399.
33. Evans, R.W., Chen, T.J., Hendry, W.J. III and Leavitt, W.W. (1980). Progesterone regulation of the estrogen receptor in the hamster uterus during the estrous cycle. Endocrinology 107, 383-390.
34. Ewing, L.L. and Zirkin, B. (1983). Leydig cell structure and steroidogenic function. Rec. Prog. Horm. Res. 39, 599-636.
35. Feldman, D., Do, Y., Burshell, A., Stathis, P. and Loore, D.S. (1982). An estrogen-binding protein and endogenous ligand in *Saccharamyces cerevisiae*: possible hormone receptor system. Science 213, 297-298.
36. Fine, M.L., Keefer, D.A., and Leichnetz, G.R. (1982). Testosterone uptake in the brainstem of a sound producing fish. Science 215, 1265-1267.
37. Fostier, A. and Breton, B. (1975). Binding of steroids by plasma of a teleost: the rainbow trout, *Salmo gairdnerii*. J. Steroid Biochem. 6, 345-351.
38. Fox, T.O. (1975). Androgen and estrogen binding macromolecules in developing mouse brain: biochemical and genetic evidence. Proc. Natl. Acad. Sci. ((USA) 72, 4303-4307.
39. Fox, T.O. and Pardee, A.B. (1971). Proteins made in the mammalian cell cycle. J. Biol. Chem. 245, 6159-6165.
40. Freeman, H.C. and Idler, D.R. (1969). Sex hormone binding proteins II. Isolation from serum of an elasmobranch (*Raja radiata*). Gen. Comp. Endocrinol. 13, 83-91.

41. Freeman, H.C. and Idler, D.R. (1971). Binding affinities of blood proteins for sex hormones and corticosteroids in fish. Steroids 17, 233-250.

42. Haussler, M.R. and Norman, A.W. (1967). The subcellular distribution of physiological doses of vitamin D_3. Arch. Biochem. Biophys. 118, 145-153.

43. Hayward, M.A., Mitchell, T.A. and Shapiro, D.J. (1980). Induction of estrogen receptor and reversal of the nuclear/cytoplasmic receptor ratio during vitellogenin synthesis and withdrawal in *Xenopus laevis*. J. Biol. Chem. 255, 11308-11312.

44. Hayward, M.A. and Shapiro, D.J. (1981). A middle-affinity estrogen-specific binding protein in livers of vitellogenic and nonvitellogenic *Xenopus laevis*. Devel. Biol. 88, 333-340.

45. Ho, S.M. and Callard, I.P. (1984). High affinity binding of [^{3}H]R5020 and [^{3}H]progesterone by putative progesterone receptors in cytosol and nuclear extract of turtle oviduct. Endocrinology 114, 70-79.

46. Ho, S.M., Tsang, P and Callard, I.P. (1980). Some properties of a steroid-binding protein in the plasma of an ovoviviparous dogfish, *Squalus acanthias*, at different stages of the life cycle. Biol. Reprod. 23, 281-289.

47. Horgen, P.A. (1977). Cytosol-hormone stimulation of transcription in the aquatic fungus, *Achyla ambisexualis*. Biochem. Biophys. Res. Comm. 75, 1022-1028.

48. Idler, D.R. and Freeman, H.C. (1968). Binding of testosterone 1α-hydroxycorticosterone and cortisol by plasma proteins of fish. Gen. Comp. Endocrinol. 11, 366-372.

49. Idler, D.R. and Freeman, H.C. (1979). Sex hormone binding proteins I. Binding of steroids by serum of elasmobranchs (*Raja radiata*). Gen. Comp. Endocrinol. 13, 83-91.

50. Jenkins, N., Joss, J.P., and Dodd, J.M. (1980). Biochemical and autoradiographic studies on the oestradiol-concentrating cells in the diencephalon and pituitary gland of the female dogfish (*Scyliorhinus canicula* L.). Gen. Comp. Endocrinol. 40, 211-219.

51. Jensen, E.V., Greene, G.L., Class, L.E., DeSombre, E. and Nadji, M. (1982). Receptors reconsidered: A 20 year perspective. Rec. Prog. Horm. Res. 38, 1-40

51a. Kelly, M.I., Kuhnt, U. and Wuttke, W. (1980). Hyperpolarization of hypothalamic parvocellular neurons by 17β-estradiol and their identification through intercellular staining with Procion yellow. Expt. Brain Res. 40, 440-447.

52. Kelley, D.B., and Pfaff, D.W. (1978). Generalizations from comparative studies on neuro-anatomical and endocrine mechanisms of sexual behavior. *In* "Biological Determinants of Sexual Behavior" (J.B. Hutchison, ed.), pp. 225-254.

53. Kim, Y.S., Stumpf, W.E. and Sar, M. (1978). Topography of estrogen target cells in the forebrain of goldfish, *Carassius auratus*. J. Comp. Neur. 182, 611-620.

54. Kim, Y.S., Stumpf, W.E. and Sar, M. (1981). Anatomical distribution of estrogen target neurons in turtle brain. Brain Res. 230 (1-2), 195-204.

55. King, W.J., and Green, G.L. (1984). Monoclonal antibodies localize oestrogen receptor in the nuclei of target cells. Nature (London) 307, 747-749.

56. Kleis-San Francisco, S.M. and Callard, I.P. (1986). Identification of a putative progesterone receptor in the oviduct of a viviparous water-snake, *Nerodia*. Gen. Comp. Endocrinol. (in press).

57. Kleis-San Francisco, S.M. and Callard, I.P. (1986). Progesterone receptors in the oviduct of a viviparous snake (*Nerodia*): Correlations with ovarian function and plasma steroid levels. Gen. Comp. Endocrinol. (in press).

58. Koch, B., Lutz, B., Briaud, B. and Mialhe, C. (1976). Heterogeneity of pituitary glucocorticoid binding: evidence for a transcortin-like compound. Biochim. Biophys. Acta. 444, 497-507.
59. Kon, O.L., Webster, R.A. and Spelsberg, T.C. (1980). Isolation and characterization of the estrogen receptor in hen oviduct: Evidence for two molecular species. Endocrinology 107, 1182-1191.
60. Lazier, C.B., Lonergan, K. and Mommsen, T.P. (1985). Hepatic estrogen receptors and plasma estrogen-binding activity in the Atlantic salmon. Gen. Comp. Endocrinol. 57, 234-245.
61. Le Menn, F., Rochefort, H., and Garcia, M. (1980). Effect of androgen mediated by the estrogen receptor of fish liver: vitellogenin accumulation. Steroids 35, 315-328.
62. Leung, K. and Munck, A. (1975). Peripheral actions of glucocorticoids. Ann. Rev. Physiol. 37, 245-272.
63. Lyttle, C.R., Medloch, K.L. and Sheehan, D.M. (1984). Eosinophils as the source of uterine nuclear type II estrogen binding sites. J. Biol. Chem. 259, 2697-2700.
64. Mak, P. (1984). Estrogen receptors in the brain-testicular axis of some non-mammalian species. Ph.D. Dissertation, Boston University.
65. Mak, P. and Callard, G.V. (1985). Characterization of estrogen receptors in hamster brain. J. Steroid Biochem. 22, 355-361.
66. Mak, P., Callard, I.P. and Callard, G.V. (1983). Characterization of an estrogen receptor in the testis of the urodele amphibian _Necturus maculosus_. Biol. Reprod. 28,261-270.
67. Mak, P., Ho, S.-M. and Callard, I.P. (1982). Estrogen receptors in the turtle brain. Brain Res. 231, 63-74.
68. Mak, P., Ho, S-M., and Callard, I.P. (1983). Characterization of an estrogen receptor in the turtle testis. Gen. Comp. Endocrinol. 52, 182-189.
69. Markaverich, B.M., Roberts, R.R., Alejandro, M.A. and Clark, J.H. (1985). Type II estrogen binding sites in the rat uterus are not of eosinophil origin. Proc. Endocrine Soc. 67th Ann. Mtg., Abs. #334, p. 84.
70. Martin, B. (1975). Steroid-protein interaction in non-mammalian vertebrates. I. Elasmobranch steroid-binding protein (ESBP) in dogfish serum (_Scyliorhinus canicula_). Gen. Comp. Endocrinol. 25, 42-51.
71. Martin, B. (1980). Steroid-protein interactions in non-mammalian vertebrates: Distribution, origin, regulation, and physiological significance of plasma steroid binding proteins. _In_ "Steroids and their Mechanisms of Action in Nonmammalian Vertebrates" (G. Delrio and J. Bracket, eds.), pp. 63-83. Raven Press, New York.
72. Martin, B. and Ozon, R. (1975). Steroid protein interactions in nonmammalian vertebrates. II. Steroid binding protein in the serum of amphibians: a physiological approach. Biol. Reprod. 13, 371-380.
73. Martin, B. and Xavier, F. (1981). High-affinity binding of progesterone, estradiol-17β and testosterone by plasma proteins of the reptile _Lacerta vivipara_, J. Gen. Comp. Endocrinol. 43, 299-307.
74. May, F.E.B. and Knowland, J. (1981). Oestrogen receptor levels and vitellogenin synthesis during development of _Xenopus laevis_. Nature 292, 853-855.
75. May, F.E.B., Westley, B.R. and Knowland, J. (1981). Vitellogenin synthesis and characterisation of the liver estrogen receptor in the neotenous salamander _Ambystoma mexicanum_. Devel. Biol. 82, 350-357.
76. McEwen, B.S. (1978). Sexual maturation and differentiation: the value of the gonadal steroids. _In_ "Maturation of the Nervous System" (M.A. Corner et al., eds.), pp. 291-307. Elsevier, Amsterdam.

77. Mester, J. and E.-E. Baulieu. (1972). Nuclear estrogen receptor of chick liver. Biochim. Biophys. Acta 261, 236-244.
78. Moger, W.H. (1980). Direct effects of estrogens on the endocrine function of the mammalian testis. Can. J. Physiol. Pharmacol. 58, 1011-1022.
79. Moore, F.L., Spielvogel, S.P., Zoeller, R.I. and Wingfield, J. (1983). Testosterone-binding protein in a seasonally breeding amphibian. Gen. Comp. Endocrinol. 49, 15-21.
80. Murphy, B.E.P. (1968). Binding of testosterone and estradiol in plasma. Can. J. Biochem. 46, 299-302.
81. Nakhla, A.M., Matshu, J.P., Janne, O.A., and Bardin, C.W. (1984). Estrogen and androgen receptors in Sertoli, Leydig, myoid and epithelial cells: effects of time in culture and cell density. Endocrinology 115, 121-128.
82. Norman, A.W., Roth, J. and Orci, L. (1982). The vitamin D endocrine system: Steroid metabolism, hormone receptors, and biological response (calcium, binding proteins). Endocr. Rev. 3 (4), 331-366.
83. Pardridge, W.M. (1981). Transport of protein-bound hormones into tissues _in vivo_. Endocr. Rev. 2, 103-123.
84. Parikh, I., Anderson, W.L., and Neame, P. (1980). Identification of high affinity estrogen binding sites in calf uterine microsomal membranes. J. Biol. Chem. 255, 10266-10270.
85. Pasmanik, M. and Callard, G.V. (1985). Identification and characterization of an androgen receptor in the brain of goldfish (_Carassius auratus_). Amer. Zool. 25, 115A. (Abstract)
86. Pasmanik, M. and Callard, G.V. (1986). Identification of sex hormine binding protein in goldfish serum. Biol. Reprod. (in press).
87. Pearlman, W.H., Crepy, O. and Murphy, M. (1967). Testosterone binding levels in the serum of women during the normal mestrual cycle, pregnancy and post-partum period. J. Clin. Endocrinol. Metab. 27, 1012-1018.
88. Pudney, J. and Callard, G.V. (1984). Identification of Leydig-like cells in the testis of the dogfish _Squalus acanthias_. Anat. Rec. 209, 323-330.
89. Pudney, J., Canick, J.A., Mak, P. and Callard, G.V. (1983). The differentiation of Leydig cells, steroidogenesis, and the spermatogenetic wave in the testis of _Necturus maculosus_. Gen. Comp. Endocrinol. 50, 43-66.
90. Richards, J.S. (1979). Hormonal control of ovarian follicular development: A 1978 perspective. Rec. Prog. Horm. Res. 35, 343-373.
91. Riehl, R.M., Toft, D.O., Meyer, M.D., Carlson, G.L. and McManus, T. C. (1984). Detection of a pheromone-binding protein in the aquatic fungus _Achlya ambisexualis_. Exp. Cell Res. 153, 544-549.
92. Riley, D. and Callard, I.P. (1984). Estrogen binding proteins in the liver of the viviparous snake, _Nerodia_. Amer. Zool. 24, 117A. (Abstract)
93. Riley, D., McPherson, R., and Callard, I.P. (1983). Hepatic estrogen receptor in the turtle _Chrysemys picta_. Partial characterization, seasonal changes and pituitary dependence. J. Steroid Biochem. (in press).
93a. Riley, D., Kleis-San Francisco, S.M. and Callard, I.P. (1986). A steroid hormone binding protein in the plasma of the viviparous snake, _Nerodia_. Biol. Reprod. 34, Suppl. 1, p. 71.
94. Ruh, M.F., Sing, P.K., Mak, P. and Callard, G.V. (1986). Tissue and species specificity unmasked nuclear acceptor sites for the estrogen receptor of _Squalus_ testes. Endocrinology, 118, 811-818.
95. Ruh, M.F., and Toft, D.O. (1984). Characterization of unusual

sex steroid binding component from chicken oviduct. J. Steroid Biochem. 21, 1-8.

96. Salhanick, A.R. and Callard, I.P. (1979). Sex steroid binding proteins in non-mammalian vertebrates. In "Steroid Hormone Receptor Systems" (W. W. Leavitt and J.H. Clark, eds), pp. 441-459. Plenum Press, New York.

97. Salhanick, A.R. and Callard, I.P. (1980). A sex-steroid binding protein in the plasma of fresh-water turtle, *Chrysemys picta*. Gen. Comp. Endocrinol. 42, 163-166.

98. Salhanick, A.R., Vito, C.C., Fox, T.O. and Callard, I.P. (1979). Estrogen-binding proteins in the oviduct of the turtle, *Chrysemys picta*: evidence for a receptor species. Endocrinology 105, 1388-1395.

99. Sandor, T. and A.Z. Mehdi. (1979). Steroids and evolution. In "Hormones and Evolution," Vol. I (E.J.W. Barrington, ed.), pp. 1-72. Academic Press, New York.

100. Schlatterer, B. and Roth, U. (1984). Cell fractionation of lamb uterus in non-aqueous media and nuclear compartmentalization of the oestrogen receptor. Vol. Vet. Med. A, 31, 431-436.

100a. Schimke, R.T., McKnight, G.S., Shapiro, D.J., Sullivan, D., and Palacios, R. (1975). Hormonal regulation of ovalbumin synthesis in the chick oviduct. Rec. Progr. Horm. Res. 31, 175-208.

101. Seal, U.S. and Doe, R.P. (1965). Vertebrate distribution of corticosteroid-binding globulin and some endocrine effects on concentrations. Steroids, 5, 827-841.

102. Sernia, C. (1978). Steroid binding proteins of the Echidna *Tachyglossus aculeatus*, with comparative data for some marsupials and reptiles. Aust. Zool. 20, 87-98.

103. Siiteri, P.K., Murai, J.T., Hammond, G.L., Nisker, J.A., Raymore, W.J., and Kuhn, R.W. (1982). The serum transport of steroid hormones. Rec. Prog. Horm. Res. 38, 457-510.

104. Sloop, T.C., Sherman, B. and Lucier, G.W. (1984). Characterization of cytosolic steroid-binding protein in the liver of the winter flounder, *Pseudopleuronectes americanus*. Gen. Comp. Endocrinol. 55, 157-162.

105. Smirnova, O.V., Smirnov, A.N., Shoshina, S.V. and Rozen, V.B. (1978). Some properties of sex hormone-binding protein from blood serum of the frog *Xenopus laevis* and the detection of estradiol-binding component differing from sex hormone-binding protein in frog liver cytosol. Biokhimiya 43 (8), 1444-1451.

106. Smith, J.S., and Thomas, P. (1984). Cytosolic estrogen receptors in the liver of spotted seatrout, *Cynoscion nebulosus*. Amer. Zool. 24, 116A. (Abstract)

107. Sone, S. and Ishikawa, K. (1982). Progesterone-binding components in cytosol fraction from mature female bullfrog (*Rana catesbeiana*) oviduct. Reports of Faculty of Science, Shizuoka University 16, 81-91.

108. Spelsberg, T.C., Littlefield, B.A., Seelke, R., Martin-Dani, G., Toyoda, H., Boyde-Leinen, P., Thrall, C., and Kon, O. L. (1983). Role of specific chromosomal receptors. Rec. Prog. Horm. Res. 39, 463-513.

109. Sutherland, R.B. and Horgen, P.A. (1977). Effects of steroid sex hormone, antheridiol, on the initiation of RNA synthesis in the simple eukaryote, *Achlya ambisexualis*. J. Biol. Chem. 252, 8812-8820.

110. Szego, C.M. (1984). Mechanisms of hormone actions: Parallels in receptor-mediated signal propagation for steroid and peptide effectors. Life Sciences 35, 2383-2396.

111. Thieulant, M.L., Samperez, S. and Jowan, P. (1981). Evidence for 5α-androstan-$3\beta,17\beta$-diol binding to the estrogen receptor in the cytosol from male rat pituitary. Endocrinology 108, 1552-1560.

112. Tokarz, R.R., Crews, D. and McEwen, B.S. (1981). Estrogen-sensitive progestin binding sites in the brain of the lizard, Anolis carolinensis. Brain Res. 220, 95-105.

113. Tsai, Y.H., Sanborn, B.M., Steinberger, A. and Steinberger, E. (1977). The interaction of testicular androgen receptor complex with rat germ cell and Sertoli cell chromatin. Biochem. Biophys. Res. Comm. 75, 366-372.

114. Turner, R.T., Dickhoff, W.D. and Gorbman, A. (1981). Estrogen binding to hepatic nuclei of Pacific hagfish, Eptatretus stouti. Gen. Comp. Endocrinol. 45, 26-29.

115. Walters, M. (1985). Steroid hormone receptors and the nucleus. Endocr. Rev. 6, 512-543.

116. Watanabe, M., Phillips, K. and Chen, T. (1973). Steroid-receptor in Pseudomonas testosteroni released by osmotic shock. J. Steroid. Biochem. 4, 613-621.

117. Watanabe, M. and Watanabe, H. (1974). Periplasmic steroid binding proteins and steroid transforming enzymes of Pseudomonas testeroni. J. Steroid Biochem. 5, 439-446.

118. Welshons, W.V., Lieberman, M.E. and Gorski, J. (1984). Nuclear localization of unoccupied oestrogen receptors. Nature (London) 307, 745-747.

119. Westley, B. and Knowland. (1978). An estrogen receptor from Xenopus laevis possibly connected with vitellogenin synthesis. Cell 15, 367-374.

120. Westphal, V. (1971). "Steroid Protein Interactions." (F. Gros, ed.). Springer-Verlag, Berlin.

121. Williams, D. and Gorski, J. (1972). Kinetic and equilibrium analysis of estradiol in uterus: A model of binding site distribution in uterine cells. Proc. Natl. Acad. Sci. (USA) 69, 3464-3468.

122. Wingfield, J.C. (1980). Sex steroid-binding proteins in vertebrate blood. In "Hormones, Adaptation and Evolution" (S. Ishii, T. Hirano and M. Wada, eds.), pp. 135-144. Springer-Verlag, Berlin and New York.

123. Wright, W. S. and Frankel, A.E. (1980). An androgen receptor in the nuclei of late spermatids in testes of male rats. Endocrinology 107, 314-318.

124. Xavier, F. (1982). Progesterone in the viviparous lizard Lacerta vivipara: Ovarian biosynthesis, plasma levels, and binding to transcortin-type protein during the sexual cycle. Herpetologica 38, 62-70.

125. Yamada, M., and Miyaji, H. (1982). Binding of sex hormones by male rat liver microsomes. J. Steroid Biochem. 16, 437-446.

126. Yamamoto, K.R., and Alberts, B. (1976). Steroid receptors: Elements for modulation of eukaryotic transcription. Ann. Rev. Biochem. 45, 722-746.

127. Yu, J.Y.-L., Dickhoff, W.D., Swanson, P. and Gorbman, A. (1981). Vitellogenesis and its hormonal regulation in the Pacific hagfish Eptatretus stouti L. Gen. Comp. Endocrinol. 43, 492-502.

128. Yund, M.A., King, D.S. and Fristrom, J.W. (1978). Ecdysteroid receptors in imaginal discs of Drosophila melanogaster. Proc. Natl. Acad. Sci. (USA) 75, 6039-6043.

SEASONAL REPRODUCTION: PATTERNS AND CONTROL

Joan M. Whittier and David Crews

Institute of Reproductive Biology
Department of Zoology
University of Texas at Austin
Austin, TX 78712

I. ADAPTIVENESS OF SEASONALITY

The great diversity of reproductive tactics among and within vertebrate species is a testament to the many levels of biological organization at which selection acts to shape the reproductive process. In this chapter, we define seasonal reproduction as the occurrence of breeding in an individual at a specific time of year. Populations exhibit seasonal reproduction insofar as the individuals of the population breed synchronously. Seasonal reproductive cycles are not necessarily synonymous with annual reproductive cycles; for example, an individual that breeds in the spring every two or three years has a seasonal cycle, but not an annual one. Aseasonal (or nonseasonal) reproduction includes reproductive patterns in which individuals breed continuously or year around. Aseasonal reproduction may also include reproductive patterns in which individuals of species breed in response to specific yet erratic environmental cues; for example, species living in the deserts of mid-western Australia may breed only when occasional and unpredictable rains inundate an area.

Not all aspects of reproduction, including gonadal growth, sex steroid hormone secretion and sexual behavior, are expressed at the same time. In most species that have been studied with respect to the environmental cues and physiological mechanisms influencing reproduction, these three events are functionally associated (Fig. 1). However, many species do not fit this pattern and have a dissociated reproductive tactic in which gonadal activity and sexual behavior are expressed at different times.[23] Other species exhibit a constant reproductive pattern in which gonads are maintained in a

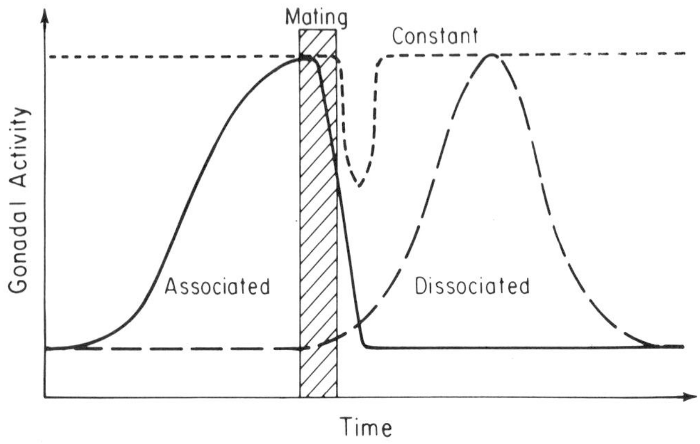

Fig. 1. Three reproductive patterns exhibited in vertebrates. Gonadal activity is defined in terms of the maturation and shedding of gametes and/or the secretion of sex steroid hormones. In individuals exhibiting the associated reproductive pattern, gonadal activity increases immediately prior to mating. In species exhibiting the dissociated reproductive pattern, gonadal activity is minimal at mating. In species exhibiting the constant reproductive pattern, the gonads are maintained at or near maximum activity.

near ready state and sexual behavior is expressed in response to particular environmental cues. Such differing patterns of reproductive activity indicate that a wide variety of physiological mechanisms have evolved to meet the demands of reproduction in different environments.

Patterns of reproduction in animals have evolved to maximize an individual's contribution of genetic information to the next generation[65] (see also review by Angelini and Ghiara[5]). Timing of reproduction in a population is determined by (1) when the most offspring survive and (2) when parents, most often females, are capable of energetically supporting the production of viable young at the least cost to themselves. Further, timing of reproduction not only reflects current conditions but also is influenced by factors regulating the total lifetime production of offspring.[83,92,94,102] Thus, the pattern of reproduction may be influenced by generation time, age to reproduction, life expectancy and age-specific mortality as well as by the predictability of the environment, ecological niche, and body shape and size.

Reproduction involves associated costs, including energetic demands and risks of mortality from predation during courtship or mating,[22] complications of pregnancy and parturition, and nutritional depletion.[16,44,95] Thus, all patterns of reproduction that evolve in some manner maximize the "benefits" of reproduction while minimizing its "costs."

A. Costs of Reproduction

Reproduction by a female requires an investment of a large percentage of her total energy budget. Reproductive effort, the metabolic expense of reproduction (Williams, 1966), has been measured directly in relatively few species, although a large body of literature reports indirect estimates of reproductive effort. When coupled with data on metabolic rates and activity patterns, indirect measures approximate directly measured values. Since most of the energetic cost of producing young appears to be incurred by females, many studies address only the costs of reproduction in the female. However, costs to males may also be high, including expenditures in the production of sperm, but most frequently in the behaviors associated with reproduction.[72,73]

1. Fishes. In fishes, reproductive effort ranges from 9-18% of the female's food intake, with variance due to temperature (see Hirshfield[44]). Most of these costs are incurred during the yolking of eggs.

2. Amphibians. Plethodontid salamanders have high costs of reproduction, primarily devoted to the production of large eggs per unit body mass. Forty-eight percent of the annual energy budget of female *Desmognathus ochrophaeus* (Plethodontidae) is contributed to reproduction; 69% of this annual energy available for reproduction and growth is devoted to the deposition of egg yolk.[35] In frogs and toads the costs of reproduction are probably less than these figures, but no data are available.

3. Reptiles. Indirect measures of reproductive effort are available primarily for lizards, with a few observations on snakes.[17,95,97] A direct measurement of reproductive effort of a tropical anole lizard (*Anolis limifrons*) indicates that 23% of the females' net metabolizable energy is directed to egg production.[4] When coupled with metabolic data, it is estimated that females of three North American lizard species (*Sceloporus jarrovi*, *Uta stansburiana*, and *Sceloporus graciosus*) expend 10-24% of their total energy budget in the production of a single clutch; the energetic costs of reproduction increase tremendously when more than one clutch is produced per season.[95]

B. Relation of Reproductive Cycles to Other Seasonal Cycles

Seasonal cycles in reproduction are often functionally associated with other seasonal cycles. For example, seasonal cycles in fat deposition, migration and hibernation have profound influences on animals and may directly relate to the reproductive cycle (see Gregory[42]). Thus, when attempting to understand reproductive patterns and control mechanisms, it is important to keep in mind that reproduction is inexorably linked to other seasonal cycles.

II. PATTERNS OF REPRODUCTION

A. Timing of Reproduction

Different environmental situations provide unique cues that can be exploited to predict the time when reproduction is most likely to succeed. In benign or predictably harsh (extreme latitude and altitude) environments, photoperiod and temperature are reliable cues associated with the prediction of spring or summer-like conditions. While tropical environments do not undergo seasonal extremes in photoperiod and temperature, often there are regularly recurring wet and dry periods. Some desert environments similarly may have seasonal rainfall. Aquatic environments exhibit seasonal change in illumination, temperature, and salinity, and also may be subject to less well-defined, but predictable changes in parameters associated with seasonal flooding. Other environments appear to lack regularly occurring external cues. These would include areas that have favorable conditions year around, as in some tropical environments. Finally there are environments where favorable conditions for reproduction are unpredictable, including some deserts and certain high montane areas.

In addition to environmental cues, some species utilize endogenous cues to regulate timing of reproduction. Endogenous circannual cycles are defined by the occurrence of complete reproductive cycles on a near annual basis under constant conditions (e.g., 24 h of darkness or nonvarying 10L:14D photoregimens and constant temperature[43]). Interactions between endogenous cycles in sensitivity to environmental stimuli and the regular progression of changes in the environment regulate the period (length) of the cycles.

1. Fishes. Timing of reproduction in fishes may be influenced by a number of factors including photoperiod, temperature, salinity, ionic content of water, water flow, food abundance, nest site availability, and social conditions (reviewed by de Vlaming,[96] Lam,[52] Lowe-McConnell,[61] Bye[13]). Temperate-zone fishes breed in response to short daylengths and cool temperatures (e.g., salmonids) or long daylengths and warm temperatures (e.g., sticklebacks and goldfish). However, temperature can act as a cue independent of photoperiod, as seen in *Fundulus*, the medaka (*Cyprinodontidae*), and in the guppy (*Poecilidae*); see Lam.[52]

In tropical and subtropical fishes, as one might expect, cues other than photoperiod or temperature are utilized to predict optimal timing for reproduction. However, even in areas where annual temperature variation is slight, there exist fishes that rely primarily on temperature as a cue. For example, Asian catfish (Claridae) respond to increasing temperature as well as long photoperiod.[93]

More commonly, however, tropical and subtropical fish may exhibit seasonal reproduction in association with the onset of the rainy season, which is correlated with increases in food availability as well as with habitat.[61] In many species, gonadal maturation often precedes the onset of the rainy season, possibly in response to increasing ionic concentrations in drying pools.[52] Spawning is thought to be triggered by critical substances entering the water from runoff, changes in salinity, or, in some cases, the physical perturbation of the surface water from raindrops.[52] In tropical species, breeding can be found throughout the year although peaks in reproduction occur in association with the rainy season.

Among fishes, reproduction may occur in relation to specific cues available in the environment, making phylogenetic generalizations meaningless. For example, in some tropical teleosts in Venezuela (*Hoplias malebaricus*, *Aequidens pulcher*), populations inhabiting lowland streams breed in association with the rainy season, whereas many populations of the same species in higher gradient streams breed during the dry season (K. Winemiller, personal communication). Evidently, selection pressures in such completely different situations are sufficient so that local populations of the same species can respond to entirely different cues. Such plasticity of response to local conditions even within the same species may explain much of the variation in reproductive patterns among tropical fishes.

2. Amphibians. Since there are consistent differences between anuran and urodelean reproductive cycles, they will be discussed separately. Temperate zone anurans primarily breed in the early spring and summer.[85] In most species that breed in early spring, oogenesis and spermatogenesis occur in summer and the gonads are quiescent in winter (see Rastogi[85]). Anurans breeding in late spring or summer undergo gonadal maturation in winter and early spring. At least two species of anurans (*Hyla raddiana andina* and *Telmatobius schreiteri*)[85] are capable of spermatogenesis in the winter and apparently are not affected by cold temperatures in this regard. Both species occur in the Andes and are subject to unpredictable temperature extremes.

Tropical anurans reproduce seasonally in environments that have a seasonal wet/dry period. For example, Phyllomedusa callidryas and Phyllomedusa dacnicolor in Southern Veracruz, Mexico, breed in May, June and July after sufficient rain has collected in pools.[84] Similarly, Old World anurans Phyrnobatrachus natalensis and Bufo regularis (from Katanga Province, Congo) breed during the rainy season.[48] All of these species anticipate the onset of the wet season with complete gonadal development, but the cues to which they are responding are not known. There is evidence for erratic reproduction in several species of tropical anurans inhabiting environments with little seasonal change (e.g., Rana erythraea[49]).

Reproductive cycles of a number of temperate zone, urodelean amphibians have been described (see Houck[45]). Breeding occurs in the spring in some species (e.g., Notophthalmus,[1] Taricha[67]) or in the fall (e.g., Pseudotriton montanus[12]).

All tropical species of salamanders belong to the New World family Plethodontidae, making a tropical/temperate comparison appropriate. The reproductive cycle of Pseudotriton montanus is typical of a northern temperate zone plethodontid salamander.[12] Males undergo spermatogenesis from April until June, and evacuation of sperm occurs in late July. Sperm are stored in the vasa deferentia from August to September. Although actual observations of mating are rare in these fossorial salamanders, it is believed that breeding occurs in late fall. Gravid females are found from July to November, and females oviposit in late autumn. Females tend their large eggs during winter; the eggs hatch the following spring.

In tropical plethodontid salamanders, reproduction is associated with the rainy season. In Bolitoglossa rostrata, eggs are laid before the dry season and are tended during the dry season. Hatching occurs in early parts of the following wet season.[45] However, males of this species exhibit continuous spermatogenesis, unlike other temperate and tropical plethodontids. The continuous nature of spermatogenesis may indicate that mating can occur at any time and that females store sperm until oviposition.[45]

3. Reptiles. Temperate reptiles generally are restricted to a seasonal pattern of reproduction, in which birth of offspring occurs in late spring or summer. Perhaps most interesting is the variety of ways in which this pattern of reproduction is achieved among reptiles. Many species of snakes and freshwater turtles exhibit dissociated reproductive cycles in which gonadal activity and sexual behavior peak at different times of year.[23]

Many species of tropical lizards and snakes are known to exhibit seasonal cycles of reproduction, including *Anolis*, *Iguana*, *Ctenosaura*, and *Cyclura*, as well as *Agama agama*, *Varanus sp.*, and *Chamaeleo milleri*.[33,34,41] As in fish, seasonal breeding may be superimposed on a low level of reproduction that occurs in the population during the rest of the year. Some tropical lizards appear to be particularly dependent on precipitation. Ovarian development in *Anolis lineatopus* is maximal during the wet season and ceases in most females during the dry season[58] (see also Sexton and Turner,[88] Stamps and Crews[91]). Similarly, the number of hatchlings is maximal in this species in the wet season, although hatchlings are found in every month of the year.[98] Juvenile lizards that hatch during the dry season exhibit significantly reduced growth rates compared to juveniles hatched during the wet season; however, reduced survivorship of hatchlings emerging in the dry season was not established.[98] In contrast, *Anolis* lizard species that inhabit warm, humid environments with slight seasonal change reproduce asynchronously. Thus, in any population reproduction occurs at all times (see Fitch[34]).

B. Synchrony and Regularity of Reproduction

Seasonal patterns of reproduction in a population of animals are a result of a synchronous response of individuals' physiological mechanisms influencing reproduction. Synchrony in a population may result from endogenous circannual controls initiated at birth or puberty or perhaps more commonly, from the influence of exogenous factors on reproductive controlling mechanisms. At the individual level, however, seasonal reproductive patterns may exist that are not necessarily synchronous with reproductive responses of other individuals. Indirect evidence for asynchronous seasonal breeding may come from observations of frequencies of adult individuals breeding at any one time in a population where extended breeding seasons occur. If only a fraction of the population is breeding or in reproductive condition at any one time, then it is possible that individuals are breeding on a seasonal, cyclic basis but not synchronously. This is an important point in regard to extended breeding cycles in tropical species. Do extended cycles result from asynchronous, seasonally predictable cycles in particular individuals or are individuals in fact continuous breeders? Answers to such questions can be obtained only from studies using observations on individuals for time periods of over several years, and to our knowledge no such studies are available.

Tropical species have traditionally been considered to be reproductively acyclic. However, the categorization of a species as acyclic implies that each individual of that species is constantly reproductively active even though it has been emphasized that 'acyclic breeding' refers to the species or population as a whole, and not to the individual animal.[47] In addition, it has been suggested that although fewer species inhabit temperate regions, greater numbers of individuals per species share larger geographical ranges. In tropical and equatorial regions, however, there may be a greater number of species with more restricted geographical ranges, greater inter-specific competition, and fewer individuals of each species.[30] Finally, there are individuals of some species that may breed during other than the maximal period. Thus, in a study designed to determine the reproductive cycle of a temperate-zone species, the probability that a significant number of these aberrant individuals will be included in the sample will be quite low, thereby giving an impression of a sharp onset and decline of the reproductive season. In tropical and equatorial regions, however, the chances of collecting a significant number of these individuals in any given sample is much greater, thereby giving the impression of acyclic reproductive activity.

Not all adult members of populations breed every year, and there is evidence that absence of reproduction in some individuals during particular years is related to the energetic cost of reproduction. The phenomenon is most often observed in females, where the energetic cost of vitellogenesis is great. Two factors may help explain such observations which usually are confined to longer-lived species. First, large body size is correlated with an increase in fecundity in nonmammalian vertebrates. Second, nonmammalian vertebrates exhibit indeterminate growth so that mature individuals continue to grow throughout their lifespan. Assuming high year-to-year survivorship, an individual that does not reproduce in a given year can divert energy into growth and may possibly produce a larger number of offspring the following year. However, an increase in clutch size would be offset by the loss of a complete, albeit smaller, clutch of the previous year.

Bruce[12] and Murphy[71] have offered alternative evolutionary arguments for the occurrence of irregular, nonannual breeding among some females. Bruce[12] suggested that, in addition to energetic savings, selection may favor spreading reproduction over a number of years under certain conditions. Murphy[71] has provided a mathematical model of these ideas, in which it is assumed that females that skip reproduction live longer (see Hirshfield[44]), that there is a long time to maturity, and that mortality of young may be high or intermittent. The model predicts that a female producing young every few years over a long period of time will contribute more sexually reproducing offspring than a female that reproduces every year. Unfortunately, data of these kind are difficult to collect, again involving observations of reproductive activity of individuals over periods of several years.

1. Fishes. Regularity and synchrony of reproduction in fish varies with productivity of the environment. In less productive environments, fewer spawnings occur over a shorter period of time. In highly productive environments, multiple spawning over an extended time period may occur. Transplants of fish from one population to another indicate that, at least in some species, these trends are largely determined by the environment (see Mann et al.[62]).

2. Amphibians. Anuran amphibians are known for their "explosive" breeding habits, particularly those species that occur at high latitudes or altitudes. For example, *Rana cascadae* breeds in pools in the Cascade Mountains of western Oregon as soon as open water is available after spring thawing. At particular ponds, all egg laying occurs in a matter of days.[75] Such synchronous breeding indicates precise regulation of the reproductive cycle.

Many temperate and tropical female plethodontid salamanders are believed to reproduce biennially[45] (see also Bruce[12]). These animals produce large eggs and invest a considerable portion of their annual energy budget into reproduction. While energetic costs may be cited to explain the absence of reproduction in every year, Bruce[12] suggests they are not sufficient. In *Pseudotriton montanus*, 25% of the adult females, as determined by ovarian condition, are nonreproductive every year. However, in a sympatric species, *Pseudotriton ruber*, all females reproduce every year. The factors that regulate frequency of breeding in females of these two species remain enigmatic.

3. Reptiles. The observation of nonreproductive, sexually mature individuals (based on body size) during the breeding season has often been noted in snakes and lizards. Species exhibiting this include *Vipera berus*, *Agkistrodon* sp., *Dipsosaurus dorsalis*, *Crotaphytus wislizenii*, *Sauromalus* sp. and *Eumeces obsoletus* (see Fitch[33]). These observations may be related to energetic restrictions of breeding or possibly to the influence of environmental factors suggested by Murphy.[71] In some species, including *Sceloporus orcutti*,[63] *Uma scoparia*[64] and *Xantusia vigilis*,[66] low reproductive success in populations has been noted in response to low precipitation, and

presumably, limited food resources. Under these conditions, ovarian follicles do not develop (*Uma* and *Xantusia*) and males fail to undergo testicular recrudescence (*Uma*). We recently gathered evidence of year-to-year variation in reproduction of female red-sided garter snakes (*Thamnophis sirtalis parietalis*) from the Interlake region of Manitoba, Canada. Only 10% of females collected in the spring of 1983 underwent ovarian growth, whereas 70% of females collected in the spring of 1984 became vitellogenic.[100] The environmental conditions influencing ovarian growth in this species have yet to be determined, but possibly relate to reduced snow cover in the winter of 1982.

Energetic constraints on particular age classes may influence the frequency of reproduction within a season as well. In some species that produce more than one clutch per year, the number of clutches produced increases with age. For example, *Sceloporus olivaceous* in their first year produce one clutch, in the second year produce two clutches, and in their third year produce three or more. The few individuals that survived four years exhibited a decline in clutch frequency.[10]

III. CONTROL OF SEASONAL REPRODUCTION

A. Pre-programmed (Closed) vs. Labile (Open) Control

Mechanisms that control seasonal reproduction may be viewed as lying on a continuum between two extremes. At one extreme, in pre-programmed or closed control mechanisms, seasonal reproduction is entirely determined by endogenous cycles and is not influenced by external cues. At the other extreme of this continuum lies labile control, in which seasonal reproduction occurs as a result of responses to exogenous cues. In predictable environments, individuals of a population that rely on pre-programmed control mechanisms may have an advantage, because they are ready to reproduce and need not rely on external influences to initiate reproduction. In less predictable environments, pre-programmed control may be disadvantageous if reproductive responses occur under inappropriate conditions. Labile control mechanisms, although they may retard a reproductive response until conditions are favorable, result in fewer risks in unpredictable environments.

Many organisms exhibit intermediate control mechanisms between these two extremes. For example, preparatory events leading to reproduction, including gonadal growth, can be influenced by endogenous factors, but final stages of maturation as well as reproductive behavior can be controlled by integration of specific cues. Another intermediate stage of control is represented by the "hour glass" mechanism, in which external events initiate but have no further influence on an endogenously controlled cycle.[43]

B. Endogenous Cycles

Evidence for endogenous cycles of seasonal reproduction come primarily from the mammalian and avian literature.[43,103] An endogenous, circannual cycle of reproduction is indicated by the presence of reproduction on an approximately annual basis while the organism is maintained on constant conditions (e.g., OL:24D). Several models have been proposed to account for the occurrence of endogenous circannual cycles. In most species that exhibit circannual cycles, external influences (Zeitgebers) serve to entrain these cycles.[43]

1. Fishes. Endogenous rhythms in gonadal cycles have been found in several species of trout maintained on constant, nonvarying temperatures and photoperiods.[8] Other species known to exhibit annual reproductive cycles under constant conditions include the killifish, *Fundulus heteroclitus*, and the Asian catfish, *Heteropneustes fossilis* (see Lam[52]).

Several species of fishes exhibit seasonal variation in sensitivity to exogenous cues such as photoperiod and temperature. In the Asian catfish and stickleback, a seasonal endogenous cycle of photosensitivity is related to circadian rhythms of photosensitivity, although temperature may modify the cycle.[6,93]

2. Amphibians. The best evidence for the presence of endogenous circannual rhythms in amphibians comes from a study of the American toad, *Bufo americanus*.[78] In this species, an annual rhythm in plasma levels of corticosterone occurs when the toads are maintained on constant conditions (12L:12D, 18°C). There is also a daily rhythm in corticosterone levels, with a peak associated with peak levels of locomotor activity. Further, in this species a naturally-occurring seasonal rhythm in corticosterone levels is present.[78] While this suggests there may be an endogenous circannual cycle in the American toad, the relationship of this cycle to seasonal reproduction has not been established.

Refractory periods, or times when an organism is unresponsive to stimulatory environmental cues, have been described in amphibians. For example, several ranid frogs exhibit an endogenously controlled cessation of spermatogenesis during winter months. During this time, long photoperiods and warm temperatures that stimulate spermatogenesis in the spring are ineffective (see Rastogi[85]). A similar refractory period to the effects of photoperiod and temperature occurs in *Plethodon cinereus*.[99] A refractory period may be an endogenous component of an annual reproductive cycle. However, evidence of a refractory period is not sufficient to conclude that the organism has an endogenous, free-running circannual rhythm (see Gwinner[43] for a discussion).

3. Reptiles. Few long-term studies have investigated endogenous circannual cycles in reptiles. What has been cited as the best evidence of such in reptiles (*Cnemidophorus uniparens*[29]) does not rule out alternative hypotheses.[70] Several species of reptiles do exhibit seasonal variation in sensitivity to exogenous cues (see Duvall et al.[32]). This is primarily manifested by a refractory period following the breeding season and extending to mid-winter. This has been documented most extensively in the green anole (*Anolis carolinensis*). Females of this species become unresponsive to long photoperiods and warm temperatures after the breeding season. The lack of response may be due, at least partly, to the presence of large corpora atretica.[27] Male anoles also exhibit refractoriness to exogenous cues during late summer.[53] Again, circadian rhythmicity to exogenous cues have been implicated as mediating the response.[74]

C. Environmental Control

As seen in the discussion of endogenous cycles, there often are interactions between endogenous events and seasonal cues. Thus, environmental factors can alter reproduction during particular gonadal phases. Such considerations complicate experimental approaches to environmental influences on reproduction (see Licht[55] for a discussion of these problems.)

1. Fishes. Fishes respond to a wider variety of cues than any other group of vertebrates. Rather than attempt a comprehensive review of this literature, we cite examples of particular interest. Reproductive cycles of many species of temperate fishes can be influenced by experimental manipulation of photoperiod or temperature. In species that breed during long days, long photoperiods and warm temperatures stimulate gonadal maturation and spawning.[6,9] Similarly, species breeding during short days respond to short photoperiods and cool temperatures.[93]

Fine-tuning of reproductive responses in relation to the environment is seen in fishes. In some species, different parts of the gonadal cycle respond to the same cue in opposite ways. Thus, in the killifish cool temperatures stimulate gametogenesis, but warm temperatures are required for spawning.[60] In other species, both sexes may not respond to the same cues.

For example, female seaperch (*Cymatogaster aggregata*) undergo fol
maturation in response to low temperatures while spermatogenesis in
cued by long photoperiod.[101]

2. *Amphibians*. Temperature is the primary cue that regulates seasonal reproduction in most amphibians.[76,93] Exposing *Rana esculenta* or *Notophthalmus viridescens* to warm temperatures at any time of year induces spermatogenesis.[37,46] Similarly, cool temperatures during summer halt spermatogenesis. Photoperiod is not of primary importance in the regulation of amphibian reproductive cycles; e.g., an increasing (nature-simulating) photoperiod or constant light has no effect on spermatogenesis in *N. viridescens*.[46] Constant darkness also does not influence spermatogenesis in *N. viridescens* or *Rana temporaria*.[15,46,77]

Although rainfall may regulate breeding behavior and the presence of breeding males may stimulate reproductive response in female anurans, studies employing experimental manipulations of these factors in reference to their effects on the reproductive cycle are lacking. Social factors (possibly pheromones) may be important in the control of final stages of reproduction in the rough-skinned newt, *Taricha granulosa*. Ovulation in this species may in fact be induced by the prolonged courtship and mating activity of males.[68]

Nutritional factors also are important in the control of reproduction in Amphibia, particularly in reference to storage of fat.[36] Removal of fat bodies induces testicular regression in *R. esculenta* and *N. viridescens*; replacement of fat tissue (implanted in body cavity) restores gonadal development (see Rastogi[85] and Adams and Rae[2])..

3. *Reptiles*. The role of environmental information in reptilian reproduction has been reviewed extensively.[25,26,32,57] In general, photoperiod and temperature both serve as proximate cues for gonadal maturation. In the green anole, temperature is of primary importance in the induction of gonadal recrudescence whereas photoperiod influences testicular regression.[54,56]

Social cues are important regulators of reptilian reproductive cycles. Male courtship behavior (in particular, extension of the dewlap) facilitates ovarian development in *A. carolinensis*.[18,19] In the red-sided garter snake ovarian recrudescence is correlated with mating; however, a causal effect of mating on ovarian recrudescence has yet to be established.[11,40,100] The stimuli associated with mating that may play an inductive role in ovarian growth can occur either in the spring or in the previous fall. In the an growth is not initiated until the following spring.[100] ain in the form of stored fat, has a profound influence on tiles.[31] Many species of northern temperate snakes vely, often every two or three years rather than on a s. Such facultative reproduction may occur in part as a ce of nutritional reserves on reproduction. For ed-sided garter snake, females of less than average body weights at a given snout-vent length are less likely to reproduce than heavier individuals (J. Whittier and D. Crews, unpublished data). Although low body weight significantly influences ovarian recrudescence in this species, body weight (within normal ranges) does not influence sexual attractivity or receptivity of females (J. Whittier and D. Crews, unpublished data). Since female red-sided garter snakes can store viable sperm for long periods of time (commonly for 9 months), perhaps it is not surprising that ovarian responses and sexual behavior are independently regulated by factors such as body weight. Thus it is conceivable that, although a female would not undergo ovarian recrudescence due to energetic constraints, she could mate with males and maintain a reserve of stored sperm for future reproduction. This indicates that sexual attractivity, receptivity, and ovarian development in this species are differentially affected by body weight and related internal mechanisms of assessment of body condition (see Beach[7]). In a related species, the Mexican garter snake, *Thamnophis melanogaster*, body condition more directly influences the reproductive cycle; nutritional condition of females regulates female attractivity and the incidence of male sexual behavior and testicular cycles.[39]

IV. NEUROENDOCRINE CONTROL OF REPRODUCTION

The specific cues utilized for regulating reproduction, be they internal or external, must be integrated in such a way so that gonadal products are produced and sexual behavior expressed at the appropriate time(s). The neuroendocrine system has evolved to perform this function in a precise, and often complex, way. Redundancy often is found in neuroendocrinological controlling mechanisms, so that removal or addition of particular substances may not always result in a single predictable result. In other words, the notion that one mechanism regulates all reproductive events or even a single reproductive event is overly simplistic.

Basic neuroendocrinological mechanisms regulate reproduction in nearly all groups of vertebrates that have been studied. Sensory systems and specific brain areas serve to integrate external and internal events to bring about changes in the reproductive system. The eyes and the pineal appear to be involved in the influence of photoperiod on the reproductive system. The integration of other cues is less well understood, but may be mediated through the eyes, olfactory, vomeronasal or auditory sensory systems with input to specific brain nuclei. In the brain, specific nuclei in the limbic system that concentrate steroid hormones receive input from these sensory areas, and in turn project to the hypothalamus where both internal and external cues are integrated. The hypothalamus secretes gonadotropin-releasing hormone (GnRH) that influences release of gonadotropin (Gn) from the anterior pituitary. Gonadotropins in turn stimulate gonadal maturation and steroid hormone production. Steroid hormones feed back on the hypothalamus and the pituitary to modulate subsequent function of these organs.

Of particular interest in terms of seasonal control of reproduction is at what level in the brain, pituitary, and gonadal axis do cues impinge to bring about an effect. Does seasonal gonadal recrudescence or regression result from alterations in stimuli provided by neural substrates on the brain, or by a change in the sensitivity of the gonad to unvarying signals? Do both mechanisms play a role?

A. Mechanisms of Endogenous Control

As difficult as it is to establish that an endogenous circannual cycle occurs in any vertebrate, it is not surprising that there is little evidence of the mechanisms of such in fishes, amphibians and reptiles. Both the pineal and melatonin have been implicated in the regulation of gonadal events in fishes,[79] amphibians,[85] and reptiles[57] (see also Chapter 15). Long-term systematic investigations will be required at several levels of reproductive organization before the mechanisms of endogenous control will be understood.

B. Mechanisms of Environmental Control

1. Fishes. A number of studies have examined long term seasonal changes in the neuroendocrine system in fishes.[8,9,79-81] Brain areas that have been implicated in the control of seasonal reproduction in fishes include the nucleus lateralis tuberis (NLT). Lesions in this area induce regression of the gonad in early phases of gonadal growth; however, in mature females, lesions stimulate ovulation.[80]

In all species of teleosts that have been studied, hypophysectomy inhibits annual gonadal cycles. Seasonal increases in Gn as well as daily surges of Gn in fishes maintained on stimulatory conditions for breeding indicate that Gn is responsible for annual gonadal cycles.[9,79-81] Daily peaks in Gn appear to be more important than elevated baseline levels in the stimulation of gonadal function in goldfish.[9]

In addition to Gn being implicated in gonadal maturation, marked changes in Gn occur during spawning (see Billard et al.[9]). Few studies of these phenomenon have been conducted on fishes in the field. However, recently the endocrine changes associated with mass spawning in the white sucker (*Catostomus commersoni*) from central Alberta have been documented.[87,90] In a two-year study, variation in spring conditions resulted in different patterns of the onset and duration of spawning; however, hormonal changes in association with spawning were consistent in both years. In both sexes of white suckers, Gn increases during spawning and decreases afterwards. Unlike goldfish[79] and salmonids,[9] no daily cycle in Gn secretion was found in white suckers. Finally, Gn levels in male white suckers may rise in response to the presence of females, a phenomenon that also occurs in goldfish.[51]

Gonadal steroids have been implicated directly in the induction of late ovarian maturation, ovulation and spermiation in teleosts. A prespawning rise in Gn is correlated with gonadal steroid secretion in rainbow trout and white suckers.[86,87] In female white suckers, plasma testosterone and androstenedione peak prior to ovulation and decline thereafter. At ovulation, the steroid biosynthetic pathway dramatically shifts from testosterone and androstenedione production to the production of progestogens. Similar shifts occur in males, although testosterone does not show a prespawning peak, as seen in females. In males, 11-ketotestosterone (the primary androgenic hormone in many male teleosts) is elevated before spawning and drops afterwards.[87]

2. Amphibians. Neuroendocrinological events in relation to seasonal reproduction have been intensively studied in the rough-skinned newt, *Taricha granulosa*.[69,89,105] In males there is an annual breeding cycle in which spermatogenesis occurs during late summer and early fall. Sexual behavior is initiated during late fall and continues into early spring. Plasma androgens are highest in males during the early breeding season but decline in the late spring even though sexual activity at this time is maximal.[89] Plasma corticosterone levels rise in summer but are depressed during the breeding season.[105] Gonadotropin-releasing hormone appears to be an important regulator of seasonal reproduction in this species.[69,105] A decline of GnRH in specific brain areas is associated with decreasing plasma androgen concentrations in late spring.[105] However, pituitary-gonadal responsiveness to exogenous GnRH is present at this time of year.

A seasonal change in the responsiveness of the brain to the effects of negative feedback of gonadal androgen occurs in the rough-skinned newt

(R.T. Zoeller and F.L. Moore, unpublished data). During early phases of the breeding season, immunoreactive GnRH in specific areas of the newt brain decreases in response to castration; exogenous testosterone restores brain levels of GnRH. However, GnRH in the brain is not affected by castration or testosterone treatment after the end of the breeding season when testicular growth is initiated (R.T. Zoeller and F.L. Moore, unpublished data).

Seasonal changes in arginine vasotocin (AVT) also have been observed in areas of the hypothalamus regulating sexual function of male newts.[106] In particular, AVT in the optic tectum peaks during a time when sexual activity is maximal and brain GnRH and plasma androgens have declined. The significance of this to sexual behavior of male newts is discussed elsewhere in this volume (Chapter 17).

3. Reptiles. Specific brain areas, including the hypothalamus and preoptic areas, are thought to be involved in the regulation of reproduction in reptiles.[21,22] These areas are sites for sex steroid binding, and lesions in these sites interrupt gonadal recrudescence and sexual behavior. Administration of exogenous GnRH stimulates release of Gn in the turtle, the only reptile to our knowledge for which this has been demonstrated.[59] In *Iguana iguana*, repeated pulses of GnRH (chicken II; see Chapter 2) administered to females increases the frequency of aggressive and reproductive behaviors of conspecific males.[82] Injection of GnRH increases the expression of receptive behavior in female lizards (*A. carolinensis*).[3] The concept of neuroendocrinological control of reproductive behavior in the green anole is supported further by evidence that hypophysectomy eliminates both sexual behavior and gonadal function.

In contrast, in the red-sided garter snake (*T. s. parietalis*), an animal that exhibits a dissociated reproductive cycle (see Fig. 1), neither hypophysectomy nor administration of exogenous GnRH or a number of other neuroendocrine substances affects male sexual behavior.[14,24,38] In fact, in this species an eight-week pulse of cold temperature is the only known stimulus that activates sexual behavior in adult males.[14] The effect of hypophysectomy and hypothalamic hormones on gonadal recrudescence in this species is not known, however.

Although there is a large amount of information available on neuroendocrinological control of reproduction in reptiles, little is known of the interactions responsible for normal seasonal changes. For example, female *A. carolinensis* undergo a seasonal cycle in which their ovaries recrudesce in the spring, produce eggs for approximately twelve weeks and then regress in late summer. Sexual behavior is linked to the ovarian cycle and can be stimulated by exogenous hormones (estrogen and progesterone; see Crews,[20,22] Jones et al.,[50] Wu et al.[104]). There is seasonal change in sensitivity of the brain of ovariectomized female green anoles to the stimulatory effects of estrogen and progesterone on sexual receptivity[104] (Table 1). Paralleling this is the seasonal refractoriness of ovarian responses to exogenous FSH.[27] Thus, in the green anole, while there is a suggestion of seasonal changes at the brain and gonad, potential interactions between these events over an entire reproductive season are not understood.

V. SUMMARY AND DIRECTIONS FOR FUTURE RESEARCH

In the 19th century, biology consisted primarily of two separate endeavors: (1) description and cataloguing of biological diversity, and (2) research on basic physiological mechanisms. This dichotomy continued in the first part of the 20th century but expanded to include study of the adaptive significance of life history in a wide variety of organisms and more complex levels of physiological organization. The challenge of the next half century will be the merging of these two traditionally separate disciplines. The study of reproduction offers excellent opportunities to contribute to this end.

In fishes, amphibians and reptiles, a wide variety of seasonal reproductive patterns and controls exist both within and among species. For a selected number of species within these taxa, the specific endogenous and environmental cues that influence reproduction are becoming better understood. Similarly, we are gaining insight into the physiological mechanisms underlying reproductive responses to particular environmental and endogenous cues. Certainly much more information on a wider array of taxa is needed. However, we believe that, in addition to cataloguing and describing reproductive patterns and underlying physiological mechanisms, we must begin to establish an evolutionary context in which to view these problems. Thus, the questions of interest become (1) what patterns of reproduction exist and how are they controlled, (2) under what conditions are particular patterns and control mechanisms expressed,and (3) how do new patterns and control mechanisms arise as a result of natural selection?

Variation in mechanisms controlling reproduction from the taxonomic level to the individual level is an important factor to consider in the context of evolution of reproductive patterns. Variation has traditionally been considered problematic in experimental situations rather than critical information for the understanding of reproductive function in natural

Table 1. Influence of Time of Year on Sexual Receptivity of Female *Anolis carolinensis* Treated with Estradiol Benzoate (EB) Followed by Progesterone (P) or a Control Injection (SSV, Steroid Suspension Vehicle)

Time of year	Treatment	N	% Receptive
Summer	0.4 μg EB/SSV	6	17
	0.4 μg EB/P	8	85
Winter	0.4 μg EB/SSV	26	0
	0.4 μg EB/P	28	0

species. Short-term studies often cannot incorporate long-term variation and thus may color interpretations of the adaptiveness of particular reproductive processes or controls. For example, it is conceivable that in longer-lived species that breed more than once in a lifetime, the relative influence of particular cues on reproduction may vary from year to year, just as they vary from season to season, between sexes and among individuals.

More importantly, since ongoing evolutionary processes are difficult if not impossible to detect, comparative data offer the best opportunity to study evolutionary mechanisms. An appropriate question to be addressed is under what conditions are particular patterns and control mechanisms favored by natural selection? For example, what are the relative benefits of closed versus open control mechanisms in an evolutionary context? These two extremes in mechanisms controlling reproduction may be likened to the familiar specialist/generalist dichotomy of evolutionary biology. As specialists, organisms relying on endogenous mechanisms for the control of reproduction may be adapted to a particular or extreme environment but limited in the capacity to utilize less predictable environmental situations. On the other hand, generalists, represented by organisms relying on environmental cues, may accommodate well to unpredictable or clinal conditions. Similar comparisons can be made between species that respond to a single cue and those that integrate several cues to control reproduction.

Constraints that shape reproductive mechanisms are yet to be identified. Further, little is understood regarding the degree of genetic or developmental plasticity governing mechanisms that control reproduction. While the basic system of communication, involving the brain-pituitary-gonadal axis, appears to be shared by all vertebrate groups (except perhaps some cyclostomes), the specifics of particular components of this system may differ within a species and between closely related species. These issues raise fundamental questions about the evolution of reproductive patterns. Do patterns and controls common among taxa represent evolutionary conservation of ancestral mechanisms, indicating that it is a static system, or is it a case of repeated convergent evolution, indicating a dynamic system? The study of reproductive processes in an evolutionary context will provide insight about diversity and variation in reproduction and at the same time contribute to knowledge of natural selection and evolution of complex processes.

Finally, data gathered on reproduction from free-living populations will be useful in planning potential population and captive management programs of the ever more numerous threatened or rare species of all classes.

VI. ACKNOWLEDGMENTS

We thank K. Winemiller, R.T. Zoeller, and F.L. Moore for allowing our use of their unpublished data. We would also like to thank F.L. Moore and M.C. Moore for discussions on seasonal reproduction.

VII. REFERENCES

1. Adams, E. (1940). Sexual condition in *Triturus viridescens*. III. The reproductive cycle of the adult aquatic form of both sexes. Am. Jour. Anat. 66, 235-275.
2. Adams, E. and E.E. Rae. (1929). An experimental study of the fat bodies in *Triturus (Diemyctylus) viridescens*. Anat. Rec. 41, 181-203.
3. Alderete, M.R., Tokarz, R.R. and D. Crews. (1980). Luteinizing hormone-releasing hormone and thyrotropin-releasing hormone induction of female sexual receptivity in the lizard, *Anolis carolinensis*. Neuroendocrinology 30, 200-205.
4. Andrews, R.M. (1979). Reproductive effort of female *Anolis limifrons* (Sauria, Iguanidae). Copeia, 1979, 620-626.
5. Angelini, F. and Ghiara, G. (1984). Reproductive modes and strategies in vertebrate evolution. Boll. Zool. 51, 121-203.
6. Baggerman, B. (1980). Photoperiodic and endogenous control of the annual reproductive cycle in teleost fishes. *In* "Environmental Physiology of Fishes" (M.M. Ali, ed.), pp. 533-567. Plenum Press, New York.
7. Beach, F.A. (1976). Sexual attractivity, proceptivity, and receptivity in female mammals. Horm. Behav. 7, 105-138.
8. Billard, R., Breton, B., Fostier, A. Jalabert, B. and Weil, C. (1978). Endocrine control of the teleost reproductive cycle and its relation to external factors. *In* "Comparative Endocrinology" (P.J. Gaillard and H.H. Boer, eds.), pp. 37-48. Elsevier/North Holland Biomedical Press, Amsterdam.
9. Billard, R., Fostier, A., Weil, C., and Breton, B. (1982). Endocrine control of spermatogenesis in teleost fish. Canad. J. Fish. Aquat. Sci. 39, 65-79.

10. Blair, W.F. (1960). The rusty lizard, a population study. Univ. of Texas Press, Austin xvi. pp. 1-185.
11. Bona-Gallo, A. and Licht, P. (1983). Effect of temperature on sexual receptivity and ovarian recrudescence in the garter snake, Thamnophis sirtalis parietalis. Herpetologica 39, 173-182.
12. Bruce, R.C. (1975). Reproductive biology of the mud salamander, Pseudotriton montanus, in western South Carolina. Copeia, 1975, 129-137.
13. Bye, V.J. (1984). The role of environmental factors in the timing of reproductive cycles. In "Fish Reproduction: strategies and tactics" (G.W. Potts and R.J. Wooton, eds.), pp. 187-202. Academic Press, London.
14. Camazine, B., Garstka, W., Tokarz, R., and Crews, D. (1980). Effects of castration and androgen replacement on male courtship behavior in the red-sided garter snake (Thamnophis sirtalis parietalis). Horm. Behav. 14, 358-372.
15. Cei, G. (1944). Analisi biogeografica e richerche biologiche e sperminentali sul ciclo sessuale annuo della rane rose d'Europa. Monit. Zool. Ital. 54 (Suppl. 1), 1-117.
16. Clutton-Brock, T.H., Guinness, F.E., and Albon, S.D. (1982). Red Deer Behavior and ecology of two sexes. p. 77 Univ. Chicago Press.
17. Congdon, J.D., Vitt, L.J. and Hadley, N.F. (1978). Parental investment: comparative reproductive energetics in bisexual and unisexual lizards. Genus Cnemidophorus. Amer. Nat. 112, 509-519.
18. Crews, D. (1974). Effects of group stability, male-male aggression, and male courtship behavior on environmentally-induced ovarian recrudescence in the lizard, Anolis carolinensis. J. Zool. 172, 419-441.
19. Crews, D. (1975). Effects of different components of male courtship behavior on environmentally induced ovarian recrudescence and mating preferences in the lizard, Anolis carolinensis. Animal Behav. 23, 349-356.
20. Crews, D. (1979). Endocrine control of reptilian reproductive behavior. In "Endocrine control of sexual behavior" (C. Beyer, ed.), pp. 167-222. Raven Press, New York.
21. Crews, D. (1979). Neuroendocrinology of lizard reproduction. Biol. Reprod. 20, 51-73.
22. Crews, D. (1980). Interrelationships among ecological, behavioral, and neuroendocrine processes in the reproductive cycle of Anolis carolinensis. Adv. Study Behav. 11, 1-74.
23. Crews, D. (1984). Gamete production, sex hormone secretion, and mating behavior uncoupled. Horm. Behav. 18, 22-28.
24. Crews, D., Camazine, B., Diamond, M., Mason, R., Tokarz, R.R., and Garstka, W.R. (1984). Hormonal independence of courtship behavior in the male garter snake. Horm. Behav. 18, 29-41.
25. Crews, D. and Garrick, L.D. (1980). Methods of inducing reproduction in captive reptiles. In "The Reproductive Biology and Diseases of Captive Reptiles" (J. Murphy and J.T. Collins, eds.), pp. 49-70. Society for the Study of Amphibians and Reptiles, Lawrence, Kansas.
26. Crews, D. and Greenberg, N. (1981). Function and causation of social signals in lizards. Amer. Zool. 21, 273-294.
27. Crews, D. and Licht, P. (1974). Inhibition by corpora atretica of ovarian sensitivity to environmental and hormonal stimulation in the lizard, Anolis carolinensis. Endocrinology 95, 102-106.
28. Cuellar, H.S. and Cuellar, O. (1977). Absence of gonadal refractoriness in the lizards Cnemidophorus uniparens and Sceloporus graciosus. Copeia 1977, 185-188.

29. Cuellar, H.S. and Cuellar, O. (1977). Evidence for endogenous rhythmicity in the reproductive cycle of the parthenogenetic lizard, Cnemidophorus uniparens (Reptilia: Teiidae). Copeia 1977, 554-557.
30. Darlington, P.J. (1957). Zoogeography. The geographical distribution of animals. Wiley, New York, p. 179.
31. Derickson, W.K. (1976). Lipid storage and utilization in reptiles. Amer. Zool. 16, 711-723.
32. Duvall, D., Guillette, L.J., Jr., and Jones, R.E. (1982). Environmental control of reptilian reproductive cycles. In "Biology of the Reptilia" (C. Gans and F.H. Pough, eds.), Vol. 13, pp. 201-231. Academic Press, London.
33. Fitch, H.S. (1970). Reproductive cycles of lizards and snakes. Univ. Kansas Mus. Natl. Hist. Misc. Publ. No. 52. pp. 1-247.
34. Fitch, H.S. (1982). Reproductive cycles in tropical reptiles. Occ. Pap. Mus. Nat. Hist. Univ. Kansas. No. 96, 1-53.
35. Fitzpatrick, L. (1973). Energy allocation in the Allegheny Mountain Salamander, Desmognathus ochrophaeus. Ecol. Mongr. 43, 43-58.
36. Fitzpatrick, L.C. (1976). Life history patterns of storage and utiliztion of lipids for energy in amphibians. Amer. Zool. 16, 725-732.
37. Galgano, M. (1932). Prime richerche intorno all influenza della temperatura sui processi spermatogenetici normale ed abberanti di Rana esculenta L. Monit. Zool. Ital. 43, 156-160.
38. Garstka, W.R., Camazine, B., Crews, D. (1982). Interactions of behavior and physiology during the annual reproductive cycle of the red-sided garter snake (Thamnophis sirtalis parietalis). Herpetologica 38, 104-123.
39. Garstka, W.R. and Crews, D. (1982). Female control of male reproductive function in a Mexican snake. Science 217, 1159-1160.
40. Garstka, W.R., Tokarz, R.R., Diamond, M., Halpert, A., and Crews, D. (1985). Behavioral and physiological control of yolk synthesis and deposition in the female red-sided garter snake (Thamnophis sirtalis parietalis). Horm. Behav. 19, 137-153.
41. Gorman, G.C. and Licht, P. (1974). Seasonality in ovarian cycles among tropical Anolis lizards. Ecol. 55, 360-369.
42. Gregory, P.T. (1982). Reptilian hibernation. In "Biology of the Reptilia" (C. Gans and F.H. Pough, eds.), Vol 13, pp. 53-154. Academic Press, London.
43. Gwinner, E. (1971). Circannual systems. In "Biological rhythms, Handbook of behavioral neurobiology" (J. Aschoff, ed.), vol. 4, pp. 391-408. Plenum Press, New York.
44. Hirshfield, M.F. (1980). An experimental analysis of reproductive effort and cost in the Japanese Medaka Oryzias latipes. Ecology 6, 282-292.
45. Houck, L.D. (1977). Reproductive biology of a Neotropical salamander, Bolitoglossa rostrata. Copeia 1977, 70-83.
46. Ifft, J.D. (1942). The effect of environmental factors on the sperm cycle of Triturus viridescens. Biol. Bull. 33, 111-128.
47. Immelmann, K. (1971). Ecological aspects of periodic reproduction. In "Avian Biology Vol. I" (D.S. Farner and J.R. King, eds), pp. 121-157. Academic Press, New York.
48. Inger, R.F. and Greenberg, B. (1956). Morphology and seasonal development of sex characters in two sympatric African toads. J. Morphol. 99, 549-574.
49. Inger, R.F. and Greenberg, B. (1963). The annual reproductive pattern of the frog Rana erythraea in Sarawak. Physiol. Zool. 36, 21-33.
50. Jones, R.E., Guillette, L.J., Jr., Summers, C.H., Tokarz, R.R. and Crews, D. (1983). The relationship among ovarian condition, steroid hormones, and estrous behavior in Anolis carolinensis. J. Exp. Zool. 227, 145-154.

51. Kyle, A.L., Stacey, N.E., and Peter, R.E., (1982). Sexual stimuli associated with increased gonadotropin and milt levels in goldfish. In "Proceedings of the International Symposium on Reproductive Physiology of Fish" (C.J.J. Richter and H.J. Th. Goos, eds.), p. 240. Pudoc: Wageningen, The Netherlands.
52. Lam. T.J. (1983). Environmental influences on gonadal activity in fish. In "Fish Physiology, vol.IX, Reproduction" (W.S. Hoar, D.J. Randall and E.M. Donaldson, eds.), pp. 65-116. Academic Press, New York.
53. Licht, P. (1967). Environmental control of annual testicular cycles in the lizard, Anolis carolinensis. I. Interaction of light and temperature in the initiation of testicular recrudescence. J. Exp. Zool. 165, 505-516.
54. Licht, P. (1971). Regulation of the annual testis cycle by photoperiod and temperature in the lizard, Anolis carolinensis. Ecology 52, 240-252.
55. Licht, P. (1972). Problems in experimentation on annual physiological cycles in reptiles. In "Proceedings of the IVth Symposium on Hibernation and Hypothermia" (F. South, ed.), pp. 681-712. Elsevier Publ. Co., Amsterdam.
56. Licht, P. (1973). Influence of temperature and photoperiod on the annual ovarian cycle in the lizard, Anolis carolinensis. Copeia 1973, 465-472.
57. Licht, P. (1984). Seasonal cycles in reptilian reproductive physiology. In "Marshall's Physiology of Reproduction" (E. Lamming, ed.), pp. 206-282. Churchill Livingstone, Edinborough, U.K.
58. Licht, P. and Gorman, G.C. (1970). Reproductive and fat cycles in Caribbean Anolis lizards. Univ. Calif. Publ. Zool. 95, 1-52.
59. Licht, P., Millar, R., King, J.A., McCreery, B.R., Mendonca, M.T., Bona-Gallo, A., and Lofts, B.. (1984). Effects of chicken and mammalian gonadotropin-releasing hormones (GnRH) on in vivo pituitary gonadotropin release in amphibians and reptiles. Gen. Comp. Endocrinol. 54:89-96.
60. Lofts, B., Pickford, G.E., and Atz, J.W.. (1968). The effects of low temperature and cortisol on testicular regression in the hypophysectomized cyprinodont fish Fundulus heteroclitus. Biol. Bull. 134, 74-86.
61. Lowe-McConnell, R.H. (1979). Ecological aspects of seasonality in fishes of tropical waters. Symp. Zool. Soc. Lond. 44, 219-241.
62. Mann, R.H.K., Mills, C.A. and Crisp, D.T. (1984). Geographical variation in the life-history tactics of some species of freshwater fish. In "Fish Reproduction: strategies and tactics" (G.W. Potts and R.J. Wooton, eds.), pp. 171-185. Academic Press, London.
63. Mayhew, W.W. (1963). Reproduction in the granite spiny lizard, Sceloporus orcutti. Copeia 1963, 144-152.
64. Mayhew, W.W. (1966). Reproduction in the psammophilous lizard, Uma scoparia. Copeia 1966, 114-122.
65. Mayr, E. (1974). Behavior programs and evolutionary strategies. Amer. Sci. 62, 650-659.
66. Miller, M.R. (1954). Further observations on reproduction in the lizard Xantusia vigilis. Copeia 1954, 38-40.
67. Miller, M.R. and Robbins, M.E. (1954). The reproductive cycle in Taricha torosa (Triturus torosus). J. Exp. Zool. 125, 415-445.
68. Moore, F.L., McCormack, C. and Swanson, L. (1979). Induced ovulation: Effects of sexual behavior and insemination on ovulation and progesterone levels in Taricha granulosa. Gen. Comp. Endocrinol. 39, 262-269.
69. Moore, F.L., Miller, L.J., Spielvogel, S.P., Kubiak, T. and Folkers, K. (1982). Luteinizing hormone-releasing hormone involvement in the reproductive behavior of a male amphibian. Neuroendocrinology 35, 212-216.

70. Moore, M.C., Whittier, J.M. and Crews, D. (1984). Environmental control of seasonal reproduction in a parthenogenetic lizard Cnemidophorus uniparens. Physiol. Zool. 57, 544-549.
71. Murphy, G.I. (1968). Pattern in life history and the environment. Amer. Nat. 102, 391-403.
72. Nagy, K.A. (1983). Ecological energetics. In "Lizard Ecology Studies of a Model Organism" (R. Huey, E. Pianka, and T. Schoener, eds.), pp. 24-54. Harvard Univ. Press, Cambridge, Mass.
73. Nakatasuru, K. and Kramer, D.L. (1982). Is sperm cheap? Limited male fertility and female choice in the lemon tetra (Pisces, Characidae). Science 216, 753-754.
74. Noeske, T.A. and Meier, A.H. (1977). Photoperiodic and thermoperiodic interaction affecting fat stores and reproductive indexes in the green anole, Anolis carolinensis. J. Exp. Zool. 202, 97-102.
75. Nussbaum, R.A., Brodie, E.D., Jr. and Storm, R.M. (1983). Amphibians and reptiles of the Pacific Northwest. Univ. Idaho Press, Moscow, pp. 166-170.
76. van Oordt, G.J. (1960). The influence of internal and external factors in the regulation of the spermatogenic cycle in Amphibia. Zool. Soc. London Symp. 2, 29-52.
77. van Oordt, P.G.W.J. (1956). The role of temperature in regulating the spermatogenic cycle in the common frog, (Rana temporaria). Acta Endocrinol. 23, 251-264.
78. Pancak, M.K. and Taylor, D.H. (1983). Seasonal and daily plasma corticosterone rhythms in American toads. Bufo americanus. Gen. Comp. Endocrinol. 50, 490-497.
79. Peter, R.E. (1981). Gonadotropin secretion during reproductive cycles in teleosts: Influences of environmental factors. Gen. Comp. Endocrinol. 45, 294-305.
80. Peter, R.E. and Crim, L.W. (1979). Reproductive endocrinology of fishes: Gonadal cycles and gonadotropin in teleosts. Ann. Rev. Physiol. 41, 323-335.
81. Peter, R.E. and Honatela, A. (1978). Annual gonadal cycles in teleosts: Environmental factors and gonadotropin levels in the blood. In "Environmental Endocrinology" (I. Assenmacher and D.S. Farner, eds.), pp. 20-25. Springer-Verlag, Berlin and New York.
82. Phillips, A., Alexander, N., Karesh, W.B., Millar, R. and Lasley, B.L. (1985). Stimulating male sexual behavior with repetitive pulses of Gn-RH in female Green iguana, Iguana iguana. J. Exp. Zool. 234: 481-484.
83. Pianka, E.R. (1976). Natural selection of optimal reproductive tactics. Amer. Zool. 16, 775-784.
84. Pyburn, W.F. (1970). Breeding behavior of the Leaf-frogs Phyllomedusa callidryas and Phyllomedusa dacnicolor in Mexico. Copeia 1970, 209-218.
85. Rastogi, R.K. (1976). Seasonal cycle in anuran (Amphibia) testis: the endocrine and environmental controls. Boll. Zool. 43, 151-172.
86. Scott, A.P., Sumpter, J.P. and Hardimann, P.A. (1983). Hormone changes during ovulation in the rainbow trout (Salmo gairdneri Richardson). Gen. Comp. Endocrinol. 49, 128-134.
87. Scott, A.P., MacKenzie, D.S. and Stacey, N.E. (1984). Endocrine changes during natural spawning in the white sucker Catostomus commersoni. II. Steroid hormones. Gen. Comp. Endocrinol. 56, 349-359.
88. Sexton, O.J. and Turner, O. (1971). The reproductive cycle of a neotropical lizard. Ecology 52, 159-164.
89. Specker, J.L. and Moore, F.L. (1980). Annual cycle of plasma androgens and testicular composition in the rough-skinned newt, Taricha granulosa. Gen. Comp. Endocrinol. 42, 297-303.

90. Stacey, N.E., MacKenzie, D.S., Marchant, T.A., Kyle, A.L. and Peter, R.E. (1984). Endocrine changes during natural spawning in the white sucker, Catostomus commersoni. I. Gonadotropin, growth hormone, and thyroid hormones. Gen. Comp. Endocrinol. 56, 333-348.
91. Stamps, J.A. and Crews, D. (1976). Seasonal changes in reproduction and social behavior in the lizard Anolis aeneus. Copeia 1976, 467-476.
92. Sterns, S.C. (1976). Life-history tactics: a review of the ideas. Quarterly Rev. Biol. 51, 3-47.
93. Sundararaj, B.I., and Vasal, S. (1976). Photoperiod and temperature control in the regulation of reproduction in the female catfish Heteropneustes fossilis. J. Fish. Res. Board Can. 33, 959-973.
94. Tinkle, D.W. (1969). The concept of reproductive effort and its relation to the evolution of life histories of lizards. Amer. Nat. 103, 501-516.
95. Tinkle, D.W. and N.F. Hadley. (1975). Lizard reproductive effort: caloric estimates and comments on its evolution. Ecology 56, 427-434.
96. de Vlaming, V.L. (1974). Environmental and endocrine control of teleost reproduction. In "Control of Sex in Fishes" (C. Schreck, ed.), pp. 13-83. Sea Grant Publ., Virginia Polytech Inst. and State University, Blacksburg, Va.
97. Vitt, L.J. (1977). Caloric content of lizard and snake (Reptilia) eggs and bodies and the conversion of weight to caloric data. J. Herpetol. 12, 65-72.
98. Vogel, P. (1984). Seasonal hatchling recruitment and juvenile growth of the lizard Anolis lineatopus. Copeia 1984, 747-757.
99. Werner, J.K. (1969). Temperature-photoperiod effects on spermatogenesis in the salamander Plethodon cinereus. Copeia 1969, 592-601.
100. Whittier, J.M. and Crews, D. (1986). Ovarian development in red-sided garter snakes, Thamnophis sirtalis parietalis: relationship to mating. Gen. Comp. Endocrinol. 61, 5-12.
101. Wiebe, J.B. (1968). The effects of temperature and daylength on the reproductive physiology of the viviparous seaperch, Cymatogaster aggregata Gibbons. Can. J. Zool. 46, 1207-1219.
102. Williams, G. (1966). Natural selection, the costs of reproduction, and a refinement of Lack's principle. Amer. Nat. 100, 687-692.
103. Wingfield, J.C. and Farner, D.S. (1980). Control of seasonal reproduction in temperate zone birds. In "Seasonal Reproduction in Higher Vertebrates" (R.J. Reiter and B.K. Follett, eds.), pp. 62-101. Karger, Basel.
104. Wu, J., Whittier, J.M., and Crews, D. (1985). Role of progesterone in the control of female sexual receptivity in Anolis carolinensis. Gen. Comp. Endocrinol. 58, 402-406.
105. Zoeller, R.T. and Moore, F.L. (1985). Seasonal changes in LHRH concentrations in microdissected brain regions of male rough-skinned newts (Taricha granulosa). Gen. Comp. Endocrinol. 58, 222-230.
106. Zoeller, R.T. and Moore, F.L. (1986). Correlation between immunoreactive vasotocin in optic tectum and seasonal changes in reproductive behaviors of male rough-skinned newts. Horm. Behav. 20, 148-154.

THYROID HORMONES AND REPRODUCTION

John F. Leatherland

Department of Zoology
University of Guelph
Guelph, Ontario
Canada N1G 2W1

INTRODUCTION

A. Control of Thyroid Function

The regulation of the hypothalamus-pituitary-thyroid (HPT) axis shown schematically in Figure 1 is best understood in mammals, but appears to be relatively conservative throughout vertebrates. Details of regulation in mammals have been described extensively in several reviews[17,21,29,94] and need not be dealt with at length here.

Briefly, thyrotropin releasing hormone (TRH), a modified tripeptide (L-pyroglutanyl-L-histidyl-L-proline amide), is secreted in the hypothalamus, released into the blood (probably into the vascular bed in the median eminence) and carried into the adenohypophysis by way of the portal system where it binds to receptors on the thyrotropic cells and increases intracellular cyclic AMP levels. TRH stimulates first the release and then the synthesis of the glycoprotein hormone, thyrotropin (TSH). Somatostatin, a cyclic tetradecapeptide synthesized in, and secreted by the hypothalamus, also plays a role in the regulation of thyrotropic cell function by inhibiting TSH secretion although, at least in mammals, the stimulatory role of TRH is considered to be the dominant factor in thyrotropic cell regulation. TSH stimulates the synthesis and release of thyroid hormones (THs) (principally L-thyroxine (T_4)) by the thyroid gland, probably acting <u>via</u> the production of cyclic AMP. Several of the metabolic processes involved in TH synthesis are affected, including iodide transport, synthesis of thyroglobulin, the iodination of the thyroglobulin as it is released into the follicle lumen, and the subsequent release of the THs from the luminal colloid. Although TSH appears to be the most important regulator of thyroid gland activity,

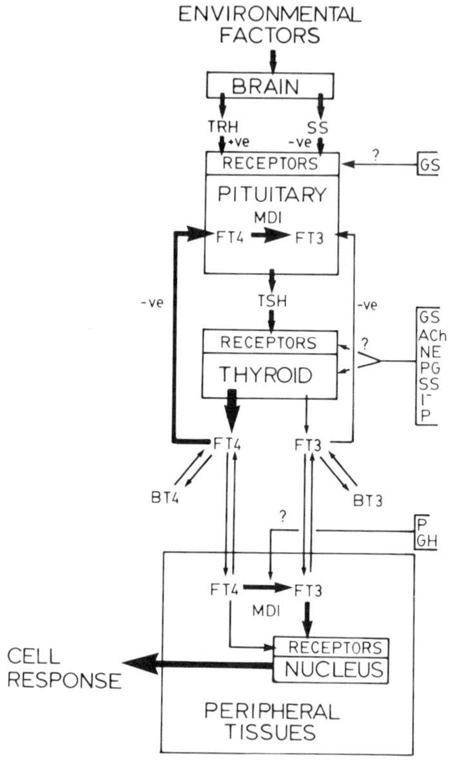

Fig. 1. Generalized scheme of the regulation of the brain-pituitary-thyroid axis. Details of the regulatory mechanism differ among vertebrate classes and full details are not available for most classes. Abbreviations: ACh, acetyl choline; BT_3 and BT_4, protein-bound T_3 and T_4; FT_3 and FT_4, free (unbound) T_3 and T_4; GH, growth hormone; GS, gonadal steroid; I^-, iodide; MDI, enzymatic monodeiodination; NE, norepinephrine; P, prolactin; PG, prostaglandin; SS, somatostatin; TRH, thyrotropin-releasing hormone; TSH, thyroglobulin. Stimulatory (+ve) or inhibitory (-ve) effects are indicated for the main brain-pituitary-thyroid axis only; see original papers for description of proposed actions of ACh, GH, GS, I^-, P and PG.

other factors have been shown to alter thyroid activity *in vivo* and *in vitr* including acetylcholine, catecholamines and prostaglandins.

The major secretory product of the thyroid in mammals is T_4, although some triiodo-L-thyronine (T_3) is released also. Most of the T_3 present in the blood is derived from peripheral enzymatic monodeiodination of the 'pro-hormone' T_4 in tissues such as the liver and kidney. The peripheral monodeiodination rate is finely regulated, but the details of the control are as yet not known; in some fishes, prolactin and growth hormone are thought to alter peripheral T_4 monodeiodination rates.[46,52]

Both T_4 and T_3 appear to be involved in the negative feedback control of pituitary TSH secretion; the thyrotropic cells convert T_4 to T_3. The negative feedback appears to operate mainly at the level of the pituitary gland rather than via the hypothalamus, although a local feedback between TSH and TRH is also suspected.

In primates, the thyroid hormones in the blood are largely (more than 99%) bound to thyroid hormone-binding globulins (TBG), and only the unbound or free hormone is considered to have biological activity. The degree of T_4 and T_3 binding varies considerably in different species and is dependent on the physiological state of the animal. For example, estrogens increase plasma TBG levels and thus alter the ratio between bound and unbound thyroid hormones. The significance of the binding has yet to be fully understood, but it probably provides a relatively stable reserve of hormone which can be supplied as free hormone to the tissue as required.

B. Interpretations of Observations

Human clinical studies have shown that an accurate assessment of thyroid economy is dependent on understanding the dynamic interactions of the several parts of the HPT axis. When a single investigational technique is used to evaluate the activity of the HPT axis (as is the case in the majority of nonhuman studies), the interpretation of results may be controversial.

Estimates of 'thyroid activity' based on thyroid histology or radioiodide uptake rate (RIR) alone are questionable. Increased thyroid epithelial cell height (usually interpreted to indicate increased 'thyroid activity') and RIR are associated with both hypo- and hyperthyroid states; conversely, clinical hypo- and hyperthyroidism are not necessarily associated with alterations in either thyroid histology or RIR. Moreover, the remarkable heterogeneity of the follicle cells both within and between follicles[38] (which probably reflects the histophysiological cycle of follicular activity[23]) has serious implications in both histological and RIR assessment of 'thyroid activity.'[12,36] Moreover, the handling of the animals to administer radioiodide may itself affect RIR measurements.

Competitive binding assays enabling the measurement of plasma or serum thyroid hormone levels have provided a useful means of assessing thyroid economy. However, most assays measure the total T_4 or T_3 concentration, and since only the unbound ('free') hormone fraction appears to have physiological significance, some measure of protein-bound:free hormone has to be made to enable accurate interpretation of blood TH levels. Moreover, blood TH concentrations represent a measure, at a single point in time, of the balance between secretory rate on the one hand and the rate of removal (by receptor binding and subsequent catabolism and by excretion) of the hormone from the circulation on the other. Studies of the kinetics of TH turnover are clearly desirable, but the increased handling of experimental animals inherent in such studies may interfere with 'normal' thyroid economy and thus alter TH kinetics.

Measurement of blood TSH levels provides an important means of monitoring the activity of the HPT axis. Unfortunately, there are currently no validated radioimmunoassays for measurement of TSH in fishes, amphibians or reptiles. Teleostean TSH has been partially characterized in separate laboratories by H. Kawauchi, P. Swanson (unpublished data) and D. Idler and coworkers.[64] Thyrotropins also have been isolated from sea turtles, snapping turtles[56c] and bullfrogs[56a,56b] and have been tested for thyroid-stimulating activity in a number of vertebrate species (see MacKenzie and Licht[56a]).

Except for the role of the THs in anuran metamorphosis and the silvering of the integument in some salmonid teleosts, there are relatively few unequivocal examples of cause-effect relationships between the THs and physiological processes in ectothermic animals. Alterations in thyroid activity which are correlated with a specific physiological event do not *ipso facto* indicate a causative role of the THs; the apparent correlation may simply reflect a response of the thyroid to the altered physiological state.

This is particularly problematical in studies of seasonal cycles of the HPT axis. Seasonal alterations in physiology are associated with changes in many biotic and abiotic factors including ambient temperature, ambient salinity, photoperiod, reproductive status, dietary status, activity (including varied behavioral processes such as mate selection, migration, hibernation), growth rate and energy expenditure, and it is often impossible to unequivocally separate these related events. Thus, cause-effect relationships based on studies of the cyclic activity of the thyroid gland are at best questionable.

Much of the data pertaining to the role of the THs has been derived from studies of hypo- and hyperthyroid animals. Experimentally-induced hypothyroidism (induced experimentally by surgical removal of the thyroid, by chemical thyroidectomy or by radiothyroidectomy) and hyperthyroidism (induced by the administration of THs *via* the diet, by injection, or by immersion of aquatic vertebrates in solutions of TH hormones) have limited value as experimental tools. Surgical thyroidectomy must be complete to prevent compensatory enlargement of any fragments left behind, but parathyroids must be left intact for maintenance of normal calcium homeostasis. Some of the reported effects of goitrogen-induced hypothyroidism on gonadal function have been attributed to the toxic action of these chemicals on the gonads.[46]

Similarly, radiothyroidectomy, used as a replacement for surgical thyroidectomy, in those species (e.g., teleostean fishes) in which the thyroid tissue is not discrete, is limited by the fact that the ovary and other tissues compete with the thyroid for available iodide.[23,46,65a] Consequently, changes in gonadal function may be related to radiation damage *per se* rather than to alterations in thyroid hormone levels.[65a]

The results of experimental hyperthyroidism studies are equally equivocal. Administration of exogenous TH does not necessarily alter the free T_4 and free T_3 levels in the circulation. Target tissues can modify their receptivity to the thyroid hormones, thus effectively 'buffering' the physiological response to the high hormone levels. Moreover, although several studies show a dose-dependent response to the THs, with different responses to 'physiological' compared with 'pharmacological' doses of THs, it is still not clear what constitutes a 'physiological' TH value.

C. Thyroid Function and Reproduction in Mammals

Evidence of an involvement of TH in mammalian reproduction is substantive, but largely indirect. The evidence is derived from three types of studies; 1) reports of sexual dysfunction in patients suffering disturbances in thyroid economy, 2) reports of alterations in gonadal steroid metabolism in response to increased or decreased levels of thyroid hormone, and 3) reports of alteration in thyroid hormone economy in response to increased or decreased levels of gonadal steroid hormones.

The results of clinical studies of hypo- and hyperthyroid patients suggest that an optimal level of THs are required for normal gonadal maturation and maintenance in both men and women.[8,38,45,63] Whether the gonadal dysfunction in hypo- and hyperthyroid subjects is a result of inappropriate metabolic rate of the gonads themselves rather than a role of the THs in other aspects of gonadal function, is not yet clear. However, well-documented changes in gonadal steroid metabolism associated with hypo- and hyperthyroid states (Table 1) cannot be explained on the basis of a metabolic rate effect.

Alterations in TH economy are associated with sexual maturation, pregnancy and lactation in humans, and periods of the life cycle associated with pronounced changes in gonadal steroid hormone secretion (particularly in women; e.g., puberty, pregnancy and menopause) are accompanied by an apparent increased activity of the thyroid gland and an increased incidence of goiters.[8,38,39,45,63] Despite changes in plasma total TH levels during pregnancy and lactation in humans, laboratory and domesticated animals, there is relatively little change in the 'free' TH fraction; the changes in total

Table 1. Some Major Reported Effects of Hypothyroidism and Hyperthyroidism on Gonadal Steroid Metabolism in Humans

	Hyperthyroid	Hypothyroid
Sex-steroid binding protein levels	+	0
Affinity of sex steroid-binding protein for T	+	0
Total serum T, DHT and E levels	+	-
Total AD levels	+	0
Free serum T, DHT and E levels	-	+
Androgen secretion rates	0	+
Conversion rates:		
T to DHT	+	0
T to E	+	0
T to AD	-	+
AD to T	+	0
AD to E	+	0
Metabolic clearance rates:		
T	-	0
DHT	-	0
E	-	0

Data are taken from review papers by Chopra (1975), Gorden and Southren (1977), Mornex and Orgiazzi (1980) and Ingbar and Woeber (1981). Abbreviations: AD, androgens; DHT, dihydrotestosterone; E, estrogens, T, testosterone, +, increase; -, decrease; 0, no data.

TH levels are largely attributable to changes in the TH-binding properties of the plasma.[8,28,38,45]

Estrogens seems to be the most potent steroids affecting thyroid state, although androgens lower serum TBG levels in humans and progesterone reduces hypertrophy of the thyroid in rats.[38] High doses of estrogen given over prolonged periods exert a generally depressive effect on the HPT axis in humans by decreasing TRH secretion with a consequent reduction in TSH and TH secretions. Conversely, estrogen doses which are closer to physiological levels when administered to intact humans or intact or ovariectomized rats increase the sensitivity of the thyrotropic cells to TRH stimulation, probably by increasing the number of TRH receptors. In rats this results in a lowered pituitary TSH content, and in both estradiol-treated humans and rats, serum TSH levels are elevated by estradiol. In humans serum TBG, total T_4 and total T_3 levels are elevated after estrogen treatment. In rats serum total and free T_4 levels are reduced, serum total and free T_3 levels are elevated, and peripheral monodeiodination of T_4 to T_3 is enhanced by estrogen.[4,9,14,17,31,38,39,41,41,83,100,101]

II. THYROID HORMONES AND GONADAL FUNCTIONS

A. Fishes

There is no direct evidence for a role of THs in gonadal development of cyclostomes. Hypophysectomy, which may remove the source of the glycoprotein hormone(s) involved in both gonadal and thyroid regulation, has little[25] or no apparent effect on the reproduction of cyclostomes.[44,56] Indeed, the pituitary control of thyroid function and of reproduction (see Chapters 5 and 10) in cyclostomes is questionable.[19]

In the acipenserines (sturgeons) the THs appear to be essential for gonadal growth and maturation.[2,55] Responsiveness of ovarian follicles to gonadotropins deteriorates if females are retained at low temperatures or for significant periods of time in captive situations. Dettlaf and Davydova[18] showed that T_3 restored the ability of the ovarian follicles to respond to gonadotropin stimulation.

Dodd[20] reviewed the evidence in support of a requirement of TH for gonadal maturation in female elasmobranchs. Surgical thyroidectomy of dogfish (*Scyliorhinus canicula*) in May prevented the annual ovarian recrudescence in the autumn. Ovaries of thyroidectomized fish contained no vitellogenic oocytes, suggesting that either vitellogenin incorporation into the oocytes or hepatic vitellogenesis is TH-dependent.

In many species of teleosts, chemically-induced hypothyroidism led to an inhibition of gonadal recrudescence, or induced gonadal atrophy in both male and female (see reviews).[1,2,23,46,55] In some studies these responses appeared to be correlated with a reduction in gonadotropic cell activity,[57,71,86] suggesting an indirect effect of the THs on gonadal function.

Some of the drawbacks of chemically-induced hypothyroidism discussed above (Section I B) might be overcome by induction of hypothyroidism by other means such as iodide-insufficiency. However, this has proved particularly difficult to bring about in teleosts (J.F. Leatherland and J.W. Hilton, unpublished data). Endemic hypothyroidism of salmon in the Great Lakes of North America is associated with low fecundity of at least one stock (Lake Erie[61]) which exhibits impaired gonadotropic function[51] and low gonadal steroid levels.[48,62]

Evidence of a stimulatory effect of the THs on gonadal function in teleosts is less abundant. Most reports describe either no apparent effect on immature fish or a mild enhancement of gonadal maturation (see reviews).[1,75] Since administration of T_4 can reverse the effect of goitrogens or radiothyroidectomy on gonadal function,[1] the response of the gonad to the two chemical thyroidectomy procedures does not appear to be due solely to radiation or chemical change of the gonadal tissues.

Hurlburt[37] and Sen and Bhattacharya[85] show that THs enhance gonadotropin action on ovarian development *in vivo* in *Carassius auratus* and on ovarian cholesterol mobilization *in vivo* and *in vitro* in *Anabas testudineus*.

B. Amphibians

The evidence of a direct involvement of the THs in gonadal development in amphibians is limited to the work of Sarkar and Rao,[80,81] who showed that chemical and surgical thyroidectomy prevent ovulation of *Rana cyanophlyctis*; T_4 administered to thyroidectomized frogs permitted ovulation *in vivo*, but had no effect on oocytes *in vitro*, suggesting that the THs interact with other factors, possibly gonadotropins. Conversely, T_4 appears to alter the sensitivity of the oocytes in *Ambystoma tigrinum* to other hormones; prolactin increases the LH or progesterone-induced ovarian changes (polar body formation and ovulation) seen in *in-vitro* preparations.[66] When T_4 was administered to the salamanders before removing the oocytes for *in-vitro* incubation, the effect of prolactin *in vitro* was eliminated, thus suggesting a T_4-induced reduction in the sensitivity of the oocyte to prolactin.

C. Reptiles

Hypo- and hyperthyroidism give rise to impaired reproductive function (atresia of ovarian follicles and impaired ovarian growth in females, and a loss of testicular Leydig cells and impaired testicular growth in males) in several reptilian species, which suggests that optimal levels of the THs are required for normal gonadal function.[13,16,26,32,58,65,74,89,90,92] Haldar-Misra and Thapliyal[32] noted seasonal changes in testicular response to thyroid manipulation in the lizard, *Calotes versicolor*. A greater inhibition of testicular size by T_4 was evident in animals tested during the peak of their reproductive cycle, but Leydig cells were atrophied in T_4-treated *C. versicolor* regardless of the time of the reproductive cycle when treatment was applied. Similarly, Nilson[65] showed that Leydig cell numbers were decreased in adders (*Vipera berus*) treated with T_4 at most times of the year except for the time of seasonally maximal thyroid follicle epithelial cell height, when there were no apparent changes in testicular structure.

These observations suggest that the THs exert their effects on gonadal function by interacting with other hormones. The nature of this interaction and the hormones involved are as yet unknown, but Haldar-Misra and Thapliyal[32] suggest an interaction between the thyroid and pituitary somatotropic and gonadotropic hormones.

Lance and Sawin[43] found that ovine TSH increased plasma testosterone levels within one to four hours of administration to *Chrysemys picta* suggesting a possible role of the THs in steroid hormone secretion, although the response also could be attributed to heterogonadotropic activity of the ovine TSH.

III. SEASONAL CHANGES IN THYROID ACTIVITY

A. Fishes

Plitsetskaya et al.[73] reported higher plasma T_4 levels in *Petromyzon marinus* sampled during spermiation and at the beginning of ovulation than in pre-migratory animals; in the females plasma T_4 levels fell and plasma T_3 levels rose at the end of ovulation. Hornsey[33] also reported higher levels of T_4 in the serum of spawning male and female *P. marinus* compared with pre- and post-spawning animals. These data contradict the earlier histological[76] and radioiodide uptake[27] data which were interpreted to indicate a low activity of the thyroid during spawning of *P. marinus*.

As well as sex differences in thyroid gland size in elasmobranchs, there are several reports of apparent increases in thyroid 'activity' (assessed using histological evidence) as the females reach 'puberty' and become sexually mature.[20]

An increase in the activity of the HPT axis (based on histological evidence) has been reported during the time of vitellogenic oocyte growth in several oviparous teleost fishes.[23,45,55,99] Whether these histological changes reflect an increased need of TH during the reproductive event (i.e., a function of the hormones) or simply reflect a gonadal steroid-induced alteration in TH economy has yet to be determined.

In migratory salmonid fishes the histological and radioiodide-uptake data suggest a peak of thyroid activity shortly after the fish migrate into the rivers, and a subsequent decline of activity during migration and spawning.[46] Leatherland and Sonstegard,[49,50] working with *Oncorhynchus kisutch*, were the first to show the marked decline of total and free T_4 and T_3 blood levels during this period. The findings were subsequently confirmed by other authors[3,87,93] and suggest a reduced need for the THs after gonadal maturation. However, the low TH levels do not necessarily mean that T_4 and T_3 are not required at that time.

One possible explanation for the decreases in blood TH levels during the period of ovarian maturation is the binding of THs to vitellogenin, and the incorporation of vitellogenin-TH complexes into developing oocytes.[6a] In hormonally-induced infertile female coho salmon (which do not produce yolky oocytes), the fall in plasma TH levels associated with gonadal maturation of fertile females is not found (W.W. Dickhoff, personal communication).

The TH content of the eggs of several teleostean species is relatively high; TH levels in the yolk decrease throughout the developmental period of the yolksac fry.[6a] Plasma T_4 levels in developing embryos peak at the time of maximal yolk reabsorption, whereas T_3 levels are non-detectable for most of the developmental period, indicating that peripheral monodeiodination of T_4 is not well developed at that stage of development.[6a,88a,88b] The changes in TH content of the yolk may indicate the importance of maternal TH in the development of embryonic/larval stages. Experimental evidence for such a role of THs in teleosts has been reported in several studies.[39a,42a]

In non-migratory species, White and Henderson[97] and Burke and Leatherland[7] reported a fall in plasma T_4 and T_3 levels during the spawning period of *Salvelinus fontinalis* and *Ictalurus nebulosus*, respectively. Eales and Fletcher[24] reported an increasing plasma T_4 level in *Pseudopleuronectes americanus* between January and May when the testes were increasing in size. A similar spring 'surge' of plasma T_4 was found in *Salmo trutta*[72] and *I. nebulosus*.[7] However, this equally could be attributed to, or associated with, the increasing water temperature and/or increased feeding activity occurring at this period. The marked seasonal changes in serum T_3 levels in *I. nebulosis* for the most part paralleled changes in ambient water temperature and were not correlated with ovarian or testicular changes. Osborn et al.[70] found that the seasonal parallel increases in gonadosomatic index (GSI) and plasma T_4 and T_3 levels in sexually mature *S. gairdneri* were evident also in sexually immature animals. However, the increase in plasma T_3 in *P. americanus* in the autumn associated with increasing GSI values may reflect a TH-gonadal relationship.

B. Amphibians

Seasonal changes in thyroid histology, thyroid radioiodide uptake or plasma T_4 and T_3 levels have been reported for *Bufo bufo*, *B. viridis*[78,79] and *Ambystoma tigrinum*.[67] There is little evidence to indicate a correlation between thyroid changes and gonadal function in any of the species, with the possible exception of *A. tigrinum* in which minimum plasma T_4 levels were correlated with the peak GSI in both males and females.[67]

C. Reptiles

Although there are several reports of seasonal changes in thyroid activity of reptiles, for the most part these are closely correlated with fluctuations in environmental temperature.[53] Some studies (e.g., Eakin et al.,[22] Wilhoft[98]) suggest alterations in thyroid activity associated with reproduction rather than ambient temperature. However, because of the basking habit of reptiles, ambient temperatures do not necessarily correlate with core body temperature. Nevertheless, there is evidence of an increased thyroid activity associated with yolk deposition, mating and/or ovulation in several species.[53] Similarly, the males of some species (*Takydromus* sp., *Xantusia vigilis*, *Leiolopisma rhomboidalis*, *Naja naja*, *Natrix maura*, *Clemmys caspica leprosa*) exhibit an apparently increased thyroid activity associated with spermatogenesis and mating.[53] Conversely, Nilson[65] found the lowest thyroid epithelial cell height in *Vipera verus* during the reproductive period, and in *Cnemidophorus sexlineatus* and *Dipsosaurus dorsalis* there was no apparent correlation between plasma TH levels and reproductive events.[40,84] Bona-Gallo et al.[5] reported a seasonal peak in plasma T_4 levels concomitant with vitellogenesis and ovulation in *N. naja*, but there was evidence in the male that plasma T4 levels are inversely correlated with plasma androgen levels. A similar inverse correlation between plasma T_4 and testosterone levels, well-known in some birds,[77] also is suspected in other reptiles (see Section IV C).

There was no evidence of a change in thyroid activity during pregnancy and parturition of live-bearing reptilian species.[53]

IV. THYROID HORMONES AND VITELLOGENESIS

The topic of vitellogenesis is reviewed in Chapter 6, and only the interactions of THs with vitellogenic events are discussed here.

A. Fishes

There are no data pertaining to the possible role of THs in vitellogenesis in cyclostomes.

In elasmobranchs the effect of thyroidectomy on the inhibition of the formation of yolky eggs is suggestive of a role for THs either in hepatic vitellogenesis or in the incorporation of vitellogenin into the oocyte.[20]

The limited available data regarding a possible role of the THs in the process of vitellogenesis in teleostean fishes are contradictory. Saxena and Mani[82] found that thiourea retards gonadal recrudescence and that the production of yolky oocytes is blocked in *Channa punctatus*, and Hurlburt[37] showed that T_4 enhances the gonadotropin-stimulated enlargement of oocytes and enlargement of the hepatosomatic index (HSI) in *Carassius auratus*. Thyroid hormones are thought to enhance vitellogenesis and accelerate gonadal maturation in the guppy, *Poecilia reticulata*.[42a] However, Ng et al.[64] reported that elevation of endogenous thyrotropin, T_3 or T_4 levels, and treatment with thiourea had no effect on yolk incorporation into the ovary of hypophysectomized *Pseudopleuronectes americanus*.

There is evidence of TH effects at the level of the oocyte, but not necessarily associated with vitellogenesis. Dettlaf and Davydova[18] found tht T_3 increased the sensitivity of oocytes of *Acipenser stellatus* to gonadotropic hormones. Hurlburt[37] found a similar permissive action of T_4 on oocyte response to gonadotropic hormones in *C. auratus*.

B. Amphibians

THs appear to be necessary as co-hormones in the estrogen-induced synthesis of vitellogenin by the liver of *Xenopus laevis*.[34,95,96] No other data are available.

C. Reptiles

There is no direct evidence of a TH role in vitellogenesis. However, since hypo- and hyperthyroidism induced a loss of the yolky oocytes from the ovary of *Agama agama*,[26] an involvement of THs in normal oocyte function, possibly vitellogenin synthesis or uptake, is likely.

V. GONADOTROPINS, GONADAL STEROIDS AND THYROID ECONOMY

A. Fishes

Many tetrapod pituitary and chorionic glycoprotein hormones appear to have thyrotropic potency in teleostean fishes,[46,54] possibly reflecting the common molecular origin of these hormones.[19] Thus, some of the apparent changes in the HPT axis during the reproductive process might be a consequence of alterations in blood gonadotropin levels.

In addition, androgens and estrogens have been shown to alter thyroid activity. Estradiol stimulated activity of pituitary thyrotropes, depressed thyroid epithelial cell height and lowered plasma T_4 and T_3 levels in *Anguilla anguilla*.[68,69] Essentially similar results were found in *Salmo gairdneri*, but in addition, the responsiveness of the thyroid to exogenous TSH was reduced in estrogen-treated fish.[47] Olivereau et al.[68] attributes the apparent paradox between depressed thyroid function and increased thyrotrope function in estradiol-treated eels to the estradiol-induced hypercalcemia; a similar hypercalcemia is found in estradiol-treated trout.[47] Whether the decreased sensitivity of the thyroid to TSH stimulation in estradiol-treated trout in due to the steroid effecting changes in the thyroid itself, or the result of membrane changes caused by elevated blood calcium levels remains to be determined.

A depressive effect of estrogens on thyroid economy is by no means universal in teleosts. Reports of increased thyroid activity in estrogen-treated fish have been reported in *Mystus vittatus*, *Poecilia reticulata*, and *Heteropneustes fossilis* (see reviews[23,46,55]). Recent studies by Chakraborti and co-workers indicate a stimulatory effect of gonadotropins (SG-G100 and LH) and estrogens on thyroidal RNA content, thyroid peroxidase activity and thyroid epithelial cell height in *Anabas testudineus*. Ovariectomy effected a lowering of thyroidal RNA content and thyroid peroxidase activity which could be reversed by treatment with estrogen but not by treatment with gonadotropins, suggesting that gonadotropins may exert their effect on the thyroid *via* an effect on gonadal steroid secretion.[10,11]

Androgens appear to stimulate thyroid activity in teleosts (see reviews[23,46,55]). Only one study has been made of the effects of androgens on thyroid hormone kinetics. Hunt and Eales[35,36] showed that testosterone propionate increases T_4 degradation rates, T_4 deiodination rates, T_4 monodeiodination to T_3, thyroidal radioiodide uptake and plasma T_3, iodide and protein levels, but has no effect on the rate of plasma iodide turnover. Reported effects of androgens on thyroid histology and plasma T_4, T_3 levels and thyroid responsiveness to TSH stimulation in rainbow trout have provided no evidence of a marked effect, or conflicting data.[47,59]

B. Amphibians

There are few data regarding the effects of gonadotropin or gonadal steroids on the thyroid economy in amphibians. In *Ambystoma tigrinum* the minimum seasonal levels are associated with peak GSIs in both males and females.[67]

Since peak GSIs are likely associated with high gonadal steroid levels, a possible inverse correlation between TH secretion and sex steroid levels is indicated. The same authors[67] found that the thyroid gland is least sensitive to exogenous TSH in September when the gonads are at or near maximum size and presumably the sex steroid levels are also high. This finding is similar to the antithyroid effects of estrogens seen in some teleosts (Section V A).

C. Reptiles

An inverse relationship between androgens and TH levels, similar to that documented in some birds,[77] has been reported also in some reptiles. Thapliyal et al.[91] found a decreased thyroid activity in testosterone-treated *Natrix piscator*, and Nilson[65] found the lowest thyroid activity (based on histological evidence) during the reproductive period of the adder (*V. berus*). Seasonal changes in plasma or serum TH levels also indicate a suppression of TH levels by androgen. John-Alder[40] found lowest plasma T_4 levels in an iguanid lizard *Dipsosaurus* during the months of peak reproductive activity, and Sellers et al.[84] showed evidence of decreases in plasma T_4 in a hibernating lizard, *Cnemidophorus sexlineatus*, at the time of testicular recrudescence. There is also evidence of a T_4 'rebound' during testicular regression and associated decreases in plasma testosterone levels in both *C. sexlineatus*[84] and *Naja naja*.[5]

VI. SUMMARY AND FUTURE RESEARCH DIRECTIONS

For the most part, evidence of TH involvement in gonadal function of fishes, amphibians and reptiles is fragmentary and speculative. For the most part, the data are restricted to studies of females, and relatively little is known of the TH involvement in testicular physiology. Optimal levels of TH appear to be required for normal growth and maturation of the ovary and testis in all vertebrate classes, with the possible exception of cyclostomes; this apparent need for THs by the gonad is unlikely to be due to metabolic effects of the hormones *per se*. The THs, probably acting in concert with gonadotropins, are implicated in vitellogenesis and/or incorporation of vitellogenin into the oocytes of several fishes, amphibians and reptiles as well as birds. This permissive or interactive role of the THs with other endocrines is suggested in several studies, and it is possible that the principal function of the THs (at least as regards their role in reproduction) is to moderate gonadal responses to other hormones. This may explain why unequivocal demonstration of definitive roles for THs in many classes of vertebrates is rare.

Studies of seasonal thyroid gland cycles, whilst providing signposts of possible involvement of THs in reproduction, are fraught with interpretive problems and deductions of the possible role of the THs in the reproductive process based on such studies have to be made with caution.

Much more needs to be known of the TSH and TH kinetics in all classes of vertebrates. To this end, validated TSH assays for species of fishes, amphibians and reptiles are needed. Some of the apparent cause-effect relationships between THs and gonadal function (e.g., changes in TH economy associated with alterations in gonadal steroid secretion) may be changes which are concomitant with, but not necessary related to, the gonadal events. These alterations in the activity of the HPT axis may be simple compensatory responses necessary for maintenance of the 'euthyroid' state. Investigations of TSH and TH kinetics will undoubtedly shed necessary light on these processes.

VII. ACKNOWLEDGMENT

I am indebted to Ms. Mary Anne Finkbeiner for her help in the preparation of the manuscript.

VIII. REFERENCES

1. Baker-Cohen, K.F. (1961). The role of the thyroid in the development of the platyfish. Zoologica 46, 181-222.
2. Ball, J.N. (1960). Reproduction in female bony fishes. Symp. Zool. Soc. Lond. 1, 105-135.
3. Biddiscombe, S. and Idler, D.R. (1983). Plasma levels of thyroid hormones in sockeye salmon (Oncorhynchus nerka) decrease before spawning. Gen. Comp. Endocrinol. 52, 467-470.
4. Boado, R., Ulloa, E., and Zaninovich, A. (1983). Effects of oestradiol benzoate on the pituitary-thyroid axis of male and female rats. Acta Endocrinol. 102, 386-391.
5. Bona-Gallo, A., Licht, P., Mackenzie, D.S. and Lofts, B. (1980). Annual cycles in levels of pituitary and plasma gonadotropin, gonadal steroids, and thyroid activity in the Chinese cobra (Naja naja). Gen. Comp. Endocrinol. 42,477-493.
6. Bromage, N.R. and Sage, M. (1968). The activity of the thyroid gland of Poecilia during the gestation cycle. Endocrinology 41, 303-311.

6a. Brown, C.L., Sullivan, C.V., Bern, H.A. and Dickhoff, W.W. (1986). Occurrence of thyroid hormones in early developmental stages of teleost fish. Abstr. Tenth Ann. Larval Fish Conf., Miami, May 18-23, 1986.

7. Burke, M. and Leatherland, J.F. (1983). Seasonal changes in serum thyroid hormone levels of the feral brown bullhead, Ictalurus nebulosus Lesueur. J. Fish. Biol. 23, 585-593.
8. Burrow, G.N. (1980). Thyroid function in relation to age and pregnancy. In "The Thyroid Gland" (M. De Visscher, ed.), pp. 215-230. Raven press, New York.
9. Carlson, H.E., Jacobs, L.S. and Daughaday, W.H. (1973). Growth hormone, thyrotropin, and prolactin responses to thyrotropin-releasing hormone following diethylstilbestrol pretreatment. J. Clin. Endocrinol. Metab. 37, 488-490.
10. Charkaborti, P. and Bhattacharya, S. (1982). Influence of gonadotropins and gonadal hormones on climbing perch thyroid nucleic acids. Endokrinologie 80, 213-219.
11. Charkaborti, P., Rakshit, D.K. and Bhattacharya, S. (1983). Influence of season, gonadotropins, and gonadal hormones on the thyroid activity of freshwater perch, Anabas testudineus (Bloch). Can. J. Zool. 61, 359-364.
12. Chan, H.H. and Eales, J.G. (1976). Influence of bovine TSH on plasma thyroxine levels and thyroid function in brook trout, Salvelinus fontinalis (Mitchill). Gen. Comp. Endocrinol. 28, 461-472.

13. Chandola, A., Thapliyal, J.P. and Gupta, S.C. (1973). Male hormone and thyroid function in the Indian garden lizard, *Calotes versicolor*. J. Endocrinol. 57, 325-326.
14. Chen, J.J. and Walfish, P.G. (1978). Effects of estradiol benzoate on thyroid-pituitary function in female rats. Endocrinology 103, 1023-1030.
15. Chopra, I.J. (1975). Gonadal steroids and gonadotropins in hyperthyroidism. Symp. Curr. Conc. Thyroid Disease. Medical Clinics of North America 59, 1109-1121.
16. Choubey, B.J. (1977). Effect of thyroidectomy on sex and accessory organs in a common Indian garden lizard, *Calotes versicolor*. Ind. J. Exp. Biol. 15, 598-600.
17. Demeester-Mirkine, N. and Dumont, J.E. (1980). The hypothalamo-pituitary-thyroid axis. *In* "The Thyroid Gland" (M. De Visscher, ed.), pp. 145-152. Raven Press, New York.
18. Dettlaff, T.A. and Davydova, S.I. (1979). Differential sensitivity of cells of follicular epithelium and oocytes in the stellate sturgeon to unfavorable conditions, and correlating influence of triiodothyronine. Gen. Comp. Endocrinol. 39, 236-243.
19. Dickhoff, W.W. and Darling., D.S. (1983). Evolution of thyroid function and its control in lower vertebrates. Amer. Zool. 23, 697-707.
20. Dodd, J.M. (1983). Reproduction in cartilaginous fishes (Chondrichthyes). *In* "Fish Physiology" Vol. 9A (W.S. Hoar, D.J. Randall and E.M. Donaldson, eds.), pp. 31-95. Academic Press, New York.
21. Dumont, J.E. and Lamy, F. (1980). The regulation of thyroid cell metabolism, function, growth, and differentiation. *In* "The Thyroid Gland" (M. De Visscher, ed.), pp. 153-168. Raven Press, New York.
22. Eakin, R.M., Stebbins, R.C. and Wilhnoft, D.C. (1959). Effects of parietalectomy and sustained temperatures on thyroid of lizard, *Sceloporus occidentalis*. Proc. Soc. Exp. Biol. Med. 101, 162-164.
23. Eales, J.G. (1979). Thyroid functions in cyclostomes and fishes. *In* "Hormones and Evolution," Vol. 1 (E.J.W. Barrington, ed.), pp. 341-436. Academic Press, London.
24. Eales, J.G. and Fletcher, G.L. (1982). Circannual cycles of thyroid hormones in plasma of winter flounder (*Pseudopleuronectes americanus* Walbaum). Can. J. Zool. 60, 304-304.
25. Evennett, P. and Dodd, J.M. (1963). Endocrinology of reproduction in the river lamprey. Nature, Lond. 197, 715-716.
26. Eyeson, K.N. (1970). The role of the thyroid in reproduction of the West African lizard, *Agama agama*. Gen. Comp. Endocrinol. 15, 1-5.
27. Fontaine, M. and Leloup, J. (1950). L'iodemie d'un cyclostome marin (*Petromyzon marinus* L.) au moment de sa migration reproductrice. C.R. Acad. Sci., Paris 230, 1538-1539.
28. Fukuda, H., Ohshima, K., Mori, M., Kobayashi, I. and Greer, M.A. (1980). Sequential changes in the pituitary-thyroid axis during pregnancy and lactation in the rat. Endocrinology 107,1711-1716.
29. Gorbman, A., Dickhoff, W.W., Vigna, S.R., Clark, N.B. and Ralph, C.L. (1983). "Comparative Endocrinology", John Wiley and Sons, New York.
30. Gordon, G.G. and Southren, A.L. (1977). Thyroid-hormone effects on steroid-hormone metabolism. Bull. N.Y. Acad. Med. 53, 241-259.
31. Gross, H.A., Appleman, M.D., Jr., and Nicoloff, J.T. (1971). Effect of biologically active steroids on thyroid function in man. J. Clin. Endocrinol. Metab. 33, 242-253.
32. Haldar-Misra, C. and Thapliyal, J.P. (1981). Thyroid in reproduction of reptiles. Gen. Comp. Endocrinol. 43, 537-542.
33. Hornsey, D.J. (1977). Triiodothyronine and thyroxine levels in thyroid and serum of sea lamprey, *Petromyzon marinus* L. Gen. Comp. Endocrinol. 31, 381-383.

34. Huber, S., Ryffel, G.U. and Weber, R. (1979). Thyroid hormone induces competence for oestrogen-dependent vitellogenin synthesis in developing Xenopus laevis liver. Nature 278, 65-67.
35. Hunt, D.W.C. and Eales, J.G. (1979). Iodine balance in rainbow trout (Salmo gairdneri) and effects of testosterone propionate. J. Fish. Res. Board Can. 36, 282-285.
36. Hunt, D.W.C. and Eales, J.G. (1979). The influence of testosterone propionate on thyroid function of immature rainbow trout, Salmo gairdneri Richardson. Gen. Comp. Endocrinol. 37, 115-121.
37. Hurlburt, M.E. (1977). Role of the thyroid gland in ovarian maturation of the goldfish, Carassius auratus L. Can. J. Zool. 55, 1906-1913.
38. Ingbar, S.H. and Woeber, K.A. (1981). The thyroid gland. In "Textbook of Endocrinology" (R.H. Williams, ed.), pp. 117-248. W.B. Saunders Co., London.
39. Ingenbleek, Y. (1980). Thyroid function in nonthyroid illness. In "The Thyroid Gland" (M. De Visscher, ed.), pp. 499-527. Raven Press, New York.
39a. Inui, Y. and Miwa, S. (1985). Thyroid hormone induces metamorphosis of flounder larvae. Gen. Comp. Endocrinol. 60, 450-454.
40. John-Alder, H.B. (1984). Seasonal variations in activity, aerobic energetic capacities, and plasma thyroid hormones (T_3 and T_4) in an iguanid lizard. J. Comp. Physiol. 154B, 409-419.
41. Kuhn, E.R., Bollen, M. and Darras, V. (1982). Fetal growth inhibition and decreased thyroid activity after injection of oestradiol benzoate into pregnant rats. J. Endocrinol. 93, 55-63.
42. Labrie, F., Borgeat, P., Drovin, J., Beaulieu, M., Lagace, L., Ferland, L. and Raymond, V. (1979). Mechanism of action of hypothalamic hormones in the adenohypophysis. Ann. Rev. Physiol. 41, 555-569.
42a. Lam, T.J. and Loy, G.L. (1985). Effect of L-thyroxine on ovarian development and gestation in the viviparous guppy, Poecilia reticulata. Gen. Comp. Endocrinol. 60, 324-330.
43. Lance, V. and Sawin, C.T. (1979). Effects of TSH and TRH in the turtle, Chrysemys picta. Am. Zool. 19, 964. (Abstract)
44. Larsen, L.O. and Rothwell, B. (1972). Adenohypophysis. In "The Biology of Lampreys", Vol. 2 (M.W. Haridsty and I.C. Potter, eds.), pp. 1-67. Academic Press, London.
45. Leathem, H. (1972). Role of the thyroid. In "Reproductive Biology" (H. Balin and S. Glasser, eds.), pp. 857-876. Excerta Medica, Amsterdam.
46. Leatherland, J.F. (1982). Environmental physiology of the teleostean thyroid gland: a review. Env. Biol. Fish. 7, 83-110.
47. Leatherland, J.F. (1985). Effects of 17β-estradiol and methyl testosterone on the activity of the thyroid gland in rainbow trout, Salmo gairdneri Richardson. Gen. Comp. Endocrinol. (in press).
48. Leatherland, J.F., Copeland, P., Sumpter, J.P. and Sonstegard, R.A. (1982). Hormonal control of gonadal maturation and development of secondary sexual characteristics in coho salmon, Oncorhynchus kisutch, from Lakes Ontario, Erie and Michigan. Gen. Comp. Endocrinol. 48, 196-204.
49. Leatherland, J.F. and Sonstegard, R.A. (1980). Seasonal changes in thyroid hyperplasia, serum thyroid hormone and lipid concentrations, and pituitary gland structure in Lake Ontario coho salmon, Oncorhynchus kisutch Walbaum and a comparison with coho salmon from Lakes Michigan and Erie. J. Fish. Biol. 16, 539-562.
50. Leatherland, J.F. and Sonstegard, R.A. (1981). Thyroid dysfunction in Great Lakes coho salmon Oncorhynchus kisutch (Walbaum): seasonal and interlake differences in serum T_3 uptake and serum total and free T_4 and T_3 levels. J. Fish. Dis. 4, 413-423.

51. Leatherland, J.F. and Sonstegard, R.A. (1984). Pathobiological responses of feral teleosts to environmental stressors: interlake studies of the physiology of Great Lakes salmon. In "Contaminant Effects on Fisheries" (V.W. Cairns, P.V. Hodson and J.O. Nriagu, eds.), pp. 116-149. John Wiley & Sons, New York.
52. Leloup, J. and de Luze, A. (1980). Prolactine et hormones thyroidiennes chez l'anguille (Anguilla anguilla L.). C.R. Acad. Sci., Paris 291, 87-90.
53. Lynn, W.G. (1970). The thyroid. In "Biology of the Reptilia" (C. Gans and T.S. Parsons, eds.), pp. 201-234. Academic Press, New York.
54. Mackenzie, D.S. (1982). Stimulation of the thyroid gland of a teleost fish, Gillichthys mirabilis, by tetrapod pituitary glycoprotein hormones. Comp. Biochem. Physiol. 72A, 477-482.
55. Matty, A.J. (1960). Thyroid cycles in fish. Symp. Zool. Soc. Lond. 2,1-15.
56. Matty, A.J., Tsuneki, K., Dickhoff, W.W. and Gorbman, A. (1976). Thyroid and gonadal function in hypophysectomized hagfish, Eptatretus stouti. Gen. Comp. Endocrinol. 30, 500-516.
56a. MacKenzie, D.S. and Licht, P. (1984). Studies on the specificity of thyroid responses to pituitary glycoprotein hormones. Gen. Comp. Endocrinol. 56, 156-166.
56b. MacKenzie, D.S., Licht, P. and Papkoff, H. (1978). Thyrotropin from amphibian (Rana catesbeiana) pituitaries and evidence for heterothyrotropic activity of bullfrog luteinizing hormone in reptiles. Gen. Comp. Endocrinol. 36, 566-574.
56c. MacKenzie, D.S., Licht, P. and Papkoff, H. (1981). Purification of thyrotropin from the pituitaries of two turtles: the green sea turtle and the snapping turtle. Gen. Comp. Endocrinol. 45, 39-48.
57. McBride, J.R. and Van Overbeeke, A.P. (1975). Effects of thiourea treatment of sexually maturing and gonadectomized male sockeye salmon (Oncorhynchus nerka). J. Fish. Res. Board Can. 32, 11-19.
58. Mellish, C.H. and Meyer, R.K. (1937). The effects of various gonadotropic substances and thyroxine on the ovaries of horned lizards (Phrynosoma cornutum). Anat. Rec. 69, 179-189.
59. Milne, R.S. and Leatherland, J.F. (1980). Changes in plasma thyroid hormones following administration of exogenous pituitary hormones and steroid hormones to rainbow trout (Salmo gairdneri). Comp. Biochem. Physiol. 66A, 679-686.
60. Mornex, R. and Orgiazzi, J.J. (1980). Hyperthyroidism. In "The Thyroid Gland" (M. De Visscher, ed.), pp. 279-362. Raven Press, New York.
61. Morrison, P.F., Leatherland, J.F. and Sonstegard, R.A. (1985). Comparative fecundity and egg survival of coho salmon, (Oncorhynchus kistuch Walbaum) from Lakes Ontario, Erie and Michigan. Can. J. Zool. 63, 1096-1100.
62. Morrison, P.F., Leatherland, J.F. and Sonstegard, R.A. (1985). Plasma cortisol and sex steroid levels in Great Lakes coho salmon (Oncorhynchus kisutch Walbaum) in relation to fecundity and egg survival. Comp. Biochem. Physiol. 80A, 61-68.
63. Myant, N.B. (1964). The thyroid and reproduction in mammals. In "The Thyroid Gland" (R. Pitt-Rivers and W.R. Trotter, eds.), pp. 283-302. Butterworths, London.
64. Ng, T.B., Idler, D.R. and Eales, J.G. (1982). Pituitary hormones that stimulate the thyroidal system in teleost fishes. Gen. Comp. Endocrinol. 48, 372-389.
65. Nilson, G. (1982). Thyroid activity and experimental evidence for its role in reproduction in the adder Vipera berus. Gen. Comp. Endocrinol. 47, 148-158.

65a. Norris, D.O. (1985). "Vertebrate Endocrinology," 2nd Ed., p. 167. Lea and Febiger, Philadelphia.
66. Norris, D.O. and Duvall, D. (1981). Hormone-induced ovulation in *Ambystoma tigrinum*: influence of prolactin and thyroxine. J. Exper. Zool. 216, 175-180.
67. Norris, D.O., Duvall, D., Greendale, K. and Gern, W.A. (1977). Thyroid function in pre- and postspawning neotenic tiger salamanders (*Ambystoma tigrinum*). Gen. Comp. Endocrinol. 33, 512-517.
68. Olivereau, M., Leloup, J., De Luze, A. and Olivereau, J. (1981). Effet de l'oestradiol sur l'axe hypophyso-thyroidien de l'anguille. Gen. Comp. Endocrinol. 43, 352-363.
69. Olivereau, M. and Olivereau, J. (1979). Effect of estradiol-17β on the cytology of the liver, gonads and pituitary, and on plasma electrolytes in the female freshwater eel. Cell Tiss. Res. 199, 431-454.
70. Osborn, R.H., Simpson, T.H. and Youngson, A.F. (1978). Seasonal and diurnal rhythms of thyroidal status in the rainbow trout, *Salmo gairdneri* Richardson. J. Fish Biol. 12, 531-540.
71. Pandey, S. and Leatherland, J.F. (1970). Comparison of the effects of Methallibure and thiourea on the testis thyroid, and adenohypophysis of the adult and juvenile guppy, *Poecilia reticulata* Peters. Can. J. Zool. 48, 445-450.
72. Pickering, A.D. and Christie, P. (1981). Changes in the concentrations of plasma cortisol and thyroxine during sexual maturation of the hatchery-reared brown trout, *Salmo trutta* L. Gen. Comp. Endocrinol. 44, 487-496.
73. Pisetskaya, E., Bower, S.A. and Gorbman, A. (1984). Profiles of plasma thyroid hormones and insulin during the upstream spawning migration of the sea lamprey, *Petromyzon marinus*. Gen. Comp. Endocrinol. 53, 451.
74. Plowman, M.M. and Lynn, W.G. (1973). The role of the thyroid in testicular function in the Gecko, *Coleonyx variegatus*. Gen. Comp. Endocrinol. 20, 342-346.
75. Rangneker, P.V. and Latey, A.N. (1977). Influence of the thyroid gland on gonadal activity in the teleost, *Tilapia mossambica* (Peters). J. Anim. Morphol. Physiol. 24, 344-352.
76. Reese, A.M. (1902). Structure and development of the thyroid gland in *Petromyzon*. Proc. Acad. Nat. Sci., Phil. 54, 85-112.
77. Ringer, R.K. (1976). Thyroids. *In* "Avian Physiology" (P.D. Sturkle, ed.), pp. 349-358. Springer-Verlag, New York.
78. Rosenkilde, P. (1982). The role of thyroid hormone in adult amphibians. *In* "Phylogenic Aspects of Thyroid Hormone Actions" (Institute of Endocrinology, Gunna University, ed.), pp. 91-106. Centre for Academic Publ., Tokyo.
79. Rosenkilde, P. and Jorgensen, I. (1977). Determination of serum thyroxine in two species of toads: variation with season. Gen. Comp. Endocrinol. 33, 566-573.
80. Sarkar, H.B.D. and Rao, M.A. (1979). Effect of thiourea on ovulation *in vitro* in the skipper frog, *Rana cyanophlyctis* (Schn.). Acta Endocrinol. 60, 173-176.
81. Sarkar, H.B.D. and Rao, M.A. (1971). Effect of thyroidectomy and and administration of thyroxine on ovulation and spawning *in vivo*, *in vitro* and in transplantation in the skipper frog, *Rana cyanophlyctis* (Schn.). Gen. Comp. Endocrinol. 16, 594-598.
82. Saxena, P.K. and Mani, K. (1979). Ovarian recrudescence in freshwater teleost *Channa punctatus* (Bl.), during thiourea treatment. Ind. J. Exp. Biol. 17, 1301-1304.
83. Segal, J., Troen, B.R. and Ingbar, S.H. (1982). Influence of age and sex on the concentrations of thyroid hormone in serum in the rat. J. Endocrinol. 93, 177-181.

84. Sellers, J.C., Wit, L.C., Ganjam, V.K., Etheridge, K.A. and Ragland, I.M. (1982). Seasonal plasma T_4 titers in the hibernating lizard Cnemidophorus sexlineatus. Gen. Comp. Endocrinol. 46, 24-28.
85. Sen, S. and Bhattacharya, S. (1981). Role of thyroxine and gonadotropins on the mobilization of ovarian cholesterol in a teleost Anabas testudineus (Bloch). Ind. J. Exp. Biol. 19, 408-412.
86. Singh, R., Raizada, R.B. and Singh, T.P. (1977). Effects of some antithyroid drugs on the pituitary-thyroid-gonad axis in a freshwater catfish, Heteropneustes fossilis (Bloch). Gen. Comp. Endocrinol. 31, 451-456.
87. Sower, S.A. and Schreck, C.B. (1982). Steroid and thyroid hormones during sexual maturation of coho salmon (Oncorhynchus kisutch) in seawater or fresh water. Gen. Comp. Endocrinol. 47, 42-53.
88. Stolk, A. (1951). Histo-endocrinological analysis of gestation phenomena in the cyprinodont Lebistes reticulatus Peters. I. Thyroid activity during pregnancy. Acta Morph. Neerlands-Scand. 1, 30-40.
88a. Sullivan, C.V. and Dickhoff, W.W. (1986). Thyroid hormones, gill (Na^+-K^+)-ATPase and seawater tolerance in salmon embryos. Amer. Zool. 25, Abstract 430.
88b. Sullivan, C.V., Dickhoff, W.W., Brewer, S.D. and Johnston, G.P. (1983). Plasma thyroid-hormone concentrations and gill (Na^+-K^+)-ATPase activities in postemergent pink salmon. Trans. Am. Fish. Soc. 112, 825-289.
89. Thapliyal, J.P. (1980). Thyroid in reptiles and birds. In "Hormones, Adaptation and Evolution" (S. Ishii, T. Hirano and M. Wada, eds.), pp. 241-250. Springer-Verlag, New York.
90. Thapliyal, J.P. and Chandola, A. (1973). Seasonal variation in thyroid hormonogenesis in the Indian garden lizard, Calotes versicolor. J. Endocrinol. 56, 451-462.
91. Thapliyal, J.P., Chandola, A., Kumar, D.S. and Oommen, O.V. (1974). Testosterone versus thyroid hormone in the regulation of oxidative metabolism in two species of reptiles. J. Endocrinol. 63, 419-420.
92. Thapliyal, J.P., Gupta, S.C. and Garg, R.K. (1974). Effects of thyroidectomy on the chequered water-snake, Natrix piscator. J. Endocrinol. 60, 517-524.
93. Ueda, H., Hiroi, O., Hara, A., Yamauchi, K. and Nagahama, Y. (1984). Changes in serum concentrations of steroid hormones, thyroxine, and vitellogenin during spawning migration of the chum salmon, Oncorhynchus keta. Gen. Comp. Endocrinol. 53, 203-211.
94. Walfish, P.G. (1981). Thyroid physiology and pathology. In "Pediatric Endocrinology" (R. Collu, J. Ducharme and H. Guyda, eds.), pp. 357-431. Raven Press, New York.
95. Wangh, L.J. (1982). Glucocorticoids act together with estrogens and thyroid hormones in regulating the synthesis and secretion of Xenopus vitellogenin, serum albumin, and fibrinogen. Dev. Biol. 89, 294-298.
96. Wangh, L.J. and Schneider, W. (1982). Thyroid hormones are corequisites for estradiol-17β in vitro induction of Xenopus vitellogenin synthesis and secretion. Devel. Biol. 89, 287-293.
97. White, B.A. and Henderson, N.E. (1977). Annual variations in circulating levels of thyroid hormones in brook trout, Salvelinus fontinalis, as measured by radioimmunoassy. Can. J. Zool. 55, 475-481.
98. Wilhoft, D.C. (1963). Seasonal changes in the gonads and thyroid in a tropical lizard, Leiolopisma rhomboidalis. Diss. Abstr. 24, 1304-1305.
99. Young, G. and Ball, J.N. (1983). Ultrastructural changes in the adenohypophysis during the ovarian cycle of the viviparous teleost Poecilia latipinna. II. The thyrotrophic cells and the thyroid gland. Gen. Comp. Endocrinol. 51, 24-38.

100. Zaninovich, A.A. (1981). Direct and indirect effects of estrogens on thyroidal economy of man. In "Physiopathology of Endocrine Diseases and Mechanisms of Hormone Action" (R.J. Soto, A. DeNicola and J. Blauier, eds.), pp. 95-102. Alan R. Liss, Inc., New York.

101. Zaninovich, A.A., Boado, R., Ulloa, E., Bromage, N.R. and Matty, A.J. (1982). Inhibition of thyroidal iodine release by oestrogens in euthyroid subjects. Acta Endocrinol. 99, 386-392.

PINEAL INVOLVEMENT IN SEASONALITY OF REPRODUCTION

William A. Gern, Jeanne M. Nervina and
Shelley S. Greenhouse

Department of Zoology and Physiology
University of Wyoming
Laramie, Wyoming 82071

I. INTRODUCTION

The pineal organ responds to changes in illumination electrophysiologically (in most nonmammalian vertebrates) and hormonally (in nearly all vertebrates). Very little is known about the electrophysiological information originating in the pineal organ, but pineal physiologists are beginning to understand how the hormonal messages are used by animals to regulate physiological events like maintenance of rhythms and seasonal reproduction.

In this article we will review the literature concerning the pineal and reproduction in nonmammalian vertebrates. In keeping with other chapters in this volume, an overview of the mammalian system is presented followed by a class-by-class review. Other valuable review articles on this topic have appeared, and the reader is referred to them.[16,44,46,47,70]

II. PINEAL AND REPRODUCTION

A. Mammalian Overview

Several extensive review articles examining pineal-gonadal interactions in mammals have appeared over the past 15 years. From these it is apparent that the wealth of knowledge currently existing on this topic, most of which accrued since 1975, stems from the initial work of Hoffman et al.[28] concerning photoperiodic control of Syrian hamster gonadal cyclicity. Our view of the history of this literature stems mostly from the comprehensive review by Reiter,[50] which provides an excellent synopsis of the development in this field. The first work implicating the pineal directly in seasonal control of reproduction stems from Hoffman and Reiter[29,30] and Reiter et al.[51] who demonstrated that pinealectomy blocks gonadal collapse in hamsters (both male and female) exposed to short-photophase conditions. While a considerable amount of information detailing hormonal response to pineal and photic manipulation is available, we will discuss the more generalized action of the pineal and its hormones; these are the antigonadotropic and progonadal effects. Several recent review articles outlining the pineal-gonadal axis have appeared, and the reader is referred to these;[3,23,31,39,49,50,56] however, most of the information available is from species of hamsters, laboratory rats and sheep.

In their initial studies, Hoffman and Reiter[29,30] and Reiter et al.[51] demonstrated that pinealectomy would block short-photophase-induced gonadal collapse. Thus, when present (the normal condition), the pineal is antigonadotropic in function during short photophases. This basic experiment has been repeated numerous times by many laboratories using a variety of mammalian species (see reviews cited above). A search for the antigonadotropic pineal product ensued. Melatonin, a secretory product of the pineal, was an excellent candidate for the antigonadotropic compound (c.f., Cardinali,[5] Reiter[50]). Initially, data confirming such a role for melatonin were engimatic (c.f., Reiter[50]). This was due to the time of day melatonin was administered, the dosage given, and because of the mode of application (either injection or implantation of beeswax or silastic capsules). Once it was determined, however, that reproducible gonadal collapse could be attained in long-photophase-maintained, intact hamsters by an afternoon injection of melatonin,[58] interest in melatonin as an anti-gonadotropic agent increased. Currently, melatonin is considered the prime candidate for the antigonadotropic pineal hormone. Similar studies have been carried out in other photoperiodic rodents and in sheep.[3] While afternoon injections of melatonin elicit gonadal collapse in Syrian hamsters, they induce the onset of seasonal polyestrus in sheep--a short-photophase breeder (c.f., Bittman[3]). Thus, melatonin, if present in the latter part of the day, elicits opposite effects in two species of mammals.

In hamsters, melatonin has, in addition to an antigonadotropic action, a progonadal (also called counter-antigonadotropic) action. Reiter et al.[51] demonstrated that hamsters bearing beeswax implants that continuously release melatonin do not undergo gonadal collapse when exposed to continuous darkness. Also, melatonin implants releasing only a few micrograms per day could cause this progonadal effect. Several investigations have confirmed these initial studies (c.f., Reiter,[50] Goldman,[23] Stetson and Watson-Whitmyre,[56] Bittman[3]).

In order to explain these seemingly divergent actions for melatonin,

Reiter[49,50] proposed that melatonin regulates the state of its own receptors in a sensitive reproductive center such as the hypothalamus. According to this hypothesis melatonin, normally synthesized and secreted at night, binds to its receptors and renders them insensitive to melatonin for a period of time. Thus, it down-regulates its own receptor. Reiter's hypothesis explains why morning melatonin injections are ineffective antigonadotropically; the receptors are down-regulated and not responsive. By late afternoon, the down-regulation normally caused by a nocturnal melatonin pulse has disappeared, and the receptors are responsive to melatonin. Thus, afternoon injections of melatonin elicit the antigonadotropic effect. Implants constantly releasing melatonin act to continually down-regulate melatonin receptors. Therefore, they are insensitive to either afternoon hormone injection or the natural nocturnal melatonin pulse. This hypothesis does explain both antigonadotropic and progonadal actions of melatonin in mammals, and it has withstood some testing (c.f., Reiter[50]). Clear demonstration of actual melatonin receptor down-regulation, however, is needed.

We find the above hypothesis appealing for another reason. Release of most hormones is controlled by negative feedback either by the hormone itself or by another elicited product. The vast majority of information indicates, however, that this is not the case for melatonin. That is, melatonin does not feed back to inhibit further melatonin synthesis. We think this is because melatonin's exapted actions (c.f., Gern and Karn[22]) are the result of its "ability" to time either night onset or length or both. Thus, negative feedback would interfere with the timing function that also is the product of natural selection. Therefore, physiological mechanisms may have developed to turn off the information available in a nocturnal pulse of melatonin. That is, the cellular receptors are down-regulated rather than the production of the hormone itself. This, too, awaits testing, but it can explain some of our present knowledge of melatonin biology.

B. Agnathans

The pineal organ appears to influence gonadal development in male and female lamprey (*Lampetra fluviatilis*). Joss[34] observed the effects of pinealectomizing groups of juvenile animals during November, December or January on testicular and ovarian development. In both males and females, pinealectomy delayed gonadal maturation. In the case of ovarian development, the observed delay due to pinealectomy was slight, but nonetheless significant. Spermatogenesis was affected similarly. In April, when all animals in the experiment were sacrificed, pinealectomized lamprey had

Table 1. Pinealectomy, Photoperiod, Temperature, Season and Reproductive Parameters in Teleosts

Species	Photoperiod	Temperature	Time of Year/Cycle	Pinealectomy Effects*	Reference
Goldfish (*Carassius auratus*)	8L/16D	14°	May	Increased ovarian and testicular GSI	19
	8L/16D	13°	April; June-August; October-December	No effect on ovarian or testicular GSI	
	8L/16D	23°	February	No effect on ovarian or testicular GSI	16
	8L/16D	23°	March	No effect on ovarian GSI	
	8L/16D	23°	April	Decreased ovarian and testicular GSI	
	8L/16D	23°	May	Increased testicular GSI; no effect on ovarian GSI	
	8L/16D	23°	June-July	No effect on ovarian GSI	
	16L/8D	23°	February	Decreased ovarian and testicular GSI	
	16L/8D	23°	May	Increased ovarian GSI; no effect on testicular GSI	
	16L/8D	23°	June-July	Decreased ovarian GSI	
	15.5L/8.5D	24°	November	No effect on pituitary or plasma Gn; no effect on ovarian GSI	71
	15.5L/8.5D	22°	March	Increased pituitary Gn; no effect on plasma Gn; decreased ovarian GSI	

Species	Photoperiod	Temperature	Time of Year/Cycle	Pinealectomy Effects	Reference
Goldfish (continued)	8L/16D	20°	March	Increased pituitary Gn; decreased plasma Gn; no effect on ovarian GSI	
	16L/8D	21°	November (regressed-early recrudescence)	No effect on GSI; no effect on serum Gn	33
	16L/8D	21°	Early March (mature)	Decreased GSI	
	16L/8D	21°	Late April (mature)	Decreased GSI and serum Gn	
	8L/16D	21°	Mid-March (mature)	No effect on GSI or serum Gn	
Golden Shiner (*Notemigonus crysoleucas*)	15.5L/8.5D	25°	January (late preparatory)	Decreased testicular GSI; no effect on ovarian GSI	12,15
	15.5L/8.5D	25°	Late February (preparatory)	Increased Gn activity	
	15.5L/8.5D	25°	March (early pre-spawning)	Decreased ovarian and testicular GSI	
	15.5L/8.5D	25°	Early April (mid-pre-spawning)	Decreased ovarian GSI	
	15.5L/8.5D	25°	Late April (early spawning)	Decreased ovarian and testicular GSI	
	15.5L/8.5D	12°	January (late preparatory)	Decreased testicular GSI; no effect on ovarian GSI	

Table 1 (Continued)

Species	Photoperiod	Temperature	Time of Year/Cycle	Pinealectomy Effects	Reference
Golden Shiner (continued)	15.5L/8.5D	15°	Late February (preparatory)	Decreased Gn activity	12, 15
	15.5L/8.5D	12°	March (early prespawning)	No effect on ovarian or testicular GSI	
	15.5L/8.5D	15°	Late April (early spawning)	Decreased testicular GSI; no effect on ovarian GSI	
	9L/15D	25°	January (late preparatory)	Decreased ovarian and testicular GSI	
	9L/15D	25°	Late February (preparatory)	No effect on Gn activity; no effect on ovarian or testicular GSI	
	9L/15D	25°	March (early prespawning)	Decreased ovarian and testicular GSI	
	9L/15D	25°	October-February (vitellogenesis)	Increased ovarian GSI	
	9L/15D	12°	January (late preparatory)	No effect on ovarian or testicular GSI	
	9L/15D	12°	Late February (preparatory)	Increased Gn activity; no effect on ovarian or testicular GSI	
	9L/15D	12°	March (early prespawning)	Decreased ovarian and testicular GSI	
	9L/15D	15°	October-February (vitellogenesis)	Increased ovarian GSI	

Table 1 (Continued)

Species	Photoperiod	Temperature	Time of Year/Cycle	Pinealectomy Effects	Reference
Medaka (*Oryzias latipes*)	14L/10D	23°	October-March	Decreased ovarian GSI; no effect on testicular GSI	68
	8L/16D	23°	October-March	No effect on ovarian or testicular GSI	
Gray Mullet (*Liza ramada*)	16L/8D	Not reported	December (spawning)	No effect on oocyte development; no effect on GSI	53
	8L/16D	Not reported	December (spawning)	Increased oocyte development	
Medaka (*Oryzias latipes*)	16L/8D	20°	November	Decreased ovarian GSI; no effect on testicular GSI	66
	8L/16D	20°	November	No effect on ovarian or testicular GSI	

*Gn = gonadotropin; GSI = gonadosomatic index

heavier testes compared to sham-operated controls. This difference was attributed to the natural breakdown and release of sperm-containing fluid into the body cavity in sham-operated animals, a process not yet started in the pinealectomized group. No effect due to time-of-year of pinealectomy (November, December or January) was observed. Joss concluded that this was probably due to the slow growth of the gonads occurring at this time. Also, a retardation in the development of secondary sexual characteristics was observed.

C. Teleosts

Two major environmental factors affect teleostean reproductive cycles: photoperiod and temperature. Manipulation of these factors can have profound effects on the reproductive status of teleosts, yet we do not know the mechanism(s) by which these effects are mediated. No doubt, both neural and hormonal controls are involved. This being the case, the role of the pineal organ (a known photoneuroendocrine transducer in fishes[41]) in control of teleostean reproductive cycles must be significant.

The pineal organ has been shown to be either progonadal or antigonadotropic depending upon the photoperiod-temperature regimen and season (Table 1). Pinealectomized goldfish exposed to long-photophase, warm regimens (16L/8D and $>$ 20°C) during February-April (prespawning period), June-July (mid-late spawning) or November (early recrudescence) will undergo gonadal regression.[16,33,71] Interestingly, pinealectomized females show an increase in their gonadosomatic index (GSI) when placed under long-photophase, warm conditions during May which coincides with the beginning of the spawning period; males show no response under these conditions.[16] This would argue for a second control system acting independently of the pineal organ, or one that perhaps "starts" at a time pre-set by the pineal. Although more data are needed, one study supports the latter situation. Fenwick[19] has found that ovarian and testicular GSIs increase in pinealectomized goldfish during May under short-photophase, cool-temperature conditions (8L/16D, 13°C). The implication from this work is that even under adverse environmental conditions, goldfish will spawn in May because the photoperiodic history of the animal (as detected and transduced by the pineal organ) has initiated a program of time-dependent events which lead to spawning.

Golden shiners, *Notemogonus chrysoleucas*, respond to pinealectomy in a different manner than goldfish. Removal of the pineal organ from golden shiners (held under long-photophase, warm conditions) at any point of the reproductive cycle decreases GSIs.[12,15] Increased ovarian GSI is observed only when animals are held on short-photophase photoperiods between October and February (this is a temperature-independent event). This is not surprising because the natural photophase between October and February is short, and this also is the time when vitellogenesis is occurring in these fish. While the increasing GSI is temperature-independent between October and February, by late February temperature begins to affect reproduction. During this time, pinealectomized shiners held on short-photophase, low-temperature regimens have significantly higher gonadotropin activity as compared to control animals. When held at short-photophase, warm conditions, these same animals show no change in gonadotropin activity.[15]

From the previous discussion, it is clear that there are photoperiod-dependent and temperature-dependent components that make up the control system for teleostean reproduction. It is not clear, however, which component is the more influential. Most researchers have assumed it is photoperiod that drives the reproductive cycles of teleosts. While this may be a correct assumption, caution must be taken because, as in the case of golden shiners, temperature may be more influential than photoperiod in regulating some reproductive parameters. Studies on medaka _Oryzias latipes_[66-69] and gray mullet _Liza ramada_[53] have measured effects of pinealectomy and photoperiod on reproductive status. Pinealectomy decreased ovarian but not testicular GSI under long photophases and warm temperatures during recrudescence in medaka. Changing the photophase from long to short while holding all other conditions the same produced no significant effects on the GSI of either species.[66,68] Interestingly, short photophases are stimulatory to gonadal function in the gray mullet.[53] An 8L/16D photoregimen was found to increase oocyte development, whereas a 16L/8D photoperiod had no effect on oocyte development. This may be due to the fact that gray mullet spawn in December, during the short-photophase portion of the year.

Given that the pineal organ has significant effects on reproductive cycles in some teleosts, it is not surprising to find that melatonin also has significant effects (Table 2). Injections of melatonin into goldfish,[18] medaka,[66] killifish,[14] and catfish (_Heteropneustus fossilis_)[55,57] cause pronounced antigonadotropic effects in mature fish held in long-photophase, warm-temperature environments. This would seem to contradict the work done on pinealectomized fish, but it should be pointed out that melatonin was given in high doses (3-100 μg/fish) over extended periods of time (10-60 days). The data may reflect, therefore, a pharmacological response rather than true physiological phenomena. In fact, Borg and Ekstrom[4] found that three-spined sticklebacks (_Gasterosteus aculeatus_) will only respond to high doses of melatonin (4 μg/g for 21 days) under short-photophase cold conditions. In this same study, low dosages of melatonin (0.9 μg/g) accelerated

Table 2. Effect of Melatonin, Photoperiod, Temperature and Season on Reproductive Parameters in Teleosts

Species	Photoperiod	Temperature	Time of Year/Cycle	Melatonin Treatment	Effect	Reference
Killifish (*Fundulus similis*)	13L/11D	9°-12°	January (prespawning)	3 μg given 2 h after photophase onset	Decreased GSI	14
	13L/11D	9°-12°	January (prespawning)	3 μg given 8 h after photophase onset	Decreased GSI	
	13L/11D	15.5°-18°	May (spawning)	5 μg given 2h after photophase onset	Decreased GSI	
	13L/11D	15.5°-18°	May (spawning)	5 μg given 8 h after photophase onset	Decreased GSI	
	10L/14D	9°-12°	January (prespawning)	5 μg given 8 h after photophase onset	No effect on GSI	
	10L/14D	15.5°-18°	May (spawning)	2 μg given 8 h after photophase onset	Decreased GSI	
Indian Catfish (*Heteropneustes fossilis*)	12L/12D	25°	May-June (prespawning)	20 μg (20 days)	No effect on gonadotropes; decreased ovarian GSI	57
	12L/12D	25°	May-June (prespawning)	100 μg (20 days)	Decreased gonadotropes; decreased ovarian GSI	
	12L/12D	25°	May-June (prespawning)	20 days after melatonin injections (20 μg or 100 μg)	Increased gonadotropes compared to control; increased ovarian GSI compared to controls	

gonadal development in animals held under long photophases and stimulated oocyte development in animals held under short photophasess. Obviously the data on melatonin effects are far from conclusive and still leave open the question of the mechanism of pineal regulation of teleostean reproductive cycles.

D. Amphibians

As in other vertebrates, amphibians display seasonal cyclicity in reproductive system function. Both intrinsic and extrinsic factors are involved in regulating reproductive cycles. Of the extrinsic cues, temperature, rainfall and light have been implicated (c.f., Salthe and Mecham,[54] see also Chapter 13). Considerable variation exists among the extrinsic factors most strongly correlated with reproduction. Also, difficulty in determining which factors may facilitate proximal events like oviposition or mating and which factors regulate annual cyclicity in gamete formation is encountered. Of the extrinsic factors, seasonal temperature and rainfall appear to be more important than photoperiod. However, few laboratory studies have been reported in which extrinsic factors were controlled rigorously while subsequent effects of temperature or light on gonadal conditions were recorded.[48] Werner[72] observed that, when males of *Plethodon cinereus* were held at 20°C under long-duration photophases (12L/12D or 18L/6D) during the latter part of the resting phase and during spermatogenesis (Dec.-March; Northern latitude), the maturational state of the testes increased. Placing *Bufo melanostictus* in continuous darkness decreased spermatogenesis and Leydig cell function (measured as Leydig cell nuclear area). Continuous light decreased spermatogenesis as well, but Leydig cell activity increased (Biswas et al.;[2] Table 3).

Just as little is known about the role of temperature and light in regulating seasonal changes in amphibian gonads, and only a few studies reporting a role for the pineal organ have been advanced (Table 4). Juszkiewicz and Rakalska[35,36] reported that injections of female bovine pineal homogenates into male *Rana esculenta* primed with human chorionic gonadotropin (hCG) inhibited spermatozoa excretion (sic. spermiation). Injections of pineal glands from male calves into hCG-primed frogs yielded no effect; female calf glands were as effective as cow glands in eliciting the spermatozoa-excretion effect. In similar studies using dosages of melatonin from 10-500 μg/frog, no effect on spermatozoa excretion was noted although the cow pineal extract inhibited spermatozoa excretion as noted previously.

O'Connor[40] investigated the effect of melatonin on *in-vitro* ovulation. In these investigations, *Rana pipiens* ovarian sections cultured with

Table 3. Effect of Photoperiod, Temperature and Time of Year on Reproductive Status in Amphibians

Species	Photoperiod	Temperature	Time of Year/Cycle	Effect	Reference
Plethodon cinereus	8L/16D	10°	April-July Spermatogenesis	Retarded testicular development	72
	12L/12D	10°	April-July Spermatogenesis	Retarded testicular development	
	16L/8D	10°	April-July Spermatogenesis	Regarded testicular development	
	8L/16D	20°	April-July Spermatogenesis	Accelerated testicular development	
	12L/12D	20°	April-July Spermatogenesis	Accelerated testicular development	
	16L/8D	20°	April-July Spermatogenesis	Accelerated testicular development	
	8L/16D	10°	December-March Quiescence	Normal testicular development	
	12L/12D	10°	December-March Quiescence	Normal testicular development	
	16L/8D	10°	December-March Quiescence	Normal testicular development	
	8L/16D	20°	December-March Quiescence	Normal testicular development	
	12L/12D	20°	December-March Quiescence	Accelerated testicular development	

Species	Photoperiod	Temperature	Time of Year/Cycle	Effect	Reference-
Plethodon cinereus (continued)	16L/8D	20°	December-March Quiescence	Accelerated testicular development	72
Bufo melanostictus	24D	-	-	Decreased spermatogenesis; decreased Leydig cell function	2
	24L	-	-	Decreased spermatogenesis; increased Leydig cell function.	

Table 4. Effect of Melatonin, Photoperiod, Temperature and Time of Year on Reproductive Status in Amphibians

Species	Photoperiod	Temperature	Time of Year/Cycle	Treatment[a]	Effect	Reference
Bufo melanostictus	-	-	Breeding	40 µg mel for 7 days	Decreased spermatogenetic cell next number; decreased Leydig cell nuclear size.	2
	10.5L/13.5D	18°	Dec.-Jan. (hibernating)	Px	Increased number of germ cell nests	6
	13.5L/10.5D	32°	May-June (breeding)	Px	No effect on number of germ cell nests	
	10.5L/13.5D	19°	January (hibernating)	40 µg mel during late photophase for 7 days	Increased spermatogenetic cell nest number	
	10.5L/13.5D	19°	January (hibernating)	40 µg mel during late photophase for 14 days	Increased spermatogenetic cell nest number	
	10.5L/13.5D	19°	April (breeding)	40 µg mel during late photophase for 7 days	Decreased spermatogenetic cell nest number	
	10.5L/13.5D	19°	April (breeding)	40 µg mel during late photophase for 14 days	Decreased spermatogenetic cell nest number	
Rana esculenta	Not reported	Not reported	Not reported	500 µg	Decreased spermatogenetic response	36
	"	"	"	100 µg mel + 20 I.U. chorionic gonadotropin	No effect	
	"	"	"	100 µg mel + 20 I.U. chorionic gonadotropin	No effect	

	Not reported	Not reported	Not reported	500 μg mel + 20 I.U. chorionic gonadotropin	No effect	36
	"	"	"	Homogenized bovine pineal gland + 20 I.U. chorionic gonadotropin	Decreased spermatogenetic response	
	Not reported	Not reported	Dec.-Jan	Homogenized adult bovine pineal gland	Inhibited spermatozoa secretion (spermiation)	35
	"	"	"	Homogenized juvenile bovine pineal gland	No effect	
	"	"	"	Extracts of bovine pineal gland	No effect	
Rana ridibunda	18L/6D	24°	Not reported	100 μg mel during scotophase (14 days)	Ovarian regression	11
Hyla cinerea	16L/8D	3C°	January	4 μg mel for 28 days	Decreased testicular and ovarian GSI	14
Alytes obstetricians				Px	No effect on gonadal development	10

[a]Px = pinealectomy,

mel = melatonin

gonadotropin released more than 70% of their oocytes in a 48-h period. Addition of melatonin in microgram amounts significantly reduced *in-vitro* ovulation. Effects of melatonin on gonadal growth in amphibians also have been reported.[14] A group of *Hyla cinerea* were maintained on a 16L/8D diel cycle at 30°C for 42 days. Following this acclimation period, frogs were injected with melatonin (4 μg/animal) or saline 10 h after the onset of the photophase. After 28 daily injections, GSIs were recorded; the data indicated melatonin-treated animals had reduced gonadal weights compared to control animals, and this was true for both female and male frogs in the treatment group.

Melatonin effects on testicular activity in *B. melanostictus* were examined during the breeding and hibernating phase of an annual cycle.[7] Here, identical experiments were performed at each phase of the cycle. At specific times, groups of animals were injected daily either with vehicle for two weeks, the vehicle for the first week and melatonin (40 μg/animal) for the second week or melatonin for two weeks. Following the two-week injection period, testes were removed and prepared for histological examination. During the breeding season, melatonin treatment daily for one week inhibited spermatogenesis and maturation of spermatocytes and spermatids within nests contained in the seminiferous tubules. Leydig cell nuclear areas also were increased compared to intact controls. The group receiving melatonin for two weeks was similar to the control group and different from the group receiving melatonin for one week. The authors attributed this lack of effect after two weeks of melatonin injection to testicular refractoriness to the hormone. During the hibernating season, toads receiving melatonin for one or two weeks displayed significant increases in the number of spermatogenetic nests within seminiferous tubules at each stage as well as increases in the Leydig-cell nuclear diameter. These authors describe this as a progonadal effect. Thus, during the breeding phase of an annual cycle, melatonin was antigonadotropic; during the hibernating phase, it was progonadal.

The effects of pinealectomy on the breeding and hibernating phases of testicular cycles also were examined. During the hibernation phase, toads pinealectomized 1, 2 and 3 weeks before sacrifice displayed increased gonadal maturation compared to sham-operated control animals. In an identical experiment during the breeding season, no effect of pinealectomy was noted in any group. Pineal organs taken from toads during hibernation were larger, and the cells had larger nuclei, whereas pineal organs from breeding toads were smaller.[6]

Effects of daily melatonin injection and photoperiodic gonadal response in female *Rana ridibunda* have been reported.[11] Ovarian biopsies were removed from frogs and scored for oocyte development. After acclimation to a 18L/6D photocycle (lights on at 1500h) for one week, groups of frogs were injected with saline or melatonin (100 µg/animal/day) at the onset of the scotophase. Following two weeks of treatment, frogs were sacrificed, and the ovaries were removed and scored for oocyte maturation. Animals receiving melatonin had significantly smaller oocytes (less advanced) compared to controls. Ovarian weights and GSIs for melatonin-treated frogs were significantly lower than those measured in control animals.

E. Reptiles

Studies of the pineal-gonadal axis in reptiles are scarce. In general, photothermal regimens are important in regulation of the annual reproductive cycle.[17,38] In addition, the circadian organization of those reptiles studied is very responsive to light[59,63] and involves the pineal organ and melatonin.[20,59,61,64] This indicates a role for the pineal organ in the reproductive cycles of reptiles (Table 5).

1. Males. In a series of experiments, Underwood[62,65] has carefully examined the role of the pineal organ and melatonin on testicular cyclicity in *Anolis carolinensis* throughout the year. In September, groups of animals were either pinealectomized, sham-pinealectomized or pinealectomized and implanted with a silastic capsule containing melatonin.[62] Other groups of animals were left intact but were implanted with either blank or melatonin-containing capsules. These groups were split into two subgroups, one maintained on an 8L/16D, 32°C photothermal regimen, the other on a 14L/10D, 32°C regimen. A final group was killed on the day of operation, the testes weighed and prepared for histological examination to stage testicular function. Twenty-eight days later, the remaining animals were killed. Results indicated that the 14L/10D, 32°C regimen elicited testicular growth and, in general, animals under these conditions were more advanced in testicular maturation compared to those under the 8L/16D, 32°C regimen. Pinealectomy caused a significant increase in testicular weight compared to any other group in the 14L/10D, 32°C regimen, and the testes of pinealectomized animals were in more advanced stages of development than the other groups. On the other hand, melatonin implantation decreased testicular weights in intact animals compared to blank-implanted animals held on a 14L/10D, 32°C regimen. Also, melatonin-treated, intact animals displayed retarded testicular development compared to blank-implanted controls.

Table 5. Effects of Pinealectomy (Px), Melatonin (mel), Photoperiod and Time of Year on Reproductive Status in Reptiles

Species	Photoperiod	Temperature	Time of Year/Cycle	Treatment	Effect	Reference
Anolis carolinensis	14L/10D	32°	September	Px	Increased testicular weight compared to Sham PX	65
	"	"	"	Px + 10 μg mel during early photophase, 20 days	Px effect only, no mel effect	
	"	"	"	Sham Px + 10 μg mel during early photophase; 20 days	No effect	
	"	"	"	Px + 10 μg mel during late photophase; 20 days	Px effect only, no mel effect	
	"	"	"	Sham Px + 10 μg mel during late photophase; 20 days	No effect	
	"	"	"	Px + 5 μg mel during early and late photo-phase; 20 days	Px effect only, no mel effect	
	"	"	"	Sham Px + 5 μg mel during early and late photo-phase; 20 days	No effect	
	10L/14D	32°	February-March	Px + mel silastic implant (dose unknown)	Decreased testicular weight compared to Px + blank implant group; and sham Px group	65

Calotes versicolor	Natural	Natural	December	Px	Increased testicular weight and activity	24
	"	"	March	"	Increased testicular weight and activity	
	"	"	June	"	No effect	
	"	"	September	"	Increased testicular weight and spermatogenetic activity	
Collisaurus draconoides	Natural	Natural	May	5 μg mel for 20 days	Testicular regression	43
A. carolinensis	6L/18D	31°	January	Px	Increased ovarian GSI	37
	"	"	"	Px + 10 μg mel 14 days	Decreased ovarian GSI	
	"	"	June	Px	No effect	
	"	"	"	Px + 10 μg mel	No effect	
	"	"	September	Px	No effect	
	"	"	"	Px + 10 μg mel 14 days	No effect	
	"	"	November	Px	Increased ovarian GSI	
	10L/14D	32°	September	Px	Increased testicular weight compared to sham Px	65
	"	"	"	Px + 10 μg mel early or late photophase; 20 days	Px effect only, no mel effect	

Table 5 (Continued)

Species	Photoperiod	Temperature	Time of Year/Cycle	Treatment	Effect	Reference
Anolis carolinensis (continued)	"	"	"	Px + 5 μg mel early and late photophase; 20 days	Px effect only, no mel effect	
	"	"	"	Sham Px + 10 μg mel early or late in photophase; 20 days	No effect	
	"	"	"	Sham Px + 5 μg mel early and late in photophase; 20 days	No effect	
	10L/14D	32°	December	Px + blank silastic implant	Increased testicular weight compared to control	
	"	"	"	Px + mel silastic implant (dose unknown)	Blocked testicular weight increase compared to control	
	10L/14D	32°	February-March	Px + blank silasrix implant	Increased testicular weight compared to Px + mel implant group; not different from sham Px group	
Emys leprosa	-	-	-	Px	Depresses testicular activity	8
	-	-	-	Px	Increased neurosecretory activity	1

Finally, no statistically significant differences in testicular weight were observed between any groups held under the 8L/16D, 32°C regimen. Histological examination indicated that lizards in the pinealectomy group had more mature testes than animals in the other groups.[62]

In a more recent study, Underwood[65] again examined the effect of pinealectomy and melatonin on reproductive cyclicity in male _A. carolinensis_. In these investigations, groups of animals were pinealectomized or sham-pinealectomized during September. These two groups were split into subgroups which received daily injections of melatonin (10 μg) or vehicle either during the AM or PM or both AM and PM. Thus a total of 12 experimental groups were formed. Another group of animals, serving as initial controls, were killed; one testis was weighed and the other prepared for histological examination. The remaining groups were exposed to a 14L/10D, 32°C photothermal cycle for 20 days, after which all animals were killed and their testes weighed or prepared for histological examination. Significantly increased testicular weights were observed in all groups of pinealectomized anoles regardless of the type (vehicle or melatonin) or time of injection.

Two final identical experiments examining the effects of pinealectomy and continuous melatonin application via silastic implants were conducted during December and again during March (when testicular recrudescence is reaching its maximum).[65] Here, groups of anoles were pinealectomized or sham-pinealectomized. The pinealectomy group was divided into two subgroups: one received melatonin implants and the other blank implants. An additional group of initial control animals were killed and their testes examined as before. Animals were killed after 20 days of exposure to a 10L/14D, 32°C regimen. In December, pinealectomized animals bearing a blank implant had increased testicular weights compared to intact controls, sham-pinealectomized or pinealectomized melatonin-implanted animals. Also, a significant melatonin effect was noted because testicular weights in pinealectomized, melatonin-implanted animals were not different from sham-operated controls. Pinealectomized animals also displayed greater testicular maturation compared to the other groups. In March, testicular weights in pinealectomized animals bearing blank implants were significantly greater than pinealectomized animals with melatonin implants. However, pinealectomized, blank-implanted animals did not differ from sham-operated animals, making statements concerning an antigonadal effect of melatonin tenuous.[65]

It is apparent that the pineal organ and melatonin are involved in the testicular cycle of _A. carolinensis_ but in a complicated way. As has been the case in mammals, careful experimental design must be used to determine involvement of the pineal organ and melatonin. Such experimental designs are by nature complicated because they must unambiguously sort out the actions of the pineal and melatonin superimposed on a continuously changing seasonal cycle. Examples of such designs can be found in Underwood's work which has, to date, yielded important information about this convoluted system. Clearly, pinealectomy is progonadal, with melatonin being antigonadal if applied in the correct manner or at the correct time. As discussed,[62] melatonin's antigonadal effect is apparent in pinealectomized anoles when environmental photostimulation is not acting too heavily in the system. Finally, while melatonin does have antigonadal action, it may not be the only antigonadal factor of pineal origin.[62,65]

Effects of pinealectomy and melatonin replacement on the annual testicular cycle in _Calotes versicolor_ have been studied. Pinealectomy induced earlier recrudescence and later regression of the testes. Spermatozoa were present for approximately 11 months in pinealectomized animals, while they were present for only five months in the sham-operated group.[24] Haldar-Misra and Thapliyal[26,27] observed that melatonin injections during the afternoon caused significant decreases in testicular weight. In this study, dosages of 0.15, 0.3, 0.6 and 1.2 μg melatonin/day were given for up to 44 days. In the highest dosage groups, significant gonadal weight reduction was observed after 22 days of injection. By 44 days, only the control group and those animals receiving the lowest dosage of melatonin had maintained testicular weight and activity. The other groups had gonads that were regressed, with no differences in gonadal weight among groups observable. Melatonin injection decreased weight of kidneys, largely due to regression of renal sexual segments (sex accessory structure; see Chapter 11).

In another study, Haldar-Misra and Thapliyal[25] examined the effects of melatonin injection on testicular weights during the active phase of the testicular cycle. In this study melatonin (10 μg/day) reduced testicular weight no matter the time of day injected. However, early-morning and late-evening injections appeared to be more effective. Finally, the effect of pinealectomy and melatonin injection on testicular weight and spermatogenetic activity was examined during the recrudescence and early regression phases of the cycle. Melatonin was without effect when given during the inactive phase. Pinealectomy increased testicular weight in all phases of the testicular cycle. Melatonin blocked the progonadal effect of pinealectomy, indicating melatonin is antigonadotropic in _C. versicolor_.

Finally, in a brief study Packard and Packard[43] observed melatonin-induced regression in testicular activity of _Callisaurus draconoides_. In this study, lizards were injected with 5 μg melatonin/day or with saline for 20 days. These authors did not observe a difference in testicular weight between groups; when examined histologically, however, testes from melatonin-treated animals had fewer sperm and displayed other signs of testicular regression.

2. Females. Levey[37] examined the effect of pinealectomy and melatonin replacement on the seasonal ovarian cycle of *A. carolinensis*. In four separate experiments done in January, June, September, and November, lizards were maintained under a 6L/18D light regimen (non-stimulatory) at 31°C. In January, pinealectomized lizards receiving daily saline injections had significantly increased ovarian GSI compared to sham operants receiving saline. Pinealectomized animals receiving 10 μg melatonin daily were not different from sham-operated animals receiving saline injections, indicating that melatonin replacement counteracts the effect of pinealectomy on ovarian GSI. Also, the pinealectomized saline-treated group had significantly larger primary follicle diameters. In June and September, no significant differences appeared in ovarian GSI or follicular diameters. In November, pinealectomized, saline-injected animals displayed significantly increased ovarian diameters compared to sham-operated, saline-injected lizards. Again, 10 g melatonin given daily to pinealectomized animals counteracted the effect of pinealectomy.

Cuellar[9] reported that pinealectomy did not alter the seasonal cycle in reproduction in *Cnemidophorus uniparens*. Normally this species is refractory to photothermal stimulation from July to mid-December. In this study, 12 control and 13 pinealectomized animals were exposed for one year to a thermal gradient from 27°-42°C while under a 10L/14D photoregimen. During this time, two pinealectomized animals displayed oocyte growth (determined by visual or at times, surgical examination) by mid-November. The remaining animals in both groups became reproductively active starting in December, the normal time. Thus in this species of lizard, pinealectomy did not affect ovarian maturation.

Owens et al.[42] reported levels of melatonin in serum and cerebrospinal fluid of green sea turtles *Chelonia mydas*. These animals displayed the normal, nocturnal increase in circulating melatonin observed in many vertebrates. Female green sea turtles sampled during the nesting and mating portions of a reproductive cycle had significantly reduced serum melatonin titers compared to male turtles or female turtles sampled during the inter-nesting period.

III. SUMMARY AND DIRECTIONS FOR FUTURE RESEARCH

A role for the pineal gland and reproduction in species of photoperiod-sensitive mammals has been demonstrated. In hamsters, the pineal gland influences episodic release of GnRH (see Stetson and Watson-Whitmyer[56]). Also, a clearer understanding of pineal involvement in regulation of reproduction in a short-photophase breeder, the sheep, has been obtained (see Bittman[3]). But besides species of hamsters, sheep and laboratory rats, relatively little is known about pineal involvement in the reproductive physiology of other mammalian species.

In nonmammals, only a few species from each class have been investigated. Because of this, generalities probably are misleading. There is a great need to investigate antigonadotropic and progonadal effects of the pineal organ and melatonin in nonmammals. At present, it appears as though the pineal organ plays an antigonadotropic role in nonmammals. Temperature effects on seasonality of reproduction in ectothermic vertebrates probably are great. The interactions between temperature and day-length are poorly understood, but of critical importance. Clearly, carefully controlled experiments investigating these interactions and the role of melatonin are needed badly.

Finally, as discussed by Ralph,[47] the pineal interacts with several endocrine axes. Like its role in reproduction, the pineal effects in these systems also are hard to distinguish clearly, and, like its reproductive functions, time-of-day and time-of-year contribute to the robustness observed in a pineal effect. This prompted Ralph[45,47] to hypothesize that the pineal organ acts as a seasonal _Zietgeber_ used in cueing annual cycles in endocrine function. The primary hormonal agent involved in this function is melatonin. Thus "success of a species may depend upon the correct interpretation of the seasonally-modulated, diurnally rhythmic pattern of melatonin elaboration." Finally he hypothesizes that "melatonin has been selected through evolution to invoke seasonally appropriate 'programs' within an animal's repertoire of adaptations."[47] We find this hypothesis appealing because it accounts in general for the vast variety of melatonin effects observed in vertebrates, many of which are seasonal. Finally, this hypothesis takes into account an important feature of melatonin secretion, namely its nightly appearance in the plasma of vertebrates. The evolutionary significance of this secretory profile has been discussed in terms of the primordial melatonin function,[21,22] and this hypothesis supports that of Ralph. Namely, melatonin's hormonal effects are hypothesized to have occurred at a later date compared to its other functions, but these hormonal effects nonetheless serve to increase fitness.

IV. ACKNOWLEDGMENTS

We thank Mary David for technical assistance. Partial support through the Office of Research, University of Wyoming and NSF - PCM-83-08738.

V. REFERENCES

1. Aron, E., Combescot, C., Demaret, J. and Guyon, C. (1960). Neurosecretion chez la tortue d'eau Emys leprosa apres destruction de la region epiphysaire. C. R. Acad. Sci. 251, 1914.
2. Biswas, N.M., Chakraborty, J., Chanda, S. and Sanyal, S. (1978). Effects of continuous darkness on the testicular histology of the toad (Bufo melanostictus). Endocrinol. Jpn. 25, 177-180.
3. Bittman, E.L. (1984). Melatonin and photoperiodic time measurement: evidence from rodents and ruminants. In: "Comprehensive Endocrinology. The Pineal Gland." (R.J. Reiter, ed.), pp. 155-192. Raven Press, New York.
4. Borg, B. and Ekstrom, P. (1981). Gonadal effects of melatonin in the three-spined stickleback, Gasterosteus aculeatus L., during different seasons and photoperiods. Reprod. Nutr. Dev. 21, 919-927.
5. Cardinali, D.P. (1980). Melatonin. A mammalian pineal hormone. Endocr. Rev. 2, 327-346.
6. Chanda, S. and Biswas, N.M. (1984). Influence of pineal gland on spermatogenesis of toad (Bufo melanostictus). Hormone Res. 19, 200-204.
7. Chanda, S., Chanda, R. and Biswas, N.M. (1983). Effect of melatonin on the testis of toad during breeding and hibernating seasons. IRCS Med. Sci. 11, 337-338.
8. Combescot, C. and Demaret, J. (1963). Hisophysiologie de l'epiphyse chez la tortue d'eau Emys leprosa (Schw.). Ann. Endocrinol. (Paris) 24, 204.
9. Cuellar, H. (1978). Continuance of circannual reproductive refractoriness in pinealectomized parthenogenetic whiptails (Cnemidophorus uniparens). J. Exp. Zool., 206, 207-214.
10. Declos, P. (1964). Epiphysectomie chez le tetard d'Alytes, C. R. Acad. Sci., 258, 3101.
11. Delgado, M.J., Gutierrez, P. and Alanso-Bedate, M. (1983). Effects of daily melatonin injections on the photoperiodic gonadal response of the female frog Rana ridibunda. Comp. Biochem. Physiol. 76A, 389-392.
12. deVlaming, V.L. (1975). Effects of pinealectomy on gonadal activity in the cyprinid teleost, Notemigonus crysoleucas. Gen. Comp. Endocrinol. 26, 36-49.
13. deVlaming, V.L. and Olscse, J. (1981). The pineal and reproduction in fish, amphibians and reptiles. In "The Pineal Gland. Vol. II Reproductive Effects" (R.J. Reiter, ed.), Vol. 2, pp. 1-29. CRC Press, Boca Raton.
14. deVlaming, V.L., Sage, M. and Charlton, C.B. (1974). The effects of melatonin treatment on gonosomatic index in the teleost, Fundulus similis and the tree frog Hyla cinerea. Gen. Comp. Endocrinol. 22, 433-438.
15. deVlaming, V.L. and Vodicnik, M.J. (1977). Effects of pinealectomy on pituitary gonadotrophs, pituitary gonadotropin potency and hypothalamic gonadotropin releasing activity in the teleost Notemigonus crysoleucas. J. Fish Biol. 10, 73-86.

16. deVlaming, V.L. and Vodicnik, M.J. (1978). Seasonal effects of pinealectomy on gonadal activity in the goldfish, Carassius auratus. Biol. Reprod. 19, 57-63.
17. Duvall, D., Guillette, L.J., Jr. and Jones, R.E. (1982). Environmental control of reptilian reproductive cycles. In "Biology of the Reptilia" (C. Gans and H. Pough, eds.), Vol. 13, pp. 201-232. Academic Press, New York.
18. Fenwick, J.C. (1970). Demonstration and effect of melatonin in fish. Gen. Comp. Endocrinol. 14, 86-97.
19. Fenwick, J.C. (1970). The pineal organ: Photoperiod and reproductive cycles in the goldfish, Carassius auratus L. J. Endocrinol. 46, 101-111.
20. Firth, B.T., Kennaway, D.J. and Rosenbilds, M.A.M. (1979). Plasma melatonin in a scincid lizard, Trachydosaurus rugosus: diel rhythm, seasonality and the effect of constant light and constant darkness. Gen. Comp. Endocrinol. 37, 493-500.
21. Gern, W.A. (1980). Evolution of melatonin function: An hypothesis. In "Melatonin: Current Status and Perspectives" (N. Biaru and W. Schloot, eds.) Adv. Biosci. 29, 85-87. Pergamon, Oxford.
22. Gern, W.A. and Karn, C.M. (1983). Evolution of melatonin's functions and effects. In "Pineal Research Reviews" (R.J. Reiter, ed.), Vol. 1, pp. 145-177. Alan R. Liss, New York.
24. Haldar, C. and Thapliyal, J.P. (1977). Effects of pinealectomy on the annual testicular cycle of Calotes versicolor. Gen. Comp. Endocrinol. 32, 395-399.
25. Haldar-Misra, C. and Thapliyal, J.P. (1979). Influence of time factor on administration of melatonin and serotonin to Indian garden lizard, Calotes versicolor. Indian J. Exp. Biol. 17, 1383-1385.
26. Haldar-Misra, C. and Thapliyal, J.P. (1981). Effect of melatonin on the testis and the renal sex-segment in the garden lizard, Calotes veriscolor. Can. J. Zool. 59, 70-119.
27. Haldar-Misra, N. and Thapliyal, J.P. (1981). Response of reptilian gonad to melatonin. Neuroendocrinology. 33, 328-332.
28. Hoffman, R.A., Hester, R.J. and Townes, C. (1965). Effect of light and temperature on the endocrine system of the golden hamster (Mesocricetus auratus Waterhouse). Comp. Biochem. Physiol. 15, 225-236.
29. Hoffman, R.A. and Reiter, R.J. (1965). Pineal gland: Influence on gonads of male hamsters. Science 148, 1609-1611.
30. Hoffman, R.A. and Reiter, R.J. (1966). Response of some endocrine organs of female hamsters to pinealectomy and light. Life Sci. 5, 1147-115.
31. Hoffman, K. (1981). Pineal involvement in the photoperiodic control of reproduction and other functions in the Djungarian hamster Phodopus sungorus. In "The Pineal Gland. Vol. II. Reproductive Effects" (R.J. Reiter, ed.), Vol. 2, pp. 83-102. CRC Press, Boca Raton.
32. Hontela, A. (1984). Daily cycles of serum gonadotropin hormone in fish. Trans. Am. Fish Soc. 113, 458-466.
33. Hontela, A. and Peter, R.E. (1980). Effects of pinealectomy, blinding, and sexual condition on serum gonadotropin levels in the goldfish. Gen. Comp. Endocrinol. 40, 168-172.
34. Joss, J.M.P. (1973). Pineal-gonadal relationships in the lamprey Lampetra fluviatilis. Gen. Comp. Endocrinol. 21, 118-125.
35. Juszkiewicz, T. and Rakalska, Z. (1963). Anti-estrogenic effects of bovine pineal glands. Nature (London) 200, 1329-1330.
36. Juszkiewicz, T. and Rakalska, Z. (1965). Lack of effect of melatonin on the frog spermatogenic reaction. J. Pharm. Pharmac. 17, 189-190.
37. Levey, L.I. (1973). Effects of pinealectomy and melatonin injections at different seasons on ovarian activity in the lizard Anolis carolinensis. J. Exp. Zool. 185, 169-174.

38. Licht, P. (1972). Environmental physiology of reptilian breeding cycles: role of temperature. Gen. Comp. Endocrinol. Supp. 3, 477-488.
39. Lynch, H.J. and Wurtman, R.J. (1981). Melatonin levels as they relate to reproductive physiology. In "The Pineal Gland. Vol. II. Reproductive Effects" (R.J. Reiter, ed.), Vol. 2, pp. 103-123. CRC Press, Boca Raton.
40. O'Connor, J.M. (1969). Effect of melatonin on *in vitro* ovulation of frog oocytes. Am. Zool. 9, 577 (Abstract).
41. Oksche, A. and Hartwig, H.G. (1979). Pineal sense organs - Components of photoneuroendocrine systems. In "Progress in Brain Research" (A. Kappers and P. Pevet, eds.), Vol. 52, pp. 113-129. Elsevier/North Holland, Amsterdam.
42. Owens, D.W., Gern, W.A. and Ralph, C.L. (1980). Melatonin in the blood and cerebrospinal fluid of the green sea turtle (*Chelonia mydas*). Gen. Comp. Endocrinol. 40, 180-187.
43. Packard, M.J. and Packard, G.C. (1977). Antigonadotropic effect of melatonin in male lizards (*Callisaurus draconoides*). Experientia, 33 1665-1666.
44. Peter, R.E. (1983). The brain and neurohormones in teleost reproduction. In "Fish Physiology" (W.S. Hoar, D. J. Randall and E.M. Donaldson, eds.), Vol. 9A, pp. 97-135. Academic Press, New York.
45. Ralph, C.L. (1975). The pineal complex: A retrospective view. Amer. Zool. 15 (Suppl. 1), 105-116.
46. Ralph, C.L. (1978). Pineal control of reproduction: Nonmammalian vertebrates. In "The Pineal and Reproduction" (R.J. Reiter, ed.), Prog. Reprod. Biol. 4, 30-50. Karger, Basal.
47. Ralph, C.L. (1983). Evolution of pineal control of endocrine function in lower vertebrates. Amer. Zool. 23, 597-602.
48. Rastogi, R.K., Iela, L., Saxena, P.K. and Chieffi, G. (1976). The control of spermatogenesis in the green frog, *Rana esculenta*. J. Exp. Zool. 196, 151-165.
49. Reiter, R.J. (1980). The pineal and its hormones in the control of reproduction in mammals. Endocr. Rev. 1, 109-131.
50. Reiter, R.J. (1981). Reproductive effects of the pineal gland and pineal indoles in the Syrian hamster and the albino rat. In "The Pineal Gland. Vol. II Reproductive Effects" (R.J. Reiter, ed.), Vol. 2, pp.45-81. CRC Press, Boca Raton.
51. Reiter, R.J., Hoffman, R.A. and Hester, R.J. (1966). The effects of thiourea, photoperiod and the pineal gland on the thyroid, adrenals and reproductive organs of female hamsters. J. Exp. Zool. 162, 263-271.
52. Reiter, R.J., Vaughan, M.K., Blask, D.C. and Johnson, L.Y. (1974). Melatonin, its inhibition of pineal antigonadotropic activity in male hamsters. Science 185,1169-1171.
53. Sagi, G., Abraham, M., and Ige, V.H. (1983). Pinealectomy and ovarian development in the gray mullet, *Liza ramada*. J. Fish Biol. 23, 339-346.
54. Salthe, S.N. and Mecham, J.S. (1974). Reproductive and courtship patterns. In "Physiology of the Amphibia" (B. Lofts, ed.), pp. 309-521. Academic Press, New York.
55. Saxena, P.K. and Anand, K. (1977). A comparison of ovarian recrudescence in the catfish, *Mystus tengara* (Ham.), exposed to short photoperiods, to long photoperiods and to melatonin. Gen. Comp. Endocrinol. 33, 506-511.
56. Stetson, M.H. and Watson-Whitmyre, M. (1984). Physiology of the pineal and its hormone melatonin in annual reproduction in rodents. In "Comprehensive Endocrinology. The Pineal Gland" (R.J. Reiter, vol. ed.), pp. 109-153. Raven Press, New York.
57. Sundararaj, B.J. and Keshavanath, P. (1976). Effects of melatonin and prolactin treatment on the hypophyseal-ovarian system in the catfish, *Heteropneustes fossilis* (Block). Gen. Comp. Endocrinol. 29, 84-96.

58. Tamarkin, L., Westram, W.K., Hamill, A.I. and Goldman, B.D. (1976). Effect of melatonin on the reproductive systems of male and female Syrian hamsters: a diurnal rhythm in sensitivity to melatonin. Endocrinology 99, 1534-1541.
59. Underwood, H. (1977). Circadian organization in lizards: The role of the pineal organ. Science, 195, 587-589.
60. Underwood, H. (1979). Melatonin affects circadian rhythmicity in lizards. J. Comp. Physiol. 130, 317-323.
61. Underwood, H. (1981). Circadian organization in the lizard *Sceloporus occidentalis*. The effects of pinealectomy, blinding and melatonin. J. Comp. Physiol. 141, 537-547.
62. Underwood, H. (1981). Effects of pinealectomy and melatonin on the photoperiodic gonadal response of the male lizard *Anolis carolinensis*. J. Exp. Zool. 217, 417-422.
63. Underwood, H. (1983). Circadian pacemakers in lizards: Phase response curves and effects of pinealectomy. Am. J. Physiol. 244, R857-R864.
64. Underwood, H. (1984). The pineal and circadian rhythms. *In* "Comprehensive Endocrinology. The Pineal Gland" (R.J. Reiter, vol. ed.), pp. 221-251. Raven Press, New York.
65. Underwood, H. (1985). Annual testicular cycle of the lizard *Anolis carolinensis*: Effects of pinealectomy and melatonin. J. Exp. Zool. 233, 235-242.
66. Urasaki, H. (1972). Effects of restricted photoperiod and melatonin administration on gonadal weight in the Japanese killifish. J. Endocrinol. 55, 619-620.
67. Urasaki, H. (1972). Role of the pineal gland in gonadal development in the fish *Oryzias latipes*. Annot. Zool. Jpn. 45, 152-158.
68. Urasaki, H. (1973). Effect of pinealectomy and photoperiod on oviposition and gonadal development in the fish, *Oryzias latipes*. J. Exp. Zool. 185, 241-246.
69. Urasaki, H. (1976). The role of pineal and eyes in the photoperiodic effect on the gonad of the medaka, *Oryzias latipes*. *Chronobiologia*, 3, 228-234.
70. Vivien-Roels, B. (1981). Pineal control of reproduction in nonmammalian vertebrates. *In* "The Pineal Organ: Photobiology - Biochronometry - Endocrinology" (A. Oksche and P. Pevet, eds.), pp. 315-334. Elsevier/North Holland Biomedical Press, Amsterdam.
71. Vodicnik, M.U., Kral, R.E., deVlaming, V.L. and Crim, L.W. (1978). The effects of pinealectomy on pituitary and plasma gonadotropin levels in *Carassius auratus* exposed to various photoperiod-temperature regimes. J. Fish Biol. 12, 187-196.
72. Werner, J.K. (1979). Temperature-photoperiod effects on spermatogenesis in the salamander *Plethodon cinereus*. Copeia 1969, 592-602.

STRESS AND REPRODUCTION: RECIPROCAL RELATIONSHIPS

Neil Greenberg

Department of Zoology
University of Tennessee
Knoxville, TN 37996

John C. Wingfield

Field Research Center
Rockefeller University
Millbrook, NY 12545

I. INTRODUCTION

A. Background

The sensitivity of reproductive function to the influence of environmental stressors in vertebrates has been clear since the late 1930's, as have the stressing effects of reproductive activity itself.[209,210] However, the agenda for the investigation of interactions between physiological stress and reproductive function did not emerge until after the significance of their mutual influences on population density came to light in the 1950's. Instrumental was the early work of Christian[42] in which activation of the pituitary-adrenocortical axis was associated with high population densities.

As Christian and Davis pointed out,[44] this attitude was fostered by the growing awareness of the importance of both intrinsic and extrinsic factors in explaining population regulation. Using mice as a model, they found that

> . . . within broad limits set by the environment, density-dependent mechanisms have evolved within the animals themselves to regulate population growth and curtail it short of suicidal destruction of the environment. (p. 1550)

This insight, coupled with the apparent generality of stress-related, density-dependent population regulation and Mason's [151,152] demonstration of the potency of psychogenic stressors, has engendered a new appreciation of the power of the comparative perspective and the diversity of possible mechanisms natural selection has revealed.

The nature of the adrenal and gonadal interactions in lower taxa also serves as an important reminder of the limits to which physiological systems can be adapted in the service of reproductive efficiency, dispelling the assumption that adrenal activation is necessarily deleterious. The mission of this brief review is to provide a representative, rather than a comprehensive, outline of several of the known and putative interactions of stress-axis activation and reproductive function.

B. Definitions

It is more than an empty tradition when dealing with stress to engage the problem of definition (see Levine[136a]). The stress concept means too many things to too many workers to do otherwise, and clearly definition can set the research agenda. Some definitions seek consensus in their vagueness ("stress is the nonspecific response of the body to any demand"; Selye,[212] p. 15), while others unmistakably define stress as nonspecific activation of the hypothalamic-pituitary-adrenocortical axis[146] or in terms of "coping" with danger.[5] Recalling Beer's[11] wise injunction that "rigour of definition risks *rigor mortis* of concept" (p. 155), we will, converging with Pickering,[182] employ the idea of "stress response" as physiological compensation that is found toward one end of a continuum of adaptive responses to internal or external stimuli. Any physiological, developmental, or environmental stimulus could then constitute a "stressor," the stimulus-value of which can vary greatly.

The highly adaptive nature of alterations in reproductive function of individuals within populations experiencing stressors is little doubted, but the reciprocal phenomenon, the influence of reproductive activity on responsiveness to stressors, has received little attention and is almost certainly comparably adaptive. The reciprocal and complementary activities of the hypothalamic-pituitary-adrenocortical system, hypothalamic-sympathetic-adrenomedullary system, and hypothalamic-pituitary-gonadal system, while the least studied, are especially likely to provide insightful surprises.

C. Stress Hormones

The major "stress-related" hormones produced in or released from chromaffin tissue (adrenal gland, adrenal medulla), the interrenal gland (adrenal cortex), and the pituitary gland have been identified in every vertebrate class (Table 1). In all cases, the secretion of these hormones increases in response to stressors. Acute stressors elicit an alarm ("emergency" or "flight or fight") reaction by means of hypothalamic activation of the sympathetic nervous system; attendant adrenomedullary (chromaffin) activation

Table 1. References for Dominant Stress-Related Hormones in Vertebrates

Hormone	Fishes: Agnatha	Fishes: Gnathostomes	Amphibians	Reptiles	Birds	Mammals
Epinephrine	168	168	168	168	168	168
Norepinephrine	168	168	168	168	168	168
Glucocorticosteroids* (CS)	B/F 104	B/F 104	B 104	B 104	B 104	B/F 104
Pro-opiomelanocortin	?	115,198	231	59,60	108	75
Corticotropin	28	207	231	59,60	108	6
β-endorphin	173	115,198	231	59,60	137	87
α-MSH	7	115,198	231	127	108	75

*B = corticosterone, F = cortisol

releases the catecholamines epinephrine (E) and norepinephrine (NE) into the circulation to intensify the response. Varying ratios of E to NE are associated with activation of specific neural sites in the hypothalamus[73] as well as higher neural centers.[66,153,154]

Corticotropin (ACTH) is released from the corticotropic cells of the pars distalis of the pituitary gland. ACTH is derived from the prohormone proopiomelanocortin (POMC) which is synthesized in the pars distalis and in the melanotropes of the pars intermedia (neurointermediate lobe). It is released mainly under the influence of the hypothalamic polypeptide, corticotropin-releasing factor (CRF). ACTH then stimulates release of adrenocortical (interrenal) steroids (corticosteroid, CS), most notably glucocorticoids, although it is also known to have non-corticotropic functions, including effects on the metabolism of cortisol and cholesterol (reviewed by Berliner et al.,[16]) and on behavior, particular "fear-mediated responding" (reviewed by Leshner[133]). The adrenocortical response is associated with the general adaptation syndrome first characterized by Selye[211], in which a rapid initial reaction ("alarm") is followed by sustained glucocorticoid secretion ("phase of resistance"), and eventually a dangerous debility when corticoid output cannot be sustained or target substrates for CS are exhausted ("phase of exhaustion").

During prolonged stress, adrenocorticotropin, glucocorticoids, and neuronal activity alter the adrenomedullary response by inducing changes in the activity of enzymes associated with catecholamine synthesis. Conversely, catecholamines can stimulate pituitary ACTH release.[6] The complexity of catecholamine effects on stress responses is underscored by observations, such as those of Berkenbosch et al.,[14,15] that the adrenergic beta-blocker propranolol attenuates the neurointermediate lobe response to some stressors but not to others.

Melanotropin (melanocyte-stimulating hormone, MSH), along with endorphin, is also derived from POMC, but is synthesized in the melanotropic cells of the neurointermediate lobe of the pituitary. Melanotropin has general tropic properties in the rat and is associated with stimulation of fetal growth, protein synthesis, and healing of wounds; it is also known to reverse CS-induced retardation of growth in chick embryos.[227] Birds do not have a detectable pars intermedia, but it should be noted that POMC and its components are found in the pars distalis of the pituitary. Avian pituitary cells that contain ACTH also have been found to contain MSH.[108]

Melanotropin was identified as a possible stress-responsive hormone in the mid-1970's.[75,203] It parallels the release of ACTH in response to CRF,[192] and it has diverse interactions with other stress-related hormones. Plasma adrenocorticoid levels may rise as a consequence of MSH treatment over several days,[50] possibly as a result of a stimulatory effect on ACTH release.[140] Melanotropin inhibits melatonin secretion[178] which, interestingly, has been associated with stress-response inhibition (reviewed by Datta and King[51]). Melanotropin, like ACTH, can affect behavior by means of extratropic interactive effects as well as through more direct neurobehavioral actions, probably due to an enhancement of attentive processes.[113,122] It has a synergy with testosterone[172] which also is associated with enhanced attention.[4]

Recent evidence that endogenous opioid peptides (endorphins) and ACTH are released concomitantly from the pituitary gland[87] has converged with biochemical and behavioral evidence to indicate that endorphins also play an important role in the stress response, possibly by mediating the emotional and analgesic components of stress (reviewed by Amir et al.[3]). Beta-endorphin can be released from either corticotropes or melanotropes which are differentially sensitive to effects of adrenal catecholamine on beta-adrenergic receptors. Thus, the beta-blocker propranolol prevents the melanotropic response but not the corticotropic response.[14,15] In this regard it is interesting that, while corticotropes apparently respond to all stressors, the melanotropic response may be selectively elicited by physiological stressors.

It also should be pointed out that a variety of stressors are known to influence secretion of prolactin (PRL), growth hormone (GH), thyroid-stimulating hormone (TSH), thyroxine, and triiodothyronine (e.g., El-Halawani et al.,[63] Aidara et al.[1]). However, the vast majority of these investigations have been conducted on birds and mammals, and virtually none on other classes (with the exception of Pickering and Christie,[184] see below). Since assays for GH and PRL are becoming available for use in nonmammals, significant advances concerning the responses of these hormones to stress can be expected in the future.

II. STRESS EFFECTS ON REPRODUCTIVE PHYSIOLOGY AND BEHAVIOR

A. Overview of Mammals

Clearly, inhibition of reproduction is a graded response that may involve multiple levels of reproductive function from maturation and gametogenesis through lactation and behavior.[43] Indeed, reproductive behavior can be altered at various critical points in fetal development, causing a variety of reproductive dysfunctions in female offspring[2,98] or impairing future copulatory behavior in males and enhancing lordotic behavior in females.[233] In addition to the effects of social-stress on reproductive behavior, there are less obvious effects on, and interactions between, adrenal and gonadal hormones, neonatal and adult experience, enhanced or redirected sensory function, and locomotor activity (reviewed by Nowel.[171]).

While the gradation of stress-induced reproductive effects may be clear from the perspective of population biology, the levels of response are discrete, frequently very specific, and quite complex in themselves. Researchers thus find themselves confronting an apparently stereotyped response ("physiological stress-response") to a change in the internal or external environment requiring physiological compensation ("stressor") that may manifest its regulatory influence through any one or a combination of individual components, most of which can affect reproductive success. In other words, the physiological stress responses, while seemingly stereotypes when compared to the diversity of stressors that can evoke them, are nevertheless sufficiently complex to engender, if not permit, an impressive number of adaptive responses. The more closely we examine these responses from the ethological perspective, the more it appears that we are really dealing with a close "family" of stress responses, with each response pattern reflecting differential activity of the several stress-related hormones.

Mechanisms affecting reproduction identified by Christian[43] include cross-system feedback inhibition of reproductive function, inhibition of maturation, and changes in hypothalamic sensitivity to inhibitory feedback associated with maturity, increased density, photoperiod/pineal gland, pheromones, aggression, and territoriality.

1. Effects of Stress Upon Social and Reproductive Activity. The physiological interrelations of the central nervous system and gonadal function are well-known,[149] and the effects of one individual's behavior on the gonadal activity of another are well-documented;[47,129] however, the

interrelatedness of social behavior, adrenal-mediated stress responses, and reproductive function are relatively unexplored. Reviews such as that by Kime et al.,[117] emphasizing mutual morphological and physiological effects of mammalian andrenocortical and gonadal hormones, point to a long-delayed appreciation of the reciprocal, complementary, and mutually modulating functions of these hormones.

A social environment is stressful to the extent that it engenders real or apparently imminent aggressive interactions. It has been clear since the work of Bronson et al.[25] that increased adrenocortical activity and decreased gonadotropin (Gn) secretion result from aggressive activities. But further, as Christian[43] emphasizes, the extent of reproductive inhibition can vary with social rank, sex, season, and maturity, as well as other factors. In other cases, agonistic interactions may even elevate circulating androgen levels for brief periods during establishment of a territory or formation of a dominance hierarchy.[249]

Dominance relationships, often subsequent to agonistic interactions, are generally regarded as a significant element in differential reproductive activity in a social group. Rowell[201,202] felt that the intense subordinance of low-ranking primates, indeed the social hierarchy itself, might best be thought of as a symptom of stress, particularly in captivity. But while captive conditions may intensify stress due to subjects coping with an alien or resource-limited situation, the generality and spontaneity of social dominance cannot be disregarded.[204,251] Bernstein[17] points out that dyadic dominance relationships (of which dominance hierarchies are constituted) must involve benefits to both dominants and subordinates. The ability of many species in every vertebrate class to shift between hierarchical and territorial behavior is probably related to efficient resource utilization.[141] The idea of dominance rank can, in fact, help explain an assortment of population dynamics that ultimately converge on reproductive success.[79] While relative rank does not preordain corresponding access to mates, there is a positive correlation between high status and mating success in many taxa (Wittenberger,[251] p. 588).

The relative sensitivity of females to stress is reasonable considering the complexity of reproductive endocrine events and their timing. Beyond the effects of climatic (heat) stress on the estrous cycle, social stimuli are prominent, particularly in species with dominance relationships. For example, female talapoin and marmoset monkeys of high and low social status have high and low luteinizing hormone (LH) responses to estrogen challenge, respectively (see Moberg[163]). Estrous abnormalities also are common in socially stressed rodents and domesticated animals.

There is clear association of androgen levels with aggressive behavior in males of several mammalian species and also with the availability of females in primates.[200,204] Sexual activity appears to contribute significantly to androgen fluctuations, but in a study of pigtail macaques, Bernstein et al.[18] found that members of all-male groups had mean testosterone concentrations equivalent to males living with females. Although these values rose following the addition of females to the group, the effect was not dependent on the receptivity of the females, suggesting that social realignment of the group was more important than the fact that the introduced animals were females. In other words, dominance over a female might be more significant than sexual activity per se.

In mice, submissiveness is not directly related to testosterone levels.[150] Indeed, hormones of the pituitary-adrenal axis are more important: The effects of acute and chronic stressors on the behavioral response of mice to hypothalamic-pituitary-adrenal activation may be different owing to an extra-adrenal effect of sustained secretion of ACTH associated with increased submissiveness while relatively brief ACTH activation will increase aggressiveness through its tropic action on the glucocorticoids (Leshner, 1978). Supporting this separation of casual elements in the otherwise apparent continuum of aggressiveness and submissiveness is the fact that, whereas elevated androgens are associated with aggressiveness, reduced androgens are not a necessary correlate of submissiveness.[135] ACTH treatment can reduce gonadal secretion of androgens,[133] but it also is associated with an action on the brain inducing "sexual excitement."[19] Once dominance relationships are established, different components of the neuroendocrine response to the stress of social interaction are selectively activated: an adrenomedullary response predominates in dominants while in subordinates an adrenocortical pattern is elicited.[64]

In staged two-mouse aggressive interactions, once an individual is defeated, continued combat is <u>not</u> necessary for continued ACTH secretion in the loser--the presence of the winner is sufficient.[24] In long-term cohabitants, subordinate mice may show only marginal signs of stress, indicating some adaptation to the state.[38] Considerations of past experience among other ethological factors will undoubtedly be significant in reconciling the various, sometimes seemingly contradictory effects of subordination on hormone levels in mammals (e.g., Eaton and Resko[62]) and birds.[199] For example, in some studies of birds and of mammals, the failure to find positive correlations between dominance and elevated testosterone whereas other studies do find such a correlation may be simply a matter of timing of sampling with respect to stabilization of the status relationship. Indeed, both early in the process of status-stabilization and episodically throughout the long-term maintenance of status relationships, levels of aggression are elevated in dominants. In fact several studies have found an association between elevated rank, increased frequency of aggression, and elevated testosterone (e.g., Coe et al.,[45] Rose et al.,[200] Steklis et al.,[218] and Sapolsky[204]).

Furthermore, corticosterone levels also may fail to show a relationship to status. If, however, a group is challenged by introduction of a novel individual, or by increased competition over (for example) food, then circulating testosterone increases concomitant with rising frequencies of agonistic interactions. Corticosterone also may rise if the challenge is severe, especially in the loser of a fight. When stability is gained once more, hormone titers decline, and any relationship with dominance or aggressiveness is lost (see Wingfield and Rameofsky[249]).

Aggressiveness may be reduced in some species by chronic ACTH treatment or high glucocorticoid levels,[36] whereas short-term treatment[23] or lower doses may stimulate aggressiveness.[120] On the other hand, aggressive rhesus monkeys have a larger glucocorticoid response to stress than less aggressive animals[136] indicating possible species differences. Sapolsky[204] found that subordinate baboons had increased plasma cortisol following anesthesia stress and had lower levels of testosterone than did dominants.

2. Stress Axis Effects on Gonadal Function. Gonado-suppression has been associated in several taxa with adrenal activation following exposure to physical or psychological environmental challenges to homeostasis. However, both as logicians and as researchers digging for underlying mechanisms, we have become properly alert to the fact that association does not confirm cause.

Stressful environmental conditions can affect the onset of puberty[194] (see also Moltz[164]), probably by means of hypothalamic-pituitary-adrenal axis activation in a similar way as adult fertility is affected by stress.[193] But the effect is not simple, and while stress can delay puberty and reduce fertility in juvenile mice, it can increase incidence of conception and lead to increased litter sizes in adults. This difference is associated with differential response of the rhythm of adrenocortical activity to stress: prepubertal mice shift the peak of their daily rhythm, while in normal cycling adult females, stress causes a reversal of the corticosterone rhythm.[176] Chronic stressors apparently can delay reproductive maturation by means of a little understood effect of ACTH on LH release. The adrenal glands are essential for this effect in rats, while in mice ACTH is effective even if the animals are adrenalectomized (see Mann et al.[145]). The delaying effect of corticosterone on puberty may, however, be more one of general health and rate of growth than of a more direct effect on the hypothalamic-pituitary gonadal axis, particularly if weight at puberty is considered instead of age.[161].

Adrenal androgen appears necessary for sexual receptivity in the female rhesus monkey.[68] After adrenalectomy and cortisol replacement treatments, androstenedione administration would restore receptivity.[69] A synchrony between adrenal and ovarian progesterone production was implicated in the control of sexual receptivity in the rat.[191]

Pituitary-adrenal modulation of imprinting, the process by which newly hatched precocial birds develop social attachments, was indicated by the work of Weiss et al.[236] Martin[147] demonstrated that CS reduced imprinting trailing latency and following time in ducks, whereas ACTH increased them. Interestingly, small amounts of ACTH injected embryonically stimulated following behavior, but larger amounts that presumably stimulated the embryonic adrenal cortex had no effect. The opposing effects of CS and ACTH were related by Martin to DeWied's[55] finding of similarly opposite effects in extinction of conditioned avoidance behavior.[148]

Clearly, stress affects Gn secretion, but the mechanisms are still obscure and findings contradictory. Hypothalamic Gn-releasing hormone (GnRH) release or its action on pituitary gonadotropes may be blocked. Evidence at present is tipped toward reduced sensitivity of gonadotropes to GnRH in adrenal-axis activated animals.[163] A possible mechanism for gonadal suppression by an active adrenal gland is feedback inhibition of Gn by adrenal sex steroids[117] (see also Moberg,[163]) although a direct inhibitory effect of glucocorticoids on testicular LH receptors also is known.[8]

Copulation or aggressive confrontation causes an increase in plasma CS, and an immediately subsequent increase in plasma testosterone in the boar. This has been interpreted by Liptrap and Raeside[139] to indicate a temporally biphasic effect of CS on plasma testosterone levels: an initial stimulatory effect on testes is followed, after prolonged corticosterone stimulation, by suppression of testosterone production.

In domesticated birds, stress-induced increases in ACTH and corticosterone can also delay sexual maturation or interrupt egg-laying.[97,213] This appears to be a direct effect of corticosterone on LH and follicle-stimulating hormone (FSH) secretion. However, wild avian species appear to be more resistant to several stressor stimuli. Although severe food deprivation blocks testicular growth and depresses LH and testosterone secretion, other stressors (such as low environmental temperature and handling) are without effect in some species.[246] Indeed it appears that during breeding, females, and to a lesser extent males, become less sensitive to handling stress than during the winter months.[250]

Corticosteroids can act both at the testicular level and at hypothalamic-pituitary level in mammals. However, the story in birds is less clear. Injections of CS into the hypothalamus of tree sparrows depressed plasma levels of LH and blocked testicular growth, and in ducks CS decreased plasma LH and FSH.[54,239] However, neither injections of ACTH into ducks nor both ACTH and CS into white-crowned sparrows had any effect on plasma LH and testosterone.[53,246] Also, increased secretion of corticosterone induced by severe storms did not depress plasma levels of LH and testosterone in naturally breeding sparrows.[244,245,248] This mechanism, by which plasma corticosterone increases (as body weight and fat decline) during stressful episodes, leaves LH and testosterone secretion unaffected. This is thought to provide the "stress-response" and thus the "adaptation" to the stressor, but leaves the gonad in a near functional state. Individuals can then begin breeding again immediately after the environmental conditions improve. If the stress response did result in a decline in LH and testosterone, as is found in domesticated species and some mammals, the gonads would regress, and the ability to rapidly resume reproductive activity would then be delayed considerably.

Opioid peptides affect Gn in rodents,[159] primates, and humans. Gonadotropins can induce menstrual abnormalities in these species (reviewed by Ferin[72]). A significant regulatory role for endogenous opiates is indicated by a study in which a synthetic enkephalin selectively inhibited copulatory behavior in otherwise sexually vigorous male rats, an effect blocked by the opioid antagonist naloxone. Further, the opioid antagonist naloxone induced copulatory behavior in otherwise sexually inactive males,[80] and naloxone also increased serum LH in prepubertal female rats but not in males. This finding, and the fact that estradiol blocked the response in females indicated to Blank et al.[20] that both opioids and gonadal steroids influence LH secretion and can affect the onset of puberty.

Numerous examples of delayed parturition in cattle, goats, and sheep, collated by van Tienhoven,[230] were characterized by fetal adrenal insufficiency. Administration of ACTH or corticosteroid, the fetal levels of which normally increase markedly in the period before parturition, appears to constitute a fetal signal in many species. These hormones decrease maternal progesterone and increase prostaglandin levels which then participate in parturition. In some species, fetal adrenal hyperplasia leads to abortion.

The Bruce effect, interruption of pregnancy in mice due to the female's exposure to a strange male (or his urine) in the first five days post-coitus,[26] is probably due to an increase in LH secretion.[39] In the hen, a small sharp corticosterone peak is noted about the time of oviposition and is probably associated with the recently ruptured follicle.[230] Recent evidence also suggests a role for corticosterone in the ovulatory process itself (e.g., Cunningham et al.[48]). In hens, treatment with ACTH or corticosterone can induce premature ovulation, and some authors provide evidence that adrenal steroids control the temporal phasing of responsiveness to various hormones during the ovulatory cycle (see van Tienhoven[230]). In other avian species it is not uncommon to see large increases in plasma levels of corticosterone as breeding commences. High levels have been associated with egg-laying in females of several species.[250] In general these elevated corticosterone levels are below the very high range induced by stress. After breeding, circulating corticosterone levels drop to relatively low basal levels (e.g., Wingfield[243]).

B. Fishes

A relationship between interrenal activity and social dominance hierarchies was indicated in sunfish (*Lepomis cyanellus*) by Erikson[65] and in trout (*Salmo irideus*) by Boddingius,[21] but their methodologies could not establish what social attributes were relevant. Similarly in the European eel (*Anguilla anguilla*) crowding, confinement and consequent stress resulted in an increase in circulating levels of cortisol in subordinate fish.[180] Noakes and Leatherland[169] attempted to clarify the situation by contrasting interrenal cell activity in isolated, crowded, and status-stabilized rainbow trout, *Salmo gairdneri*, in a linear hierarchy. They found that isolates had significantly lower interrenal cell activity than did fish from crowded holding tanks and that interrenal cell activity generally was correlated inversely with dominance rank. Unexpectedly, however, top-ranking fish had activity levels comparable to those in crowded tanks or fourth- and fifth-ranking animals, a result the authors tentatively attribute to their relatively greater general or agonistic activity.

Hannes[93] observed that, when male and female swordtails (*Xiphophorus helleri*) mature together, baseline androgen and corticoid hormonal states of dominant and subordinate males are similar, but when animals unknown to each other aggressively establish a hierarchy, a positive correlation between androgen and dominance is observed.[94] Similar seemingly contradictory findings also are observed in primates and birds, and, as in the swordtails, may be reconciled partly by a consideration of past experience and differences between stable and unstable social conditions.

Leloup-Hatey[130] examined the CS levels in the blood of the carp, *Cyprinus carpio*, after chasing, catching them in a hand-net, and transferring them to other aquarium tanks. Fish which had been handled and rapidly sampled showed relatively low CS levels, but immediately after half-an-hour of chasing the levels had risen significantly. These elevated levels persisted for up to 2 h.

Stirring the water of an aquarium every 5 minutes significantly raised the blood levels of CS in coho salmon, *Oncorhynchus kisutch*, within one hour.[235] Similarly in chinook salmon (*Oncorhynchus tshawytscha*) caught at sea and held in tanks on board ship, a significant rise in plasma CS was observed.[92] However, in resting, adapted sockeye salmon (*Oncorhynchus nerka*) low CS levels are found[70] which are similar to levels observed in the plaice, *Pleuronectes platessa*, 1-4 days post-capture in summer.[242]

Stimulation of the interrenal gland and an increase in plasma CS after handling was observed in the carp and Atlantic salmon (*Salmo salar*),[130,131,156,196,219] chinook salmon,[92] sockeye salmon,[70] and plaice[238,242] (see also Chester Jones and Mosley,[41] Donaldson[56]). Pollutants also can influence interrenal activity. Stress by exposure of rainbow trout to hexavalent chromium elevated plasma corticosteroids after one week, although after two to three weeks these levels were not different from controls.[99] Crude oil elevated cortisol in carp, trout, and eel (e.g., Leloup-Hatey[132]). Forced swimming also raised plasma corticosteroid levels within a few minutes in the rainbow trout, but these levels were back to normal within 24 h.[99] Similar recovery of cortisol levels after stress occurred in brown trout, *Salmo trutta*.[186] Further evidence for adaptation comes from Wedemeyer[235] who demonstrated in coho salmon and rainbow trout that interrenal ascorbate levels decreased during stress and increased to normal if adaptation occurred. Serum cortisol, however, was elevated by non-specific stress and remained high after the interrenal ascorbate level had returned to its initial level.

The response of catecholamines to stressor stimuli also has been described in fishes.[167] Plasma concentrations of E, NE, and glucose increased in response to severe physical disturbance in the rainbow trout, but anterior kidney concentrations of E and NE did not change detectably.

Leucocytosis, which accompanies stress, was found to depend on the release of ACTH and an increase in corticosteroid production in *Mollienesia latipinna* and *Fundulus heteroclitus*.[9,215] The pituitary-interrenal stress syndrome that was evident in the cold shock-leucocyte stress sequence was described in *F. heteroclitus*. It was shown that intraperitoneal injection of high physiological doses of E could stimulate the entire cold-shock response (leucocytosis after 2 h). Injection of ACTH and cortisol were also effective in inducing the cold-shock response.[187-189] In brown trout, temperature shock elevated cortisol levels,[220] and cortisol treatment reduced the number of circulating lymphocytes. These may have been linked to increased susceptibility of stressed fish to disease.[183] Moreover, fungal infections can be induced by cortisol treatment,[184,185] and Mazeaud et al.[156] found that salmonids with fungal infections had extremely high levels of circulating cortisol.

Rasquin and Atz[195] have shown that in the Mexican cave fish (*Astynax mexicanus*) an increase in ACTH secretion in response to stress was accompanied by a decrease in thyroid-stimulating hormone (TSH) secretion, although the techniques used were indirect and need to be confirmed. Osborne[174] reported that plasma thyroid hormone levels fall in the plaice after capture, although the extent to which they remain depressed is uncertain. Triiodothyronine (T_3) showed a consistent fall, while thyroxine (T_4) was more variable, at first rising, and then falling after three days. Also, Wardle[234] demonstrated that plasma glucose levels rise following capture, but after two or three days these levels return to levels found at capture. In brown trout, however, Pickering and Christie[184] found no effect of stress on T_4 levels in blood.

In a more detailed study of capture stress in plaice, plasma levels of cortisol were high at capture, declined after two days, and then slowly increased to chronically high levels 5-10 days later. Further, handling and disturbance of aquarium-adapted fish increased plasma cortisol within 2 h. The circulating levels of cortisol observed after capture appear to be lower in summer than in winter, possibly the result of a seasonal variation in the activity of the plaice interrenal tissue.[242] Also, a fall in plasma cortisol 48 hours post-capture was observed in May, and no change was observed in January. Whether or not this represents a change in resistance to stress when breeding, as shown in birds, has yet to be determined. However, recent evidence from coho salmon indicates that there is an increased responsiveness to stress (demonstrated by elevated secretion of cortisol) during smoltification.[10]

Environmental temperature can exert profound influences on secretion of cortisol. Increasing water temperature from 1°C-20°C results in an elevation of circulating levels of cortisol in goldfish. Similarly, plasma levels of cortisol were low in killifish acclimated to seawater at -1°C and high in those exposed to 20°C.[179] Cortisol is known to enhance hepatic glycogen content and mobilization in both species. In the saltwater killifish at -1°C, increased cortisol may induce hyperglycemia which acts as an antifreeze in blood. The hypotonic tissues of the killifish could possibly freeze during winter. In contrast, the totally freshwater goldfish survives by moving into deeper water beneath the ice in winter, thus avoiding the possibility of freezing.[179] Increasing temperature also elevates cortisol levels in brown trout.[220] In the European eel, there is no change in plasma level of cortisol as temperature increases from 13-20°C. However, the metabolic clearance rate for cortisol increases two-fold indicating that the secretion rate also must have increased to maintain a constant plasma level.[132] However, this does not appear to be a general trend since another freshwater fish, *Tilapia aurea*, has high levels of cortisol in blood during the cool months of winter. High cortisol levels may be due to decreased metabolic clearance rate, although it also should be noted that these fish do not feed when in cool water and thus hyperadrenocortical activity could be induced by fasting.[229]

It is questionable whether the effects of temperature, within the limits normally experienced by an individual, constitute a "true stress" or merely acclimation and adaptation to the environment. Similarly the well-known responses of cortisol to osmotic challenge (fresh water to sea water and vice versa) fall under the same category. However, individuals can be exposed to unanticipated challenges by temperature and electrolytes that could constitute a "true" stressor. This division between predicted acclimation and unpredictable challenge clearly needs further investigation.

Many studies indicate a profound involvement of interrenal tissue in the activation or synchronization of sexual maturation and gonadal development in various teleosts (see Kime et al.[117]). Inhibition of maturation in platyfish by the presence of a socially dominant individual has been documented,[216] but no physiological mechanism has been suggested. We know that the prohormone, POMC, is present in salmonids,[116] that its products can delay maturation in other taxa, and that evidence has been presented for the presence of MSH in carp and goldfish.[75] Differential response of brown trout to stressors was documented by Sumpter et al.;[220] while handling increases circulating cortisol levels, handling combined with thermal shock also increases circulating MSH and endorphin levels.

The literature on the effects of stress on non-teleostean fish is sparse indeed. This may be due in part to a certain degree of controversy as to the endocrine status of "lower fish," especially with regard to steroid hormones (e.g., Idler[104]). Some information has come to light, however, concerning the effects of stress on agnathan fishes. In the river lamprey,

Lampetra fluviatilis, handling and anesthesia caused hyperglycemia even in hypophysectomized fish.[128] Whether or not this response is accompanied by increased CS levels has yet to be verified.

In ammocoete larvae of the sea lamprey, *Petromyzon marinus*, crowding resulted in an increase in blood levels of cortisol and corticosterone, as well as in elevations of dihydrotestosterone (DHT), androstenedione, estrone, estradiol and progesterone. Curiously, testosterone did not increase and was undetectable in most individuals.[49] These authors found that decreasing temperature from 12 to 3°C resulted in higher levels of corticosterone, whereas increasing temperature from 12 to 25°C had no effect. Temperature changes had no effect on cortisol or sex steroids. Dashow et al.[49] speculate that in primitive fish, sex steroids also may be involved in the stress response, although they also point out that the biological effects of these steroids in agnathans are still largely unknown.

The effects of stress on mature, landlocked *P. marinus* also have been investigated. Androstenendione increased in response to chronic surgery (heart cannulation), acute stress (agitation) and anesthesia, compared with fish killed quickly by decapitation. Progesterone levels increased only after chronic stress, and estradiol levels decreased after fish were exposed to anesthesia.[114] Clearly the responses of petromyzontids to stress differ markedly from other vertebrate taxa studied thus far. However, until the biological functions of steroid hormones in this group are understood, it is unlikely that the significance of increasing androstenedione levels in response to stress will be clarified.

Finally, blood levels of prolactin increase in tilapia, *Oreochromis* (*Sarotherodon*, *Tilapia*) *mossambicus*, exposed to acid stress.[237] This effect is probably linked to well-known effects of osmotic challenge on prolactin (see Chapter 4) and cortisol secretion in fishes (e.g., Leloup-Hatey[132]).

C. Amphibians

Immobilization stress in the frog, *Rana esculenta* resulted in an increase of plasma corticosterone and in the concentration of NE in the adrenal tissue. There was no change in adrenal content of E.[112] In addition, temperatures below 20°C had no effect on corticosterone, but when ambient temperature rose above 20°C, there was a significant elevation of plasma corticosterone.[112]

Plasma corticosterone and cortisol levels were determined in larval anurans by Krug et al.[121] who suggested that CS are important to the development and metamorphosis of amphibians. They proposed that thyroid hormone stimulates the maturation of the pituitary-interrenal axis, possibly by sensitizing the interrenal to ACTH. Capture of tiger salamander larvae (*Ambystoma tigrinum*) and subsequent confinement in tap water causes the abrupt onset of metamorphosis.[169a,191a,191b] The observed increase in plasma thyroid hormones and apparent reduction in PRL secretion[169a] may be a consequence of interrenal activation. Recently Stiffler et al.[218a] showed that osmotic stress does elevate circulating CS in *A. tigrinum* larvae.

Following a presumably stressful capture, bullfrogs (*Rana catesbeiana*), have reduced plasma Gn and androgen coincident with elevated plasma corticosterone.[138] In the field androgen levels of mature male bullfrogs show a trend to increase prior to establishment of territories and chorusing. Plasma androgen levels do not correlate with the behavior of individuals but do show a transitory depression when chorusing begins. Corticosterone levels, however, peak at the beginning of chorusing and are correlated (weakly) with the decline in androgen, which Mendonca et al.[160] associate with an inhibitory influence of corticosterone.

The most extensive program of research into stress-related gonado-suppression in amphibians is that of Moore and his colleagues. Using rough-skinned newts (*Taricha granulosa*) as a model, a correlation between exposure to presumed stressors and reduced incidence of courtship was observed. Courtship was suppressed by CS injections in a dose-dependent manner as well as by introcerebroventricular infusion with synthetic CRF. Metyrapone, which inhibits CS synthesis, reduces the suppressive effects of stress or CRF.[165] The decrease in circulating androgen seems to be associated with increases in infundibular and rostral hypothalamic GnRH, indicating a suppression of release.[166] The opioid peptide dynorphin was found to have a delayed (24 h) inhibitory effect on sexual behavior in *T. granulosa*, and there was some evidence of a possible synergism with CS.[162]

D. Reptiles

Adrenal involvement in the social stress responses of lizards was suggested by work on *Sceloporus*.[32,33] Brackin[22] observed that enlarged adrenal glands were associated with low social status in *Cnemidophorus*. Following agonistic interactions, subordinate status is presumably reinforced and maintained by a conspicuous morphological or behavioral attribute of high status in the dominant animal. Hierarchical status relationships are readily elicited in many species of lizards when grouped in captivity, but there is also a wide and apparently species-typical range of tolerance for the presence of one or more subordinates in different species. The establishment of dominance relationships in captivity can be responsible for impaired thermoregulation, access to shelter, and resistance to disease in various species of Anolis lizards.[13] Some reptilian species, when maintained in relatively small habitats, establish clear dominant-subordinate relationships which are upset when the animals are moved to larger habitats.[96,155] In these new habitats the subordinate presumably no longer is subject to the continual influence of some signal-quality of the dominant.[84] The tendency for many territorial species to form dominance hierarchies under special conditions in nature, such as a sudden local abundance of food,[67,170] indicates that the behavioral pattern is fixed firmly in the animal's biology.

Repetitive daily handling disturbances (small subcutaneous injections of normal saline) altered fat stores and male reproductive organs in the lizard, Anolis carolinensis. The stimulatory or inhibitory direction of the effects of handling was associated with the time of the disturbance: lizards injected with saline 4 h after the onset of their 16-h photoperiod had fat bodies ten times heavier and mean testes weights twice that of animals 20 h after onset of the photoperiod.[157]

In the painted turtle, Chrysemys picta, stress due to hemorrhaging (cardiac puncture followed by removal of 1-ml blood samples) increased plasma levels of corticosterone, whereas cardiac puncture alone had no effect. Further, neither hypophysectomy nor dexamethasone treatment blocked this response, suggesting that stress-induced increases in corticosterone are only partially mediated by the pituitary.[30,31] This is in contrast to the findings of Gist and Kaplan[81] in the crocodilian, Caiman crocodilus, in which dexamethasone did block the increase in plasma corticosterone that normally follows hemorrhage. American alligators, Alligator mississippiensis, stressed by handling and blood sampling showed elevations of corticosterone after 1, 4 and 24 h.[126]

The effect of any stressor that mobilizes an adrenal catecholamine response may inhibit oviposition in female A. carolinensis by means of beta-adrenoreceptor stimulation: Jones et al.[110] observed that arginine vasotocin (AVT), which stimulates contraction of uterine smooth muscle and induces oviposition in some lizards, does not do so in A. carolinensis females unless a beta-adrenoreceptor antagonist is administered shortly before the AVT injection. Jones et al.[110] suggest that this stress-sensitive control of uterine sensitivity in Anolis is related to its approximately 14-day alternate-ovary, single-egg ovulatory pattern.

Direct evidence associating relative social status with the adrenal stress response in cohabiting male lizards, A. carolinensis, was provided by Greenberg et al.[86] who measured levels of circulating corticosterone in the dominant and subordinate member of each pair. A significant elevation in circulating corticosterone was found in subordinates relative to dominants and was paralleled by a darkening of body color. The spermatogenetic stage of dominants and subordinates were not, however, significantly different.

In a preliminary study of the effects of corticosterone implants (Greenberg and Brown, unpublished data), male *A. carolinensis* with subcutaneous corticosterone pellets were not different from controls in their susceptibility to adrenal (chromaffin) activation as indicated by the appearance of the eyespot, a positive indication of epinephrine release.[88] Skin color, on the other hand, was significantly darker, suggesting a corticosterone-induced facilitation of MSH action or secretion. During bouts of aggression, animals implanted with corticosterone were typically first to display during a territorial encounter, but they invariably "lost" the interaction.

The skin darkening associated with reduced status is a striking attribute of *A. carolinensis*, and it occurs during fights as well as during the subsequent stabilization of the dominance relationship. Circulating E and NE are known to be affected by such acute social stress: after aggressive bouts, winners have 100 percent higher E levels and 50 percent higher NE levels than do losers, suggesting that losers have exhausted their reserve of these adaptive hormones.[85] It is not likely that corticosterone would affect chromatophores directly, although ACTH, which is structurally similar to MSH, probably will. Curiously, however, corticosterone-implanted animals with presumably inhibited secretion of ACTH did get darker. It is known that the dermal chromatophores of this species are, perhaps uniquely, devoid of direct sympathetic innervation,[119] and therefore affected only by blood-borne agents. Corticosterone could act by facilitating a key enzyme in the biosynthetic pathway for E,[256,257], known to stimulate the melanin-dispersing (skin-darkening) beta-adrenergic receptors on melanophores.[88] Elevated circulating corticosterone also might be related to increased MSH release from the pars intermedia of the pituitary gland. Circulating MSH levels in dominants and subordinates were measured by radioimmunoassay and found to be elevated significantly in subordinates relative to dominants (Greenberg and Chen, unpublished data).

III. EFFECTS OF REPRODUCTIVE PHYSIOLOGY AND BEHAVIOR ON STRESS RESPONSES

A. Overview of Mammals

The review by Parkes and Deanesly[177] from which much of the following background review is derived, took direction from Zuckerman's[258] observation that the embryologically related gonads and adrenals are more closely related than any other two pituitary-dependent glands. However, sex differences in the adrenal gland are known for humans and some rodents. The adrenal of the female is usually larger and heavier although in the golden hamster the opposite is the case. In addition, there are also numerous reports that adrenals are activated during pregnancy and that this activation continues through lactation.

Parkes and Deanesly[177] also reviewed effects of gonadectomy. A key observation of great interest is that castration leads to the appearance of an adrenal cortical X-zone in castrated male mice. This X-zone disappears from normal females given androgen (Parkes and Deanesly[177], p. 1068). Further, when castrated rats with uninhibited secretion are parabiotically joined to intact males, their gonads are stimulated and their adrenal size is decreased. However, when the parabiotic partner is a female, adrenals are stimulated to enlarge.[190] Adrenal enlargement in castrated males is common, and Parkes and Deanesley[177] believe that androgens retard adrenal growth while estrogens stimulate it. In certain strains of mice castration creates a disposition to develop adrenocortical tumors capable of secreting sufficient sex steroid to replace the lost gonads.[158] In rats, the effect was thought attributable to the lack of gonadal feedback inhibition of pituitary Gn, hypophysectomy caused regression of castration-induced tumors.[102]

Gonad-sensitive adrenal function, as indicated by correlations between adrenal activity and sex as well as by changes following manipulation of sex hormone by castration or hormone administration, are well-established (reviewed by Parkes and Deanesly[177]). More recently it has become clear that estrogen and androgen can affect hypothalamic-pituitary-adrenal (H-P-A) axis activity at several levels by means of different mechanisms and to different extents in males and females (reviewed by Kitay[118]). Both pituitary ACTH secretion and metabolism of CS by the liver apparently are stimulated by estradiol and inhibited by testosterone. Interestingly, however, small amounts of estradiol and testosterone enhance whereas larger amounts depress CS production. Although estradiol stimulates ACTH secretion, CS output, and hepatic clearance, testosterone inhibits ACTH release but stimulates CS output and prolongs hepatic clearance time. The controlling roles of steroid feedback and independent effects at different sites in the two sexes remain to be determined, and Kitay[118] was unable to provide a simple unitary hypothesis to deal with the effects of estradiol and testosterone on the H-P-A axis.

An increase in the density of opioid receptors in brain (homogenates) of castrated relative to intact rats[89,90] has drawn attention to the gonadal hormone modulation of possible feedback control of pituitary Gns by endogenous opiates (see Meietes and Briski[158]).

B. Fishes

Increases in levels of sex steroids during development of rainbow, steelhead, and brown trout are paralleled by interrenal hyperplasia and increased corticosteroid levels.[27,184,205] A causal relationship is indicated by a suppression of interrenal activity after castration and its reversal by sex steroid replacement.[111] Similarly in the goldfish,[46] white sucker (*Catostomus commersoni*),[208] common whitefish (*Coregonus lavaretus*),[78] and plaice,[238,242,247]) high levels of cortisol were measured in spawning or recently post-spawned fish. In the winter flounder (*Pseudopleuronectes americanus*), there was no cycle of plasma cortisol in females, but males had highest cortisol levels just after spawning.[35]

Glucocorticoid cycles are indicated in salmonid fishes by phenomena such as seasonal interrenal hyperplasia and increases in plasma 17-hydroxycorticosteroids including several gluco- and mineralocorticosteroids measured by sulphuric acid fluorescence after prior purification by silica-gel thin-layer and paper chromatography. These changes accompany migration and gonadal maturation in the Pacific species of salmon and in steelhead trout, reaching a peak at spawning.[91,106,197,206] Idler and Truscott[107] suggested that the high concentrations of 17-hydroxycorticosteroids in sexually mature and spawned Pacific salmon were caused by impaired clearance of plasma steroids associated with the moribund state of the fish rather than by hyperactivity of the adrenal tissue. An impaired metabolic clearance rate of cortisol has been demonstrated in moribund cod, *Gadus morhua*.[105] The concentrations of plasma cortisol in maturing and mature sockeye salmon during the freshwater stage of the life cycle were quite low, however, when the fish were resting in captivity.[70] The same is true of resting salmon in their natural habitat. Handling of fish, physical exercise, and disease increase cortisol concentrations markedly. Sexual maturation and spawning per se do not necessarily induce increases in healthy mature and maturing fish, indicating that the up-river migration rather than maturation may be responsible for elevated plasma corticosteroid levels in Pacific salmon. Plasma steroid concentrations by themselves, however, do not provide any information on the level of activity of the steroid-producing tissues. The metabolic clearance rate (MCR), resting secretion rate and plasma concentration of cortisol were all significantly higher in estradiol-treated female *O. nerka*. No significant differences between cortisol-injected fish and controls were found.[57] Cortisol secretion rates were significantly higher in *O. nerka* treated with

17-methyltestosterone than in controls, although cortisol concentrations in plasma during anesthetic stress were significantly lower in groups receiving androgens. The MCR of cortisol also increased with androgen administration.[71] From these results it is apparent that sex steroids can enhance cortisol production in _O. nerka_. In addition, sexual maturation in both sexes increased the volume of distribution of cortisol in both intravascular and extravascular pools as well as MCR and secretion rate of cortisol.[58] These changes were reversed if fish were gonadectomized just before reaching functional maturity. A peak of plasma cortisol observed in adult female plaice on spawning also may be a result of the preceding peak in sex steroids, but this would not explain a similar peak in plasma cortisol observed at the same time in immature fish in which there was no preceding increase in sex steroids.[247]

The Pacific salmon die after spawning, and a large proportion of steelhead trout also die, unlike most other teleosts which can spawn for many years. However, the rainbow trout and the nikko-iwana (_Salvelinus leucomaenis_), land-locked forms of migratory salmonids, do spawn repeatedly for several years. Adrenal hyperplasia and increased plasma CS levels accompany sexual maturation, reaching a peak at spawning and a low soon after spawning in these non-migratory species.[100,197] An exception is provided by the non-migratory and freshwater _T. aurea_. This species has low levels of cortisol when spawning and higher levels in winter associated with low water temperature and a period of no feeding. Furthermore, ovariectomy and injection of estradiol has no effect on plasma levels of cortisol.[229]

The function of elevated cortisol levels during breeding in most fish investigated is rather obscure. It has been shown that ovine LH and catfish (_Heteropneustes fossilis_) pituitary homogenate are potent ovulating agents when injected into gravid catfish.[221,223] However, they are virtually ineffective at equivalent or even higher doses under _in-vitro_ conditions.[83] Gonadal steroids also are ineffective _in vivo_ or _in vitro_. However, deoxycorticosterone (DOC) and cortisol and their acetates were found to be potent ovulating agents _in vivo_[221] as well as _in vitro_[225] although the doses of DOC and cortisol used to induce oocyte maturation in the catfish were extremely high.[225] Cortisol and DOC also were found to be the natural _in-vitro_ secretory products to be produced by endogenous precursors of catfish interrenals.[22,224] Cortisol and DOC in combination were more potent than either alone in inducing _in-vitro_ maturation of catfish oocytes.[225] A role for cortisol in the final maturation of oocytes also has been postulated for whitefish, rainbow trout, and northern pike (_Esox lucius_), possibly by lowering the threshold to maturation-inducing factors.[78,109] In the catfish cortisol appears to potentiate the effects of estradiol on vitellogenesis,[226] and plasma levels of cortisol peak during both vitellogenesis and ovulation.[125]

Corticotropin added to an incubation medium containing catfish interrenal slices resulted in production of cortisol and DOC in a ratio of 1.4:1, whereas addition of ovine LH altered this ratio to 1:3.[224]. These results suggest that gonadotropins may alter interrenal function and control ovulation through elevation of CS.

The cortisol peak observed in immature as well as adult plaice[247] is rather interesting. Changes in CS are associated with migration in fishes and additionally, Woodhead[252] reported that changes in the activity of the thyroid gland of the cod (*Gadus callarias*) during its migrations in the Barent's Sea also occurred during the "dummy run" of the immature cod. Later, Woodhead and Woodhead[254] showed that the interrenal hypertrophy accompanying the spawning migration of Barent's Sea cod also occurred to a lesser extent in immature fish. Plaice do undergo a spawning migration into deeper water, and although immature plaice also migrate into deeper water in the spring in the North Sea, it is not known whether similar migrations take place in the Irish Sea[241] where Wingfield and Grimm[247] sampled their fish. As Irish Sea migrations are likely to be rather limited in distance, it is perhaps unlikely that the observed cortisol peak is concerned primarily with migration.

It is possible that the rise in cortisol to a peak in March may simply be coincidental with spawning time. March usually marks the end of the winter period when food is in short supply. Mackinnon[144] has shown that in the American plaice (*Hippoglossoides platessoides*) energy is stored in the summer months for metabolism and gonadal maturation in the winter months. The closely related plaice has a similar reproductive cycle. Liver weight and liver glycogen content were maximal in October and coincided with the onset of vitellogenesis in the female. Insufficient data were collected for the male.[247] Cortisol levels are just beginning to rise at this time and, as gonadal maturation progresses and cortisol levels increase, liver weight and glycogen content decrease to a low at spawning. Thus, cortisol may be involved in the mobilization of glycogen, lipid and protein stores at this time and especially in vitellogenic females. In the Barent's Sea cod, increased interrenal activity in immatures and adults may be related to the winter period of poor feeding, and the greater activity in adult fish may be associated with the pronounced physiological demands of gonadal maturation.[254]

Similarly, in *T. aurea*, high levels of cortisol were measured in the blood of fish sampled in winter when most individuals are aphagic.[229]

In contrast, liver glycogen content of the male was consistently greater than that of the female throughout the anadromous river migration of *O. nerka*. In both sexes it decreases during the early stage of migration, but increases during the later stages, and levels at the spawning grounds are approximately twice those at the mouth of the river.[37] The increase in plasma corticosteroids during the river migration of *O. nerka* is possibly linked with liver glycogen deposition at the expense of protein.[206] Cortisol has been linked with tissue breakdown, particularly of skeletal muscle,[12,40] and the resultant amino acids possibly could be used in the synthesis of yolk protein during vitellogenesis. Cortisol also has been shown to increase liver glycogen in eels[29] and in *O. mossambicus*. Many fish cease to feed during gonadal maturation despite the extra nutritional requirements.[101] Fat and protein reserves are withdrawn from the liver as well as the muscles. This also may occur in the plaice as the body muscles in post-spawned fish appear very translucent and watery compared to October fish where the same muscles are firm and opaque. Thus cortisol could have a catabolic role in mobilizing muscle protein, liver glycogen, and lipid stores for vitellogenesis in the female and to a lesser extent for general metabolism in adult males and immatures over the winter period. In contrast to some mammals and all birds, many species of fish continue to grow after sexual maturity has been attained. Growth and reproduction are thus alternate activities in such species, and they show a seasonal decline in growth at the onset of maturation, the extent of which varies among species and often between sexes.[253]

In the Indian catfish, goldfish, pike, and rainbow trout, the effect of Gn on the ovary probably is regulated by interrenal CS; however, in the Japanese rice fish, *Oryzias latipes*, Gn appears to stimulate CS production by ovarian tissue, an effect observed in at least three other teleostean species. Catecholamines seem to have a role in this since they can induce ovulation *in vitro*, and this effect can be blocked by an alpha-aminergic receptor site antagonist (reviewed by van Tienhoven[230]). In the guppy, *Poecilia reticulata* (aplacental and viviparous), deoxycortisol often accelerates parturition; van Tienhoven[230] speculates that embryonic CS secretion would be an effective mechanism for the young to control their own destiny.

There is evidence that changes in sex steroids during development may influence catecholamine synthesis in the chromaffin tissue of rainbow trout. Both androgen and estrogen administration decreased *in-vitro* activities of dopamine-hydroxylase and phenylethanolamine-N-methyl transferase.[111]

C. Amphibians

Effects of the gonadal axis on chromaffin ("adrenomedullary") function were indicated by the work of Wilson et al.[240] who observed that GnRH elicits a modest (3X) NE response and large (20X) E response. They concluded that GnRH mobilizes catecholamines. The catecholamine pressor response, they suggest, is part of the coordinated neuroendocrine and cardiovascular preparation for seasonal reproductive activity.

Plasma testosterone levels were correlated positively with the immobility reaction (to skin pressure) in the crested newt (*Triturus cristatus carnifex*).[143] In reproductively active male toads (*Bufo bufo*), testosterone also was found to increase the duration of immobility responses, while response latencies were unaffected.[142]

Seasonal rhythms in circulating levels of glucocorticosteroids also have been investigated. Plasma levels of corticosterone in the American toad, *Bufo americanus*, were basal during winter peaking in April during breeding. Levels declined during the summer but peaked again in autumn just prior to hibernation.[175] A maximum of plasma corticosterone level also was observed in April in the edible frog, *Rana esculenta*. Further, distinct diel rhythms of corticosterone levels were observed in summer and autumn, but not in winter during hibernation.[61] In contrast, there were no annual changes in corticosterone in the blood of the giant swamp frog (*Discoglossus occipitalis*) sampled in an essentially seasonal environment in equatorial Africa.[122] Interestingly, this species appears to breed year round and does not hibernate, unlike anuran species in the north temperate latitudes.

D. Reptiles

Androgen may influence the CS response to enforced subordination in *A. carolinensis*. In one study, CS levels of castrates, who generally become subordinate despite frequently intense aggressive bouts, were examined.[86] Circulating CS levels of castrates were found to be essentially unaffected by reduced status, as was body color. The effect of castration on the MSH response to stress also is marked; castrates when stressed by chasing to exhaustion had circulating MSH levels 75% of control castrates while subjects with testosterone implants had only 34% of control levels. Also, castrates that won aggressive encounters (against other castrates) had 125% of control levels while testosterone-implanted winners had 56% of control levels (Greenberg and Chen, unpublished data).

Information on seasonal corticosteroid cycles and interactions with the reproductive system is sparse for reptiles. Corticosterone levels were generally low in the blood of _Sceloporus cyanogenys_ from January to April and then rose suddenly as breeding commenced in May.[52]

In the painted turtle, FSH, LH, PMSG (pregnant mare's serum gonadotropin) and estradiol stimulate increased corticosterone levels in females but not males. Both LH and PMSG increased corticosterone levels in female _Dipsosaurus dorsalis_, whereas estradiol was effective in males. Also, only estradiol increased corticosterone levels in male _S. cyanogenys_.[31]

IV. THE ADAPTIVE SIGNIFICANCE OF GONADAL-ADRENAL RECIPROCAL INTERACTIONS

From the literature cited above it is clear that a wide variety of stressors result in an increase in secretion of CS throughout the vertebrate classes. Similar effects on CRF, ACTH, β-endorphin, and MSH also appear to be widespread, although more comparative work is needed to strengthen this statement. These hormones also may be elicited differentially depending on the individual's history and the nature of the stressor. It is also clear that plasma levels of CS increase concomitantly with environmental change, reproduction, and other seasonal phenomena such as migration. With very few exceptions, plasma CS levels are elevated during, or just after, ovulation and oviposition, and may even play an important role in the maturation and ovulation process itself (at least in birds and fishes). Increases in adrenocortical activity also tend to accompany migration in birds and fishes, and smoltification in salmonids (a morphologic change) associated with down-river migration). In these species, CS may be involved in both the deposition of energy stores (e.g., fat) for migration and the mobilation of those stores during migratory movement itself. Temperature fluctuations, and transitions from fresh to salt water, or _vice versa_, also are accompanied by rises in blood levels of CS.

Thus, it appears that even normal and predictable transitions in physiologic state are associated with increased adrenocortical activity. Since such events as reproduction, migration, seasonal changes in temperature, and movements between fresh and salt water are "normal" processes that are usually predicted well in advance by an individual, it is perhaps not appropriate to refer to these phenomena as stressful (e.g., the "stress of reproduction" or the "stress of migration"). They are certainly energetically demanding and may even result in an individual being more susceptible to true stressors that are in general unpredictable (e.g., harassment by predators, reduced food availability, habitat destruction) or in some cases predictable in the short term (e.g., severe storms accompanied by marked temperature change, transient reduction in food, etc.). In these cases plasma levels of CS increase even further and often result in disruption of reproductive function.

Numerous laboratory experiments have demonstrated clearly that application of severe stressors can have profound influences on reproduction, sometimes resulting in complete suppression of reproductive function. This effect clearly appears to be mediated by a direct effect of CS on the hypothalamic-pituitary-gonadal axis. However, a few observations of effects of natural stressors on free-living individuals contradict the laboratory studies. For example, inclement weather during the breeding season of birds results in disruption of the reproductive cycle if the storm occurs after the actual nesting cycle has begun. Storms earlier in the year (i.e., before egg-laying commenced) are less likely to be stressful and may have only marginal effects on gonadal development. Abandonment of the breeding attempt during these periods of inclement weather is accompanied by increases in CS but no declines in LH and testosterone in males. This mechanism may be highly adaptive in allowing the normal stress-response (an increase in CS and declines in body mass and fat stores) but leaving the reproductive system unaltered so that when conditions improve a renesting attempt can begin immediately. If LH and testosterone were to decline, gonadal collapse would follow. It would then take several weeks for the gonads to return to a functional state by which time the breeding season would be over.[243] Clearly these mechanisms allowing a stress response, redirecting reproductive behavior temporarily but leaving the gonad intact, would increase reproductive success greatly. To date, this phenomenon has been demonstrated only in the class Aves. However, it is highly likely that similar or additional responses to stressors in free-living organisms of other vertebrate classes exist and simply await discovery.

It is also pertinent that endocrine responses to stress in captive individuals can be highly artificial. While individuals often accommodate stressors by means of a particular endocrine response pattern, an animal in the laboratory may be unable to manifest possible alternative mechanisms because of the necessarily restrictive nature of the captive environment. Thus the resultant decrease in reproductive function may be an artifact of captivity and not characteristic of the animal in the field. Under natural conditions a stress response (increase in CS) occurs, but the individual also may respond by changing behavior or temporarily halting the reproductive effort before irreparable damage is done. Such responses are highly adaptive but have not, as far as we are aware, been studied in any detail under laboratory conditions. A compromising investigative strategy was used by Greenberg[84] who replicated field conditions in the laboratory and then gradually reduced their complexity until only the relevant cues for the behavior of interest remained. However, this approach can accommodate only a few potentially relevant variables at best. Such efforts must be predicated on detailed field studies and are enormously time-consuming.

Changes in reproductive activity appear to influence the responses to at least acute stressors. Once again, in birds, females appear to delay the increase in CS that occurs in response to acute stressors such as capture and handling. More chronic stressors will eventually evoke a response, however. The delayed response to acute stressors during breeding also may be most marked in those species that breed at high latitudes and altitudes.[243] This may be adaptive since the reproductive season in these more extreme environments is very short. Resistance to disruptions due to acute stress at this time may increase the likelihood of reproductive success and counterbalance possible detrimental effects of the stressors themselves. Whether or not similar mechanisms exist in other vertebrate classes has yet to be determined.

Alternatively, in the lizard _A. carolinensis_, data suggest that the experimental reduction of androgen ameliorates the response to stress.[86] If subsequent work confirms that androgen is reduced spontaneously in subordinate lizards as it is in many mammals (e.g., Harding[95]) and birds (e.g., Wingfield and Ramenofsky[249]), we could posit a homeostatic mechanism by which the organism could avoid the potentially deleterious consequences of further stress responses caused by behavior that is activated or facilitated by androgen. In nature, such an individual may retain the capacity to respond relatively quickly to a future circumstance in which androgen levels may rise once again. A mechanism such as the inhibition of testicular activity sustained by elevating circulating CS[139] could contribute to this as could feedback of adrenal sex steroids on pituitary Gn release.[117,145]

Tentative conclusions suggest that increased secretion of CS accompanies all environmental change whether it be a normal seasonal or developmental progression or an environmental challenge beyond an animal's normal adaptive scope. In fact, CS secretion may be highly advantageous during many of these processes and only when the animal is confronted with unpredictable, or abnormal, stimuli is a more extreme or prolonged response evoked. However, even this extreme response is not necessarily deleterious since the individual in nature may have adaptive strategies not available in the laboratory. For example, some animals may hide, become torpid, or move away until conditions improve. Future investigators must consider this when interpreting laboratory data.

V. FUTURE RESEARCH DIRECTIONS

The great diversity in types of social organization (including mating systems) and the varying means by which specific organizations are established and maintained, as well as an animal's developmental (including experiential) history must caution us in our interpretation of apparent causal links between hormonal levels (or endocrine responsiveness) and reproductive condition. A broader appreciation of context and pattern will contribute much to liberate our understanding of the neuroendocrine responses to environmental challenge. In this regard, Giuliani[32] believes that "much confusion has risen as the result of a too naive acceptance of single, definable responses following noxious stimuli rather than heterogeneous attempts by the whole organism of achieving homeostatic balance" (p. 303).

Research on stress also is plagued by teleological language. Terms such as CRF, ACTH, MSH, now known to have functions beyond those of the targets for which they were named, and the false but common assumption that stress and adrenal activation is necessarily deleterious, have laid the "dead hand of tradition" on the imaginations of countless students. This is by no means a unique problem; misleading language was one of the false "idols" that Francis Bacon knew impeded the "Advancement of Learning."

Stress has been treated as part of a more general adaptive "arousal" system by Frankenhaeuser[76] who concluded that positive aspects of psychological arousal ("effort") and negative aspects ("distress") are associated differentially with catecholamine secretion and corticosteroid secretion, respectively. In this sense, the work of Ely and Henry[64] indicating a predominantly sympatho-adrenomedullary response pattern in dominant animals and a pituitary-adrenocortical pattern in subordinates, is interesting. The categories, if not the response, recall Selye's distinction between "eustress" and "distress," associated with desirable and undesirable effects of stressors, respectively.[212]

It is reasonable to propose stages in adaptation to social and psychological stressors not unlike the "phases" of alarm, resistance and exhaustion proposed by Selye (1956) for some somatic insults. We might term such stages "challenge, resolution, and stabilization." In the "challenge" phase, all participants in a stressful social interaction, in which each individual is unknown to the others, may respond with an initial surge of catecholamine, ACTH, CS, and testosterone regardless of subsequent rank (see Wingfield and Ramenofsky[249]). In contrast, when social relationships have already been established, individuals tend to show increases in catecholamines, ACTH and CS only (e.g., Sapolsky,[204] Wingfield and Ramenofsky[249]). Subsequent to "challenge," a phase of "resolution" can be identified in which levels of androgens and adrenomedullary hormones remain elevated in the prevailing animal while defeated individuals retain only the elevated adrenocortical hormones. In the final phase of "stabilization," individuals adapt species-typical adjusted basal rates of hormone secretion or endocrine response patterns characteristic of their changed circumstances (e.g., Ely and Henry[64]).

In many respects it is premature for future research to take any specific direction from reviews of stress and reproduction. Rather, a broader advance along all present frontiers is in order, but perhaps one more circumspect than we have often witnessed. Advance along more reductionist paths with its attendent increased discrimination and resolution must parallel integrative efforts, particularly now that we can appreciate the complexity of interactions and the diversity of effects of once thought-to-be function-dedicated tropic agents. In terms of reductionist work, clarification of mechanisms in diverse taxa seem most needed, while integrative-minded workers could fruitfully clarify the relationship of various classes of endocrine mechanisms to social organization and ecology, also in diverse taxa. Perhaps most importantly researchers along these two fronts must remain in mutual dialogue.

VI. ACKNOWLEDGMENTS

We are grateful for the support of the National Science Foundation (BNS-8406028 to Greenberg, DCB-8316155 to Wingfield), a University of Tennessee Faculty Research Award (Greenberg) and a Charles H. Revson Foundation Fellowship in Biomedical Research (Wingfield). We appreciate the helpful suggestions of Dr. Richard Strange. Present address of J.C. Wingfield is Department of Zoology NJ-15, University of Washington, Seattle WN 98195.

VII. REFERENCES

1. Aidara, D., Tahiri-Zagret, C. and Robyn, C. (1981). Serum prolactin concentration in mangabey (Cercobusatys lunulotus, and patas (Erythrocebus patas) monkeys in response to stress, ketamine, TRH, sulpiride and levodopa. J. Reprod. Fertil 62, 165-172.
2. Allen, T.O. and Haggett, B.N. (1977). Group housing of pregnant mice reduces copulatory receptivity of female progeny. Physiol. Behav. 19, 61-68.
3. Amir, S., Brown, Z.W. and Amit, Z. (1980). The role of endorphin in stress: evidence and speculations. Neurosc. Biobehav. Rev. 4, 77-86.
4. Andrew, R.J. (1978). Increased persistence of attention produced by testosterone, and its implications for the study of sexual behavior. In "Biological Determinants of Sexual Behavior" (J. R. Hutchison, ed.), pp. 254-275. Wiley, New York.

5. Archer, J. (1979). "Animals Under Stress." University Park Press, Baltimore.
6. Axelrod, J. and Reisine, T.D. (1984). Stress hormones: their interaction and regulation. Science 224,452-459.
7. Baker, B.I. and Buckingham, J.C. (1983). A study of corticotrophic and melanotrophic activities in the pituitary and brain of the lamprey *Lampetra fluviatilis*. Gen. Comp. Endocrinol. 52, 283-290.
8. Bambino, T.H. and Hsueh, A.J.W. (1981). Direct inhibitory effect of glucocorticoids upon testicular luteinizing hormone receptor and steroidogenesis *in vivo* and *in vitro*. Endocrinology 108, 2142-2148.
9. Ball, J.N. and Slicher, A.M. (1962). Influence of hypophysectomy and of an adrenocortical inhibitor (SU 4885) on the stress response of the white blood cells in the teleost fish (*Mollissima latipinna*). Nature 196, 1331-1332.
10. Barton, B.A., Schreck, C.B., Ewing, R.D., Hemmingsen, A.R. and Patino, R. (1985). Changes in plasma cortisol during stress and smoltification in coho salmon, *Oncorhynchus kisutch*. Gen. Comp. Endocrinol. 59, 468-571.
11. Beer, C.G. (1977). What is a display? Amer. Zool., 17, 155-165.
12. Bellamy, D., Leonard, R.A., Dulieu, K. and Stevenson, A. (1968). Starvation metabolism and plasma corticosterone with reference to the actions of metopirone and propylthiouracil. Gen. Comp. Endocrinol. 10, 119-125.
13. Bels, V.L. (1984). Ethological problems of the anoline lizards in captivity. Acta Zool. Pathol. Antverpiensa. 78, 85-100.
14. Berkenbosch, F., Tilders, F.J.H. and Vermes, I. (1983). Beta-adrenoreceptor activation mediates stress-induced secretion of beta-endorphin related peptides from the intermediate but not from the anterior pituitary. Nature 305, 237-239.
15. Berkenbosch, F., Vermes, I. and Tilders, F.J.H. (1984). The β-adrenoceptor-blocking drug propranolol prevents secretion of immunoreactive β-endorphin and α-melanocyte-stimulating hormone in response to certain stress stimuli. Endocrinology 115, 1051-1059.
16. Berliner, D.L., Keller, N. and Dougherty, T.F. (1969). Mediated and direct effects of ACTH and corticosteroids in stress. *In* "The Physiology and Pathology of Adaptation Mechanisms" (E. Bajusz, ed.), pp. 204-213. Pergamon Press, New York.
17. Bernstein, I.S. (1981). Dominance: The baby and the bathwater. Behav. Brain. Sci. 4, 419-457.
18. Bernstein, I.S., Gordon, T.P., Rose, R.M. and Peterson, M.S. (1978). Influences of sexual and social stimuli upon circulating levels of testosterone in male pigtail macaques. Behav. Biol. 24, 400-404.
19. Bertolini, A., Vergoni, W., Gessa, G.L. and Ferrari, W. (1969). Induction of sexual excitement by the action of adrenocorticotropic hormone in brain. Nature 221, 667-669.
20. Blank, M.S., Panerai, A.E. and Friesen, H.G. (1979). Opioid peptides modulate luteinizing hormone secretion during sexual maturation. Science 203, 1129-113.
21. Boddingius, J. (1976). The influence of social rank on adenohypophysial cell activity in *Salmo irideus*, with special reference to basophil ACTH-cells of the proximal pars distalis. Cell Tiss. Res. 170, 383-414.
22. Brackin, M.F. (1978). The reaction of rank to physiological state in *Cnemidophorus sexlineatus* dominance hierarchies. Herpetologica 34, 185-19.
23. Brain, P.F. and Evans, C.M. (1973). Some recent studies on the effects of corticotrophin on agonistic behavior in the house mouse and the golden hamster. J. Endocrinol. 57, 39-40.

24. Bronson, F.H. and Eleftheriou, B.E. (1965). Adrenal response to fighting in mice: separation of physical and psychological causes. Science 146, 627-628.
25. Bronson, F.H., Stetson, M.H. and Stiff, M.E. (1973). Serum FSH and LH in male mice following aggressive and nonaggressive interaction. Physiol. Behav. 10, 369-372.
26. Bruce, H.M. (1959). An exteroceptive block to pregnancy in the mouse. Nature 184, 105.
27. Bry, C. (1985). Plasma cortisol levels of female rainbow trout (Salmo gairdnerii) at the end of the reproductive cycle: Relationship with oocyte stages. Gen. Comp. Endocrinol. 57, 47-52.
28. Buckingham, J.C., Leach, J.H., Plisetskaya, E., Sower, S. and Gorbman, A. (1985). Corticotropin-like bioactivity in the pituitary gland and brain of the Pacific hagfish, Eptatretus stouti. Gen. Comp. Endocrinol. 57, 434-437.
29. Butler, D.G. (1968). Hormonal control of gluconeogenesis in the North American eel (Anguilla rostrata). Gen. Comp. Endocrinol. 10, 85-91.
30. Callard, G.V. (1975). Control of the interrenal gland of the freshwater turtle, Chrysemys picta, in vivo and in vitro. Gen. Comp. Endocrinol. 25, 323-331.
31. Callard, I.P. and Callard, G.V. (1978). The adrenal gland in reptilia. Part 2. Physiology. In "General, Comparative, and Clinical Endocrinology of the Adrenal Cortex" (I. Chester Jones, and I.W. Henderson, eds.), pp. 370-418. Academic Press, New York and London.
32. Callard, I.P. and Chester Jones, I. (1971). The effect of hypothalamic lesions and hypophysectomy on adrenal weight in Sceloporus cyanogenys. Gen. Comp. Endocrinol. 17, 194-202.
33. Callard, I.P., Chan, S.W.C. and Callard, G.V. (1973). Hypothalamic-pituitary-adrenal relationships in reptiles. In "Brain-Pituitary-Adrenal Relationships," pp. 270-212. Karger, Basel.
34. Callard, I.P. and Willard, E. III. (1969). Effects of intrahypothamic betamethazone implants on adrenal function in male Sceloporus cyanogenys. Gen. and Comp. Endocrinol. 13, 460-467.
35. Campbell, C.M., Walsh, J.M. and Idler, D.R. (1976). Steroids in the plasma of the winter flounder (Pseudopleuronectes americanus Walbaum). A seasonal study and investigation of steroid involvement in oocyte maturation. Gen. Comp. Endocrinol. 29, 14-20.
36. Candland, D.K. and Leshner, A.I. (1974). A model for agonistic behavior: endocrine and automatic correlates. In "Limbic and autonomic nervous systems research" (L.V. DiCara, ed.), pp. 137-164. Plenum Press, New York.
37. Chang, V.M. and Idler, D.R. (1960). Biochemical studies on sockeye salmon during spawning migration, XXI. Liver glycogen. Can. J. Biochem. Physiol. 38, 553-558.
38. Chapman, V.M., Desjardins, C. and Bronson, F.H. (1969). Social rank in male mice and adrenocortical response to open field exposure. Proc. Soc. Exp. Biol. Med. 130, 624-627.
39. Chapman, V.M., Desjardins, C. and Whiten, W.K. (1970). Pregnancy block in mice: changes in pituitary LH and LTH and plasma progestin levels. J. Reprod. Fertil. 21, 333-337.
40. Chester Jones, I., Bellamy, D., Chan, D.K.)., Follett, B.K., Henderson, I.W., Phillips, J.G. and Snart, R.S. (1972). Biological action of steroid hormones in non-mammalian vertebrates. In "Steroids in Non-Mammalian Vertebrates" (D.R. Idler, ed.), pp. 414-480. Academic Press, New York.
41. Chester Jones, I. and Mosley, W. (1980). The interrenal gland in pisces. Part 1. Structure. In "General, Comparative, and Clinical Endocrinology of the Adrenal Cortex." (I. Chester Jones, and I.W. Henderson, eds.), pp. 395-472. Academic Press, New York and London.

42. Christian, J.J. (1950). The adreno-pituitary system and population cycles in mammals. J. Mammal. 31, 247-259.

43. Christian, J.J. (1980). Endocrine factors in population regulation. In "Biosocial Mechanisms of Population Regulation" (M.N. Cohen, R.S. Malpass, H.G. Klein, eds.), pp. 55-115. Yale University Press, New Haven.

44. Christian, J.J. and Davis, D.E. (1964). Endocrines, behavior and populations. Science 146, 1550-1560.

45. Coe, C.L., Franklin, D., Smith, E.G. and Levine, S. (1982). Hormonal responses accompanying fear and agitation in the squirrel monkey. Physiol. Behav. 29, 1051-1059.

46. Cook, A.F., Stacey, N.E. and Peter, R.E. (1980). Periovulatory changes in serum cortisol levels in the goldfish, Carassius auratus. Gen. Comp. Endocrinol. 40, 507-510.

47. Crews, D.(1975). Psychobiology of reptilian reproduction. Science 189, 1059-1065.

48. Cunningham, F.C., Wilson, S.C., Knight, P.G. and Gladwell, R.T. (1984). Chicken ovulatory cycle. J. Exp. Zool. 232, 485-494.

49. Dashow, L., Katz, Y., Trachtman, M.S. and Epple, A. (1984). Plasma steroids in the ammocoete of Petromyzon marinus. Gen. Comp. Endocrinol. 55, 361-366.

50. Datta, P.C. and King, M.G. (1978). Effects of MIF-I and melatonin on novelty-induced defecation and associated plasma 11-OHCS and brain catecholamines. Pharmacol. Biochem. Behav. 11, 173-181.

51. Datta, P.C. and M.G. King (1980). Melatonin: effects on brain and behavior. Neurosci. Biobehav. Rev. 4, 451-458.

52. Daugherty, D.R. and Callard, I.P. (1972). Plasma corticosterone levels in the male iguanid lizard Sceloporus cyanogenys under the various physiological conditions. Gen. Comp. Endocrinol. 19, 69-79.

53. Deviche, P., Balthazart, J., Heyns, W. and Hendrick, J.-C. (1980). Endocrine effects of castration followed by androgen replacement and ACTH injections in the male domestic duck (Anas platyrhynchos L.). Gen. Comp. Endocrinol. 41, 53-61.

54. Deviche, P., Heyns, W., Balthazart, J. and Hendrick, J.-C. (1979). Inhibition of LH plasma levels by corticosterone administration in the male duckling (Anas platyrhynchos). I.R.C.S. Med. Sci. 7, 622.

55. DeWied, D. (1974). Pituitary-adrenal system hormones and behavior. In "The Neurosciences, Third Study Program" (F.O. Schmidt and F.G. Wordon, eds.), pp.653-663. Rockefeller University Press, New York.

56. Donaldson, E.M. (1981). The pituitary-interrenal axis as an indicator of stress in fish. In "Stress and Fish" (A.D. Pickering, ed.), pp. 11-47. Academic Press, New York.

57. Donaldson, E.M. and Fagerlund, U.H.M. (1969). Cortisol secretion rate in gonadectomized female sockeye salmon; effects of estrogen and cortisol treatment. J. Fish. Res. Bd. Can. 26, 1789-1799.

58. Donaldson, E.M. and U.H.M. Fagerlund. (1970). Effect of sexual maturation and gonadectomy at sexual maturity on cortisol secretion rate in sockeye salmon. J. Fish. Res. Bd. Can. 27, 2287-2296.

59. Dores, R.M. (1983). Further characterization of the major forms of reptile beta-endorphin. Peptides, 4, 897-905.

60. Dores, R.M. and Suprenant, A. (1983). Biosynthesis of multiple forms of β-endorphin in the reptile intermediate pituitary. Peptides 4, 889-896.

61. Dupont, W., Bourgeois, P., Reinberg, A. and Vaillant, R. (1979). Circannual and circadian rhythms in the concentration of corticosterone in the plasma of the edible frog (Rana esculenta). J. Endocrinol. 80, 117-126.

62. Eaton, G.G. and Resko, J.A. (1974). Plasma testosterone and male

dominance in a Japanese macaque (Macada fuscata) troop compared with repeated measures of testosterone in laboratory males. Horm. Behav. 5, 251-259.
63. El-Halawani, M.E., Silsby, J.L., Fehrer, S.C. and Behnke, E.J. (1985). The influence of acute or repeated immobilization on plasma prolactin levels in the turkey (Meleagris gallopavo). Gen. Comp. Endocrinol. 59, 410-415.
64. Ely, D.L. and Henry, J.P. (1978). Neuroendocrine response patterns in dominant and subordinate mice. Horm. Behav. 10, 156-169.
65. Erikson, J.G. (1967). Social hierarchy, territoriality, and stress reactions in sunfish. Physiol. Zool. 40, 40-47.
66. Euler, U.S.V. and Folkow, B. (1958). The effect of stimulation of autonomic areas in the cerebral cortex upon the adrenaline and noradrenaline secretion from the adrenal gland in the cat. Acta Physiol. Scand. 42, 313-320.
67. Evans, L.T. (1951). A field study of the social behavior of the black lizard Ctenosaura pectinata. Amer. mus. notivates, 1493, 1-26.
68. Everitt, B.J. and Herbert, J. (1969). Adrenal glands and sexual receptivity in female rhesus monkeys. Nature, 222, 1065-1066.
69. Everitt, B.J., Herbert, J. and Hamer, J.D. (1972). Sexual receptivity of bilaterally adrenalectomized female rhesus monkeys. Physiol. Behav. 8, 409-415.
70. Fagerlund, U.H.M. (1967). Plasma cortisol concentration in relation to stress in adult sockeye salmon during the freshwater stage of their life cycle. Gen. Comp. Endocrinol. 8, 197-207.
71. Fagerlund, U.H.M. and Donaldson, E.M. (1969). The effect of androgens on the distribution and secretion of cortisol in gonadectomized male sockeye salmon (Oncorhynchus nerka). Gen. Comp. Endocrinol. 12, 438-448.
72. Ferin, M. (1984). Endogenous opiod peptides and the menstrual cycle. Trends in Neurosci. 7, 194-196.
73. Folkow, B. and von Euler, U.S. (1954). Selective activation of noradrenaline and adrenaline producing cells in the cat's adrenal gland by hypothalamic stimulation. Circ. Res. 2, 191-195.
74. Follenius, E., van Dorsselaer, N. and Meunier, A. (1985). Separation and partial characterization by high performance liquid chromatography and radioimmunoassay of different forms of melanocyte-stimulating hormone from fish (Cyprinidae) neurointermediate lobes. Gen. Comp. Endocrinol. 57, 198-207.
75. Francis, M.G. and Peaslee, M.H. (1974). Effects of social stress on pituitary melanocyte-stimulating hormone activity in male mice. Neuroendocrinology 16, 1-7.
76. Frankenhaeuser, M. (1982). Challenge-control interaction as reflected in sympathetic-adrenal and pituitary-adrenal activity: comparison between the sexes. Scand. J. Psychol. Suppl. 1, 158-164.
77. Fryer, J.M. (1975). Stress and adrenocorticosteroid dynamics in the goldfish, Carassius auratus. Can. J. Zool. 53, 1012-1020.
78. Fuller, J.D., Scott, D.B.C. and Fraser, R. (1976). The reproductive cycle of Coregonus lavaretus (L) in Loch Lomond, Scotland, in relation to seasonal changes in plasma cortisol concentration. J. Fish. Biol. 9, 105-117.
79. Gauthreaux, S.A., Jr. (1978). The ecological significance of behavioral dominance. In "Perspectives in Ethology," Vol. 3 "Social Behavior" (P.P.G. Bateson and P.H. Klopfer, eds.), pp. 17-54. Plenum Press, New York.
80. Gessa, G.L., Paglietti, E., Pellegrina Quarantotti, B. (1979). Induction of copulatory behavior in sexually inactive rats by naloxone. Science, 204, 203-205.

81. Gist, D.H. and Kaplan, M.L. (1976). Effects of stress and ACTH on plasma corticosterone levels in the caiman, Caiman crocodilus. Gen. Comp. Endocrinol. 28, 413-419.
82. Giuliani, G. (1969). Studies on gonadotropin release during stressful situations and the role of the central nervous system. In "Physiology and Pathology of Adaptive Mechanisms" (E. Bajusz, ed.), pp. 293-309. Pergamon Press, New York.
83. Goswami, S.V. and Sundararaj, B.I. (1971). In vitro maturation and ovulation of oocytes of the catfish, Heteropneustes fossilis (Bloch): Effects of mammalian hypophyseal hormones, catfish pituitary homogenate, steroid precursors and metabolites, and gonadal and adrenocortical steroids. J. Exp. Zool. 178, 467-478.
84. Greenberg, N. (1978). Ethological considerations in the experimental study of lizard behavior. In "Behavior and Neurology of Lizards" (N. Greenberg and P.D. MacLean, eds.), pp. 203-224. NIMH, Rockville, MD.
85. Greenberg, N. (1983). Central and autonomic aspects of aggression and dominance in reptiles. In "Advances in Vertebrate Neuroethology" (J.-P. Ewert, R. R. Capranica, D.J. Ingle, eds.), pp. 1135-1143. Plenum Press, New York.
86. Greenberg, N., Chen, T. and Crews, D. (1984). Social status, gonadal state, and the adrenal stress response in the lizard, Anolis carolinensis. Hormones and Behavior, 18, 1-11.
87. Guillemin, R., Vargo, T.M., Rossier, J., Minick, S., Ling, N., Rivier, C., Vale, W. and Bloom, F. (1977). β-Endorphin and adrenocorticotrophin are secreted concomitantly by the pituitary gland. Science 197, 1367-1369.
88. Hadley, M.E. and Goldman, J.M. (1969). Physiological color changes in reptiles. Amer. Zool. 9, 489-504.
89. Hahn, E.F. and Fishman, J. (1979). Changes in rat brain opiate receptor content upon castration and testosterone replacement. Biochem. Biophys. Res. Commun. 90, 819-823.
90. Hahn, E.F. and Fishman, J. (1985). Castration affects male rat brain opiate receptor content. Neuroendocrinology 41, 60-63.
91. Hane, S. and Robertson, D.H. (1959). Changes in the plasma 17-hydroxycorticosteroids accompanying sexual maturation and spawning of the Pacific salmon (Oncorhynchus nerka) and rainbow trout (Salmo (gairdnerii). Proc. Nat. Acad. Sci. U.S.A. 45, 886-893.
92. Hane, S., Robertson, O.H., Wexler, B.C. and Krupp, M.A. (1966). Adrenocortical response to stress and ACTH in Pacific salmon and rainbow trout at successive states in the sexual cycle. Endocrinology 78, 791-800.
93. Hannes, R.-P. (1984). Androgen and corticoid levels in blood and body extracts of high- and low-ranking swordtail males (Xiphophorus helleri) before amd after social isolation. Z. Tierpsychol. 66, 70-76.
94. Hannes, R.-P., Franck, D. and Liemann, F. (1984). Effects of rank-order fight on whole-body and blood concentrations of androgens and corticosteroids in the male swordtail (Xiphophorus belleri). Z. Tierpsychol. 65, 53-65.
95. Harding, C.F. (1981). Social modulation of circulating hormone levels in the male. Amer. Zool. 21, 223-232.
96. Harris, V.A. (1964). "The life of the rainbow lizard." Hutchinson Tropical Monographs. Hutchinson and Co., London.
97. Harvey, S., Phillips, J.G., Rees, A. and Hall, T.R. (1984). Stress and adrenal function. J. Exp. Zool. 232, 633-646.
98. Herrenkohl, L.R. (1979). Prenatal stress reduces fertility and fecundity in female offspring. Science 206, 1097-1099.
99. Hill, C.W. and Fromm, P.O. (1968). Response of the interrenal gland of rainbow trout to stress. Gen. Comp. Endocrinol. 11, 69-77.

100. Honma, Y. and Tamura, E.1. (1965). Studies on the Japanese char, the iwana (Genus Salvelinus). I. Seasonal changes in the endocrine glands of the Nikko-Iwana, Salvelinus leucomaenis pluvius (Hilgendorf). Jpn. Soc. Sci. Fish. 31, 867-877.
101. Honmans, R.E.S. and Vladykov, V.D. (1954). Relation between feeding and the sexual cycle of the haddock. J. Fish. Res. Bd. Can. 11, 535.
102. Houssay, A.B., Cardega, A.F., Foglia, V.G. and Pinto, R.M. (1953). Adrenal tumors in gonadectomized rats. Acta Physiol. Latinoamer. 3, 125. (In Parkes and Deanesly, 1966.)
103. van den Hurk, R., Gielen, J.T. and Terlou, M. (1984). Accumulation of glycoprotein gonadotropin in the pituitary of juvenile rainbow trout in response to androgens and C-21 steroids, including 11-steroids. Cell Tiss. Res. 235, 636-642.
104. Idler, D.R. (1972). "Steroids in Non-Mammalian Vertebrates." Academic Press, New York and London.
105. Idler, D.R. and Freeman, H.C. (1965). A demonstration of impaired hormone metabolism in moribund Atlantic cod (Gadus morhua). Can. J. Biochem. Physiol. 43, 620-623.
106. Idler, D.R., Ronald, A.P., and Schmidt, P.J. (1959). Biochemical studies on sockeye salmon during spawning migration. VII. Steroid hormones in plasma. Can. J. Biochem. Physiol. 37, 1227-1238.
107. Idler, D.R. and Truscott, B. (1963). In vivo metabolism of steroid hormones, by sockeye salmon. (a) Impaired hormone clearance in mature and spawned Pacific salmon (Oncorhynchus nerka). (b) Precursors of 11-ketotestosterone. Can. J. Biochem. Physiol. 41, 875-887.
108. Iturriza, F.C., Estivariz, F.E. and Levitin, H.P. (1980). Co-existence of melanocyte-stimulating hormone, and adrenocorticotrophin in all cells containing either of the two hormones in the duck pituitary. Gen. Comp. Endocrinol. 42, 110-115.
109. Jalabert, B. (1976). In vitro oocyte maturation and ovulation in rainbow trout (Salmo gairdnerii), northern pike (Esox lucius), and goldfish (Carassius auratus). J. Fish. Res. Bd. Can. 33, 974-988.
110. Jones, R.E., Summers, C.H. and Lopez, K.H. (1983). Adrenergic inhibition of uterine contractions and oviposition in the lizard Anolis carolinensis. Gen. Comp. Endocrinol., 51, 77-83.
111. Jonsson, A-C. and Hansson, T. (1984). Influence of sex steroids on the catecholamine biosynthesis in chromaffin tissue from rainbow trout, Salmo gairdneri. Gen. Comp. Endocrinol. 55, 200-207.
112. Jurani, M., Murgas, K., Mikulaj, L. and Babusikova, F. (1973). Effects of stress and environmental temperature on adrenal function in Rana esculenta. J. Endocrinol. 57, 385-291.
113. Kastin, A.J., Miller, L.H., Nockton, R., Sandman, C.A., Schally, A.V. and Stratton, L.O. (1973). Behavioral aspects of melanocyte-stimulating hormone. Progr. Brain Res. 39, 461-470.
114. Katz, Y., Dashow, L., and Epple, A. (1982). Circulating steroid hormones of anadromous sea lampreys under various experimental conditions. Gen. Comp. Endocrinol. 48, 261-268.
115. Kawauchi, H. (1985). Salmon pituitary hormones: Evolutionary aspects of pro-opiocortin and related hormones. In "Current trends in Comparative Endocrinology" (B. Lofts and W.N. Holmes, eds.), pp. 155-157. Univ. Hong Kong Press, Hong Kong.
116. Kawauchi, H., Kawazol, I., Adachi, Y., Buckley, D.I. and Ramachandran, J. (1984). Chemical and biological characterization of salmon melanocyte-stimulating hormones. Gen. Comp. Endocrinol. 53, 37-48.
117. Kime, D.E., Vinson, G.P., Major, P.W. and Kilpatrick, R. (1980). Adrenal-gonadal relationships. In "General, Comparative and Clinical Endocrinology of the Adrenal Cortex," Volume 3 (I. Chester Jones and I.W. Henderson, eds.), pp. 133-264. Academic Press, New York.

118. Kitay, J.I. (1968). Effects of estrogen and androgen on the adrenal cortex of the rat. In "Functions of the Adrenal Cortex," Vol. 2 (K. McKerns, ed.), pp. 775-811. Appleton-Century-Corfts, New York.
119. Kleinholz, L.H. (1938). Studies in reptilian color changes. III. Control of light phase and behavior of isolated skin. J. Exp. Biol. 15, 492-499.
120. Kostowski, W., Rewerski, W. and Piechocki, T. (1970). Effects of some steroids on aggressive behavior in mice and rats. Neuroendocrinology 6, 311-318.
121. Krug, E.C., Honn, K.V., Battista, J. and Nicoll, C.S. (1983). Corticosteroids in serum of Rana catesbeiana during development and metamorphosis. Gen. Comp. Endocrinol. 52, 232-241.
122. Kuhn, E.R., Darras, V.M. and Gevaerts, H. (1985). Circadian and annual hormonal rhythms in amphibians. In "The Endocrine System and the Environment" (B.K. Follett, S. Ishii and A. Chandola, eds.), pp. 55-69. Japan. Sci. Soc. Press., Tokyo and Springer-Verlag, Berlin.
123. LaHoste, G.J., Olson, G.A., Kastin, A.J. and Olson, R.D. (1980). Behavioral effects of melanocyte stimulating hormone. Neurosci. Bio-Behav. Rev. 4, 9-16.
124. Lamba, V.J., Goswami, S.V. and Sundararaj, B.I. (1982). Radioimmunoassay for plasma cortisol, testosterone, estradiol-17β, and estrone in the catfish Heteropneustes fossilis (Bloch): Development and validation. Gen. Comp. Endocrinol. 47, 170-181.
125. Lamba, V.J., Goswami, S.V. and Sundararaj, B.I. (1983). Circannual and circadian variations in plasma levels of steroids (cortisol, estradiol-17β, estrone, and testosterone) correlated with the annual gonadal cycle in the catfish, Heteropneustes fossilis (Bloch). Gen. Comp. Endocrinol. 50, 205-225.
126. Lance, V.A. and Lauren, D. (1984). Circadian variation in plasma corticosterone in the American alligator, Alligator mississippiensis, and the effects of ACTH injections. Gen. Comp. Endocrinol. 54, 1-7.
127. LaPointe, J.L., Salas, M.A. and Hemer, J.H. (1985). Release of MSH from the pituitary gland of the lizard, Anolis carolinensis. In "Current Trends in Comparative Endocrinology" (B. Loft and W. N. Holmes, eds.), pp. 117-122. Univ. Hong Kong Press, Hong Kong.
128. Larsen, L.O. (1976). Blood glucose levels in intact and hypophysectomized river lampreys (Lampetra fluviatilis) treated with insulin, "stress," or glucose before and during the period of sexual maturation. Gen. Comp. Endocrinol. 29, 1-12.
129. Lehrman, D.S. (1965). Interaction of internal and external environments in the regulation of the reproductive cycle of the ring dove. In "Sex and Behavior" (F.A. Beach, ed.), pp. 355-380. Wiley, New York.
130. Leloup-Hatey, J. (1958). Influence de l'agitation motrice sur la teneur du plasma en 17-hydroxycorticosteroids d'un teleosteen la carpe. C.R. Acad. Sci. 246, 1088-1091.
131. Leloup-Hatey, J. (1964). Influence d'un accroissement de la salinite du milieu exterieur sur le fonctionnement d'interrenal anterieur de divers teleosteens. Arch. Sci. Physiol. 18, 293.
132. Leloup-Hatey, J. (1985). Environmental effects on the fish interrenal gland. In "The Endocrine System and the Environment" (B.K. Follett, S. Ishii and A. Chandola, eds.), pp. 13-22. Japan. Sci. Soc. Press, Tokyo and Springer-Verlag, Berlin.
133. Leshner, A.I. (1978). "An Introduction to Behavioral Endocrinology." Oxford University Press, New York.
134. Leshner, A.I. (1979). Kinds of hormonal effects on behavior: a new view. Neurosci. and Biobeh. Rev. 3, 69-73.
135. Leshner, A.I. and Politch, J.A. (1979). Hormonal control of submissiveness in mice: irrelevance of the androgens and relevance of the pituitary adrenal hormones. Physiol. Behav. 22, 531-534.

136. Levine, M.D., Gordon, T.P., Peterson, R.H. and Rose, R.M. (1970). Urinary 17-OHCS response of high- and low-aggressive rhesus monkeys to shock avoidance. Physiol. Behav. 5, 919-924.
136a. Levine, S. (1985). A definition of stress? In "Animal Stress" (G.P. Moberg, ed.), pp. 51-67. American Physiological Society, Bethesda, MD.
137. Li, C.H., Oelossen, W. and Naude, R.J. (1978). Adrenocorticotrophin 53. The amino acid sequence of the hormone from the ostrich pituitary gland. Biochem. Biophys. Res. Commun. 81, 900-906.
138. Licht, P., McCreery, B.R., Barnes, R. and Pang, R. (1983). Seasonal and stress related changes in plasma gonadotropins, sex steroids, and corticosterone in the bullfrog, *Rana catesbeiana*. Gen. Comp. Endocrinol. 50, 124-145.
139. Liptrap, R.M. and Raeside, J.I. (1978). A relationship between plasma concentrations of testosterone and corticosteroids during sexual and aggressive behaviour in the boar. J. Endocrinol. 76, 75-85.
140. Lis, M., Janos, J., Gutkowska, J., Genest, J. (1982). Corticotropin-releasing activity of α-melanotropin. Science, 215, 675-677.
141. Lott, D.F. (1984). Intraspecific variation in the social system of wild vertebrates. Behaviour 88, 266-325.
142. Lupo, C., Lodi, L. and Paluffi, G. (1985). Immobility reaction, plasma sex steroids and testosterone metabolism in the brain of mating and non-mating toads. Boll. Zool. 52, 177-180.
143. Lupo, C., Paluffi, G. and Carli, G. (1983). Behavioral and hormonal effects of captivity on the immobility reaction in crested newts (*Triturus cristatus carnifex*). Boll. Zool 50, 47-50.
144. MacKinnon, J.C. (1972). Summer storage of energy and its use for winter metabolism and gonad maturation in the American plaice (*Hippoglossoides platessoides*). J. Fish. Res. Bd. Can. 29, 1749-1759.
145. Mann, D.R., Jackson, G.G. and Blank, M.S. (1982). Influence of adrenocorticotropin and adrenalectomy on gonadotropin secretion in immature rats. Neuroendocrinology 34, 20-26.
146. Makara, G.B., Palkovits, M., and Szentogothai, J. (1980). The endocrine hypothalamus and the hormonal response to stress. In "Selye's Guide to Stress Research" (Hans Selye, ed.), Vol. I, pp. 280-337. Van Nostrand Reinhold, New York.
147. Martin, J.T. (1978). Imprinting behavior: pituitary-adrenocortical modulation of the approach response. Science 200, 565-567.
148. Martin, J.T. (1978). Embryonic pituitary adrenal axis, behavior development and domestication in birds. Amer. Zool. 18, 489-499.
149. Martini, L. and Ganong, W.G. (1966). "Neuroendocrinology." Academic Press, New York.
150. Maruniak, J.A., Desjardins, C. and Bronson, F.H. (1977). Dominant-subordinate relationships in castrated male mice bearing testosterone implants. Amer. J. Physiol. 233, E495-E499.
151. Mason, J.W. (1959). Psychological influences on the pituitary-adrenal cortical system. Rec. Progr. Horm. Res. 15, 345-378.
152. Mason, J.W. (1968). A review of psychoendocrine research on the pituitary-adrenal system. Psychosom. Med. 30, 576-607.
153. Matsui, H. (1981). Adrenal medullary secretory response to pontine and mesencephalic stimulation in the rat. Neuroendocrinology 33, 84-87.
154. Matsui, H. (1984). Adrenal medullary secretion in response to diencephalic stimulation in the rat. Neuroendocrinology 38, 164-168.
155. Mayhew, W.W. (1963). Observations on captive *Amphibolorus pictus*, an Australian Agamid lizard. Herpetologica 19 (2), 81-88.
156. Mazeaud, M.M., Mazeaud, F. and Donaldson, E.M. (1977). Primary and secondary effects of stress in fish: Some new data with a general review. Trans. Amer. Fish. Soc. 106, 201-212.

157. Meier, A.H., Trobed, T.N., Haymaker, H.G., MacGregor, R. III, and Russo, A.C. (1973). Daily variations in the effects of handling on fat storage and testicular weights in several vertebrates. J. Exp. Zool. 184, 281-288.
158. Meites, J. and Brisk, K. (1984). Relation of endogenous opiates to secretion of gonadotropins. In "Hormonal Control of the Hypothalamic-Pituitary-Gonadal Axis" (K.W. McKerns and Z. Naor, eds), pp. 509-519. Plenum Press, New York.
159. Meites, J., Bruni, J.F. and Van Vugt, D.A. (1979). Effects of endogenous opiate peptides on release of anterior pituitary hormones. In "Central Nervous System Effects of Hypothalamic Hormones and Other Peptides" (R. Collu, A. Barbeau, D.R. Ducharmie, G.G. Rochefort, eds.), pp. 261-268. Raven Press, New York.
160. Mendonca, M.T., Licht, P., Ryan, M.J. and Barnes, R. (1985). Changes in hormone levels in relation to breeding behavior in male bullfrogs (Rana catesbeiana) at the individual and population levels. Gen. Comp. Endocrinol. 58, 270-279.
161. Meys-Reolofs, H.M.A. and Moll, J. (1978). Sexual maturation and the adrenal glands. J. Reprod. Fertil. 52, 413-418.
162. Miller, L.J. and Moore, F.L. (1983). Intracranial administration of corticotropin-like peptides increases incidence of amphibian reproductive behavior. Peptides 4, 729-733.
163. Moberg, G. P. (1985). Influence of stress on reproduction: Measure of well-being. In "Animal Stress" (G.P. Moberg, ed.), pp. 246-267. American Physiological Society, Bethesda, MD.
164. Moltz, H. (1975). The search for the determinants of puberty in the rat. In "Hormonal Correlates of Behavior" (B.E. Eleftheriou and R.L. Spratt, eds.), Vol. 1, pp. 35-154. Plenum Press, New York.
165. Moore, F.L. and Miller, L.J. (1984). Stress-induced inhibition of sexual behavior: Corticosterone inhibits courtship behaviors of male amphibian. Horm. Behav. 18, 400-410.
166. Moore, F.L. and Zoeller, R.T. (1985). Stress-induced inhibition of reproduction: evidence of suppressed secretion of LHRH in an amphibian. Gen. Comp. Endocrinol. 60, 252-258.
167. Nakano, T. and Tomlinson, N. (1967). Catecholamine and carbohydrate concentrations in the rainbow trout (Salmo gairdnerii) in relation to physical disturbance. J. Fish. Res. Bd. Can. 24, 1701-1715.
168. Narasimham, C. and Kamat, D.N. (1970). A comparative study on the distribution of medullary (chromaffin) cell in the vertebrate adrenal glands. Cytologia 35, 409-414.
169. Noakes, D.L.G. and Leatherland, J.F. (1977). Social dominance and interrenal cell activity in rainbow trout, Salmo gairdneri (Pisces, Salmonidae). Env. Biol. Fish. 2, 131-136.
169a. Norris, D.O. (1978). Hormonal and environmental factors involved in the determination of neoteny in urodeles. In "Comparative Endocrinology" (P.J. Gaillard and H.H. Boer, eds.), pp. 109-112. Elsevier/North-Holland Biomedical Press, Amsterdam.
170. Norris, K.S. (1953). The ecology of the desert iguana, Dipsosaurus dorsalis. Ecology 34, 265-287.
171. Nowell, N.W. (1980). Adrenocortical function in relation to mammalian population densities and hierarchies. In "General, Comparative and Clinical Endocrinology of the Adrenal Cortex" (I. Chester Jones and I.W. Henderson, eds.), Vol. 3, pp. 349-393. Academic Press, New York.
172. Nowell, N.W., Thody, A.J. and Woodley, R. (1980). α-MSH and aggressive behavior in the male mouse. Physiol. Behav. 24, 5-9.
173. Nozaki, M. and Gorbman, A. (1985). Immunoreactivity for met-enkephalin and substance P in cells of the adenohypophysis of larval and adult sea lampreys, Petromyzon marinus. Gen. Comp. Endocrinol. 57, 172-183.
174. Osborne, R.H. (1971). Thyroid function in the plaice, Pleuronectes platessa L. Ph.D. Dissertation, Univ. Aberdeen, 187 pp.

175. Pancak, M.K. and Taylor, D.H. (1983). Seasonal and daily plasma corticosterone rhythms in American toads, Bufo americanus. Gen. Comp. Endocrinol. 50, 490-498.
176. Paris, A.L. and Ramaley, J.A. (1974). Adrenal-gonadal rhythms and fertility: The effects of repeated stress upon the adrenal rhythm. Neuroendocrinology 15,126-136.
177. Parkes, A.S. and Deanesly, R. (1966). Relation between the gonads and the adrenal glands. In "Marshall's Physiology of Reproduction" (A.S. Parkes, ed.), Vol. III, pp. 1064-1111. Little Brown and Co., Boston.
178. Patterson, A.T., Rickerby, J., Simpson, J. and Vickers, C. (1980). Possible interaction of melanocyte-stimulating hormone and the pineal in the control of territorial aggression in mice. Physiol. Behav. 24, 843-848.
179. Paxton, R., Gist, D.H. and Umminger, B.L. (1984). Serum cortisol levels in thermally acclimated goldfish (Carassius auratus) and killifish (Fundulus heteroclitus): Implications in control of hepatic glycogen metabolism. Comp. Biochem. Physiol. B 78, 813-816.
180. Peters, G., Delventhal, H. and Klinger, H. (1980). Physiological and morphological effects of social stress in the eel (Anquilla anquilla L). Arch. Firsch. Wiss. 30, 157-180.
181. Phillips, J.G. and Wright, A. (1966). Adrenocorticosteroids and fish migration. Proc. XI Pac. Sci. Cong. 7 (Symp. 55), 3. Abstract.
182. Pickering, A.D. (1981). Introduction: the concept of biological stress. In "Stress and Fish" (A.D. Pickering, ed.), pp. 1-9. Academic Press, New York.
183. Pickering, A.D. (1984). Cortisol-induced lymphocytopenia in brown trout, Salmo trutta L. Gen. Comp. Endocrinol. 53, 252-259.
184. Pickering, A.D. and Christie, P. (1981). Changes in the concentrations of plasma cortisol and thyroxine during sexual maturation of the hatchery reared brown trout, Salmo trutta L. Gen. Comp. Endocrinol. 44, 487-496.
185. Pickering, A.D. and Duston, J. (1983). Administration of cortisol to brown trout, Salmo trutta L., and its effects on the susceptibility to Saprolegnia infection and furunculosis. J. Fish Biol. 23, 163-175.
186. Pickering, A.D., Pottinger, T.G. and Christie, P. (1982). Recovery of brown trout, Salmo trutta L., from acute handling stress. J. Fish Biol. 20, 229-244.
187. Pickford, G.E., Srivastava, A.K., Slicher, A.M. and Pang, P.K.T. (1971). The stress response in the abundance of circulating leucocytes in the killifish, Fundulus heteroclitus. I. The cold shock sequence and the effects of hypophysectomy. J. Exp. Zool. 177, 89-96.
188. Pickford, G.E., Srivastava, A.K., Slicher, A.M. and Pang, P.K.T. (1971). The stress response in the abundance of circulating leucocytes in the killifish, Fundulus heteroclitus. II. The role of catecholamines. J. Exp. Zool. 177, 97-108.
189. Pickford, G.E., Srivastava, A.K., Slicher, A.M. and Pang, P.K.T. (1971). The stress response in abundance of circulating leucocytes in the killifish, Fundulus heteroclitus. III. The role of the adrenal cortex and a concluding discussion of the leucocyte-stress syndrome. J. Exp. Zool. 177, 109-118.
190. Pinto, R.M. (1945). Interrelations of adrenals and sex glands in parabiotic rats. Amer. J. Physiol. 144, 652. (In Parkes and Deanesly, 1966.)
191. Plas-Roser, S. and Aron, C. (1977). New data concerning the control by the adrenals of sexual receptivity in the rat. Physiol. Behav. 19, 57-60.

191a. Platt, J.E. (1974). The role of prolactin in neoteny as found in Colorado populations of the tiger salamander (Ambystoma tigrinum. I. responses to alterations in external electrolyte concentrations. Gen. Comp. Endocrinol. 62, 290-297.
191b. Platt, J.E. (1976). The effects of ergocornine on tail height, spontaneous and T_4-induced metamorphosis and thyroidal uptake of radioiodide in neotenic Ambystoma tigrinum. Gen. Comp. Endocrinol. 28, 71-81.
192. Proulx-Ferland, L., Labrie, F., Dumont, D. and Cote, J. (1982). Corticotropin-releasing factor stimulates secretion of melanocyte-stimulating hormone from the rat pituitary. Science 217, 62-64.
193. Ramalay, J.A. (1974). Adrenal-gonadal interactions at puberty. Life Sci. 14, 1623-1633.
194. Ramalay, J.A. (1979). Development of gonadotropin regulation in the prepubertal mammal. Biol. Reprod. 20, 1-31.
195. Rasquin, P. and Atz, E.H. (1952). Effects of ACTH and cortisone on the pituitary, thyroid, and gonads of the teleost Astynax mexicanus. Zoologica New York 37, 77-87.
196. Redgate, E.S. (1974). Neural control pituitary adrenal activity in Cyprinus carpio. Gen. Comp. Endocrinol. 22,35-41.
197. Robertson, O.H., Krupp, M.A., Thomas, S.F., Favour, C.B., Hane, S. and Wexler, B.C. (1961). Hyperadrenocorticism in spawning migratory and non-migratory rainbow trout (Salmo gairdnerii); comparison with Pacific salmon (Oncorhynchus sp.). Gen. Comp. Endocrinol. 1, 473-484.
198. Rodrigues, K.T., Jenks, B.G. and Sumpter, J.P. (1983). Biosynthesis of pro-opiomelanocortin-related peptides in the neurointermediate lobe of the pituitary gland of the rainbow trout (Salmo gairdnerii). J. Endocrinol. 98, 271-282.
199. Rohwer, W. and Wingfield, C. (1981). A field study of social dominance, plasma levels of luteinizing hormone, and steroid hormones in wintering Harris' sparrows. Z. Tierpsychol. 57, 173-183.
200. Rose, R.M., Gordon, T.P. and Berstein, I.S. (1972). Plasma testosterone levels in the male rhesus: influences of sexual and social stimuli. Science 178, 643-645.
201. Rowell, T.E. (1972). "The Social Behavior of Monkeys." Penguin, Baltimore, Md.
202. Rosell, T.E. (1974). The concept of social dominance. Behav. Biol. 11, 131-154.
203. Sandman, C.A., Kastin, A.J., Shally, A.V., Kendall, J.W. and Miller, L.H. (1973). Neuroendocrine responses to physical and psychological stress. J. Comp. Physiol. Psychol. 34, 386-390.
204. Sapolsky, R.M. (1982). The endocrine stress response and social status in the wild baboon. Horm. Behav. 16, 279-292.
205. Schmidt, S.P. and House, E.H. (1982). Plasma cortisol measurements in precocious male and non-precocious male juvenile steelhead trout (Salmo gairdnerii). J. Comp. Physiology B 146, 123-128.
206. Schmidt, P.J. and Idler, D.R. (1962). Steroid hormones in the plasma of salmon at various stages of maturation. Gen. Comp. Endocrinol. 2, 204-214.
207. Scott, A.P. and Baker, B.I. (1975). ACTH production by the pars intermedia of the rainbow trout pituitary. Gen. Comp. Endocrinol. 27, 193-202.
208. Scott, A.P., MacKenzie, D.S. and Stacey, N.E. (1984). Endocrine changes during natural spawning in the white sucker, Catostomus commersoni. II. Steroid hormones. Gen. Comp. Endocrinol. 56, 349-359.
209. Selye, H. (1937). Studies on adaptation. Endocrinology 21, 169-188.
210. Selye, H. (1939). The effect of adaptation to various damaging agents on the female sex organs in the rat. Endocrinology 25, 615-624.
211. Selye, H. (1956). "The Stress of Life." McGraw-Hill, New York.

212. Selye, H. (1976). "Stress in Health and Disease." Butterworth's, Boston.

213. Siegel, H.S. (1980). Physiological stress in birds. BioScience 30, 529-534.

214. Siegel, H.S. and Siegel, P.B. (1959). The relationship of social competition with endocrine weights and activity in male chickens. Anim. Behav. 9, 151-158.

215. Slicher, A.M. and Ball, J.N. (1962). Evidence for pituitary-adrenal mediation of leucocytosis induced by cold shock in fishes. Amer. Zool. 2, 449. (Abstract)

216. Sohn, J.J. (1977). Socially induced inhibition of genetically determined maturation in the platyfish, Xiphophorus maculatus. Science 195, 199-201.

217. Speiler, R.E. (1974). Short term serum cortisol concentrations in goldfish (Carassius auratus) subjected to serial sampling and restraint. J. Fish. Res. Bd. Can. 31, 1240-1242.

218. Steklis, H.D., Brammer, G.L., Raleigh, M.J. and McGuire, M.T. (1985). Serum testosterone, male dominance, and aggression in captive groups of vervet monkeys (Cercopithecus aethiops sabaeus). Horm. Behav. 19, 154-163.

218a. Stiffler, D.F., DeRuyter, M.L., Hanson, P.B. and Marshall, M. (1986). Interrenal function in larval Ambystoma tigrinum. I. Responses to alterations in external electrolyte concentrations. Gen. Comp. Endocrinol. 62, 290-297.

219. Strange, R.J., Schreck, C.B. and Golden, J.T. (1977). Corticoid responses to handling and temperature in salmonids. Trans. Amer. Fish. Soc. 106, 213-218.

220. Sumpter, J.P., Pickering, A.D. and Pottinger, T.G. (1985). Stress-induced elevation of plasma α-MSH and endorphin in brown trout, Salmo trutta. Gen. Comp. Endocrinol., 59, 257-265.

221. Sundararaj, B.I. and Goswami, S.V. (1966). Effects of mammalian hypophysial hormones, placental gonadotropins, gonadal hormones and adrenal corticosteroids on ovulation and spawning in hypophysectomized catfish (Heteropneustes fossilis). J. Exp. Zool. 16, 287-296.

222. Sundararaj, B.I. and Goswami, S.V. (1966). Effect of metopirone (SU-4885) on LH and corticosteroid-induced ovulation and spawning in hypophysectomized catfish, Heteropneustes fossilis (Bloch). J. Exp. Zool. 163, 49-54.

223. Sundararaj, B.I. and Goswami, S.V. (1968). Some aspects of induced spawning in the hypophysectomized catfish, Heteropneustes fossilis. Bull. Nat. Inst. Sci. India No. 36, 189-193.

224. Sundararaj, B.I. and Goswami, S.V. (1969). Role of interrenal in luteinizing hormone induced ovulation and spawning in the catfish, Heteropneustes fossilis (Bloch). Gen. Comp. Endocrinol. Suppl. 2, 374-384.

225. Sundararaj, B.I. and Goswami, S.V. (1971). Effects of deoxycorticosterone and hydrocortisone singly and in various combinations on in vitro maturation of oocytes of the catfish, Heteropneustes fossilis (Bloch). Gen. Comp. Endocrinol. 17, 570-575.

226. Sundararaj, B.I., Goswami, S.V. and Lamba, V.J. (1982). Role of testosterone, estradiol-17β, and cortisol during vitellogenin synthesis in the catfish, Heteropneustes fossilis (Bloch). Gen. Comp. Endocrinol. 48, 390-397.

227. Swaab, D.F. and Martin, J.T. (1981). Functions of α-melanotropin and other opiomelanocortin peptides in labour, intrauterine growth and brain development. In "Peptides of the Pars Intermedia," pp. 196-217. Ciba Foundation Symposium 81, Pitman Medical, London.

228. Swallow, R.L. and Flemming, W.R. (1970). The effect of oxaloacetate, ACTH, and cortisol on the liver glycogen levels of Tilapia mossambica. Comp. Biochem. Physiol. 36, 93-98.

229. Terkatin-Shimony, A., Ilan, Z., Yaron, Z. and Johnson, D.W. (1980). Relationship between temperature, ovarian recrudescence, and plasma cortisol level in *Tilapia aurea* (Cichlidae, Teleostei). Gen. Comp. Endocrinol. 40, 143-148.
230. van Tienhoven, A. (1983). "Reproductive Physiology of Vertebrates," 2nd edition. Cornell University Press, Ithaca, New York.
231. Vaudry, H., Jenks, B.G. and van Overbeeke, A.P. (1984). Biosynthesis, processing, and release of pro-opiomelanocortin related peptides in the intermediate lobe of the pituitary gland of the frog (*Rana ridibunda*). Peptides 5, 905-912.
232. Veldhuis, H.D. and DeWied, D. (1984). Differential behavioral actions of corticotropin-releasing factor (CRF). Pharmacol. Biochem. Behav. 21, 707-713.
233. Ward, I.L. and Weisz, J. (1980). Maternal stress alters plasma testosterone in fetal males. Science 207, 328-329.
234. Wardle, C.S. (1972). Some aspects of the physiology of the stress reaction of the marine teleost following capture from the wild. Ph.D. Dissertation, Univ. Aberdeen, 146 pp.
235. Wedemeyer, M. (1969). Stress-induced ascorbic acid depletion and cortisol production in two salmonid fishes. Comp. Biochem. Physiol. 29, 1247-1251.
236. Weiss, J., Kohler, W., Landsberg, J.W. (1977). Increase in the corticosterone level in ducklings during the sensitive period of the following response. Developmental Psychobiology 10, 59-64.
237. Wendelaar Bonga, S.E., van der Meij, J.C.A. and Flik, G. (1984). Prolactin and acid stress in the teleost *Oreochromis* (formerly *Sarotherodon*) *mossambicus*. Gen. Comp. Endocrinol. 55, 323-332.
238. White, A. and Fletcher, T.C. (1984). Radioimmunoassay of serum cortisol in the plaice (*Pleuronectes platessa*). Gen. Comp. Endocrinol. 53, 410-417.
239. Wilson, F.E. and Follett, B.K. (1975). Corticosterone-induced gonadosuppression in photostimulated tree sparrows. Life Sci. 17, 1451-1456.
240. Wilson, J.X., Van Vliet, B.N., and West, N.H. (1984). Gonadotropin-releasing hormone increases plasma catecholamines and blood pressure in toads. Neuroendocrinology 39, 437-441.
241. Wimpenny, R.S. (1953). The Plaice. Buckland Lectures for 1949, Arnold, London.
242. Wingfield, J.C. (1973). Pituitary-steroid interrelationships in the plaice, *Pleuronectes platessa* L. Ph.D. Dissertation, Univ. Wales, U.K., 163 pp.
243. Wingfield, J.C. (1984). Influences of weather on reproduction. J. Exp. Zool. 232, 589-594.
244. Wingfield, J.C. (1985). Influences of weather on reproductive function in male song sparrows, *Melospiza melodia*. J. Zool., Lond. (A) 205, 525-544.
245. Wingfield, J.C. (1985). Influences of weather on reproductive function in female song sparrows, *Melospiza melodia*. J. Zool. Lond. (A), 545-558.
246. Wingfield, J.C. and Farner, D.S. (1980). Endocrinologic and reproductive states of bird populations under environmental stress. Report cc 699095 to U.S. Environ. Prot. Ag., Washington, D.C., 122 pp.
247. Wingfield, J.C. and Grimm, A.S. (1977). Seasonal changes in plasma cortisol, testosterone, and oestradiol-17β in the plaice, *Pleuronectes platessa* L. Gen. Comp. Endocrinol. 31, 1-11.
248. Wingfield, J.C., Moore, M.C. and Farner, D.S. (1983). Endocrine responses to inclement weather in naturally breeding populations of white-crowned sparrows (*Zonotrichia leucophrys pugetensis*). Auk 100, 56-62.

249. Wingfield, J.C. and Ramenofsky. (1985). Testosterone and aggression in birds. *In* "Neurobiology and Behaviour" (R. Gilles and J. Balthazart, eds.), pp. 92-104. Springer-Verlag, Berlin.
250. Wingfield, J.C., Smith, J.P. and Farner, D.S. (1982). Endocrine responses of white-crowned sparrows to environmental stress. Condor 84, 399-409.
251. Wittenberger, J.F. (1981). "Animal Social Behavior," Duxbury Press, Boston.
252. Woodhead, A.D. (1959). Variations in the activity of the thyroid gland of the cod, *Gadus morhua*, L. in relation to its migrations in the Barent's Sea. II. The 'dummy run' of the immature fish. J. Mar. Biol. Assoc. U.K. 38, 407-415.
253. Woodhead, A.D. (1960). Presence of ACTH in the pituitary of the Arctic cod (*Gadus morhua*). J. Endocrinol. 21, 295-301.
254. Woodhead, A.D. and Woodhead, P.M.J. (1965). Seasonal changes in the physiology of the Barent's Sea cod, *Gadus morhua*, in relation to its environment. I. Endocrine changes particularly affecting migration and maturation. Int. Commun. N.W. Atl. Fish. Spec. Publ. No. 6, 691-716.
255. Wooley, G. (1958). Tumors of the adrenal cortex. "Ciba Foundation Coll. Endocrinol." 12, 122. (In Parks and Deanesly, 1966.)
256. Wurtman, R.J. and Axelrod, J. (1965). Adrenalin synthesis: control by the pituitary gland and adrenal glucocorticoids. Science 150, 1464-1465.
257. Wurtman, R.J., Axelrod, J. and Tramezzani, J. (1967). Distribution of the adrenalin-forming enzyme in the adrenal gland of a snake, *Xenodon merremii*. Nature 215, 879-880.
258. Zuckerman, S. (1952). The adreno-genital relationship. "Proc. Colston. Res. Soc." 5, 69. (J.M. Yoffey, ed.). (In Parkes and Deansely, 1966.)

REGULATION OF REPRODUCTIVE BEHAVIORS

Frank L. Moore

Department of Zoology
Oregon State University
Corvallis, Oregon 97331

Behavioral endocrinology is an emerging discipline dedicated to understanding hormone-behavior interactions. Because nonmammalian vertebrates have proven to be excellent research models for these studies, the comparative approach to behavioral endocrinology is a fast-growing subdiscipline.

Studies of hormone-behavior interactions take many approaches to answer two central questions: How do hormones modulate behaviors? How do behaviors modulate hormones? The first step in investigating hormone-behavior interactions is to determine which hormones and which anatomical sites are involved. Most of behavioral endocrinology is still at this stage.

There are surprisingly few examples in behavioral endocrinology where experiments provide convincing evidence of a causal relationship between a specific hormone and a particular behavior. To be convinced that Hormone A causes an increase in Behavior B, for example, researchers should demonstrate that 1) the experimental elevation of Hormone A increases the incidence of Behavior B, 2) the experimental suppression of Hormone A decreases the incidence of Behavior B, and 3) the endogenous concentrations of Hormone A are correlated positively with the natural occurrence of Behavior B. Most studies fail to satisfy conditions #2 and #3 because of the technical difficulties associated with suppressing just one hormone and with measuring specific hormone concentrations at the site of action.

The anatomical site of hormone action is difficult to determine, often because hormones appear to have multiple sites of action. In male South African clawed frogs (*Xenopus laevis*), for example, dihydrotestosterone (DHT) appears to control mate calling by acting at several sites, including a motor nucleus for the laryngeal motor neurons and a midbrain nucleus that responds to auditory stimuli.[31] These two sites are of particular interest because one is associated with sensory input and the other with motor output during mate calling. In *X. laevis* and other vertebrates, there are specific brain areas that appear to mediate the behavioral actions of gonadal steroids, such as the preoptic area, medial hypothalamus, limbic structures, and an area in the mesencephalon (reviewed by Kelley and Pfaff,[34] Morrell and Pfaff[61,62]). From these observations, Kelley[31] hypothesizes that gonadal steroid hormones act at all stations of the neural pathway controlling the behavior.

In reviewing research about the hormonal control of behavior in teleostean fishes, amphibians, and reptiles, this chapter emphasizes the short-term effects of hormones on reproductive behaviors, as opposed to the long-term developmental effects, the latter being covered in Chapter 1. Data are not available for cyclostomes and cartilaginous fishes.

I. FEMALE REPRODUCTIVE BEHAVIORS

To organize this review of the hormonal basis of female reproductive behaviors, behaviors are grouped into three categories: sexual attractivity, proceptivity, and receptivity. These terms are defined below to encompass reproductive behaviors of poikilotherms, adapting the definitions of Beach[4] that originally applied to estrus in female mammals. Sexual attractivity refers to those stimulus qualities of the female (visual, tactile, olfactory, auditory) which enhance sexual responses in conspecific males. Proceptivity refers to those behavioral reactions to a conspecific male, such as solicitation or affiliation, which facilitate sexual interactions prior to actual mating or spawning. Receptivity refers to those responses which are necessary and sufficient for mating or spawning.

A. Teleostean Fishes: Females

1. Sexual Attractivity. Morphological structures and pheromones influence the sexual attractivity of teleostean females. In some species, males respond to specific visual cues and are most likely to court gravid females rather than sexually immature females or conspecific males. The hormonal basis of the morphological component of attractivity is largely unknown, because female fish often have the appearance of sexually immature males, a morphology which generally results from the absence of testicular hormones rather than from the presence of feminizing hormones.

Some female teleosts secrete pheromones that attract and/or sexually excite male conspecifics (reviewed by Liley[39]). A pheromone in female goldfish (*Carassius auratus*), for example attracts sexually active males and is produced when estrogen levels are high or when exogenous estrogen is administered.[86] In female guppies (Poecilia reticulata), a pheromone is synthesized in the ovaries in response to estrogen administration.[46]

2. Proceptive and Receptive Behaviors. The hormonal basis of proceptive behaviors in female fish is poorly understood, primarily because most female fish do not exhibit pronounced proceptive behaviors. Perhaps this apparent lack of proceptive behaviors results from the sexual activity of the males, giving females such attention that proceptivity is rarely expressed. A notable exception is the paradise fish (Macropodus opercularis), in which males are highly territorial. Female paradise fish approach and solicit males with tailwags, flank bites, and flank butts.[81,82] Prostaglandin ($PGF_{2\alpha}$), if injected within a few days of ovulation, stimulates these behaviors indicating a relationship between the behavioral actions of $PGF_{2\alpha}$ and the gonadal state of the female.

Information about the hormonal control of receptive behaviors in female fishes comes from studies of only a few species (reviewed by Liley and Stacey[40], Stacey[74]). These studies can be divided according to the reproductive mode of the subject: ovoviviparous or oviparous (see Chapter 18 for discussion of reproductive modes).

The most studied ovoviviparous species is the guppy, P. reticulata.[38,41] Because female guppies are internally fertilized and can store sperm, mating behaviors often precede ovulation. Virgin females exhibit receptive behaviors from the time of ovarian development until mated. In virgin guppies, receptivity appears to be independent of ovarian steroids, because receptive behaviors are unaffected by ovariectomy. In multiparous females, in contrast, postpartum cycles of receptivity correspond with maximal ovarian steroidogenesis. These postpartum receptive behaviors are abolished by ovariectomy and restored by estrogen replacement.[38]

Goldfish (C. auratus) are the most studied of the oviparous teleosts, at least in terms of the hormonal basis of receptive behaviors (reviewed by Stacey[73,74]). In an aquarium, female goldfish with mature ovarian follicles ovulate in response to rising water temperature and to the presence of green vegetation, the substrate for egg deposition. Shortly after ovulation, female goldfish become receptive and spawn if sexually active males are present. Their receptive behaviors appear to be controlled by prostaglandins. In nonovulated females, injection of prostaglandins ($PGF_{2\alpha}$ and PGE_2) stimulates receptivity, whereas administration of indomethacin, which blocks prostaglandin synthesis, inhibits receptivity in females which have had ovulated oocytes injected into the reproductive tract.[72] It has been suggested that, in recently ovulated female goldfish, oocytes in the oviducts trigger prostaglandin synthesis; the prostaglandins then act on brain cells to enhance receptivity.[76,77]

Prostaglandins apparently regulate receptive behaviors of other female teleosts as well. Spawning behaviors of Pacific herring (*Clupea harengus*) are inhibited by indomethacin.[75] Oviposition behaviors of a cichlid fish (*Cichlasoma bimaculatum*) are stimulated by an injection of $PGF_{2\alpha}$.[7] In paradise fish, exogenous prostaglandins stimulate receptive behaviors if injected within a few days of ovulation,[81] indicating that ovarian steroids may modulate the behavioral actions of prostaglandins. In summary, the action of prostaglandins on receptive behaviors appears to be a widespread phenomenon in oviparous teleosts.

Neurohypophysial hormones such as arginine vasotocin (AVT) may regulate spawning behaviors of killfish (*Fundulus heteroclitus*).[42,46] An injection of AVT results in the activation of S-shaped postures, a behavior displayed by both male and female killifish when spawning. Because the spawning behaviors induced by exogenous AVT are not directed toward a sexual partner, this response should not be equated with the stimulation of receptivity.

B. Amphibians: Females

1. *Sexual Attractivity*. Studying sexual attractivity in amphibians is difficult because males generally are not very discriminating. Sexually active toads, for example, have been observed to attempt to clasp a person's rubber boot. Despite the fact that some male amphibians are not discriminating sexual partners, female amphibians do vary in sexual attractivity.

By measuring the percentage of males attempting to clasp particular females, Kelley[32] demonstrated that sexual attractivity in female *X. laevis* is enhanced by ovarian hormones and gonadotropin-releasing hormone (GnRH). Ovariectomized *X. laevis* were most attractive if treated with estradiol and progesterone plus GnRH. Moore[49] found that an injection of progesterone enhances sexual attractivity in intact female newts (*Taricha granulosa*). Plasma progesterone levels in female newts are correlated with sexual attractivity; when female newts first enter breeding ponds, both progesterone levels and sexual attractivity are highest.[52] Other hormones probably act with progesterone to enhance attractivity in *T. granulosa* females. Interestingly, neither Kelley[32] nor Moore[49] has identified the stimuli which enhance sexual attractivity.

2. Proceptive and Receptive Behaviors. The hormonal basis of receptive behaviors has not been studied in any urodele and, in anurans, has been studied in only *Rana pipiens* and *X. laevis.*

When clasped by a male, sexually unreceptive female *R. pipiens* emit release calls, a particular vocalization with abdominal vibrations. Conversely, receptive females remain quiescent when clasped. Thus, experimental inhibition of release calling can be interpreted as the induction of receptivity. Experimentally, an injection of AVT or prostaglandin (PGE_2 or $PGF_{2\alpha}$) inhibits release calls.[18] Because indomethacin antagonizes this inhibitory effect of AVT, the action of AVT on receptivity appears to involve prostaglandins.[17] In addition, there is evidence that receptivity in *R. pipiens* is independent of ovarian steroids; ovariectomy does not alter the inhibitory effect of AVT on release calls,[19] nor does the administration of estradiol and/or progesterone to ovariectomized females influence release calling.[20]

Diakow[17] hypothesized that the behavioral effect of AVT on release calling results from the uptake of water and the subsequent increase in internal pressure. Considerable evidence supports this hypothesis (reviewed in Diakow and Raimondi[19]). However, Diakow and Nemiroff[18] report that PGE_2 or $PGF_{2\alpha}$ injections inhibit release calls without stimulating water uptake and that indomethacin suppresses the behavioral effect of AVT without suppressing the water uptake effect of AVT. These observations appear to weaken the water uptake/internal pressure hypothesis and should stimulate further research.

Female *X. laevis*, if receptive when clasped by a male, contract their hindlegs (adduct the thighs and flex the knees) and do not emit ticking vocalizations.[32] These receptive behaviors can be stimulated in intact, but not ovariectomized, females by injections of gonadotropin (hCG) or GnRH.[32] In ovariectomized *X. laevis* replacement treatments with progesterone plus estradiol, but neither steroid alone, can maintain the receptive behaviors. This synergism between progesterone and estrogen apparently results from estradiol stimulating the formation of progesterone receptors in the hypothalamus-preoptic area of female frogs.[67a] Furthermore, because GnRH injections can enhance receptivity in ovariectomized females treated with progesterone plus estradiol, receptive behaviors in this amphibian appear to be controlled by progesterone, estrogen, and GnRH.[32]

C. Reptiles: Females

1. Sexual Attractivity. Female reptiles, as do other vertebrates, exhibit morphological, behavioral and olfactory cues that influence their

sexual attractivity. In two lizard species, *Holbrookia propiqua* and *Crotaphytus collaris*, females that are sexually active have bright lateral colorations which are controlled by gonadal steroids.[9,10] The hormonal control of olfactory cues in sexual attractivity of reptiles is best documented for garter snakes, *Thamnophis sirtalis* and *Thamnophis melanogaster*.[24] Males identify females by pheromonal cues and initiate courtship in response to a lipid-soluble substance, perhaps vitellogenin, that is on the skin of certain females (reviewed by Garstka et al.[23]). In *T. sirtalis* females, ovariectomy reduces sexual attractivity, and estrogen treatment enhances it. In both *T. sirtalis* and *Thamnophis radix*, females become sexually unattractive when mated. The seminal plug that is deposited during copulation contains a pheromone which inhibits courtship by other males.[66,85]

2. Proceptive and Receptive Behavior. To date, the hormonal control of proceptive behaviors in reptiles has not been investigated, and information about the control of receptive behaviors is limited to the lizard *Anolis carolinensis*.

In *A. carolinensis*, receptive females remain in the vicinity of courting males and, when approached, exhibit neck-bending behaviors. These receptive behaviors are correlated with the ovarian cycle; females are receptive from the period of late follicular development until ovulation.[11] Predictably, ovariectomy abolishes and estradiol treatment restores receptivity. Although high doses of estrogen can completely activate receptive behaviors in *A. carolinensis*, at low doses of estradiol, progesterone acts synergistically and enhances receptive behaviors.[45] These findings are consistent with the observed profiles of ovarian steroids in the plasma; when female lizards are receptive, there are low estradiol and high progesterone concentrations in the blood.[29] There is evidence that neuropeptides might also be involved in the regulation of receptivity in anoline lizards;[2] the administration of thyrotropin-releasing hormone (TRH) or GnRH stimulates receptivity in ovariectomized, estrogen-treated *A. carolinensis*.

In *A. carolinensis*, female receptivity is terminated within minutes of copulation. This post-copulatory decrease in receptivity can be prevented in intact, but not in ovariectomized, females by an injection of estrogen,[80] indicating that the ovaries are involved. Since an injection of prostaglandin (e.g., $PGF_{2\alpha}$) rapidly terminates receptivity in *A. carolinensis* females,[79] it is possible that copulatory stimuli trigger the release of prostaglandins from the ovaries and that the prostaglandins mediate the post-copulatory inhibition of receptivity. Likewise, female *T. sirtalis* become unreceptive when injected with $PGF_{2\alpha}$, indicating that the post-copulatory decrease in sexual receptivity in female snakes also may be mediated by prostaglandins.[84a]

II. MALE REPRODUCTIVE BEHAVIORS

The terms proceptive and receptive, which were defined originally for females, can be applied as well to males. Male courtship behaviors, in most cases, are proceptive behaviors because the stereotyped displays and solicitations maintain the sexual interactions with the females. Copulatory, mating, and spawning behaviors generally are receptive behaviors. However, some reproductive behaviors--nest building and territorial defense, for example--do not fit either category.

A. Teleostean Fishes: Males

1. Sexual Attractivity. Testicular androgens control many masculine traits among teleosts, including nuptial coloration, morphological shape, specialized intromittent organs, and mucus-secreting glands.[40] Some of these androgen-dependent traits undoubtedly enhance the attractiveness of the male to conspecific females. Female sticklebacks (*Gasterosteus aculeatus*), for example, select mates on the basis of color, an androgen-dependent trait.[68]

In certain male teleosts, sexual pheromones function to attract females or to induce ovulation and spawning (reviewed by Liley[39]). Male black gobies (*Gobius jozo*) have testicular glands that produce a sexual pheromone.[8] This pheromone, a 5β-reduced conjugated androgen, attracts female gobies. Moreover, injections of mammalian luteinizing hormone (LH) stimulate the production of pheromones in male *Blennius pavo*.[36] It seems likely, considering the temporal relationship between production of pheromones and reproductive cycles, that secretion of other pheromones also is under hormonal control.[40]

2. Proceptive and Receptive Behaviors. Teleosts exhibit a wide range of reproductive behaviors and patterns. The endocrine basis of reproductive behaviors has been investigated in a variety of species (reviewed by Liley and Stacey,[40] Stacey[74]). With surprisingly few exceptions, studies of teleosts provide evidence that testicular androgens are involved in the expression of male behaviors. Evidence supporting this conclusion includes the following observations: plasma androgen levels are correlated with reproductive behaviors, castration or the administration of antiandrogens suppresses the behaviors, and the administration of androgens (usually testosterone) increases the occurrence of male reproductive behaviors.

Despite evidence that testicular androgens are involved in masculine behaviors of teleosts, a number of important questions remain unanswered. For example, it has not been resolved which specific behaviors are androgen-dependent. There are inconsistencies in the literature regarding whether nest-building and aggressive behaviors (territorial and nonterritorial) are dependent on testicular androgens. Even for those reproductive behaviors that are androgen-dependent, there remains the issue of which testicular androgen, testosterone or 11-ketotestosterone, is the primary regulator. In rainbow trout (*Salmo gairdneri*) and several other species, spawning behaviors are correlated with plasma concentrations of 11-ketotestosterone, not testosterone.[67] Unfortunately, there are no published reports that compare the effectiveness of different testicular steroids in stimulating male behaviors in teleosts.

Another unresolved issue is whether gonadotropins can activate reproductive behaviors in teleosts, independent of the testes. In some teleosts, exogenous gonadotropin administration can stimulate reproductive behaviors in castrated males. However, these data may be explained by testicular regeneration in the castrates (see discussion by Stacey[74]).

Other hormones besides androgen may be involved in the control of reproductive behaviors in teleosts. For example, Hoar[28] found that the stimulatory effects of androgens on reproductive behaviors in castrated sticklebacks depend on the photoperiod. It is striking that neuropeptides have not been implicated in the reproductive behaviors of teleosts.

B. Amphibians: Males

1. Sexual Attractivity. Variations in male attractivity have been documented for a number of species of amphibians, particularly anurans.[78,84] No studies have demonstrated hormonal involvement in male sexual attractivity.

2. Proceptive and Receptive Behaviors. A number of studies demonstrate that castration causes a drop in reproductive behaviors of male amphibians (reviewed by Moore[51]). Kelley and Pfaff,[33] using male *X. laevis*, were the first to demonstrate that androgens can restore reproductive behaviors in a castrated amphibian. Subsequent studies with *T. granulosa* and *Triturus cristatus* confirm that androgens can maintain amplectic clasping and other courtship behaviors in castrates.[3,50] Recently, calling was restored in castrated male *X. laevis* by androgen administration.[83] Paradoxically, many studies found that the administration of androgen has no effect on amphibian sexual behaviors.[56,57,63]

The main difference between these studies with positive and negative results is in the use of sexually active males. Moore[50] found that castration abolishes reproductive behaviors and that implants of testosterone and DHT maintain sexual behaviors, but only in those males which are sexually active at the time of castration. These results support the conclusion that testicular androgens are necessary, but not sufficient, for the expression of amphibian sexual behaviors.[50]

This conclusion is consistent with studies of plasma androgen levels. In *T. granulosa*, *Ambystoma tigrinum*, *Rana catesbeiana*, and *Rana esculenta*, plasma androgen (testosterone plus DHT) concentrations change seasonally such that the lowest levels occur when the males are not breeding.[21,37,62a,71] But on any day, plasma testosterone (or DHT) concentrations are not significantly different in behaviorally active versus inactive male *T. granulosa*.[57]

It seems likely that hormones besides testicular androgens also are involved as regulators of reproductive behaviors, and evidence points to AVT as one such regulator. In *T. granulosa*, an injection of AVT can stimulate amplectic clasping in behaviorally inactive males during the breeding season.[63,58] Furthermore, clasping behaviors of *T. granulosa* are suppressed by intracranial injections of anti-AVT immune serum or injections of a synthetic AVT antagonist,[53] indicating that the site of action for AVT is in the brain, not in peripheral tissues that control water balance.

There is evidence that, in *T. granulosa*, testicular androgens maintain the behavioral response to AVT.[53,58,87] In castrated newts, the behavioral response to AVT persists for several weeks and then disappears completely, unless the castrates receive androgen treatment.[87] This gradual decline in responsiveness to AVT injections is reminiscent of the gradual postcastration decline in reproductive behaviors that occurs in most vertebrates.

Concentrations of AVT in specific brain areas of *T. granulosa* are correlated with reproductive behaviors (Zoeller and Moore, unpublished). The specific brain areas, in which AVT concentrations are significantly higher in behaviorally active than in behaviorally inactive males, include the dorsal preoptic areas, optic tectum, cerebrospinal fluid, and ventral infundibular nucleus. Another study of *T. granulosa* males found that AVT concentrations in the optic tectum change seasonally. Concentrations of AVT are low throughout the summer, fall, and winter, but rise over fivefold in the spring when sexual behaviors are most prevalent in male newts.[89]

The above studies support the conclusion that AVT regulates the initiation of courtship behavior in *T. granulosa*. Other neuropeptides also appear to be involved. An intracranial injection of hypothalamic GnRH stimulates courtship behaviors in _T. granulosa_ males.[55] Conversely, injections of synthetic GnRH antagonists inhibit reproductive behaviors. These responses to exogenous GnRH only occurred early in the breeding season, not during the months of March and April. Coincidentally, immunoreactive GnRH concentrations in newt hypothalami drop to undetectable levels during March and April.[88] The exact role of GnRH in amphibian behavior must be investigated further, particularly because there appears to be at least three different endogenous analogs of GnRH in _T. granulosa_.[70] (See also Chapters 2 and 3.)

Prolactin (PRL) also may be involved; injections of PRL into intact or hypophysectomized male newts (_T. cristatus_, _T. vulgaris_, _T. helveticus_) can increase the incidence of courtship behaviors.[43] For more information about PRL in amphibian behaviors see Chapter 4.

Opioid and corticotropin-like peptides also have been implicated in newt reproductive behaviors. Injections of corticotropin-like peptides (ACTH 4-10, MSH, or ACTH 1-39) can stimulate courtship behaviors in _T granulosa_, particularly if low doses are used.[48] In contrast, an intracranial injection of dynorphin, a potent mammalian opioid peptide, decreases the incidence of newt sexual behaviors.[47] However, the behavioral responses to opioid and corticotropin-like peptides are not nearly as consistent as the behavioral responses to AVT.

Exposure to stressful stimuli can interfere with reproduction and reproductive behaviors of _T. granulosa_.[54] When male newts are exposed to harsh stimuli--invasive surgery, extensive handling, high temperatures, repeated injections, etc.--there is an increase in plasma corticosterone and a decrease in reproductive behaviors. An injection of corticosterone inhibits reproductive behaviors within minutes. This and other evidence indicates that corticosterone mediates the stress-induced inhibition of courtship in _T. granulosa_.[54] Furthermore, it appears that, when stress causes plasma androgen levels to drop, corticosterone mediates this stress response by blocking the release of GnRH in _T. granulosa_.[59]

What is important here is the fact that, in male amphibians, there are two components in the control of reproductive behaviors. One component activates the behaviors; the other component inhibits the behaviors. Testicular androgens, PRL, GnRH, and AVT appear to function as activators; whereas corticosterone appears to function as an inhibitor.

C. Reptiles: Males

1. Sexual Attractivity. The hormonal basis of sexual attractivity in male reptiles has not been studied.[15]

2. Proceptive and Receptive Behaviors. The hormonal control of reproductive behaviors in male reptiles has been studied most extensively in the green anoline lizard (A. carolinensis) and garter snakes (Thamnophis spp) (reviewed by Crews,[9,10] Garstka et al.[23]). Anoline lizards are like other vertebrates in that testicular androgens are involved in regulating courtship and copulatory behaviors. When A. carolinensis are castrated, the frequency of courtship decreases significantly.[1,44] Predictably, the administration of testosterone or DHT to castrated males restores reproductive behaviors to precastration levels.[1,16,44]

In contrast to anoles, there is no evidence that garter snakes utilize testicular androgens (or any other gonadal steroid for that matter) to control male reproductive behaviors.[5] When T. sirtalis males are castrated, reproductive behaviors are unaffected. In both long-term and short-term castrates, courtship and copulatory behaviors occur during the normal breeding season. For example, males that are castrated at the end of one breeding season, prior to testicular recrudescence, exhibit normal courtship and copulatory behaviors during the next breeding season.[13] The administration of testosterone and other sex steroids to intact or castrated snakes has no effect, either positive or negative, on reproductive behaviors. Furthermore, adrenalectomy, hypophysectomy, castration, or the administration of a variety of hormones are all without effects on reproductive behaviors. Thus there are no hints, to date, as to which hormones are involved in activating reproductive behaviors in male garter snakes.

III. BEHAVIORAL ACTIVATION OF ENDOCRINE SYSTEM

Thus far, this chapter has discussed the hormonal control of reproductive behaviors. But this is only half the picture; the other half is the behavioral activation of the endocrine system by stimuli (olfactory or behavioral) emanating from the sexual partner(s). Keep in mind that in this section the emphasis is on the activation of the endocrine system, not on the stimuli which cause this response. (See the preceding sections on Sexual Attractivity for information about the endocrine control of the sexual stimuli which elicit behavioral activation.)

A. Fishes: Behavioral Activation

Examples of behavioral activation in teleosts are varied in form and apparent function. The first examples indicate that androgen concentrations can be influenced by feminine behavioral cues. When male sticklebacks (G. aculeatus) are exposed to conspecific females or motorized models of females, the androgen-dependent coloration in the males becomes more pronounced.[65] When male cichlids (Haplochromis burtoni) and swordtails (Xiphophorus helleri) are exposed to females, plasma androgen levels are elevated.[27]

Other studies indicate that masculine behavioral cues can activate reproductive hormones in female teleosts. Female angelfish (Pterophyllum scalare), for example, spawn more frequently when exposed to olfactory and visual stimuli from conspecific males.[6] Female zebrafish (Brachydanio rerio) frequently ovulate when sexually active males are placed in the same tank.[22]

There also is evidence that gonadal steroid levels increase during spawning. In Tilapia aurea, spawning females have elevated levels of testosterone, 11-ketotestosterone, and deoxycorticosterone.[30] Likewise, the concentrations of gonadotropin and testosterone increase in the plasma of male goldfish (C. auratus) that are exposed to spawning pairs.[35]

Perhaps the most extreme example of behavioral activation among vertebrates is that of sex change in coral reef fish. In Anthias squamipinnis, for example, sexually mature females can be induced to become males by removing the male from their social group.[69]

B. Amphibians: Behavioral Activation

It would seem that, with examples of chorusing in anurans and lengthy courtship in urodeles, there is ample opportunity for behavioral activation. however, only one species, T. granulosa, has been studied to date.

Female T. granulosa enter breeding ponds prior to ovulation, and only those females that are courted and inseminated ovulate.[52] If female newts are courted only, without insemination, progesterone concentrations are elevated without any signs of ovulation. When female newts are exposed to vasectomized males, normal courtship behaviors occur and an aspermic spermathophore is produced. But females do not ovulate when mated with vasectomized males (Spielvogel and Moore, unpublished data). These studies indicate that female newts are induced ovulators and that the trigger for ovulation is contained in the spermatophore and comes from the sperm duct.

C. Reptiles: Behavioral Activation

Crews[11] provides evidence that ovarian recrudescence in female A. carolinensis is enhanced by masculine behavioral cues. The presence of intact males in cages with females that are undergoing ovarian recrudescence enhances the rate of ovarian development. Without males, anoline females produce unshelled eggs. When males are prevented from extending their dewlap, by hyoidectomy, females no longer exhibit enhanced ovarian development, indicating that the dewlap display is a critical component of male courtship in this species.

The whiptail lizard, *Cnemidophorus uniparens*, is a parthenogenetic species that provides an interesting example of behavioral activation. Individual females alternate between exhibiting male and female sexual behaviors, depending on ovarian state.[14,60] The pseudomale mounting behavior has been shown to increase reproductive output in the females that are mounted.[26]

Another interesting example of behavioral activation occurs in the semiaquatic garter snake, *T. melanogaster*.[25] Male courtship behaviors were not observed in a wild population of these snakes until a group of females were treated with estrogen and released. The estrogen-treated females are sexually attractive and induce courtship in the males. These males, after 6 to 8 days of sexual activity, exhibit marked testicular development and elevated androgen concentrations. These observations support the hypothesis that in *T. melanogaster*, the reproductive state of the female is the proximate factor controlling the reproductive cycle in the male.

IV. SUMMARY AND FUTURE RESEARCH

Comparative behavioral endocrinology is a young field, replete with unanswered questions about the hormonal control of reproductive behaviors in fishes, amphibians, and reptiles. In general, testicular steroid hormones (testosterone, dihydrotestosterone, and estradiol) activate male reproductive behaviors and ovarian steroid hormones (estradiol and progesterone) activate female reproductive behaviors. In addition, neuropeptides (particularly GnRH, PRL, and AVT) and prostaglandins appear to activate reproductive behaviors in some fishes, amphibians, and reptiles. However, other species apparently control reproductive behaviors with other hormones (courtship behaviors in male garter snakes, for example, appear to be independent of testicular steroid hormones). These results, particularly when compared to results with birds and mammals, indicate that there is some evolutionary conservatism as well as species specificity in the control of reproductive behaviors of vertebrates.

Any given reproductive behavior is probably controlled by a suite of hormones, not by a single hormone. Studies of amplectic clasping in male rough-skinned newts, for example, indicate that this behavior is activated by several steroids and neuropeptides (testosterone, dihydrotestosterone, estradiol, AVT, GnRH and MSH) and is inactivated by several stress-related hormones (corticosterone, corticotropin-releasing hormone, and opioid peptides). Such complexity probably imparts better control of the behaviors. There are adaptive advantages to those animals in which fine-tuned regulatory mechanisms activate reproductive behaviors only at appropriate times and in appropriate places.

The need for additional research is apparent. Most investigations cited in this chapter are injection type of studies--inject the hormone and then record the behaviors. This type of study, alone, is not adequate to prove a causal relationship. To date, unequivocal evidence for a causal relationship between any hormone and a particular reproductive behavior is absent in fishes, amphibians, and reptiles. Furthermore, there is a paucity of information about the functional interactions among the hormones as well as the sites of action and mechanisms of actions of the hormones. Although there are many unanswered questions about the effects of hormones on behaviors, there is even less known about the effects of behaviors on hormones.

V. REFERENCES

1. Adkins, E. and Schlesinger, L. (1979). Androgens and the social behavior of male and female lizards (Anolis carolinensis). Horm. Behav. 13, 139-152.
2. Alderete, M.R., Tokarz, R.R. and Crews, D. (1980). Luteinizing hormone-releasing hormone and thyrotropin-releasing hormone induction of female sexual receptivity in the lizard, Anolis carolinensis. Neuroendocrinology 30, 200-295.
3. Andreoletti, G.E., Malacarne, G. and Vellano, C. (1983). Androgen control of male sex behavior in the crested newt (Triturus cristatus carnifex Laur.): Castration and sex steroid administration. Horm. Behav. 17, 103-110.
4. Beach, F.A. (1976). Sexual attractivity, proceptivity, and receptivity in female mammals. Horm. Behav. 7, 105-138.
5. Camazine, B., Garstka, W., Tokarz, R. and Crews, D. (1980). Effects of castration and androgen replacement on male courtship behavior in the red-sided garter snake (Thamnophis sirtalis parietalis). Horm. Behav. 14, 358-372.
6. Chien, A.K. (1973). Reproductive behaviour of the angelfish Pterophyllum scalare (Pisces: Cichilidae) II. Influence of male stimuli upon the spawning rate of females. Anim. Behav. 21, 456-463.
7. Cole, K.S. and Stacey, N.E. (1984). Prostaglandin induction of spawning behavior in Cichlasoma bimaculatum (Pisces Cichlidae). Horm. Behav. 18, 235-248.
8. Colombo, L., Marconato, A., Belvedere, P.C. and Fiso, C. (1980). Endocrinology of teleost reproduction: A testicular steroid pheromone in the black goby, Gobius jozo L. Boll. Zool. 47, 355-364.
9. Cooper, W.E., Jr. (1984). Female secondary sexual coloration and sex recognition in the keeled earless lizard, Holbrookia propinqua. Anim. Behav. 32, 1142-1150.
10. Cooper, W.E., Jr. and Ferguson, G.W. (1972). Steroids and color change during gravity in the lizard Crotophytus collaris. Gen. Comp. Endocrinol. 18, 69-72.
11. Crews, D. (1975). Psychobiology of reptilian reproduction. Science 189, 1059-1065.

11. Crews, D. (1975). Psychobiology of reptilian reproduction. Science 189, 1059-1065.
12. Crews, D. (1984). Gamete production, sex hormone secretion, and mating behavior uncoupled. Horm. Behav. 18, 22-28.
13. Crews, D., Camazine, B., Diamond, M., Mason, R., Tokarz, R.R. and Garstka, W.R. (1984). Hormonal independence of courtship behavior in the male garter snake. Horm. Behav. 18, 29-41.
14. Crews, D. and Fitzgerald, K. (1980). "Sexual" behavior in parthenogenetic lizards. Proc. Natl. Acad. Sci. USA 77, 499-502.
15. Crews, D. and Greenberg, N. (1981). Function and causation of social signals in lizards. Amer. Zool. 21, 273-294.
16. Crews, D., Traina, V., Wetzel, F.T. and Muller, C. (1978). Hormonal control of male reproductive behavior in the lizard, *Anolis carolinensis*: Role of testosterone, dihydrotestosterone, and estradiol. Endocrinology 103, 1814-1821.
17. Diakow, C. (1978). A hormonal basis for breeding behavior in female frogs: Vasotocin inhibits the release call of *Rana pipiens*. Science 199, 1456-1457.
18. Diakow, C. and Nemiroff, A. (1981). Vasotocin, prostaglandin, and female reproductive behavior in the frog, *Rana pipiens*. Horm Behav. 15, 86-93.
19. Diakow, C. and Raimondi, D. (1981). Physiology of *Rana pipiens* reproductive behavior: A proposed mechanism for inhibition of release call. Amer. Zool. 21, 295-304.
20. Diakow, D., Wilcox, J.N. and Woltmann, R. (1978). Female frog reproductive behavior elicited in the absence of the ovaries. Horm. Behav. 11, 183-189.
21. D'Istria, M., Delrio, G., and Chieffi, G. (1974). Radioimmunoassay of testosterone, 17β-oestradiol and oestrone in the male and female plasma of *Rana esculenta* during sexual cycle. Steroids Lipids Res. 5, 42-48.
22. Eaton, R.C. and Farley, R.D. (1974). Spawning cycle and egg production in zebrafish, *Brachydanio rerio*, in the laboratory. Copeia 1974, 195-204.
23. Garstka, W.R. and Crews, D. (1981). Female sex pheromone in the skin and circulation of a garter snake. Science 214, 681-683.
24. Garstka, W.R. and Crews, D. (1982). Female control of male reproductive function in a Mexican Snake. Science 217, 1159-1160.
25. Garstka, W.R., Camazine, B. and Crews, D. (1982). Interactions of behavior and physiology during the annual reproductive cycle of the red-sided garter snake (*Thamnophis sirtalis parietalis*). Herpetologica 38, 104-123.
26. Gustafson, J. and Crews, D. (1981). Effects of group size and physiological state of a cagemate on reproductive effort in the parthenogenetic lizard, *Cnemidophorus uniparens*. Behav. Ecol. Sociobiol. 8, 267-272.
27. Hannes, R. and Franck, D. (1983). The effect of social isolation on androgen and corticosteroid levels in a cichlid fish (*Haplochromis burtoni*) and in swordtails (*Xiphophorus helleri*). Horm. Behav. 17, 292-301.
28. Hoar, W.S. (1962). Hormones and the reproductive behaviour of the male three-spined stickleback (*Gasterosteus aculeatus*). Anim. Behav. 10, 247-266.
29. Jones, R.E., Guillette, L.J., Summers, C.H., Tokarz, R.R. and Crews, D. (1983). Ovarian follicular and luteal condition, and related steroid hormone levels, during the estrous cycle of the lizard, *Anolis carolinensis*. J. Exp. Zool. 227, 145-154.
30. Katz, Y. and Eckstein, B. (1974). Changes in steroid concentration in blood of female *Tilapia aurea* (Teleostei, Cichlidae) during initiation of spawning. Endocrinology 95, 963-968.

31. Kelley, D.B. (1980). Auditory and vocal nuclei in the frog brain concentrate sex hormones. Science 207, 553-555.
32. Kelley, D.B. (1982). Female sex behaviors in the South African clawed frog, Xenopus laevis: Gonadotropin-releasing, gonadotropic, and steroid hormones. Horm. Behav. 16, 158-174.
33. Kelley, D. B. and Pfaff, D.W. (1976). Hormone effects on male sex behavior in adult South African clawed frogs, Xenopus laevis. Horm. Behav. 7, 159-182.
34. Kelley, D.B. and Pfaff, D.W. (1978). Generalizations from comparative studies on neuroanatomical and endocrine mechanisms of sexual behaviour. In "Biological Determinants of Sexual Behaviour" (J. Hutchison, ed.), pp. 226-254. J. Wiley and Sons, Chichester.
35. Kyle, A.L., Stacey, N.E., Peter, R.E. and Billard, R. (1985). Elevations in gonadotropin concentrations and milt volumes as a result of spawning behavior in the goldfish. Gen. Comp. Endocrinol. 57, 10-22.
36. Laumen, J., Pern, U. and Blum, V. (1974). Investigations on the function and hormonal regulation of the anal appendices in Blennus pavo (Risso). J. Exp. Zool. 190, 47-56.
37. Licht, P., McCreery, B.R., Barnes, R. and Pang, R. (1983). Seasonal and stress related changes in plasma gonadotropins, sex steroids, and corticosterone in the bullfrog, Rana catesbeiana. Gen. Comp. Endocrinol. 50, 124-145.
38. Liley, N.B. (1972). The effects of estrogens and other steroids on the sexual behavior of the female guppy, Poecilia reticulata. Gen. Comp. Endocrinol. 3, 542-552.
39. Liley, N.B. (1982). Chemical communication in fish. Can. J. Fish. Aquat. Sci. 39,22-35, 1982.
40. Liley, N.R. and Stacey, N.E. (1983). Hormones, pheromones and reproductive behavior in fish. In "Fish Physiology, Volume IXB" (Hoar, Randall, and Donaldson, eds.),pp. 1-63. Academic Press, New York.
41. Liley, N.R. and Wishlow, W.P. (1974). The interaction of endocrine and experiential factors in the regulation of sexual behaviour in the female guppy, Poecilia reticulata. Behaviour 48, 185-214.
42. Macey, M.J., Pickford, G.E. and Peter, R.E. (1974). Forebrain localization of the spawning reflex response to exogenous hypophyseal hormones in the killifish, Fundulus heteroclitus. J. Exp. Zool. 190, 269-280.
43. Malacarne, G., Giacomo, C., Vellano, C. and Mazzi, V. (1982). Prolactin and sexual behaviour in the crested newt (Triturus cristatus carnifex Laur). Gen. Comp. Endocrinol. 47, 139-147.
44. Mason, P. and Adkins, E.K. (1976). Hormones and social behavior in the lizard, Anolis carolinensis. Horm. Behav. 7, 75-86.
45. McNicol, D. and Crews, D. (1979). Estrogen/progesterone synergy in the control of female sexual receptivity in the lizard, Anolis carolinensis. Horm. Behav. 38, 68-74.
46. Meyer, J.H. and Liley, N.R. (1982). Hormonal control of pheromone production in the guppy (Poecilia reticulata Peters). Can. J. Zool. 60, 1505-1510.
47. Miller, L.J. and Moore, F.L. (1981). Evidence that an opioid peptide inhibits sexual behavior in rough-skinned newts. Amer. Zool. 22, 92. (Abstract)
48. Miller, L.J. and Moore, F.L. (1983). Intracranial administration of corticotropin-like peptides increases incidence of amphibian reproductive behavior. Peptides 4, 729-733.
49. Moore, F.L. (1978). Effects of progesterone on sexual attractivity in female rough-skinned newts, Taricha granulosa. Copeia 1978, 530-532.
50. Moore, F.L. (1978). Differential effects of testosterone plus dihydrotestosterone on male courtship of castrated newts, Taricha granulosa. Horm. Behav. 11, 202-208.
51. Moore, F.L. (1983). Behavioral endocrinology of amphibian reproduction. BioScience 33, 557-561.

52. Moore, F.L., McCormack, C. and Swanson, L. (1979). Induced ovulation: Effects of sexual behavior and insemination on ovulation and progesterone levels in Taricha granulosa. Gen. Comp. Endocrinol. 39, 262-269.
53. Moore, F.L. and Miller, L.J. (1983). Arginine vasotocin induces sexual behavior of newts by acting on cells in the brain. Peptides 4, 97-102.
54. Moore, F.L. and Miller, L.J. (1984). Stress-induced inhibition of sexual behavior: corticosterone inhibits courtship behaviors of a male amphibian (Taricha granulosa). Horm. Behav. 18, 400-410.
55. Moore, F.L., Miller, L.J., Spielvogel, S.P., Kubiak, T., and Folkers, K. (1982). Luteinizing hormone-releasing hormone involvement in the reproductive behavior of a male amphibian. Neuroendocrinology 35, 212-216.
56. Moore, F.L. and Muller, C.H. (1977). Androgens and male mating behavior in rough-skinned newts, Taricha granulosa. Horm. Behav. 9, 309-320.
57. Moore, F.L., Specker, J.L. and Swanson, L. (1978). Effects of testosterone and methallibure on courtship and plasma androgen concentrations of male newts, Taricha granulosa. Gen. Comp. Endocrinol. 34, 259-264.
58. Moore, F.L. and Zoeller, R.T. (1979). Endocrine control of amphibian sexual behavior: Evidence for a neurohormone-androgen interaction. Horm. Behav. 13, 207-213.
59. Moore, F.L. and Zoeller, R.T. (1985). Stress-induced inhibition of reproduction: Evidence of suppressed secretion of luteinizing-hormone releasing hormone (LHRH) in an amphibian. Gen. Comp. Endocrinol. 60, 252-258.
60. Moore, M.C., Whittier, J.M., Billy, A.J. and Crews, D. (1985). Male-like behaviour in an all-female lizard, relationship to ovarian cycle. Anim. Behav. 33, 284-289.
61. Morrell, J.I. and Pfaff, D.W. (1978). A neuroendocrine approach to brain function: Localization of sex steroid concentrating cells in vertebrate brains. Amer. Zool. 18, 447-460.
62. Morrell, J.I. and Pfaff, D.W. (1981). Autoradiographic technique for steroid hormone localization: Application to the vertebrate brain. In "Neuroendocrinology of Reproduction Physiology and Behavior" (N.T. Adler, ed.), pp. 519-531. Plenum Press, New York.
62a. Norris, D.O., Norman, M.F., Pancak, M.K. and Duvall, D. (1985). Seasonal variations in spermatogenesis, testicular weights, vasa deferentia and androgen levels in neotenic male tiger salamanders, Ambystoma tigrinum. Gen. Comp. Endocrinol. 60, 51-57.
63. Palka, Y.S. and Gorbman, A. (1973). Pituitary and testicular influenced sexual behavior in male frogs, Rana pipiens. Gen. Comp. Endocrinol. 148-151.
64. Pickford, G.E. and Strecker, E.L. (1977). The spawning reflex response of the killifish, Fundulus heteroclitus: Isotocin is relatively inactive in comparison to arginine vasotocin. Gen. Comp. Endocrinol. 32, 132-137.
65. Reisman, H.M. (1968). Effects of social stimuli on the secondary sex characters of male three-spined sticklebacks, Gasterosteus aculeatus. Copeia 1968, 816-826.
66. Ross, P. and Crews, D. (1977). Influence of the seminal plug on mating behaviour in the garter snake. Nature 267, 344-345.
67. Scott, A.P., Bye, V.J., Baynes, S.M. and Springate, J.R.C. (1980). Seasonal variations in plasma concentrations of 11-ketotestosterone and testosterone in male rainbow trout, Salmo gairdneri richardson. J. Fish Biol. 17, 495-505.
67a. Roy, E.J., Wilson, M.A. and Kelley, D.B. (1986). Estrogen-induced progestin receptors in the brain and pituitary of the South African clawed frog, Xenopus laevis. Neuroendocrinology 42, 51-56.
68. Semler, D.E. (1971). Some aspects of adaptation in the polymorphism of breeding color in the threespine stickleback (Gasterosteus aculeatus). J. Zool. 165, 291-302.

69. Shapiro, D.Y. and Boulon, R.H. (1982). The influence of females on the initiation of female-to-male sex change in a coral reef fish. Horm. Behav. 16, 66-75.
70. Sherwood, N.M., Zoeller, R.T. and Moore, F.L. (1986). Multiple forms of gonadotropin-releasing hormone in amphibian brains. Gen. Comp. Endocrinol. 61, 313-322.
71. Specker, J.L. and Moore, F.L. (1980). Annual cycle of plasma androgens and testicular composition in the rough-skinned newt, _Taricha granulosa_. Gen. Comp. Endocrinol. 42,297-393.
72. Stacey, N.E. (1976). Effects of indomethacin and prostaglandins on spawning behavior of female goldfish. Prostaglandins 12, 113-126.
73. Stacey, N.E. (1981). Hormonal regulation of female sexual behavior in teleosts. Amer. Zool. 21, 305-316.
74. Stacey, N.E. (1983). Hormones and pheromones in fish sexual behavior. BioScience 33, 552-556.
75. Stacey, N.E. and Hourston, A.S. (1982). Spawning and feeding behavior of captive Pacific herring, _Clupea harengus pallasi_. Can. J. Fish Aquat. Sci. 39, 489-498.
76. Stacey, N.E. and Liley, N.R. (1974). Regulation of spawning behavior in the female goldfish. Nature 247, 71-72.
77. Stacey, N.E. and Peter, R.E. (1979). Central action of prostaglandins in spawning behavior of female goldfish. Physiol. Behav. 22, 1191-1196.
78. Sullivan, B.K. (1983). Sexual selection in Woodhouse's toad (_Bufo woodhousei_) II. Female choice. Anim. Behav. 31, 1011-1017.
79. Tokarz, R.R. and Crews, D. (1981). Effects of prostaglandins on sexual receptivity in the female lizard, _Anolis carolinensis_. Endocrinology 108, 451-457.
80. Valenstein, P. and Crews, D. (1977). Mating-induced termination of behavioral estrus in the female lizard, _Anolis carolinensis_. Horm. Behav. 9, 362-370.
81. Villars, T.A. and Burdick, M. (1986). Rapid decline in the behavioral response of female paradise fish (_Macropodus opercularis_) to prostaglandin $F_{2\alpha}$ treatment. Gen. Comp. Endocrinol., in press
82. Villars, T.A., Hale, N., and Chapnick, D. (1985). Prostaglandin-$F_{2\alpha}$ stimulates reproductive behavior of female paradise fish (_Macropodus opercularis_). Horm. Behav. 19, 21-35.
83. Wetzel, D.M. and Kelley, D.B. (1983). Androgen and gonadotropin effects on male mate calls in South African clawed frogs, _Xenopus laevis_. Horm. Behav. 17, 388-404.
84. Whitney, C.L. and Krebs, J.R. (1975). Mate selection in Pacific tree frogs. Nature 255, 325-326.
84a. Whittier, J.M. and Crews, D. (1985). Prostaglandins and mating behavior in female garter snakes. Amer. Zool. 25, 117A. (Abstract)
85. Whittier, J.M. and Crews, D. (1986). Ovarian development in red-sided garter snakes, _Thamnophis sirtalis parietalis_: Relationship to mating. Gen. Comp. Endocrinol. 61, 5-12.
86. Yamazaki, F. and Watanabe, K. (1979). The role of sex hormones in sex recognition during spawning behaviour of the goldfish, _Carassius auratus_. L. Proc. Indian Natl. Sci. Acad. Part B-45, 505-511.
87. Zoeller, R.T. and Moore, F.L. (1982). Duration of androgen treatment modifies behavioral response to arginine vasotocin in _Taricha granulosa_. Horm. Behav. 16, 23-30.
88. Zoeller, R.T. and Moore, F.L. (1985). Seasonal changes in LHRH concentrations in microdissected brain regions of male rough-skinned newts (_Taricha granulosa_). Gen. Comp. Endocrinol. 58, 222-230.
89. Zoeller, R.T. and Moore, F.L. (1986). Correlation between immunoreactive vasotocin in optic tectum and seasonal changes in reproductive behaviors of male rough-skinned newts. Horm. Behav. 20, 148-154.

THE EVOLUTION OF VIVIPARITY IN FISHES, AMPHIBIANS AND REPTILES: AN ENDOCRINE APPROACH

Louis J. Guillette, Jr.

Department of Zoology
University of Florida
Gainesville, FL 32611

I. INTRODUCTION AND TERMINOLOGY

Vertebrate viviparity has been of interest to biologists for many years. In fishes, amphibians and reptiles, studies have focused primarily on the anatomical and ecological aspects of this parity mode (for recent reviews, see Wourms,[170] Wake,[174,175] Shine[150]). Historically, few studies have examined the physiological adaptations required for the change from oviparity to viviparity in these vertebrate groups. In recent years, a resurgence of interest has occurred, and many studies have begun to focus on physiological and endocrine adaptations related to the evolution of viviparity.

Before I begin the discussion of vertebrate viviparity, I need to define some of the terms to be used. One aspect of the reproductive strategy of a species is its parity type. The terminology describing parity types is varied and recently has been reviewed.[6] In this paper, I have chosen the most general terms and definitions. That is, females in which complete embryonic development occurs within the reproductive tract are defined as viviparous. Likewise, the terms pregnancy or pregnant will be used to define viviparous females possessing developing young in their oviducts. In contrast, the term gravid will be used for oviparous females. Those animals exhibiting complete embryonic development within the female's body but outside of the reproductive tract are termed live-bearers. Oviparous females are those that lay eggs in which much of embryonic development is completed outside the female. Ovoviviparity is defined in numerous ways by different workers (see Bertin,[15] Blackburn,[16] Guillette,[61] and Wake[175] for discussions of these definitions). However, most authors agree that it implies an evolutionary stage intermediate between oviparity and viviparity, one in which the young are born completely developed but receive no nurturing from the mother except from the initial yolk stores laid down in the egg before ovulation and/or fertilization. Species employing this ovoviviparous strategy will be included with the viviparous species in this chapter.

Many viviparous species possess placental structures, but the female provides little nutritive support for the developing young through these structures. A problem develops in the definition of what constitutes nurturing. Does providing 5% or 25% of the nutritive needs of the young via a placenta constitute 'true viviparity' or does 50%? These are problems which will not be discussed further here but must be considered. Suffice it to say that those species having complete embryonic development within the female's reproductive tract are viviparous. Broader definitions are more useful until we have a better understanding of the degree of placentation and nutrient exchange in more species. As with other classification systems, it is impossible to fit all species into exactly defined boxes, and the concepts of oviparity and viviparity are different extremes along a spectrum of reproductive adaptations.

Finally, a note on anatomical terminology. I will use the term 'oviducts' as a designation for the entire system derived from the mullerian ducts (see Chapter 1). Since the oviducts of mammals, birds, reptiles, amphibians and some fishes are homologous, the terms myometrium and endometrium can be used to refer to the muscular and mucosal components, respectively.

II. VIVIPARITY IN FISHES, AMPHIBIANS AND REPTILES

Viviparity has evolved repeatedly in fishes,[180] amphibians,[174] and reptiles.[16,150] In fact, in reptiles alone, approximately 100 separate origins may be calculated using reproductive and taxonomic data.[150] Similar calculations have not been performed for amphibians and fishes, but examination of the appropriate data suggest multiple evolutionary events for these groups as well (see Wourms,[180] Wake[174]). Thus, viviparity has been an adaptive strategy in a large number of species in many environmental settings. Although many researchers have favored cold environments as being a major factor in the evolution of viviparity,[126,147,151,178] enough viviparous species exist in tropical environments to suggest that no single environmental setting alone selects for viviparity. Likewise, there is probably no single anatomical or physiological pathway from oviparity to viviparity. One need only examine the differences among viviparous chondrichthyan, teleostean and anuran species.[174,179,180] However, for most vertebrates, a basic set of anatomical and physiological constraints exist upon which selection must work. Thus, similarities among species will exist. Furthermore, as in all evolutionary processes, the intermediate stages in the evolution of viviparity from oviparity must be advantageous or at the least neutral to the selection process. The need for advantageous intermediate stages is a powerful constraint on most species that forces changes in reproductive strategies to be conservative.

Several intermediate stages have been hypothesized for the evolution of viviparity,[71,126,151,159,179] including: (1) egg retention, whereby an organism achieves longer and longer periods of *in-utero* development and (2) placentation, whereby structures develop to maximize exchange between the mother and developing young. Most studies have examined only the anatomical and ecological correlates of these stages. However, the physiological control of viviparity presents an excellent model system with which to examine the evolution of endocrine adaptations; that is, how has selection modified and adapted the endocrine control of reproduction during the evolution of viviparity?

As mentioned above, viviparity involves changes in gestation length as well as modification of the maternal and fetal tissues for exchange. The change in gestation length can be achieved by several pathways. That is, the female (or embryo) may inhibit expulsion of the embryo (egg); i.e., delay oviposition. This can happen as an obligate phenomenon, or it may be facultative. It has been hypothesized[63] that the evolution of viviparity involves selection of females from a population that exhibits facultative egg retention. In these animals, retention is stimulated by an environmental factor, is adaptive, and has a genetic basis. Differential reproductive success selects for egg-retaining females because the young produced accrue a selective advantage, and thus facultative retention gradually becomes obligate. Obligate egg retention does not imply retention of eggs *in utero* until development is complete (i.e., viviparity), but instead denotes a condition in which obligate retention of eggs occurs for a specific period of time during development. Those species having such a mechanism for egg retention would possess a possible preadaptation for the evolution of viviparity. For example, many species of squamate reptiles exhibit egg retention for as much as half of the developmental period.[148] This ability to exhibit extensive egg retention may help explain the large number of separate origins of viviparity in some reptilian groups (lizards, snakes) and the lack of viviparity in others (turtles, crocodiles). In addition to being able to retain eggs, retention of eggs for the complete developmental period may require formation of a placenta. Many viviparous vertebrates develop structures used for gaseous and/or nutrient exchange and waste removal during embryonic development. That is, during embryonic development the embryo requires two factors from the surrounding environment: water and oxygen. If an egg is retained within the mother's body, the same requirements must be met. Procurement of water may not be a problem, as the estrogen/progesterone-dominated uterus is edematous and very vascular.[70] Gaseous exchange, however, would be problematic if retention of an egg occurred. The egg, surrounded by a relatively thick shell membrane(s) or coat(s) and remaining in the oviduct which has large glandular and muscular layers, would encounter problems due to the diffusion barriers these tissues impose. This problem can be solved in viviparous species by the formation of a placenta. Thus, for this discussion a structure for the exchange of any or all materials (not just nutrients) between the embryo and mother is termed a placenta. This definition of placentation is similar to that proposed by Mossman,[12] but is different from that used by other workers (see Amoroso,[3] Bauchot,[13] Morriss,[121]). However, placentae with nutritive functions do not evolve *de novo* but probably begin in many species as organs for exchange of gas, water and waste. Thus, for this discussion a structure for the exchange of any or all materials (not just nutrients) between the embryo and mother is termed a placenta. These structures may be as 'simple' as the apposition of two vascular structures, one maternal and the other embryonic. They may, however, be highly specialized structures such as the trophotaeniae of teleosts or the trophoblastic placentae of eutherian mammals, both of which are used to supply nutrients during embryonic development.

Viviparous species are thought to evolve from oviparous ones (for discussion, see Hogarth[85]); thus, an assumption that can be made at the onset is that species evolving viviparity have eggs containing all of the nutrients required within them for complete embryonic development. Therefore, a placenta evolves initially not as a nutrient organ but as a structure for the exchange of other factors. Additionally, the placental structures observed do not occur *de novo* but are at the onset modifications of systems already in existence and probably used for a similar function during oviparous development. For example, the yolk-sac placenta may be an elaboration of a structure which forms normally in an oviposited egg and may function in the egg by maximizing uptake of water and/or gases from the environment.

Formation of the early placenta would require several developments: (1) a maternal and embryonic tissue must become (or already be) apposed,

(2) these structures must be capable of extensive vascularization, (3) a reduction in tissue mass must occur to maximize diffusion by decreasing diffusion distance and (4) an increase in surface area must occur to support exchange.

Few studies have examined the endocrine aspects of placental formation in non-mammalian vertebrates. In fact, very few have even addressed this question. The purpose of this chapter is to examine the phenomenon of viviparity in fishes, amphibians and reptiles from an endocrine perspective. Specifically, this paper will address: (a) the maintenance of gestation and its influence on viviparity, (b) the phenomenon of egg-coat secretion and its relationship to the evolution of a placenta, and (c) unanswered questions and possible models for the evolution of egg retention and placentation. It is not my goal to review all the specific ecological, anatomical, and physiological data available on viviparous species. Pertinent observations will be used throughout, however, to clarify environmental pressures and anatomical constraints and the influences these phenomena have had on endocrine adaptations for viviparity.

III. GESTATION MAINTENANCE AND EARLY PLACENTATION

A. Chondrichthyan Fishes

1. Previous Research. Fishes exhibit an extensive array of reproductive adaptations, and it is in the chondrichthyans that viviparity is first observed. This parity mode is prominent among the sharks and rays, with 48 families, 99 genera and 428 species exhibiting viviparity.[180] In the chondrichthyan fishes, viviparity is characterized by a fairly extensive period of embryonic development in the oviduct when compared to oviparous forms. The eggs are ovulated, enter the oviduct via the ostium, and then pass through the shell (or nidamental) gland where a case is secreted around the egg immediately after fertilization. The control of egg case secretion is not understood. In fact, very little is known concerning the factors controlling function of the shell gland or other portions of the oviduct. In the skate *Raja erinacea*, shell gland development and growth is correlated with follicular development and plasma levels of estradiol.[107] These data support earlier observations that estrogens stimulate increased oviductal vascularity and nidamental gland activity in *Squalus acanthias* and *Mustelus canis*.[472,81,82] It is interesting to note that the shell case is reduced or vestigial in viviparous species.[179,180] Reduction in shell case thickness is correlated with changes in the uterine portion of the oviduct, which involve an extensive increase in surface area of the luminal mucosa, a reduction in diffusion distance by a thinning of the tissues, and an increase in

oviductal vascularity.[93] (See also reviews by Hogarth[85] and Wourms[179]). These changes modify the pregnant oviduct into a respiratory membrane. No information is available regarding the physiological control of these changes. Vascular changes, especially those occurring at the capillary level, may not require a direct hormonal stimulus but may be brought about by local hypoxia caused by fetal uptake of oxygen. The mechanism by which extensive reduction in oviductal thickness and makeup occurs, however, is not clear as we do not understand the process by which these glands are stimulated.

In addition to changes in the oviductal tissues, ovarian modifications are noted as well. The primary change occurs in the postovulatory follicle as it becomes reorganized and forms a structure homologous with the mammalian corpus luteum[83,136] (see also Chapter 9). This structure appears to be a temporary endocrine organ capable of secreting steroids. The enzyme 3β-hydroxysteroid dehydrogenase (3β-HSD) has been demonstrated in these tissues (see Lance and Callard[109]). Additionally, Lupo di Prisco (cited by Saidapur[136]) observed that postovulatory follicles of several chondrichthyan species secrete steroids. Tissue from _Torpedo marmorata_ synthesized estrogens, whereas progesterone but no estrogens were produced by corpora lutea from _Scyliorhinus stellaris_. Formation and secretion of steroids from luteal tissue may be gonadotropin independent. That is, hypophysectomy in _M. canis_ does not affect luteal function, and pregnancy continues for 3 months after surgery.[34,83]

In addition to steroid hormones, the corpora lutea of some sharks (e.g., _Odontaspis taurus_) are capable of secreting the peptide hormone relaxin.[56,141] Shark relaxin appears to have many structural features in common with mammalian relaxin.[56] Koob et al.[106] and Callard et al.[29] have demonstrated that porcine relaxin stimulates relaxation of oviductal smooth muscle and subsequent premature parturition in the sand shark, _Squalus acanthias_. Relaxin, however, did not increase the cross-sectional diameter of the anterior regions of the oviduct in this species. These data support the earlier suggestion of Koob and Callard[105] that relaxin may play an important role in normal parturition of the dogfish.

Studies of the mechanisms for gestation maintenance in elasmobranchs are rare. Hypophysectomy of _Mustelus canis_ at the beginning of gestation does not effect gestation or development of the young as indicated above. LaPointe[112] reviewed the research on effects of neurotransmitters and neurohypophysial (NH) peptides on the uterus of chondrichthyan species. Isolated oviducts of all species examined exhibit smooth muscle contraction in response to treatment with acetylcholine, oxytocin, arginine vasotocin, aspargtocin, valitocin, glumitocin, and elasmobranch neural lobe extracts. Interestingly, isolated oviducts from immature and pregnant leopard sharks, *Triakis semifasciata*, do not respond to NH hormones, whereas oviducts from post-partum females do.[112] These data imply that sex steroids may influence the contractile response of the oviduct to these hormones as they do in other vertebrates.[76] Dodd[46] suggested that neural and hormonal factors may control oviposition/parturition in elasmobranchs. Data are still scarce, however, and one can only speculate as to the mechanisms regulating gestation and parturition. Available data indicate that (1) neurotransmitters can stimulate uterine contractions, (2) NH peptides stimulate uterine contractions, (3) reproductive condition (steroid hormones environment?) influences the response of the uterus to these factors, and (4) mammalian relaxin relaxes the posterior oviductal smooth muscles and allows premature parturition. These factors, however, characterize oviparous as well as viviparous species. Anoroso et al. (1979) and Hogarth (1976) concluded that the transition from oviparity to viviparity in elasmobranchs was not associated directly with fundamental changes in endocrine regulation. Fundamental changes must occur, however, if one is to accept that this trait evolved through progressive stages of egg retention and did not appear *de novo*. That is, the duration of gestation increased substantially in viviparous elasmobranchs. For example, pregnancy length may be as much as 18-22 months in the spiny dogfish, *S. acanthias*. Thus, although at this stage one cannot determine the endocrine adaptations which have evolved, we do know that they are not identical to the mechanisms observed in mammalian species. That is, the corpus luteum does not appear to require hormones from the adenohypophysis to function. However, this does not imply that endocrine adaptations for viviparity have not occurred.

2. Future Research Needs. Future studies on the control of parturition and the maintenance of gestation in chondrichthyans are needed. Specifically, what controls the synthesis and release of relaxin? What controls the release of the NH peptides? How do steroid hormones mediate the response of the oviduct to these factors? Luteal progesterone may not be required to maintain pregnancy but may influence NH peptide receptor formation. Moreover, progesterone may only be important during the first few hours of pregnancy following ovulation. Prostaglandins are important factors during parturition in other vertebrate species, yet nothing is known about the function or synthesis of these factors in the reproductive tract of chondrichthyans. Does the reproductive tract of chondrichthyans synthesize prostaglandins? What are their functions? That is, are they involved with

oviductal contractions or luteolysis? If the corpus luteum is not required for the maintenance of gestation, what then is its role? Is it just a means of cleaning up the postovulatory follicle or does it influence an as yet unknown aspect of chondrichthyan reproduction?

A fundamental problem with our research efforts in lower vertebrate reproduction may be a desire to demonstrate conservatism in the controlling mechanisms among vertebrate groups. Thus, we utilize mammals as the model for all other vertebrates and in doing so we miss the primary function of the structure under observation. This may be our present problem in interpreting elasmobranch reproduction. The problems and questions presented above need to be answered before any hypothesis concerning the physiological mechanisms associated with the evolution of viviparity can be developed for these fishes.

B. Bony Fishes

1. Previous Research. Viviparity also is exhibited by species of the class Osteichthyes, but at a lower frequency than observed in chondrichthyan fishes. Wourms[180] reported that this parity mode occurs in 13 families, 122 genera and 510 species of teleostean fishes. In addition, the coelacanth (Sarcopterygii) is viviparous.[152]

Viviparity in teleostean fishes is very different in one aspect when compared to the "normal" pattern of vertebrate viviparity. That is, viviparous teleosts have no mullerian-derived duct system (see Chapter 1), and gestation occurs within the ovary. Most teleosts have a centrally located ovary, the lumen of which is covered with a surface epithelium.[163] Gestation within viviparous species occurs in two forms. First is the phenomenon of intraovarian gestation. The eggs are ovulated into the lumen of the hollow ovary, and development occurs subsequent to fertilization. In the second type, gestation may take place within the follicle itself, so-called intrafollicular gestation. Fertilization occurs prior to ovulation, and development takes place within the follicle. The blenny (*Zoarces viviparus*) exemplifies ovarian gestation, whereas the poeciliids (e.g., *Poecilia reticulata*; *Xiphophorus hellerii*; *Xiphophorus maculatus*) exhibit intrafollicular gestation.[22] Thus, the sequence of reproductive events (e.g., ovulation, fertilization, parturition) are varied depending on the species examined (for a complete discussion, see Wourms[180]). As with chondrichthyan species, physiological and anatomical adaptations for the prolongation of gestation occur in viviparous forms.

The anatomical adaptations have been studied extensively in several viviparous species (see Wourms, 1981). Early placentation in teleosts is characterized by (1) loss or reduction in the fibrous fertilization membrane in the region of the placenta, (2) greatly increased capillary density in the follicular wall in those exhibiting intrafollicular gestation, and (3) an increase in embryonic surface area.[45,91,92,104,165] As for early placentation in elasmobranchs, no information is available for bony fishes concerning the role of hormones in controlling the modifications described above. Progesterone may be involved with the maintenance of follicular vascularity,[5] but this suggestion has not been tested.

Extensive data indicate that ovarian development in teleosts is dependent upon pituitary hormones (see Goetz,[55] Nagahama[123]). Furthermore, teleostean ovarian tissue possesses the enzymes required for steroidogenesis and synthesizes and secretes all classes of steroid hormones (progestogens, androgens, estrogens, corticosteroids).[109] Postovulatory follicles form corpora lutea in oviparous species (see Khoo, 1975, and Saidapur, 1982) and in all species examined having intraovarian gestation[136] (see also Chapter 9). However, the role of luteal hormones in gestation maintenance is unknown. Hypophysectomy does not affect gestation or embryonic mortality in *Poecilia latipinna*,[9] whereas this operation does increase embryonic mortality in freshwater *Gambusia* unless replacement therapy with salt solutions and/or prolactin (PRL) is performed.[32] Similar observations concerning the need for PRL following hypophysectomy were reported by Hogarth[85] for the seahorse, *Hippocampus*. He noted that gestation was not dependent on gonadal hormones (stimulated by the hypophysis), but PRL was needed for maintenance of the ionic environment in the pouch. This is not at all surprising given the fact that PRL is essential for normal ionic balance and life in a number of freshwater fishes. Ball and Ensor[10,11] observed that freshwater-adapted *Poecilia latipinna* die following hypophysectomy due to a considerable loss of sodium; death could be prevented by administration of PRL. Interestingly, the pituitary of *P. latipinna* contains six times the PRL when freshwater-adapted than when saltwater-adapted.[12] Thus, the experimental environment may be very important when examining the role of PRL in gestational maintenance by freshwater viviparous fish. Prolactin may not be a primary gestational hormone per se, as this hormone is essential for normal ionic balance in many freshwater fish (see Chapter 4).

For obvious reasons, ovariectomy and deluteinization have not been performed on viviparous teleosts, so no data exist concerning the requirement of corpora lutea for normal gestation in viviparous fish. As mentioned, corpora lutea are capable of secreting progesterone. The functional significance of this, however, is unknown in teleostean species although progesterone may play a role in placentation as discussed above.

Although little is known concerning the endocrine control of gestation in teleosts, data have been compiled on ovulation and spawning in these animals (for recent reviews, see Donaldson and Hunter,[49] Goetz,[55] Liley and Stacey,[115] Dodd and Sumpter;[48] see also Chapter 8). Almost all of these data, however, are from oviparous fishes. In many oviparous fishes and in viviparous fishes exhibiting intrafollicular development, the ovulation and spawning/parturition events are related closely. Hogarth[85] presented a model for the evolution of intrafollicular gestation, suggesting that the follicle loses its sensitivity to the "ovulation-inducing effects of gonadotropin or estrogen." Thus, ovulation does not occur, and fertilization occurs within the follicle. Development ensues, and "by the time the second batch of follicles is fully mature, the first batch (now containing advanced embryos) has belatedly become sensitive to gonadotropin (or estrogen)." He reports further that LH and estrogen administration in late-pregnant females does in fact induce parturition in a way that is similar to ovulation in oviparous species.

A more complex picture exists concerning the phenomenon of ovulation/parturition in those viviparous fish exhibiting intrafollicular gestation. Kujala,[108] studying the guppy Poecilia reticulata, observed that injections of deoxycortisol induced premature parturition. He further noted that deoxycortisol stimulated the breakdown of follicular membranes. These observations are similar to those made during the final maturational process of oocytes from other piscine species.[55] Several different steroids have been implicated in germinal vescicle breakdown and final oocyte maturation, the most important being $17\alpha,20\beta$-dihydroxy-4-pregnen-3-one as well as the 11-deoxygenated or 11-oxygenated corticosteroids[55] (see also Chapter 7). The presence of high plasma concentrations of DOC and other corticosteroids suggests that the adrenal gland has an important functional role in teleostean ovulation (see Chapter 8).

In addition to deoxycortisol, Kujala[108] observed that NH hormones (oxytocin, vasotocin, carp neurohypophysial extract) also stimulate premature parturition. Heller[76] observed that smooth muscle surrounds the follicles and ovary of viviparous fishes (e.g., P. reticulata, Molliensia sphenops, X. hellerii, Limnia melanogaster) and that this musculature is sensitive to NH peptide stimulation. Furthermore, he observed that the sensitivity of this tissue to these peptides increases as gestation progresses. The change in sensitivity is correlated with sex steroid secretion: estrogens enhance NH-peptide induced contractions whereas progesterone inhibits them.[76]

2. Future Research Needs. As discussed above, little is known concern-

ing the maintenance of gestation in viviparous bony fishes. Corpora lutea form from postovulatory follicles, and appear capable of steroidogenesis. However, data available suggest that luteal steroidogenesis is not under pituitary control. Furthermore, the role of corpora lutea during gestation is unknown. Ovulation and parturition are related events in those species exhibiting intrafollicular gestation, and are complex phenomena with neural, endocrine and environmental factors influencing their outcome. So little comparative data exist that, at this time, a definitive understanding of gestation maintenance in viviparous forms is not possible. At least one factor which is important in chondrichthyan species has not been examined in teleosts; that is, the occurrence or role of the hormone relaxin. Relaxin acts primarily on the mullerian-derived reproductive tract in other vertebrates. As this system is missing in many bony fishes, it might be concluded that relaxin has a different role in teleosts, if it is present. The data presented indicate that parturition in viviparous fishes is a complex event. It is certainly not as simplistic as the model Hogarth[85] presented suggesting that intrafollicular development evolves simply due to loss of sensitivity to the ovulation-inducing effects of gonadotropin and/or estrogen. Nevertheless, Hogarth's model outlines some important features and provides a plausible hypothesis which can be tested.

Lance and Callard[109] suggested that ovulation in fishes may be stimulated by a 'stress' response. This hypothesis is based on the observations that corticosteroid hormones are associated with in-vivo and in-vitro ovulation. As reported above, Kujala[108] observed that deoxycortisol induced premature parturition as did the stress of immersion of pregnant females in distilled water. Could a 'stress response' then be part of a normal parturition event in viviparous teleosts? Stress as used here does not imply a pathological condition but, as defined by Selye,[145] it is "the nonspecific response of the body to any demand made upon it." Exposure to a stressor elicits an immediate response, the alarm phase, during which catecholamines are released from the chromaffin tissue associated with the adrenal gland.[143-146] This initial response is followed soon by the resistance phase if the stressor becomes chronic. The resistance phase is characterized by synthesis and release of corticosteroids from the adenocortical (interrenal) tissue.

Stressors can stimulate a number of pathological conditions. However, the stress response is an adaptation to environmental and/or physical stimuli. Could an embryonic stress response stimulate the parturition event in those species exhibiting intrafollicular gestation? I suggest the following hypothesis by which a stress response is adaptive. At the end of development, oxygen consumption by the embryo is high and the demand must be met by the mother. Adaptations of the ovarian follicle and embryo occur to maximize gaseous exchange, but as the embryo continues to grow, metabolic mass increases and problems related to gaseous exchange eventually will occur. The embryo will encounter a hypoxic environment which stimulates an embryonic stress response. This response initiates the synthesis and release of adrenal corticosteroids and catecholamines which cross the "placenta" and initiate ovulation. This reaction would be coupled during late gestation with the steroid hormone-induced increase in sensitivity to NH peptides which would initiate contractility of the smooth muscle of the follicle. Mechanical stimulation of the muscle would provide the stimulus for release of ovarian prostaglandins. An alternative model would suggest that the stress response was initiated in the mother, with the same outcome. I suggest that this hypothesis concerning stress activation of parturition may explain the situation in viviparous teleosts with intrafollicular or intraovarian gestation.

It is much more difficult to explain the initial loss of sensitivity to the "primary" ovulatory event. Hogarth[85] suggests that intrafollicular gestation might evolve by loss of sensitivity to the ovulatory stimulus. As yet, the interrelationship of the events occurring during teleostean ovulation are not well understood, so a model is difficult to develop (see Chapter 8). Prostaglandins are active agents in the ovulation and spawning events of oviparous teleosts. What role do they play in viviparous forms? Could inhibition of prostaglandin synthesis influence the ovulatory event? What are the selective forces for intrafollicular fertilization? What are the intermediate stages, if any? Future studies examining these questions and the relationship among the adrenal, ovary and pituitary may provide the data necessary to develop a testable hypothesis for the evolution of viviparity in teleosts.

C. Amphibians

1. Previous Research. Amphibians exhibit a diversity of reproductive strategies as extensive as the fishes. All three amphibian orders exhibit viviparity.[174,175] In each, there are excellent examples that have been well-studied. For this discussion, I would like to briefly examine the best understood examples of amphibian viviparity to illustrate the characteristics of this parity mode. This overview, however, is by no means an exhaustive review of amphibian viviparity (for more complete discussion see reviews by Xavier[182] and Wake[170,171,173-175]).

Two anuran species exhibiting viviparity have been studied; these are the toad Nectophrynoides occidentalis[182] and the frog Eleutherodactylus jasperi.[172] These species are characterized by a reduction in the size of the ovary and the number of eggs the female produces (N. occidentalis; 2-10 eggs/clutch; E. jasperi: 1-6 eggs/clutch); E. jasperi exhibits increased egg size (3.3 mm diameter) compared to other anurans, whereas N. occidentalis exhibits very small eggs (500-600 μm diameter). In addition to the above traits, these species have complete embryonic development in the oviduct (including metamorphosis) and a reduction in egg coat thickness. Eggs are ovulated and enter the oviduct where they are fertilized. Internal fertilization is common in viviparous anurans but rare in oviparous forms.[160] During development the embryo is surrounded by an egg coat secreted by the oviduct (at least in anurans) as the eggs are transported through the reproductive tract.[138] The number of coats secreted is species-specific, and may vary intraspecifically.[14,138,140] The composition also is different among species, but in general is composed of mucopolysaccarides.[14,54,138] Data concerning the control of egg jelly secretion is contradictory, as some authors suggest it is controlled by estrogens and others, by progestogens (reviewed by Saidapur,[136] Lofts[116]). Oviductal development at the onset of the reproductive cycle is controlled by estrogens, but steroid hormones from postovulatory follicles may influence secretion of jelly by the oviductal glands. Additional studies are needed to determine the control of this phenomenon, as a reduction in the thickness of this coat occurs in N. occidentalis.[181] Wake[172] did not describe the egg coat in E. jasperi; however, this species has a relatively small egg tooth (which is common in species with thin egg coats) implying that a reduction in egg coat is correlated with viviparity. Once an understanding of the process of jelly secretion is obtained in related oviparous species, we may be able to construct an hypothesis concerning the reduction or loss of these structures in viviparous forms during the evolution of viviparity.

During embryonic development, corpora lutea form from the postovulatory follicles of N. occidentalis. Xavier et al.[184] observed high activity for the enzyme 3β-HSD in the luteal granulosa cells of N. occidentalis, indicating that these cells are steroidogenic. This conclusion is supported by in-vitro studies demonstrating that ovarian tissue from pregnant N. occidentalis converts pregnenolone to progesterone.[183] Furthermore, the conversion rate decreases as pregnancy progresses, so that progesterone synthesis is maximal at the beginning of gestation.[125,183,184]

Wake[172] reported no corpora lutea in E. jasperi, and no additional information is available concerning ovarian steroidogenesis during pregnancy; this does not mean that it does not occur. Historically, anuran postovulatory follicles were thought to have little or no function, as they are transitory structures (for discussion, see Saidapur[136]). Recent evidence, however, suggests that corpora lutea are steroidogenic in oviparous amphibians, if only for a short time[127,136] (see also Chapter 9); the hormone produced, progesterone, may actively stimulate the oviductal glands and oviductal secretion in oviparous forms.[75,137] In the viviparous N. occidentalis, ovariectomy stimulated spontaneous abortion in primigravid females,[184] whereas in older females experiencing their second or third pregnancy, ovariectomy had no effect on maintenance of gestation. Although gonadectomy did not cause spontaneous abortion in older females, it did elicit a more rapid embryonic developmental rate and shorter gestational period.[184] This change in gestational period and developmental rate appears to be in response to removal of progesterone, because exogenous progesterone slows embryonic development. Xavier[184] concluded that progesterone does not maintain gestation per se but influences the oviductal environment by stimulating development of vascularized oviductal pouches in which the young develop. Furthermore, luteal progesterone affects the developmental rate of the embryos by an undetermined mechanism. In addition to progesterone, estrogens are secreted by the ovary, but their role in gestation is unknown. Thus, steroid hormones influence anuran gestation, but an understanding of the maintenance of gestation is not readily available. Could maintenance of gestation then be passive and the termination of gestation be the initiation of parturition? That is to say, gestation may be lengthened by inhibition of parturition and not by actual "maintenance."

Anuran oviposition and parturition are not well understood. As with teleosts, several patterns exist; i.e., a female may oviposit all eggs at one time (e.g., Rana) or may lay the eggs singly and sequentially over a period of time (e.g., Xenopus). These differences suggest several patterns of control. Early studies showed that Xenopus females do not shed all eggs simultaneously and that ovulated ova are moved down the oviduct by ciliary movement.[176] Heller et al.[77] reported that the oviduct of Xenopus was responsive to NH peptides (arginine vasotocin, arginine vasopressin, lysine vasopressin) as were oviducts from other anurans (Bufo marinus, B. bufo). Apparently, studies have not been performed on oviducts from viviparous anuran species. In addition to NH peptides, other smooth muscle factors such as prostaglandins and relaxin have not been examined for their action on the anuran reproductive tract.

In addition to the 'normal' in-utero forms of viviparity, anurans have experimented with alternative forms of live-bearing. These forms have been termed 'pseudoviviparous'; however, I feel as does Wake[174] that they represent a true form of live-bearing since development takes place "within" the female (or male) parent's body. Several forms exist, including (1) development of fertilized eggs in or on the dorsal integument (e.g., Gastrotheca riobambae,[43] Pipa pipa,[130,131] (2) gastric brooding, Rheobatrachus silus,[38,161] (3) brooding in vocal sacs, Rhinoderma darwini,[24,139] (4) brooding in bilateral inguinal canals, Assa (Crinia) darlingtoni.[88,157] In all these strategies, anatomical and physiological adaptations of the parent and embryo occur although incubation and embryonic development do not occur within the parent's reproductive tract.

The adaptations occurring in the marsupial frog, Gastrotheca riobambae, have been well studied. In this species, development occurs in a dorsal pouch formed from the integumentary tissues under the influence of estrogen.[51,99] During pregnancy, the mouth of the pouch closes, the pouch becomes very vascular, and vascular outgrowths develop that partially envelop the embryos.[43] The jelly coat is reduced to less than 10 μm thickness.[43] As with previous discussions of oviductal adaptations for placentation in fish, the reduction of the egg coat in Gastrotheca is correlated with changes in the maternal tissues, in this case the integumentary pouch, not the oviduct. The pouch of Gastrotheca has been defined as a specialized respiratory organ due to its great vascularization and relative decrease in tissue mass which minimize the diffusion distance.[43,124,154] In fact, del Pino et al.[43] reported that reduction of maternal tissue is so extensive that maternal and fetal blood vessels are separated by only 10 μm of tissue. The development of specialized "bell gills" occur in the embryo very early in development, probably related to an increased need for gaseous and/or water exchange at this time. All these adaptations mimic the developments occurring in the oviduct during pregnancy in other vertebrates. In addition, the pouch modifications occurring during early pregnancy are correlated with luteal activity.[42] That is, postovulatory follicles (corpora lutea) develop after ovulation and remain active (histologically) for approximately 25-30 days of the 100-125 day incubation period.[43,44] Neither circulating plasma steroid concentrations nor pituitary peptide hormones have been measured during the reproductive cycle. Thus, at this time, the factors supporting pouch incubation during the last two-thirds of gestation are not known. Preliminary data obtained by del Pino and Sanchez[44] indicate that ovariectomy during the first 20 days of incubation (or that period during which the corpora lutea are present) stimulates spontaneous abortion. Ovariectomy after day 40, however, has no effect on embryonic developmental rate or time of birth. Furthermore, del Pino[42] observed that, under progesterone treatment, the pouch develops incubatory chambers around plastic beads. These bead-induced incubatory chambers remain active for only about two weeks, suggesting possible synergistic factors from the parent and/or embryo are necessary for normal gestation. Parturition from the pouch does not appear to be caused by muscular contractions of the pouch as in the seahorse[85] but is associated with distinct maternal behaviors; what triggers these maternal behaviors is unknown.

A similar situation as described above occurs in the Surinam toad *Pipa pipa*.[3,130,131] Development occurs in depressions which form on the dorsal integument of the female. The endocrine control of pouch formation and maintenance of gestation are unknown. As in *Gastrotheca*, corpora lutea develop and persist in the ovary; they may contribute to the formation of the pouch although no conclusive data exist.

Another example of a live-bearing anuran is a leptodactylid frog of the genus *Rheobatrachus*. Female *R. silus* incubate embryos in the stomach, with parturition of juveniles occurring via the mouth.[38,161] During gastric brooding, the stomach wall distends and becomes highly vascular.[50] The inhibition of gastric secretion is controlled by a prostaglandin (PGE_2), which is secreted by the developing young.[162] What controls vascularity is unknown, although PGE_2 is a potent vasodilator.[86,118] Likewise, the inhibition of peristaltic contractions of the stomach musculature is not understood. Ovarian morphology and histology have not been described during gestation, so no data are available concerning the fate of postovulatory follicles. In the original paper describing this phenomenon,[38] the authors stated that "the control mechanism for the inhibition [of gastric secretion] is under the influence of reproductive hormones, for an intimate interaction is required." However, with the recent data indicating that inhibition is controlled by PGE_2 derived from the young, a reproductive hormone (gonadotropin or steroid) influence may not be required. A remaining question involves the triggering stimulus for oral parturition. Reports indicate that birth in this species is a propulsive ejection of the juveniles following compression of the lateral musculature.[38,161] Moreover, birth occurred in many instances of stress (handling; placing in new aquaria; removal of refuge materials from aquarium). Could parturition involve an adrenal response to the stress of many young in the stomach? Unfortunately, this species is very rare, and future answers to physiological questions may not be forthcoming.

Within the salamanders, only five or six of the 348 described species exhibit viviparity.[89] The best studied examples are the species *Salamandra* atra[166-168] and Salamandra salamandra.[57-59,94-98] Similar adaptations occur in these species as those observed in the anurans. The clutch size is unusually small (one or two young), gestation is long (as much as four years), and numerous changes occur within the reproductive system. Most salamanders have internal fertilization (except the Cryptobranchidae and Hynobiidae) and many have developed terrestrial eggs (including species closely related to S. atra). As with chondrichthyan fishes and anurans, the egg is ovulated and enters the oviduct where it is fertilized and covered with an egg coat secreted by the oviduct. Vilter[166] reported that only the first egg in the oviduct of S. atra becomes coated with the species-specific number of jelly layers, whereas the following eggs are not coated completely. Only one embryo forms in each oviduct, usually the first one. Data now available suggest that the egg jelly, especially the inner coats, are required for amphibian fertilization.[140] This may indicate why the egg coat still remains during early development in viviparous anurans and urodeles. Control of jelly secretion in urodeles is poorly understood; the stimulus may be as simple as direct contact of the egg with the oviductal gland, or jelly secretion may have elaborate endocrine control.[116]

Embryonic development occurs within the oviduct of viviparous urodeles, and gestation is correlated with the formation of postovulatory structures similar to corpora lutea[136] (see also Chapter 9). Fine structure of the corpora lutea of S. salamandra indicates a steroidogenic capability.[98] Histochemical studies also have suggested that postovulatory follicles from urodelean species are capable of steroidogenesis.[95,98] The role of luteal steroids in pregnancy, however, is not well understood. Estradiol stimulates hypertrophy of the oviduct in the oviparous urodele Ambystoma tigrinum, as does testosterone.[73] These findings are similar to studies on other urodeles and other vertebrate species (see Chapter 11). Although progesterone does not stimulate further hypertrophy in Ambystoma over that initiated by estrogens, it is required for the oviduct to respond to the NH peptide, arginine vasotocin.[73] These observations suggest that steroid hormones from the follicles and corpora lutea may act to sensitize the oviduct. Heller[76] suggested that the sensitivity of the vertebrate oviduct to NH peptides is dependent on the physiological state of the female. Oviductal tissues from several anuran and urodelean species exhibit pronounced seasonal differences in sensitivity to oxytocin and AVT.[77] Steroid hormones may influence the sensitivity of the oviduct and thus the strength of the oviductal contractions in urodeles. This in turn would determine gestation length. What role prostaglandins, relaxin or other hormonal factors might play are unknown, for no data are available concerning their presence or action in urodeles.

Wake[171,174,175] has reviewed the phenomenon of viviparity in the Apoda. Viviparity is exhibited by species in three of the four families of caecilians. In fact, as many as 50% of the caecilian species may be viviparous.[171] The taxonomy of this group suggests that viviparity has evolved at least twice in the Apoda, once in the South American branch and once in the South African families. More evolutionary events may have occurred, however.

Viviparity is characterized in apodans by direct development in the oviduct through metamorphosis and the formation of several specialized structures including: (1) specialized secretory structures in the oviduct that produce "uterine milk," (2) specialized fetal dentition used to obtain this nutritive source, (3) expansion of external gills, possibly to serve in gaseous and/or water exchange during development in the oviduct, (4) formation of 'active' corpora lutea, and (5) reduction and/or loss of the egg coats. Although extensive data on the anatomical adaptations for viviparity are available, almost nothing is known concerning the physiological adaptations for viviparity. As with other viviparous amphibians species, apodan postovulatory follicles form structures resembling corpora lutea.[169] These structures remain histologically distinct throughout pregnancy. No histochemical or *in-vitro* techniques have been used to examine ovarian material, so the secretory nature of these structures is unknown. However, high concentrations of plasma progesterone have been obtained from pregnant female *Dermophis mexicanus* (McCreery and Wake, unpublished findings, cited in Wake[174]). Wake[174] suggests that progesterone functions in a manner similar to that observed in *N. occidentalis* and *S. atra*; that is, in the maintenance of pregnancy. No studies examining the role of NH peptides, prostaglandins or relaxin have been performed. Furthermore, the effect of hypophysectomy or ovariectomy (or deluteinization) have not been determined. Thus, no hypothesis concerning the role of hormones in the maintenance of gestation or the evolution of viviparity in apodans can be presented now, due to an almost total lack of endocrine data.

2. Future Research Needs. As described, amphibians exhibit some interesting forms of viviparity and live-bearing. Many basic features of the endocrine influence in these systems, however, are unknown. Estrogens stimulate proliferation and vascularity of both the oviduct and skin in preparation for pregnancy, as they do in other vertebrates.[116] Progesterone is implicated in oviductal function in that it appears to maintain oviductal

vascularity and influence egg jelly secretion. Progesterone also affects developmental rate in at least one species (Nectophrynoides). However, progesterone and luteal function do not appear essential for normal gestation or development. What endocrine factor(s) inhibit oviposition and thus lead to egg retention is unknown. No studies are available demonstrating the effects of NH peptides on the oviduct of viviparous amphibians. In addition to NH peptides, other factors possibly influencing smooth muscle contraction, such as prostaglandins and relaxin, have not been examined. Anurans synthesize prostaglandins,[78] and future studies must examine the role of these paracrines in the smooth muscle of the oviduct. Also, relaxin has not yet been reported from any amphibian ovarian tissues.

The anuran live-bearing forms present many interesting questions concerning the endocrine control of reproduction outside the reproductive system. For example, Gastrotheca presents some very interesting paradoxes. What maintains the corpora lutea after oviposition? In other anurans the corpora lutea are short-lived structures (except in N. occidentalis) which regress rapidly after oviposition.[136] What then is the signal in Gastrotheca for maintaining the corpora lutea? Pancharatna and Saidapur[127] observed that homoplastic pituitary pars distalis homogenate, PMSG and hCG all are capable of extending the life of postovulatory luteal tissue in the ovary of the frog Rana cyanophlyctis. Studies examining vertebrates at all levels (elasmobranchs, reptiles, mammals) have reported that luteal formation can occur in the absence of luteotropic factors from the pituitary or other sources (for discussion, see Rothchild,[134] Rowlands and Weir[135]). How then do such long-lived corpora lutea evolve? One possible hypothesis suggests that during the ovulatory event, the follicle is luteinized and forms a short-lived (in some cases steroidogenic) structure. By exposure to a luteotropic factor, the corpora lutea remain "active" for longer periods of time, secrete progesterone and stimulate the development of progesterone-dependent structures. In the case of Gastrotheca, the incubatory pouches are such progesterone-dependent tissues. Maintenance of these pouches during the non-luteal phase of pregnancy must require some other factor possibly supplied by the developing embryo (see discussion above). Future studies examining the phenomenon of pouch maintenance during the non-luteal (late pregnancy) phase are needed as are studies examining luteal formation after oviposition in this species.

It is interesting to note that, as in the fishes, viviparous and live-bearing amphibians exhibit a reduction in the thickness of the egg coat. What role this reduction may have is not known; it may serve to reduce the diffusion distance between the female and the embryo. The endocrine control (if any) of this phenomenon is unknown. How does the reduction in the egg coats take place? Is it a gradual thinning or is the secretion of certain layers blocked? These questions can be answered with morphological studies. However, to understand the endocrine control of this reduction will require studies examining the mechanism of normal egg jelly secretion, for without an understanding of the normal jelly secretion process, it will be difficult to determine how or why reduction occurs.

D. Reptiles

1. Previous Research. Fossil evidence suggests that viviparity has been an amniote reproductive strategy for millions of years.[67] Viviparity in living reptiles is observed in approximately 20% of the squamate reptiles (lizards and snakes) whereas it does not occur in the turtles, crocodilians or the tuatara.[150] Extensive studies examining the ecology, anatomy and physiology of viviparous reptiles have been performed (see the following reviews: Tinkle and Gibbons,[159] Packard et al.,[126] Shine and Bull,[151] Guillette et al.,[71] Blackburn,[16] Shine,[150] Yaron [186]).

Most reptiles are characterized by having paired ovaries, paired oviducts, large yolky eggs and internal fertilization. Following ovulation and fertilization, the eggs pass through the oviducts and accumulate a species-specific number of albumen layers, shell membranes and calcium deposits. The shell membrane in viviparous reptilian species is reduced and in many is resorbed during late pregnancy.[52,63] The shell is secreted by glands of the oviduct in reptiles,[53] but the hormonal control of these glands is unknown. Initial development of the oviductal shell glands during the reproductive cycle is correlated with plasma estrogen levels, whereas the shelling process itself is correlated with high levels of plasma progesterone.[53] Mead et al.[119] observed that estrogen treatment of the snake *Thamnophis elegans* stimulates glandular development, and progesterone maintains the glands. However, the appearance of steroid-treated glands is never identical to that seen during the natural cycle, indicating that additional factors may be required.

As with other vertebrates, the changes in shell thickness are correlated with an extensive reduction of maternal tissue and an increase in oviductal vascularity.[63,156] Oviductal vascularity is stimulated by estrogens and maintained by progesterone.[164,186] Again, as with glandular development, exogenous steroid treatment never produces results exactly like those occurring naturally, although the vascular pattern resembles the natural pattern.[186]

The loss of oviductal glands and endometrial tissue during the evolution of viviparity is documented for several species with "newly" evolved viviparity (*Sceloporus bicanthalis*,[70] *Sphenomorphus fragile*[62]). Although the anatomical changes are documented, the physiological changes are still poorly understood. Recent studies[53,117] examining the natural anatomical and endocrine changes occurring in oviductal tissue during shelling may help to determine what changes take place during the evolutionary process.

The amount of time the eggs remain in the oviduct is species-specific. Shine[148] noted that egg retention occurs regularly in many oviparous species, and eggs are retained in the oviduct of many oviparous lizards and snakes for at least half of the normal embryonic developmental period. Thus, an initial step in the evolution of viviparity already occurs in many oviparous squamates, and understanding the normal processes of egg retention and egg shelling could provide clues to the mechanism by which gestation could be extended for the entire developmental period.

The literature concerning the endocrine control of egg retention and maintenance of gestation in reptiles has been reviewed recently.[63,186] To summarize, squamates form postovulatory follicles which are similar in structure and function to mammalian corpora lutea[136] (see also Chapter 9). In most squamates, particularly viviparous species, corpora lutea give a positive reaction for the enzyme 3ß-HSD.[109,136,186] This enzyme is usually located in the granulosa (luteal) cells but also may be found in the theca interna of the corpus luteum of many squamates.[17,25,35,120,142,164,185] Thus, the corpus luteum is capable of steroidogenesis and is believed to be the main site of progesterone synthesis in gravid or pregnant squamates. This conclusion is supported by studies[25-27,30,33,52,79,164] which have shown a positive correlation between plasma progesterone levels and luteal activity. Furthermore, removal of the corpora lutea in the gravid oviparous lizard *Anolis carolinensis*[65] or the pregnant viviparous lizard *Sceloporus cyanogenys*[25,27] lowers circulating levels of progesterone. Moreover, prostaglandin-induced luteolysis in the lizard *A. carolinensis* initiates an abrupt decline in plasma progesterone levels.[72] Arslan et al.[7] found that gravid oviparous lizards (*Uromastix hardwicki*) had higher plasma progesterone concentrations after ovulation than during the preovulatory phase. Similar observations are available for other oviparous and viviparous squamate species (Table 1).

It is unclear whether luteal progesterone is essential for the maintenance of gravidity in squamates and, if so, how this hormone exerts its effects.[63,80,100] The effects of ovariectomy and deluteinization during pregnancy in squamates are summarized in Table 2. In viviparous snakes

Table 1. Plasma Progesterone Levels (ng/jl)[b] During the Reproductive Cycle of Squamates. (See also Table 2, Chapter 9).

Species	Follicular Growth	Preovulatory	Early Pregnancy	Mid Pregnancy	Late Pregnancy	Postpartum	References
OVIPAROUS							
Uromastix hardwicki	---c	1.66 ± 0.30	13.41 ± 1.43	---	---	---	7
Crotophytus collaris	0.18 ± 0.08	---	21.25 ± 2.11	---	10.97 ± 0.46	0.30 ± 0.12	53,117
Eumeces obsoletus	0.14 ± 0.05	---	11.41 ± 6.24	---	0.42 ± 0.07	0.52 ± 0.08	53,117
Sceloporus undulatus	0.27 ± 0.18	25.81 ± 1.22	12.51 ± 2.73	3.78 ± 1.19	0.32 ± 0.14	NC[d]	66,117
Agama atra	2.14 ± 0.50	2.97 ± 0.99	7.45 ± 1.72	---	---	---	163a
Cnemidophorus uniparens	<1.0	∿2.50	1.24 - 70.0[e]	---	---	∿2.5	120a
Naja naja	∿1.4	∿3.00	10.5 ± 1.12	---	---	---	16a
VIVIPAROUS							
Lacerta vivipara	---	57.90 ± 4.81	132.65 ± 16.79	216.22 ± 17.92	347.20 ± 17.38	3.71 ± 0.56	182a
*Sceloporus cyanogenys**	0.70 ± 0.15	0.90 ± 0.38	3.30 ± 0.48	---	3.5 ± 0.34	1.6 ± 0.22	27,109
Sceloporus jarrovi	---	---	0.75 ± 0.11	0.95 ± 0.05	4.1 ± 0.94	0.43 ± 0.06	74
Bradypodion (Chameleo) pumilis	0.86	0.95 ± 0.71	4.95 ± 3.9	2.30 ± 0.34	---	---	164

Species	Follicular Growth	Preovulatory	Early Pregnancy	Mid Pregnancy	Late Pregnancy	Postpartum	References
Nerodia taxispilota	0.44 ± 0.04	0.91 ± 0.08	1.93 ± 0.24	---	---	---	33
Nerodia sipedon	1.27 ± 0.19	3.93 ± 0.83	4.95 ± 1.41	6.94 ± 0.78	2.81 ± 0.44	---	33
Thamnophis elegans	---	---	1.70 ± 0.30	6.20 ± 1.0	---	1.0	79

[b] $\bar{X}$ ± SE

[c] not measured

[d] not detectable

[e] values for entire period of gravidity

Table 2. Effects of Removal of Corpora Lutea or Entire Ovary on Pregnancy or Parturition of Viviparous Squamates

Species	Structure Removed	Stage of Pregnancy	Results	Source
SNAKES				
Thamnophis and *Nerodia*	Ovaries	All stages	Two females aborted	132
		All stages	Normal gestation for 25 day post-operation	
Nerodia	Ovaries	Early	Resorption of all embryos	37
		Mid	Resorption or abortion of all embryos	
		Late	Normal parturition.	
Thamnophis sirtalis	Ovaries	Late	Normal parturition	37
Storeria dekayi	Ovaries	Mid	Resorption or abortion of all embryos.	37
Bothrops	CL	Unknown	All embryos died, females died 10 days after surgery.	53a
Nerodia sipedon	Ovaries	Mid	Development and parturition normal in 1 female of 3. Two females died.	21
Thamnophis elegans	CL	Early	Development normal, parturition delayed.	80
LIZARDS				
Lacerta vivipara	Ovaries and CL (Unilateral)	All stages	Normal development and parturition.	129
	Ovaries	All stages	Parturition delayed in most females.	129
	CL	All stages	Parturition delayed in 2 of 3 females.	129

	Structure Removed	Stage of Pregnancy	Results	Source
Chalcides ocellatus	Sham-operated	All stages	Normal development and parturition.	8
	CL	All stages	Normal parturition	
Sceloporus cyanogenys	Ovaries	Unknown	Normal development, parturition delayed.	114
Xantusia vigilis	Sham-operated	Early & mid	Majority of embryos aborted.	185, 185a
	Ovaries	Early & mid	Majority of embryos aborted.	
	CL	Early & mid	Majority of embryos aborted.	
Trachydosaurus rugosus	Ovaries	Early	Embryonic resorption, stillbirth and parturition delayed.	18
Bradypodion pumilis	CL	Mid-late	Late embryos normal, early embryos resorbed.	
	Ovaries	Early & mid	Majority of embryos aborted or resorbed	164
Mabuya carinata	CL	Early	Normal development and parturition.	142
Sceloporus jarrovi	CL	Early	Embryos aborted or resorbed.	b
Mabuya capensis	CL	Early-mid	Normal development for 4 weeks post-surgery	c

[b] Guillette, unpublished

[c] Yaron and Tai, unpublished

(*Natrix*, *Storeria*, *Thamnophis*), ovariectomy during early gestation elicits resorption of embryos, whereas during mid-gestation this operation causes birth of dead embryos.[36,37] The ovary is not required, however, for normal gestation during the last third of development in these species. Hypophysectomy during early or mid-gestation has yielded similar results.[36,37] Clausen[36,37] suggested that corpora lutea are important during the first half of pregnancy in the above species, whereas during late gestation other tissues supply the necessary endocrine factors. He concluded that reptilian gestation is controlled by a series of hormones, with the corpora lutea involved only during early gestation. These results were not confirmed, however, by other studies. Rahn[132] and Bragdon[21] observed that early ovariectomy, or hypophysectomy and ovariectomy, respectively, had no effect on gestation but did delay parturition. Highfill and Mead[80] also observed that early deluteinization did not prevent normal embryonic development or affect the rate of embryonic development in the viviparous snake, *Thamnophis elegans*. However (as observed by Bragdon), parturition was significantly delayed in deluteinized females. They concluded that the corpora lutea function through the secretion of a hormone which ensures that parturition occurs at the normal time.

Deluteinization induces early oviposition and disruption of shelling in the oviparous lizards *Sceloporus undulatus*[133] and *Cnemidophorus uniparens*.[39] In viviparous lizards, ovariectomy or deluteinization during gestation appears to have varying effects as observed in snakes. Bilateral ovariectomy or deluteinization caused stillbirth and delayed parturition in the lizard *Lacerta vivipara* when compared with controls.[128,129] Lien and Callard[114] and Callard et al.[25], studying the viviparous lizard *S. cyanogenys*, observed similar results. In contrast, Badir[8] observed normal gestation and parturition in *Chalcides ocellatus* after ovariectomy. Likewise, Sekharappa and Devaraj Sarkar[142] observed that deluteinization had no effect on gestation length in the viviparous lizard *Mabuya carinata*. Corpora lutea, however, appear to be important in the maintenance of early and mid-gestation in the viviparous lizards *B. pumilus*,[164] *S. jarrovi* (Guillette, unpubl. findings), and *T. rugosa*.[18] Replacement therapy using progesterone does not always alleviate the effects of deluteinization, however,[37,114,129] but an increase in embryonic survivorship is observed in deluteinized *B. pumilus* following treatment with this hormone.[164]

In addition to luteal factors, other hormones may influence the maintenance of gestation and the timing of parturition in reptiles. Panigel[129] and LaPointe[110] demonstrated that oxytocin induced premature birth in lizards. Although this hormone is not found in the neurohypophysis of reptiles, it is closely related to the hormone which is, arginine vasotocin (AVT),[1,2] and AVT is a potent stimulator of both oviposition and parturition.[60,69] In at least one species (*A. carolinensis*), however, β-adrenergic blockage must be overcome if AVT is to be effective.[102,158] Data on plasma AVT concentration during pregnancy were not available until recently. Fergusson and Bradshaw (personal communication) observed that plasma AVT rises slightly approximately 20-30 days before parturition in the viviparous lizard, *T. rugosa*. The significance of this pattern is unknown at present, but does imply that oviposition/parturition in reptiles may not be due to the release of AVT but to the appearance of AVT receptors or the release of a block on oviductal contractions.

In-vitro tests demonstrate that AVT is a potent stimulator of rhythmic contractions of smooth muscle in the reptilian oviduct.[28,68,101,111,112] However, as Heller[76] observed in fishes and amphibians, the reproductive state of the organism influences the responsiveness of the oviduct to NH peptide hormones. Lemus et al.[113] observed that spontaneous contractions occur more readily during late pregnancy than during early pregnancy in the lizard *Liolaemus gravenhorstii*. Furthermore, they noted that oviducts were more sensitive to oxytocin at this time. Callard and Hirsch[28] demonstrated that, in the turtle *Chrysemys picta*, sex steroids affect the *in-vitro* response of oviducts to AVT; estrogens enhance and progesterone inhibits this phenomenon. LaPointe[111] found no effect of pretreatment with sex steroid hormones on *in-vitro* oviductal contractions in the lizard, *Xantusia riversiana*. To further confuse this discussion, Guillette and Jones[68] observed that progesterone enhanced, and estradiol inhibited, AVT-induced *in-vitro* contractions in the lizard *A. carolinensis*. Thus, although these studies indicate a sex steroid influence on the contractile response of some species to AVT, the effect is by no means clear or conclusive, and additional species need to be examined.

In addition to NH peptides, prostaglandins influence parturition in reptiles. Blockage of prostaglandin synthesis with indomethacin has no effect on the ability of AVT to induce oviposition in the lizard *S. undulatus* (Guillette, Pitts, and Daniel, unpublished findings). The reproductive tract of this lizard synthesizes prostaglandins as does the reproductive tract of the viviparous lizard *S. jarrovi*.[66] Moreover, the pattern of synthesis changes with the reproductive state of the female. *In-vitro* studies have demonstrated that prostaglandins influence smooth muscle contractions in the reproductive tract of the lizard *A. carolinensis* (Jones, R.E., personal communication). Additionally, prostaglandin $F_{2\alpha}$ potentiates neurotransmitter-induced striated muscle contractions in the lizard *Crotaphytus collaris*, and these contractions may contribute to the parturition event in reptiles.[64] That is, oviposition and parturition in reptiles is achieved by contraction of the oviductal myometrium and abdominal body wall musculature, and prostaglandin $F_{2\alpha}$ influences the abdominal wall but may or may not have a major effect on the oviduct.

2. Future Research Needs. The role of corpora lutea and luteal progesterone during gestation in viviparous squamates is still in question. The data presented indicate that in most species corpora lutea are required during early pregnancy but are not essential during the last half of pregnancy. These data suggest that an additional source of hormones may be available during late pregnancy as suggested by Clausen.[36,37] Guillette et al.[74] observed that the major peak in plasma progesterone concentration occurred in the viviparous lizard *S. jarrovi* after the corpora lutea became inactive (determined by histological observation); they suggested that the chorioallantoic placenta may be an endocrine structure, but additional data are not available to test this hypothesis. An alternate hypothesis implies that the adrenal glands secrete progesterone. A recent paper by Dauphin-Villemant and Xavier[40] demonstrated that adrenals from pregnant *L. vivipara* secrete large amounts of progesterone *in vitro*. Additional studies examining the function of the adrenal glands during gravidity in oviparous and viviparous species are needed to determine the role of these glands in reptilian gestation.

Relaxin has not been isolated from reptiles nor has its role (if any) been ascertained. However, the physiological role of this hormone in viviparous reptiles has not been determined. Future studies examining its occurrence and function in reptiles are needed, especially in light of the data obtained from deluteinization studies demonstrating a delay in parturition. One factor involved may be the removal of the source of relaxin (corpora lutea), and thus the normal changes associated with parturition may not occur. Koob et al.[106] observed that relaxin stimulated premature parturition in the shark *Squalus acanthias*. The subsequent delay in parturition following removal of the corpus luteum may be due partly to the absence of relaxin. This hypothesis needs to be tested.

The control of gestation in reptiles is complex. However, strong similarities exist in the endocrine control of gestation and oviposition/parturition in oviparous and viviparous squamates. One can reason that: (1) corpora lutea appear to function in the maintenance of early pregnancy in viviparous reptiles, although in some species their role is still questionable, (2) the survival of embryos in the oviduct of viviparous species is enhanced by the presence of corpora lutea, and (3) NH peptides and prostaglandins stimulate or enhance uterine or body-wall contractions associated with birth. If luteolysis and uterine contractions were inhibited, gestation possibly would be prolonged, and egg retention would result. An hypothesis for the evolution of egg retention in reptiles could be generated from this assumption. This hypothesis has been developed[63] and can be summarized as follows: an environmentally-induced stress response stimulates the release of adrenal steroids, which are capable of blocking prostaglandin synthesis and thus luteolysis. Active corpora lutea would inhibit the release of NH hormones associated with oviductal contractions. Selection for this response in oviparous females living in areas with environmentally poor nesting conditions would lead to egg retention.

Previous researchers have suggested that reptilian viviparity is an adaptation to extreme environmental conditions that would decrease the survivorship of oviposited eggs (see Weekes,[178] Packard et al.,[126] Tinkle and Gibbons,[159] Shine and Bull,[151] Guillette et al., [71] Guillette[61]). The conditions suggested include cold climates with short growing/activity seasons, excessively mesic or xeric environments, and marine environments. Several lizard species exhibit facultative egg retention during periods of adverse environmental conditions.[71,87,155] It is not known whether stress-induced egg retention is due to the presence of inhibitory stimuli or the absence of positive stimuli. However, either stimulus situation would register as a stressor.[145] The major problem with this hypothesis is that normal luteal function may require a luteotropic substance for maximal activity. Interestingly, some stressors appear capable of stimulating hyperactive luteal tissue.[153] Circumstantial evidence indicates that reptilian luteal tissue also is capable of stimulation by a stressful event.[19,65] However, little is known concerning the ability of corticotropin (ACTH) to stimulate luteal activity in mammals or reptiles. Callard et al.[31] demonstrated that mammalian ACTH has no effect on progesterone secretion by turtle corpora lutea. Thus, maintenance of luteal activity by stress may not provide a plausible pathway for the evolution of egg retention in reptiles. However, an alternative source of progesterone may be available to reptiles (and other vertebrates). A recent study indicates that the adrenal gland of a viviparous lizard is capable of synthesizing and releasing large amounts of progesterone.[40] In fact, maximal _in-vitro_ synthesis occurs during the pregnancy period. Maximal plasma progesterone in the viviparous lizard _S. jarrovi_ occurred during late pregnancy after luteolysis had occurred.[74] Could then, selection for hyperactivity of the adrenal occur? No comparative information is available to test this hypothesis, as adrenal activity and secretion in oviparous forms is poorly understood.

Finally, additional research is needed on the endocrine control of oviductal shell gland function. Early placentation is correlated not only with the loss of the fibrous component of the shell but with a significant regression in the shell glands present in the oviductal endometrium. Estrogen and progesterone treatment stimulates glandular function, but not to maximal levels. What other factor(s) are needed? As with fishes and amphibians, an understanding of the normal secretory process is required before an understanding of the reduction process can be obtained. Loss of the shell and shell glands appears to be a requirement for placentation, as placentation is necessary for complete development of the embryo within the oviduct. Studies examining the endocrine control of placentation in viviparous reptiles may provide important insights into the evolution of viviparity in mammals as well as the absence of viviparity in birds.

IV. ACKNOWLEDGMENTS

I would like to thank the editors, D.O. Norris and R.E. Jones, for their very helpful comments and their patience while I prepared this chapter. I also would like to thank Drs. B. Fergusson and S.D. Bradshaw for allowing me to present portions of their unpublished research findings. Unpublished data from my laboratory concerning prostaglandins comes from studies supported by NSF grant DCB 84-16707.

V. REFERENCES

1. Archer, R., Chauvet, J. and Chauvet, M.T. (1969). The neurohypophyseal hormones of reptiles: Comparison of the viper, cobra, and elaphe active principles. Gen. Comp. Endocrinol. 13, 357-360.
2. Archer, R., Chauvet, J. and Chauvet, M.T. (1972). Reptilian neurohypophyseal hormones: The active peptides of a Saurian *Iguana iguana*. Gen. Comp. Endocrinol. 19, 345-348.
3. Amoroso, E.C. (1952). Placentation. *In* "Marshall's Physiology of Reproduction" (A.S. Parkes, ed.), pp. 127-311. Longman's Green, London.
4. Amoroso, E.C. (1981). Viviparity. *In* "Cellular and Molecular Aspects of Implantation" (S.R. Glasser and D.W. Bullock, eds.), pp. 3-25. Plenum Press, New York.
5. Amoroso, E.C., Heap, R.B. and Renfree, M.B. (1979). Hormones and the evolution of viviparity. *In* "Hormones and Evolution" (E.J.W. Barrington, ed.), pp . 925-989. Academic Press, New York.
6. Angelini, F. and Ghiara, G. (1984). Reproductive modes and strategies in vertebrate evolution. Boll. Zool. 51, 121-203.
7. Arslan, M., Zaidi, P., Lobo, J., Zaidi, A.A. and Quzi, M.H. (1978). Steroid levels in preovulatory and gravid lizards (*Uromastix hardwicki*). Gen. Comp. Endocrinol. 34, 300-303.

8. Badir, N. (1968). Structure and function of corpus luteum during gestation in the viviparous lizard, Chalcides ocellatus. Anat. Anz. 122, 1-10.
9. Ball, J.N. (1962). Brood-production after hypophysectomy in the viviparous teleost Mollienesia latipinna Le Sueur. Nature 194, 787.
10. Ball, J.N. and Ensor, D.M. (1965). Effect of prolactin on plasma sodium in the teleost, Poecilia latipinna. J. Endocrinol. 32, 269-270.
11. Ball, J.N. and Ensor, D.M. (1967). Specific action of prolactin on plasma sodium levels in hypophysectomized Poecilia latipinna (Teleostei). Gen. Comp. Endocrinol. 8, 443-440.
12. Ball, J.N. and Ingleton, P.M. (1973). Adaptive variations in prolactin secretion in relation to external salinity in the teleost Poecilia latipinna. Gen. Comp. Endocrinol. 20, 312-325.
13. Bauchot, R. (1965). La Placentation chez les Reptiles. Ann. Biol. 4, 547-575.
14. Beattie, R.C. (1980). A physico-chemical investigation of the jelly capsules surrounding eggs of the common frog (Rana temporaria temporaria). J. Zool. (London) 190, 1-25.
15. Bertin, L. (1952). Oviparite, Ovoviviparite, viviparite. Bull. Soc. Zool. Fr. 77, 84-88.
16. Blackburn, D.G. (1982). Evolutionary origins of viviparity in the Reptilia. 1. Sauria. Amphibia-Reptilia 3, 185-205.
16a. Bona-Gallo, A., Licht, P., MacKenzie, D., and Lofts, B. (1980). Annual cycles in levels of pituitary and plasma gonadotropins, gonadal steroids and thyroid activity in the Chinese cobra, Naja naja. Gen. Comp. Endocrinol. 42, 477-493.
17. Botte, V. and Delrio, G. (1965). Richerche istochimiche sulla distribuzione dei 3α-17-chetosteroidi e di alcuni enzimi della steroid ogenesi nell' ovario de Lacerta sicula. Boll. Zool. 32, 191-195.
18. Bourne, A.R. (1972). Reproductive endocrinology of the viviparous lizard, Tiliqua rugosa. PhD Dissertation, University of Adelaide, pp. 1-171.
19. Bourne, A.R. (1981). Progesterone-like activity in the plasma of the viviparous skink Trachydosaurus rugosus (Stump-tailed lizard). In "Proceedings of the Melbourne Herpetological Symposium" (C.B. Banks and A.A. Martin, eds.), Zoological Board of Victoria, Victoria.
20. Boyd, M.M. (1942). The oviduct, foetal membranes and placentation of Hoplodactylus maculatus (Gray). Proc. Zool. Soc. London Ser. A. 112, 65-104.
20a. Bragdon, D.E. (1950). Hormonal control of the reproductive cycle of ovoviviparous snakes as related to the evolution of viviparity. Virginia J. Sci. 1:391-392.
21. Bragdon, D.E. (1951). The non-essentiality of the corpora lutea for the maintenance of gestation in certain live-bearing snakes. J. Exp. Biol. 118, 419-435.
22. Breder, C.M. and Rosen, D.E. (1966). Modes of Reproduction in Fishes. Am. Mus. Nat. Hist., N.Y.
23. Browning, H.C. (1973). The evolutionary history of the corpus luteum. Biol. Reprod. 8, 128-157.
23a. Burrage, B.R. (1973). Comparative ecology and behaviour of Chamaeleo pumilus pumilus (Gmelin) and C. namaquensis A. Smith. Ann. South Afr. Mus. 61, 1-158.
24. Busse, K. (1970). Care of the young by male Rhinoderma darwini. Copeia 1970, 395.
25. Callard, I.P., Bayne, C.G. and McConnell, W.F. (1972). Hormones and reproduction in the female lizard, Sceloporus cyanogenys. Gen. Comp. Endocrinol. 18, 175-195.

26. Callard, I.P., Chan, S.W.C. and Poots, M.A. (1972). The control of the reptilian gonad. Amer. Zool. 12, 273-287.
27. Callard, I.P., Doolittle, J., Banks, W.I., and Chan, W.C. (1972). Recent studies on the control of the reptilian ovarian cycle. Gen. Comp. Endocrinol. Suppl. 3, 65-75.
28. Callard, I.P. and Hirsch, M. (1976). The influence of oestradiol-17β and progesterone on the contractility of the oviduct of the turtle, *Chrysemys picta*, *in vitro*. J. Endocrinol. 68, 147-152.
29. Callard, I.P., Koob, T., Tsang, P. and Riley, D. (1985). Hormone-receptor interactions and the evolution of viviparity. Paper presented at 10th International Symposium on Comparative Endocrinology, Copper Mountain, Colorado, July 21-26.
30. Callard, I.P. and Lance, V. (1977). The control of reptilian follicular cycles. *In* "Reproduction and Evolution" (C.H. Tyndale-Biscoe, ed.). Proc. 4th Int. Symp. Comp. Biol. Reprod., Australian Acad. Sci., Canberra.
31. Callard, I.P., McChesney, I., Scanes, C. and Callard, G.V. (1976). The influence of mammalian and avian gonadotropins on *in vitro* ovarian steroid synthesis in the turtle (*Chrysemys picta*). Gen. Comp. Endocrinol. 28, 2-9.
32. Chambolle, P. (1964). Influence de l'hypophysectomie sur la gestation de *Gambusia sp*. *C. R. Acad. Sci.*, (Paris) Ser. D. 259, 3855-3857.
33. Chan, S.W.C., Ziegel, S. and Callard, I.P. (1973). Plasma progesterone in snakes. Comp. Biochem. Physiol. 44, 631-637.
34. Chieffi, G. (1962). Integration of reproductive functions. II. Endocrine aspects of reproduction in elasmobranch fishes. Gen. Comp. Endocrinol. Suppl. 1, 275-285.
35. Chieffi, G. and Botte, V. (1970). The problem of luteogenesis in non-mammalian vertebrates. Bull. Zool. 37, 85-102.
36. Clausen, H.J. (1935). The effects of ovariotomy and hypophysectomy on parturition in snakes. Anat. Rec., 64; suppl. 88.
37. Clausen, H.J. (1940). Studies on the effect of ovariotomy and hypophysectomy on gestation in snakes. Endocrinology 27, 700-704.
38. Corben, C.J., Ingram, G.J. and Tyler, M.J. (1974). Gastric brooding: unique form of parental care in an Australian frog. Science 186, 946-947.
39. Cuellar, H. (1979). Disruption of gestation and egg shelling in deluteinized oviparous whiptail lizards *Cnemidophorus uniparens*. Gen. Comp. Endocrinol. 39, 150-157.
40. Dauphin-Villemant, C. and Xavier, F. (1985). *In vitro* steroid biosynthesis by the adrenal gland of the female *Lacerta vivipara* Jacquin: The metabolism of exogenous precursors. Gen. Comp. Endocrinol. 58, 1-9.
41. del Pino, E.M. (1980). Morphology of the pouch and incubatory integument in marsupial frogs (Hylidae). Copeia 1980, 10-17.
42. del Pino, E.M. (1983). Progesterone induces incubatory changes in the brooding pouch of the frog *Gastrotheca riobambae* (Fowler). J. Exp. Zool. 227, 159-163.
43. del Pino, E.M., Galarza, M.L., de Albuja, C.M. and Humphries, A.A., Jr. (1975). The maternal pouch and development in the marsupial frog *Gastrotheca riobambae* (Fowler). Biol. Bull. 149, 480-491.
44. del Pino, E.M. and Sanchez, G. (1977). Ovarian structure of the marsupial frog *Gastrotheca riobambae* (Fowler). J. Morphol. 153, 153-162.
45. Dobbs, G.H. (1975). Scanning electron microscopy of intraovarian embryos of the viviparous teleost, *Micrometrus minimus* (Gibbons), (Perciformes: Embiotocidae) J. Fish Biol. 7, 209-214.
46. Dodd, J.M. (1975). The hormones of sex and reproduction and their effect in fish and lower chordates: twenty years on. Am. Zool. 15 (Suppl. 1), 137-171.

47. Dodd, J.M., Evennett, P.J. and Goddard, C.K. (1960). Reproductive endocrinology in cyclostomes and elasmobranchs. Symp. Zool. Soc. Lond. 1, 77-103.
48. Dodd, J.M. and Sumpter, J.P. (1984). Fishes. In "Marshall's Physiology of Reproduction, 4th Ed." Vol. 1: Reproductive Cycles of Vertebrates. (G.E. Lamming, ed.), pp. 1-126. Churchill Livingstone, Edinburgh.
49. Donaldson, E.M. and Hunter, G.A. (1983). Induced final maturation, ovulation and spermiation in cultured fish. In "Fish Physiology" Vol. 9, Part B (W.S. Hoar, D.J. Randall, and E.M. Donaldson, eds.), pp. 351-404. Academic Press, Inc., New York.
50. Fanning, J.C., Tyler, M.J. and Shearman, D.J.C. (1982). Converting a stomach to a uterus: The microscopic structure of the stomach of the gastric brooding frog Rheobatrachus silus. Gastroenterology 82, 62-70.
51. Fitzgerald, K.T., Guillette, L.J. Jr., and Duvall, D. (1979). Notes on birth, development and care of Gastrotheca riobambae tadpoles in the laboratory (Amphibia, Anura, Hylidae). J. Herpetol. 13, 457-460.
52. Fox, H. (1977). The urinogenital system of reptiles. In "Biology of the Reptilia" Vol. 6 (C. Gans, ed.), pp. 1-157. Academic Press, New York.
53. Fox, S.L. (1984). Ovarian and oviductal morphology during the reproductive cycle of two lizard species, Crotaphytus collaris and Eumeces obsoletus. MS Thesis, Wichita State University, Wichita, KS.
53a. Fraenkel, L., Martins, T. and Mello, R.F. (1940). Studies on the pregnancy of viviparous snakes. Endocrinology 27, 836-387.
54. Freeman, S.B. (1968). A study of the jelly envelopes surrounding the egg of the amphibian, Xenopus laevis. Biol. Bull. 135, 501-513.
55. Goetz, F.W. (1983). Hormonal control of oocyte final maturation and ovulation in fishes. In "Fish Physiology" Vol. 9, Part B (W.S. Hoar, D.J. Randall, and E.M. Donaldson, eds.), pp. 117-170. Academic Press Inc., New York.
56. Gowan, L.K., Reinig, J.W., Schwabe, C., Bedarker, S. and Blundell, T.L. (1981). On the primary and tertiary structure of relaxin from the sand tiger shark (Odontaspis taurus). FEBS (Fed. Eur. Biochem. Soc.) Lett. 129, 80-82.
57. Greven, H. (1976). Notizen zum Geburtsvorgang be im Feuersalamander, Salamandra salamandra (L.). Salamandra 12, 87-93.
58. Greven, H. (1977). Comparative ultrastructure investigations of the uterine epithelium in the viviparous Salamandra atra Laur. and the ovoviviparous Salamandra salamandra (L.) (Amphibia, Urodela). Cell. Tiss. Res. 181, 215-237.
59. Greven, H., Kuhlmann, D. and Reineck, U. (1975). Anatomie und Histologie des oviductes von Salamandra salamandra (Amphibia: Urodela). Zool. Beitr. N.F. 21, 325-345.
60. Guillette, L.J., Jr. (1979). Stimulation of parturition in a viviparous lizard (Sceloporus jarrovi) by arginine vasotocin. Gen. Comp. Endocrinol. 38, 457-460.
61. Guillette, L.J., Jr. (1982). The evolution of viviparity and placentation in the high elevation, Mexican lizard Sceloporus aeneus. Herpetologica 38, 94-103.
62. Guillette, L.J., Jr. (1984). Ovarian and oviductal morphology of the New Guinea scincid lizard, Sphenomorphus fragilis. Paper presented at the annual meeting of the Australian Society of Herpetologists. Springwood, Australia, September 1-2.
63. Guillette, L.J., Jr. (1985). The evolution of egg retention in lizards: A physiological model. In "Biology of Australasian Frogs and Reptiles" (G. Grigg, R. Shine and H. Ehmann, eds.), pp. 379-386. Royal Zoological Society of New South Wales, Sydney.

64. Guillette, L.J., Jr. and Dickey, D. (1984). Potentiation of acetylcholine-induced skeletal muscle contractions in vitro by Prostaglandin $F_{2\alpha}$ in the lizard, Crotaphytus collaris. Can. J. Zool. 62, 2188-2191.
65. Guillette, L.J. Jr. and Fox, S.L. (1985). Effect of deluteinization on plasma progesterone concentration and gestation in the lizard, Anolis carolinensis. Comp. Biochem. Physiol. 80A, 303-306.
66. Guillette, L.J., Jr. and Herman, C.A. (1985). Synthesis of series two prostaglandins by the reptilian reproductive tract. Paper presented at the 10th International Symposium on Comparative Endocrinology, Copper Mountain, Colorado; July 21-26.
67. Guillette, L.J., Jr. and Hotton, N. (1985). The evolution of mammalian reproductive characteristics in therapsid reptiles. In "The Biology of the Mammal-like Reptiles" (Roth, J.J. et al., eds.), in press. Smithsonian Press, Washington, D.C.
68. Guillette, L.J., Jr. and Jones, R.E. (1980). Arginine vasotocin-induced in vitro oviductal contractions in Anolis carolinensis: Effects of steroid hormone pretreatment in vivo. J. Exp. Zool. 212, 147-152.
69. Guillette, L.J. Jr., and Jones, R.E. (1982). Further observations on arginine vasotocin-induced oviposition and parturition in lizards. J. Herpetol. 16, 140-144.
70. Guillette, L.J., Jr. and Jones, R.E. (1985). Ovarian, oviductal and placental morphology of the reproductively bimodal lizard species, Sceloporus aeneus. J. Morphol. 184, 85-98.
71. Guillette, L.J., Jr., Jones, R.E., Fitzgerald, K.T. and Smith, H.M. (1980). Evolution of viviparity in the lizard genus Sceloporus. Herpetologica 36, 201-215.
72. Guillette, L.J., Jr., Lavia, L.A., Walker, N. and Roberts D. (1984). Luteolysis induced by Prostaglandin $F_{2\alpha}$ in the lizard, Anolis carolinensis. Gen. Comp. Endocrinol. 56, 271-277.
73. Guillette, L.J. Jr., Norris, D.O., Norman, M. (1985). Response of amphibian (Ambystoma tigrinum) oviduct to arginine vasotocin and acetylcholine in vitro: Influence of steroid hormone pretreatment in vivo. Comp. Biochem. Physiol. 80C, 151-154.
74. Guillette, L.J. Jr., Spielvogel, S. and Moore, F.L. (1981). Luteal development, placentation and plasma progesterone concentration in the viviparous lizard, Sceloporus jarrovi. Gen. Comp. Endocrinol. 43, 20-29.
75. Guraya, S.S. (1968). Histochemical study of granulosa (follicular) cells in the preovulatory and postovulatory follicles of amphibian ovary (bullfrog). Gen. Comp. Endocrinol. 10, 138-146.
76. Heller, H. (1972). The effect of neurohypophyseal hormones on the female reproductive tract of lower vertebrates. Gen. Comp. Endocrinol. Suppl. 3, 703-714.
77. Heller, H., Ferreri, E., and Leathers, D.H.G. (1970). The effect of neurohypophyseal hormones on the amphibian oviduct in vitro, with some remarks on the histology of this organ. J. Endocrinol. 47, 495-509.
78. Herman, C.A., Gonzales, D.V., Doolittle, K., and Jackson, L. (1981). Frogs produce a physiologically active compound from eicosapentaenoic acid. Prostaglandins 21, 297-306.
79. Highfill, D.R. and Mead, R.A. (1975). Function of corpora lutea of pregnancy in the viviparous garter snake, Thamnophis elegans. Gen. Comp. Endocrinol. 27, 401-407.
80. Highfill, D.R. and Mead, R.A. (1975). Sources and levels of progesterone during pregnancy in the garter snake, Thamnophis elegans. Gen. Comp. Endocrinol. 27, 389-400.
81. Hisaw, F.L. and Abramowitz, A.A. (1938). The physiology of reproduction in the dogfish, Mustelus canis. Rep. Woods Hole Oceanogr. Inst. 1937, 21-22.

82. Hisaw, F.L. and Abramowitz, A.A. (1939). Physiology of reproduction in the dogfishes, Mustelus canis and Squalus acanthias. Rep. Woods Hole Oceanogr. Inst. 1938, 22.
83. Hisaw, F.L., Jr and Hisaw, F.L. (1959). The corpora lutea of elasmobranch fishes. Anat. Rec. 135, 269-278.
84. Hoar, W.S. (1969). Reproduction. In "Fish Physiology" Vol. 3 (W. A. Hoar and D.J. Randall, eds.), pp. 1-72. Academic Press, Inc., New York.
85. Hogarth, P.J. (1976). Viviparity. Studies in Biology #75. Arnold, London.
86. Horrobin, D.F. (1978). Prostaglandins: physiology, pharmacology and clinical significance. Eden Press Inc., Montreal, Que.
87. Huey, R.B. (1977). Egg retention in some high altitude Anolis lizards. Copeia 1977, 373-375.
88. Ingram, G.J., Anstis, M. and Corben, C.J. (1975). Observations on the Australian Leptodactylid frog, Assa darlingtoni. Herpetologica 31, 425-49.
89. Iverson, J.G. (1978). Amphibia and Reptilia. In "Vertebrates of the World, A Preliminary List." Gainesville Field Station, U.S. Fish and Wildlife Laboratory, Gainesville.
90. Jalabert, B. (1976). In vitro oocyte maturation and ovulation in rainbow trout (Salmo gairdneri), northern pike (Esox lucius) and goldfish (Carassius auratus). J. Fish. Res. Board Can. 33, 974-988.
91. Jollie, W. P. and Jollie, L. G. (1964). The fine structure of the ovarian follicle of the ovoviviparous poeciliid fish, Lebistes reticulatus. I. Maturation of follicular epithelium. J. Morphol. 114, 479-502.
92. Jollie, W.P. and Jollie, L.G. (1964). The fine structure of the ovarian follicle of the ovoviviparous poeciliid fish, Lebistes reticulatus. II. Formation of follicular pseudoplacenta. J. Morphol. 114, 503-526.
93. Jollie, W.P. and Jollie, L.G. (1967). Electron microscopic observations on accommodations to pregnancy in the uterus of the spiny dogfish, Squalus acanthias. J. Ultrastruc. Res. 20, 161-178.
94. Joly, J. (1960). Le cycle sexuel de la Salamandre tachettee, Salamandra salamandra quadrivirgata, dans l'ouest de la France. C. R. Acad. Sci. (Paris) Ser. D. 251, 2594-2596.
95. Joly, J. (1965). Mise en evidence histochemique d'une Δ^5,3β-hydroxysteroide - deshydrogenase dans l'ovaire de l'urodele Salamandra salamandra (L.) a differents stages du cycle sexuel. C. R. Acad. Sci. (Paris) Ser. D. 261, 1569-1571.
96. Joly, J. (1968). Donnees ecologiques sur la Salamandre tachetee Salamandra salamandra (L.). Ann. Sc. Nat. Zool. 10, 301-366.
97. Joly, J. and Boisseau, C. (1973). Localisation des spermatozoides dans l'oviducte de la salamandra terrestre, Salamandra salamandra (L.) (amphibien urodele) au moment de la fecondation. C. R. Acad. Sci. (Paris) Ser. D. 277, 2537-2540.
98. Joly, J. and Picheral, B. (1972). Ultrastructure, histochimie et physiologie du follicule pre-ovulatorie et du corps jaune de l'urodele ovo-vivipare Salamandra salamandra (L.). Gen. Comp. Endocrinol. 18, 235-259.
99. Jones, R.E., Gerrard, A.M., and Roth, J.J. (1973). Estrogen and brood pouch formation in the marsupial frog, Gastrotheca riobambae. J. Exp. Zool. 184, 177-184.
100. Jones, R.E. and Guillette, L.J., Jr. (1982). Hormonal control of oviposition and parturition in lizards. Herpetologica 38, 80-93.
101. Jones, R.E., Guillette, L.J., Jr., Norman, M. and Roth, J.J. (1982). Corpus luteum-uterine relationships in the control of uterine contraction in the lizard Anolis carolinensis. Gen. Comp. Endocrinol. 48, 104-112.

102. Jones, R.E., Summers, C. and Lopez, K. (1983). Adrenergic inhibition of uterine contractions and oviposition in the lizard Anolis carolinensis. Gen. Comp. Endocrinol. 51, 77-83.
103. Khoo, K.H. (1975). The corpus luteum of goldfish (Carassius auratus L.) and its functions. Can. J. Zool. 53, 1306-1323.
104. Knight, F.M., Lombardi, J., Wourms, J.P. and Burns, J.R. (1985). Follicular placenta and embryonic growth of the viviparous four-eyed fish (Anableps). J. Morphol. 185, 131-142.
105. Koob, T.J. and Callard, I.P. (1982). Relaxin: speculations on its physiological importance in some non-mammalian species. Ann. N.Y. Acad. Sci. 380,163-173.
106. Koob, T.J., Laffan, J.J. and Callard, I.P. (1984). Effects of relaxin and insulin on reproductive tract size and early fetal loss in Squalus acanthias. Biol. Reprod. 31, 231-238.
107. Koob, T.J., Tsang, P., Laffan, J.J. and Callard, I.P. (1982). Possible role for estradiol in nidamental gland function. Bull. Mt. Desert Isl. Biol. Lab 22,97-99.
108. Kujalȧ, G.A. (1978). Corticosteroid and neurohypophyseal hormone control of parturition in the guppy, Poecilia reticulata. Gen. Comp. Endocrinol. 36, 286-296.
109. Lance, V. and Callard, I.P. (1978). Hormonal control of ovarian steroidogenesis in nonmammalian vertebrates. In "The Vertebrate Ovary: Comparative Biology and Evolution" (R.E. Jones, ed.), pp. 361-407. Plenum Press, New York.
110. LaPointe, J.L. (1964). Induction of oviposition in lizards with the hormone oxytocin. Copeia 1964, 541-542.
111. LaPointe, J.L. (1979). Effects of ovarian steroids and neurohypophysial hormones on the oviduct of the viviparous lizard, Klauberina riversiana. J. Endocrinol. 43, 197-205.
112. LaPointe, J.L. (1977). Comparative physiology of neurohypophysial hormone action on the vertebrate oviduct-uterus. Am. Zool. 17, 763-773.
113. Lemus, D., Zurich, L., Paz de la Vega-Lemus, Y. and Wacyk, J. (1968). Actividad espontanea y effecto de oxitocina en el utero aislado de Liolaemus gravenhorti y Liolaemus tenuis T. Arch. Biol. Med. Exp. 7, 11-13.
114. Lien, J.P. and Callard, I.P. (1968). The effect of gonadectomy and hormonal replacement therapy on reproduction and the reproductive tract of the ovoviviparous lizard, Sceloporus cyanogenys. Va. J. Sci. 19, 179.
115. Liley, N.R. and Stacey, N.E. (1983). Hormones, pheromones, and reproductive behavior in fish. In "Fish Physiology" Vol. 9, Part B (W.S. Hoar, D.J. Randall, and E.M. Donaldson, eds.), pp. 1-64. Academic Press Inc., New York.
116. Lofts, B. (1984). Amphibians. In "Marshall's Physiology of Reproduction, 4th ed." Vol. 1: Reproductive cycles of Vertebrates. (G.E. Lamming, ed.), pp. 127-205. Churchill Livingstone, Edinburgh.
117. Masson, G.R. (1985). Changes in uterine vascularity during the reproductive cycle of three oviparous lizards: Eumeces obsoletus, Sceloporus undulatus, Crotaphytus collaris. MS Thesis, Wichita State University, Wichita, KS.
118. McGill, J.C. (1981). Prostaglandins, prostacyclin and thromboxanes. Ann. Rev. Pharmacol. Toxicol. 21, 479-509.
119. Mead, R.A., Eroschenko, V.P., and Highfill, D.R. (1981). Effects of progesterone and estrogen on the histology of the oviduct of the garter snake, Thamnophis elegans. Gen. Comp. Endocrinol. 45, 345-354.
120. Morat, M. (1969). Contributions a l'etude de l'activite Δ^5-3β-hydroxysteroide dehydrogenasicue chez chelques reptiles de massif central. Ann. Sta. Biol. Beese, Chandesse 4, 1-55.

120a. Moore, M.C., Whittier, J.M. and Crews, D. (1985). Sex steroid hormones during the ovarian cycle of an all-female, parthenogenetic lizard and their correlation with pseudo-sexual behavior. Gen. Comp. Endocrinol. 60, 144-153.

121. Morriss, G. (1975). Placental evolution and embryonic nutrition. In "Comparative Placentation: Essays in Structure and Function" (D.H. Steven, ed.), pp. 87-107. Academic Press, New York.

122. Mossman, H.W. (1937). Comparative morphogenesis of the fetal membranes and accessory uterine structures. Cont. Embryol. 158, 130-193.

123. Nagahama, Y. (1983). The functional morphology of teleost gonads. In "Fish Physiology" Vol. 9, Part A (W.S. Hoar, D.J. Randall, and E.M. Donaldson, eds.). Academic Press, New York.

124. Noble, G.K. (1925). The integumentary, pulmonary and cardiac modifications correlated with increased cutaneous respiration in the amphibia: a solution of the "hairy frog" problem. J. Morphol. 40, 3410 416.

125. Ozon, R. and Xavier, F. (1968). Biosynthese in vitro des steroides par l'ovaire de l'Anoure vivipare Nectophrynoides occidentalis au cours du cycle sexuel. C.R. Acad. Sci. (Paris) Ser. D. 266, 1173-1175.

126. Packard, G.C., Tracy, C.R. and Roth, J.J. (1977). The physiological ecology of reptilian eggs and embryos, and the evolution of viviparity within the class Reptilia. Biol. Rev. 52, 71-105.

127. Pancharatna, M. and Saidapur, S.K. (1984). The luteotrophic effect of homoplastic pituitary pars distalis homogenate, PMSG, and HCG on the corpora lutea of the hypophysectomized frog, Rana cyanophlyctis (SCHN). Gen. Comp. Endocrinol. 53, 375-380.

128. Panigel, M. (1953). Role des corps jaunes au cours de la gestation chez le lezard vivipare, Zootoca vivipara. C. R. Acad. Sci. (Paris) Ser D. 236, 849-851.

129. Panigel, M. (1956). Contribution a l'etude de l'ovoviviparite chez les reptiles: Gestation et parturition chez le lezard vivipare, Zootoca vivipara. Ann. Sci. Nat. Zool. Biol. Anim. XI 18, 569-668.

130. Rabb, G.S. and Rabb, M.S. (1960). On the mating and egg-laying behavior of the Surinam toad, Pipa pipa. Copeia 1960, 271-276.

131. Rabb, G.S. and Rabb, M.S. (1963). Additional observations on the breeding behavior of the Surinam toad, Pipa pipa. Copeia 1963, 636-642.

132. Rahn, H. (1938). Structure and function of placenta and corpus luteum in viviparous snakes. Proc. Soc. Exp. Biol. Med. 40, 381-382.

133. Roth, J.J., Jones, R.E. and Gerrard, A. (1973). Corpora lutea and oviposition in the lizard Sceloporus undulatus. Gen. Comp. Endocrinol. 21, 596-572.

134. Rothchild, I. (1981). The regulation of the mammalian corpus luteum. Rec. Prog. Horm. Res. 37, 183-298.

135. Rownads, I.W. and Weir, B.J. (1984). Mammals: Non-primate eutherians. In "Marshall's Physiology of Reproduction, 4th ed." Vol. 1: Reproductive Cycles of Vertebrates (G.E. Lamming, ed.), pp. 455-658. Churchill Livingstone, Edinburgh.

136. Saidapur, S.K. (1982). Structure and function of postovulatory follicles (Corpora Lutea) in the ovaries of nonmammalian vertebrates. Int. Rev. Cytol. 75, 243-285.

137. Saidapur, S.K. and Nadkarni, V.B. (1974). Steroid synthesizing cellular sites in the amphibian ovary. A histochemical study. Gen. Comp. Endocrinol. 22, 459-462.

138. Salthe, S.N. (1963). The egg capsules in the amphibia. J. Morphol. 113, 161-171.

139. Salthe, S.E. and Duellman, W.E. (1973). Quantitative constraints associated with reproductive mode in anurans. In "Evolutionary Biology of the Anurans: Contemporary Research on Major Problems" (J.L. Vial, ed.), pp. 229-249. Univ. Missouri Press, Columbia, MO.

140. Salthe, S.N. and Mecham, J.S. (1974). Reproductive and courtship patterns. In "Physiology of the Amphibia" (B. Lofts, ed.), pp. 309-501. Academic Press, New York.

141. Schwabe, C., Steinetz, B., Weiss, G., Segaloff, A., McDonald, J.K., O'Byrne, E.M., Hochman, J., Carriere, B. and Goldsmith, L. (1978). Relaxin. Recent Prog. Horm. Res. 34, 123-199.

142. Sekharappa, B.M. and Devaraj Sarkar, H.B. (1978). Role of corpora lutea in gestation in the skink *Mabuya carinata*. Indian J. Exp. Biol. 16, 1097-1098

143. Selye, H. (1936). A syndrome produced by diverse noxuous agents. Nature 138, 32.

144. Selye, H. (1936). Thymus, adrenals and thyroid in the response of the organism to certain drugs. Am. J. Physiol. 116, 141.

145. Selye, H. (1937). Studies on Adaptation. Endocrinology 21, 169-188.

146. Selye, H. (1950). Stress. Acta Inc., Montreal, 1005 pp,

147. Sergeev, A.M. (1940). Researches on the viviparity of reptiles. Moscow Soc. Nat. 1-34.

148. Shine, R. (1983). Reptilian reproductive modes: the oviparity-viviparity continuum. Herpetologica 39, 1-8.

149. Shine, R. (1984). Physiological and ecological questions on the evolution of reptilian viviparity. In "Respiration and Metabolism of Embryonic Vertebrates" (R. Seymour, ed.), pp.147-154. Junk, The Hague.

150. Shine, R. (1985). The evolution of viviparity in reptiles: An ecological analysis. In "Biology of the Reptilia" Vol. 15 (C. Gans and F. Billett, eds.), pp. 604-694. Wiley Interscience, St. Louis.

151. Shine, R. and Bull, J.J. (1979). The evolution of live-bearing in lizards and snakes. Am. Nat. 113, 905-923.

152. Smith, C.L., Rand, C.S., Schaeffer, B. and Atz, J.W. (1975). *Latimeria*, the living coelecanth, is ovoviviparous. Science 190, 1105-1106.

153. Sod-Moria, U.S. and Bedrak, E. (1973). Corpora lutea activity in heat-acclimated (HA) female rats. Acta endocrinol., Copenh. (Suppl) 177, 337.

154. Spannhof, I. and Spannhof, L. (1972). Beobachtungen zur brutbiologie und larvenetwicklung von *Gastrotheca marsupiata*. Wiss. Z. Univ. Rostock Math.-Naturwiss. Reihe 20,97-104.

155. Stamps, J.A. (1976). Egg retention, rainfall and egg laying in a tropical lizard *Anolis aeneus*. Copeia 1976, 759-764.

156. Stewart, J.R. (1985). Placentation in the lizard *Gerrhonotus caeruleus* with a comparison to the extraembryonic membranes of the oviparous *Gerrhonotus multicarinatus* (Sauria, Anguidae). J. Morphol. 185, 101-114.

157. Straughan, I.R. and Main, A.R. (1966). Speciation and polymorphism in the genus *Crinia* Tschudi (Anura, Leptodactylidae) in Queensland. Proc. Roy. Soc. Queensland 78, 11-28.

158. Summers, C.H., Austin, H.B. and Jones, R.E. (1985). Induction of oviposition in cycling *Anolis carolinensis* requires a β-adrenoreceptor blocker and a high dosage of arginine vasotocin. Gen. Comp. Endocrinol. 57, 389-392.

159. Tinkle, D.W. and Gibbons, J.W. (1977). The distribution and evolution of viviparity in reptiles. Misc. Publ. Mus. Zool. Univ. Michigan #154.

160. Townsend, D.S., Stewart, M.M., Pough, F.H., and Brussard, P.F. (1981). Internal fertilization in an oviparous frog. Science 212, 469-470.

161. Tyler, M.J. and Carter, D.B. (1981). Oral birth of the young of the gastric brooding frog *Rheobatrachus silus*. Anim. Beh. 29, 280-282.

162. Tyler, M.J., Shearman, D.J.C., Franco, R., O'Brien, P., Seamark, R.F., Kelly, R. (1983). Inhibition of gastric acid secretion in the gastric brooding frog, *Rheobatrachus silus*. Science 220, 609-610.

163. van Tienhoven, A. (1983). Reproductive Physiology of Vertebrates, 2nd Ed. Cornell University Press, Ithaca, N.Y.

163a. van Wyk, J.H. (1984). Physiological changes during the ovarian cycle of the female rock lizard, _Agama atra_ (Sauria: Agamidae). S.-Afr. Tydskr. Dierk. 19, 253-260.

164. Veith, W.J. (1974). Reproductive biology of _Chameleo pumilus pumilus_ with special reference to the role of the corpus luteum and progesterone. Zool. Afr. 9, 161-183.

165. Veith, W.J. (1980). Viviparity and embryonic adaptations in the teleost _Clinus superciliosus_. Can. J. Zool. 58, 1-12.

166. Vilter, V. (1967) Histologie de l'oviducte chez _Salamandra atra_ mature, urodele totalement vivipare de haute montaigne. C.R. Soc. Biol. (Paris) 161, 260-264.

167. Vilter, A. and Vilter, V. (1962). Role de l'oviducte dans l'uniparite uterine de la salamandre noire des Alpes orientales (_Salamandra atra_ Laur.). C.R. Soc. Biol. (Paris) 156, 49-51.

168. Vilter, V. and Vilter, A. (1964). Sur l'evolution des corps jaunes ovariens chez _Salamandra atra_ Laur. des Alpes Vaudoises. C.R. Soc. Biol. (Paris) 158, 457-461.

169. Wake, M.H. (1968). Evolutionary morphology of the caecilian urogenital system. I. The gonads and the fat bodies. J. Morphol. 126, 291-332.

170. Wake, M.H. (1977). Fetal maintenance and its evolutionary significance in the Amphibia: Gymnophiona. J. Herpetol. 11, 379-386.

171. Wake, M.H. (1977). The reproductive biology of caecilians: an evolutionary perspective. _In_ "Reproductive Biology of the Amphibia" (S. Guttman and D. Taylor, eds.), pp. 73-101. Plenum Press, New York.

172. Wake, M.H. (1978). An ovoviviparous _Eleutherodactylus_, with comments on the evolution of live-bearing systems. J. Herpetol. 12, 121-133.

173. Wake, M.H. (1980). The reproductive biology of _Nectophrynoides malcolmi_ (Amphibia: Bufonidae), with comments on the evolution of reproductive modes in the genus _Nectophrynoides_. Copeia 1980, 192-209.

174. Wake, M.H. (1982). Diversity within a framework of constraints. Amphibian Reproductive Modes. _In_ "Environmental Adaptation and Evolution" (D. Mossakowski and G. Roth, eds.), pp. 87-107. Gustav Fischer, Stuttgart, New York.

175. Wake, M.W. (1985). Oviduct structure and function in non-mammalian vertebrates. Fortsch. Zool. 30, 427-435.

176. Waring, H., Landgrebe, F.W. and Neill, R.M. (1941). Ovulation and oviposition in anura. J. Exp. Biol. 18, 11-25.

177. Welsch, U., Muller, W. and Schubert C. (1977). Electron microscopical and histochemical observations on the reproductive biology of viviparous caecilians (_Chtonerpetron indistinctum_). Zool. Jb. Anat. 97, 532-549.

178. Weekes, H.C. (1935). A review of placentation among reptiles with particular regard to the function and evolution of the placenta. Proc. Zool. Soc. London 1935, 625-645.

179. Wourms, J.P. (1977). Reproduction and development in chondrichthyan fishes. Am. Zool. 17, 379-410.

180. Wourms, J.P. (1981). Viviparity: The maternal-fetal relationship in fishes. Am. Zool. 21, 473-515.

181. Xavier, F. (1973). Le cycle des voies genitales femelles de _Nectophrynoides occidentalis_ Angel, amphibien anoure vivipare. Z. Zellforsch, 140, 509-534.

182. Xavier, F. (1977). An exceptional reproductive strategy in Anura: _Nectophyrnoides occidentalis_ Angel (Bufonidae), an example of adaptation to terrestrial life by viviparity. _In_ "Major Patterns in Vertebrate Evolution" (M.K. Hecht, P.C. Goody, and B.M. Hecht), pp. 545-552. Plenum Press, New York.

182a. Xavier, F. (1982). Progesterone in the viviparous lizard Lacerta vivipara: ovarian biosynthesis, plasma levels, and binding to transcortin-type protein during the sexual cycle. Herpetologica 38, 62-70.

183. Xavier, F. and Ozon, R. (1971). Recherches sur l'activite endocrine de l'ovaire de Nectophrynoides occidentalis Angel (Amphibien Anoure vivipare). II. Synthese in vitro des steroides. Gen. Comp. Endocrinol. 16, 30-40.

184. Xavier, F., Zuber-Vogeli, M. and LeQuang Trong, Y. (1970). Recerches sur l'activite endocrine de l'ovaire de Nectophrynoides occidentalis Angel (Amphibien anoure vivipare). Gen. Comp. Endocrinol. 15, 425-431.

185. Yaron, Z. (1972). Endocrine aspects of gestation in viviparous reptiles. Gen. Comp. Endocrinol. Suppl. 3, 668-674.

185a. Yaron, Z. (1977). Embryo-maternal interrelations in the lizard Xantusia vigilis. In. "Reproduction and Evolution" (J.H. Calaby and C.H. Tyndale-Biscoe, eds.), 271-277. Australian Academy of Sci. Canberra.

186. Yaron, Z. (1985). Reptilian placentation and gestation: Structure, function and endocrine control. In "Biology of the Reptilia" Vol. 15 Development B. (C. Gans and F. Billett, eds.), pp. 527-603. John Wiley and Sons, N.Y.

AGING OF THE NEUROENDOCRINE SYSTEM

Martin P. Schreibman,
Henrietta Margolis-Nunno and
Leslie Halpern-Sebold

Department of Biology
Brooklyn College
Brooklyn, New York 11210

I. INTRODUCTION

A. Background

The study of senescence is, essentially, a science in its infancy. As a result, there is a paucity of fundamental information available on the aging of basic physiological processes, including reproductive phenomena. Concepts about what leads to the faltering of neuroendocrine mechanisms and subsequent demise of the individual are, for the most, hypotheses that are only beginning to be tested.

The preparation of this review on the aging of the neuroendocrine system in poikilotherms was facilitated by the observation that the majority of studies on this subject have been restricted to mammals, and these have been summarized in several recent reviews.[1,2,32,40] Essentially, the limited amount of information that is available on the aging of physiological systems in fishes, amphibians and reptiles is based on either field studies or isolated reports that compare a group of "young" animals with a group of "old" ones. Our investigations of the brain-pituitary-gonad (BPG) axis in platyfish, *Xiphophorus maculatus*, constitute the most extensive aging study performed in any poikilotherm. The information we have gathered, although far from being complete, will serve as the foundation of this treatise, which is formulated under the premise that the process of aging begins at birth (if not earlier) and differs from, but includes, senescence, the processes of growing old. Thus, the sections of the chapter that follow which deal with the changes that occur in the brain, pituitary and gonad from the onset of sexual maturation into the senescence of platyfish, will be built upon the foundation of available knowledge of the events that occur from birth to puberty.

B. The Platyfish (*Xiphophorus maculatus*; Poeciliidae): a Model to Study Aging

The platyfish, a small viviparous freshwater teleost, has been studied intensively for more than 60 years.[19] Sufficient information has been amassed on the structure and function of its reproductive system, and the hereditary mechanisms that control its age at sexual maturation, to make the platyfish invaluable for studies involving the mechanisms of aging of the BPG axis. The structure and function of the BPG axis from birth to senescence has been under intensive investigation in our laboratory.

Variation in the time of onset of sexual maturation of male and female platyfish is due to genetic factors.[19,21,22,45] A sex-linked gene, *P*, controls the age at which the BPG axis becomes activated and sexual maturation ensues. Five *P* alleles (P^1 through P^5) have been identified in natural populations and laboratory stocks[20] and, depending upon genotype, puberty occurs at a specific predictable age between 8 and 104 weeks. There is no overlap in maturation time when different genotypes are raised under identical conditions.[21,22,45] The various *P* factors are carried on the sex chromosomes (X and Y) and are closely linked to a number of genes for body pigmentation that serve as their phenotypic markers.

The life span of platyfish is two to three years; however, fish raised in isolation and under optimal conditions have lived for as long as five years. Unlike other vertebrates, the platyfish presents us with a unique model system with which to study the interaction of genetic and environmental factors, and the effect of these factors on the senescence of integral parts of the nervous and endocrine systems and on the general phenomenon of aging. By using known genetic stocks, some with large degrees of homozygosity, we can eliminate the variability in animal populations that plague the analysis of so many aging studies.

Metamorphosis of the anal fin into a gonopodium (the intromittent organ) is invaluable for ascertaining the state of gonadal maturation by gross inspection.[8,21] All male platyfish regardless of genotype enter into stage 1 of anal fin transformation at five weeks of age. Stage 2 commences with the initiation of sexual maturation (early maturers, 11 weeks; late, 26 weeks). At stage 6 fish are fully mature and have functional testes (early, 18 weeks; late, 52 weeks). Females follow a time table for maturation that is comparable to males of similar *P* allele constitution.[45]

II. BRAIN

A. Birth to Sexual Maturity

1. Gonadotropin-Releasing Hormone (GnRH). Using antisera generated against synthetic GnRH (see Schreibman et al.[43]) we have localized immunoreactive (ir-) material in three brain centers. There is a sequential development of the three ir-GnRH containing areas that is directly related to the stage of sexual development and not to chronological age. The nucleus olfactoretinalis (NOR) is the first region to contain ir-GnRH which then appears in the nucleus preopticus periventricularis (NPP) followed by the nucleus lateralis tuberis pars posterioris (NLT). Anatomically, the sequence of this pattern of development proceeds from anterior to posterior and is essentially similar in both early- and late-maturing genotypes. This phenomenon has been termed the "cascade effect," because of the apparent continuum in the pattern of development of the NOR, NPP, NLT and pituitary. Numerous fibers containing ir-GnRH material emanate in all directions from the NOR, and processes can be clearly delineated between this nucleus and the other nuclei, the pineal gland (habenular region), the olfactory tract and the optic nerve.[3,10,33,48]

In the NOR of early and late maturers cellular and nuclear areas increase from one (neonatal) to five (stage 1 gonopodial development) weeks of age (Fig. 1). Ir-GnRH first appears in NOR perikarya at five weeks of age in early maturers and six weeks later (still in stage 1), in late maturers (Fig. 2). Ir-NOR perikarya are at their maximal number when they

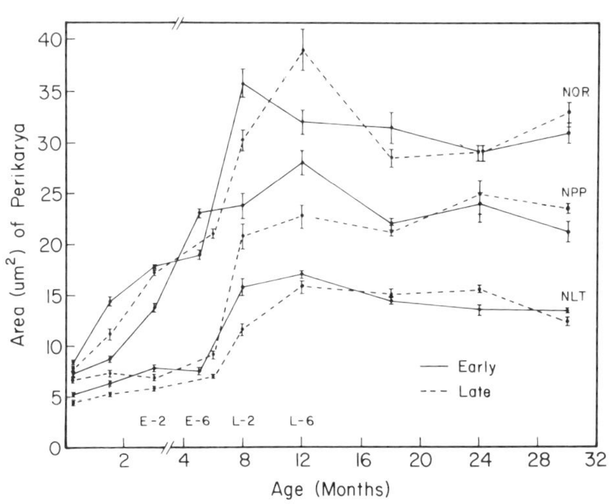

Fig. 1. Age-related changes in the area of perikarya in the NOR, NPP, and NLT of early- and late-maturing male platyfish. Age at gonopodial stages 2 and 6 indicated for early (E) and late (L) genotypes.

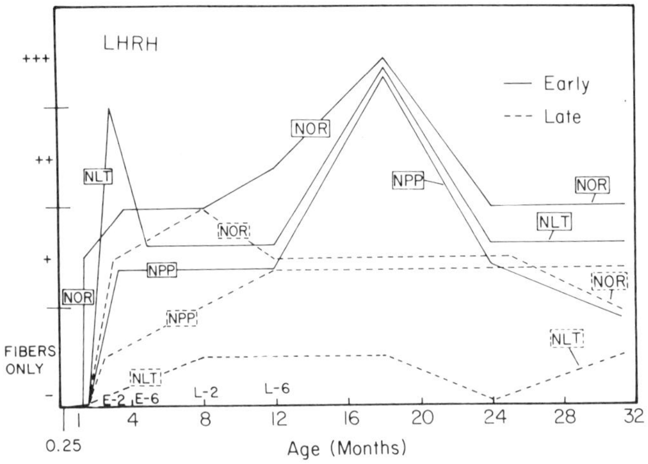

Fig. 2. Age-related changes in relative intensities (- to +++) of ir-GnRH in perikarya of the NOR, NPP and NLT of early- and late-maturing male platyfish (see text for detail). Age at gonopodial stages 2 and 6 indicated for early (E) and late (L) genotypes.

first appear, and this number, which is similar in both genotypes, remains constant into adulthood (Fig. 3). There are, however, significant increases in the dimensions of NOR perikarya in both genotypes from stage 1 to stage 2, a period that is characterized by the appearance of ir-fibers between the NOR and the NPP and NLT, and a slight increase in immunoreactivity of NOR perikarya in early maturers. These observations suggest increased cellular activity, a concept supported by cytological observations of increased basophilia, paler nuclei and more prominent nucleoli.

In the NPP, the dimensions of perikarya show substantial increases in both genotypes subsequent to those shown by the NOR (Fig. 1). The NPP enlargement seen at stage 2 corresponds with the initial appearance of ir-NPP perikarya. By stage 6 (sexual maturity) the NPP of both genotypes attains its maximal number of ir-perikarya, with early maturers containing approximately fifty percent more than late-maturing siblings (Fig. 3).

The dimensions of NLT perikarya follow a similar pattern of change in both genotypes, and they increase up to stage 2 and then decrease from stage 2 to 6 (Fig. 1). Ir-GnRH fibers first appear in stage 2 in both genotypes; however, although ir-perikarya begin to appear shortly thereafter (at their maximal number) in early maturers, ir-GnRH cell bodies are never seen in the NLT of late maturers, whose NLT perikarya also display less basophilia than early maturers at stage 2. The intensity and number of ir-cell bodies in the NLT of early maturers decrease dramatically from stage 2 to stage 6

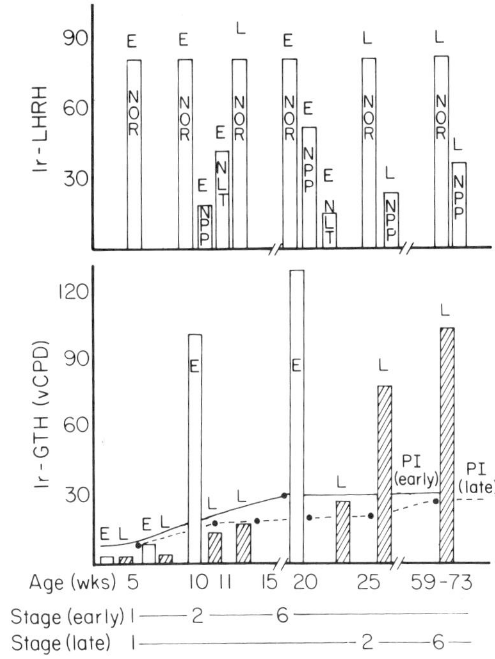

Fig. 3. A correlation between the number of ir-GnRH (LHRH) containing perikarya in the NOR, NPP and NLT (upper panel) and ir-Gn-containing cells (Ir-GTH) in the vCPD (bars) and PI (lines) from birth to sexual maturity in early (E) and late (L) genotypes.

(Fig. 3) and although late maturers continue to lack NLT ir-perikarya, ir-fibers persist at stage 6 (Fig. 2).

Other essential differences also exist between genotypes. In late maturers, specific steps of the "çascade effect" take place at similar developmental stages, but at different ages (older animals) and require more time to complete. This delay in time creates significant differences in the cytometric, cytologic, and immunologic characteristics of the three brain regions in fish of the same age but different genotype (Figs. 1, 2 and 3).

B. Maturity to Senescence (8 to 30 months)

1. GnRH in NOR. Ir-GnRH-containing NOR perikarya retain their fusiform shape through life in platyfish of both early- and late-maturing genotypes. In early-maturing males, the average area of NOR cells decreases gradually between 8 and 24 months and then shows a minor increase at 30 months (Fig. 1). The nuclear area of these cells also decreases between 8 and 30 months. Although cytoplasmic basophilia is somewhat elevated in early males at 12 and 18 months, the NOR nuclei are consistently pale and their nucleoli prominent at all ages.

Late-maturing males differ from early maturers in several parameters. In both genotypes the areas of NOR cell bodies are similar at 24 and 30 months of age; they are significantly smaller in late maturers at 8 and 18 months but very much larger at 12 months (Fig. 1). Nuclear area is also significantly larger in late fish at 12 months while other values for the two genotypes are comparable. Cytoplasmic basophilia at 8 months is somewhat darker in late (stage 2) than in early maturers (sexually mature); however, as in early maturers, nuclei are consistently round and pale staining, and nucleoli are prominent at all ages.

In early-maturing males the intensity of ir-GnRH in NOR perikarya and fibers is similar at 8, 24 and 30 months (Fig. 2). At 12 months ir-GnRH is only slightly more intense; however, at 18 months there is a significant increase in the immunoreactive response. By contrast, in early-maturing females immunoreactivity is consistently light at all ages between 8 and 30 months.[11,44a]

In late-maturing males, NOR ir-GnRH is most intense at 8 months; however, late maturers are still in stage 2 at this age. Similar amounts of ir-GnRH, considerably less than found at 8 months, are present in the NOR cell bodies at all other ages (unpublished); the increase of NOR-ir-GnRH which occurs in early-maturing males at 18 months is not present in late males (Fig. 2).

2. GnRH in NPP. Perikarya in the NPP are generally ovoid to fusiform in shape throughout life in both early- and late-maturing genotypes. In early maturers, the area of NPP perikarya is maximal at 12 months and lowest at 18 and 30 months (Fig. 1). Differences between the NPP cell areas of early and late genotypes are greatest at 12 months of age. Similarly, the nuclei of NPP perikarya are markedly larger in early maturers at 8 and 12 months (same chronological but different physiological age than late maturers) but are similar in both genotypes at older ages. Generally, in late genotypes cytoplasmic basophilia is light, nuclei pale and nucleoli prominent. At 18 months in early maturers there is a perceptible increase in basophilia and nuclei are more dense in appearance.

There is a noticeable increase of ir-GnRH in NPP perikarya of early-maturing males at 18 months; the degree of immunoreactivity then decreases at 24 months and is at its lowest level (since 8 months) at 30 months (Fig. 2). Early-maturing females show fluctuations in NPP immunoreactivity similar to those observed in early-maturing males; however, at 18 months the intensity of the ir-response is less in females, and their number of ir-perikarya is fewer.[11,44a] Immunoreactivity in the NPP of late-maturing males remains at a constant level between 12 and 30 months and is lowest at 8 months (unpublished). The increase of NPP ir-GnRH present in early males at 18 months does not occur in late males.

3. GnRH in NLT. The typical unipolar shape of cell bodies in the NLT of younger fish tends to become more variable in older (24 and 30 months) males of both genotypes. Perikaryal area shows the greatest difference between genotypes at 8 and 24 months (Fig. 1) while the nuclear area differs at 8, and somewhat less at 12, months. Cytoplasmic basophilia is never intense in NLT cells of either genotype at any age; however, small perceptible increases in its intensity are seen at 18 and 30 months in early maturers, while in late males basophilia gradually intensifies between 18 and 30 months. Nuclei are pale staining at most ages in both genotypes. In early maturers, nucleoli are prominent except at 30 months; in late-maturing males, nucleoli are prominent only at 18 and 24 months.

As in the NOR and NPP, ir-GnRH in the NLT is at its greatest intensity since stage 2, at 18 months in early males (Fig. 2). In early-maturing females between 8 and 30 months, NLT ir-GnRH is restricted to a few fibers and immunoreactivity is never seen in NLT perikarya[11,44a] In late males, ir-perikarya are never seen in the NLT and pale immunoreactivity of some NLT neuronal processes is present only at 12, 18 and 30 months of age (unpublished).

As far as can be ascertained, our studies in platyfish are the only ones available in any nonmammalian vertebrate which deal with age-related changes in immumocytochemical GnRH. The distribution of ir-GnRH in mature animals representing each class of vertebrates recently has been comprehensively reviewed.[3]

4. Serotonin (5HT). In early-maturing platyfish of all ages studied, immunoreactivity to anti-serotonin (Immunonuclear, Inc.) is found in the forebrain in the wall of the third ventricle and its lateral (NRL) and posterior (NRP) recesses (paraventricular organ, PVO), and, if fish are sacrificed before 1 P.M., in the pineal gland[28,29,31] (Fig. 4a). In addition, when cross-sections of brains from 8-month-old early-maturing males were examined, ir-5HT was observed in perikarya and processes of the raphe region and in perikarya in the valvular of the cerebellum.[29] In early-maturing males ir-5HT first appears in perikarya of the nucleus preopticus (NPO) at 18 months, and PVO-immunoreactivity is most intense at 24 months.[31]

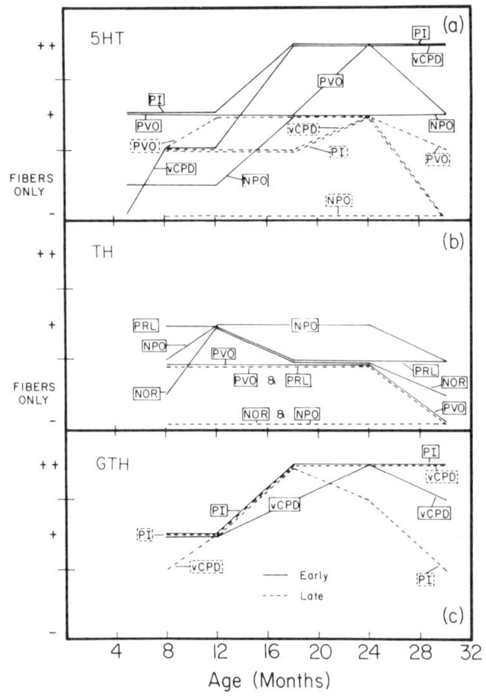

Fig. 4. Age-related changes during adulthood in relative intensities (- to +++) of (a) ir-serotonin (5HT), (b) ir-tyrosine hydroxylase (TH), and (c) ir-Gn (GTH) in indicated areas of the brain and/or pituitary in early (E) and late (L) male genotypes. See text for details.

In late-maturing males at all ages between 8 and 30 months (unpublished) results show patterns of immunoreactive staining which are significantly paler than, but similar in pattern of distribution to, early males. The most obvious exceptions are that in late maturers ir-5HT is never found in NPO perikarya, and PVO-immunoreactivity is unchanged at 24, and decreased at 30, months (Fig. 4a; unpublished).

Ir-5HT localization in 5- and 8-month-old early maturers shows no sexual dimorphism.[29] Aging studies on early females show 5HT immunoreactivity that is similar to that of early males at all ages except that ir-5HT containing NPO perikarya were never found in females (unpublished).

Serotonin has been histochemically and biochemically identified in nervous tissue of most vertebrate classes;[34] however, there is no information available on aging changes in ir-5HT distribution in poikilotherms. The most common characteristics of monoaminergic systems in the hypothalamus of nonmammalian vertebrates are the cerebrospinal fluid-contacting serotonergic cells which form the paraventricular organ. These cells are numerous in anamniotes and have been documented in several species (*Rana temporaria*,[36]

Xenopus laevis,[50] *Carassius auratus*[17,59]) as well as in platyfish. It is believed that the PVO is involved in the regulation of various endocrine events, including ovulation (cf. Parent[34]).

5. Tyrosine Hydroxylase (TH). Tyrosine hydroxylase is an enzyme necessary for catecholamine synthesis. In 12-month-old early-maturing males, ir-TH[16] is localized in perikarya and processes of the NOR and NPO, and paler immunoreactivity is found in perikarya in the wall of the third ventricle and the PVO.[14] In the NOR, ir-TH is greatest at 12 months and lowest (limited to fibers only) at 8 and 30 months; in the NPO it is at its highest levels between 12 and 24 months. In the PVO it is at low levels between 8 and 24 months and is absent at 30 months. In late-maturing males, low levels of ir-TH are found in the PVO between 8 and 24 months only; no ir-TH perikarya are localized in the NOR or NPO (unpublished; Fig. 4b).

There are no other reports available for age-related changes in ir-TH distribution in cold-blooded vertebrates; however, ir-TH recently has been localized in adult goldfish,[18,59] and Parent[34] has reviewed catecholamine localizations in all vertebrate classes.

III. PITUITARY GLAND

Most aging changes in the pituitary gland discussed below will deal with those occurring in gonadotropin- (Gn) containing cells. In platyfish these include gonadotropes of the caudal pars distalis (CPD) and of the pars intermedia (PI; GTH^P cells, see Schreibman[41]).

A. Birth to Sexual Maturity

1. Gn and GnRH. Cells containing both ir-Gn (beta subunit, see Schreibman et al.[46]) and ir-GnRH can be identified in neonates in the PAS-positive cells of the PI and in clusters of cells in lateral pockets of the CPD; gonadotropes of the ventral (v) CPD are represented by a few scattered ir-cells.[47] The gonadotropes of the PI are numerous at 1 week of age while those of the vCPD begin to appear just prior to the onset of sexual maturation and then increase markedly in number, size and activity[9,12,13] (Figs. 3,5). There is a direct correlation between the number of ir-GnRH-containing neurons in the brain and the number of ir-Gn-containing CPD and PI cells in the pituitary of both early and late maturers from one week to adulthood.[9,12,13] In adult early and late maturers, the number of ir-cells in the PI are similar, but late maturers have significantly fewer ir-cells in the CPD (Fig. 3). In both genotypes, the initial appearance of ir-GnRH in

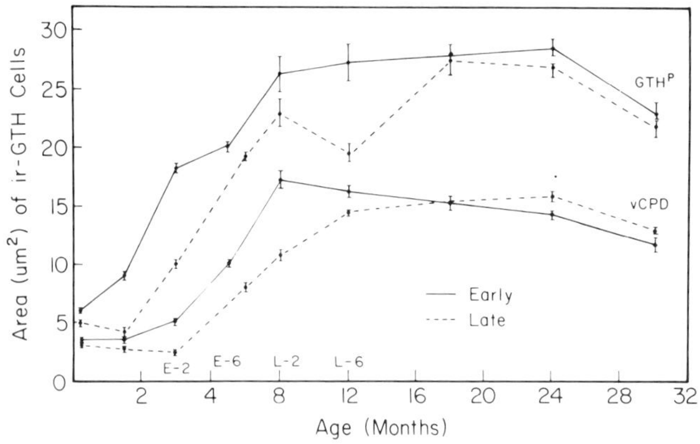

Fig. 5. Age-related changes in the area of ir-Gn-containing (IR-GTH) cells in the PI (GTHP) and vCPD of early- and late-maturing male platyfish. Age at gonopodial stages 2 and 6 indicated for early (E) and late (L) genotypes.

the NPP and NLT occurs concomitantly with the proliferation of gonadotropes in the CPD.

B. Maturity to Senescence

1. Gn and GnRH: Ventral CPD. Gonadotropes in the vCPD are consistently irregular in shape in both early- and late-maturing platyfish at all ages.[40a] In both male genotypes basophilia is maximal, nuclei pale staining, and nucleoli most prominent at 18 and/or 24 months. Cellular area in early maturers declines steadily from 8 to 30 months. By comparison, in late-maturing males cellular area increases steadily from 8 to 24 months and then decreases at 30 months (Fig. 5).

Ir-Gn in the vCPD is at its maximal intensity in 24-month-old early-maturing males (Fig. 4c). The intensity of the immunoreactive response using anti-GnRH in the pituitary is similar to that obtained with anti-Gn at the various ages. In late-maturing males, immunoreactivity to Gn, as well as to GnRH, is highest at 18, 24, and 30 months (Fig. 4c).

There are sexually dimorphic, age-related changes in the distribution of pituitary gonadotropes in platyfish.[30] In females shortly after reaching sexual maturity cells containing ir-Gn begin to appear among the prolactin cells in the rostral pars distalis, and they become more numerous with age. In males, if they do appear, it is almost a year later, and they are much fewer in number than in females.

2. Gn and GnRH: Pars Intermedia. GTHP cells in the PI vary in shape at all ages, and their cytoplasm displays nonspecific fluctuations in basophilia

within a narrow range. In early-maturing males, nuclei in GTH^p cells become smaller at 30 months and contain difficult-to-locate nucleoli. In late maturers, nuclei are generally pale-staining and contain nucleoli that are most prominent at 18 and 24 months.

In early-maturing males, PI cell area increases steadily from 8 to 24 months and then decreases at 30 months. In late-maturing males, area decreases markedly from 8 to 12 months and then increases dramatically at 18 months. A moderate decrease at 24 months, and a more definitive one at 30, has been noted. The largest difference between genotypes in cell area occurs at 12 months (Fig. 5).

The immunoreactive response to Gn antiserum is greatest at 18 months. This intensity is maintained at 24 and 30 months in early males whereas late maturers show declines in immumoreactivity at 24 and 30 months (Fig. 4c). In both male genotypes, ir-GnRH levels in the PI show fluctuations similar to those seen for ir-Gn. In early-maturing females the pattern of immunoreactivity for ir-Gn and ir-GnRH is similar to that seen in early males except at 18 and 30 months where females appear to have somewhat less ir-Gn in their PI cells (unpublished). Even in the oldest male platyfish (54 months) examined, some almost twice the average life span for platyfish, ir-Gn and ir-GnRH are localized in the same pituitary cell types (i.e., gonadotropes of the vCPD and PI) as they are in young sexually mature fish.[46]

3. Serotonin. Ir-5HT is localized in the GTH^p cells of the PI, in fish at all ages examined beginning at 5 months. In early-maturing males, it is most intense at 18, 24 and 30 months.[28,29,31] In some fish, beginning at 8 months, low levels of ir-5HT are localized in gonadotropes of the vCPD. In early maturers the greatest ir-response in vCPD cells occurs at 18, 24 and 30 months of age. In late maturers, in both the vCPD and PI, ir-5HT is elevated only at 24 months, is always less than is seen at the same age in early males, and is absent at 30 months (Fig. 4a).

In early-maturing females, pale 5HT immunoreactivity first appears in vCPD cells at 18 months, and it remains unchanged to 30 months. Ir-5HT intensity in both the vCPD and PI is always much paler in females than males.

In other immunocytochemical studies of the pituitary, ir-5HT containing cells have been found in the pars distalis of stingray,[38] bullfrog[23] and rat;[54] however, no aging studies are available.

4. Tyrosine Hydroxylase. In male and female platyfish of early-maturing genotype, prolactin cells of the rostral pars distalis contain ir-TH at all ages (Fig. 4b). In addition, paler ir-TH can be seen in certain cells

of the PI of some fish at 12 and 18 months of age[14] (also unpublished). Age-related changes in ir-TH in the pituitary have not been investigated in other vertebrates.

IV. GONADS

A. Birth to Sexual Maturity

1. Testis. In neonatal platyfish the testis consists of 2 lobes comprised of small clusters (cysts) of spermatogonia surrounding a central stroma. As testicular development continues, spermatogonia proliferate, and efferent ducts appear.[42] All immature fish which are still in stage 1 of gonopodial development have testes that have progressed in spermatogenesis up to, but not including, the primary spermatocyte stage.[21,45]

In the youngest males examined (3-day-old), Δ^5-3β-hydroxysteroid dehydrogenase (3β-HSD) and glucose-6-phosphate dehydrogenase (G6PD), enzymes, which when present together suggest the potential for sex steroid formation, were identified histochemically in Leydig cells located in the central stroma and among the spermatogonia in the two lateral regions of the testes.[42] With the initiation of the process of sexual maturation, 3β-HSD- and G6PD-positive cells become more numerous and display greater activity at the periphery of the testes and in the stroma around the efferent duct system. his pattern of enzyme distribution is also characteristic of the fully developed testis where, in addition, testicular duct epithelium shows a positive G6PD-reaction. Sertoli cells are negative for these enzymes at all stages of development.

2. Ovary. In the absence of functional gonadotropes in the vCPD, oogenesis in platyfish proceeds only up to the yolk deposition stage. In both early- and late-maturing females oil droplets appear in oocytes at approximately the same age.[45] Oocytes at this stage are surrounded by active follicle cells and a prominent zona pellucida.

In 3-day-old females 3β-HSD and G6PD activity are found in the stroma of gonads composed exclusively of oogonia. These cells increase in number and activity with subsequent ovarian development.[42] By comparison, granulosa cells are negative for the same enzymes at all stages of development.

B. Maturity to Senescence

1. Testis. The testes from males aged 18 months and older can be recognized on dissection by the presence of white lobules in a somewhat translucent matrix and by an increase in melanophores on the gonadal

surface. At the histological level, the testes of sexually mature fish are composed of cysts, each containing cells representative of a particular stage of spermatogenesis. Cysts at more advanced stages of gametogenesis are closest to the center of the testis, in proximity to the main testicular ducts. Mature sperm arranged in an almost perfect sphere (spermatozeugmata) are found within the ducts.[21,52]

In older fish, all stages of spermatogenesis continue to be present in the testes. However, beginning at about 18 months, the pattern of organization becomes distorted, especially in the cysts furthest from the periphery of the testis, and the degree of distortion increases with advancing age. In the oldest fish, proportionately more than the normal complement of spermatozeugmata appears to be present; however, they lack the definitive organization characteristic of the younger gonad, and frequently large clusters of loose sperm are seen, particularly near and in the main ducts. Ducts are laden with spermatozeugmata in various stages of (dis)organization. This suggests a decrease in sperm release which may be related to an age-dependent decline in the reproductive capacity of live-bearing poeciliids[7] or to celibacy imposed by experimental design.[46]

There is an age-related increase in connective tissue in the gonad which begins at about 18 months. In the oldest fish, collagenous fibers are present throughout, and large deposits of adipose tissue are located between the main ducts with smaller amounts interspersed among the tubules. The testes of aged platyfish show no apparent changes in efferent duct columnar or main duct cuboidal epithelia or in the histological appearance of Leydig cells.[46]

Our observations on testes correlate with most of those found for other teleosts. With advancing age in guppies,[57] _Astyanax mexicanus_,[37] and bitterlings,[15] the production of mature sperm continues, as in platyfish, despite the conspicuous accumulation of connective tissue in the testes of old fish. In aged _Betta splendens_,[55] however, in addition to the observed increase in connective tissue infiltration, there is an almost total regression of the gonads and absence of germinal tissue similar to that seen in hypophysectomized fish.[35]

Preliminary histochemical studies of male platyfish suggest that there is a general decrease in 3β-HSD, 3α-HSD, G6PD and lactic dehydrogenase activity suggesting hampered cellular metabolic and steroidogenic activity, and cellular degeneration with increasing age (unpublished with R. van den Hurk).

The endocrinology of reproduction in amphibians and reptiles has been summarized recently[24,25,27] (see also Chapter 10); however, age-related changes in the testes are not detailed in these reviews.

2. Ovary. Age-related changes in the gross morphology of the platyfish ovary are less consistent than those observed in the testis. Most frequently one notes that with increasing age yolky oocytes become depleted and/or their physical characteristics change (e.g., they become very hard). It frequently appears as if the ovary is atrophied, and there is an increase in brown or bright yellow "bodies" observed on its surface (unpublished). Nevertheless, yolky oocytes may be observed even in the oldest females.

In preliminary histochemical studies of ovaries from aging platyfish we have observed marked decreases in 3β-HSD, 3α-HSD, and G6PD activity with increasing age, suggesting that metabolic and steroidogenic activity is reduced with age (unpublished with R. van den Hurk).

In many teleosts, resorption of yolky eggs, thickening of the ovarian wall, increased connective tissue and general atrophy are characteristics of aging ovaries (cf. Woodhead and Setlow[58]). These observations do not hold for all teleosts, however, since few age-related changes occur in ovaries of guppies and bitterlings (cf. Haranghy et al.[15]).

Studies in reptiles have shown that there are significant differences in the numbers and quality of eggs produced in young, mid-aged and old female alligators.[4] These relationships correlate well with the reproductive and behavioral biology of alligators in these age groups.[4,5] In addition, "reproductively senescent" females have few growing oocytes, no corpora lutea, resorbing ovarian follicles, and eggs in the oviduct that are undergoing resorption (cf. Ferguson[5]). Although there is a limit on the number of oocytes that a female can produce, it appears that oognial proliferation continues through much of adulthood.[4]

V. SYNTHESIS, SUMMARY AND SUGGESTIONS FOR FUTURE RESEARCH DIRECTIONS

This chapter details the changes that occur from birth to senescence in the neuroendocrine axis which regulates reproductive function in platyfish of known genetic makeup. As such, the information presented represents the most complete "longitudinal" study available for any cold-blooded vertebrate. Yet, it is clear that there are still wide abysses in our knowledge, even for the platyfish. The platyfish is a unique research model in that it permits the analysis of the interplay of environmental, genetic, and neuroendocrine components and their concomitant influences on the aging of the reproductive system. However, it was the prime intent of this chapter to present this information in order for it to serve as a framework, and to instigate such studies in other organisms, including other fishes.

Despite the incomplete state of our knowledge of the aging process in platyfish, we have observed some striking trends which suggest possible lines for future investigation.

1. There are distinct changes in the distribution and relative quantity of neuropeptides, neurotransmitters and pituitary and gonadal hormones, as well as in the structure of the cells that manufacture, transport or are target organs for them, during the lifespan of platyfish.

During the maturation of the reproductive system, there is a sequence in the initial appearance of the three ir-GnRH-containing regions in the brain that is related directly to the stage of sexual maturation and not to chronological age. The appearance of these centers precedes, and is essential for, the ontogeny of pituitary gonadotropes in the caudal pars distalis (CPD). The presence of functional gonadotropes is necessary for complete maturation of both testis and ovary.[21,45] It appears, therefore, that the early and later stages of gonadal development are under separate control. Early stages of testicular and ovarian development may be regulated by gonadotropins originating in the GTHP cells of the PI[41] and perhaps from cells in the lateral CPD; completion of gonadal development and subsequent maintenance may be the function served by vCPD gonadotropes.

With increasing age (8 to 30 months), distinct fluctuations in the distribution and intensity of ir-GnRH in the brain are noticed with major changes occurring at 18 months (Section II B). In early-maturing males, the intensity of the response to anti-GnRH at 18 months markedly increases in the nucleus olfactoretinalis (NOR), nucleus preopticus periventricularis (NPP) and nucleus lateralis tuberis (NLT). In both early and late maturers levels of ir-GTH and ir-GnRH in the pituitary are elevated at 18 and/or 24 months. These aging shifts follow, or occur concomitantly with, distinct changes in the morphometry and cytology of ir-GnRH-containing perikarya and pituitary gonadotropes (Section III). Serotonin (5HT), a neurotransmitter important in reproductive physiology and behavior (cf. Timiras,[51] Vitale et al.[53]) makes its initial appearance in the nucleus preopticus (NPO) at 18 months. At this age, ir-5HT increases in intensity in GTHP cells and first appears consistently in vCPD gonadotropes. All these changes which have been recorded for the brain and pituitary conceivably could be translated into the "degenerative" transformations which are noted in the testes of early maturers at 18 months (Section IV B).

Immunoreactivity in the NPO to anti-tyrosine hydroxylase (TH) does not increase, and may even decrease somewhat, during aging. This, coupled with our findings of ir-5HT in the NPO only in aging platyfish, offers grounds for speculating that new interplays of the different neurotransmitters, caused by changes in their relative concentrations in specific areas, may occur during, and thereby also contribute to, senescence.

The interdigitating cyclical events which occur during platyfish aging suggest the exchange of information among components of the BPG axis. They need to be examined at shorter time intervals in order to determine the specific sequence of events and the direct cause and effect relationships that may contribute to the process of sensescence. Eighteen to 24 months seems to be a significant period to study in greater detail because marked changes occur in the brain, pituitary and gonads. However, it may be even more enlightening to study the ages immediately preceding this period.

2. Our studies indicate that each ir-GnRH-containing region of the brain plays a significant, and perhaps unique, role during specific phases of the life span of platyfish. The NOR (presumably homologous with the nervus terminalis[3]), may serve to initiate and orchestrate the activity in the other two GnRH-containing centers which culminates in the development and functioning of pituitary gonadotropes (Section II A1). The NOR appears to function when Gn levels are low, as for example, in immature platyfish and in mature fish following hypophysectomy.[44] The NOR may serve to "communicate" and "translate" environmental cues into neuroendocrine activity.[10,44,48] The close functional relationship between the NOR and the NPP and NLT may be demonstrated by the elevated levels of GnRH immunoreactivity observed concomitantly in all three regions at specific times during development and aging (early-maturing males at stage 2 and 18 months; late-maturing males at stage 2). The decreases observed in the 3 regions at 24 months and then again in the NOR and NPP at 30 months, further illustrate that these areas, although unique in their structural and functional qualities, may also act in concert to affect neuroendocrine activity.

Of the three centers, it is the NLT which displays the greatest age-, sex- and genotype-related variability in its GnRH immunoreactivity. Ir-GnRH is present in NLT perikarya in early-maturing males at all ages after stage 2. Its intensity, however, is high only at 11 weeks (stage 2) and 18 months. Although ir-GnRH can be seen in their cell processes, we have been unable to identify ir-GnRH in NLT perikarya in early-maturing females after an age comparable to male sibs in stage 2, or in male and female late maturers at all ages examined (Sections II A, II B). These observations of variability related to age, sex, and physiological state, as well as those showing correlations between ir-NLT cell number and the number of vCPD gonadotropes, lead us to speculate that the NLT may play a role in "fine tuning" or modulating gonadotropin activity in the pituitary gland.

3. Although data on females are not as yet available for all phases of our aging study, it is clear that the process of senescence exhibits sexual dimorphism. There are variations with increasing age between the sexes in NLT activity, as well as in the distribution of neurotransmitters and gonadotropes. Also, the degree of age-related modifications in the gross morphology is not similar in testes and ovaries; ovarian changes take place later and are less apparent than those of the testes (Sections IV, B1 and 2). Histological analysis is necessary to clarify these observations. Are there clues in the sexual dimorphic pattern of age-related changes observed that signal differential mechanisms of aging of these basically similar systems? Is there any correlation in our observations that may explain the higher natural mortality rate of males in all classes of vertebrates?[6,49,56]

4. It appears that in the oldest animals examined, gonads retain the capacity for gametogenesis, and ir-Gn and ir-GnRH are present in the pituitary gland. Thus, the age-dependent decrease in reproductive capacity found in some teleosts[7] need not be an obligatory phenomenon. Since Gn production and the potential for GnRH stimulation of Gn release persist in aged platyfish, the observed age-related changes in the platyfish gonad may be, as suggested for the guppy,[57] associated with general endocrine dysfunction rather than a "physiological hypophysectomy" that results from a decline in Gn with advancing age.

It is obvious, however, that the gonads display numerous signs of demise which become apparent in males at 18 months, an age that is coincident with obvious changes taking place in the brain and pituitary gland. It would be useful to determine how gonadal structure and function in aging fish is related to changing levels of circulating pituitary hormones and to our immunocytochemical observations. Our data suggest that age-related differences are quantitative (e.g., number of ir-cells in brain and pituitary, intensity of immunoreactivity in these cells, and proportion of mature gametes), and the research we suggest would help to test this hypothesis.

5. In platyfish genetic factors determine the timing and sequence of events in the brain and pituitary gland that culminate in puberty (Sections I, II A). The pattern of development of the BPG axis is similar in early- and late-maturing genotypes except that in late maturers specific steps take place in older animals and require more time to complete. This delay creates differences between genotypes in cytometric, cytological and immunological characteristics. There is a positive correlation between the number of ir-GnRH neurons and the number of ir-gonadotropes in the vCPD in early- and late-maturing fish from one week to adulthood (Fig. 3). In addition, the number of ir-GnRH cells in the NOR as well as the PI are similar in adult

early and late maturers, but late maturers have significantly fewer ir-cells in the NPP, NLT and CPD than early maturing sibs (Section III A).

Genetic regulation in platyfish of the structure and function of the BPG axis is carried into senescence. In addition to the variations observed in the aging of the BPG axis which are related to genotype and discussed above (Section V, 2.), there are several others that should receive special attention. In late maturers we do not observe the sharp increase in intensity of ir-GnRH that is seen in early genotypes at 18 months. In addition, the ir-5HT and ir-TH that is noted in the NPO of early maturers is absent (5HT), or present only at minimal levels (TH), in late maturers. The significance of these observations requires clarification by additional study. However, it leads us to suggest that further studies concentrating on the development and interactivity of neurotransmitters and neuropeptides in animals of different genotypes will be rewarding in our general comprehension of aging.

Future research in platyfish will help us to determine whether a clock for aging resides in the genotype of the organism in the same way that one exists for puberty. The basic question that needs to be resolved, of course, is where and how genes exert their control over the development and aging of the BPG axis. Is the P allele also involved in orchestrating senescence in platyfish, or are there separate alleles for this process which differ from those related to sexual maturation? We need to know more about the relationship between development, aging and longevity and how these processes are influenced by genotype.[26,39] These difficult problems are certain to be made easier to probe with the advent of newer, sophisticated methods for gene manipulation. In the immediate future, however, our studies should soon provide information for the platyfish that will clarify the relationship between the age at sexual maturation and the age at death. In other words, do early- and late-maturing platyfish differ in their life expectancies?

We are on the frontier of studying the phenomenon of aging. The years to come are certain to provide fundamental information that will enable us to better understand and, hopefully, control senescence.

VI. ACKNOWLEDGMENTS

This chapter is dedicated to Professor Louis G. Moriber, teacher, colleague and dear friend, on the occasion of his retirement from the Department of Biology at Brooklyn College.

The research of our laboratory has been supported by the National Science Foundation (PCM 77-15981), the National Institute on Aging (AGO 1938)

and the City University of New York (PSC-CUNY). We sincerely appreciate the technical assistance of Ethel Suben and Sandra L. Shaatal-Weiner. We thank Drs. H.J.Th. Goos, T.H. Joh, E. Burzawa-Gerard and Y.A. Fontaine for gifts of antisera, and Dr. K.D. Kallman for stocks of platyfish.

VII. REFERENCES

1. Adelman, R.C. and Roth, G.S., eds. (1982). "Endocrine and Neuroendocrine Mechanisms of Aging." CRC Press, Florida.
2. Cristofalo, V.J., Baker, G.T., Adelman, R.C. and Roberts, J., eds. (1984). "Altered Endocrine Status During Aging." Alan R. Liss, Inc., New York.
3. Demski, L.S. (1984). The evolution of neuroanatomical substrates of reproductive behavior: sex steroid and LHRH-specific pathways including the terminal nerve. Amer. Zool. 24, 809-830.
4. Ferguson, M.W.J. (1984). Craniofacial development in Alligator mississippiensis. In "The Structure, Development and Evolution of Reptiles" (M.W.J. Ferguson, ed.), pp. 223-273. Academic Press, Orlando, Fla.
5. Ferguson, M.W.J. (1985). Reproductive biology and embryology of the crocodilians. In "Biology of the Reptilia" (C. Gans, F. Billet and P.F.A. Maderson, eds.) Vol. 14, pp. 329-491. John Wiley & Sons, Inc., New York.
6. Finch, C.E. and Hayflick, L., eds. (1977). "Handbook of the Biology of Aging." Van Nostrand Reinhold Company, New York.
7. Gerking, S.D. (1959). Physiological changes accompanying ageing in fishes. In "The Lifespan of Animals (Ciba Foundation Colloquia on Ageing)" (G.E.W. Wolstenholme and M. O'Connor, eds.), pp.181-208. Little, Brown & Co., Boston.
8. Grobstein, D. (1948). Endocrine and developmental studies of gonopod differentiation in certain poeciliid fishes. I. The structure and development of the gonopod in Platypoecilus maculatus. Univ. of Calif. Publ. Zool. 47, 1-22.
9. Halpern-Sebold, L., Schreibman, M.P. and Margolis-Nunno, H. (1986) Differences between early and late maturing genotypes of the platyfish (Xiphophorus maculatus) in the morphometry of their immunoreactive LHRH containing cells: A developmental study. J. Exp. Zool. 840, in press.
10. Halpern-Sebold, L.R. and Schreibman, M.P. (1983). Ontogeny of centers containing luteinizing hormone-releasing hormone in the brain of platyfish (Xiphophorus maculatus) as determined by immunocytochemistry. Cell Tissue Res. 229, 75-84.
11. Halpern-Sebold, L.R., Schreibman, M.P. and Margolis-Kazan, H. (1984). Sexually dimorphic age-related changes in the distribution of immunoreactive LHRH. Amer. Zool. 24, 71A. (Abstract)
12. Halpern-Sebold, L., Schreibman, M.P. and Margolis-Kazan, H. (1984). Ontogeny of luteinizing hormone releasing hormone (ir-LHRH) containing centers in the brain and their relationship to sexual maturation. Neuroscience 10, 456. (Abstract)
13. Halpern-Sebold, L., Schreibman, M.P., Margolis-Kazan, H. and Goos, H.J.Th. (1983). A relationship between the number of luteinizing hormone releasing hormone (LHRH) containing neurons and gonadotrops in platyfish. Amer. Zool. 23, 882. (Abstract)
14. Halpern-Sebold, L.R., Margolis-Kazan, H., Schreibman, M.P. and Joh, T. (1985). Immunoreactive tyrosine hydroxylase in the brain and pituitary gland of the platyfish. Proc. Soc. Exper. Biol. Med 178, 486-489.
15. Haranghy, L., Penzes, L., Kerenyi, T. and Penzes, B. (1977). Sexual activity and ageing in the teleost fish, Rhodeus sericeus amarus. Exper. Gerontol. 12, 17-25.

16. Joh, T.H., Gregham, G. and Reis, D.J.C. (1973). Immunochemical demonstration of increased tyrosine hydroxylase protein in sympathetic ganglia and adrenal medulla elicited by reserpine. Proc. Natl. Acad. Sci. (U.S.A.) 70, 2767-2771.
17. Kah, O. and Chambolle, P. (1983). Serotonin in the brain of the goldfish, *Carassius auratus*. Cell Tissue Res. 234, 319-333.
18. Kah, O., Chambolle, P. and Thibault, J. (1984). Existence of dopaminergic neurons in the preoptic region of the goldfish. Neuroscience Letters 48, 293-298.
19. Kallman, K.D. (1975). The Platyfish *Xiphophorus maculatus*. *In* "Handbook of Genetics" (R.C. King, ed.), pp. 81-132. Plenum Publ. Corp., New York.
20. Kallman, K.D. and Borkoski, V. (1978). A sex-linked gene controlling the onset of sexual maturation in female and male platyfish (*Xiphophorus maculatus*), fecundity in females and adult size in males. Genetics 89, 79-119.
21. Kallman, K.D. and Schreibman, M.P. (1973). A sex-linked gene controlling gonadotrop differentiation and its significance in determining the age of sexual maturation and size of the platyfish, *Xiphophorus maculatus*. Gen. Comp. Endocrinol. 21, 287-304.
22. Kallman, K.D., Schreibman, M.P. and Borkoski, V. (1973). Genetic control of gonadotrop differentiation in the platyfish, *Xiphophorus maculatus* (Poeciliidae). Science 58, 678-680.
23. Kondo, Y., Nagatsu, I., Yoshida, M., Karasawa, N. and Nagatsu, T. (1982). Existence of noradrenalin cells and serotonin cells in the pituitary gland of *Rana catesbeiana*. Cell Tissue Res. 228, 405-408.
24. Lance, V. (1984). Endocrinology of reproduction in male reptiles. *In* "The Structure, Development and Evolution of Reptiles" (M. W. J. Ferguson, ed.), pp. 357-383. Academic Press, Orlando, Fla.
25. Licht, P. (1984). Reptiles. *In* "Marshall's Physiology of Reproduction" (G.E. Lamming, ed.), Vol. I, pp. 206-282. Churchill Livingstone, London.
26. Lints, F.A. (1981). Genetics and aging. Experientia 37, 1046-1050.
27. Lofts, B. (1984). Amphibians. *In* "Marshall's Physiology of Reproduction" (G.E. Lamming, ed.), Vol. I, pp. 127-205. Churchill Livingstone, London.
28. Margolis-Kazan, H., Halpern-Sebold, L.R. and Schreibman, M.P. (1984). Aging changes in the distribution of immunoreactive serotonin (ir-5HT) in platyfish. Amer. Zool. 24, 70A. (Abstract)
29. Margolis-Kazan, H., Halpern-Sebold, L.R. and Schreibman, M.P. (1985). Immunocytochemical localization of serotonin in the brain and pituitary gland of the platyfish (*Xiphophorus maculatus*). Cell Tissue Res. 240, 311-314.
30. Margolis-Kazan, H. and Schreibman, M.P. (1984) Sexually dimorphic, age-related changes in pituitary gonadotrop distribution. Mech. Ageing Devel. 24, 325-333.
31. Margolis-Nunno, H., Halpern-Sebold, L. and Schreibman, M.P. (1985). Immunocytochemical changes in serotonin in the forebrain and pituitary of aging fish. Neurobiol. Aging, 7, 17-21.
32. Meites, J., ed. (1983). "Neuroendocrinology of Aging." Plenum Press, New York.
33. Münz, H., Stumpf, W.E. and Jennes, L. (1981). LHRH systems in the brain of platyfish. Brain Res. 221, 1-13.
34. Parent, A. (1984). Functional anatomy and evolution of monoaminergic systems. Amer. Zool. 24, 783-790.
35. Pickford, G.E. and Atz, J.W. (1957). "The Physiology of the Pituitary Gland of Fishes." New York Zoological Society, New York.
36. Prasada Rao, P.D. and Hartwig, H.G. (1974). Monoaminergic tracts of

the diencephalon and innervation of the pars intermedia in *Rana temporaria*. A fluorescence and microspectrofluorometric study. Cell Tissue Res. 151, 1-26.

37. Rasquin, P. and Hafter, E. (1951). Age changes in the testes in the teleost, *Astyanax mexicanus*. J. Morphol. 89, 397-408.
38. Ritchie, T. C., Livingston, C.A., Hughes, M.G., McAdoo, D.J. and Leonard, R.B. (1983). The distribution of serotonin in the CNS of an elasmobranch fish: Immunocytochemical and biochemical studies in the Atlantic stingray, *Dasyatis sabina*. J. Comp. Neurol. 221, 429-443.
39. Rose, M.R. (1984). The evolution of animal senescence. Can. J. Zool. 62, 1661-1667.
40. Samuel, D., Algeri, S., Gershon, S. and Grimm, V. (eds.). (1983). "Aging of the brain." Raven Press, New York.

40a. Schreibman, M.P. (1964). Studies on the pituitary gland of *Xiphophorus maculatus* (the platyfish). Zoologica 49, 217-243.

41. Schreibman, M.P. (1986). Pituitary gland morphology. *In* "Vertebrate Endocrinology: Fundamentals and Biomedical Implications" (P.K.T. Pang and M.P. Schreibman, eds.), in press. Academic Press, New York.
42. Schreibman, M.P., Berkowitz, E.J. and van den Hurk, R. (1982). The histology and histochemistry of the platyfish (*Xiphophorus maculatus*) testis and ovary from birth to sexual maturity. Cell Tissue Res. 224, 81-88.
43. Schreibman, M.P., Halpern, L.R., Goos, H.T.Th. and Margolis-Kazan, H. (1979). Identification of luteinizing hormone-releasing hormone (LH-RH) in the brain and pituitary gland of a fish by immunocytochemistry. J. Exp. Zool. 210, 153-160.
44. Schreibman, M.P., Halpern-Sebold, L.R., Ferin, M., Margolis-Kazan, H. and Goos, H.J.Th. (1983). The effect of hypophysectomy and gonadotropin administration on the distribution and quantity of LHRH in the brains of platyfish: A combined immunocytochemistry and radioimmunoassay study. Brain Res. 267, 293-300.

44a. Schreibman, M.P., Halpern-Sebold, L. and Margolis-Nunno, H. (1985). Sexually dimorphic age-related changes in the distribution of immunoreactive luteinizing hormone releasing hormone in the platyfish. Mech. Ageing Devel. 33, 29-37.

45. Schreibman, M.P. and Kallman, K.D. (1977). The genetic control of the pituitary-gonadal axis in the platyfish, *Xiphophorus maculatus*. J. Exp. Zool. 200, 277-294.
46. Schreibman, M.P., Margolis-Kazan, H., Bloom, J.L. and Kallman, K. D. (1983). Continued reproductive potential in aging platyfish as demonstrated by the persistence of gonadotropin, luteinizing hormone releasing hormone and spermatogenesis. Mech. Ageing. Devel. 22, 105-112.
47. Schreibman, M.P., Margolis-Kazan, H. and Halpern-Sebold, L.R. (1982). Immunoreactive gonadotropin and luteinizing hormone-releasing hormone in the pituitary gland of neonatal fish. Gen. Comp. Endocrinol. 47, 385-391.
48. Schreibman, M.P., Margolis-Kazan, H., Halpern-Sebold, L. and Goos, H.J.Th. (1982). The functional significance of the nucleus olfactoretinalis in the platyfish, *Xiphophorus maculatus*. *In* "Proceedings of the International Symposium on Reproductive Physiology of Fishes" (C.J.J. Richter and H.J. Th. Goos, eds.), p. 59. Pudoc, Wacheningen.
49. Stini, W.A. (1981). Association of early growth patterns with the process of aging. Federation Proceedings 40, 2588-2594.
50. Terlou, M. and Ploemacher, R.E. (1973). The distribution of monoamines in the tel-, di- and mesencephalon of *Xenopus laevis* tadpoles, with special reference to the hypothalamo-hypophysial system. Z. Zellforsch. 137, 521-540.

51. Timiras, P.O. (1983). Neuroendocrinology of aging: retrospective, current, and prospective views. In "Neuroendocrinology of Aging" (J. Meites, ed.), pp. 5-30. Plenum Press, New York.
52. van den Hurk, R. (1974). Steroidogenesis in the testes and gonadotropic activity in the pituitary during postnatal development of the black molly, *Molliensia latipinna*. Proc. Kon. Ned. Akad. Wetensch., Ser. C 77, 193-200.
53. Vitale, M.L., de las Nieves Parasi, M., Chiocchio, S.R. and Tramezzani, J.H. (1984). Median eminence serotonin involved in the proestrus gonadotropin release. Neuroendocrinology 39, 136-141.
54. Westlund, K.N. and Childs, G.V. (1983). Double label immunocytochemical localization of neuronal fibers and cells in the rat adenophypophysis. J. Histochem. Cytochem. 31, 1066.
55. Woodhead, A.D. (1974). Ageing changes in the Siamese fighting fish, *Betta splendens*. I. The testis. Exper. Geront. 9, 75-81.
56. Woodhead, A.D. (1978). Fish in studies of aging. Exper. Geront. 13, 125-140.
57. Woodhead, A.D. and Ellett, S. (1969). Endocrine aspects of ageing in the guppy, *Lebistes reticulatus* (Peters). III. The testis. Exper. Geront. 4, 17-25.
58. Woodhead, A.D. and Setlow, R.B. (1979). Aging changes in the ovary of the amazon molly, *Poecilia formosa*. Exper. Geront. 14, 205-209.
59. Yoshida, M., Nagatsu, I., Kawakami-Kondo, Y., Karasawa, N., Spatz, M. and Nagatsu, T. (1983). Monoaminergic neurons in the brain of goldfish as observed by immunohistochemical techniques. Experientia 39, 1171-1174.

CONTRIBUTORS

ELIZABETH ADKINS-REGAN	Department of Psychology and Section of Neurobiology and Behavior Cornell University Ithaca NY 14853
GLORIA V. CALLARD	Department of Biology Boston University 2 Cummington St. Boston MA 02115
IAN P. CALLARD	Department of Biology Boston University 2 Cummington St. Boston MA 02115
GIOVANNI CHIEFFI	Instituto di Biologia Generale Universita Degli di Napoli 80138 Napoli, Italy
DAVID CREWS	Institute of Reproductive Biology Department of Zoology University of Texas at Austin Austin TX 78712
WILLIAM A. GERN	Department of Zoology and Physiology University of Wyoming Laramie WY 82071
NEIL GREENBERG	Department of Zoology University of Tennessee Knoxville TN 37996
SHELLEY S. GREENHOUSE	Department of Zoology and Physiology University of Wyoming Laramie WY 80271
LOUIS J. GUILLETTE, JR.	Department of Zoology University of Florida Gainesville FL 32611
LESLIE HALPERN-SEBOLD	Department of Biology Brooklyn College Brooklyn NY 11210

SHUK-MEI HO
Department of Biology
Tufts University
Medford MA 02155

RICHARD E. JONES
Laboratory of Comparative Reproduction
Department of Environmental, Population and Organismic Biology
University of Colorado
Boulder CO 80309-0334

JOHN F. LEATHERLAND
Department of Zoology
University of Guelph
Guelph, Ontario
Canada N1G 2W1

PAUL LICHT
Department of Zoology
University of California
Berkeley CA 94720

BRIAN LOFTS
Department of Zoology
University of Hong Kong
Hong Kong

HENRIETTA MARGOLIS-NUNNO
Department of Biology
Brooklyn College
Brooklyn NY 11210

VALDO MAZZI
Dipartimento di Biologia Animale
Universita di Torino
Via Giolitti 34
10123 Torino, Italy

FRANK L. MOORE
Department of Zoology
Oregon State University
Corvallis OR 97331

YOSHITAKA NAGAHAMA
Laboratory of Reproductive Biology
Institute of Basic Biology
Okazaki 444, Japan

JEANNE M. NERVINA
Department of Zoology and Physiology
University of Wyoming
Laramie WY 80271

DAVID O. NORRIS
Laboratory of Comparative Endocrinology
Department of Environmental, Population and Organismic Biology
University of Colorado
Boulder CO 80309-0334

RICCARDO PIERANTONI
Stazione Zoologica
Universita Degli di Napoli
80121 Napoli, Italy

DAVID PORTER
Department of Zoology
University of California
Berkeley CA 94720

NANCY SHERWOOD
Department of Biology
University of Victoria
Victoria B.C., Canada V8W 2Y2

MARTIN P. SCHREIBMAN
Department of Biology
Brooklyn College
Brooklyn NY 11210

CAMILLO VELLANO
Dipartimento di Biologia Animale
Universita di Torino
Via Giolotti 34
10123 Torino, Italy

JOAN W. WHITTIER
Institute of Reproductive Biology
Department of Zoology
University of Texas at Austin
Austin TX 78712

JOHN WINGFIELD
Department of Zoology
University of Washington
Seattle WN 98195

FRANCOISE XAVIER
Ecole Normale Superieure
CNRS U.A. 686
Biochimie et Physiologie du Developpement
46 rue d'Ulm, F-75230
Paris Cedex 05, France

SPECIES INDEX

Species are arranged according to major taxonomic groups. Within a group, species are listed alphabetically. Latin names are used for wild species, and common names (followed by Latin names) for domestic or laboratory species.

SUBJECT INDEX

References to tables are indicated by t; references to figures are indicated by f.